T0074900

CAMBRIDGE STUDIES IN ADVANCED MATHEMATICS 189

Editorial Board
B. BOLLOBÁS, W. FULTON, F. KIRWAN,
P. SARNAK, B. SIMON, B. TOTARO

HIGHER INDEX THEORY

Index theory studies the solutions to differential equations on geometric spaces, their relation to the underlying geometry and topology, and applications to physics. If the space of solutions is infinite dimensional, it becomes necessary to generalise the classical Fredholm index using tools from the K-theory of operator algebras. This leads to higher index theory, a rapidly developing subject with connections to noncommutative geometry, large-scale geometry, manifold topology and geometry, and operator algebras.

Aimed at geometers, topologists and operator algebraists, this book takes a friendly and concrete approach to this exciting theory, focusing on the main conjectures in the area and their applications outside of it. A well-balanced combination of detailed introductory material (with exercises), cutting-edge developments and references to the wider literature make this a valuable guide to this active area for graduate students and experts alike.

Rufus Willett is Professor of Mathematics at the University of Hawai'i, Mānoa. He has interdisciplinary research interests across large-scale geometry, K-theory, index theory, manifold topology and geometry, and operator algebras.

Guoliang Yu is the Powell Chair in Mathematics and University Distinguished Professor at Texas A&M University. He was an invited speaker at the International Congress of Mathematicians in 2006, is a Fellow of the American Mathematical Society and a Simons Fellow in Mathematics. His research interests include large-scale geometry, K-theory, index theory, manifold topology and geometry, and operator algebras.

CAMBRIDGE STUDIES IN ADVANCED MATHEMATICS

Editorial Board
B. Bollobás, W. Fulton, F. Kirwan, P. Sarnak, B. Simon, B. Totaro

All the titles listed below can be obtained from good booksellers or from Cambridge University Press. For a complete series listing, visit www.cambridge.org/mathematics.

Already Published
149 C. Godsil & K. Meagher *Erdős-Ko-Rado Theorems: Algebraic Approaches*
150 P. Mattila *Fourier Analysis and Hausdorff Dimension*
151 M. Viana & K. Oliveira *Foundations of Ergodic Theory*
152 V. I. Paulsen & M. Raghupathi *An Introduction to the Theory of Reproducing Kernel Hilbert Spaces*
153 R. Beals & R. Wong *Special Functions and Orthogonal Polynomials*
154 V. Jurdjevic *Optimal Control and Geometry: Integrable Systems*
155 G. Pisier *Martingales in Banach Spaces*
156 C. T. C. Wall *Differential Topology*
157 J. C. Robinson, J. L. Rodrigo & W. Sadowski *The Three-Dimensional Navier-Stokes Equations*
158 D. Huybrechts *Lectures on K3 Surfaces*
159 H. Matsumoto & S. Taniguchi *Stochastic Analysis*
160 A. Borodin & G. Olshanski *Representations of the Infinite Symmetric Group*
161 P. Webb *Finite Group Representations for the Pure Mathematician*
162 C. J. Bishop & Y. Peres *Fractals in Probability and Analysis*
163 A. Bovier *Gaussian Processes on Trees*
164 P. Schneider *Galois Representations and (φ, Γ)-Modules*
165 P. Gille & T. Szamuely *Central Simple Algebras and Galois Cohomology (2nd Edition)*
166 D. Li & H. Queffelec *Introduction to Banach Spaces, I*
167 D. Li & H. Queffelec *Introduction to Banach Spaces, II*
168 J. Carlson, S. Müller-Stach & C. Peters *Period Mappings and Period Domains (2nd Edition)*
169 J. M. Landsberg *Geometry and Complexity Theory*
170 J. S. Milne *Algebraic Groups*
171 J. Gough & J. Kupsch *Quantum Fields and Processes*
172 T. Ceccherini-Silberstein, F. Scarabotti & F. Tolli *Discrete Harmonic Analysis*
173 P. Garrett *Modern Analysis of Automorphic Forms by Example, I*
174 P. Garrett *Modern Analysis of Automorphic Forms by Example, II*
175 G. Navarro *Character Theory and the McKay Conjecture*
176 P. Fleig, H. P. A. Gustafsson, A. Kleinschmidt & D. Persson *Eisenstein Series and Automorphic Representations*
177 E. Peterson *Formal Geometry and Bordism Operators*
178 A. Ogus *Lectures on Logarithmic Algebraic Geometry*
179 N. Nikolski *Hardy Spaces*
180 D.-C. Cisinski *Higher Categories and Homotopical Algebra*
181 A. Agrachev, D. Barilari & U. Boscain *A Comprehensive Introduction to Sub-Riemannian Geometry*
182 N. Nikolski *Toeplitz Matrices and Operators*
183 A. Yekutieli *Derived Categories*
184 C. Demeter *Fourier Restriction, Decoupling and Applications*
185 D. Barnes & C. Roitzheim *Foundations of Stable Homotopy Theory*
186 V. Vasyunin & A. Volberg *The Bellman Function Technique in Harmonic Analysis*
187 M. Geck & G. Malle *The Character Theory of Finite Groups of Lie Type*
188 B. Richter *Category Theory for Homotopy Theory*

Higher Index Theory

RUFUS WILLETT
University of Hawai'i, Mānoa

GUOLIANG YU
Texas A&M University

CAMBRIDGE
UNIVERSITY PRESS

University Printing House, Cambridge CB2 8BS, United Kingdom

One Liberty Plaza, 20th Floor, New York, NY 10006, USA

477 Williamstown Road, Port Melbourne, VIC 3207, Australia

314–321, 3rd Floor, Plot 3, Splendor Forum, Jasola District Centre,
New Delhi – 110025, India

79 Anson Road, #06–04/06, Singapore 079906

Cambridge University Press is part of the University of Cambridge.

It furthers the University's mission by disseminating knowledge in the pursuit of education, learning, and research at the highest international levels of excellence.

www.cambridge.org
Information on this title: www.cambridge.org/9781108491068
DOI: 10.1017/9781108867351

© Rufus Willett and Guoliang Yu 2020

This publication is in copyright. Subject to statutory exception
and to the provisions of relevant collective licensing agreements,
no reproduction of any part may take place without the written
permission of Cambridge University Press.

First published 2020

Printed in the United Kingdom by TJ International Ltd. Padstow Cornwall

A catalogue record for this publication is available from the British Library.

Library of Congress Cataloging-in-Publication Data
Names: Willett, Rufus, 1983– author. | Yu, Guoliang, 1963– author.
Title: Higher index theory / Rufus Willett, Guoliang Yu.
Description: Cambridge ; New York, NY : Cambridge University Press, 2020. |
Series: Cambridge studies in advanced mathematics ; 189 |
Includes bibliographical references and index.
Identifiers: LCCN 2019060048 (print) | LCCN 2019060049 (ebook) |
ISBN 9781108491068 (hardback) | ISBN 9781108867351 (epub)
Subjects: LCSH: Index theory (Mathematics) | C*-algebras. | K-theory.
Classification: LCC QA614.92 .W55 2020 (print) | LCC QA614.92 (ebook) |
DDC 512/.556–dc23
LC record available at https://lccn.loc.gov/2019060048
LC ebook record available at https://lccn.loc.gov/2019060049

ISBN 978-1-108-49106-8 Hardback

Cambridge University Press has no responsibility for the persistence or accuracy of URLs for external or third-party internet websites referred to in this publication and does not guarantee that any content on such websites is, or will remain, accurate or appropriate.

Dedicated to our mathematical mentors.
献给我们的老师.

For RW:

Sue Barnes
Glenys Luke
John Roe

For GY:

刘文虎 (Liu Wenhu)
孙顺华 (Sun Shunhua)
Ronald G. Douglas

Contents

Introduction

One of the greatest discoveries in mathematics is the Fredholm index. This measures the size of the solution space for a linear system. The beauty of this index is that it is invariant under small perturbations of the linear system.

The topology and geometry of a smooth closed manifold M is governed by certain natural elliptic differential operators. These operators have Fredholm indices that are computed by the famous Atiyah–Singer index formula. The work underlying this formula was one of the foremost mathematical achievements of the last century, and has important applications in geometry, topology, and mathematical physics.

A central question in mathematics is to extend the Atiyah–Singer index theory to non-compact manifolds. In the non-compact case, the classic Fredholm index is not well defined since the solution spaces of natural elliptic differential operators can be infinite-dimensional. A vast generalisation of the Fredholm index, called the higher index, can be defined for differential operators within the framework of Alain Connes' noncommutative geometry. A key idea in the definition of higher index is to develop a notion of dimension for possibly infinite-dimensional spaces using operator algebras. This dimension theory has its root in John von Neumann's theory of continuous geometry and is formalised using K-theory of operator algebras. Crucial features are that the higher index is invariant under small perturbations of the differential operator, and that it is an obstruction to invertibility of the operator.

Higher index theory has been developed in the work of many mathematicians over the last 40 years. It has found fundamental applications to geometry and topology, such as to the Novikov conjecture on topological rigidity and the Gromov–Lawson conjecture on scalar curvature.

The purpose of the book is to give a friendly exposition of this exciting subject!

Structure

This book is split into four parts.

Part ONE summarises background on C^*-algebras and K-theory, often only including full proofs for non-standard material. The reader should not expect to have to understand all of this before approaching the rest of the book. Part ONE ends with a section motivating some of the techniques we will study, based on the problem of existence of positive scalar curvature metrics. We also give a more detailed summary of the book's contents here.

Part TWO discusses Roe algebras, localisation algebras and the assembly maps connecting them. Roe algebras and localisation algebras are C^*-algebras associated to the large- and small-scale structures of a space, and assembly is a map between them. Assembly is closely related to taking higher indices. This section finishes with a description of the Baum–Connes conjectures, which posit that a certain universal assembly map is an isomorphism.

Part THREE moves into the theory of differential operators on manifolds, which is where the main applications of the theory developed in Part TWO lie; in the earlier parts, we do not really discuss manifolds at all. We discuss how elliptic operators naturally give rise to K-theory classes, and the flavour thus becomes more explicitly index-theoretic. We also discuss how Poincaré duality in K-theory relates to differential operators, and summarise some of the most important applications to geometry and topology.

Part FOUR looks at the (Baum–Connes) assembly maps in more detail. We give an elementary approach to some results in the case of almost constant bundles. We spend some time giving a new and relatively elementary proof of the coarse Baum–Connes conjecture for spaces that admit a coarse embedding into Hilbert space, a particularly important theorem for applications. We also discuss some counterexamples.

Finally, the book closes with several appendices summarising an ad hoc collection of material from general topology and coarse geometry, from representation theory, from the theory of unbounded operators, and about graded C^*-algebras and Hilbert spaces.

Intended Audience and Prerequisites

The prerequisites are something like a first course in C^*-algebra K-theory, some of which is summarised in Part ONE. For Part THREE, it will also help to have some background in manifold topology and geometry, although this is generally kept to a minimum.

The intended audience consists of either operator algebraists who are interested in applications of their field to topology and geometry, or topologists and geometers who want to use tools from operator algebras and index theory.

We have done our best to keep the exposition as concrete and direct as possible.

Acknowledgements

We have benefitted from direct and indirect contact with many workers in this area. Many mathematicians have had a deep influence over our view of the subject: we would particularly like to acknowledge Michael Atiyah, Arthur Bartels, Paul Baum, Xiaoman Chen, Alain Connes, Joachim Cuntz, Ron Douglas, Siegfried Echterhoff, Steve Ferry, Guihua Gong, Sherry Gong, Erik Guentner, Nigel Higson, Xinhui Jiang, Gennadi Kasparov, Wolfgang Lück, Ralf Meyer, Henri Moscovici, Ryszard Nest, Hervé Oyono-Oyono, John Roe, Thomas Schick, Georges Skandalis, Ján Špakula, Xiang Tang, Romain Tessera, Jean-Louis Tu, Shmuel Weinberger, Jianchao Wu, Zhizhang Xie and Rudolf Zeidler in this regard.

The following people have made appreciated comments on earlier drafts: Arthur Bartels, Xiaoman Chen, Jintao Deng, Chun Ding, Johannes Ebert, Hao Guo, Amanda Hoisington, Baojie Jiang, Matthew Lorentz, Amine Marrakchi, Lisa Ritter, Xiaoyu Su, Xianjin Wang, Alexander Weygandt and Rudolf Zeidler.

We are also grateful to the US NSF (RW and GY) and the International Exchange Program of the Chinese NSF (GY) for support during the writing of this book.

Finally, we would like to thank Christopher Chien and Saritha Srinivasan for their great work on the proofs.

PART ONE

Background

1

C^*-algebras

Our goal in this chapter is to summarise enough of the theory of C^*-algebras for what we need in the rest of the book. We are not aiming for a self-contained introduction: we do sketch some proofs where we think these are illuminating and do not take us too far afield, but leave many important results unjustified. We are also not attempting a comprehensive overview of basic C^*-algebra theory: for us, in this text, the theory of C^*-algebras is generally a means to an end, rather than an end in its own right, and we have tried to limit this overview to what we actually use.

The most concrete way to think of a C^*-algebra is as a norm-closed and adjoint-closed subalgebra of the collection $\mathcal{B}(H)$ of bounded operators on some Hilbert space H; one could define 'C^*-algebra' to mean exactly this, with no loss of generality. Moreover, many of the C^*-algebras we consider in this text arise naturally in this form, and thus thinking of C^*-algebras in this way will suffice for many of our purposes.

However, it is a remarkable fact that a quotient of a C^*-algebra by a closed ideal is still a C^*-algebra, and this would not be at all obvious if one just defined 'C^*-algebra' to mean 'norm and adjoint closed subalgebra of $\mathcal{B}(H)$'. Largely motivated by the need to explain this, we develop some basic theory and use the standard definition of C^*-algebra in terms of certain natural axioms.

The chapter is structured as follows. In Section 1.1 we start the chapter proper, and give the axiomatic definition of a C^*-algebra and some basic examples.

In Section 1.2 we discuss basic facts about spectrum and invertible elements, and in Section 1.3 we go through most of a proof of the celebrated Gelfand–Naimark theory, which characterises commutative C^*-algebras as all being of the form $C_0(X)$ for some locally compact Hausdorff space X. As this theory is so fundamental (and beautiful!), we give more exposition and proofs than is strictly necessary for the rest of the text. We also give a more detailed discussion of functoriality than is standard in the literature, and having a relatively detailed exposition helps with this.

Having done with this, Section 1.4 uses the commutative theory to construct the continuous functional calculus, a very powerful tool for general C^*-algebras. Already in this section we start to introduce some ideas without proof, in particular the holomorphic functional calculus; this will not be used much in the main text, but is occasionally important for K-theoretic arguments. Section 1.5 completes what one might think of as the 'basic abstract theory', discussing ideals and quotients. Our main goal is the fundamental fact that a quotient of a C^*-algebra by a closed ideal is still a C^*-algebra.

Section 1.6 then completes the general theory by discussing the relation between the concrete (i.e. as norm closed $*$-subalgebras of $\mathcal{B}(H)$) treatment of C^*-algebras and the axiomatic one. Perhaps the most glaring of our omissions in terms of basic theory is that we do not really justify this fact (nor do we discuss 'states', the key ingredient in its proof). We also discuss the relationship between representations of $C_0(X)$ and of the C^*-algebra $B(X)$ of bounded Borel functions on X; this is maybe less standard in C^*-algebra theory, but will be used over and over again in the main text.

Finally, Sections 1.7 and 1.8 discuss material that is again a little less standard, but that will occasionally be important to use. Section 1.7 discusses multiplier algebras. We do this in a very concrete (and somewhat ad hoc) way that is convenient for our purposes. We also give a treatment of Morita equivalence in terms of full corners; again this is ad hoc, but very convenient for our applications. Finally, Section 1.8 gives a quick overview of tensor products of Hilbert spaces and (spatial) tensor products of C^*-algebras, with particular emphasis on what happens in the commutative case.

1.1 Definition and Examples

Definition 1.1.1 A *C^*-algebra* is an algebra over $\mathbb{C}$ (i.e. simultaneously a ring and a vector space over $\mathbb{C}$, with the two structures being compatible), that is also equipped with a $*$-operation $*: A \to A$ and a complete norm with the following properties. The $*$-operation should be involutive, i.e. $(a^*)^* = a$ for all $a \in A$, and compatible with the algebra structure, meaning that

$$(ab)^* = b^*a^* \quad \text{and} \quad (\lambda a + \mu b)^* = \overline{\lambda}a^* + \overline{\mu}b^*$$

for all $a, b \in A$ and $\lambda, \mu \in \mathbb{C}$, where $\bar{\cdot}$ denotes complex conjugation. The norm should interact in the usual way with the linear structure on A and be submultiplicative, meaning that $\|ab\| \leqslant \|a\|\|b\|$ for all $a, b \in A$.

Finally, all three structures – norm, $*$ and algebra – should be compatible via the *C^*-identity*:

$$\|a^*a\| = \|a\|^2$$

for all $a \in A$.

A *C^*-subalgebra* of a C^*-algebra is a norm-closed, $*$-closed, subalgebra.

The zero algebra will be allowed as a C^*-algebra, unless doing so makes some statement obviously false. We leave deciding whether it should be allowed or not in any given context to the good judgement of the reader; similar comments apply to the empty topological space.

Remark 1.1.2 We will occasionally need to work with more general types of (normed, $*$) algebras; sometimes we will also state results in more generality than we actually need, mainly as this sometimes makes what is going on in a particular proof clearer. Here is the relevant terminology.

(i) A *$*$-algebra* will mean a complex algebra equipped with a $*$-operation satisfying the conditions in Definition 1.1.1, but not necessarily with a norm.
(ii) A *Banach algebra* is a complex algebra equipped with a complete norm that satisfies $\|ab\| \leqslant \|a\|\|b\|$ for all elements a and b, but that does not necessarily also have a $*$-operation.
(iii) A *Banach $*$-algebra* is a complex algebra equipped with an isometric $*$-operation and a complete norm satisfying all the conditions in Definition 1.1.1 except for possibly the C^*-identity.

We will always assume that the unit in a unital Banach algebra has norm one (this is automatic in a unital C^*-algebra).

Remark 1.1.3 The C^*-identity and submultiplicativity imply that the $*$ operation is isometric. Indeed, for any non-zero $a \in A$

$$\|a\|^2 = \|a^*a\| \leqslant \|a\|\|a^*\|,$$

and thus $\|a\| \leqslant \|a^*\|$. The reverse inequality follows by symmetry. Thus in particular, a C^*-algebra is a special type of Banach $*$-algebra.

Example 1.1.4 The fundamental example of a C^*-algebra is the complex numbers $\mathbb{C}$, with its usual absolute value for a norm, and complex conjugation for the $*$-operation.

Example 1.1.5 Let H be a Hilbert space. Then the collection $\mathcal{B}(H)$ of all bounded operators on H is a C^*-algebra: the linear operations are the pointwise operations inherited from the linear structure on H, multiplication is composition of operators, the $*$ is the adjoint of an operator, and the norm is the usual operator norm

$$\|T\| := \sup_{v \in H, \|v\| \leqslant 1} \|Tv\|_H.$$

In particular, if H is the finite-dimensional Hilbert space $\mathbb{C}^n$ with its usual inner product, then the $n \times n$ matrices $M_n(\mathbb{C})$ form a C^*-algebra.

Example 1.1.6 A subalgebra of a C^*-algebra that is stable under the $*$-operation and norm-closed is a C^*-algebra. As a special case, recall that an operator T on a Hilbert space H is *compact* if the image of the unit ball in H under T has compact closure, or equivalently if T is a norm limit of finite-rank operators. The collection of all compact operators, denoted $\mathcal{K}(H)$, is a C^*-algebra. If H is separable and infinite-dimensional, then we will often just write $\mathcal{K}$ for this C^*-algebra: it plays an important role in the theory.

Note that the example above will not contain a unit if H is infinite-dimensional. In particular, a C^*-algebra does not have to have a unit.

Example 1.1.7 One can form the direct sum of two C^*-algebras A and B: the algebra operations and $*$ are defined pointwise, and the norm is given by

$$\|(a,b)\| := \max\{\|a\|, \|b\|\};$$

it is straightforward (exercise!) to check that this is a C^*-algebra. More generally, if $(A_i)_{i \in I}$ is any collection of C^*-algebras, then their *product* is the collection

$$\prod_{i \in I} A_i := \left\{(a_i)_{i \in I} \mid a_i \in A_i \text{ and } \sup_{i \in I} \|a_i\| < \infty\right\}$$

of bounded sequences from the collection, equipped with pointwise operations and the norm

$$\|(a_i)\| := \sup_{i \in I} \|a_i\|.$$

The *direct sum* $\bigoplus_{i \in I} A_i$ is the C^*-subalgebra of $\prod_{i \in I} A_i$ consisting of all sequences $(a_i)_{i \in I}$ such that $\|a_i\| \to 0$ as $i \to \infty$ (or in other words, such that for any $\epsilon > 0$, there is a finite subset $F \subseteq I$ such that $\|a_i\| < \epsilon$ for all $i \notin F$).

If the index set I is finite, $\bigoplus_{i \in I} A_i$ and $\prod_{i \in I} A_i$ are the same, and we will generally use the sum notation. As a special case of the above example, one can form finite direct sums of matrix algebras, say $A = \bigoplus_{k=1}^{N} M_{n_k}(\mathbb{C})$. In fact, any finite-dimensional C^*-algebra turns out to be of this form.

We finish this collection of examples with something a little more involved.

Example 1.1.8 Let G be a discrete group. The *group algebra* $\mathbb{C}[G]$ is the collection of all formal linear combinations

$$a = \sum_{g \in G} a_g g,$$

where the coefficients a_g are in $\mathbb{C}$, and only finitely many of them are non-zero. The group algebra is equipped with pointwise linear structure, with the $*$-operation defined by

$$\Big(\sum_{g \in G} a_g g\Big)^* := \sum_{g \in G} \overline{a_g} g^{-1},$$

and with multiplication induced by multiplication in the group in the natural way, i.e.

$$\Big(\sum_{g \in G} a_g g\Big)\Big(\sum_{h \in G} b_h h\Big) := \sum_{g,h \in G} a_g b_h (gh).$$

Collecting terms, note that we can write this product in the standard form of an element of $\mathbb{C}[G]$: indeed,

$$\sum_{g,h \in G} a_g b_h (gh) = \sum_{k \in G} \Big(\sum_{g \in G} a_g b_{g^{-1}k}\Big) k.$$

A representation

$$\pi : G \to \mathcal{B}(H), \quad g \mapsto \pi_g$$

(i.e. a homomorphism into the invertible elements of $\mathcal{B}(H)$) is *unitary* if for all $g \in G$, π_g is a unitary operator, i.e. a norm preserving bijection. Such a representation extends to a representation of $\mathbb{C}[G]$ by linearity. Define a norm $\|\cdot\|_{\max}$ on $\mathbb{C}[G]$ to be the supremum over all semi-norms pulled back from unitary representations of G; in symbols

$$\Big\|\sum_{g \in G} a_g g\Big\|_{\max} := \sup\Big\{\Big\|\sum_{g \in G} a_g \pi_g\Big\|_{\mathcal{B}(H)} \ \Big| \ \pi : G \to \mathcal{B}(H) \text{ a unitary representation}\Big\}.$$

This is finite: indeed, the fact that π_g is unitary implies that $\|\pi_g\|_{\mathcal{B}(H)} = 1$ for all $g \in G$, and so

$$\left\| \sum_{g \in G} a_g \pi_g \right\|_{\mathcal{B}(H)} \leqslant \sum_{g \in G} |a_g|,$$

i.e. there is a uniform bound of the numbers that we are taking the supremum of. One can check that $\| \cdot \|_{\max}$ is a norm rather than just a semi-norm using that the left-translation action of G on $\ell^2(G)$ induces a faithful representation $\mathbb{C}[G] \to \mathcal{B}(\ell^2(G))$.

The *group C^*-algebra* of G, denoted $C^*_{\max}(G)$, is defined to be the completion of $\mathbb{C}[G]$ for the norm above. Using that $\mathcal{B}(H)$ is a C^*-algebra for any Hilbert space H, it is not too difficult to check that $C^*_{\max}(G)$ is indeed a C^*-algebra.

As C^*-algebras are not always unital, it is often convenient to adjoin a unit as in the following definition (we justify it below).

Definition 1.1.9 Let A be a complex algebra. Its *unitisation* is the complex algebra denoted[1] A^+ with underlying vector space $A \oplus \mathbb{C}$ and multiplication defined by

$$(a,\lambda)(b,\mu) := (ab + \lambda b + \mu a, \lambda\mu).$$

If A is a $*$-algebra, then A^+ is made a $*$-algebra via the definition by $(a,\lambda)^* := (a^*, \overline{\lambda})$.

If A is a non-unital C^*-algebra, then A^+ is equipped with the norm defined by

$$\|(a,\lambda)\| := \sup_{b \in A, \|b\| \leqslant 1} \|ab + \lambda b\|_A.$$

If A is already unital, then A^+ can also be equipped with a (unique) C^*-algebra norm for which it is isomorphic to $A \oplus \mathbb{C}$ as a C^*-algebra: see Exercise 1.9.1.

Lemma 1.1.10 *If A is a non-unital C^*-algebra, then the unitisation A^+ is a unital C^*-algebra, and the natural inclusion*

$$A \to A^+, \quad a \mapsto (a,0)$$

identifies A isometrically with a $*$-subalgebra of A^+.

[1] We use the symbol A^+ by analogy with the one point compactification X^+ of a space. It should not be confused with the collection of positive elements of A as in Definition 1.1.11 below!

Proof It is clear that A^+ is a $*$-algebra. Note that for any non-zero $a \in A$ we have

$$\|(a,0)\| = \sup_{b \in A,\, \|b\| \leqslant 1} \|ab\| \leqslant \|a\|$$

and also using that $\|a\| = \|a^*\|$ and the C^*-identity

$$\|(a,0)\|| = \sup_{b \in A,\, \|b\| \leqslant 1} \|ab\| \geqslant \left\| a\frac{a^*}{\|a\|} \right\| = \|a\|.$$

It follows that A identifies isometrically with the collection $\{(a,0) \in A^+ | a \in A\}$ in the obvious way; clearly this identification also preserves the $*$-algebra operations.

We claim now that the given formula defines a submultiplicative norm on A^+. Almost all of this just involves writing down the definitions. The trickiest point is seeing why any element of zero norm must be zero. For this, note that if $\|(a,\lambda)\| = 0$, then for any $b \in A$, $ab + \lambda b = 0$; if $\lambda = 0$, this is impossible unless $a = 0$ on taking $b = a^*$ by the argument above. On the other hand, if $\lambda \neq 0$, it follows that $(-\frac{1}{\lambda}a)b = b$ for all $b \in A$ and taking adjoints gives $b(-\frac{1}{\lambda}a^*) = b$ for all $b \in A$. Hence $-\frac{1}{\lambda}a = -\frac{1}{\lambda}a^*$, and this element is a unit for A, contradicting that A is non-unital.

Note moreover that the norm on A^+ must also be complete: indeed, we have seen that it identifies with the norm of A on the subspace $\{(a,0) \in A^+ \mid a \in A\}$, and is therefore complete when restricted to this subspace. As this subspace has finite codimension (actually, codimension one), it is complete on all of A^+.

Finally, to check that A^+ is a C^*-algebra, we need to check the C^*-identity: this follows as if we think of $(a,\lambda) \in A^+$ as the operator $\mathrm{Op}_{a,\lambda}$ of left multiplication by $a + \lambda$ on A, then

$$\|(a,\lambda)\|^2 = \sup_{b \in A,\, \|b\| \leqslant 1} \|ab + \lambda b\|^2 = \sup_{b \in A,\, \|b\| \leqslant 1} \|b^*(a^*a + \lambda a^* + \overline{\lambda} a + |\lambda|^2)b\|,$$

where the second equality is the C^*-identity; the element inside the norm on the right should be thought of as shorthand for the element of A one gets on multiplying it out. Continuing, this is bounded above by

$$\sup_{b \in A,\, \|b\| \leqslant 1} \|(aa^* + \lambda + \overline{\lambda} + |\lambda|^2)b\| = \|(a,\lambda)^*(a,\lambda)\|;$$

as the inequality $\|(a,\lambda)\|^2 \geqslant \|(a,\lambda)^*(a,\lambda)\|$ follows from submultiplicativity of the operator norm and the fact that $*$ is an isometry, we are done. □

One should think of the element (a,λ) of A^+ defined above as the sum $a + \lambda$, and we will typically just write it like that; having identified A and $\mathbb{C}$

with the subalgebras $\{(a,0) \in A^+ \mid a \in A\}$ and $\{(0,\lambda) \in A^+ \mid \lambda \in \mathbb{C}\}$ of A^+ respectively, this makes sense.

There are many types of elements of C^*-algebras that play special roles and have special names. The most important are as follows.

Definition 1.1.11 An element of a C^*-algebra A is:

(i) *self-adjoint* if $a = a^*$;
(ii) *normal* if $a^*a = aa^*$;
(iii) *positive* if it equals b^*b for some $b \in A$;
(iv) a *contraction* if $\|a\| \leqslant 1$;
(v) an *idempotent* if $a^2 = a$;
(vi) a *projection* if $p^2 = p$ and $p = p^*$;
(vii) a *partial isometry* if $vv^*v = v$ and $v^*vv^* = v^*$.

If, in addition, A has an identity element 1, an element of A is:

(viii) *invertible* if it has a multiplicative inverse;
(ix) *unitary* if $uu^* = 1 = u^*u$;
(x) an *isometry* if $v^*v = 1$;
(xi) a *co-isometry* if $vv^* = 1$.

The terminology is mainly motivated by the roles that operators satisfying these conditions play in the special case that $A = \mathcal{B}(H)$ (or just $A = M_n(\mathbb{C})$): for example, an operator in $\mathcal{B}(H)$ is a projection if and only if it is the orthogonal projection onto a closed subspace of H; it is an isometry if and only if it preserves norms; and it is unitary if and only if it is bijective and preserves norms.

Our final task in this section is to introduce the notion of morphisms most appropriate to C^*-algebras.

Definition 1.1.12 A $*$-*homomorphism* between two C^*-algebras (or more generally, two $*$-algebras) is an algebra homomorphism $\phi\colon A \to B$ that satisfies $\phi(a^*) = \phi(a)^*$ for all $a \in A$. A $*$-*isomorphism* is an invertible $*$-homomorphism.

In the special case that $B = \mathcal{B}(H)$ is the bounded operators on some Hilbert space, a $*$-homomorphism $\pi : A \to B$ will typically be called a $*$-*representation*, or just a *representation*.

Throughout this book, we will work mainly in the category with objects C^*-algebras, and morphisms being $*$-homomorphisms. By default, maps between C^*-algebras will generally be assumed to be $*$-homomorphisms, and we will often say 'homomorphism' rather than '$*$-homomorphism' for brevity.

There are two natural conditions that are not demanded by the definition: firstly, that ϕ be unital when A and B have units; secondly, that ϕ be continuous for the norm. The first of these is omitted simply because the extra generality is useful. The second is omitted as it turns out to be automatic: indeed we will see below (see Corollary 1.2.8) that $*$-homomorphisms between C^*-algebras are always contractive, i.e. satisfy $\|\phi(a)\| \leqslant \|a\|$ for all a in the domain.

Remark 1.1.13 Any $*$-homomorphism $\phi\colon A \to B$ between possibly non-unital C^*-algebras extends uniquely to a unital $*$-homomorphism $\phi^+\colon A^+ \to B^+$ between their unitisations via the formula $\phi^+(a + \lambda) = \phi(a) + \lambda$.

1.2 Invertible Elements and Spectrum

We start with a fact relating the invertible elements in a unital C^*-algebra to the topology. A key use of this result is that it allows one to do approximation arguments in K-theory: it is thus (arguably!) the main reason that K-theory for C^*-algebras is more tractable than its purely algebraic cousin. Although we are almost exclusively interested in C^*-algebras, we sometimes work in more generality to highlight exactly what assumptions are involved.

Theorem 1.2.1 *Let A be a unital Banach algebra. Then if $a \in A$ satisfies $\|a - 1\| < 1$, a is invertible with inverse given by the norm convergent series*[2]

$$a^{-1} = \sum_{n=0}^{\infty} (1 - a)^n.$$

Moreover, the set of invertible elements in a C^*-algebra is open, and the inverse operation is continuous on this set.

Proof Write $a = 1 - b$, so $\|b\| < 1$. The sum

$$c := \sum_{n=0}^{\infty} b^n$$

converges absolutely in norm, and one has that

$$c(1 - b) = \lim_{N \to \infty} \sum_{n=0}^{N} b^n(1 - b) = \lim_{N \to \infty} 1 - b^N = 1,$$

[2] Sometimes called a *Neumann series* in this context.

and, similarly, that $(1-b)c = 1$. Hence $a = 1-b$ is invertible with inverse c. Note for later use that whenever $\|1-a\| < 1$, we have the estimate

$$\|a^{-1}\| \leqslant \sum_{n=0}^{\infty} \|b\|^n = \frac{1}{1-\|1-a\|}. \tag{1.1}$$

To see that the collection of all invertible elements is open, let $a \in A$ be invertible, and say $\|a-b\| < \|a^{-1}\|^{-1}$. Then

$$\|1-a^{-1}b\| = \|(a^{-1}(a-b)\| \leqslant \|a^{-1}\|\|a-b\| < 1,$$

whence $a^{-1}b$ is invertible, and so b is invertible.

Finally, to check continuity of the inverse operation, say (a_n) is a sequence of invertible elements converging to some invertible $a \in A$. Then we have

$$\begin{aligned} a_n^{-1} - a^{-1} &= a^{-1}(aa_n^{-1} - 1) = a^{-1}((a_n a^{-1})^{-1} - 1) \\ &= a^{-1}((1-(a-a_n)a^{-1})^{-1} - 1). \end{aligned}$$

For simplicity of notation, set $d_n := 1-(a-a_n)a^{-1}$, and so we get from the above that

$$\|a_n^{-1} - a^{-1}\| \leqslant \|a^{-1}\|\|1-d_n^{-1}\|. \tag{1.2}$$

Line (1.1) gives that for all n large enough so that $\|(a-a_n)a^{-1}\| < 1/2$ we have that

$$\|d_n^{-1}\| \leqslant \frac{1}{1-\|(a-a_n)a^{-1}\|} \leqslant 2.$$

Hence as soon as n is large enough we get from line (1.2) that

$$\begin{aligned} \|a_n^{-1} - a^{-1}\| &\leqslant \|a^{-1}\|\|1-d_n^{-1}\| \\ &\leqslant \|a^{-1}\|\|d_n\|^{-1}\|d_n - 1\| \\ &\leqslant 2\|a^{-1}\|\|(a-a_n)a^{-1}\| \\ &\leqslant 2\|a^{-1}\|^2\|a-a_n\| \end{aligned}$$

and we are done. □

We now define the spectrum of an element of a C^*-algebra.

Definition 1.2.2 Let A be a unital complex algebra and let a be an element of A. The *spectrum* of a, denoted by spec(a), is the collection of all $\lambda \in \mathbb{C}$ such that $a - \lambda$ (which is shorthand for $a - \lambda 1_A$, where 1_A is the unit of A) is not invertible.

If A is non-unital, the spectrum of $a \in A$ is defined to be its spectrum in the unitisation A^+.

Remark 1.2.3 There is an ambiguity in the notation above: if $A \subseteq B$ is a nested pair of algebras with the same unit, it is not obvious that the spectrum of an element of A is the same when it is considered as an element of A as when it is considered as an element of B. Indeed, for general (Banach) algebras, the 'spectrum relative to B' and the 'spectrum relative to A' can be different (see Exercise 1.9.2 below). Fortunately, however, for C^*-algebras spectrum cannot change under unital inclusions in this sense. This is one of the many ways in which C^*-algebras behave better than arbitrary Banach algebras: Exercise 1.9.3 leads you through a proof of this.

The following result is the first fundamental fact about the spectrum.

Theorem 1.2.4 *Let A be a unital Banach algebra, and a an element in A. Then the spectrum of a is a non-empty, compact subset of $\mathbb{C}$.*

Proof To see that the spectrum is closed, we show that its complement is open. Let then λ be such that $a - \lambda$ is invertible, and note that if $|\mu - \lambda|$ is suitably small, Theorem 1.2.1 shows that $a - \mu$ is also invertible. The spectrum of a is also bounded (in fact by $\|a\|$) as if $|\lambda| > \|a\|$ then

$$\left\| 1 - \frac{1}{\lambda}(\lambda - a) \right\| = \frac{1}{|\lambda|}\|a\| < 1,$$

and thus $\lambda - a$ is invertible by Theorem 1.2.1 again. Hence the spectrum is compact.

To show the spectrum is not empty, assume for contradiction that it is. Then we have a well-defined function

$$f_a \colon \mathbb{C} \to A, \quad \lambda \mapsto (a - \lambda)^{-1}.$$

For any bounded linear functional $\phi \colon A \to \mathbb{C}$, one can check that $\phi \circ f_a \colon \mathbb{C} \to \mathbb{C}$ is holomorphic in the usual sense for functions on $\mathbb{C}$: indeed,

$$f_a(z) - f_a(w) = (a - z)^{-1} - (a - w)^{-1} = (a - z)^{-1}(z - w)(a - w)^{-1}$$

whence

$$\frac{\phi(f_a(z)) - \phi(f_a(w))}{z - w} = \frac{z - w}{z - w}\phi((a - z)^{-1}(a - w)^{-1}),$$

and this converges to $\phi((a - z)^{-2})$ as $w \to z$ by continuity of the inverse map (Theorem 1.2.1) and of ϕ. On the other hand, whenever $|\lambda| > \|a\|$ we have

$$|\phi(f_a(\lambda))| \leqslant \|\phi\|\|(a - \lambda)^{-1}\| \leqslant \|\phi\||\lambda|^{-1}\left\|\left(\frac{1}{\lambda}a - 1\right)^{-1}\right\|.$$

This tends to zero as λ goes to infinity by continuity of the inverse operation (Theorem 1.2.1 yet again). Hence in particular, $\phi \circ f_a$ is bounded, and so constant by Liouville's theorem, so constantly equal to zero as it tends to zero at infinity. As this holds for every bounded linear functional on A, the Hahn–Banach theorem forces f_a to be constantly zero, which is impossible. □

Definition 1.2.5 For a complex algebra A and $a \in A$, define the *spectral radius* of a to be

$$r(a) := \max\{|\lambda| \mid \lambda \in \operatorname{spec}(a)\}.$$

The next fundamental result about the spectrum is called the *spectral radius formula*.

Theorem 1.2.6 *Let A be a Banach algebra. Then for any $a \in A$*

$$r(a) = \lim_{n\to\infty} \|a^n\|^{1/n}.$$

Proof Observe first that if $|\lambda| > \|a\|$, then $a-\lambda = \lambda(\frac{a}{\lambda}-1)$, which is invertible by Theorem 1.2.1. Moreover, by the spectral mapping theorem for polynomials (see Exercise 1.9.4), if $\lambda \in \operatorname{spec}(a)$, then $\lambda^n \in \operatorname{spec}(a^n)$. Hence by the first observation, $|\lambda|^n \leqslant \|a^n\|$, or in other words, we have that $|\lambda| \leqslant \|a^n\|^{1/n}$ for all n and all $\lambda \in \operatorname{spec}(a)$. It thus suffices to prove that $\limsup_{n\to\infty} \|a^n\|^{1/n} \leqslant r(a)$.

We may assume that $a \neq 0$. Set $R = 1/r(a)$ (interpreted as ∞ if $r(a) = 0$), and let D be the disk in $\mathbb{C}$ centred at 0 and of radius R.

We claim that for all $\lambda \in D$, the sequence $((\lambda a)^n)$ is bounded in A. For this, from the uniform boundedness principle it suffices to show that for any bounded linear functional ϕ on A the sequence $(\lambda^n \phi(a^n))$ is bounded (with a bound depending a priori on ϕ). Consider the function

$$f : D \to \mathbb{C}, \quad \lambda \mapsto \phi\big((1-\lambda a)^{-1}\big).$$

For $z, w \in D$, we have

$$(1-za)^{-1} - (1-wa)^{-1} = (1-za)^{-1}(z-w)a(1-wa)^{-1}$$

whence

$$\frac{f(z)-f(w)}{z-w} = \frac{z-w}{z-w}\phi((1-za)^{-1}a(1-wa)^{-1}).$$

Combined with continuity of inversion (Theorem 1.2.1) and of ϕ, this gives that

$$\frac{f(z)-f(w)}{z-w} \to \phi(a(1-za)^{-2}) \quad \text{as} \quad w \to z.$$

Hence f is holomorphic on D. On the other hand, note that if $|\lambda| < 1/\|a\|$, we have by Theorem 1.2.1 that

$$(1-\lambda a)^{-1} = \sum_{n=0}^{\infty} \lambda^n a^n.$$

Hence on the disk $D_{\|\cdot\|} := \{\lambda \in \mathbb{C} \mid |\lambda| < 1/\|a\|\}$ (which is potentially a proper subset of D) we have that

$$f(\lambda) = \sum_{n=0}^{\infty} \lambda^n \phi(a^n). \tag{1.3}$$

Hence f is given by the power series in line (1.3) for all $\lambda \in D$, by uniqueness of power series expansions. Convergence of this power series implies that $(\lambda^n \phi(a^n))$ is bounded as claimed.

Now, from the claim we have that there is $M \geqslant 0$ such that $|\lambda^n|\|a^n\| \leqslant M$ for all n, and all $\lambda \in D$. Rearranging and taking nth roots,

$$\|a^n\|^{1/n} \leqslant \frac{M^{1/n}}{|\lambda|}$$

for all n. Taking the limsup in n gives then that

$$\limsup_{n\to\infty} \|a^n\|^{1/n} \leqslant \frac{1}{|\lambda|}$$

for any $\lambda \in D$. However, D has radius $1/r(a)$, so letting λ converge to $1/r(a)$ gives the required inequality, completing the proof. □

In the remainder of this section, we specialise back to C^*-algebras.

Corollary 1.2.7 *If $a \in A$ is a normal element of a C^*-algebra then $r(a) = \|a\|$.*

Proof Working in the unitisation if necessary, we may assume that A is unital. First assume that a is self-adjoint. Then the C^*-identity from Definition 1.1.1 reduces to $\|a^2\| = \|a\|^2$. Hence Theorem 1.2.6 gives

$$r(a) = \lim_{n\to\infty} \|a^{2^n}\|^{2^{-n}} = \|a\|.$$

If a is normal, we get

$$r(a) = \lim_{n\to\infty} \|a^{2^n}\|^{2^{-n}} = \lim_{n\to\infty} \|(a^*)^{2^n} a^{2^n}\|^{2^{-n-1}} = \lim_{n\to\infty} \|(a^*a)^{2^n}\|^{2^{-n-1}},$$

where the first equality is Theorem 1.2.6, the second is the C^*-identity and the third uses normality of a to rearrange inside the norm. Applying Theorem 1.2.6 again, we get

$$\lim_{n\to\infty} \|(a^*a)^{2^n}\|^{2^{-n-1}} = r(a^*a)^{1/2}.$$

As a^*a is self-adjoint, we may apply what we proved already to get

$$r(a^*a)^{1/2} = \|a^*a\|^{1/2},$$

and the C^*-identity now finishes the proof. □

The following corollary is very important; it gets used all the time (without explicit reference).

Corollary 1.2.8 *Let $\phi\colon A \to B$ be a $*$-homomorphism between C^*-algebras. Then ϕ is contractive.*

Proof Using Remark 1.1.13, we may assume that A and B are unital, and that ϕ takes unit to unit. Writing $r(a) = \max\{|\lambda| \mid \lambda \in \operatorname{spec}(a)\}$ again, we have for any $a \in A$,

$$\|\phi(a)\|^2 = \|\phi(a)^*\phi(a)\| = \|\phi(a^*a)\| = r(\phi(a^*a)), \tag{1.4}$$

where the last step uses Corollary 1.2.7. As ϕ is a unital $*$-homomorphism, we have that the spectrum of $\phi(a^*a)$ is contained in that of a^*a, and thus $r(\phi(a^*a)) \leqslant r(a^*a)$. Combining this with line (1.4) above and another use of Corollary 1.2.7 gives

$$\|\phi(a)\|^2 \leqslant r(a^*a) = \|a^*a\| = \|a\|^2,$$

completing the proof. □

1.3 Commutative C^*-algebras

In this section, we discuss the structure of commutative C^*-algebras. Here is the basic example.

Example 1.3.1 Let X be a compact Hausdorff topological space, and let $C(X)$ denote the collection of all continuous functions $f\colon X \to \mathbb{C}$. Equipped with pointwise operations (the $*$ is pointwise complex conjugation) and the supremum norm

$$\|f\| := \sup_{x\in X} |f(x)|,$$

$C(X)$ is a C^*-algebra.

More generally, say X is locally compact and Hausdorff, and X^+ is its one point compactification (see Definition A.1.4). Then $C_0(X)$ is defined to consist of all continuous functions $f: X^+ \to \mathbb{C}$ such that f vanishes at the point at infinity; if X is not compact we can equivalently say that $C_0(X)$ consists of all continuous functions $f: X \to \mathbb{C}$ such that for all $\epsilon > 0$, the set $\{x \in X \mid |f(x)| \geqslant \epsilon\}$ is compact.

Note that if X is compact, then $C(X)$ and $C_0(X)$ are canonically isomorphic, as X^+ is just the disjoint union of X and an isolated point ∞ in this case. We sometimes use the notation '$C_0(X)$' when X might be compact, even though this is a little non-standard.

Remark 1.3.2 For any locally compact Hausdorff space X, the unitisation $C_0(X)^+$ identifies canonically with $C(X^+)$: see Exercise 1.9.6 below.

It turns out that any commutative C^*-algebra is of the form $C_0(X)$ for some canonically associated locally compact Hausdorff space X. Our goal in this section is to prove this fact, and discuss how the correspondence $X \leftrightarrow C_0(X)$ behaves as a functor.

The following result is called the *Gelfand–Mazur theorem*: it says there are no non-trivial division algebras that are also Banach algebras.

Theorem 1.3.3 *Let A be a unital Banach algebra in which every non-zero element is invertible. Then there is a unique unital isometric isomorphism from A to $\mathbb{C}$.*

Proof Let a be an element of A. Theorem 1.2.4 implies that there is some λ in the spectrum of a, i.e. so that $a - \lambda$ is not invertible. Hence $a - \lambda = 0$ by assumption, i.e. $a = \lambda$. This says that the canonical unital algebra homomorphism $\mathbb{C} \to A$ sending a scalar to that multiple of the identity is an isomorphism. It is isometric as we always assume (see Remark 1.1.2) that the identity in a Banach algebra has norm one. □

Definition 1.3.4 Let A be a unital commutative Banach algebra. The *spectrum of* A, denoted $\widehat{A}$, is the collection of non-zero multiplicative linear functionals $\phi: A \to \mathbb{C}$. We equip $\widehat{A}$ with the subspace topology that it inherits from the weak-$*$ topology on the dual.[3] A^*.

Example 1.3.5 If $A = C(X)$ for some compact Hausdorff space, then any point $x \in X$ defines an element of $\widehat{A}$ by the evaluation map $\phi_x : f \mapsto f(x)$. In fact, the natural map

$$X \to \widehat{A}, \quad x \mapsto \phi_x$$

is a homeomorphism. The reader is led through a proof of this in Exercise 1.9.7 below.

[3] We will see shortly that elements of $\widehat{A}$ are automatically bounded, so this makes sense.

In the next lemma (and also later in this section), we will need the usual notion of an *ideal* I in an algebra A: recall that an ideal is a subset that is closed under the algebra operations, and is such that if $a \in A$ and $b \in I$, then ab and ba are also in I. Associated to an ideal is the *quotient algebra* A/I, whose elements are cosets $a + I$, and with operations induced by those of A. If in addition A is a Banach algebra and I is closed in the norm topology, we may define a norm on A/I by the formula

$$\|a + I\|_{A/I} := \inf_{b \in I} \|a - b\|_A.$$

One can check that this is indeed a norm, and that it makes A/I into a Banach algebra in its own right.

Lemma 1.3.6 *A (non-zero) multiplicative linear functional $\phi\colon A \to \mathbb{C}$ on a unital Banach algebra is automatically unital and contractive.*

Proof Unitality follows as ϕ is non-zero, and as the only non-zero idempotent in $\mathbb{C}$ is the identity.

For boundedness, let $I \subseteq A$ be the kernel of ϕ, which is an ideal, and is not all of A as ϕ is non-zero. As the Banach algebra operations are continuous, the closure $\overline{I}$ is also an ideal, and as the invertible elements of A are open (Theorem 1.2.1), $\overline{I}$ is also not all of A. However, I is codimension one, whence it is a maximal ideal, and so $\overline{I}$ must equal I by maximality. As I is closed, the quotient A/I is a Banach algebra.

Moreover, the map $A/I \to \mathbb{C}$ induced by ϕ is an algebra isomorphism. Hence A/I has no non-trivial zero divisors, whence the Gelfand–Mazur theorem (Theorem 1.3.3) implies that there is a unique isometric isomorphism from A/I to $\mathbb{C}$, which must be the map induced by ϕ. As ϕ is the composition

$$A \to A/I \to \mathbb{C}$$

of the canonical quotient map and the map induced on the quotient by ϕ, and as the first of these is contractive and the second an isometry, we see that ϕ is contractive. □

Now, the above corollary says in particular that $\widehat{A}$ is a subset of the closed unit ball of A^*. It is moreover straightforward to check that it is a closed subset when the latter space is equipped with the weak-$*$ topology. Hence the following corollary is immediate from the Banach–Alaoglu theorem.

Corollary 1.3.7 *The spectrum of a unital commutative Banach algebra is a compact Hausdorff space.* □

Definition 1.3.8 Let A be a unital commutative Banach algebra. Then the *Gelfand transform* is the map

$$A \to C(\widehat{A}), \quad a \mapsto \widehat{a},$$

where $\widehat{a}(\phi) := \phi(a)$.

We leave it to the reader to check that the Gelfand transform is a well-defined unital contractive homomorphism of Banach algebras. In general, the Gelfand transform need not be injective or surjective. Nonetheless, for general unital commutative Banach algebras, the Gelfand transform still provides a powerful tool: the following corollary of the Gelfand–Mazur theorem (which we will need below) gives some evidence of why this is the case.

Corollary 1.3.9 *Let A be a unital commutative Banach algebra. Then for any $a \in A$,*

$$spec(a) = \{\phi(a) \mid \phi \in \widehat{A}\}.$$

Proof Say first that $\lambda = \phi(a)$ for some $\phi \in \widehat{A}$. Then $\phi(a - \lambda) = 0$, whence $a - \lambda$ is contained in the proper ideal Kernel(ϕ) of A, and so in particular is not invertible and so λ is contained in spec(a).

Conversely, say λ is in spec(a). As $a - \lambda$ is not invertible (and A is commutative) there is a maximal proper ideal $I \subseteq A$ containing $a - \lambda$. As I is maximal, the quotient A/I is a field, whence isomorphic to $\mathbb{C}$ by the Gelfand–Mazur theorem (see Theorem 1.3.3). Let $\phi : A \to A/I \cong \mathbb{C}$ be the composition of the quotient map and the isomorphism $A/I \cong \mathbb{C}$. Then ϕ is a multiplicative linear functional, and $\phi(a - \lambda) = 0$, whence $\phi(a) = \lambda$. □

We have now gone as far as we want to with general Banach algebras, and specialise to C^*-algebras. Here the Gelfand transform turns out to be an isometric $*$-isomorphism. The remaining key ingredient that we need is the following lemma, which will be used to show that the Gelfand transform is a $*$-homomorphism; this lemma is not true for general commutative Banach *-algebras.

Lemma 1.3.10 *Let A be a C^*-algebra, and let $a \in A$ be self-adjoint. Then spec(a) is contained in $\mathbb{R}$.*

Proof Let us assume first that A is commutative and unital. We claim that if $u \in A$ is unitary, then spec(u) is contained in $\mathbb{T} := \{z \in \mathbb{C} \mid |z| = 1\}$. Indeed, the spectral radius of u is bounded above by its norm; as the norm of a unitary is one (from the C^* identity), this gives that spec(u) is contained in

$\{z \in \mathbb{C} \mid |z| \leqslant 1\}$. As u^{-1} also has norm one, we also get that $\mathrm{spec}(u^{-1})$ is also contained in $\{z \in \mathbb{C} \mid |z| \leqslant 1\}$. However, it is straightforward to check that

$$\mathrm{spec}(u^{-1}) = \{\lambda^{-1} \mid \lambda \in \mathrm{spec}(u)\},$$

so we have that $\mathrm{spec}(u)$ is contained in

$$\{z \in \mathbb{C} \mid |z| \leqslant 1\} \cap \{z \in \mathbb{C} \mid |z^{-1}| \leqslant 1\} = \mathbb{T}$$

as required.

Let now $a \in A$ be self-adjoint. Then the power series

$$e^{ia} := \sum_{n=0}^{\infty} \frac{(ia)^n}{n!}$$

converges in norm, and elementary manipulations show that $(e^{ia})^* = e^{-ia}$, and that $e^{ia}e^{-ia} = e^{-ia}e^{ia} = 1$. Combining the first claim with Corollary 1.3.9 gives that

$$\mathbb{T} \supseteq \mathrm{spec}(e^{ia}) = \{\phi(e^{ia}) \mid \phi \in \widehat{A}\} = \{e^{i\phi(a)} \mid \phi \in \widehat{A}\} = \{e^{i\lambda} \mid \lambda \in \mathrm{spec}(a)\}$$

(the first and third equalities are from Corollary 1.3.9, and the second follows as ϕ is a multiplicative linear functional, whence continuous by Lemma 1.3.6). This is impossible unless $\mathrm{spec}(a)$ is contained in $\mathbb{R}$, so we are done in the commutative unital case.

In the general case, let A be an arbitrary C^*-algebra. As the spectrum of a is by definition the spectrum of A considered as an element of A^+, we may assume that A is unital. Let B be the C^*-subalgebra of A^+ generated by a and the unit. As a is self-adjoint, B is commutative, whence the first part of the proof gives that the spectrum of a considered as an element of B is contained in the reals. However, the spectrum of a considered as an element of A^+ is contained in the spectrum of a considered as an element of B (as it is easier to be invertible in the larger algebra A^+), so we are done. □

Here is the fundamental result about commutative C^*-algebras.

Theorem 1.3.11 *Let A be a commutative unital C^*-algebra. Then the Gelfand transform is an isometric $*$-isomorphism.*

Proof We first claim that the Gelfand transform preserves adjoints. Let then a be an element of A. Write $a_r := \frac{1}{2}(a + a^*)$ and $a_i := \frac{1}{2i}(a - a^*)$. Then a_r and a_i are self-adjoint, and $a = a_r + ia_i$. For any $\phi \in \widehat{A}$, $\phi(a) = \phi(a_r) + i\phi(a_i)$. On the other hand, using that $a^* = a_r - ia_i$ and the fact that $\phi(a_r)$ and $\phi(a_i)$ are both real (Corollary 1.3.9 and Lemma 1.3.10), we get that

$$\phi(a^*) = \phi(a_r) - i\phi(a_i) = \overline{\phi(a_r) + i\phi(a_i)},$$

which is exactly the statement that the Gelfand transform is $*$-preserving.

We next claim that the Gelfand transform is an isometry. Note that every element of A is normal in the sense of Definition 1.1.11. Hence Corollary 1.2.7 and Corollary 1.3.9 imply that

$$\|a\| = r(a) = \sup\{|\lambda| \mid \lambda \in \operatorname{spec}(a)\} = \sup\{|\phi(a)| \mid \phi \in \widehat{A}\}, \tag{1.5}$$

which is exactly the statement that the Gelfand transform is an isometry.

We are left to show that the Gelfand transform is surjective. This is now immediate from the Stone–Weierstrass theorem, however: indeed, the image of the Gelfand transform is a $*$-subalgebra of $C(X)$ that tautologically separates points, and thus is dense by Stone–Weierstrass; however, as we already know that the Gelfand transform is an isometry, its image is closed, and we are done. □

The following result can be deduced without too much difficulty from Theorem 1.3.11 and Exercise 1.9.7. We leave the details to the reader.

Theorem 1.3.12 *There is a well-defined contravariant functor from the category of compact Hausdorff topological spaces and continuous maps to the category of commutative unital C^*-algebras and unital $*$-homomorphisms defined as follows.*

(i) On objects, the functor takes X to $C(X)$.
(ii) On morphisms, the functor takes $f: X \to Y$ to the map $C(Y) \to C(X)$ defined by precomposition with f.

There is a well-defined contravariant functor from the category of commutative unital C^*-algebras and unital $*$-homomorphisms to the category of compact Hausdorff topological spaces and continuous maps defined as follows.

(i) On objects, the functor takes A to $\widehat{A}$.
(ii) On morphisms, the functor takes $\phi: A \to B$ to the map $\widehat{B} \to \widehat{A}$ defined by precomposition with ϕ. □

Moreover, these functors define a contravariant equivalence of categories. □

It will be important to us that this extends to the non-unital case. For a non-unital commutative C^*-algebra A, define the spectrum to be

$$\widehat{A} := \{\phi \in \widehat{A^+} \mid \phi(A) \neq \{0\}\}.$$

Thus $\widehat{A}$ is $\widehat{A^+}$ with the single point corresponding to the canonical quotient map $A^+ \to A^+/A \cong \mathbb{C}$ removed. In particular, $\widehat{A}$ is locally compact and Hausdorff.

Using Remark 1.3.2, it is straightforward to see that Theorem 1.3.11 implies the following.

Theorem 1.3.13 *Let A be a commutative non-unital C^*-algebra. Then the Gelfand transform for the unitisation A^+ restricts to an isometric $*$-isomorphism between A and $C_0(\widehat{A})$.* □

Using Remarks 1.1.13 and 1.3.2, Theorem 1.3.12 can be bootstrapped up to a non-unital result as follows. For the statement, consider the category $\mathcal{LC}$ with objects all locally compact Hausdorff spaces, and morphisms from X to Y being continuous maps $f\colon X^+ \to Y^+$ that send the point at infinity to the point at infinity. Again, the proof consists of direct checks that we leave to the reader.

Theorem 1.3.14 *There is a well-defined contravariant functor from the category $\mathcal{LC}$ to the category of commutative C^*-algebras and $*$-homomorphisms defined as follows.*

(i) On objects, the functor takes X to $C_0(X)$.
(ii) On morphisms, the functor takes $f\colon X^+ \to Y^+$ to the map $C_0(Y) \to C_0(X)$ defined by precomposition with f.

There is a well-defined contravariant functor from the category of commutative C^*-algebras and $*$-homomorphisms to the category $\mathcal{LC}$ defined as follows.

(i) On objects, the functor takes A to $\widehat{A}$.
(ii) On morphisms, the functor takes $\phi\colon A \to B$ to the map $\widehat{B} \to \widehat{A}$ defined by precomposition with ϕ. □

Moreover, these functors define a contravariant equivalence of categories. □

The category $\mathcal{LC}$ may seem a little strange at first, but it is a useful place to work. See Proposition A.1.8 for an alternative description.

We finish this section with some important consequences that develop Corollary 1.2.8 a bit.

Corollary 1.3.15 *Say $\phi\colon A \to B$ is an injective $*$-homomorphism between C^*-algebras. Then ϕ is isometric.*

Proof Using Remark 1.1.13, we may assume that A and B are unital, and that ϕ preserves the units. Using the C^*-identity, it suffices to prove that

$\|\phi(a)\| = \|a\|$ when $a \in A$ is self-adjoint. We then have that the C^*-algebras $C^*(a, 1)$ and $C^*(\phi(a), 1)$ generated by a and $\phi(a)$ respectively and the units are commutative; restricting to $C^*(a, 1)$, it suffices to show that an injection between commutative unital C^*-algebras is isometric.

Assume then that $\phi\colon A \to B$ is an injective $*$-homomorphism between commutative and unital C^*-algebras. As we have a contravariant equivalence of categories in Theorem 1.3.12, the injection $\phi\colon A \to B$ corresponds to a surjection $\phi^*\colon Y \to X$ of compact Hausdorff spaces: a fancy way to deduce this is to use that a contravariant equivalence of categories takes monomorphisms to epimorphisms in the categorical sense. The result follows from this as for any $f \in A = C(X)$, we have

$$\|\phi(f)\| = \sup_{y \in Y} |f(\phi^*(y))| = \sup_{x \in X} |f(x)| = \|f\|,$$

where the middle equality uses surjectivity. □

Corollary 1.3.16 *Say A is a $*$-algebra, and $\|\cdot\|_1$ and $\|\cdot\|_2$ are two norms on A, satisfying all the conditions so that $(A, \|\cdot\|_1)$ and $(A, \|\cdot\|_2)$ are C^*-algebras, except that $\|\cdot\|_2$ might not be complete. Then $\|\cdot\|_1 = \|\cdot\|_2$.*

Proof Equip A with the C^*-algebra norm $\|\cdot\|_1$, and let B denote the completion of A for the potentially non-complete $\|\cdot\|_2$, so B is also a C^*-algebra. The identity map on A can then be thought of as an injective $*$-homomorphism from $A \to B$; Corollary 1.3.15 implies it is an isometry, so we are done. □

1.4 Functional Calculus

Let A be a unital C^*-algebra, and let $a \in A$ be a normal element as in Definition 1.1.11. The C^*-subalgebra of A generated by a and the unit 1, denoted $C^*(a, 1)$, is commutative, and so by Theorem 1.3.11, it is canonically isomorphic to $C(X)$, where X is the spectrum of $C^*(a, 1)$. Note that by Corollary 1.3.9, we have a natural continuous surjective map

$$X \to \mathrm{spec}(a), \quad \phi \mapsto \phi(a).$$

As a and the unit together generate $C^*(a, 1)$, this map is also injective, whence a homeomorphism as both spaces are compact and Hausdorff. In other words, we have a canonical $*$-isomorphism

$$C^*(a, 1) \cong C(\mathrm{spec}(a)).$$

The inverse of this map is denoted

$$C(\mathrm{spec}(a)) \to C^*(a,1), \quad f \mapsto f(a). \tag{1.6}$$

The notation $f \mapsto f(a)$ is motivated by the fact that if $f = f(z,\overline{z})$ is a polynomial in the standard complex coordinate z and its conjugate $\overline{z}$, then $f(a)$ as defined above agrees with the naive notion of $f(a)$ one gets by just substituting in a for z and a^* for $\overline{z}$ in the formula for f; we leave this as an exercise for the reader.

More generally, if a is a normal element in a not-necessarily-unital C^*-algebra, then the same ideas show that the C^*-algebra $C^*(a)$ generated by a is canonically isomorphic to $C_0(\mathrm{spec}(a) \setminus \{0\})$, and we again get a $*$-isomorphism

$$C_0(\mathrm{spec}(a) \setminus \{0\}) \to C^*(a), \quad f \mapsto f(a). \tag{1.7}$$

Again, this agrees with the naive notion of $f(a)$ if f is a polynomial in z and $\overline{z}$ with no constant term: the assumption that f has no constant term ensures that $f(0) = 0$ and so, as $\mathrm{spec}(a)$ is compact, that f is an element of $C_0(\mathrm{spec}(a) \setminus \{0\})$.

Definition 1.4.1 If A is a commutative C^*-algebra, then either of the $*$-homomorphisms in lines (1.6) and (1.7) above are called the (continuous) *functional calculus* for a.

The functional calculus has the following useful continuity property.

Proposition 1.4.2 *Let A be a C^*-algebra, and let K be a compact subset of $\mathbb{C}$. Let A_K denote those normal elements of A with spectrum contained in K. Let f be any function in $C(K)$. Then the function*

$$A_K \to A, \quad a \mapsto f(a)$$

is uniformly norm continuous.

Proof Let $\epsilon > 0$. Let $p \in C(K)$ be a polynomial such that $\sup_{x \in K} |p(x) - f(x)| < \epsilon/3$. Note that the norm of any element of A_K equals its spectral radius by Corollary 1.2.7, and in particular all these norms are uniformly bounded. We may thus choose $\delta > 0$ (depending only on K and p) such that for all $a,b \in A_K$, if $\|a-b\| < \delta$, then $\|p(a)-p(b)\| < \epsilon/3$. Combining this, if $\|a-b\| < \delta$ for $a,b \in A_K$ then

$$\|f(a)-f(b)\| \leqslant \|f(a)-p(a)\| + \|p(a)-p(b)\| + \|p(b)-f(b)\| < \epsilon$$

and we are done. □

In order to apply the functional calculus, it is sometimes useful to know a little more than we currently do about the spectra of elements in a C^*-algebra. The following results are very useful in this regard.

Theorem 1.4.3 *A normal element in a C^*-algebra (unital as necessary for the definitions to make sense) is:*

1. *self-adjoint (i.e. $a = a^*$) if and only if its spectrum is contained in $\mathbb{R}$;*
2. *a projection (i.e. $a^2 = a$ and $a = a^*$) if and only if its spectrum is contained in $\{0, 1\}$;*
3. *unitary (i.e. $aa^* = a^*a = 1$) if and only if its spectrum is contained in the unit circle $\{z \in \mathbb{C} \mid |z| = 1\}$;*
4. *positive (i.e. $a = b^*b$ for some $b \in A$) if and only if its spectrum is contained in $[0, \infty)$.*

Sketch proof The first three parts follow from the functional calculus isomorphisms $C(\text{spec}(a)) \cong C^*(a, 1)$ and $C_0(\text{spec}(a) \setminus \{0\}) \cong C^*(a)$ and the corresponding properties for functions (recalling that the spectrum of a continuous function on a compact space is just its range).

For the fourth, assume first that a is normal, and the spectrum of a is contained in $[0, \infty)$. Then we may use the functional calculus to define $b = a^{1/2}$, which has the right property. The converse involves some clever tricks; we leave it to the reader to find this in the references provided at the end of this chapter. □

A useful application of the functional calculus is that it can be used to replace elements that are 'close to being projections' with actual projections, and invertible elements with unitaries as in the next two examples; this will come up when we come to discuss K-theory later.

Example 1.4.4 Say a is a normal element in a C^*-algebra A such that $\|a^2 - a\| < 1/4$. Thinking of a as a function on $\text{spec}(a)$ via the functional calculus isomorphism $C^*(a) \cong C_0(\text{spec}(a) \setminus \{0\})$, this is only possible if the spectrum of a (i.e. the range of the corresponding function on $\text{spec}(a)$) avoids the line $\text{Re}(z) = 1/2$. The characteristic function χ of the set $\{z \in \mathbb{C} \mid \text{Re}(z) > 1/2\}$ is thus continuous on the spectrum of a, and so we can form $\chi(a)$. Note that χ is a projection in $C_0(\text{spec}(a) \setminus 0)$; as the functional calculus is a $*$-isomorphism, $\chi(a)$ is a projection in $C^*(a)$ (and therefore also in A).

Example 1.4.5 Say a is an invertible element of a unital C^*-algebra A. Then a^*a is also invertible. Its spectrum is moreover contained in $[0, \infty)$ by Theorem 1.4.3 part (4.), and thus in $[c, \infty)$ for some $c > 0$ by invertibility and the fact that spectra are compact (Theorem 1.2.4). Hence $(a^*a)^{-1/2}$ makes sense. We claim that $u := a(a^*a)^{-1/2}$ is unitary. Computing,

$$u^*u = (a^*a)^{-1/2}a^*a(a^*a)^{-1/2} = 1.$$

As u is a product of invertible elements, it is invertible, so the fact that u^* is a one-sided inverse for u implies it is a two-sided inverse.

We give an alternative argument to show that $uu^* = 1$, as it involves a useful trick. First, note that if p is any polynomial and b an element of a C^*-algebra, then $bp(b^*b) = p(bb^*)b$ as one can check directly. Hence by an approximation argument, the same identity holds for any function p that is continuous on the spectra of b^*b and of bb^*. Hence we have

$$uu^* = a(a^*a)^{-1/2}(a^*a)^{-1/2}a^* = a(a^*a)^{-1}a^* = aa^*(aa^*)^{-1} = 1$$

as required.

There is also a notion of functional calculus, called the *holomorphic functional calculus*, that works for elements that are not necessarily normal (and indeed, in any Banach algebra). The functional calculus one gets is less powerful, but still very useful as it applies in great generality. We will not justify this here (partly as we use it very rarely in the main text): see the references at the end of the chapter as proof.

Theorem 1.4.6 *Let A be a unital Banach algebra. Let Ω be an open subset of $\mathbb{C}$, and let $\mathcal{H}(\Omega)$ denote the space of holomorphic functions of Ω, equipped with the topology of uniform convergence on compact subsets of Ω. Let a be an element of A with spectrum contained in Ω.*

Then there is a unique continuous unital algebra homomorphism

$$\mathcal{H}(\Omega) \to A, \quad h \mapsto h(a)$$

that sends the identity function to a. This also works in a non-unital Banach algebra, if one restricts to holomorphic functions that send zero to zero.

Moreover, let K be a compact subset of $\mathbb{C}$, and let A_K denote those elements of A with spectrum contained in K. Let Ω be any open set containing K, and let h be an element of $\mathcal{H}(\Omega)$. Then the function

$$A_K \to A, \quad a \mapsto h(a)$$

is norm continuous. □

The last theorem gives us a useful general version of Example 1.4.4.

Example 1.4.7 Say a is an element of a C^*-algebra (or just of a unital Banach algebra), and that $\|a^2 - a\| < 1/4$. Then Exercise 1.9.4 combined with the direct consequence

$$\max\{|\lambda| \mid \lambda \in \operatorname{spec}(a^2 - a)\} \leqslant \|a^2 - a\|$$

of Theorem 1.2.6 implies that $\operatorname{spec}(a)$ does not intersect the line $\operatorname{Re}(z) = 1/2$. Then the characteristic function χ of the set $\{z \in \mathbb{C} \mid \operatorname{Re}(z) > 1/2\}$ is holomorphic on the spectrum of a, and thus we may apply $\chi(a)$ to get an idempotent in A.

Note that Example 1.4.5 has no real analogue here: it already works perfectly well without assuming that the input element is normal.

1.5 Ideals and Quotients

Our main goal in this section is to see that quotients of C^*-algebras are again C^*-algebras. This fact is not obvious: the key technical ingredient is the notion of an approximate unit, which we introduce later.

When discussing C^*-algebras, we will follow the usual conventions in the literature and define ideals as follows.

Definition 1.5.1 Let A be a C^*-algebra. An *ideal* in A is a norm-closed, two-sided ideal that is stable under the $*$-operation.

Note that an ideal of a C^*-algebra means something more than the purely ring-theoretic ideals that appeared in some arguments earlier in this chapter. Occasionally, we will have need to speak of non-closed ideals in a C^*-algebra: in this case, we will say something like 'algebraic ideal' or 'not-necessarily-closed ideal'.

Remark 1.5.2 It turns out that any two-sided norm-closed ideal in a C^*-algebra is stable under the $*$-operation. This fact is not completely obvious, and it is generally easy to check $*$-closure in cases of interest, so we just include it in the definition.

Now, we want to show that if I is an ideal in a C^*-algebra A, then the usual quotient norm

$$\|a + I\|_{A/I} := \inf_{b \in I} \|a - b\|$$

from Banach space theory makes A/I into a C^*-algebra. For this we will need approximate units as in the next definition.

Definition 1.5.3 Let A be a C^*-algebra. An *approximate unit* for A is a net $(h_i)_{i\in I}$ of positive contractions such that

$$\lim_{i\in I} \|h_i a - a\| = \lim_{i\in I} \|a h_i - a\| = 0$$

for all $a \in A$. An approximate unit is *increasing*[4] if $i \leqslant j$ implies $h_i \leqslant h_j$.

Example 1.5.4 Let $A = C_0(\mathbb{R})$. For each n, let $f_n : \mathbb{R} \to [0,1]$ be a continuous function that is constantly equal to one on $[-n,n]$ and supported in $[-n-1, n+1]$. Then (f_n) is an approximate unit for A.

Example 1.5.5 Let $\mathcal{K}$ denote a copy of the compact operators on a separable, infinite-dimensional Hilbert space. Choose an orthonormal basis for the Hilbert space indexed by $\mathbb{N}$, and let p_n denote the orthogonal projection onto the span of the first n basis vectors. Then (p_n) is an approximate unit for $\mathcal{K}$.

Similarly, say H is a general Hilbert space and I is the net of finite-dimensional subspaces of H, ordered by inclusion. For $i \in I$ set p_i to be the orthogonal projection onto the subspace i. Then $(p_i)_{i\in I}$ is an approximate unit for $\mathcal{K}(H)$.

The following fundamental theorem is needed for many arguments involving non-unital C^*-algebras. The theorem makes use of dense two-sided, but not necessarily norm closed ideals in a C^*-algebra. Good examples to bear in mind are: the ideal $I = C_c(X)$ of compactly supported functions on X inside $A = C_0(X)$; and I the ideal of finite rank operators inside the compact operators $A = \mathcal{K}(H)$.

Theorem 1.5.6 *Let A be a C^*-algebra, and let I be a dense (not necessarily closed!) two-sided ideal in A. Then A has an approximate unit consisting of elements from I, and that may be chosen to be a sequence if A is separable.*

Proof We let

$$\Lambda_I := \{h \in I \mid h \geqslant 0, \text{ and } \|h\| < 1\}.$$

Note that if $h \in I$, then $1 - h$ is invertible in the unitisation A^+ and so if I_+ is the positive part of I, then the map

$$\Lambda_I \to I_+, \quad h \mapsto h(1-h)^{-1}$$

makes sense. It moreover preserves order, as follows from Exercise 1.9.10 and the formula $h(1-h)^{-1} = (1-h)^{-1} - 1$. In particular, Λ_I is a directed set as I_+ is: indeed, an upper bound for $a, b \in I_+$ is given by $a + b$. We claim that Λ_I (indexed by itself) is an increasing approximate unit for A.

[4] For self adjoint elements a and b in a C^* -algebra, '$a \leqslant b$' means by definition that $b - a$ is positive.

To see this, note that as the positive elements span A (Exercise 1.9.8), it suffices to show that

$$(1-h)a \to 0 \quad \text{as} \quad h \to \infty \quad \text{in} \quad \Lambda_I, \tag{1.8}$$

whenever $a \in A$ is positive and satisfies $\|a\| < 1$. Moreover, the positive elements I_+ of I are dense in the positive elements A_+ of A: this follows as we can write a general element of A_+ as a^*a, and can then approximate a by an element b from I, whence b^*b is a positive element of I approximating a^*a. Hence to show the condition in line (1.8), it suffices to show that for any $\epsilon > 0$ there exists $h \in \Lambda_A := \{a \in A_+ \mid \|a\| < 1\}$ such that $\|(1-h)a\| < \epsilon$. Using that $a \geqslant 0$, and $\|h\| < 1$, the functional calculus shows us that $h = a^{1/n}$ will work for suitably large n.

We leave the statement about separability as an exercise for the reader. □

Remark 1.5.7 It is clear from the definition of a C^*-algebra that an ideal in a C^*-algebra is a C^*-algebra in its own right. In particular, Theorem 1.5.6 shows that any ideal in a C^*-algebra contains an approximate unit for itself.

One of the most important consequences of the existence of approximate units is that they allow us to show that quotients of C^*-algebras are C^*-algebras. The following lemma gives a useful formula for the quotient norm.

Lemma 1.5.8 *Let A be a C^*-algebra and I an ideal in A. Let $(h_j)_{j\in J}$ be an approximate unit for I. Then for any $a \in A$,*

$$\|a + I\|_{A/I} = \lim_j \|a - ah_j\|.$$

Proof Working instead inside the unitisation of A, we may assume that A is unital. Then for any $a \in A$, any $b \in I$, and any $j \in J$,

$$\|a - ah_j\| \leqslant \|(a+b)(1-h_j)\| + \|b - bh_j\|.$$

As each h_j is a positive contraction, we have that $\|1 - h_j\| \leqslant 1$. Hence the above inequality implies

$$\|a - ah_j\| \leqslant \|a+b\| + \|b - bh_j\|,$$

and thus

$$\limsup_j \|a - ah_j\| \leqslant \|a+b\|.$$

Taking the infimum over all $b \in I$ gives

$$\limsup_j \|a - ah_j\| \leqslant \|a + I\|_{A/I}.$$

On the other hand, as I is an ideal, ah_j is in I for all j and so we get $\|a + I\|_{A/I} \leqslant \|a - ah_j\|$ for any j. The result follows. □

Theorem 1.5.9 *Let A be a C^*-algebra, and I an ideal in A. Then when equipped with the quotient norm*

$$\|a + I\|_{A/I} := \inf_{b \in I} \|a - b\|,$$

A/I is a C^*-algebra.[5]

Proof We leave the checks that A is a Banach *-algebra (which follow readily from general facts from algebra and Banach space theory) as an exercise for the reader; it remains to check the C^*-identity. For this, we may again assume that A is unital by working in its unitisation. Let $(h_j)_{j \in J}$ be an approximate unit for I (which exists by Remark 1.5.7). Then for any $a \in A$, Lemma 1.5.8 implies that

$$\begin{aligned} \|a + I\|^2_{A/I} &= \lim_j \|a(1 - h_j)\|^2_A = \lim_j \|(1 - h_j)a^*a(1 - h_j)\|_A \\ &\leqslant \lim_j \|a^*a(1 - h_j)\|_A = \|a^*a + I\|_{A/I} = \|(a + I)^*(a + I)\|_{A/I}. \end{aligned}$$

Using that the $*$-operation is an isometry on A as in Remark 1.1.3 and preserves I, we see that the $*$-operation is an isometry on A/I. We thus get

$$\|(a + I)^*(a + I)\|_{A/I} \leqslant \|a + I\|_{A/I}\|(a + I)^*\|_{A/I} = \|a + I\|^2_{A/I}$$

and are done. □

The following corollary gets used all the time without explicit reference.

Corollary 1.5.10 *Let $\phi: A \to B$ be a $*$-homomorphism of C^*-algebras. Then the image $\phi(A)$ is a closed C^*-subalgebra of B.*

Proof The image is isomorphic as a $*$-algebra to A/I, where I is the kernel of ϕ. A priori, it has two different norms: the quotient norm, and the norm it inherits as a sub–*-algebra of B. The quotient norm is complete, but the subalgebra norm in principle may not be. However, using Corollary 1.3.16, the two norms are the same. Hence the norm $\phi(A)$ inherits as a sub–C^*-algebra of B is also complete, and thus $\phi(A)$ is closed. □

In the commutative case, ideals and quotients can be characterised directly in terms of the associated topological space. This is crucial for applications of C^*-algebra theory to geometry and topology, so we give a proof.

[5] Called the *quotient C^*-algebra.*

Theorem 1.5.11 *Let $A = C_0(X)$ be a commutative C^*-algebra. Then for any open subset U of X, $C_0(U)$ canonically identifies with an ideal in $C_0(X)$ in such a way that if $F := X \setminus U$ is the closed complement, then there is a canonical short exact sequence*

$$0 \longrightarrow C_0(U) \longrightarrow C_0(X) \longrightarrow C_0(F) \longrightarrow 0 .$$

Moreover, for any ideal I in $C_0(X)$, there is a canonically associated open subset U of X such that I identifies with $C_0(U)$.

Proof Let U be an open subset of X, and let I be the collection of all $f \in C_0(X)$ that vanish on the complement of U in the one-point compactification X^+. It is then straightforward to see that I is a (closed, $*$-closed, two-sided) ideal, and that restriction of functions from the one-point compactification X^+ to the subspace U^+ identifies I with $C_0(U)$. Moreover, if $F \subseteq X$ is the closed complement of U, then the restriction map $C_0(X) \to C_0(F)$ is surjective by the Tietze extension theorem (applied to the closed subspace F^+ of X^+), and clearly has kernel $C_0(U)$, giving the short exact sequence in the statement.

It remains to show that any ideal I in $C_0(X)$ canonically identifies with an ideal of the form $C_0(U)$. Identify X with the collection $\widehat{C_0(X)}$ of non-zero multiplicative linear functionals $\phi \colon C_0(X) \to \mathbb{C}$ as in Theorem 1.3.14. Let $U = \{\phi \in \widehat{C_0(X)} \mid \phi(I) \neq \{0\}\}$. Then it is clear that $\widehat{C_0(X)} \setminus U$ is closed, whence U is an open subset of X. Moreover, I is a commutative C^*-algebra, so to complete the proof, it suffices by Theorem 1.3.14 to show that restrictions of elements of U to I are precisely the elements of $\widehat{I}$. For this, it suffices to show that any $\phi \in \widehat{I}$ extends uniquely to an element $\widetilde{\phi}$ of $\widehat{C_0(X)}$.

For this last statement, let $\phi \colon I \to \mathbb{C}$ be an element of $\widehat{I}$. As ϕ is non-zero and linear, it is surjective, so there is some element $a_1 \in I$ such that $\phi(a_1) = 1$. Define $\widetilde{\phi} \colon C_0(X) \to \mathbb{C}$ by $\widetilde{\phi}(a) := \phi(aa_1)$, which makes sense as a_1 is in I, and ϕ is an ideal. It is straightforward to see that $\widetilde{\phi}$ is linear, and multiplicativity follows as if $a, b \in C_0(X)$, then

$$\widetilde{\phi}(ab) = \phi(aba_1) = \phi(aba_1)\phi(a_1) = \phi(aba_1^2) = \phi(aa_1)\phi(ba_1) = \widetilde{\phi}(a)\widetilde{\phi}(b),$$

where the first and last equalities are by definition of $\widetilde{\phi}$, the second uses that $\phi(a_1) = 1$, the third and fourth use multiplicativity of ϕ, and the fourth also uses that $C_0(X)$ is commutative. It is clear that any multiplicative linear extension of ϕ satisfies the formula defining $\widetilde{\phi}$, so we are done. □

To conclude this section, we give another example of a construction one can perform on C^*-algebras.

Example 1.5.12 Let $(A_i)_{i\in I}$ be a collection of C^*-algebras, where I is a directed set (in particular, I could be the natural numbers). Assume moreover that for each $i \leqslant j$ there is a $*$-homomorphism $\phi_{ji}\colon A_i \to A_j$ such that each ϕ_{ii} is the identity, and such that if $i \leqslant j \leqslant k$, then $\phi_{kj} \circ \phi_{ji} = \phi_{ki}$. This data is called a *directed system* of C^*-algebras indexed by I. In fancy language, a directed system of C^*-algebras is a functor from I considered as a category to the category of C^*-algebras and $*$-homomorphisms.

Given a directed system $(A_i)_{i\in I}$ as above (we follow the usual convention of leaving the maps ϕ_{ij} implicit), we may form the *direct limit* C^*-algebra $\lim\limits_{i\in I} A_i$ as follows. First, note that the direct sum C^*-algebra $\bigoplus_{i\in I} A_i$ is naturally an ideal in the direct product C^*-algebra $\prod_{i\in I} A_i$ (see Example 1.1.7 for notation), so that one can take the quotient C^*-algebra

$$B := \prod_{i\in I} A_i \Big/ \bigoplus_{i\in I} A_i.$$

For each $i \in I$, there is a natural $*$-homomorphism $\phi_i\colon A_i \to B$ defined by setting the component of $\phi_i(a)$ in A_j to be $\phi_{ji}(a)$ if $j \geqslant i$, and zero otherwise. The direct limit $\lim\limits_{i\in I} A_i$ can then be defined as the C^*-subalgebra of B generated by $\phi_i(A_i)$ for all i.

The direct limit $A := \lim\limits_{i\in I} A_i$ has the following universal property: if C is any other C^*-algebra equipped with a collection of $*$-homomorphisms $\psi_i\colon A_i \to C$ such that the diagrams

$$\begin{array}{ccc} A_i & \xrightarrow{\phi_{ji}} & A_j \\ \downarrow{\scriptstyle \psi_i} & & \downarrow{\scriptstyle \psi_j} \\ C & = & C \end{array}$$

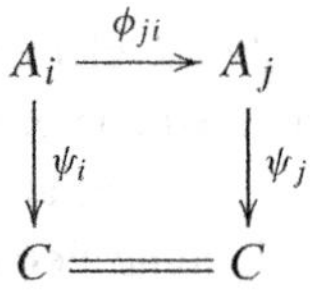

commute, then there is a unique $*$-homomorphism $\phi\colon A \to C$ such that the diagrams

$$\begin{array}{ccc} A_i & \xrightarrow{\phi_i} & A \\ & \searrow^{\psi_i} & \downarrow{\scriptstyle \phi} \\ & & C \end{array}$$

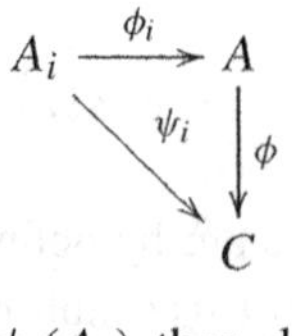

commute. Indeed, if $\phi_i(a) \in \phi_i(A_i)$ then define $\phi(\phi_i(a)) := \psi_i(a)$; it is not difficult to see that this gives a $*$-homomorphism on the densely defined $*$-subalgebra $\bigcup_{i\in I} \phi_i(A_i)$ that extends to the required map $\phi\colon A \to C$.

In the special case that the family $(A_i)_{i\in I}$ consists of C^*-subalgebras of some fixed C^*-algebra B, ordered by inclusion, then the direct limit admits a simpler description: it is just the closure of the union $\bigcup_{i \in I} A_i$ inside B. As a concrete example, consider the directed system

$$M_1(\mathbb{C}) \to M_2(\mathbb{C}) \to M_3(\mathbb{C}) \to \cdots,$$

where each map is the top left inclusion map $a \mapsto \begin{pmatrix} a & 0 \\ 0 & 0 \end{pmatrix}$. We may view each $M_n(\mathbb{C})$ as acting on $\ell^2(\mathbb{N})$ via the usual action on $\ell^2(\{1,\ldots,n\})$, and the zero action on the orthogonal complement $\ell^2(\{n+1,n+2,\ldots\})$. These representations are compatible with the inclusions, and it is not difficult to see that the union is dense in $\mathcal{K}(\ell^2(\mathbb{N}))$. Thus we have $\mathcal{K} = \lim_{n \in \mathbb{N}} M_n(\mathbb{C})$.

1.6 Spatial Theory

Our goal in this section is to discuss C^*-algebras as they arise as C^*-subalgebras of the bounded operators on some Hilbert space.

Definition 1.6.1 A *concrete C^*-algebra* is a C^*-subalgebra of the bounded operators $\mathcal{B}(H)$ on some Hilbert space.

Example 1.6.2 Let G be a group, and $\pi: G \to \mathcal{B}(H)$ be a unitary representation as in Example 1.1.8 above. Then we may take the C^*-algebra generated by the image $\{\pi_g \mid g \in G\}$ of this representation. Particularly important examples are the C^*-algebras generated by the left and right regular representations on $\ell^2(G)$ defined respectively by

$$\lambda_g : \delta_h \mapsto \delta_{gh} \quad \text{and} \quad \rho_g : \delta_h \mapsto \delta_{hg^{-1}}.$$

The C^*-algebras these generate are denoted by $C^*_\lambda(G)$ and $C^*_\rho(G)$ respectively. The unitary isomorphism

$$U : \ell^2(G) \to \ell^2(G), \quad \delta_g \mapsto \delta_{g^{-1}}$$

conjugates one into the other,[6] and thus they are $*$-isomorphic. Either of them is usually called the *reduced group C^*-algebra of G*. In the literature, $C^*_\lambda(G)$ is probably the default option when one wants to consider a concrete copy of the reduced group C^*-algebra, although our conventions in this text force us to favour $C^*_\rho(G)$.

Example 1.6.3 Say X is a locally compact Hausdorff space, and equip X with a measure μ which is positive on every non-empty open subset. Then $H = L^2(X,\mu)$ is a Hilbert space, and we can realise $C_0(X)$ as a concrete C^*-algebra

[6] In symbols: $UC^*_\lambda(G)U^* = C^*_\rho(G)$.

on $\mathcal{B}(H)$ acting by multiplication operators: the condition that the measure is positive on open sets guarantees that the norm that $f \in C_0(X)$ inherits from $\mathcal{B}(H)$ is the same as its supremum norm.

Throughout this book, we will see several more examples of C^*-algebras introduced as concrete algebras of operators on some Hilbert space in this way. In fact, *any* C^*-algebra is $*$-isomorphic to a concrete C^*-algebra of operators on some Hilbert space. We will not need any ideas from the proof of this fact in this book so do not discuss it. Before stating a precise version, we give some definitions relating to representations of C^*-algebras.

Definition 1.6.4 Let A be a $*$-algebra. A *representation* of A is a $*$-homomorphism $\pi : A \to \mathcal{B}(H)$ from A to the C^*-algebra of bounded operators on some Hilbert space. A representation $\pi : A \to \mathcal{B}(H)$ is:

(i) *non-degenerate* if whenever $v \in H$ is such that $\pi(a)v = 0$ for all non-zero $a \in A$, we have that $v = 0$;
(ii) *faithful* if π is injective;[7]
(iii) *ample* if no non-zero element of A acts as a compact operator.

Remark 1.6.5 Say A is a unital C^*-algebra. Then a representation $\pi : A \to \mathcal{B}(H)$ is non-degenerate if and only if it is unital. Indeed, if π is unital then it is clearly non-degenerate. On the other hand, if π is not unital, then $P := \pi(1)$ is a non-identity projection on H. If v is any non-zero vector in $(1-P)H$, then $\pi(a)v = \pi(a1)v = \pi(a)Pv = 0$ for all $a \in A$, so π is not non-degenerate.

Note that any representation of the zero C^*-algebra is non-degenerate by our definition (this is not just a curiosity: slightly irritatingly, we will need to use this fact at a couple of points below).

Remark 1.6.6 Let A be a C^*-algebra and $\pi : A \to \mathcal{B}(H)$ a faithful representation. Then Remark 1.3.15 implies that π is isometric. Thus $\pi(A)$ is just a copy of A with all the same algebraic and metric structure.

Remark 1.6.7 Let $\pi : A \to \mathcal{B}(H)$ be a representation of a C^*-algebra. We claim that π is non-degenerate if and only if the subspace

$$\pi(A)H := \{\pi(a)v \mid a \in A, v \in H\}$$

of H is dense. Indeed, if $\pi(A)H$ is dense, let $v \in H$ be such that $\pi(a)v = 0$ for all $a \in A$. Then by our density assumption, for any $\epsilon > 0$ we may find

[7] If A is a C^*-algebra and π is a $*$-homomorphism, this is equivalent to saying that π takes non-zero positive elements to non-zero positive elements; the latter condition is the 'correct' definition of 'faithful' for certain more general classes of maps between C^*-algebras.

$w \in H$ and $b \in A$ such that $\|v - \pi(b)w\| < \epsilon$. Let (h_i) be an approximate unit for A (see Definition 1.5.3). Then we get that

$$\epsilon > \|\pi(h_i)\|\|v - \pi(b)w\| \geqslant \|\pi(h_i)(v - \pi(b)w)\| = \|\pi(h_i b)w\| \to \|\pi(b)w\|.$$

Hence $\|v\| \leqslant \|\pi(b)w\| + \epsilon \leqslant 2\epsilon$, and as ϵ was arbitrary, $v = 0$.

Conversely, say π is non-degenerate. Let H_0 be the orthogonal complement of $\pi(A)H$. It suffices to show that H_0 is zero. Let then v be an element of H_0, and let a be an element of A. Then for any $a \in A$,

$$\|\pi(a)v\|^2 = \langle \pi(a)v, \pi(a)v \rangle = \langle v, \pi(a^*a)v \rangle,$$

which is zero as $\pi(a^*a)v$ is in $\pi(A)H$. Hence $\pi(a)v = 0$ for all $a \in A$, and by non-degeneracy this forces $v = 0$ as we wanted.

The following is the celebrated Gelfand–Naimark theorem. We do not need the ideas behind the proof in this book, so we will not go into them here.

Theorem 1.6.8 *Let A be a C^*-algebra. Then there exists a faithful (hence isometric by Remark 1.6.6) non-degenerate representation on some Hilbert space. Moreover, this Hilbert space can be chosen to be separable if A is.* □

Remark 1.6.9 Say $\pi : A \to \mathcal{B}(H)$ is a faithful non-degenerate representation. Let $H^{\oplus\infty}$ be the countably infinite direct sum of copies of H: precisely, this means one takes the algebraic direct sum of countably infinitely many copies of H, so elements are sequences $(v_n)_{n=1}^{\infty}$ with each v_n in H and only finitely many non-zero, and completes with respect to the metric induced by the inner product

$$\langle (v_n), (w_n) \rangle := \sum_{n=0}^{\infty} \langle v_n, w_n \rangle_H.$$

Equivalently, one could define $H^{\oplus\infty}$ to equal the space $\ell^2(\mathbb{N}, H)$ of square-summable functions from $\mathbb{N}$ to H with the natural inner product. Then there is a countable infinite direct sum representation $\pi^{\oplus\infty} : A \to \mathcal{B}(H^{\oplus\infty})$ defined to act by $\pi(a)$ in each component. This new representation will still be faithful and non-degenerate. It will moreover be ample, as no operator 'repeated' or 'amplified' infinitely many times in this way can be compact unless it is zero.

In summary, Theorem 1.6.8 implies that any C^*-algebra has ample non-degenerate representations.

Here is a sample application of Theorem 1.6.8: although this may seem at first like it 'should' be elementary, no way of justifying it that does not essentially go through Theorem 1.6.8 above seems to be known.

Example 1.6.10 Let A be a C^*-algebra. Then the $*$-algebra of $n \times n$ matrices $M_n(A)$ over A admits a norm (which is unique by Corollary 1.3.16) making it a C^*-algebra. Indeed, choose a faithful (non-degenerate) representation of A on some Hilbert space H. Then $M_n(A)$ is naturally represented on the direct sum $H^{\oplus n}$ of n copies of H, and so inherits a C^*-algebra norm from $\mathcal{B}(H^{\oplus n})$.

Our last goal in this section is a result about extending representations of $C_0(X)$ to the C^*-algebra of bounded Borel functions on X. This will be useful in its own right, and also to deduce the existence of the Borel functional calculus. To state it, recall that a sequence (or a net) (T_n) of bounded operators on a Hilbert space H converges *strongly* to a bounded operator T if for every $v \in H$ we have that $\|T_n v - Tv\| \to 0$ as n tends to infinity.

Proposition 1.6.11 *Let X be a second countable locally compact Hausdorff space, and let*

$$\pi : C_0(X) \to \mathcal{B}(H)$$

be a non-degenerate representation of $C_0(X)$ on some Hilbert space H. Then there exists a canonical extension of π to a unital representation

$$\pi : B(X) \to \mathcal{B}(H)$$

of the C^*-algebra of bounded Borel functions on X.

Moreover, this extension has the property that if (f_n) is a uniformly bounded sequence of Borel functions converging pointwise to a Borel function f, then $(\pi(f_n))$ converges strongly to f.

Proof Let $v \in H$ be a vector, and consider the bounded linear functional

$$\mu_v : C_0(X) \to \mathbb{C}, \quad f \mapsto \langle v, \pi(f)v \rangle.$$

The Riesz representation theorem implies that μ_v corresponds to a unique finite, positive, Borel measure on X, which we also denote by μ_v.

Fix now a bounded Borel function $f \in B(X)$. Then for each $v \in H$, the map

$$\phi_f : H \to \mathbb{C}, \quad u \mapsto \frac{1}{4}\sum_{k=0}^{3} i^k \int_X f \, d\mu_{(u+i^k v)}$$

(the formula is inspired by the polarisation identity) defines a bounded linear functional on H. Hence there is a unique vector that we call $\pi(f)v$ such that $\phi_f(u) = \langle \pi(f)v, u \rangle$ for all $u \in H$. It now follows from direct checks that this prescription defines a bounded linear operator $\pi(f) : H \to H$ that agrees with the original definition of $\pi(f)$ when $f \in C_0(X)$, and moreover that the corresponding map

$$\pi : B(X) \to \mathcal{B}(H)$$

is a unital $*$-homomorphism.

To complete the proof, it remains to check the claimed continuity property. Let then (f_n) be a uniformly bounded sequence of functions in $B(X)$ that converges pointwise to some (bounded, Borel) function $f : X \to \mathbb{C}$. Then for any $v \in H$ and any $n \in \mathbb{N}$

$$\|(\pi(f_n) - \pi(f))v\|^2 = \langle v, \pi((f_n - f)^*(f_n - f))v\rangle = \int_X |f_n - f|^2 d\mu_v.$$

The right-hand side tends to zero using the dominated convergence theorem (here we use that $\sup_n \|f_n\|$ is finite, and that μ_v is a finite measure), so we are done. □

Corollary 1.6.12 *Let $T \in \mathcal{B}(H)$ be a bounded normal operator on a Hilbert space with spectrum $X \subseteq \mathbb{C}$, and let $B(X)$ be the C^*-algebra of bounded Borel functions on X. Then there is a unique $*$-homomorphism*

$$B(X) \to \mathcal{B}(H), \quad f \mapsto f(T),$$

called the Borel functional calculus, *that takes the identity function to T, and that takes bounded pointwise convergent sequences of functions to strongly convergent sequences of operators.*

Proof The usual functional calculus gives a unique $*$-homomorphism

$$C(X) \to \mathcal{B}(H), \quad f \mapsto f(T)$$

subject to the condition that the identity function $X \to \mathbb{C}$ goes to T. Proposition 1.6.11 extends this to a $*$-homomorphism $B(X) \to \mathcal{B}(H)$ with the claimed 'pointwise-to-strong' continuity property. Uniqueness follows as for a compact subset of $\mathbb{C}$, the bounded Borel functions $B(X)$ are the smallest class of functions containing $C(X)$, and closed under pointwise limits of bounded sequences. □

1.7 Multipliers and Corners

In this section, we use the spatial theory from the last section to discuss multiplier algebras, and use these to give a definition of corners and Morita equivalence.

Definition 1.7.1 Let $A \subseteq \mathcal{B}(H)$ be a concrete C^*-algebra such that the identity representation is non-degenerate. The *multiplier algebra* of A, denoted $M(A)$, is the C^*-subalgebra of those $T \in \mathcal{B}(H)$ such that

$$Ta \in A \quad \text{and} \quad aT \in A$$

for all $a \in A$.

Note that A is an ideal in $M(A)$, and that $M(A)$ is always unital. On the other hand, if A is unital, then Remark 1.6.5 implies that any non-degenerate representation of A is unital, and thus $M(A) = A$ in this case.

Definition 1.7.2 An ideal A in a C^*-algebra B is *essential* if whenever $b \in B$ is such that $ba = 0$ for all $a \in A$, then $b = 0$.

Note that Remark 1.6.7 implies that in the situation of Definition 1.7.1, A is an essential ideal in $M(A)$.

Our first task in this section is to show that $M(A)$ can be reasonably defined for any C^*-algebra, even a non-concrete one. The key point is the next result.

Proposition 1.7.3 *Let A be a C^*-algebra, let B be a C^*-algebra containing A as an ideal. Let $\pi : A \to \mathcal{B}(H)$ be a non-degenerate representation. Then there is a unique $*$-homomorphism $\widetilde{\pi} : B \to M(\pi(A))$ extending π.*

Proof Using Remark 1.6.7, the subspace $\pi(A)H = \text{span}\{\pi(a)v \mid a \in A, v \in H\}$ is dense. We will attempt to define

$$\widetilde{\pi}(b)\left(\sum_{i=1}^{n} \pi(a_i)v_i\right) := \sum_{i=1}^{n} \pi(ba_i)v_i \tag{1.9}$$

for all elements $\pi(a)v$ in $\pi(A)H$ and show that this extends to a bounded operator on all of H. Let (h_j) be an approximate unit for A. Then

$$\left\|\sum_{i=1}^{n} \pi(ba_i)v_i\right\| = \lim_j \left\|\sum_{i=1}^{n} \pi(bh_ja_i)v_i\right\| = \lim_j \left\|\pi(bh_j)\left(\sum_{i=1}^{n} \pi(a_i)v_i\right)\right\|$$

$$\leqslant \limsup_j \|bh_j\| \left\|\sum_{i=1}^{n} \pi(a_i)v_i\right\| \leqslant \|b\| \left\|\sum_{i=1}^{n} \pi(a_i)v_i\right\|.$$

This computation shows that $\pi(b)$ as in line (1.9) is well-defined as a bounded linear operator on the dense subspace $\pi(A)H$ of H, and (therefore) also that it extends uniquely to a bounded linear operator defined on all of H. It is routine to check that $\widetilde{\pi}$ as in line (1.9) is a $*$-homomorphism extending π, that it is unique as such, and that it takes values in $M(\pi(A))$, so we are done. □

Corollary 1.7.4 *Let A and B be concrete C^*-algebras on H_A, H_B respectively, and let $\pi : A \to B$ be a $*$-homomorphism that takes an approximate unit for A to an approximate unit of B (e.g. π is unital, or surjective). Then there is a unique map $\widetilde{\pi} : M(A) \to M(B)$ extending π.*

Proof Uniqueness follows from non-degeneracy of π, considered as a representation of A. For existence, note that the fact that $\pi : A \to B$ takes an approximate unit to an approximate unit and non-degeneracy of the identity representation of B implies that $\pi : A \to B \subseteq \mathcal{B}(H_B)$ is non-degenerate when considered as a representation of A. Hence from Proposition 1.7.3 and the fact that A is an ideal in $M(A)$, we get a $*$-homomorphism $\widetilde{\pi} : M(A) \to M(\pi(A))$ extending π. The only issue is to show that $M(\pi(A))$ is contained in $M(B)$. Indeed, let (h_j) be an approximate unit for A such that $(\pi(h_j))$ is an approximate unit for B. Then for any $m \in M(\pi(A))$ and $b \in B$ we have

$$mb = \lim_j mh_j b.$$

As each mh_j is in $\pi(A) \subseteq B$, we have that mb is in B. Similarly bm is in B, and we are done. □

Corollary 1.7.5 *Let $\pi_1 : A \to \mathcal{B}(H_1)$ and $\pi_2 : A \to \mathcal{B}(H_2)$ be faithful non-degenerate representations of a C^*-algebra A. Let $M_1(A)$ and $M_2(A)$ be the multiplier algebras of $\pi_1(A)$ and $\pi_2(A)$ respectively. Then the map $\pi_2 \circ \pi_1^{-1} : \pi_1(A) \to \pi_2(A)$ extends uniquely to a $*$-isomorphism $M_1(A) \to M_2(A)$.*

Proof Using Proposition 1.7.3, we have a unique $*$-homomorphism $\sigma_{21} : M_1(A) \to M_2(A)$ extending the map $\pi_2 \circ \pi_1^{-1} : \pi_1(A) \to \pi_2(A)$, and similarly for $\sigma_{12} : M_2(A) \to M_1(A)$. The uniqueness clause from Proposition 1.7.3 implies that the map $\sigma_{12} \circ \sigma_{21} : M_1(A) \to M_1(A)$ (which extends the identity map on $\pi_1(A)$) is the identity, and similarly for the other composition. □

The following definition, although an abuse of terminology, makes sense up to canonical isomorphism (there are other ways to define $M(A)$ that do not depend on any choices: readers who dislike the approach below can see Section 1.10, Notes and References, at the end of the chapter).

Definition 1.7.6 Let A be a C^*-algebra. Its *multiplier algebra* $M(A)$ is defined to be the multiplier algebra of $\pi(A)$ for any faithful non-degenerate representation $\pi : A \to \mathcal{B}(H)$.

Example 1.7.7 Let X be a locally compact Hausdorff space. One can compute that the multiplier algebra of $C_0(X)$ naturally identifies with the C^*-algebra $C_b(X)$ of continuous, bounded functions $f: X \to \mathbb{C}$. Exercise 1.9.14 leads the reader through one approach to this.

Having defined multiplier algebras, we now use them to define a version of Morita equivalence.

Definition 1.7.8 Let A be a C^*-algebra, and $M(A)$ its multiplier algebra. A *corner* of A is any C^*-subalgebra of the form pAp where p is a projection in $M(A)$. A corner (or the projection defining it) is *full* if the ideal generated by p, i.e. the norm closure of the set

$$ApA := \text{span}\{apb \mid a,b \in A\},$$

is dense in A.

We conclude this section with one more important definition.

Definition 1.7.9 Two C^*-algebras A and B are *elementarily Morita equivalent* if one is a full corner in the other. Two C^*-algebras are *Morita equivalent* if there is a chain $A = A_0, \ldots, A_n = B$ of C^*-algebras with A_{i-1} and A_i elementarily Morita equivalent for each $i \in \{1, \ldots, n\}$.

It is a fact that if A and B are Morita equivalent, then there is a third C^*-algebra C containing both as full corners in a complementary way, but we will not prove that here.

Example 1.7.10 Let $M_n(A)$ be the $n \times n$ matrices over a C^*-algebra A, and let p be the element of the multiplier algebra of $M_n(A)$ that has a copy of the identity in the top-left entry. Then $pM_n(A)p$ identifies canonically with A, and $M_n(A)pM_n(A) = M_n(A)$ as one can easily check (recall here that $M_n(A)pM_n(A)$ is the *span* of the set $\{apb \mid a,b \in M_n(A)\}$, not just this set itself). Hence A and $M_n(A)$ are elementarily Morita equivalent.

1.8 Tensor Products

Our goal in this section is to give a brief introduction to the theory of the spatial tensor product of two C^*-algebras. We assume that the reader knows how to form the algebraic tensor product over $\mathbb{C}$ of two complex vector spaces (if not, see Section 1.10, Notes and References, at the end of the chapter for recommended background references). If V, W are complex vector spaces, we write $V \odot W$ for their algebraic tensor product over $\mathbb{C}$; we use '$\odot$' rather than '$\otimes$' to distinguish the algebraic tensor product from the completed tensor products that we will use later. As we will want to identify algebraic tensor products $V \odot W$ with a subspace of the relevant completed version $V \otimes W$, we still write elementary tensors in $V \odot W$ as $v \otimes w$ (sorry). The key universal property of $V \odot W$ is that if $\phi \colon V \times W \to U$ is a bilinear map, then there is a unique linear map $\Phi \colon V \odot W \to U$ such that $\Phi(v \otimes w) = \phi(v, w)$ for all elementary tensors $v \otimes w$.

If A and B are $*$-algebras, then $A \odot B$ is also a $*$-algebra in a natural way: the adjoint and multiplication are determined by the formulas

$$(a_1 \otimes b_1)(a_2 \otimes b_2) := a_1 a_2 \otimes b_1 b_2, \quad (a \otimes b)^* := a^* \otimes b^*$$

for elementary tensors. Using the universal property of $A \odot B$, it is not too difficult to check that these formulas do indeed determines a well-defined $*$-algebra structure on $A \odot B$; we leave this as an exercise for the reader (or see Section 1.10, Notes and References, at the end of the chapter).

Remark 1.8.1 Let A, B, and C be $*$-algebras. One can check that $A \odot B$ has the following universal property: if $\phi \colon A \to C$ and $\psi \colon B \to C$ are $*$-homomorphisms with commuting images, then there is a unique $*$-homomorphism $\phi \otimes \psi \colon A \odot B \to C$ satisfying $(\phi \otimes \psi)(a \otimes b) = \phi(a)\psi(b)$ on elementary tensors: see Exercise 1.9.19.

The following fundamental example will be used many times in the main text.

Example 1.8.2 Let A be a $*$-algebra and let $M_n(\mathbb{C})$ be the $*$-algebra of $n \times n$ matrices. Then if e_{ij} are the usual matrix units with a one in the (ij)th position and zeros everywhere else, the collection $(e_{ij})_{i,j=1}^n$ is a basis for $M_n(\mathbb{C})$, and so every element in $A \odot M_n(\mathbb{C})$ can be written uniquely as $\sum_{i,j=1}^n a_{ij} \otimes e_{ij}$ for some collection $(a_{ij})_{i,j=1}^n$ of elements of A. One checks directly that the map

$$A \odot M_n(\mathbb{C}) \to M_n(A), \quad \sum_{i,j} a_{ij} \otimes e_{ij} \mapsto \begin{pmatrix} a_{11} & a_{12} & \dots & a_{1n} \\ a_{21} & a_{22} & \dots & a_{2n} \\ \vdots & \vdots & \ddots & \vdots \\ a_{n1} & a_{n2} & \dots & a_{nn} \end{pmatrix}$$

is a $*$-isomorphism. In particular, if A is a C^*-algebra, then $A \odot M_n(\mathbb{C})$ can also be made into a C^*-algebra via this isomorphism and Example 1.6.10, and the resulting C^*-algebra norm is unique by Corollary 1.3.16.

We now turn to completed tensor products, starting with Hilbert spaces.

Lemma 1.8.3 *Let H and K be Hilbert spaces. Then the form on $H \odot K$ defined by*

$$\Big\langle \sum_i u_i \otimes w_i, \sum_j v_j \otimes x_j \Big\rangle := \sum_{i,j} \langle u_i, v_j \rangle_H \langle w_i, x_j \rangle_K$$

is a well-defined inner product.

Proof Let V be the vector space of all conjugate linear functionals from $H \odot K$ to $\mathbb{C}$. We define a map

$$\phi \colon H \oplus K \to V, \quad \phi(v,x) \colon \sum_i u_i \otimes w_i \mapsto \sum_i \langle u_i, v \rangle_H \langle w_i, x \rangle_K .$$

One checks directly that ϕ is bilinear, and so defines a linear map from $H \odot K$ to V by the universal property of the tensor product; we also denote this map by ϕ. One computes directly that with the definition in the statement

$$\Big\langle \sum_i u_i \otimes w_i, \sum_j v_j \otimes x_j \Big\rangle = \phi\Big(\sum_j v_j \otimes x_j\Big)\Big(\sum_i u_i \otimes w_i\Big),$$

and thus that the form in the statement is well-defined. The inner product properties are all straightforward to check, except possibly non-degeneracy. For this, let $\sum_{i=1}^n u_i \otimes w_i$ be an arbitrary element of $H \odot K$. Choosing an orthonormal basis for $\text{span}\{u_1, \ldots, u_n\}$ and expanding in terms of this basis, we may assume that $u_1, \ldots, u_n$ is an orthonormal collection. We then get that

$$\Big\langle \sum_i u_i \otimes w_i, \sum_i u_i \otimes w_i \Big\rangle = \sum_{i,j=1}^n \langle u_i, u_j \rangle_H \langle w_i, w_j \rangle_K = \sum_i \|w_i\|^2;$$

as the only way this can be zero is if all the w_i are zero, we are done. □

Definition 1.8.4 Let H and K be Hilbert spaces. The *Hilbert space tensor product* of H and K, denoted $H \otimes K$, is defined to be the completion associated to the inner product from Lemma 1.8.3 above.

The following example will be used many times (usually without reference).

Example 1.8.5 Let X be a set, let H be a Hilbert space, and let $\ell^2(X,H)$ denote the Hilbert space of square-summable functions from X to H. Then there is a canonical unitary isomorphism $\ell^2(X) \otimes H \to \ell^2(X,H)$ determined by the formula

$$u \otimes v \mapsto \big(x \mapsto u(x)v\big)$$

on elementary tensors. Checking the details of this is a good exercise in making sure one has understood the definitions above.

We now turn to bounded operators on tensor products of Hilbert spaces.

Lemma 1.8.6 *Let S and T be bounded operators on Hilbert spaces H and K. Then there is a unique bounded operator written $S \otimes T$ on $H \otimes K$ that satisfies*

$$S \otimes T : u \otimes v \mapsto Su \otimes Tv \tag{1.10}$$

on elementary tensors. Moreover, one has that $\|S \otimes T\| = \|S\|\|T\|$.

Proof It is clear that the formula in line (1.10) defines a bilinear map $H \oplus K \to H \otimes K$, and thus gives a unique linear operator $H \odot K \to H \otimes K$ by the universal property of the algebraic tensor product. We need to show first that this extends to $H \otimes K$.

We first consider the special case that $S = 1$. Let then $u \in H \otimes K$ be arbitrary, and write $u = \sum_{i=1}^{n} e_i \otimes v_i$, where $e_1, \ldots, e_n$ are orthonormal vectors in H, and $v_1, \ldots, v_n$ are some vectors in K. Then

$$\begin{aligned}\|(1 \otimes T)u\|^2 = \left\| \sum_{i=1}^{n} e_i \otimes T v_i \right\|^2 &= \sum_{i=1}^{n} \|T v_i\|^2 \\ &\leqslant \|T\|^2 \sum_{i=1}^{n} \|v_i\|^2 = \|T\|^2 \|u\|^2,\end{aligned}$$

whence $\|1 \otimes T\|$ is bounded by $\|T\|$ and in particular extends to a bounded operator on $H \otimes K$. A precisely analogous argument works for $S \otimes 1$. The operator $(1 \otimes T)(S \otimes 1)$ agrees with the operator in line (1.10) on elementary tensors, so this gives our desired extension. Note moreover that this argument shows that

$$\|S \otimes T\| \leqslant \|1 \otimes T\| \|S \otimes 1\| \leqslant \|S\| \|T\|.$$

To get the reverse inequality, let (u_n) and (v_n) be sequences of unit vectors in H and K respectively such that $\|T u_n\| \to \|T\|$ and $\|S v_n\| \to \|S\|$. Then $\|u_n \otimes v_n\| = 1$ for all n and

$$\|(S \otimes T)(u_n \otimes v_n)\| = \|S u_n \otimes T v_n\| = \|S u_n\| \|T v_n\| \to \|S\| \|T\|,$$

completing the proof. □

Remark 1.8.7 Let $\pi : A \to \mathcal{B}(H_A)$ be a representation of a C^*-algebra, and let H be another Hilbert space. Then the representation

$$\pi \otimes 1 : A \to \mathcal{B}(H_A \otimes H), \quad a \mapsto \pi(a) \otimes 1$$

makes sense by Lemma 1.8.6 above. We will frequently have use of this construction: the resulting representation $\pi \otimes 1$ of A is called the *amplification* of π to $H_A \otimes H$.

Now, say A and B are $*$-algebras, and that $\pi_A: A \to \mathcal{B}(H_A)$ and $\pi_B: B \to \mathcal{B}(H_B)$ are representations. Then Remark 1.8.7 gives amplified representations $\pi_A \otimes 1: A \to \mathcal{B}(H_A \otimes H_B)$ and $1 \otimes \pi_B: B \to \mathcal{B}(H_A \otimes H_B)$. Clearly these have commuting images, and so Remark 1.8.1 gives a $*$-homomorphism

$$(\pi_A \otimes 1) \otimes (1 \otimes \pi_B): A \odot B \to \mathcal{B}(H_A \otimes H_B). \tag{1.11}$$

Definition 1.8.8 With notation as above, we write $\pi_A \otimes \pi_B: A \odot B \to \mathcal{B}(H_A \otimes H_B)$ for the homomorphism in line (1.11) above, and call it the *tensor product* of π_A and π_B. Concretely, we have the formula

$$(\pi_A \otimes \pi_B)(a \otimes b) = \pi_A(a) \otimes \pi_B(b)$$

for elementary tensors in $A \odot B$, where the operator on the right-hand side is the one defined in Lemma 1.8.6.

Proposition 1.8.9 *Let $\pi_A: A \to \mathcal{B}(H)$ and $\pi_B: B \to \mathcal{B}(H_B)$ be faithful representations of C*-algebras. Then the tensor product representation*

$$\pi_A \otimes \pi_B: A \odot B \to \mathcal{B}(H_A \otimes H_B)$$

is injective. Moreover, for $c \in A \odot B$, the norm defined by $\|c\| := \|(\pi_A \otimes \pi_B)(c)\|_{\mathcal{B}(H_A \otimes H_B)}$ does not depend on the choice of π_A and π_B.

Proof For injectivity, say $c = \sum_{i=1}^n a_i \otimes b_i \in A \odot B$ is in the kernel of $\pi_A \otimes \pi_B$. Choosing a basis for $\operatorname{span}\{b_1, \ldots, b_n\}$ and rewriting c in terms of this basis, we may assume that the collection $b_1, \ldots, b_n$ is linearly independent. Now, for any $u, v \in H_A$ and $w, x \in H_B$ we have

$$0 = \langle u \otimes w, (\pi_A \otimes \pi_B)(c)(v \otimes x)\rangle = \sum_{i=1}^n \langle u, \pi_A(a_i)v\rangle \langle w, \pi_B(b_i)x\rangle$$

$$= \left\langle w, \pi_B\left(\sum_{i=1}^n \langle u, \pi_A(a_i)v\rangle b_i\right) x\right\rangle.$$

As w and x are arbitrary and π_B is injective, this forces $\sum_{i=1}^n \langle u, \pi_A(a_i)v\rangle b_i = 0$, and linear independence of $b_1, \ldots, b_n$ then forces $\langle u, \pi_A(a_i)v\rangle = 0$ for each i. As u and v are arbitrary and π_A is injective, this forces $a_i = 0$ for each i. Hence $c = 0$, so so we are done with injectivity.

We now check that the norm defined above is independent of the choice of π_A; by symmetry, this suffices. Fix an increasing net (P_i) of finite rank projections on H_B converging strongly to the identity.[8] Using Lemma 1.8.6,

[8] i.e. so that $\|P_i v - v\| \to 0$ as $i \to \infty$; for example, the net of all finite rank projection ordered by inclusion of images has this property.

define $Q_i := 1 \otimes P_i$. We claim that the net (Q_i) converges strongly to the identity on $H_A \otimes H_B$: indeed, it follows from a direct check that $\|Q_i v - v\| \to 0$ when $v \in H_A \otimes H_B$ is a finite sum of elementary tensors, and the general case follows as such finite sums are dense in $H_A \otimes H_B$, and as Lemma 1.8.6 implies that $\|Q_i\| = 1$ for all i. It follows for the claim that for any operator T on $H_A \otimes H_B$,

$$\|T\| = \lim_i \|Q_i T Q_i\|. \tag{1.12}$$

Now, let $\sum_{j=1}^n a_j \otimes b_j$ be an element of $A \odot B$. Then for any i,

$$Q_i \left(\sum_{j=1}^n \pi_A(a_j) \otimes \pi_B(b_j) \right) Q_i = \sum_{j=1}^n \pi_A(a_j) \otimes P_i \pi_B(b_j) P_i. \tag{1.13}$$

Write $n = \mathrm{rank}(P_i)$, and consider the representation $\pi : M_n(\mathbb{C}) \to \mathcal{B}(H_B)$ defined by some choice of isomorphism $M_n(\mathbb{C}) \cong \mathcal{B}(P_i H)$. Then

$$\pi_A \otimes \pi : A \odot M_n(\mathbb{C}) \to \mathcal{B}(H_A \otimes H_B)$$

is injective by the first part of the proof, and the computation in line (1.13) shows that its image contains $Q_i(\sum_{j=1}^n \pi_A(a_j) \otimes \pi_B(b_j))Q_i$. Moreover, Example 1.8.2 shows that we may identify the domain of $\pi_A \otimes \pi$ with $M_n(A)$.

If then σ_A is any other faithful representation of A, we may apply the above discussion to both π_A and σ_A. Corollary 1.3.16 implies that the C^*-algebra norm on

$$M_n(A) \cong (\pi_A \otimes \pi)(A \odot M_n(\mathbb{C})) \cong (\sigma_A \otimes \pi)(A \odot M_n(\mathbb{C}))$$

is unique. This then implies that

$$\left\| Q_i \left(\sum_{j=1}^n \pi_A(a_j) \otimes \pi_B(b_j) \right) Q_i \right\| = \left\| Q_i \left(\sum_{j=1}^n \sigma_A(a_j) \otimes \pi_B(b_j) \right) Q_i \right\|.$$

Taking the limit over i and using line (1.12), we get

$$\left\| \sum_{j=1}^n \pi_A(a_j) \otimes \pi_B(b_j) \right\| = \left\| \sum_{j=1}^n \sigma_A(a_j) \otimes \pi_B(b_j) \right\|$$

and are done. □

We are now ready for completed tensor products of C^*-algebras. Proposition 1.8.9 implies that the definition below both gives a norm, and that the resulting norm does not depend on the choices involved.

Definition 1.8.10 Let A and B be C^*-algebras. Choose faithful representations π_A and π_B of A and B respectively, and define the *spatial norm* on $A \odot B$ by

$$\|c\| := \|(\pi_A \otimes \pi_B)(c)\|.$$

The *spatial tensor product* of A and B, denoted $A \otimes B$, is defined to be the associated completion of $A \odot B$.

Example 1.8.11 Let $C_0(X)$ be a commutative C^*-algebra, and let A be any C^*-algebra. Let $C_0(X, A)$ be the collection of continuous functions from X to A that vanish at infinity, which is a C^*-algebra when equipped with pointwise operations and the supremum norm. We claim that the map determined by

$$C_0(X) \odot A \to C_0(X, A), \quad f \otimes a \mapsto (x \mapsto f(x)a) \tag{1.14}$$

extends to a $*$-isomorphism $C_0(X) \otimes A \cong C_0(X, A)$. Indeed, to build $C_0(X) \otimes A$ we may use $\ell^2(X) \otimes H_A$ for some faithful representation H_A of A, and where $C_0(X)$ acts on $\ell^2(X)$ by multiplication. Then $C_0(X, A)$ is also represented on this Hilbert space via the formula

$$f : \delta_x \otimes v \mapsto \delta_x \otimes f(x)v.$$

Identifying $C_0(X, A)$ with its image under this representation, it is clear that the representation of $C_0(X) \odot A$ on $\ell^2(X) \otimes H$ maps this $*$-algebra into $C_0(X, A)$ via the map in line (1.14). It is also straightforward to see that the image is dense using a partition of unity argument. As images of $*$-homomorphisms are closed, we are done.

Arguing quite analogously, we get canonical identifications

$$C_0(X) \otimes C_0(Y) = C_0(X, C_0(Y)) = C_0(X \times Y),$$

which will occasionally be used in the main body of the text.

Remark 1.8.12 Using the ideas in the proof of Proposition 1.8.9, it is not too difficult to see that the spatial tensor product is functorial in the following sense: for each pair $\phi : A \to C$ and $\psi : B \to D$ of $*$-homomorphisms there is a unique $*$-homomorphism $\phi \otimes \psi : A \otimes B \to C \otimes D$ that satisfies $(\phi \otimes \psi)(a \otimes b) = \phi(a) \otimes \psi(b)$ on elementary tensors. Moreover, the homomorphism $\phi \otimes \psi$ is injective if ϕ and ψ both are. We leave this as an exercise for the reader: see Exercise 1.9.16 below.

If at least one of the algebras involved in a tensor product is commutative then we have the following more general functoriality result.

Lemma 1.8.13 *Let A be a commutative C^*-algebra. Then for any C^*-algebras B and C, and any $*$-homomorphisms $\phi: A \to C$ and $\psi: B \to C$ with commuting images, there is a unique $*$-homomorphism*

$$\phi \otimes \psi: A \otimes B \to C$$

satisfying $(\phi \otimes \psi)(a \otimes b) = \phi(a)\psi(b)$ on elementary tensors.

Proof As in Exercise 1.9.19, there is a unique $*$-homomorphism $\phi \otimes \psi: A \odot B \to C$ satisfying $(\phi \otimes \psi)(a \otimes b) = \phi(a)\psi(b)$ on elementary tensors. Our task is to show that $\phi \otimes \psi$ extends to the spatial tensor product $A \otimes B$. Using the injectivity comment in Remark 1.8.12, there is a canonical inclusion $A \otimes B \subseteq A^+ \otimes B^+$ of the spatial tensor product of A and B into the spatial tensor product of their unitisations. Replacing A, B and C with their unitisations, and replacing ϕ and ψ with the corresponding unital maps between unitisations, it thus suffices to prove the lemma in the case that all the C^*-algebras and $*$-homomorphisms are unital. In particular, we may write $A = C(X)$ for some compact space X, and Example 1.8.11 gives us a canonical identification $A \otimes B = C(X, B)$.

Let now $f \in A$ and $b \in B$ be arbitrary non-zero elements. We claim that for if $\mathcal{U}$ is an open cover of X such that $|f(x) - f(y)| < \epsilon/\|b\|$ for all x, y in the same element of $\mathcal{U}$, $\{g_1, \ldots, g_m\}$ is a partition of unity on X subordinate to this cover, and $x_1, \ldots, x_m$ are points in X such that x_i is in the support of g_i, then we have

$$\left\| \phi(f)\psi(b) - \sum_{i=1}^{m} f(x_i)\phi(g_i)\psi(b) \right\|_C < \epsilon.$$

Indeed, using that $*$-homomorphisms are contractive, this is bounded above by

$$\left\| \phi\left(f - \sum_{i=1}^{m} f(x_i)g_i \right) \right\|_C \|\psi(b)\|_C \leqslant \left\| f - \sum_{i=1}^{m} f(x_i)g_i \right\|_{C(X)} \|b\|_B,$$

and the claim follows.

Now, let $c = \sum_{i=1}^{n} f_i \otimes b_i$ be an arbitrary element of the algebraic tensor product $A \odot B$. Let $g_1, \ldots, g_m$ be a partition of unity on X subordinate to an open cover $\mathcal{U}$ such that each f_i satisfies $|f_i(x) - f_i(y)| < \epsilon/(n \max \|b_i\|)$ whenever x, y are in the same element of $\mathcal{U}$. Let $x_1, \ldots, x_m \in X$ be such that x_i is in the support of g_i. Define maps

$$\alpha: C(X, B) \to \bigoplus_{i=1}^{n} B, \quad f \mapsto \left(f(x_i)\right)_{i=1}^{n}$$

and

$$\beta\colon \bigoplus_{i=1}^{n} B \to C, \quad (b_i)_{i=1}^{n} \mapsto \sum_{i=1}^{n} \phi(g_i)\psi(b_i).$$

The choice of the g_i and x_i and the claim above imply that $\|\alpha \circ \beta(c) - (\phi \otimes \psi)(c)\|_C < \epsilon$. Moreover, α is clearly a $*$-homomorphism so contractive, and β is clearly unital and takes positive elements to positive elements,[9] so has norm bounded by 4 by Exercise 1.9.17. Hence

$$\|(\phi \otimes \psi)(c)\|_C \leqslant \|\alpha \circ \beta(c)\|_C + \|\alpha \circ \beta(c) - (\phi \otimes \psi)(c)\|_C \leqslant 4\|c\|_{A\otimes B} + \epsilon.$$

As ϵ and c were arbitrary, this gives that $\phi \otimes \psi\colon A \odot B \to C$ has norm bounded above by 4, whence extends to $A \otimes B$ as required.[10] □

Remark 1.8.14 A C^*-algebra A satisfying the conclusion of Lemma 1.8.13 is said to be *nuclear*. Thus Lemma 1.8.13 can be succinctly restated as follows: commutative C^*-algebras are nuclear.

It is more common to express nuclearity in the following way. For C^*-algebras A and B, let S be the collection of all triples (ϕ, ψ, C) where C is a C^*-algebra, and $\phi\colon A \to C$ and $\psi\colon B \to C$ are $*$-homomorphisms with commuting images. Note that such a triple gives rise to a $*$-homomorphism $\phi \otimes \psi\colon A \odot B \to C$ as in the proof of Lemma 1.8.13. One then defines the *maximal* C^*-algebra norm[11] on $A \odot B$ by

$$\|c\|_{\max} := \sup\{\|(\phi \odot \psi)(c)\|_C \mid (\phi, \psi, C) \in S\}.$$

As C^*-algebra homomorphisms are contractive, one has that for any $(\phi, \psi, C) \in S$ and any $c = \sum_i a_i \otimes b_i \in A \odot B$,

$$\|\phi \odot \psi(c)\|_C \leqslant \sum_i \|a_i\|\|b_i\|,$$

and thus the supremum defining $\|c\|_{\max}$ is finite. Hence $\|\cdot\|_{\max}$ is indeed a C^*-algebra norm on $A \odot B$. The *maximal tensor product* of A and B, denoted $A \otimes_{\max} B$, is defined to be the associated completion. The identity map on $A \odot B$ can be shown to extend to a quotient map

$$A \otimes_{\max} B \to A \otimes B. \tag{1.15}$$

Indeed, if $\pi_A\colon A \to \mathcal{B}(H_A)$ and $\pi_B\colon B \to \mathcal{B}(H_B)$ are faithful representations used to define the norm on $A \otimes B$, then we get a $*$-homomorphism

[9] Actually, it satisfies the stronger property of 'complete positivity', which implies in particular that it is contractive, but we do not need this.

[10] At which point we know it is a $*$-homomorphism between C^*-algebras, so has norm one.

[11] This is not quite the usual definition, but is equivalent: see Exercise 1.9.20.

$\pi_A \otimes \pi_B \colon A \otimes_{\max} B \to \mathcal{B}(H_A \otimes H_B)$ by definition of the maximal completion, and it is not difficult to check that the image actually lies in the natural copy of $A \otimes B$ inside $\mathcal{B}(H_A \otimes H_B)$. The usual definition of nuclearity is that the quotient map in line (1.15) is an isomorphism; it is clearly equivalent to the definition we have given above.

We will need to go back to maximal tensor products in the graded setting, as this is the most convenient way for us to discuss the external product in K-theory: see Section 2.10 below.

The C^*-algebra $\mathcal{K}$ of compact operators on a separable, infinite-dimensional Hilbert space plays a special role in the theory; we make the following standard definition.

Definition 1.8.15 A C^*-algebra is *stable* if $A \cong A \otimes \mathcal{K}$.

Remark 1.8.16 For any C^*-algebra A, A is elementarily Morita equivalent to $A \otimes \mathcal{K}$ in the sense of Definition 1.7.9 above. Indeed, fix any rank one-projection q in $\mathcal{K}$. Then the operator $p := 1 \otimes q$ makes sense as an element of the multiplier algebra of $A \otimes \mathcal{K}$ (even if A is not unital). It is not too difficult to show that A is $*$-isomorphic to $p(A \otimes \mathcal{K})p$, and that p is full in $A \otimes \mathcal{K}$.

1.9 Exercises

1.9.1 Show that if A is a unital C^*-algebra, then A^+ is $*$-isomorphic to the C^*-algebra direct sum $A \oplus \mathbb{C}$.

1.9.2 Let B be the unital C^*-algebra generated by the *bilateral shift*

$$U \colon \ell^2(\mathbb{Z}) \to \ell^2(\mathbb{Z}), \quad \delta_n \mapsto \delta_{n+1}$$

on $\ell^2(\mathbb{Z})$, and let $A \subseteq B$ be the unital Banach algebra generated by U (so A does not contain U^*). Show that the spectrum of U 'relative to A' is the closed unit disk in $\mathbb{C}$, and the spectrum of U 'relative to B' is the unit circle.

1.9.3 The goal of this exercise is to show that the sort of behaviour exhibited in Exercise 1.9.2 cannot happen for C^*-algebras; this is sometimes called *spectral permanence*. Let A be a C^*-algebra, and let B be a sub-C^*-algebra of A. For $b \in B$, write $\mathrm{spec}_B(a)$ for the spectrum of a considered as an element of B, and $\mathrm{spec}_A(b)$ for the spectrum of b considered as an element of A. Replacing the algebras with their unitisations, we may assume everything is unital, and that A and B have the same unit.

(a) Observe that $\mathrm{spec}_A(b) \subseteq \mathrm{spec}_B(b)$ (this is true for general algebras).

(b) Show that the boundary of $\text{spec}_B(b)$ is contained in $\text{spec}_A(b)$ (this is true for general Banach algebras).
Hint: by translation, it suffices to show that if 0 is in the boundary of $\text{spec}_B(b)$*, then b is not invertible in A. Aiming for contradiction, assume that b is invertible in A, and that 0 is in the boundary of* $\text{spec}_B(b)$*. Then there is a sequence* (λ_n) *not in* $\text{spec}_B(b)$ *converging to zero. Now use continuity of inversion in A applied to the sequence* $((b - \lambda_n)^{-1})$ *in B to deduce a contradiction.*

(c) Show that if b is self-adjoint, then $\text{spec}_B(a) = \text{spec}_A(b)$.
Hint: Lemma 1.3.10.

(d) To complete the proof, by translation it suffices to show that if b is invertible in A, then it is invertible in B. Do this.
Hint: note that b^*b *is invertible in A, and use the previous part to deduce that it is invertible in B.*

1.9.4 The result of this exercise is called the *spectral mapping theorem for polynomials.*

Let A be a unital algebra over $\mathbb{C}$, and $a \in A$. Prove that if $p\colon \mathbb{C} \to \mathbb{C}$ is a complex polynomial, then $\text{spec}(p(a)) \subseteq p(\text{spec}(a))$.

1.9.5 Show that if a, b are elements of a complex algebra, then $\text{spec}(ab) \cup \{0\} = \text{spec}(ba) \cup \{0\}$. Give an example where $\text{spec}(ab) \neq \text{spec}(ba)$.
Hint: working in the unitisation, show that it suffices to prove that if $ab - 1$ *is invertible then* $ba - 1$ *is invertible. Let then c be the inverse of* $ab - 1$*, and show that* $bca - 1$ *is the inverse of* $ba - 1$.

1.9.6 Prove the claim of Remark 1.3.2.

1.9.7 Let $A = C(X)$, where X is a compact Hausdorff space. Prove that the map

$$X \to \widehat{A}, \quad x \mapsto \phi_x$$

of Example 1.3.5 is a homeomorphism.
Hint: first show (this is straightforward) that the map above is injective and continuous. As both X and $\widehat{A}$ *are compact and Hausdorff, to complete the proof it suffices to show that the map above is surjective. For this, let* $\phi\colon C(X) \to \mathbb{C}$ *be a multiplicative linear functional, and note that as* ϕ *is automatically bounded (Lemma 1.3.6), it is given by integration against some measure by the Riesz representation theorem. Show that this measure must be a Dirac mass in order for* ϕ *to be multiplicative.*

1.9.8 Prove that any element a of a C^*-algebra is a linear combination of four positive elements $y_1, \ldots, y_4$ satisfying $\|y_i\| \leqslant \|a\|$.
Hint: first write a as a linear combination of its real *and* imaginary *parts defined respectively by $\frac{1}{2}(a + a^*)$ and $\frac{1}{2i}(a - a^*)$. Define functions $f_+, f_- : \mathbb{R} \to \mathbb{R}$ by*

$$f_+(t) = \begin{cases} t & t \geqslant 0, \\ 0 & t < 0, \end{cases} \quad \textit{and} \quad f_+(t) = \begin{cases} -t & t \leqslant 0, \\ 0 & t > 0, \end{cases}$$

respectively. For a self-adjoint element $b \in A$, use the functional calculus to define $b_+ := f_+(b)$ and $b_- := f_-(b)$, the so-called positive *and* negative *parts of b, and note that $b = b_+ - b_-$.*

1.9.9 Any element a in a C^*-algebra A can be written $a = bb^*b$ for some $b \in A$. Here is a guided proof using a '2×2 matrix trick' (there are many such tricks in the theory of operator algebras).

Write $c = \left(\begin{smallmatrix} 0 & a^* \\ a & 0 \end{smallmatrix}\right) \in M_2(A)$, which is self-adjoint. Moreover, if $u = \left(\begin{smallmatrix} 1 & 0 \\ 0 & 1 \end{smallmatrix}\right) \in M_2(A^+)$, then $ucu^* = -c$.

(i) Show that for any normal element d of a C^*-algebra D, any $f \in C_0(\text{spec}(d) \setminus \{0\})$ and any unitary $u \in D^+$, $f(udu^*) = uf(d)u^*$.
(ii) Conclude from this that $uc^{1/3}u^* = -c^{1/3}$, and thus $c = \left(\begin{smallmatrix} 0 & b^* \\ b & 0 \end{smallmatrix}\right)$ for some $b \in A$. This b has the right property.

1.9.10 Let A be a unital C^*-algebra, and let $a, b \in A$ be invertible elements with $0 \leqslant a \leqslant b$. Show that $0 \leqslant b^{-1} \leqslant a^{-1}$.
Hint: first show that

$$b^{-1/2}ab^{-1/2} \leqslant b^{-1/2}bb^{-1/2} = 1.$$

From this and the functional calculus deduce that

$$1 \leqslant (b^{-1/2}ab^{-1/2})^{-1} = b^{1/2}a^{-1}b^{1/2};$$

now compress by $b^{-1/2}$.

1.9.11 Let A be a C^*-algebra, and let $a, b \in A$ be positive elements such that $a \leqslant b$. Show that $a^{1/2} \leqslant b^{1/2}$.
Hint: working in the unitisation of A it suffices to assume that b is invertible, as for general b, $b + \epsilon$ is invertible and $(b + \epsilon)^{1/2} \to b$ as $\epsilon \to 0$. Assuming this, check that $b^{-1/2}ab^{-1/2} \leqslant 1$, and therefore $\|b^{-1/2}a^{1/2}\| \leqslant 1$. Now, using Exercise 1.9.5, the spectral radius of $b^{-1/4}a^{1/2}b^{-1/4}$ equals that of $b^{-1/2}a^{1/2}$, so is at most 1. Hence by Corollary 1.2.7, $\|b^{-1/4}a^{1/2}b^{-1/4}\| \leqslant 1$, and so $b^{-1/4}a^{1/2}b^{-1/4} \leqslant 1$, from which the result follows. Be warned that the analogous statement '$0 \leqslant a \leqslant b \Rightarrow a^2 \leqslant b^2$' is false!

1.9.12 Show that if I and J are ideals in a C^*-algebra A, then $I \cap J = I \cdot J$, where $I \cdot J$ is the algebraic span of all products from I and J.
Hint: $I \cdot J \subseteq I \cap J$ is immediate; for the converse, you can use Exercise 1.9.9 applied to the C^*-algebra $I \cap J$.

1.9.13 Show that if I is an ideal in a C^*-algebra A, and $B \subseteq A$ is a C^*-subalgebra, then $I + B$ is a C^*-subalgebra of A.
Hint: to see that $B + I$ is closed, let $\pi : A \to A/I$ be the quotient map, note that $B+I = \pi^{-1}(\pi(B))$ and apply Corollary 1.5.10 to see that $\pi(B)$ is closed.

1.9.14 Show that if X is a locally compact Hausdorff space, then the multiplier algebra $M(C_0(X))$ naturally identifies with $C_b(X)$.
Hint: represent $C_0(X)$ by multiplication operators on $\ell^2(X)$. Considered as an X-by-X matrix, show that any multiplier must have no non-zero off-diagonal entries, and from here that the diagonal entries must define a bounded continuous function on X.

1.9.15 Show that if H and K are Hilbert spaces, then there is a natural $*$-isomorphism

$$\mathcal{K}(H) \otimes \mathcal{K}(K) \cong \mathcal{K}(H \otimes K).$$

1.9.16 Justify the comments about functoriality of the C^*-algebra tensor product in Remark 1.8.12.
Hint: given $\phi\colon A \to C$ and $\psi\colon B \to D$, fix faithful representations $\pi_C\colon C \to \mathcal{B}(H_C)$ and $\pi_D\colon D \to \mathcal{B}(H_D)$ of C and D respectively, and also π_A and π_B of A and B. Then by definition of the spatial tensor product, there is faithful representation

$$(\pi_A \oplus \pi_C \circ \phi) \otimes (\pi_B \oplus \pi_D \circ \psi)\colon A \otimes B \to \mathcal{B}((H_A \oplus H_C) \otimes (H_B \oplus H_D)).$$

Let $P_C\colon H_A \oplus H_C \to H_C$ and $P_D\colon H_B \oplus H_D \to H_D$ be the orthogonal projections, and consider the map $A \otimes B \to \mathcal{B}(H_C \otimes H_D)$ determined by

$$a \otimes b \mapsto (P_C \otimes P_D)\Big((\pi_A \oplus \pi_C \circ \phi)(a) \otimes (\pi_B \oplus \pi_D \circ \psi)(b)\Big)(P_C \otimes P_D).$$

Show that this takes image in $C \otimes D$, and gives the required map $\phi \otimes \psi$.

1.9.17 Show that if $\phi\colon A \to B$ is a unital map between C^*-algebras that takes positive elements to positive elements, then it has norm bounded by four (this is not optimal, but good enough for what we need).
Hint: if $a \in A$ is positive, then $0 \leqslant a \leqslant \|a\|$, whence positivity and unitality of ϕ give that $\phi(a) \leqslant \phi(\|a\|) = \|a\|$ in B. Now use that any element x in a C^*-algebra is a sum of four positive elements $y_1, \ldots, y_4$ that satisfy $\|y_i\| \leqslant \|x\|$ as in Exercise 1.9.8

1.9.18 Show that the compact operators $\mathcal{K}(H)$ on any Hilbert space is a nuclear C^*-algebra in the sense of Remark 1.8.14. More generally (the reader should regard the more general statement as a hint for how to do the proof in the special case!), show that any direct limit of finite-dimensional C^*-algebras is nuclear.

1.9.19 (i) Let A and B be $*$-algebras. Show that the algebraic tensor $A \odot B$ has the following universal property: whenever $\phi_A : A \to C$ and $\phi_B : B \to C$ are $*$-homomorphisms with commuting images, there is a unique $*$-homomorphism $\phi_A \otimes \phi_B : A \odot B \to C$ such that for all elementary tensors $a \otimes b \in A \odot B$, we have $(\phi_A \otimes \phi_B)(a \otimes b) = \phi_A(a)\phi_B(b)$.

(ii) Show the analogous universal property, but now assuming that A, B, and C are C^*-algebras, and with $\odot$ replaced by $\otimes_{\max}$ as in Remark 1.8.14.

1.9.20 In Remark 1.8.14, we gave a slightly non-standard definition of the maximal tensor product norm on $A \odot B$. The usual definition is

$$\|c\|_{\max} := \sup\{\|\pi(c)\|_{\mathcal{B}(H)} \mid \pi : A \odot B \to \mathcal{B}(H) \text{ a non-degenerate representation}\}.$$

Show that this defines the same norm, and then use this to show that $\otimes_{\max}$ has the analogous functoriality property to $\otimes$ from Remark 1.8.12.

Hint: the key point is to show that if $\pi : A \odot B \to \mathcal{B}(H)$ is non-degenerate, then there are non-degenerate representations $\pi_A : A \to \mathcal{B}(H)$ and $\pi_B : B \to \mathcal{B}(H)$ with commuting images, and such that $\pi(a \otimes b) = \pi_A(a)\pi_B(b)$ for all $a \in A$ and $b \in B$. This is not difficult when both A and B are unital: define $\pi_A(a) := \pi(a \otimes 1)$ and similarly for π_B. In the non-unital case, let (h_i) be an approximate unit for A, and try to define π_B by setting $\pi_B(b)$ to be the strong limit of the net $\pi(h_i \otimes b)$ (and similarly with the roles of A and B reversed).

For functoriality, given $*$-homomorphisms $\phi : A \to C$ and $\psi : B \to D$, fix a faithful non-degenerate representation $\pi : C \otimes_{\max} D \to \mathcal{B}(H)$. Use the argument above to find representations π_C and π_D with the properties there, and consider the map $(\pi_C \circ \psi) \otimes (\pi_D \circ \psi) : A \odot B \to \mathcal{B}(H)$ arising from Remark 1.8.1.

1.10 Notes and References

There are several general introductory references on C^*-algebras. Douglas [81] and Arveson [4] are excellent general introductions to spectral theory and operator theory: their main focus is not C^*-algebras, but they nonetheless cover much of the basics (and in particular, most of the material needed for this text). Murphy [189] is a good introduction specifically focused on C^*-algebra theory. Davidson [76] also establishes the basic theory as well as containing detailed studies of many interesting examples that are quite different from the subject matter of the current text; it demands more of the reader than Murphy.

In terms of more advanced C^*-algebra texts, the classic monographs of Dixmier [80] and Pedersen [204] contain a wealth of information (the former particularly on the relationship of C^*-algebra theory to representation theory); while in principal they start from the beginning of the subject, we would not strongly recommend either to the beginner (particularly not the beginner who is motivated mainly by the topics covered in this book, as both go much further than anything required by this text early on in their expositions). Blackadar [34] is a very useful modern survey: it gives an overview of much of the 'standard' theory of operator algebras.

The material on multiplier algebras and tensor products in the last two sections of this chapter is probably the least standard material that we discuss. The cleanest way to approach multiplier algebras is probably through Hilbert modules, and we recommend the expositions found in chapter 2 of Lance [163] and chapter 2 of Raeburn and Williams [209], for the Hilbert module approach. Our definition of Morita equivalence (Definition 1.7.9) using multiplier algebras is a little ad-hoc, but convenient for our applications, and equivalent to the more usual definitions thanks to the *linking algebra* interpretation of Morita equivalence. See for example chapter 3 of Raeburn and Williams [209] for a textbook discussion of the general theory here.

For the theory of C^*-algebra tensor products, we particularly recommend the exposition in chapter 3 of Brown and Ozawa [44], which also contains a detailed account of the purely algebraic theory that we skipped. A different exposition that is closer to classical representation theory can be found in chapter 6 of Murphy [189]. These references both contain proofs of the beautiful theorem of Takesaki (theorem 2 in Takesaki [238]) that the spatial tensor product norm is the smallest C^*-algebra norm on the algebraic tensor product $A \odot B$ of two C^*-algebras; for this reason, the spatial tensor product is often called the *minimal* C^*-algebra tensor product.

The fact (see Lemma 1.8.13 and Remark 1.8.14) that commutative C^*-algebras are nuclear is originally theorem 1 of Takesaki [238]. Two quite different textbook proofs can be found in theorem 6.4.15 in Murphy ([189]), and by combining propositions 2.4.2 and 3.6.12 in Brown and Ozawa [44]. Our proof is a low-tech version of that which appears in Brown and Ozawa, which is the standard modern proof.

It is not so relevant for the topics in this book, but nuclearity turns out to be one of the most important properties in the general theory of C^*-algebras: we recommend Brown and Ozawa [44] for an exposition of (some of) the theory of nuclearity, and particularly chapter 3 of Brown and Ozawa [44] for an introduction to the maximal and spatial tensor products.

2

K-theory for C^*-algebras

Our goal in this chapter is to give an overview of the facts from C^*-algebra K-theory that we will need in the rest of the book, as well as to establish notation and conventions. As in Chapter 1, the aim is not to be completely self-contained, but we do at least sketch proofs where we think this helps with intuition, or where a result is difficult to find in the literature in the form we need it.

The chapter is structured as follows. In Section 2.1 we recall the definition of the K_0 group for a complex algebra. This section involves no analysis or topology. In Section 2.2 we put a complete norm on our algebra, and discuss some powerful equivalent descriptions of K_0 in this setting. Section 2.3 adds yet more analysis, using some relatively delicate C^*-algebraic machinery to set up the theory of the maps on K-theory induced by unbounded traces. At this point we are done with facts that touch only on the K_0 group.

In Section 2.4, we go back to pure algebra, introducing the index map to measure an obstruction to the existence of long exact sequences in K-theory. At this point, one can develop the K_1 group underlying the index map in a way that is either partly topological or purely algebraic; unlike the K_0 group, the two choices give something quite different. Our applications dictate that we make the topological choice, and in Section 2.5 we discuss the topological K_1 group. Section 2.6 then ties the K_0 and K_1 groups together via the fundamental Bott periodicity theorem, which completes the basic theory.

The remaining four sections contain slightly less standard material. Section 2.7 gives a grab bag of computational tools. All of these are well known, but some are not so prominent in the literature, so we give details here. Section 2.8 discusses various index constructions of elements in K-theory in more detail, focusing on explicit formulas; this will be important for some applications. Section 2.9 discusses an alternative picture of K-theory, the so-called spectral picture, which is particularly well-suited to discussions of products and index theory. Finally, in Section 2.10, we use the spectral picture of K-theory to discuss products.

2.1 Algebraic K_0

In this section we define the K_0 group of a C^*-algebra purely algebraically. To emphasise that the theory is algebraic, we will work with arbitrary $\mathbb{C}$-algebras.[1] Even if one is only interested in C^*-algebras, knowing what can be done purely algebraically is often useful, as algebraic arguments sometimes give more precise information than analytic or topological ones. Of course, this is balanced by the fact that analytic and topological tools are often more powerful than purely algebraic ones.

Contrary to standard conventions in many algebra texts, we emphasise that algebras are not assumed unital, and morphisms between unital algebras need not preserve the units.

Definition 2.1.1 Let R be a $\mathbb{C}$-algebra. Let $M_\infty(R)$ be the (non-unital) $\mathbb{C}$-algebra of $\mathbb{N} \times \mathbb{N}$ matrices over R, all but finitely many of whose entries are zero.

In fancier language, one can equivalently define $M_\infty(R)$ to be the direct limit (in the category of not-necessarily unital $\mathbb{C}$-algebras and not-necessarily unital $\mathbb{C}$-algebra homomorphisms) of the $\mathbb{C}$-algebras $M_n(R)$ under the top left corner inclusion maps

$$M_n(R) \to M_{n+1}(R), \quad a \mapsto \begin{pmatrix} a & 0 \\ 0 & 0 \end{pmatrix}.$$

Recall now that an *idempotent* in a ring is an element e such that $e^2 = e$.

Definition 2.1.2 Two idempotents e, f in a $\mathbb{C}$-algebra S are *Murray–von Neumann equivalent*, written $e \sim_{MvN} f$, if there are $v, w \in S$ with $vw = e$ and $wv = f$.

Remark 2.1.3 We think of v and w as mutually inverse isomorphisms between e and f: schematically, one has

$$e \underset{v}{\overset{w}{\rightleftarrows}} f .$$

[1] With minor changes, the theory even applies to arbitrary rings.

Murray–von Neumann equivalence is indeed an equivalence relation. For transitivity, if schematically one has

$$e \underset{v}{\overset{w}{\rightleftarrows}} f \underset{t}{\overset{s}{\rightleftarrows}} g$$

then sw and vt induce a Murray–von Neumann equivalence between e and g: for example,

$$(sw)(vt) = s(wv)t = sft = s(ts)t = (st)(st) = gg = g$$

and similarly in the opposite order.

Definition 2.1.4 Let R be a unital $\mathbb{C}$-algebra, and let $V(R)$ denote the collection of all Murray–von Neumann equivalence classes of idempotents in $M_\infty(R)$. Equip $V(R)$ with the addition operation[2] defined by

$$[e] + [f] := \begin{bmatrix} e & 0 \\ 0 & f \end{bmatrix}.$$

To make sense of this, think of e and f as the images in $M_\infty(R)$ of finite, say $n \times n$ and $m \times m$, matrices. Let then $\begin{pmatrix} e & 0 \\ 0 & f \end{pmatrix}$ be the image in $M_\infty(R)$ of the corresponding $(n+m) \times (n+m)$ matrix. There are choices involved here: for a start, m and n are not unique. We leave it to the reader (see Exercise 2.11.2) to show that the operation on $V(R)$ is well-defined, and turns $V(R)$ into a commutative monoid (with 0 as the identity element).

Remark 2.1.5 Equivalently, $V(R)$ can be described as the collection of all isomorphism classes of finitely generated projective modules over R, with the operation induced by direct sum of modules. Indeed, to go from an idempotent $e \in M_n(R) \subseteq M_\infty(R)$ to a projective (right) module one takes $M = e(R^n)$. The other direction is a little more involved: see Exercise 2.11.3 for a full justification of this.

Example 2.1.6 Let $R = \mathbb{C}$. Then two idempotents in $M_\infty(\mathbb{C})$ are Murray–von Neumann equivalent if and only if they have the same rank, and addition of equivalence classes corresponds to adding ranks. Thus $V(\mathbb{C})$ is isomorphic as a monoid to $\mathbb{N}$.

Example 2.1.7 Let $R = \mathcal{B}(H)$ be the bounded operators on a separable infinite-dimensional Hilbert space. Similarly to Example 2.1.6, the monoid $V(R)$ is completely determined by rank, and thus $V(\mathcal{B}(H))$ is isomorphic as

[2] Sometimes called *block sum*.

a monoid to $\mathbb{N} \cup \{\infty\}$ (with the usual addition on $\mathbb{N}$, and with infinity plus anything equalling infinity).

Example 2.1.8 Let $R = C(X)$, where X is a compact Hausdorff space. Then the *Serre–Swan theorem* says that $V(R)$ consists precisely of isomorphism classes of vector bundles over X, with operation given by direct sum (also called *Whitney sum* in this context) of bundles: see Exercise 2.11.4.

Proposition 2.1.9 *Let R be a unital complex algebra. Then idempotents $e, f \in M_\infty(R)$ are Murray–von Neumann equivalent if and only if there is an invertible element u in some $M_n(R)$ such that $ueu^{-1} = f$.*

If R is a non-unital complex algebra, let R^+ be the unitisation of R as in Definition 1.1.9. Then idempotents $e, f \in M_\infty(R)$ are Murray–von Neumann equivalent if and only if there is an invertible element u in some $M_n(R^+)$ that conjugates e to f.

Proof First note that if $ueu^{-1} = f$, then we make take $v = u^{-1}f$ and $w = fu$ in the definition of Murray–von Neumann equivalence (this works either in the unital or non-unital case).

On the other hand, say $e = vw$ and $f = wv$ for some $v, w \in M_k(R) \subseteq M_\infty(R)$. From Exercise 2.11.1, we may assume that $ev = vf = v$ and $we = fw = w$. Define now for R unital

$$u = \begin{pmatrix} w & 1-f \\ 1-e & v \end{pmatrix} \in M_{2k}(R) \tag{2.1}$$

(if R is non-unital, then u is in $M_{2k}(R^+)$). Using the formulas $vwv = v$ and $wvw = w$, one checks that u is invertible, with

$$u^{-1} = \begin{pmatrix} v & 1-e \\ 1-f & w \end{pmatrix}.$$

Computing gives us

$$u \begin{pmatrix} e & 0 \\ 0 & 0 \end{pmatrix} u^{-1} = \begin{pmatrix} f & 0 \\ 0 & 0 \end{pmatrix};$$

however, by definition of $M_\infty(R)$ the expressions $\left(\begin{smallmatrix} e & 0 \\ 0 & 0 \end{smallmatrix}\right)$ and e are just different ways of writing the same element of this ring, so we are done. □

Remark 2.1.10 It may make the formula for u in line (2.1) above more conceptual if one considers it as a map between two copies of $R \oplus R$ decomposed in different ways as follows

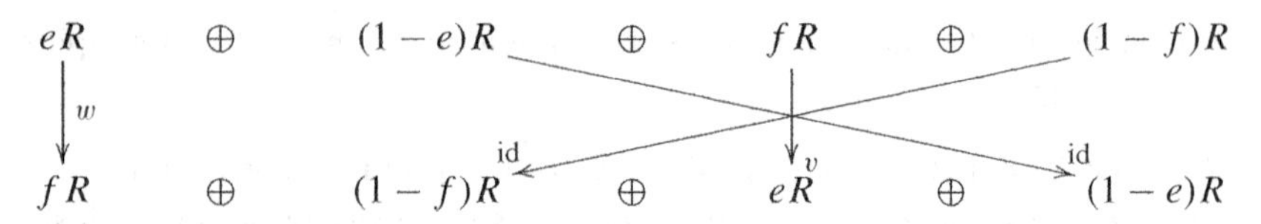

We are now ready to define the K_0-group.

Definition 2.1.11 Let R be a unital $\mathbb{C}$-algebra. The group $K_0(R)$ is determined up to canonical isomorphism by the following universal property. It is equipped with a monoid homomorphism $V(R) \to K_0(R)$ such that for any monoid homomorphism $V(R) \to A$ to an abelian group, the dashed arrow in the diagram

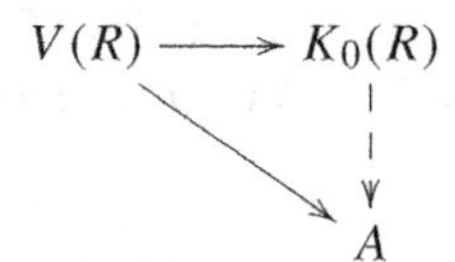

can be filled in with a unique monoid (equivalently, group) homomorphism.

There are several direct constructions of a group $K_0(R)$ satisfying the universal property above: see Exercise 2.11.5. In general, a group satisfying the above universal property for some abelian monoid M in place of $V(R)$ is called the *Grothendieck group* of M.

Example 2.1.12 The K_0 group of $\mathbb{C}$ is $\mathbb{Z}$. This follows from Example 2.1.6 above and the fact that the Grothendieck group of the monoid $\mathbb{N}$ is $\mathbb{Z}$ (as an exercise, either check the universal property or construct a concrete isomorphism using the constructions of Exercise 2.11.5).

Now, $K_0(R)$ is functorial for unital $\mathbb{C}$-algebra homomorphisms in a natural way: if $\phi\colon R \to S$ is a unital ring homomorphism, and $e \in M_\infty(R)$ is an idempotent, then $\phi(e)$ is an idempotent in $M_\infty(S)$, and the map $[e] \mapsto [\phi(e)]$ is well-defined as a map $V(R) \to V(S)$. Hence it induces a map

$$\phi_*\colon K_0(R) \to K_0(S)$$

by the universal properties of this group (or using the explicit descriptions of K_0 from Exercise 2.11.5). This also lets us make the following definition.

Definition 2.1.13 Let R be a not-necessarily unital $\mathbb{C}$-algebra, and let R^+ be its unitisation (see Definition 1.1.9), which is equipped with a canonical unital $\mathbb{C}$-algebra homomorphism $\phi\colon R^+ \to \mathbb{C}$ with kernel R. Then the group $K_0(R)$ is defined by

$$K_0(R) := \text{Kernel}(\phi_*\colon K_0(R^+) \to K_0(\mathbb{C})).$$

This is consistent with our earlier definition in the case that R is unital, in the sense that the two definitions lead to canonically isomorphic abelian groups. This follows from the following facts, which we leave as exercises for the reader: first, if R is unital, then R^+ is canonically isomorphic to the $\mathbb{C}$-algebra $R \oplus \mathbb{C}$; second, that if $R \oplus S$ is a direct product of unital rings then there is a canonical isomorphism $K_0(R \oplus S) \cong K_0(R) \oplus K_0(S)$; and third, that with respect to this isomorphism the canonical quotient map $\phi\colon R^+ \to \mathbb{C}$ induces the projection onto the second factor $K_0(R^+) \cong K_0(R) \oplus K_0(\mathbb{C}) \to K_0(\mathbb{C})$.

The new definition of K_0 for possibly non-unital $\mathbb{C}$-algebras is still functorial: indeed, say $\phi\colon R \to S$ is any algebra homomorphism (possibly, for example, a non-unital homomorphism between unital algebras). Then it induces a unital ring homomorphism

$$\phi^+\colon R^+ \to S^+, \quad (r,\lambda) \mapsto (\phi(r),\lambda)$$

that makes the diagram

$$\begin{array}{ccc} R^+ & \longrightarrow & \mathbb{C} \\ \big\downarrow{\scriptstyle \phi^+} & & \big\| \\ S^+ & \longrightarrow & \mathbb{C} \end{array}$$

commute, where the horizontal maps are the canonical quotient maps. Hence by functoriality of K_0 in the unital case, there is a commutative diagram of short exact sequences

$$\begin{array}{ccccccccc} 0 & \longrightarrow & K_0(R) & \longrightarrow & K_0(R^+) & \longrightarrow & K_0(\mathbb{C}) & \longrightarrow & 0 \\ & & \big\downarrow & & \big\downarrow{\scriptstyle \phi^+_*} & & \big\| & & \\ 0 & \longrightarrow & K_0(S) & \longrightarrow & K_0(S^+) & \longrightarrow & K_0(\mathbb{C}) & \longrightarrow & 0. \end{array}$$

The homomorphism $\phi_*\colon K_0(R) \to K_0(S)$ is then defined to be the unique dashed arrow making the diagram commute; in the unital case, this again agrees with the earlier definition.

Remark 2.1.14 It is often convenient to have a more concrete picture for elements of $K_0(R)$ in the non-unital case. Here is one such way to represent elements. Let x be an element of $K_0(R)$, so using Exercise 2.11.5 x can be represented by a formal difference $[e] - [f]$ of idempotents in some $M_n(R^+)$, with the property that if $\pi : R^+ \to \mathbb{C}$ is the natural quotient, then the induced map $\pi_*\colon K_0(R^+) \to K_0(\mathbb{C})$ takes $[e]$ and $[f]$ to the same class. Adding $[1_n - f] - [1_n - f]$ where 1_n is the unit in $M_n(R^+)$, we see that our element can be represented by $[e'] - [1_n]$, where e' is an idempotent in $M_{2n}(R^+)$, and $[\pi(e')] = [1_n]$ in $M_n(\mathbb{C})$ (here we abuse notation slightly and use the same symbol π for the map induced by π on matrices). Thanks to the ideas in Examples 2.1.6 and 2.1.12, this means there is an invertible $u \in M_{2n}(\mathbb{C})$ such that $u\pi(e')u^{-1} = 1_n$. Now, u also makes sense as an element of $M_{2n}(R^+)$, and we have that $[ue'u^{-1}] - [1_n]$ represents the same class in $K_0(R)$ by the discussion in Proposition 2.1.9.

We conclude from the above discussion that any class in $K_0(R)$ can be represented as a formal difference $[e] - [1_n]$ for some n, with the property that the map on matrices induced by the natural quotient $\pi : R^+ \to \mathbb{C}$ takes e to 1_n.

Remark 2.1.15 Let $M_n(R)$ denote the $\mathbb{C}$-algebra of $n \times n$ matrices over R. Then there is a (non-unital) homomorphism $R \to M_n(R)$ defined via 'top left corner inclusion'

$$a \mapsto \begin{pmatrix} a & 0 \\ 0 & 0 \end{pmatrix}.$$

The induced map $K_0(R) \to K_0(M_n(R))$ is an isomorphism. If R is unital, this is straightforward, as the top left corner inclusion is easily seen to induce an isomorphism $M_\infty(R) \cong M_\infty(M_n(R))$. If R is non-unital, this requires a little more thought (note that $M_n(R)^+$ is not the same as $M_n(R^+)$) and we leave the details as an exercise for the reader: compare Exercise 2.11.6 below.

Remark 2.1.16 Let R be a unital $\mathbb{C}$-algebra, and let $\tau : R \to \mathbb{C}$ be a trace, i.e. τ is a linear functional with the property that $\tau(ab) = \tau(ba)$ for all $a, b \in R$. Define

$$\tau_\infty : M_\infty(R) \to \mathbb{C}, \quad a \mapsto \sum_{n \in \mathbb{N}} \tau(a_{nn}),$$

i.e. τ_∞ sums the values of τ on all the diagonal entries of a; this makes sense, as elements of $M_\infty(R)$ have only finitely many non-zero entries. Note that τ_∞ is still a trace, whence it agrees on Murray–von Neumann equivalent elements, and descends to a well-defined map $\tau_* : V(R) \to \mathbb{C}$, which clearly respects the monoid structure. The universal property of K_0 now gives a well-defined group homomorphism $\tau_* : K_0(R) \to \mathbb{C}$.

All this adapts easily to the non-unital case: start by extending a trace $\tau : R \to \mathbb{C}$ to a unital trace $\tau : R^+ \to \mathbb{C}$, thus getting a map $K_0(R^+) \to \mathbb{C}$ as above. The induced map $\tau_* : K_0(R) \to \mathbb{C}$ is then just the restriction to the subgroup $K_0(R)$ of $K_0(R^+)$.

To summarise, traces give 'linear functionals' on K_0 groups.

2.2 Approximation and Homotopy in K_0

In this section, we specialise from $\mathbb{C}$-algebras to Banach algebras,[3] and also sometimes to C^*-algebras. We will give several different interpretations of the K_0 group: in terms of homotopy classes of idempotents, in terms of almost idempotents, and in terms of projections.

For this discussion to make sense, we also need norms on A^+ and $M_n(A)$ where A is a Banach algebra. If A is a C^*-algebra, we equip A^+ and $M_n(A)$ with the unique C^*-algebra norms (see Definition 1.1.9 and Example 1.6.10). If A is a Banach algebra, there are no really canonical choice of norms but any 'reasonable' choice will do: for concreteness, we use the norms defined in Exercise 2.11.7.

We first discuss homotopies.

Definition 2.2.1 Let A be a Banach algebra and $e_0, e_1 \in A$ be idempotents. A *homotopy* between e_0 and e_1 is a continuous map

$$[0,1] \to A, \quad t \mapsto e_t$$

that agrees with e_0 and e_1 at the endpoints, and with the property that each e_t is an idempotent. Two idempotents are *homotopic* if there is a homotopy between them. By definition, two idempotents e_0, e_1 in $M_\infty(A)$ are *homotopic* if they there is a homotopy between them in some $M_n(A)$.

We want to show that homotopic idempotents (in some matrix algebra over A) define the same element in K_0. The key point is the following result, which says that close idempotents are conjugate.

Lemma 2.2.2 *Let $e_0, e_1 \in A$ be idempotents in a Banach algebra, and assume that the inequality $\|e_0 - e_1\| < 1/\|2e_0 - 1\|$ holds in the unitisation A^+. Then there is an invertible element u in A^+ with $ue_0u^{-1} = e_1$.*

Proof Working in A^+, set

$$u = e_0e_1 + (1 - e_0)(1 - e_1). \tag{2.2}$$

Then

$$u - 1 = 2e_0e_1 - e_0 - e_1 = (2e_0 - 1)(e_0 - e_1),$$

and so our assumption implies that $\|u - 1\| < 1$. Hence u is invertible by Theorem 1.2.1. A direct computation gives that $e_0u = ue_1$, and so u has the property we want. □

[3] The main text almost always works in the context of C^*-algebras, but the Banach algebra theory will be used very occasionally, so we include the basics here.

Remark 2.2.3 To try to get a more geometric intuition for the formula defining u in line (2.2) above, let us assume that we are working in a concrete C^*-algebra and that e_0 and e_1 are not-necessarily-orthogonal projection operators on a Hilbert space. The condition in the statement of Lemma 2.2.2 implies that the operator $e_0 e_1$ takes the image of e_1 isomorphically (although not necessarily isometrically) onto the image of e_0, and similarly that $(1-e_0)(1-e_1)$ takes the complement of the image of e_1 onto the complement of the image of e_0. From this, we see that $u = e_0 e_1 + (1-e_0)(1-e_1)$ is invertible, and satisfies the formula $e_0 u = u e_1$.

Proposition 2.2.4 *Let A be a Banach algebra and e_0, e_1 be idempotents in $M_\infty(A)$. Then e_0 and e_1 are homotopic if and only if they are Murray–von Neumann equivalent (see Definition 2.1.2).*

Proof Say first that there is a homotopy $(e_t)_{t\in[0,1]}$ between e_0 and e_1. Then by Lemma 2.2.2, we may find finitely many points $t_0, \ldots, t_N$ in $[0,1]$ that are close enough so that for each $i \in \{1,\ldots,N\}$ there is an invertible element $u_i \in A^+$ with $u_i e_{t_{i-1}} u_i^{-1} = e_{t_i}$. It follows that $u := u_N \cdots u_2 u_1$ conjugates e_0 to e_1, and thus by Proposition 2.1.9 that e_0 and e_1 are Murray–von Neumann equivalent.

Conversely, say $e_0, e_1 \in M_\infty(A)$ are Murray–von Neumann equivalent. Then by Proposition 2.1.9 there is an invertible $u \in M_n(A^+)$ for some n such that $u e_0 u^{-1} = e_1$. Consider the homotopy $(v_t)_{t\in[0,\pi/2]}$ (note the unusual domain – of course, one could reparametrise it to have domain $[0,1]$ if one wants) defined by

$$v_t := \begin{pmatrix} u & 0 \\ 0 & 1 \end{pmatrix}\begin{pmatrix} \cos(t) & -\sin(t) \\ \sin(t) & \cos(t) \end{pmatrix}\begin{pmatrix} 1 & 0 \\ 0 & u^{-1} \end{pmatrix}\begin{pmatrix} \cos(t) & \sin(t) \\ -\sin(t) & \cos(t) \end{pmatrix}, \qquad (2.3)$$

so v_0 is the diagonal matrix with entries u and u^{-1}, and v_1 is the identity in $M_{2n}(A^+)$. Then the homotopy

$$\left(v_t \begin{pmatrix} e_0 & 0 \\ 0 & 0 \end{pmatrix} v_t^{-1} \right)_{t\in[0,\pi/2]}$$

is contained in $M_{2n}(A)$ (as this is an ideal in $M_{2n}(A^+)$), and connects $\begin{pmatrix} e_0 & 0 \\ 0 & 0 \end{pmatrix}$ to $\begin{pmatrix} e_1 & 0 \\ 0 & 0 \end{pmatrix}$ via idempotents. Recalling that e_i and $\begin{pmatrix} e_i & 0 \\ 0 & 0 \end{pmatrix}$ are the same element in $M_\infty(A)$, we are done. □

We now connect $K_0(A)$ to the $*$-structure on A when A is a C^*-algebra.

Proposition 2.2.5 *Let A be a C^*-algebra.*

(i) *If $e \in A$ is an idempotent, then there are a projection $p \in A$ and invertible $u \in A^+$ with $u^{-1}pu = e$.*

(ii) *If $p,q \in A$ are Murray–von Neumann equivalent projections, then there is a partial isometry $x \in A$ with $x^*x = p$ and $xx^* = q$.*

(iii) *If $p,q \in A$ are projections that are conjugate by some invertible $u \in A^+$, then they are also conjugate by a unitary element of A^+.*

Part (i) of this result does not hold for arbitrary Banach-$*$ algebras: the proof uses that $1+a^*a$ is invertible for any element a of a unital C^*-algebra, and this need not hold in an arbitrary Banach $*$-algebra. See Exercise 2.11.20 below for a counterexample.

Proof For part (i), define $x := 1+(e-e^*)(e^*-e)$, an element of the unitisation A^+. Then x is of the form $1+a^*a$, so self-adjoint and invertible. One computes that $ex = ee^*e = xe$, so e (and hence also e^*) commutes with x, Moreover,

$$ee^*x = e(e^*ee^*) = (ee^*)^2.$$

Define $p = ee^*x^{-1}$, which is in A not just A^+. Then p is self-adjoint as it is a product of the commuting self-adjoint elements ee^* and x^{-1}. Moreover,

$$p = (ee^*)x^{-1} = (ee^*x)x^{-2} = (ee^*)^2x^{-2} = (ee^*x^{-1})^2 = p^2,$$

so p is a projection. One computes that $ep = p$ and $pe = e$; setting $u = 1-p+e \in A^+$, these formulas show that u is invertible with inverse $u^{-1} = 1+p-e$ and we get that $u^{-1}pu = e$ as claimed.

For part (ii), we may assume p and q are non-zero. Using Exercise 2.11.1, we may assume that $p = vw$ and $q = wv$, where w and v satisfy $pvq = v$ and that $qwp = w$. Then $p = p^*p = w^*v^*vw \leqslant \|v\|^2w^*w$. Hence w^*w is invertible in the unital C^*-algebra pAp, so there exists $r \in pAp$ with $(w^*w)^{1/2}r = p$. One checks that $x = wr$ now works.

For part (iii), note that if $u^{-1}pu = q$ for projections p,q, then $pu = uq$, and as p and q are self-adjoint, we also get that $u^*p = qu^*$. Hence in particular

$$u^*uq = u^*pu = qu^*u,$$

so q commutes with anything in the C^*-algebra generated by u^*u. Hence if we set $v = u(u^*u)^{-1/2}$, then v is unitary, and

$$v^* p v = (u^* u)^{-1/2} u^* p u (u^* u)^{-1/2} = (u^* u)^{-1/2} u^* u u^{-1} p u (u^* u)^{-1/2}$$
$$= (u^* u)^{1/2} q (u^* u)^{-1/2} = q,$$

so v conjugates p to q, and we are done. □

Remark 2.2.6 Let us try to explain the idea of the above proof, which the algebraic formalism might obscure slightly. For notational simplicity, let e be an idempotent in A, so we may assume that e is a not-necessarily-orthogonal projection operator on some Hilbert space H: more precisely, H splits as a direct sum of closed subspaces $K \oplus E$ and e acts by sending everything in K to zero, and by the identity on E; however, as e need not be self-adjoint, it need not be true that K and E are mutually orthogonal. Let p be the orthogonal projection with range E. Then it is not too difficult to see that there is a (non-unique) invertible operator $u: H \to H$ that takes K to $E^{\perp}$, and acts as the identity on E; we therefore have $u^{-1} p u = e$. The difficulty is to show that u, which is a priori just a bounded operator on H, can be chosen in such a way that it is in A^+: this is what the formulas in the argument above achieve.

The last main topic of this section is the use of 'approximate idempotents' to define K-theory.

Definition 2.2.7 Let A be a Banach algebra. A *quasi-idempotent* is an element $a \in A$ such that $\|a^2 - a\| < 1/4$. If A is a C^*-algebra, a *quasi-projection* is a quasi-idempotent that is self-adjoint.

Construction 2.2.8 Let $a \in A$ be a quasi-idempotent in a Banach algebra. As in Example 1.4.7, one sees that the spectrum of a misses the line $\mathrm{Re}(z) = 1/2$, and so we may use the holomorphic functional calculus (see Theorem 1.4.6) to build an idempotent $e := \chi(a)$, where χ is the characteristic function of $\{z \in \mathbb{C} \mid \mathrm{Re}(z) > 1/2\}$. As a result any quasi-idempotent e in $M_\infty(A)$ defines a class $[\chi(a)]$ in $K_0(A)$ in a canonical way.

Note that if $(e_t)_{t \in [0,1]}$ is a continuous path of quasi-idempotents, then Theorem 1.4.6 implies that $\chi(e_t)$ is a continuous path of actual idempotents.

The advantage of looking at quasi-idempotents is that thinking this way allows approximation: the condition '$\|a^2 - a\| < 1/4$' defines an open subset of A, unlike the condition of actually being an idempotent which defines a closed subset.

To conclude this section, we give a summary of some different ways of describing the monoid $V(A)$ underlying the K_0 group of a C^*-algebra A. This list is by no-means exhaustive, and the reader should by no means try to memorise them all; we just aim to give a sense of some of the flexibility inherent in the definition. The proof follows by combining ideas from Proposition 2.1.9, Proposition 2.2.4, Proposition 2.2.5, and Construction 2.2.8: we leave the details to the reader.

Proposition 2.2.9 *For a Banach algebra A, equivalence classes for the following sets and equivalence relations all define naturally isomorphic monoids $V(A)$, and therefore models of K-theory. In all cases, the zero element is given by the class of zero, and the operation by setting $[x]+[y]$ to be the class of the matrix $\begin{pmatrix} x & 0 \\ 0 & y \end{pmatrix}$.*

(i) Set: idempotents in $M_\infty(A)$. Equivalence relation: Murray–von Neumann equivalence (this is our original Definition 2.1.11 above).
(ii) Set: idempotents in $M_\infty(A)$. Equivalence relation: conjugation by invertibles in $M_\infty(A^+)$ (or in $M_\infty(A)$ if A is unital).
(iii) Set: idempotents in $M_\infty(A)$. Equivalence relation: homotopy through idempotents (as in Definition 2.2.1).
(iv) Set: almost idempotents in $M_\infty(A)$. Equivalence relation: homotopy through almost idempotents (see Exercise 2.11.9).

If in addition A is a C^*-algebra, then the following also give the same monoid.

(i) Set: projections in $M_\infty(A)$. Equivalence relation: p, q are equivalent if there is a partial isometry in $M_\infty(A)$ with $v^*v = p$ and $vv^* = q$.
(ii) Set: projections in $M_\infty(A)$. Equivalence relation: conjugation by unitaries in $M_\infty(A^+)$ (or in $M_\infty(A)$ if A is unital).
(iii) Set: projections in $M_\infty(A)$. Equivalence relation: homotopy through projections. □

Remark 2.2.10 Let A and B be Banach algebras. A *homotopy* between two homomorphisms $\phi_0, \phi_1 : A \to B$ is a path $(\phi_t : A \to B)_{t\in[0,1]}$ of homomorphisms[4] that connects them, and that is point-norm continuous, i.e. for each $a \in A$, the path

$$[0,1] \to B, \quad t \mapsto \phi_t(a)$$

is continuous. Two homomorphisms are *homotopic* if there exists a homotopy between them. It is clear from the descriptions in line (2.2.1) above that the functor K_0 from the category of Banach algebras and homomorphisms to the category of abelian groups takes homotopic homomorphisms to the same group homomorphism.

[4] If A and B are C^*-algebras and ϕ_0 and ϕ_1 are $*$-homomorphisms, it is perhaps more natural to require that each ϕ_t is also a $*$-homomorphism; however, for the purposes of basic K-theory, this additional restriction does not matter.

2.3 Unbounded Traces

Recall from Remark 2.1.16 above that a trace $\tau\colon R \to \mathbb{C}$ on a complex algebra induces a map $\tau_*\colon K_0(R) \to \mathbb{C}$ on K_0 groups. In applications, however, one sometimes has to consider traces on C^*-algebras that are only defined on a dense subspace. Our goal here is to show that under suitable conditions, such traces also induce maps on K_0.

A secondary goal (which is anyway needed for our study of traces) is to introduce some sufficient conditions for the inclusion $\mathcal{A} \to A$ of a dense $*$-subalgebra in a C^*-algebra to induce an isomorphism on K-theory. Although we will not use these results much in this book, they are very important in the subject more broadly, so it seemed a useful service to the reader to at least touch on them here.

This section will not be used in the subsequent development of basic K-theory, and can safely be skipped on a first reading. We include relatively full details as this material is not included in any standard textbook that we know of. It is also of quite a different character from the rest of the chapter, involving some quite delicate C^*-algebraic arguments.

Definition 2.3.1 Let A be a C^*-algebra, and A_+ the collection of positive elements of A. A *positive trace* on A is a map $\tau\colon A_+ \to [0,\infty]$ such that:

(i) $\tau(0) = 0$;
(ii) for all $a \in A$, $\tau(a^*a) = \tau(aa^*)$;
(iii) for all $a_1, a_2 \in A_+$ and all $\lambda_1, \lambda_2 > 0$, $\tau(\lambda_1 a_1 + \lambda_2 a_2) = \lambda_1\tau(a_1) + \lambda_2\tau(a_2)$.

Remark 2.3.2 If a positive trace τ on a C^*-algebra A takes only finite values on A_+, we will say that τ is *bounded*. In this case, straightforward algebraic checks using that A is spanned by its positive elements show that τ extends uniquely to a positive linear functional $\tau\colon A \to \mathbb{C}$. Moreover the *polarisation identity*

$$ab = \frac{1}{4}\sum_{k=0}^{3} i^k (i^k a^* + b)^*(i^k a^* + b) \tag{2.4}$$

shows that the linear functional $\tau\colon A \to \mathbb{C}$ satisfies the usual trace property: $\tau(ab) = \tau(ba)$ for all $a, b \in A$.

Thus a bounded positive trace on A is the same thing as a linear functional that takes positive elements to positive[5] elements, and is also a trace in the

[5] We use 'positive' here to mean 'non-negative' as in the usual C^*-algebra conventions!

usual sense. In the bounded case, we will treat the maps $\tau\colon A_+ \to [0,\infty)$ and $\tau\colon A \to \mathbb{C}$ interchangeably.

We warn the reader, however, that the terminology is a little misleading: a 'positive trace' is not the same thing as a special type of trace in general.

Example 2.3.3 Let H be a Hilbert space, and let $\mathcal{B}(H)$ denote the bounded operators on H. Let $(e_i)_{i\in I}$ be an orthonormal basis for H. The *canonical trace* on $\mathcal{B}(H)$ is defined by

$$\mathrm{Tr}\colon \mathcal{B}(H)_+ \to [0,\infty], \quad \mathrm{Tr}(T) := \sum_{i\in I} \langle e_i, Te_i\rangle.$$

We also call the restriction of this map to $\mathcal{K}(H)_+$ the *canonical trace* on $\mathcal{K}(H)$. A straightforward computation shows that for any $T \in \mathcal{B}(H)$, if T is represented by the matrix (T_{ij}) with respect to the given basis,[6] then

$$\mathrm{Tr}(T^*T) = \sum_{i,j\in I} |T_{ij}|^2 = \mathrm{Tr}(TT^*) \tag{2.5}$$

and thus Tr is a positive trace. Note that Tr is bounded if and only if H is finite-dimensional, and in that case it agrees with the usual matrix trace.

If moreover $(f_i)_{i\in I}$ is another orthonormal basis, and $U\colon H \to H$ the change of basis unitary determined by $U(e_i) = f_i$, then we have using line (2.5) that

$$\sum_{i\in I} \langle f_i, Tf_i\rangle = \sum_{i\in I} \langle Ue_i, TUe_i\rangle = \mathrm{Tr}(U^*TU) = \mathrm{Tr}(T^{1/2}UU^*T^{1/2}) = \mathrm{Tr}(T).$$

This shows that $\mathrm{Tr}(T)$ is independent of the choice of orthonormal basis.

Example 2.3.4 There is an important generalisation of Example 2.3.3 above to stabilised C^*-algebras (this is the main example we will need in applications). Indeed, say A is a C^*-algebra, and $\tau\colon A \to \mathbb{C}$ a bounded, positive trace. Let H be a Hilbert space equipped with an orthonormal basis $(e_i)_{i\in I}$, and let $\mathcal{K} = \mathcal{K}(H)$ denote the compact operators on H. Using the orthonormal basis (e_i), we may identify elements of the spatial tensor product $A\otimes\mathcal{K}$ with I-by-I matrices[7] $(a_{ij})_{i,j\in I}$ with entries from A. Define

$$\tau\otimes\mathrm{Tr}\colon (A\otimes\mathcal{K})_+ \to [0,\infty], \quad (a_{ij})_{i,j\in I} \mapsto \sum_{i\in I} \tau(a_{ii})$$

(the notation is motivated by the fact that on elementary tensors, $(\tau\otimes\mathrm{Tr})(a\otimes T) = \tau(a)\mathrm{Tr}(T)$). Then $\tau\otimes\mathrm{Tr}$ is again a trace in the sense of Definition 2.3.1

[6] Precisely, this means that $T_{ij} := \langle e_i, Te_j\rangle$.

[7] Saying when exactly such a matrix gives an element of $A\otimes\mathcal{K}$ is difficult, but that does not matter for the current discussion.

(see Exercise 2.11.14), and one can check quite analogously to Example 2.3.3 that it does not depend on the choice of basis involved in its construction.

Associated to a positive trace on A are two important subsets of A.

Definition 2.3.5 Let τ be a trace on a C^*-algebra A. Define

$$I_{\tau,1} := \text{span}\{a \in A_+ \mid \tau(a) < \infty\}$$

and

$$I_{\tau,2} := \{a \in A \mid \tau(a^*a) < \infty\}.$$

Example 2.3.6 Let $A = \mathcal{K}(H)$ and Tr the canonical trace from Example 2.3.3. Then the elements of $I_{\tau,1}$ are called the *trace class* operators on H, and those of $I_{\tau,2}$ are called the *Hilbert–Schmidt operators*. In this special case, $I_{\tau,i}$ is more usually denoted $\mathcal{L}^i(H)$ for $i \in \{1,2\}$.

Elementary algebra shows that τ extends to a $*$-preserving linear functional $\tau\colon I_{\tau,1} \to \mathbb{C}$; from now on we will abuse notation and write τ for both the map $A_+ \to [0,\infty]$ and the associated linear functional on $I_{\tau,1}$. The spaces $I_{\tau,1}$ and $I_{\tau,2}$ and associated linear functional τ automatically have substantial algebraic structure. Here is the basic result; for the statement, recall that a subset S of a C^*-algebra is *hereditary* if whenever $0 \leqslant a \leqslant b$ and $b \in S$, we have that $a \in S$. Define also $I_{\tau,2}^2 := \{ab \in A \mid a,b \in I_{\tau,2}\}$.

Proposition 2.3.7 *Let τ be a positive trace on a C^*-algebra A, and let $I_{\tau,1}$ and $I_{\tau,2}$ be as in Definition 2.3.5.*

(i) $I_{\tau,2}$ is a $*$-closed algebraic ideal in A.
(ii) $I_{\tau,1} = I_{\tau,2}^2$.
(iii) $I_{\tau,1}$ is a hereditary $*$-closed algebraic ideal in A.
(iv) If $a,b \in I_{\tau,2}$, then $\tau(ab) = \tau(ba)$.
(v) If $a,b \in A$, and at least one of a,b is in $I_{\tau,1}$, then $\tau(ab) = \tau(ba)$. In particular, $\tau\colon I_{\tau,1} \to \mathbb{C}$ is a trace in the purely algebraic sense of Remark 2.1.16.

Proof For part (i), note first that $I_{\tau,2}$ is clearly closed under scalar multiplication and taking adjoints. To see that it is closed under taking sums, note that for $a,b \in I_{\tau,2}$,

$$(a+b)^*(a+b) \leqslant (a+b)^*(a+b) + (a-b)^*(a-b) = 2(a^*a + b^*b),$$

whence

$$\tau((a+b)^*(a+b)) \leqslant 2\tau(a^*a) + 2\tau(b^*b) < \infty.$$

Hence $a + b \in I_{\tau,2}$, so we now have that $I_{\tau,2}$ is a $*$-closed subspace of A. Moreover, for any $a \in A$ and $b \in I_{\tau,2}$,

$$(ab)^*(ab) = b^*a^*ab \leqslant \|a\|^2 b^*b,$$

and thus ab is in $I_{\tau,2}$. This says that $I_{\tau,2}$ is a left ideal in A, whence it is a two-sided ideal as $*$-closed.

For part (ii), first note that if $a, b \in I_{\tau,2}$, then as $I_{\tau,2}$ is a $*$-closed subalgebra of A for each $k \in \{0,1,2,3\}$ we have that $i^k a^* + b \in I_{\tau,2}$, and so $\tau((i^k a^* + b)^*(i^k a^* + b)) < \infty$. Hence each product $(i^k a^* + b)^*(i^k a^* + b)$ is in $I_{\tau,1}$. The polarisation identity

$$ab = \frac{1}{4}\sum_{k=0}^{3} i^k (i^k a^* + b)^*(i^k a^* + b),$$

then gives that $ab \in I_{\tau,1}$, and so we have that $I_{\tau,2}^2 \subseteq I_{\tau,1}$. For the opposite inclusion, as $I_{\tau,1}$ is spanned by the positive elements that it contains, it will suffice to show that if $a \in A_+ \cap I_{\tau,1}$, then $a \in I_{\tau,2}^2$. Note however that in this case $a^{1/2} \in I_{\tau,2}$ by definition, and hence $a = a^{1/2}a^{1/2} \in I_{\tau,2}^2$.

For part (iii), it is clear that $I_{\tau,1}$ is a $*$-closed subspace of A. It then follows from parts (i) and (ii) that $I_{\tau,1}$ is an ideal.[8] The fact that $I_{\tau,1}$ is hereditary follows as if $0 \leqslant a \leqslant b$, then $\tau(b) = \tau(a) + \tau(b - a)$; as $\tau(b - a) \geqslant 0$, this forces $\tau(a) \leqslant \tau(b)$.

For part (iv), say that $a, b \in I_{\tau,2}$. Then using the polarisation identity

$$ab = \frac{1}{4}\sum_{k=0}^{3} i^k (i^k a^* + b)^*(i^k a^* + b)$$

and the fact that $\tau(c^*c) = \tau(cc^*)$ for all $c \in A$ we have that $\tau(ab) = \tau(ba)$.

For part (v), say first that $a \in A$ is arbitrary and $b \in I_{\tau,1}$ is positive. Then $b^{1/2}$ is in $I_{\tau,2}$. Using that $I_{\tau,2}$ is an ideal and part (iv) (twice), we get

$$\tau(ab) = \tau(ab^{1/2}b^{1/2}) = \tau(b^{1/2}ab^{1/2}) = \tau(b^{1/2}b^{1/2}a) = \tau(ba).$$

The case with $a \in A$ and $b \in I_{\tau,1}$ arbitrary follows from this as $I_{\tau,1}$ is spanned by the positive elements that it contains. The case with $a \in I_{\tau,1}$ and $b \in A$ is similar. □

Remark 2.3.8 If A is unital, and if the set $\{a \in A_+ \mid \tau(a) < \infty\}$ is dense in A_+, then this set contains an invertible element. As $I_{\tau,1}$ is an ideal, it is thus all of A. Hence $\tau \colon A \to \mathbb{C}$ is a bounded positive trace as in Remark 2.3.2, and

[8] See Exercise 2.11.10 for a different argument.

we are in the situation of Remark 2.1.16. Thus the theory we are developing is only interesting when A is not unital (as in Examples 2.3.3 and 2.3.4 above).

In order to make further progress, we need to make some analytic assumptions on our positive traces. As we want to include the canonical trace on $\mathcal{K}(H)$ of Example 2.3.3 in the theory, it would be too much to assume that τ is bounded; fortunately, a significantly weaker analytic assumption will be sufficient for our purposes.

Definition 2.3.9 A positive trace $\tau : A_+ \to [0, \infty]$ on a C^*-algebra A is *lower semicontinuous* if for any norm convergent sequence (a_n) in A_+,

$$\tau(\lim_{n\to\infty} a_n) \leqslant \liminf_{n\to\infty} \tau(a_n).$$

It is *densely defined* if the collection $\{a \in A_+ \mid \tau(a) < \infty\}$ is dense in A_+.

Example 2.3.10 Let Tr be as in Example 2.3.3. It is not difficult to check that Tr is lower semicontinuous: this is essentially Fatou's lemma for the set I with counting measure. However, it is not densely defined on $\mathcal{B}(H)_+$ if H is infinite-dimensional: if it were, it would be everywhere defined by Remark 2.3.8, and it is clear that $\text{Tr}(1) = \infty$. On the other hand, Tr is clearly finite on all (positive) operators that have a finite matrix representation with respect to the basis (e_i). As such operators are dense in (the positive part of) the compact operators $\mathcal{K}(H)$, we see that Tr gives a densely defined trace on $A = \mathcal{K}(H)$ in our sense.

This all works similarly for the trace in Example 2.3.4: see Exercise 2.11.14.

Here are the key analytic properties of lower semicontinuous traces. Recall for the statement that if a is an element of a C^*-algebra, then $|a|$ denotes $(a^*a)^{1/2}$.

Proposition 2.3.11 *Let τ be a densely defined, lower semicontinuous positive trace on a C^*-algebra A with associated $*$-ideals $I_{\tau,1}$ and $I_{\tau,2}$ as in Proposition 2.3.7. Also let $\tau : I_{\tau,1} \to \mathbb{C}$ denote the associated linear functional. Then one has the following facts.*

(i) For all $a \in I_{\tau,1}$ and $b \in A$,

$$|\tau(ab)| \leqslant \tau(|a|)\|b\| \quad \text{and} \quad |\tau(ba)| \leqslant \tau(|a|)\|b\|.$$

(ii) For all $a \in A$, a is in $I_{\tau,1}$ if and only if $|a|$ is in $I_{\tau,1}$.
(iii) For all $a, b \in A$,

$$\tau(|a + b|) \leqslant \tau(|a|) + \tau(|b|).$$

(iv) The formula

$$\|a\|_\tau := \|a\|_A + \tau(|a|)$$

defines a norm on $I_{\tau,1}$ with respect to which it is a Banach $*$-algebra.

Proof We first look at part (i), although we will only get the result in the special case that a is positive at first. Using the equality $I_{\tau,1} = I_{\tau,2}^2$ from Proposition 2.3.7, we have for $a, b \in I_{\tau,2}$ that $\tau(a^*b)$ is a well-defined complex number. Hence we may define a positive semi-definite inner product on $I_{\tau,2}$ by the formula

$$\langle a, b\rangle := \tau(a^*b).$$

In particular, if $\|a\|_2 := \sqrt{\langle a, a\rangle}$ is the associated semi-norm, then this inner product must satisfy the Cauchy–Schwarz inequality

$$|\langle a, b\rangle| \leqslant \|a\|_2\|b\|_2.$$

Note also that if $a \in I_{\tau,2}$ and $b \in A$, then using part (iv) of Proposition 2.3.7, we get

$$\|ab\|_2^2 = \tau(b^*a^*ab) = \tau(abb^*a^*) \leqslant \tau(a\|bb^*\|a^*) = \|b\|^2\|a\|_2^2.$$

Combining the last two displayed inequalities, we see that if $a \in I_{\tau,1}$ is positive and $b \in A$ then

$$|\tau(ab)| = |\langle a^{1/2}, a^{1/2}b\rangle| \leqslant \|a^{1/2}\|_2\|a^{1/2}b\|_2 \leqslant \|a^{1/2}\|_2^2\|b\| = \tau(a)\|b\|.$$

Using the trace property from part (v) of Proposition 2.3.7, this gives the special case of the inequality from part (i) that we were aiming for.

Fix now $a \in I_{\tau,1}$; we will show that that $|a| \in I_{\tau,1}$, establishing half of part (ii). As $I_{\tau,1}$ is spanned by its positive elements, the special case of part (i) for positive $a \in I_{\tau,1}$ implies that for any $a \in I_{\tau,1}$ there is a constant $c > 0$ depending on a such that for all $b \in A$,

$$|\tau(ab)| \leqslant c\|b\|. \tag{2.6}$$

Let

$$w_n := a((1/n) + a^*a)^{-1/2}. \tag{2.7}$$

As $I_{\tau,1}$ is an ideal, w_n is in $I_{\tau,1}$ for all n. We have moreover that

$$w_n^*w_n = a^*a((1/n) + a^*a)^{-1},$$

so $\|w_n\|^2 = \|w_n^*w_n\| \leqslant 1$ by the functional calculus. Further, we have

$$w_n^*a = ((1/n) + a^*a)^{-1/2}a^*a,$$

whence again by the functional calculus, the sequence $(w_n^* a)_{n=1}^\infty$ converges in norm to $|a|$. Using the inequality in line (2.6) we see that

$$|\tau(w_n^* a)| \leqslant c\|w_n^*\| \leqslant c$$

for all n, i.e. the sequence $(\tau(w_n^* a))_{n=1}^\infty$ is uniformly bounded. Lower semicontinuity now gives that $\tau(|a|) < \infty$, completing the proof that if a is in I_τ, then $|a|$ is too.

We can now deduce the general case of the inequality in part (i). Note first that it suffices to prove the inequality for $b \in I_{\tau,1}$ as $I_{\tau,1}$ is norm dense in A, and both sides of the inequalities in (i) are norm continuous in b (for the left-hand side, this follows from line (2.6) above). Analogously to line (2.6), there exists some constant $c > 0$ depending on b such that

$$|\tau(ab) - \tau(w_n|a|b)| = |\tau((a - w_n|a|)b)| \leqslant c\|a - w_n|a|\|$$

for all n. Moreover one computes using the functional calculus that $(w_n|a| - a)^*(w_n|a| - a)$ converges to zero in norm and thus by the C^*-identity that $w_n|a| \to a$ in norm. Hence the inequality in the previous displayed line implies that

$$|\tau(ab)| = \lim_{n\to\infty} |\tau(w_n|a|b)|.$$

On the other hand, we have

$$\lim_{n\to\infty} |\tau(|a|bw_n)| \leqslant \limsup_{n\to\infty} \tau(|a|)\|bw_n\| \leqslant \tau(|a|)\|b\|,$$

where the first inequality follows from the version of part (i) with a positive, and the second inequality follows as $\|w_n\| \leqslant 1$ for all n. The last two displayed lines complete the proof of part (i).

We now look at part (iii). Let first $a \in A$ be arbitrary, and consider w_n as in line (2.7) above. Then the sequence $(w_n^* a)$ consists of positive elements. Moreover, using the functional calculus it is monotone increasing and converges in norm to $|a|$. Hence $\tau(|a| - w_n^* a) \geqslant 0$ for all n, and so

$$\limsup_{n\to\infty} \tau(w_n^* a) \leqslant \tau(|a|).$$

On the other hand, lower semicontinuity gives that

$$\tau(|a|) \leqslant \liminf_{n\to\infty} \tau(w_n^* a),$$

whence we have

$$\tau(|a|) = \lim_{n\to\infty} \tau(w_n^* a). \tag{2.8}$$

Let now $a, b \in I_{\tau,1}$, and let v_n be defined analogously to w_n, but starting with $a+b$ rather than a. Then combining line (2.8) above with part (i) we get

$$\begin{aligned}\tau(|a+b|) &= \lim_{n\to\infty} |\tau(v_n^*(a+b))| \leqslant \limsup_{n\to\infty}(|\tau(v_n^*b)| + |\tau(v_n^*a)|) \\ &\leqslant \limsup_{n\to\infty} \|v_n^*\|(\tau(|a|) + \tau(|b|)) \leqslant \tau(|a|) + \tau(|b|),\end{aligned}$$

Thus we get part (iii) in the special case when a, b are in $I_{\tau,1}$. Say now that $a, b \in A$ are arbitrary. Let (h_i) be an increasing approximate unit for A that is contained in $I_{\tau,1}$ (such an approximate unit exists by Theorem 1.5.6). Set $a_i := h_i a$ and $b_i := h_i b$, which are elements of the ideal $I_{\tau,1}$. Note that

$$|a_i|^2 = a^* h_i^* h_i a_i \leqslant a^* \|h_i\|^2 a \leqslant a^* a = |a|^2$$

for all i. As taking square roots preserves inequalities amongst positive elements in C^*-algebras (see Exercise 1.9.11), we get that $|a_i| \leqslant |a|$ for all i. Hence $\tau(|a_i|) \leqslant \tau(|a|)$ for all i, and similarly $\tau(|b_i|) \leqslant \tau(|b|)$ for all i. On the other hand, using lower semicontinuity and the special case of part (iii) for $a, b \in I_{\tau,1}$ that we have already proved, we get

$$\tau(|a+b|) \leqslant \liminf_{i\to\infty} \tau(|a_i + b_i|) \leqslant \liminf_{i\to\infty} \big(\tau(|a_i|) + \tau(|b_i|)\big) \leqslant \tau(|a|) + \tau(|b|)$$

completing the proof of part (iii) in general.

We now go back to the other half of part (ii), i.e. that if $|a|$ is in $I_{\tau,1}$ for some $a \in A$, then a is also in $I_{\tau,1}$. First consider the special case that $a \in A$ is self-adjoint and such that $|a|$ is in $I_{\tau,1}$. Let a_+ and a_- be the positive and negative parts of a respectively as in the hint to Exercise 1.9.8. We then have that $|a| = a_+ + a_-$, whence a_+ and a_- are in $I_{\tau,1}$ as $I_{\tau,1}$ is hereditary. Hence $a = a_+ - a_-$ is in $I_{\tau,1}$.

In general, let $a \in A$ be arbitrary and such that $|a|$ is in $I_{\tau,1}$. Note first that with w_n as in line (2.7) we have that

$$\begin{aligned}w_n |a| w_n^* &= a((1/n) + a^*a)^{-1/2}(a^*a)^{1/2}((1/n) + a^*a)^{-1/2}a^* \\ &= ((1/n) + aa^*)^{-1}(aa^*)^{3/2},\end{aligned}$$

and thus $w_n |a| w_n^*$ converges in norm to $|a^*|$ as $n \to \infty$ by the functional calculus. Hence from part (i) and lower semicontinuity, we get that $\tau(|a^*|) < \infty$, and thus that $|a^*|$ is also in $I_{\tau,1}$. Let $x = \frac{1}{2}(a + a^*)$ and $y = \frac{1}{2i}(a - a^*)$ be the real and imaginary parts of a respectively, so x and y are self-adjoint, and $a = x + iy$. Applying (iii), we see that

$$\tau(|x|) \leqslant \frac{1}{2}\tau(|x+iy|) + \frac{1}{2}\tau(|x-iy|) = \frac{1}{2}\tau(|a|) + \frac{1}{2}\tau(|a^*|) < \infty,$$

whence $|x|$ is in $I_{\tau,1}$, and so x is in $I_{\tau,1}$ by the self-adjoint case already considered. Similarly y is in $I_{\tau,1}$, and we are done with part (ii).

Finally, for part (iv) note first that $\|\cdot\|_\tau$ satisfies the triangle inequality by (iii), and the other norm axioms are straightforward to check. It is clear that the adjoint $*$ is isometric for $\|\cdot\|_\tau$, and the fact that $\|\cdot\|_\tau$ is submultiplicative follows from part (i).

Hence to show that $\|\cdot\|_\tau$ makes $I_{\tau,1}$ into a Banach $*$-algebra, it remains to check completeness. Let (a_n) be a Cauchy sequence. In particular, note that (a_n) is Cauchy for the usual norm on A, and thus has a limit in A, say a; we need to show that a is is $I_{\tau,1}$, and that $\tau(|a_n - a|) \to 0$. As (a_n) is Cauchy for $\|\cdot\|_\tau$, we have in particular that the sequence $(\tau(|a_n|))$ is bounded, and thus by lower semicontinuity, that $\tau(|a|) < \infty$. Hence $|a|$ is in $I_{\tau,1}$, and thus a is in $I_{\tau,1}$ by part (ii). To show that $\tau(|a_n - a|) \to 0$, let $\epsilon > 0$, and let N be such that for $n, m \geqslant N$, $\tau(|a_n - a_m|) < \epsilon$. Lower semicontinuity gives us that $\tau(|a_n - a|) \leqslant \epsilon$ for $n \geqslant N$, and we are done. □

We have now established the basic properties of positive traces that we need, and turn to K-theoretic applications.

Definition 2.3.12 Let A be a complex algebra, and $\mathcal{A}$ a unital dense subalgebra (with the same unit). Then $\mathcal{A}$ is *inverse closed* in A if whenever $a \in \mathcal{A}$ has an inverse a^{-1} in A, then a^{-1} is actually in $\mathcal{A}$.

If $\mathcal{A}$ is a subalgebra of a non-unital complex algebra A, then $\mathcal{A}$ is *inverse closed* in A if the subalgebra $\mathcal{A}^+$ of the unitisation A^+ generated by $\mathcal{A}$ and the unit is inverse closed in A^+.

Lemma 2.3.13 *Let A be a C^*-algebra and τ a densely defined, lower semicontinuous positive trace on A. Then $I_{\tau,1}$ is inverse closed in A.*

Proof The unital case is trivial by Remark 2.3.8, so we will assume that A is non-unital and thus work in the unitisations. Let us norm the unitisation $I_{\tau,1}^+$ of $I_{\tau,1}$ by the formula $\|a + \lambda\|_\tau := \|a\|_\tau + |\lambda|$ for $a \in I_{\tau,1}$ and $\lambda \in \mathbb{C}$ as in Exercise 2.11.7. Note that if $a + \lambda$ is an element in A^+ with $a \in A$ self-adjoint, then the functional calculus gives us that

$$\|a\| \leqslant 2\|a + \lambda\|$$

for any $\lambda \in \mathbb{R}$. Let $a \in A$ be a general element, and let $a = x + iy$ be its decomposition into real and imaginary parts as in the hint to Exercise 1.9.8. Then for $\lambda \in \mathbb{C}$ we get using the inequality above in the self-adjoint case

$$\|a\| \leqslant \|x\| + \|y\| \leqslant 2\|x + \mathrm{Re}(\lambda)\| + 2\|y + \mathrm{Im}(\lambda)\|.$$

Using that $x + \mathrm{Re}(\lambda)$ and $y + \mathrm{Im}(\lambda)$ are respectively the real and imaginary parts of $a+\lambda$, they both have norm bounded above by $\|a+\lambda\|$, and thus we get

$$\|a\| \leqslant 4\|a+\lambda\| \tag{2.9}$$

for any $a \in A$ and any $\lambda \in \mathbb{C}$.

Now, using part (iv) of Proposition 2.3.11, the norm $\|\cdot\|_\tau$ on $I_{\tau,1}$ satisfies

$$\|ab\|_\tau \leqslant \|a\|\|b\|_\tau \tag{2.10}$$

for all $a \in A$ and $b \in I_{\tau,1}$. Applying the inequality $|\lambda| \leqslant \|a+\lambda\|$ (which holds for any $a \in A$ and $\lambda \in \mathbb{C}$) and the inequalities in lines (2.9) and (2.10) we get that for any $a \in A$, $b \in I_\tau$ and $\lambda, \mu \in \mathbb{C}$

$$\begin{aligned}\|(a+\lambda)(b+\mu)\|_\tau &= \|ab+\lambda b+\mu a\| + \tau(|ab|+\lambda b+\mu a|) + |\lambda\mu| \\ &\leqslant \|a+\lambda\|\|b\| + |\mu|\|a\| + \|b\|\tau(|a|) + |\mu|\tau(|a|) + |\lambda\mu| \\ &\leqslant \|a+\lambda\|(\|b\| + \tau(|b|) + |\mu|) + 5\|b \\ &\quad + \mu\|(\|a\| + \tau(|a|) + |\lambda|).\end{aligned}$$

Hence

$$\|(a+\lambda)(b+\mu)\|_\tau \leqslant 5(\|a+\lambda\|\|b+\mu\|_\tau + \|a+\lambda\|_\tau\|b+\mu\|). \tag{2.11}$$

Let now $a \in I^+_{\tau,1}$, and let $r_\tau(a)$ and $r_A(a)$ denote its spectral radii considered as an element of $I^+_{\tau,1}$ and of A^+ respectively. We claim that $r_A(a) = r_\tau(a)$. Indeed, applying line (2.11), we see that for any n,

$$\|a^{2n}\|_\tau \leqslant 10\|a^n\|\|a^n\|_\tau.$$

Taking nth roots, taking the limit as n tends to infinity, and applying the spectral radius formula (Theorem 1.2.6) then implies that

$$r_\tau(a)^2 \leqslant r_A(a)r_\tau(a),$$

and thus that $r_\tau(a) \leqslant r_A(a)$. As the spectrum of a in $I^+_{\tau,1}$ is no smaller than the spectrum in A^+, the opposite inequality is immediate and we get $r_\tau(a) = r_A(a)$ as claimed.

Now, let $a \in I^+_{\tau,1}$ be invertible in A^+; we need to show that it is actually invertible in $I^+_{\tau,1}$. As a is invertible in A, a^*a is also invertible in A, whence its A-spectrum is contained in $[c,\infty)$ for some $c > 0$. Hence for all real λ, the A-spectrum of $\lambda - a^*a$ is contained in $(-\infty, \lambda - c]$. For λ suitably large, this operator is positive, so has spectrum contained in $[0, \lambda - c]$, and in particular has spectral radius (in A) strictly less than λ. As we know the spectral radii in A^+ and $I^+_{\tau,1}$ are the same, however, this gives $r_\tau(\lambda - a^*a) < \lambda$, which implies that 0 cannot be in the $I^+_{\tau,1}$-spectrum of a^*a, and so a^*a is invertible in $I^+_{\tau,1}$. Hence a is left invertible in $I^+_{\tau,1}$. However, the left inverse in $I^+_{\tau,1}$ must equal the actual inverse in A^+, which is thus in $I^+_{\tau,1}$, and we are done. □

Our next goal is to show that being inverse closed is preserved under taking matrix algebras, at least under suitable assumptions. We need a technical lemma.

Lemma 2.3.14 *Let A be a unital Banach algebra, and $\mathcal{A}$ a unital dense subalgebra (with the same unit). Then the following are equivalent:*

(i) $\mathcal{A}$ is inverse closed in A;

(ii) for every maximal right ideal $\mathcal{J}$ in $\mathcal{A}$, the intersection $\overline{\mathcal{J}} \cap \mathcal{A}$ of the closure of $\mathcal{J}$ and $\mathcal{A}$ equals $\mathcal{J}$;

(iii) for any irreducible (right) $\mathcal{A}$ module $\mathcal{M}$, there exists an A module M such that the restriction of M to $\mathcal{A}$ contains $\mathcal{M}$ as a submodule.

Proof For (i) implies (ii), let $\mathcal{J}$ be a maximal right ideal in $\mathcal{A}$. It cannot contain any invertible element of $\mathcal{A}$, whence by (i) it cannot contain any invertible element of A. As the collection of invertibles in A is open (Theorem 1.2.1), the closure $\overline{\mathcal{J}}$ cannot contain any invertible in A either. Hence $\overline{\mathcal{J}} \cap \mathcal{A}$ is a right ideal in $\mathcal{A}$ that contains $\mathcal{J}$; by maximality, they are equal.

For (ii) implies (iii), say $\mathcal{M}$ is an irreducible $\mathcal{A}$ module. Choose any non-zero $m \in \mathcal{M}$, and note that the map

$$\mathcal{A} \to \mathcal{M}, \quad a \mapsto ma$$

is surjective by irreducibility. Hence if $\mathcal{J}$ is the kernel of this map, then the induced module map

$$\mathcal{J} \setminus \mathcal{A} \to \mathcal{M}$$

is an isomorphism of right A modules. As $\mathcal{M}$ is irreducible, $\mathcal{J}$ is moreover maximal. Let $J = \overline{\mathcal{J}}$ be the closure of $\mathcal{J}$ in A, which is a right ideal, and let $M = J \setminus A$. Then, using (ii) for the second isomorphism in the chain below,

$$\mathcal{M} \cong \frac{\mathcal{A}}{\mathcal{J}} \cong \frac{\mathcal{A}}{J \cap \mathcal{A}} \cong \frac{J + \mathcal{A}}{J} \subseteq \frac{A}{J} = M$$

(where the isomorphisms and inclusions are all in the category of right $\mathcal{A}$-modules); this gives the result.

Finally, for (iii) implies (i), we assume that (i) does not hold, so there is a non-invertible element a of $\mathcal{A}$ that is invertible in A. The element a cannot be right-invertible in $\mathcal{A}$, and thus it is contained in a maximal right ideal, say $\mathcal{J}$, in $\mathcal{A}$. Then $\mathcal{M} := \mathcal{J} \setminus \mathcal{A}$ is an irreducible $\mathcal{A}$ module. Letting $[1] \in \mathcal{M}$ be the class of the identity of $\mathcal{A}$ we get that $[1]a = [a] = 0$ as $a \in \mathcal{J}$. It therefore cannot be true that $\mathcal{M}$ is contained in the restriction to $\mathcal{A}$ of any A module M: indeed, the fact that a is invertible in A implies that the map

$$M \to M, \quad m \mapsto ma$$

is injective, but the above shows that $[1]$ is in the kernel of this map and $[1] \neq 0$ in $\mathcal{M}$. □

Corollary 2.3.15 *Let A be a unital Banach algebra, and let $\mathcal{A}$ be an inverse closed unital dense subalgebra (with the same unit). Then for all n, $M_n(\mathcal{A})$ is inverse closed in $M_n(A)$.*

Proof For any unital algebra R, the categories of R modules and $M_n(R)$ modules are equivalent. Indeed, if M is an R module, then $M^{\oplus n}$ defines an $M_n(R)$ module in a canonical way, while if M is an $M_n(R)$ module and $e_{11} \in M_n(R)$ is the standard matrix unit, then Me_{11} identifies with an R module in a canonical way; moreover, these processes are mutually inverse up to canonical isomorphisms. It follows from this that the inclusion $\mathcal{A} \to A$ satisfies condition (iii) of Lemma 2.3.14 if and only if the inclusion $M_n(\mathcal{A}) \to M_n(A)$ does. □

Theorem 2.3.16 *Let A be a C^*-algebra, and let $\mathcal{A}$ be an inverse closed dense $*$-subalgebra. Assume moreover that $\mathcal{A}$ is a Banach algebra in its own right, and that the inclusion $\iota \colon \mathcal{A} \to A$ is continuous. Then the map $\iota_* \colon K_0(\mathcal{A}) \to K_0(A)$ induced by ι is an isomorphism.*

Once we have defined the (topological) K_1 group in Section 2.5, it will be possible to prove an analogous result for K_1: see Exercise 2.11.11. The result can also be generalised to the case that $\mathcal{A}$ is only a Fréchet algebra: see Exercise 2.11.12.

Proof Say we have proved the result in the unital case. Then in the non-unital case, the definition of K_0 gives a commutative diagram

$$\begin{array}{ccccccccc} 0 & \longrightarrow & K_0(\mathcal{A}) & \longrightarrow & K_0(\mathcal{A}^+) & \longrightarrow & K_0(\mathbb{C}) & \longrightarrow & 0 \\ & & \downarrow & & \downarrow & & \| & & \\ 0 & \longrightarrow & K_0(A) & \longrightarrow & K_0(A^+) & \longrightarrow & K_0(\mathbb{C}) & \longrightarrow & 0 \end{array}$$

of short exact sequences. The non-unital result thus follows from the unital one and the five lemma. Hence we may assume everything is unital.

Looking first at surjectivity, it suffices to show that if $p \in M_n(A)$ is a projection, then $[p]$ is in the image of ι_*. As $\mathcal{A}$ is dense in A, there exists contractive $a \in M_n(\mathcal{A})$ with $\|a - p\| < 1/12$. As $p^2 = p$, it follows that

$$\|a^2 - a\| \leqslant \|a(a-p)\| + \|(a-p)p\| + \|a - p\| < 1/4,$$

i.e. that a is a quasi-idempotent in the sense of Definition 2.2.7. Hence the spectrum of a in $M_n(A)$ misses the line $\mathrm{Re}(z) = 1/2$. Using Lemma 2.3.15, the spectra of a in $M_n(\mathcal{A})$ and in $M_n(A)$ are the same, so the spectrum of a in $M_n(\mathcal{A})$ also misses this line. Hence the characteristic function χ of $\{z \in \mathbb{C} \mid \mathrm{Re}(z) > 1/2\}$ is holomorphic on the $M_n(\mathcal{A})$-spectrum of a; as $M_n(\mathcal{A})$ is a Banach algebra (use for example the norm from Exercise 2.11.7), we may thus use the holomorphic functional calculus (Theorem 1.4.6) to form $q := \chi(a)$ in $M_n(\mathcal{A})$. As the inclusion map $\iota \colon \mathcal{A} \to A$ is continuous, we have that $\chi(a)$ is the same whether formed using the Banach algebra structure of $M_n(\mathcal{A})$, or of $M_n(A)$. Now, q is an idempotent; using the continuity statement at the end of Theorem 1.4.6, we can force q to be as close to p as we like by forcing a to be close to $p = \chi(p)$. As long as $\|q - p\| < 1$, we have $[q] = [p]$ by Lemma 2.2.2, completing the proof of surjectivity.

For injectivity, it suffices to show that if $(p_t)_{t\in[0,1]}$ in $M_n(A)$ is a homotopy of projections with $p_0, p_1 \in M_n(\mathcal{A})$, then p_0, p_1 are also homotopic through idempotents in $M_n(\mathcal{A})$. This can be achieved by approximating the homotopy by a piecewise linear path $(a_t)_{t\in[0,1]}$ in $M_n(\mathcal{A})$ (that does not necessarily consist of idempotents) with $p_t = a_t$ for $t \in \{0,1\}$ and $\|a_t - p_t\| < 1/12$ for all t. We then replace this piecewise linear path with the path $(\chi(a_t))_{t\in[0,1]}$, which now passes through idempotents in $M_n(\mathcal{A})$ by the same argument as in the surjectivity half, and that satisfies $\chi(a_t) = a_t = p_t$ for $t \in \{0,1\}$. Moreover, the path $(\chi(a_t))_{t\in[0,1]}$ is continuous using the continuity statement for the holomorphic functional calculus at the end of Theorem 1.4.6 again. We have thus shown that p_0 and p_1 are homotopic through idempotents in $M_n(\mathcal{A})$, completing the proof. □

Putting everything together, we get to the following definition.

Definition 2.3.17 Let τ be a densely defined, lower semicontinuous positive trace on a C^*-algebra A and $I_{\tau,1}$ be the $*$-ideal of Proposition 2.3.7. Let $\tau_* \colon K_0(I_{\tau,1}) \to \mathbb{C}$ be the map induced on τ on K-theory as in Remark 2.1.16, and $\iota_* \colon K_0(I_{\tau,1}) \to K_0(A)$ be the map induced on K-theory by the inclusion $\iota \colon I_{\tau,1} \to A$ (which is an isomorphism by Proposition 2.3.11, Lemma 2.3.13, and Proposition 2.3.16). Define

$$\tau_* \colon K_0(A) \to \mathbb{C}, \quad x \mapsto \tau_* \circ \iota_*^{-1}(x).$$

Remark 2.3.18 With notation as in Definition 2.3.17, the map $\tau_*\colon K_0(A) \to \mathbb{C}$ is actually real-valued. Indeed, using inverse closedness, the proof of Proposition 2.2.5 works in $I_{\tau,1}$ to show that any idempotent in the matrices $M_\infty(I^+_{\tau,1})$ over the unitisation is equivalent in K_0 to a self-adjoint idempotent. However, any self-adjoint element of $M_\infty(I^+_{\tau,1})$ can be written as a real linear combination of two elements from $M_\infty(I^+_{\tau,1}) \cap M_\infty(A^+_+)$: this follows as whenever $a \in M_\infty(I_{\tau,1})$, we get that $|a|$ is also in $M_\infty(I_{\tau,1})$ from part (ii) of Proposition 2.3.11 (applied to the natural extension of τ to some matrix algebra over A containing a), and as $a = \frac{1}{2}(|a| + a) - \frac{1}{2}(|a| - a)$.

The map in Definition 2.3.17 is difficult to compute in general. However, following through all the various definitions, we at least have the following result: we leave the elementary checks involved to the reader.

Lemma 2.3.19 *With notation as in Definition 2.3.17, let $x \in K_0(A)$ be represented by a formal difference of projections $[p] - [q]$, where p, q are elements of $M_\infty(I^+_{\tau,1})$ such that $p - q \in M_\infty(I_{\tau,1})$. Then with $\tau_\infty\colon M_\infty(I^+_{\tau,1}) \to \mathbb{C}$ as in Remark 2.1.16, we have that*

$$\tau_*([p] - [q]) = \tau_\infty(p - q). \qquad \square$$

Remark 2.3.20 Let $A = B \otimes \mathcal{K}$, equipped with a densely defined trace of the form $\tau \otimes \mathrm{Tr}$ as in Example 2.3.4. There is then a simpler way to show that $\tau \otimes \mathrm{Tr}$ induces a map $K_0(A \otimes \mathcal{K}) \to \mathbb{R}$. Indeed, identifying $M_n(\mathbb{C})$ with the top left corner in $\mathcal{K}$ in the usual way, we get a sequence of C^*-subalgebras

$$A \otimes M_1(\mathbb{C}) \subseteq A \otimes M_2(\mathbb{C}) \subseteq A \otimes M_3(\mathbb{C}) \subseteq \cdots \subseteq A \otimes \mathcal{K},$$

whose union $M_\infty(A) := \bigcup_{n=1}^\infty A \otimes M_n(\mathbb{C})$ is dense in $A \otimes \mathcal{K}$. Then it is not difficult to see that $\tau \otimes \mathrm{Tr}$ is a finite valued trace on $M_\infty(A)$. Moreover, $M_\infty(A)$ has a holomorphic functional calculus (and even a continuous functional calculus) from the functional calculus on each $A \otimes M_n(\mathbb{C})$. The argument of Proposition 2.3.16 can thus be carried out to show that the inclusion $\iota_*\colon M_\infty(A) \to A \otimes \mathcal{K}$ induces an isomorphism on K_0-groups (compare also Corollary 2.7.2 below). Thus there is a map on K-theory $K_0(A \otimes \mathcal{K}) \to \mathbb{R}$ defined as the composition

$$(\tau \circ \mathrm{Tr})_* \circ (\iota_*)^{-1}\colon K_0(A \otimes \mathcal{K}) \to K_0(M_\infty(A)) \to \mathbb{R},$$

which one can check is the same as the map of Definition 2.3.17 in this case.

This description is rather more straightforward than the one in Definition 2.3.17, as we do not need to go through the somewhat involved analysis of the domain $I_{\tau \otimes \mathrm{Tr},1}$ of $\tau \otimes \mathrm{Tr}$. However, it is less useful for applications: the problem is that in some applications one gets elements that are naturally in $K_0(I_{\tau \otimes \mathrm{Tr},1})$, but not obviously in $K_0(M_\infty(A))$. The image of such an element under our trace map can therefore be computed using the description in Lemma 2.3.19 above, but it is not obvious that it can be computed using the description in this remark.

2.4 The Algebraic Index Map

The goal of this section is to start to discuss how the K_0 functor interacts with short exact sequences. One might hope it takes short exact sequence to short exact sequences, but this is not quite right. The first step in understanding possible failures of exactness is to define the so-called index map, which we do here.

Throughout the section, we work purely algebraically with arbitrary $\mathbb{C}$-algebras, just as we did in Section 2.1. Indeed, it is a very useful fact that the index map can be defined purely algebraically and we want to emphasise this point; we will come back to the index map in more detail in Section 2.8.

Starting the formal discussion, let

$$0 \longrightarrow I \longrightarrow R \longrightarrow Q \longrightarrow 0$$

be a short exact sequence of $\mathbb{C}$-algebras. We thus get a functorially induced sequence

$$K_0(I) \longrightarrow K_0(R) \longrightarrow K_0(Q)$$

of abelian groups. This sequence will not in general be exact any more, but one does at least have the following; the proof makes a good exercise, or can be found in the references given in Section 2.12, Notes and References, at the end of the chapter.

Proposition 2.4.1 *The functor K_0 is* half exact, *meaning that if*

$$0 \longrightarrow I \longrightarrow R \longrightarrow Q \longrightarrow 0$$

is a short exact sequence of $\mathbb{C}$-algebras, then the induced sequence

$$K_0(I) \longrightarrow K_0(R) \longrightarrow K_0(Q)$$

is exact in the middle. □

In the remainder of this section, we will discuss an obstruction to the map $K_0(I) \to K_0(R)$ being injective.

Definition 2.4.2 Let S be a unital $\mathbb{C}$-algebra. Define $GL_\infty(S)$ to consist of $\mathbb{N}$-by-$\mathbb{N}$ invertible matrices over S that only differ from the identity at finitely many entries.

More abstractly, one can equivalently define $GL_\infty(S)$ as follows. For each n, let $GL_n(S)$ denote the group of invertible $n \times n$ matrices with values in S, and let $GL_\infty(S)$ be the direct limit (in the category of groups) of the sequence

$$GL_1(S) \to GL_2(S) \to GL_3(S) \cdots$$

under the connecting maps

$$u \mapsto \begin{pmatrix} u & 0 \\ 0 & 1 \end{pmatrix}.$$

Let now

$$0 \longrightarrow I \longrightarrow R \longrightarrow Q \longrightarrow 0$$

be a short exact sequence, and let $u \in GL_n(Q^+)$ for some n. In general, there need not be an element $\widetilde{u} \in GL_\infty(R^+)$ lifting Q. However, the element

$$\begin{pmatrix} 0 & -u^{-1} \\ u & 0 \end{pmatrix} \in GL_{2n}(Q^+)$$

is always liftable to $GL_{2n}(R^+)$ via the following trick of Whitehead. Indeed, we may write

$$\begin{pmatrix} 0 & -u^{-1} \\ u & 0 \end{pmatrix} = \begin{pmatrix} 1 & 0 \\ u & 1 \end{pmatrix} \begin{pmatrix} 1 & -u^{-1} \\ 0 & 1 \end{pmatrix} \begin{pmatrix} 1 & 0 \\ u & 1 \end{pmatrix}. \tag{2.12}$$

Each of the three factors on the right-hand side then lifts to a matrix in $GL_{2n}(R^+)$, using that the map $R^+ \to Q^+$ is surjective and that any upper or lower triangular matrix with ones on the diagonal is invertible.

Note now that if v is any invertible lift to $GL_{2n}(R^+)$ of $\left(\begin{smallmatrix} 0 & -u^{-1} \\ u & 0 \end{smallmatrix}\right)$, then $v \left(\begin{smallmatrix} 1 & 0 \\ 0 & 0 \end{smallmatrix}\right) v^{-1}$ is a lift of $\left(\begin{smallmatrix} 0 & 0 \\ 0 & 1 \end{smallmatrix}\right)$. It follows that the idempotent $v \left(\begin{smallmatrix} 1 & 0 \\ 0 & 0 \end{smallmatrix}\right) v^{-1}$ is in $M_{2n}(I^+)$, and that the difference

$$v \begin{pmatrix} 1 & 0 \\ 0 & 0 \end{pmatrix} v^{-1} - \begin{pmatrix} 0 & 0 \\ 0 & 1 \end{pmatrix}$$

is in $M_{2n}(I)$. Thus the following definition makes sense.

Definition 2.4.3 Let

$$0 \longrightarrow I \longrightarrow R \longrightarrow Q \longrightarrow 0$$

be a short exact sequence of $\mathbb{C}$-algebras, and let $u \in GL_\infty(Q^+)$ be invertible. Let $v \in GL_\infty(R^+)$ be a lift of $\begin{pmatrix} 0 & -u^{-1} \\ u & 0 \end{pmatrix}$, and define the *index* of u to be the formal difference of idempotents

$$\mathrm{Ind}(u) := \left[v \begin{pmatrix} 1 & 0 \\ 0 & 0 \end{pmatrix} v^{-1} \right] - \begin{bmatrix} 0 & 0 \\ 0 & 1 \end{bmatrix} \in K_0(I).$$

This is spelled out a little more concretely in Definition 2.8.1 below. The most important example comes from the classical *Fredholm index*, as in Example 2.8.3; this is also the source of the terminology. The reader is encouraged to look forward to these examples; for now, however, we will content ourselves with looking at formal properties.

The following proposition consists of fairly direct checks: we again leave it as an exercise.

Proposition 2.4.4 *With notation as in Definition 2.4.3, Ind(u) is a well-defined element of $K_0(I)$. Moreover, the map*

$$Ind\colon GL_\infty(Q^+) \to K_0(I)$$

is a group homomorphism, and fits into an exact sequence

$$GL_\infty(Q^+) \xrightarrow{Ind} K_0(I) \longrightarrow K_0(R) \longrightarrow K_0(Q) . \qquad \square$$

Remark 2.4.5 If Q happens to be unital, one can use $GL_\infty(Q)$ in place of $GL_\infty(Q^+)$ in Definition 2.4.3 and Proposition 2.4.4, and the discussion leading up to them. The index map one gets is essentially the same: see Exercise 2.11.9.

Remark 2.4.6 If $u \in GL_\infty(Q^+)$ happens to lift to an invertible in $GL_\infty(R^+)$, then one can check that Ind(u) is zero in $K_0(I)$: indeed, in this case $\begin{pmatrix} 0 & -u^{-1} \\ u & 0 \end{pmatrix}$ lifts to an invertible matrix of the same form, and using this lift to define Ind(u), the formal difference of idempotents involved is precisely zero. It follows from this that K_0 is *split exact*: this means that if a short exact sequence

$$0 \longrightarrow I \longrightarrow R \longrightarrow Q \longrightarrow 0$$

has the property that the quotient map $R \to Q$ is split, then the sequence

$$0 \longrightarrow K_0(I) \longrightarrow K_0(R) \longrightarrow K_0(Q) \longrightarrow 0$$

is exact.[9]

[9] More precisely, this argument implies exactness at the left; exactness on the right follows from the existence of a splitting and functoriality.

This is about as far as we will take the purely algebraic theory. Indeed, at this point one is led to define $K_1(Q)$ to be some quotient of $GL_\infty(Q^+)$, and similarly for $K_1(R)$, in such a way that the exact sequence from Proposition 2.4.4 can be continued to the left, getting

$$K_1(R) \longrightarrow K_1(Q) \xrightarrow{\text{Ind}} K_0(I) \longrightarrow K_0(R) \longrightarrow K_0(Q) .$$

There are two natural ways to do this: one purely algebraic, and one bringing topology into the picture. In the next section we discuss the topological method.

2.5 The Topological K_1 Group

In this section, we define the topological K_1 group of a C^*-algebra (and more generally, of a Banach algebra).

To do this, we will need topologies on $M_n(A)$ for a Banach algebra A: for a C^*-algebra we use the unique C^*-norm as in Example 1.6.10, and for a general Banach algebra we use the norm from Exercise 2.11.7. These norm topologies restrict to topologies on $GL_n(A)$ in each case.

Motivated by the discussion in Section 2.4, the idea is to define $K_1(A)$ by a suitable equivalence relation on $GL_\infty(A^+)$ (see Definition 2.4.2), and motivated by the discussion in Section 2.2, a reasonable definition is given by forcing homotopic elements to define the same class. Thus we are led to define a *homotopy* between $u_0, u_1 \in GL_\infty(A^+)$ as a continuous path

$$[0,1] \to GL_n(A^+), \quad t \mapsto u_t$$

connecting them for some n. As usual two invertibles are *homotopic* if there is a homotopy between them. We then define $K_1(A)$ to be the quotient of $GL_\infty(A^+)$ by the equivalence relation of homotopy. With this definition, it is not completely obvious that $K_1(A)$ is a group, however, so we instead give the following equivalent description. Indeed, note that u_0 and u_1 are homotopic in $GL_n(A^+)$ if and only if $u_0u_1^{-1}$ is connected by a path in $GL_n(A^+)$ to the identity.

Let $GL_{n,0}(A^+)$ denote then the collection of elements in $GL_n(A^+)$ that are path connected to the identity, which is a normal subgroup, and let $GL_{\infty,0}(A^+)$ be the direct limit of the groups $GL_{n,0}(A^+)$ with the same connecting maps as defining $GL_\infty(A)$. It follows from the above discussion that two elements u_0 and u_1 of $GL_\infty(A^+)$ are homotopic if and only if they define the same class in the quotient group.

$$GL_\infty(A^+)/GL_{\infty,0}(A^+).$$

Hence it is natural to make the following definition.

Definition 2.5.1 Let A be a Banach algebra. We define

$$K_1(A) := GL_\infty(A^+)/GL_{\infty,0}(A^+)$$

and write $[u]$ for the class of an element $u \in GL_\infty(A^+)$ in $K_1(A)$.

Lemma 2.5.2 *The group operation on $K_1(A)$ is abelian, and can equivalently be defined by*

$$[u] + [v] := \begin{bmatrix} u & 0 \\ 0 & v \end{bmatrix}.$$

As a result, we will typically write the operation on $K_1(A)$ additively, and write $0 \in K_1(A)$ for the identity element $[1]$.

Proof Let u, v be elements in some $GL_n(A)$. Then a rotation homotopy similar to that in line (2.3) above shows that if '$\sim$' denotes 'homotopic to', then

$$\begin{pmatrix} uv & 0 \\ 0 & 1 \end{pmatrix} = \begin{pmatrix} u & 0 \\ 0 & 1 \end{pmatrix}\begin{pmatrix} v & 0 \\ 0 & 1 \end{pmatrix} \sim \begin{pmatrix} u & 0 \\ 0 & v \end{pmatrix} \sim \begin{pmatrix} v & 0 \\ 0 & u \end{pmatrix} \sim \begin{pmatrix} vu & 0 \\ 0 & 1 \end{pmatrix}$$

and thus $[u][v] = [v][u]$ in $K_1(A)$. This computation also establishes the alternative form for the group operation. □

Remark 2.5.3 If A is a unital Banach algebra, one could also consider the group $GL_\infty(A)/GL_{\infty,0}(A)$ (i.e. without using unitisations). This is canonically isomorphic to $K_1(A)$ as defined above (and often rather more natural to work with): see Exercise 2.11.8 below.

Example 2.5.4 If $A = \mathbb{C}$, $GL_n(\mathbb{C})$ is connected for all n, and thus (by Remark 2.5.3) $K_1(\mathbb{C})$ is trivial.

Remark 2.5.5 Let A be a unital C^*-algebra, let $U_n(A)$ be the subgroup of $GL_n(A)$ consisting of unitary matrices, and $U_{n,0}(A)$ be those unitaries that are connected to the identity. Let $U_\infty(A)$ and $U_{\infty,0}(A)$ be the corresponding direct limit subgroups of $GL_\infty(A)$ and $GL_{\infty,0}(A)$ respectively. Any element $u \in GL_n(A)$ is homotopic to one in $U_n(A)$: indeed, u^*u is invertible, and the path

$$[0,1] \to GL_n(A), \quad u_t := u(u^*u)^{-t/2}$$

defines a homotopy between $u_0 = u$ and a unitary u_1 (compare Example 1.4.5 above). Moreover, if $u_0, u_1 \in U_n(A)$ are homotopic via a path (u_t) in

$GL_n(A)$, then the path $(u_t(u_t^* u_t)^{-1/2})$ is a path of unitaries between u_0 and u_1; note that this implies in particular that $U_{\infty,0}(A) = GL_{\infty,0}(A) \cap U_\infty(A)$. It follows from this discussion that the inclusion $U_\infty(A) \to GL_\infty(A)$ induces an isomorphism

$$\frac{U_\infty(A)}{U_{\infty,0}(A)} \to \frac{GL_\infty(A)}{GL_{\infty,0}(A)} = K_1(A).$$

In other words, when A is a C^*-algebra we can define $K_1(A)$ as consisting of homotopy classes of unitaries in $U_\infty(A)$.

An analogous statement holds in the non-unital (or not-necessarily unital) case with A replaced by A^+ throughout.

Remark 2.5.6 A continuous algebra homomorphism $\phi\colon A \to B$ induces a unital homomorphism $A^+ \to B^+$, and thus a group homomorphism $GL_\infty(A^+) \to GL_\infty(B^+)$ by applying the extension of ϕ to unitisations entrywise on each GL_n. Moreover, as ϕ is continuous, this group homomorphism takes $GL_{\infty,0}(A^+)$ to $GL_{\infty,0}(B^+)$. Hence we get an induced map $\phi_*\colon K_1(A) \to K_1(B)$. The assignment $\phi \mapsto \phi_*$ clearly respects composition of homomorphisms, and thus K_1 defines a functor from the category of Banach algebras and continuous algebra homomorphisms to the category of abelian groups. As in Remark 2.2.10 above, the functor K_1 clearly takes homotopic homomorphisms to the same group homomorphism.

Remark 2.5.7 Let $M_n(A)$ denote the C^*-algebra of $n \times n$ matrices over A. Then there is a (non-unital) homomorphism $A \to M_n(A)$ defined via 'top left corner inclusion'

$$a \mapsto \begin{pmatrix} a & 0 \\ 0 & 0 \end{pmatrix}.$$

The induced map $K_1(A) \to K_1(M_n(A))$ is an isomorphism.

We conclude this section with the following extension of the ideas in Section 2.4. This provides a strong suggestion that our definition of K_1 is a good one. Recall first that if

$$0 \longrightarrow I \longrightarrow A \longrightarrow Q \longrightarrow 0$$

is a short exact sequence of complex algebras, then we have a well-defined *index map*

$$\text{Ind}\colon GL_\infty(Q^+) \to K_0(I).$$

The next theorem summarises the basic properties of this index map: the proof makes an instructive exercise, and can also be found in any of the standard references discussed in Section 2.12 at the end of this chapter.

Proposition 2.5.8 *Let*

$$0 \longrightarrow I \longrightarrow A \longrightarrow Q \longrightarrow 0$$

be a short exact sequence of Banach algebras. The index map then descends to a well-defined group homomorphism

$$Ind\colon K_1(Q) \to K_0(I)$$

that fits into an exact sequence

$$K_1(I) \longrightarrow K_1(A) \longrightarrow K_1(Q) \xrightarrow{Ind} K_0(I) \longrightarrow K_0(A) \longrightarrow K_0(Q)$$

in which all the maps other than Ind are those functorially induced by the short exact sequence. Finally, the exact sequence above is natural for maps of exact sequences in the obvious sense. □

2.6 Bott Periodicity and the Six-Term Exact Sequence

Our goal in this section is to extend the short exact sequence of the previous section to a long exact sequence. This can be done using fairly general machinery borrowed from algebraic topology: the so-called *Puppe sequence*. To complete the long exact sequence, we then describe the Bott periodicity theorem. First, we need to introduce cones and suspensions.

Definition 2.6.1 Let A be a Banach algebra. The *cone* over A is the Banach algebra $CA := C_0((0,1], A)$, and the *suspension* of A is the Banach algebra $SA := C_0((0,1), A)$ (in both cases, the norm is the supremum norm).

Note that if A is a C^*-algebra, then both CA and SA are C^*-algebras too. Now, associated to these two Banach algebras is a short exact sequence

$$0 \longrightarrow SA \longrightarrow CA \longrightarrow A \longrightarrow 0$$

where the quotient map $CA \to A$ is defined by evaluation at one. From Proposition 2.5.8, we get an associated exact sequence

$$K_1(SA) \longrightarrow K_1(CA) \longrightarrow K_1(A) \xrightarrow{\text{Ind}} K_0(SA) \longrightarrow K_0(CA) \longrightarrow K_0(A)\ .$$

For each $t \in [0,1]$, define a homomorphism $h_t: CA \to CA$ by the formula

$$(h_t f)(x) = f(tx);$$

this family gives a homotopy between h_1, which is the identity map, and h_0, which is the zero map (a Banach algebra for which such a homotopy exists is said to be *contractible*). From the homotopy invariance of K_0 and K_1, we thus get that $K_0(CA) = K_1(CA) = 0$, and so we have an isomorphism

$$\mathrm{Ind}: K_1(A) \overset{\cong}{\to} K_0(SA). \tag{2.13}$$

This motivates the following definition.

Definition 2.6.2 For each $n \geqslant 0$, let $S^n A$ denote the result of applying the suspension operation to a Banach algebra A n-times. Define the *nth K-theory group* to be $K_n(A) := K_0(S^n A)$.

We have been a bit sloppy here: we have now defined $K_1(A)$ once using invertible elements in Definition 2.5.1, and again above to be $K_0(SA)$; thanks to the existence of the canonical isomorphism in line (2.13), this nicety does not really matter from a practical point of view.

The assignment $A \mapsto SA$ is a functor from the category of Banach algebras and continuous homomorphisms to itself: if $\phi: A \to B$ is a continuous homomorphism, then one gets an induced map $S\phi: SA \to SB$ by applying ϕ pointwise (i.e. $(S\phi f)(x) := \phi(f(x))$), and this assignment is clearly functorial. This functor also takes homotopies to homotopies. It follows that each K_n is a functor from the category of Banach algebras to the category of abelian groups that takes homotopic (continuous) homomorphisms to the same group homomorphism.

It is moreover straightforward to see that the functor S takes short exact sequences to short exact sequences. Hence for $n \geqslant 1$ applying S^n to a short exact sequence

$$0 \longrightarrow I \longrightarrow A \longrightarrow Q \longrightarrow 0,$$

we get another short exact sequence of Banach algebras, and using Proposition 2.5.8 and the isomorphism in line (2.13) an exact sequence of K-groups

$$\begin{array}{ccccccccc} K_n(I) & \longrightarrow & K_n(A) & \longrightarrow & K_n(Q) & & & & \\ & & & & \big\downarrow & & & & \\ & & & & K_{n-1}(I) & \longrightarrow & K_{n-1}(A) & \longrightarrow & K_{n-1}(Q). \end{array}$$

Splicing these exact sequences together, we get the desired long exact sequence in K-theory. This is usually called a *Puppe sequence* as it is a version of a general construction from algebraic topology.

Proposition 2.6.3 *Let*

$$0 \longrightarrow I \longrightarrow A \longrightarrow Q \longrightarrow 0$$

be a short exact sequence of Banach algebras. Then there is a long exact sequence of abelian groups

$$\begin{array}{ccccccc} \cdots \longrightarrow K_n(A) \longrightarrow & K_n(Q) & & & & \\ & \big\downarrow {\scriptstyle Ind_n} & & & & \\ & K_{n-1}(I) & \longrightarrow & K_{n-1}(A) & \longrightarrow & K_{n-1}(Q) \longrightarrow \cdots \end{array}$$

extending infinitely far to the left, and terminating at $K_0(Q)$ on the right. It is natural for maps between short exact sequences in the obvious sense. □

Up until now, most of what we have done with K-theory outside of Section 2.3 has been relatively formal: it is probably fair to say that it is not that deep.[10] On the other hand, the next theorem, the *Bott periodicity theorem* is substantial. To state it precisely, we need some notation.

Let A be a unital Banach algebra. Let $\mathbb{T} = \{z \in \mathbb{C} \mid |z| = 1\}$ be the unit circle in $\mathbb{C}$ and $1_A \in A$ be the unit, and make the identification

$$(SA)^+ = \{f \in C(\mathbb{T}, A) \mid f(1) \in \mathbb{C}1_A\}.$$

For an idempotent $e \in M_n(A)$, define

$$\beta_A(e) \in (SA)^+, \quad \beta_A(e) \colon z \mapsto zp + (1_{M_n(A)} - e),$$

and note that $\beta_A(e)$ is an invertible element of $M_n((SA)^+)$. It is not too difficult to see that β_A induces a well-defined map

$$\beta_A \colon K_0(A) \to K_1(SA).$$

Moreover, the map β_A is natural for unital continuous homomorphisms between unital Banach algebras. Hence considering the short exact sequence

$$0 \longrightarrow A \longrightarrow A^+ \longrightarrow \mathbb{C} \longrightarrow 0,$$

we see that β_{A^+} canonically induces a map

$$\beta_A \colon K_0(A) \to K_1(SA)$$

[10] This is not quite the same thing as saying that it is easy!

whether or not A is unital. Using the canonical identification $K_1(SA) = K_2(A)$, we may thus make the following definition.

Definition 2.6.4 Let A be a Banach algebra. The *Bott map* is the homomorphism

$$\beta_A : K_0(A) \to K_2(A)$$

defined above.

Here then is a version of the Bott periodicity theorem. We will prove a version of it using index theory in Section 9.3, although for now we will just use it as a black box.

Theorem 2.6.5 *For any Banach algebra A, the Bott map*

$$\beta_A : K_0(A) \to K_2(A)$$

is an isomorphism, and is natural in A. Moreover, combining this isomorphism with the long exact sequence from Proposition 2.6.3 above associated to every short exact sequence

$$0 \longrightarrow I \longrightarrow A \longrightarrow Q \longrightarrow 0$$

gives a six-term exact sequence

$$\begin{array}{ccccc} K_0(I) & \longrightarrow & K_0(A) & \longrightarrow & K_0(Q) \\ {\scriptstyle Ind}\uparrow & & & & \downarrow{\scriptstyle Ind_2 \circ \beta_Q} \\ K_1(Q) & \longleftarrow & K_1(A) & \longleftarrow & K_1(I) \end{array}$$

which is natural for maps between short exact sequences. □

It follows from the above that up to canonical isomorphism, there are only two K-groups of a Banach algebra: $K_0(A)$ and $K_1(A)$. As such, the following definition will be useful, as it encodes all the information given by the K-groups.

Definition 2.6.6 Let A be a Banach algebra, and write $K_*(A)$ for the direct sum group

$$K_*(A) := K_0(A) \oplus K_1(A).$$

We consider $K_*(A)$ as an element of the category $\mathcal{GA}$ of $\mathbb{Z}/2$-graded abelian groups and graded group homomorphisms: precisely, objects in this category are abelian groups G equipped with a direct sum composition $G = G_0 \oplus G_1$, and a morphism from G to H is a group homomorphism takes G_i into H_i for $i \in \{0, 1\}$.

Clearly K_* is a functor from the category of Banach algebras and continuous homomorphisms to the category $\mathcal{GA}$ above.

Remark 2.6.7 Restricting to commutative C^*-algebras, we may consider K_* as a contravariant functor from the category $\mathcal{LC}$ of Definition B.1.1 to $\mathcal{GA}$ using Theorem 1.3.14. Considered like this, we will sometimes write $K^*(X)$ for $K_*(C_0(X))$ and similarly with the notation $K^0(X)$ and $K^1(X)$. The upper indices are to reflect the fact that K-theory is contravariant when considered as a functor of spaces rather than C^*-algebras.

Remark 2.6.8 The map $\text{Ind}_2 \circ \beta_Q$ appearing in Theorem 2.6.5 above is often called the *exponential map*. This is because it admits the following more concrete description. Say to begin with that A (hence also Q) is unital, and let $e \in M_n(A)$ represent some class in $K_0(Q)$. We may lift e to a self-adjoint element $a \in M_n(A)$ (not necessarily an idempotent any more). Identifying I^+ with a subalgebra of A in the obvious way, one can check that the invertible element $\exp(2\pi i a)$ (defined as a convergent power series) is in I^+: indeed it maps to the element $\exp(2\pi i e)$ of Q, and as $e^2 = e$ we have

$$\exp(2\pi i e) = \sum_{k=0}^{\infty} \frac{(2\pi i)^k e^k}{k!} = 1 + e \sum_{k=1}^{\infty} \frac{(2\pi i)^k}{k!} = 1 + e(\exp(2\pi i) - 1) = 1.$$

We then have that

$$\text{Ind}_2 \circ \beta_Q[e] = -[\exp(2\pi i a)].$$

If A is non-unital, we can proceed similarly. Using Remark 2.1.14, any class in $K_0(Q)$ can be represented by a formal difference $[e] - [1_n]$, where $e \in M_m(Q^+)$ is an idempotent such that the canonical quotient map $M_m(Q^+) \to M_m(\mathbb{C})$ takes e to the idempotent 1_n with image first n of the canonical basis vectors for $\mathbb{C}^m$. Then we may again lift e to a self-adjoint element $a \in M_m(A)$. Identifying I^+ (unitally) with a subalgebra of A^+, we then have that

$$\text{Ind}_2 \circ \beta_Q([e] - [1_n]) = -[\exp(2\pi i a)].$$

See Section 2.12, Notes and References, at the end of the chapter for justifications of this.

2.7 Some Computational Tools

In this section we discuss some useful tools for doing K-theory computations. All are well known and can be found in the literature; nonetheless, not all are completely standard, so we give reasonably complete proofs for ease of reference.

Although much of this material works for general Banach algebras, we typically restrict to the case of C^*-algebras as we do not need the additional generality, and as this tends to simplify things.

Continuity

Recall the definition of a directed system $(A_i)_{i\in I}$ of C^*-algebras and the associated notation and direct limit $\lim_{i\in I} A$ from Example 1.5.12. The following result is often summarised by saying that K-theory is *continuous*.

Proposition 2.7.1 *Let $(A_i)_{i\in I}$ be a directed system of C^*-algebras, and $A = \lim_{i\in I} A_i$ the associated direct limit. Then the functorially associated directed system of K-theory groups $(K_*(A_i))_{i\in I}$ has direct limit $K_*(A)$.*

Proof Let us work with K_0 and K_1 separately; we will just look at K_0 as the case of K_1 is similar. We will show that $K_0(A)$ has the universal property required by the direct limit. Let then G be any abelian group equipped with a compatible family $\psi_i : K_0(A_i) \to G$ of maps, i.e. so that the diagrams

$$\begin{array}{ccc} K_0(A_i) & \xrightarrow{(\phi_{ji})_*} & K_0(A_j) \\ \downarrow{\scriptstyle \psi_i} & & \downarrow{\scriptstyle \psi_j} \\ G & = & G \end{array}$$

commute for each i and j. We must construct a map $\psi : K_0(A) \to G$ such that the diagram

$$\begin{array}{ccc} K_0(A_i) & \xrightarrow{(\phi_i)_*} & K_0(A) \\ & \searrow^{\psi_i} & \downarrow{\scriptstyle \psi} \\ & & G \end{array}$$

commutes. Let $[p] - [q]$ be a formal difference of projections defining a class in $K_0(A)$. Using an approximation and the functional calculus (compare Construction 2.2.8), we may find i and classes of projections $[p_i]$ and $[q_i]$ in $K_0(A_i)$ such that $(\phi_i)_*([p_i] - [q_i]) = [p] - [q]$. Define

$$\psi([p] - [q]) := \psi_i([p_i] - [q_i]).$$

To see that this is well defined, say $[p_j] - [q_j]$ defines a class in $K_0(A_j)$ such that $(\phi_j)_*([p_i]-[q_i]) = [p]-[q]$. Another approximation argument (this time approximating the 'reason', say a homotopy, that $[p_i] - [q_i]$ and $[p_j] - [q_j]$ map to the same element $[p] - [q] \in K_0(A)$) now shows that

$$(\phi_{ki})_*([p_i] - [q_i]) = (\phi_{kj})_*([p_j] - [q_j])$$

in $K_0(A_k)$ for some $k \geqslant i, j$. Hence by compatibility of the family (ψ_i) we have that

$$\begin{aligned}\psi_i([p_i] - [q_i]) &= \psi_k((\phi_{ki})_*([p_i] - [q_i])) \\ &= \psi_k((\phi_{kj})_*([p_j] - [q_j])) = \psi_j([p_j] - [q_j])\end{aligned}$$

and are done. □

The following corollary is often summarised by saying that K-theory is *stable*.

Corollary 2.7.2 *Let $\mathcal{K}$ denote the C^*-algebra of compact operators on a separable,*[11] *infinite-dimensional Hilbert space H. Choosing an orthonormal basis $(e_n)_{n\in\mathbb{N}}$ for H, for any C^*-algebra A we may identify elements of $A \otimes \mathcal{K}$ with a subset of the collection of $\mathbb{N}$-by-$\mathbb{N}$ matrices over A. Then the inclusion $A \to A \otimes \mathcal{K}$ in the top left corner induces an isomorphism on K-theory.*

Proof Using the discussion at the end of Example 1.5.12, it is not difficult to see that $A \otimes \mathcal{K}$ is the direct limit of the system

$$A \otimes M_1(\mathbb{C}) \to A \otimes M_2(\mathbb{C}) \to A \otimes M_3(\mathbb{C}) \to \cdots$$

where each arrow is induced by the top left corner inclusion of the matrix algebras, and the identity on A. All the maps in this directed system induce isomorphisms on K-theory (compare Remarks 2.1.15 and 2.5.7 above), and therefore the map $A = A \otimes M_1(\mathbb{C}) \to A \otimes \mathcal{K}$ arising from the definition of the direct limit does too; this map is just the standard top left corner inclusion, however. □

Example 2.7.3 In the special case $A = \mathbb{C}$, the above corollary gives that a choice of top left corner inclusion $\mathbb{C} \to \mathcal{K}$ induces an isomorphism on K-theory, and thus that

$$K_i(\mathcal{K}) = \begin{cases} \mathbb{Z} & i = 0, \\ 0 & i = 1. \end{cases}$$

Moreover, inspection of the proof (compare Example 2.1.12) shows that under this isomorphism if $p \in \mathcal{K}$ is a projection, then the class $[p] \in K_0(\mathcal{K})$ corresponds under this isomorphism to $\text{rank}(p) \in \mathbb{Z}$.

[11] This assumption is not necessary: just replace the inductive system in the proof by a slightly more complicated one.

We finish the discussion of continuity with one more useful consequence: the point is that if a C^*-algebra has a particularly nice approximate unit, then $K_0(A)$ can be generated by projections from matrices with values in A, rather than needing projections from matrices with values in the unitisation of A.

Corollary 2.7.4 *Let A be a C^*-algebra and assume that A has an approximate unit $(p_i)_{i \in I}$ consisting of a directed set of projections. Then $K_0(A)$ is generated by the classes of projections in matrix algebras over A, and $K_1(A)$ is generated by unitaries of the form $v + (1 - p_i \otimes 1_n)$ for some i, where 1_n is the unit of $M_n(\mathbb{C})$, and v is an element of $M_n(A) = A \otimes M_n(\mathbb{C})$ for some n such that $v^*v = vv^* = p_i \otimes 1_n$.*

Proof As (p_i) is a directed set and an approximate unit, A is the direct limit of the directed set of subalgebras $(p_i A p_i)_{i \in I}$, ordered by inclusion. Hence by Proposition 2.7.1, $K_*(A) = \lim_{i \in I} K_*(p_i A p_i)$. As each p_i is a projection, each subalgebra $p_i A p_i$ is unital (with unit p_i). Hence $K_0(p_i A p_i)$ is generated by projections in matrix algebras over $p_i A p_i$, and thus the same is true for $K_0(A)$. The statement for K_1 follows similarly on noting that the image of the class $[v] \in K_1(p_i A p_i)$ of a unitary $v \in M_n(p_i A p_i)$ under the map $K_1(p_i A p_i) \to K_1(A)$ induced by inclusion is exactly the class of the element $v + (1 - p_i \otimes 1_n)$ of $M_n(A^+)$. □

Isometries and Eilenberg Swindles

The next few results in this section will mainly be used to construct *Eilenberg swindles*, as illustrated in Corollary 2.7.7. The basic idea of an Eilenberg swindle is that if $g \in G$ is an element of an abelian group and one can make reasonable sense of an element $\infty \cdot g$ which is the 'sum of g with itself infinitely many times', then g will satisfy the equation $\infty \cdot g + g = \infty \cdot g$, and so g must be zero. What exactly '$\infty \cdot g$' actually means will depend on the context.

The following result is a useful tool for making sense of the above ideas (and more generally). For the statement, recall the notion of the multiplier algebra $M(A)$ of a C^*-algebra A from Definition 1.7.6.

Proposition 2.7.5 *Let $\alpha \colon A \to C$ be a $*$-homomorphism and $v \in M(C)$ be a partial isometry such that $\alpha(a)vv^* = \alpha(a)$ for all $a \in A$. Then the map*

$$ad_v \circ \alpha \colon A \to C, \quad a \mapsto v\alpha(a)v^*$$

is a $*$-homomorphism, and induces the same map on K-theory as α.

Proof It is straightforward to check that $\text{ad}_v \circ \alpha$ is a $*$-homomorphism: we leave this to the reader.

First we show the result in the special case that $v = u$ is a unitary. Let $\phi\colon K_*(M_2(C)) \to K_*(C)$ be the isomorphism discussed in Remarks 2.1.15 and 2.5.7; in other words, ϕ is the inverse to the map on K-theory induced by the top left corner inclusion

$$a \mapsto \begin{pmatrix} a & 0 \\ 0 & 0 \end{pmatrix}.$$

Then the map $(\mathrm{ad}_u)_*\colon K_*(C) \to K_*(C)$ is the composition of: the map on K-theory induced by the top left corner inclusion, the map on K-theory induced by conjugation by the matrix

$$\begin{pmatrix} u & 0 \\ 0 & u^* \end{pmatrix}, \tag{2.14}$$

and ϕ. The matrix in line (2.14) is homotopic to the identity however (compare line (2.3) above), so conjugation by it induces the trivial map on K-theory, and we are done in the case v is unitary.

In the general case, consider the matrix

$$u = \begin{pmatrix} v & 1 - vv^* \\ 1 - v^*v & v^* \end{pmatrix}$$

in the multiplier algebra of $M_2(C)$. This is unitary and so by the earlier discussion $\mathrm{ad}_u\colon M_2(C) \to M_2(C)$ induces the identity on K-theory. However, the map induced by $\mathrm{ad}_v \circ \alpha$ on K-theory is the composition of the maps induced on K-theory by α, the top left corner inclusion, ad_u, and $\phi\colon K_*(M_2(A)) \to K_*(A)$. Hence $\mathrm{ad}_v \circ \alpha$ induces the same map on K-theory as α. □

Lemma 2.7.6 *Let A, B be C*-algebras, and let $\alpha_1, \alpha_2\colon A \to B$ be $*$-homomorphisms with orthogonal image, meaning that $\alpha_1(a_1)\alpha_2(a_2) = 0$ for all $a_1, a_2 \in A$. Then the linear map $\alpha := \alpha_1 + \alpha_2\colon A \to B$ is a $*$-homomorphism, and as maps on K-theory, $(\alpha_1 + \alpha_2)_* = (\alpha_1)_* + (\alpha_2)_*$.*

Proof It is straightforward to check that α is a $*$-homomorphism. Let $\phi\colon K_*(M_2(B)) \to K_*(B)$ be the inverse of the stabilisation isomorphisms of Remarks 2.1.15 and 2.5.7, and note that for $i \in \{1,2\}$ the composition of the map on K-theory induced by the $*$-homomorphism

$$A \to M_2(B), \quad a \mapsto \begin{pmatrix} \alpha_i(a) & 0 \\ 0 & 0 \end{pmatrix}$$

and ϕ equals $(\alpha_i)_*$. The same is true if we replace the above with the map

$$A \to M_2(B), \quad a \mapsto \begin{pmatrix} 0 & 0 \\ 0 & \alpha_i(a) \end{pmatrix}$$

as this differs from the previous version by conjugation by the element $\left(\begin{smallmatrix} 0 & 1 \\ 1 & 0 \end{smallmatrix}\right)$ of the multiplier algebra of $M_2(A)$, and by Proposition 2.7.5, this conjugation has no effect on K-theory. As the operations on K_0 and K_1 can both be described by block sum of matrices, it follows that $(\alpha_1)_* + (\alpha_2)_*$ is the map on K-theory induced by

$$A \to M_2(B), \quad a \mapsto \begin{pmatrix} \alpha_1(a) & 0 \\ 0 & \alpha_2(a) \end{pmatrix} \tag{2.15}$$

composed with the isomorphism $\phi\colon K_*(M_2(B)) \to K_*(B)$. On the other hand, α_* is given as the composition of the map on K-theory defined by

$$A \to M_2(B), \quad a \mapsto \begin{pmatrix} \alpha(a) & 0 \\ 0 & 0 \end{pmatrix} \tag{2.16}$$

and ϕ. It thus suffices to show that the maps in line (2.15) and line (2.16) are the same on K-theory. This follows as the path

$$\psi_t : a \mapsto \begin{pmatrix} \alpha_1(a) & 0 \\ 0 & 0 \end{pmatrix} + \begin{pmatrix} \cos(t) & -\sin(t) \\ \sin(t) & \cos(t) \end{pmatrix} \begin{pmatrix} 0 & 0 \\ 0 & \alpha_2(a) \end{pmatrix} \begin{pmatrix} \cos(t) & \sin(t) \\ -\sin(t) & \cos(t) \end{pmatrix}$$

as t varies from 0 to $\pi/2$ is a homotopy between them (the fact that α_1 and α_2 have orthogonal images is used to show that each ψ_t is indeed a $*$-homomorphism). □

The following corollary is included mainly as it illustrates how the above results can be used to perform Eilenberg swindles, i.e. to construct 'infinite sums' of a given element of a K-group.[12] The result of the corollary can also be deduced directly from Example 2.1.7 (at least when H is separable, and by a minor variation of this in general).

Corollary 2.7.7 *Let H be an infinite-dimensional Hilbert space. Then $K_*(\mathcal{B}(H)) = 0$.*

Proof As H is infinite-dimensional, we may write H as a countable direct sum $H = \bigoplus_{n=1}^{\infty} H_n$ of subspaces, each with the same dimension as H itself. For each n, choose an isometry $v_n \in \mathcal{B}(H)$ with image $H_n \subseteq H$, and let $\mathrm{ad}_{v_n}\colon \mathcal{B}(H) \to \mathcal{B}(H)$ be the associated $*$-homomorphism. Set v to be the direct sum map

$$v = \sum_{n=2}^{\infty} v_n \colon H \to H.$$

[12] Although what we actually do is construct an infinite sum of the identity map with itself in the endomorphisms of K_*.

Proposition 2.7.5 also implies that $\mathrm{ad}_{v_1} + \mathrm{ad}_v$ induces the same map on K-theory as ad_v, as these two $*$-homomorphisms are conjugate via the isometry

$$w := \sum_{n=1}^{\infty} v_{n+1}v_n^*$$

(the sum converges in the strong operator topology). Hence using Lemma 2.7.6,

$$(\mathrm{ad}_{v_1})_* + (\mathrm{ad}_v)_* = (\mathrm{ad}_{v_1} + \mathrm{ad}_v)_* = (\mathrm{ad}_v)_*.$$

Cancelling off $(\mathrm{ad}_v)_*$, we get that $(\mathrm{ad}_{v_1})_* = 0$. However, Proposition 2.7.5 (applied in the special case that α is the identity map) implies that $(\mathrm{ad}_{v_1})_*$ is the identity map, so $K_*(\mathcal{B}(H)) = 0$ as required. □

Doubles and Quasi-Morphisms

Our next goal is to discuss *quasi-morphisms*. These are a more general class of maps than $*$-homomorphisms and are useful for inducing maps between K-theory groups.

Definition 2.7.8 Let A be a C^*-algebra and I an ideal in A. Then the *double* of A along I, denoted $D_A(I)$, is the C^*-algebra

$$\{(a,b) \in A \oplus A \mid a - b \in I\}.$$

Lemma 2.7.9 *The natural inclusion $I \mapsto D_A(I)$ defined by $a \mapsto (a,0)$ leads to a split short exact sequence*

$$0 \longrightarrow I \longrightarrow D_A(I) \longrightarrow A \longrightarrow 0$$

and thus a direct sum decomposition

$$K_*(D_A(I)) \cong K_*(A) \oplus K_*(I).$$

Proof The quotient map in the short exact sequence is given by evaluation on the second coordinate, and the splitting is given by the function $A \to D_A(I)$, defined by $a \mapsto (a,a)$. The existence of the direct sum decomposition follows directly from this and the six-term exact sequence (Theorem 2.6.5). □

Definition 2.7.10 Let A and I be C^*-algebras. A *quasi-morphism* from A to I consists of a C^*-algebra B containing I as an ideal, and a pair

$$\phi, \psi : A \to B$$

of $*$-homomorphisms such that $\phi(a) - \psi(a)$ is in I for all $a \in A$.

Given a quasi-morphism as above, the induced map on K-theory

$$(\phi - \psi)_* \colon K_*(A) \to K_*(I)$$

is defined as the composition of the map on K-theory induced by the $*$-homomorphism

$$\phi \oplus \psi \colon A \to D_B(I)$$

and the quotient map

$$K_*(D_B(I)) \to K_*(I)$$

arising from the direct sum decomposition in Lemma 2.7.9.

Direct Products

For our next goal we discuss the behaviour of K-theory under products. Recall from Example 1.1.7 that if $(A_i)_{i \in I}$ is a collection of C^*-algebras, then their product $\prod_{i \in I} A_i$ is the C^*-algebra of all bounded sequences $(a_i)_{i \in I}$ with $a_i \in A_i$, equipped with pointwise operations and the supremum norm. One might hope that there is a natural isomorphism

$$K_*\left(\prod_{i \in I} A_i\right) \cong \prod_{i \in I} K_*(A_i)$$

induced by the quotient maps $\prod_{i \in I} A_i \to A_j$, but this is not true in general: see Exercise 2.11.15 below.

It becomes true if one assumes some form of stability in the sense of Definition 1.8.15: for example assuming $A_i \otimes \mathcal{K} \cong A_i$ (where $\mathcal{K}$ is the compact operators) for each i would be enough. For applications, we want to get away with something a little weaker than this, so introduce the following definition.

Definition 2.7.11 A C^*-algebra A is *quasi-stable* if for all n there exists an isometry v in the multiplier algebra of $M_n(A)$ such that vv^* is the matrix unit e_{11}.

The terminology is inspired by the case of stable C^*-algebras as in Definition 1.8.15, i.e. those C^*-algebras A isomorphic to $A \otimes \mathcal{K}$, for $\mathcal{K}$ the compact operators on a separable infinite-dimensional Hilbert space. It is not too difficult to see that a stable C^*-algebra is quasi-stable, but the converse is false: $\mathcal{B}(H)$ for infinite-dimensional H is a counterexample. The point of quasi-stability is that it has many of the same K-theoretic consequences as the more commonly used stability, but is more general and easier to check.

Proposition 2.7.12 *Let $(A_i)_{i\in I}$ be a collection of quasi-stable C^*-algebras. Then the natural quotients*

$$\pi_j \colon \prod_{i\in I} A_i \to A_j$$

induce an isomorphism

$$\prod_{i\in I}(\pi_i)_* \colon K_*\left(\prod_{i\in I} A_i\right) \to \prod_{i\in I} K_*(A_i).$$

Proof We will construct an inverse map

$$\prod_{i\in I} K_*(A_i) \to K_*\left(\prod_{i\in I} A_i\right).$$

For simplicity, let us focus on the case of K_0; the case of K_1 is similar. Let then an element of $\prod_{i\in I} K_*(A_i)$ be given, which can be represented as a sequence $([p_i]-[q_i])_{i\in I}$ of formal differences of projections, where p_i, q_i are in $M_{n_i}(A^+)$ for some n_i, and $p_i - q_i \in M_{n_i}(A)$ for all i (compare Remark 2.1.14 above). The sequence $([p_i]-[q_i])_{i\in I}$ does not obviously define an element of $K_0(\prod_{i\in I} A_i)$: the problem is that there is no uniform bound on n_i as i varies.

To get around this problem, we use quasi-stability. Let $v_i \in M_{n_i}(A_i)$ be an isometry with the property that $v_i v_i^*$ is the top left matrix unit as in the definition of quasi-stability. Using Proposition 2.7.5, for each i we have

$$[v_i p_i v_i^*] - [v_i q v_i^*] = [p_i] - [q_i]$$

in $K_0(A_i)$. However, we may identify the elements $v_i p_i v_i^*$ and $v_i q_i v_i^*$ with elements of A_i^+, whose difference is in A_i. Making this identification, we may thus define our putative inverse map

$$\prod_{i\in I} K_*(A_i) \to K_*\left(\prod_{i\in I} A_i\right), \quad ([p_i]-[q_i])_{i\in I} \mapsto [(v_i p_i v_i^*)_{i\in I}] - [(v_i q_i v_i^*)]_{i\in I}.$$

To complete the proof, we need to show that this map makes sense and really is an inverse to the map in the statement: this is all routine, and we leave the details to the reader. □

Mayer–Vietoris Sequences

There are long exact Mayer–Vietoris sequences associated to pushouts and pullbacks (defined below) of C^*-algebras. These are useful variations of the basic six-term exact sequence of Theorem 2.6.5. They generalise the classical

Mayer–Vietoris sequences for decompositions of a topological space into closed and open subsets: see Example 2.7.16 below.

We will have one Mayer–Vietoris sequence for pushouts, and a different one for pullbacks, as defined below.

Definition 2.7.13 A *pushout diagram* of C^*-algebras is a diagram of the form

$$\begin{array}{ccc} I \cap J & \xrightarrow{\iota^I} & I \\ \downarrow{\scriptstyle \iota^J} & & \downarrow{\scriptstyle \kappa^I} \\ J & \xrightarrow{\kappa^J} & A, \end{array} \tag{2.17}$$

where I and J are ideals in A, the arrows are the obvious inclusions and where the sum $I + J$ is dense.[13]

Definition 2.7.14 A *pullback diagram* of C^*-algebras is a diagram of the form

$$\begin{array}{ccc} P & \xrightarrow{\rho^A} & A \\ \downarrow{\scriptstyle \rho^B} & & \downarrow{\scriptstyle \pi^A} \\ B & \xrightarrow{\pi^B} & Q, \end{array} \tag{2.18}$$

where the maps π_A, π_B are $*$-homomorphisms with at least one of them being surjective, where $P = \{(a,b) \in A \oplus B \mid \pi_A(a) = \pi_B(b)\}$ and where the maps from P to A and B are the restrictions to the summands.

The Mayer–Vietoris sequences associated to such diagrams are then as follows.

Proposition 2.7.15 *Let A, I, J be as in Definition 2.7.13 above. Then there is a six-term Mayer–Vietoris sequence*

$$\begin{array}{ccccc} K_0(I \cap J) & \longrightarrow & K_0(I) \oplus K_0(J) & \longrightarrow & K_0(A) \\ \uparrow & & & & \downarrow \\ K_1(A) & \longleftarrow & K_1(I) \oplus K_1(J) & \longleftarrow & K_1(I \cap J), \end{array}$$

which is natural for commutative diagrams of pushout diagrams. The morphisms

$$K_*(I \cap J) \to K_*(I) \oplus K_*(J) \quad \text{and} \quad K_*(I) \oplus K_*(J) \to K_*(A)$$

[13] This actually forces in A to equal $I + J$, as we will see from the proof.

in the above are given by

$$x \mapsto \iota_*^I(x) \oplus \iota_*^J(x) \quad \text{and} \quad y \oplus z \mapsto \kappa_*^I(y) - \kappa_*^J(z)$$

respectively.

Let P, A, B, Q be as in Definition 2.7.14 above. Then there is a six-term Mayer–Vietoris sequence

$$\begin{array}{ccccc} K_0(P) & \longrightarrow & K_0(A) \oplus K_0(B) & \longrightarrow & K_0(Q) \\ \uparrow & & & & \downarrow \\ K_1(Q) & \longleftarrow & K_1(A) \oplus K_1(B) & \longleftarrow & K_1(P), \end{array}$$

which is natural for commutative maps between pullback diagrams. The morphisms

$$K_*(P) \to K_*(A) \oplus K_*(B) \quad \text{and} \quad K_*(A) \oplus K_*(B) \to K_*(Q)$$

in the above are given by

$$x \mapsto \rho_*^A(x) \oplus \rho_*^B(x) \quad \text{and} \quad y \oplus z \mapsto \pi_*^A(y) - \pi_*^B(z)$$

respectively.

Proof of Proposition 2.7.15, pushout case We first note that the map

$$I \oplus J \to A/(I \cap J), \quad (a_I, a_J) \mapsto a_I + a_J$$

is a well-defined $*$-homomorphism with dense image. It is thus surjective by Corollary 1.5.10, and so we automatically have that $A = I + J$, not merely that $I + J$ is dense in A. Hence standard isomorphism theorems from pure algebra give that

$$\frac{I}{I \cap J} \cong \frac{I + J}{J} = \frac{A}{J}.$$

Consider the following commutative diagram of short exact sequences

$$\begin{array}{ccccccccc} 0 & \longrightarrow & I \cap J & \longrightarrow & I & \longrightarrow & A/J & \longrightarrow & 0 \\ & & \downarrow & & \downarrow & & \| & & \\ 0 & \longrightarrow & J & \longrightarrow & A & \longrightarrow & A/J & \longrightarrow & 0, \end{array}$$

which by Theorem 2.6.5 gives rise to a commutative diagram of six-term exact sequences

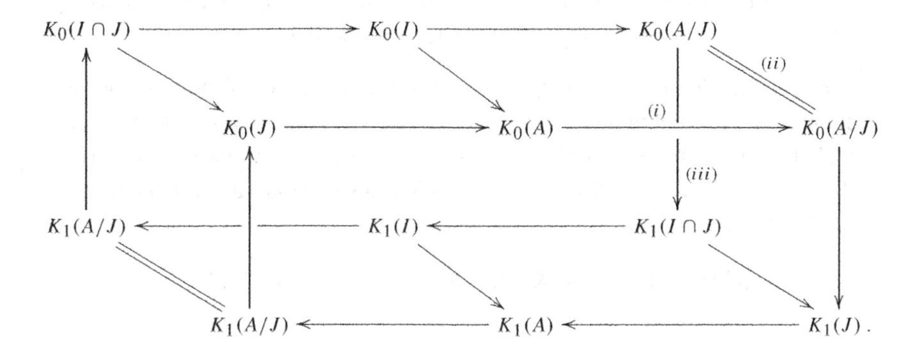

The existence of the Mayer–Vietoris sequence follows from this and some diagram chasing that we leave to the reader: for example, the map $K_0(A) \to K_1(I \cap J)$ in the Mayer–Vietoris sequence is defined as the composition of the arrows marked (i), (ii), (iii). Naturality follows from naturality of the usual six-term exact sequence. □

Proof of Proposition 2.7.15, pullback case Assume without loss of generality that the map $A \to Q$ is surjective, and let I be the kernel. Then there is a commutative diagram of short exact sequences

$$\begin{array}{ccccccccc} 0 & \longrightarrow & I & \longrightarrow & A & \xrightarrow{\pi^A} & Q & \longrightarrow & 0 \\ & & \| & & \uparrow{\scriptstyle \rho^A} & & \uparrow{\scriptstyle \pi^B} & & \\ 0 & \longrightarrow & I & \longrightarrow & P & \xrightarrow{\rho^B} & B & \longrightarrow & 0. \end{array}$$

Theorem 2.6.5 associates to this a commutative diagram of six-term exact sequences

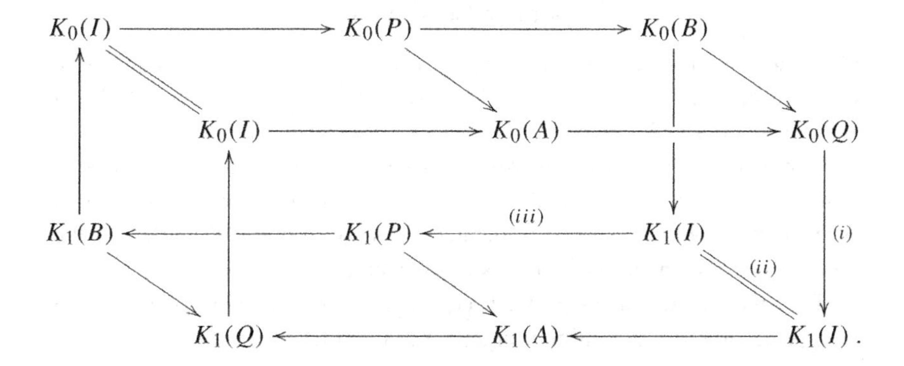

The existence of the Mayer–Vietoris sequence again follows from a diagram chase that we leave to the reader: for example, the boundary map $K_0(Q) \to K_1(P)$ is the composition of the three arrows labelled (i), (ii), (iii). Naturality again follows from naturality of the six-term exact sequence. □

Example 2.7.16 The motivating examples for these Mayer–Vietoris sequences come from the commutative case. Indeed, let X be a locally compact space, and U, V be open subsets such that $X = U \cup V$. Then $A = C_0(X)$, $I = C_0(U)$, $J = C_0(V)$ and $I \cap J = C_0(U \cap V)$ fit it into a pushout diagram as in Definition 2.7.13. The associated six-term exact sequence is

$$\begin{array}{ccccc} K^0(U \cap V) & \longrightarrow & K^0(U) \oplus K^0(V) & \longrightarrow & K^0(X) \\ \uparrow & & & & \downarrow \\ K^1(X) & \longleftarrow & K^1(U) \oplus K^1(V) & \longleftarrow & K^1(U \cap V), \end{array}$$

where we use the conventions of Remark 2.6.7. Note the functoriality in the above appears to go the 'wrong way' for a cohomology theory: this is because K-theory is a cohomology theory 'with compact supports', and is functorial in 'the wrong' direction for inclusions of open subsets.

On the other hand, if $X = E \cup F$ for closed subsets E and F, taking $P = C_0(X)$, $A = C_0(E)$, $B = C_0(F)$ and $Q = C_0(E \cap F)$, we have a pullback diagram. The associated six-term exact sequence becomes

$$\begin{array}{ccccc} K^0(X) & \longrightarrow & K^0(E) \oplus K^0(F) & \longrightarrow & K^0(E \cap F) \\ \uparrow & & & & \downarrow \\ K^1(E \cap F) & \longleftarrow & K^1(E) \oplus K^1(F) & \longleftarrow & K^1(X) \end{array}$$

(where now the functoriality is in the 'expected' direction).

Morita Invariance

Our final goal in this section is to discuss invariance of K-theory under Morita equivalence (see Definition 1.7.9). For this, it suffices to show that if B is a full corner of A as in Definition 1.7.8, then the inclusion $B \to A$ induces an isomorphism on K-theory. This result is sometimes only stated in the special case that A and B have countable approximate units, but it will be important for us that it holds in general; fortunately, it is not too difficult to deduce the general case, as long as we assume some machinery.

To introduce the machinery we need, let A be a C^*-algebra, let $p \in M(A)$ be a projection in the multiplier algebra of A and let $\mathcal{K} := \mathcal{K}(\ell^2(\mathbb{N}))$ be the compact operators. Note that by definition of the spatial tensor product we may consider $p \otimes 1$ as an element of the multiplier algebra of $A \otimes \mathcal{K}$: indeed, if A is faithfully represented on H_A, then $A \otimes \mathcal{K}$ is faithfully represented on $H_A \otimes \ell^2(\mathbb{N})$ in the natural way, and $p \otimes 1$ is clearly in the multiplier algebra for this representation.[14] Recall also from Definition 1.7.8 that a projection $p \in M(A)$ is *full* if ApA is dense in A.

We will use the following theorem as a black box: see Section 2.12, Notes and References, at the end of the chapter for more detail.

Theorem 2.7.17 *Say A is a separable*[15] *C^*-algebra and $p \in M(A)$ a full projection. Then there is an isometry $v \in M(A \otimes \mathcal{K})$ such that $v^*v = 1$ and $vv^* = p \otimes 1$.* □

Corollary 2.7.18 *Say A is a separable C^*-algebra and $B \subseteq A$ is a full corner. Then the inclusion $B \to A$ induces an isomorphism on K-theory.*

Proof Let $p \in M(A)$ be such that $pAp = B$ and ApA is dense in A as in the definition of a full corner. Let $v \in M(A \otimes \mathcal{K})$ be as in Theorem 2.7.17. Define

$$\phi\colon A \otimes \mathcal{K} \to B \otimes \mathcal{K}, \quad a \mapsto vav^*.$$

It is not difficult to check that ϕ is a $*$-isomorphism. Consider now the composition

$$A \otimes \mathcal{K} \xrightarrow{\phi} B \otimes \mathcal{K} \to A \otimes \mathcal{K},$$

of ϕ and the inclusion $B \otimes \mathcal{K} \to A \otimes \mathcal{K}$ induced by the inclusion $B \to A$ and the identity map on $\mathcal{K}$ (see Remark 1.8.12). This composition is given by $a \mapsto vav^*$, and thus induces the identity map on K-theory by Proposition 2.7.5. Hence in particular the inclusion $B \otimes \mathcal{K} \to A \otimes \mathcal{K}$ induces an isomorphism on K-theory (precisely, the inverse of the isomorphism on K-theory induced by the $*$-isomorphism ϕ). The proof is completed by considering the commutative diagram

$$\begin{array}{ccc} B & \longrightarrow & A \\ \downarrow & & \downarrow \\ B \otimes \mathcal{K} & \longrightarrow & A \otimes \mathcal{K}, \end{array}$$

where the horizontal arrows are the canonical inclusions and the vertical arrows are a choice of top left corner inclusion: indeed, we have just seen that the

[14] Underlying this, there is always a natural inclusion $M(A) \otimes M(B) \subseteq M(A \otimes B)$: exercise!

[15] More generally, σ-unital, meaning that A has a countable approximate unit.

bottom arrow induces an isomorphism on K-theory, while the two vertical arrows induce isomorphisms by Corollary 2.7.2. □

Proposition 2.7.19 *Say A is a C^*-algebra and $B \subseteq A$ is a full corner. Then the inclusion $B \to A$ induces an isomorphism on K-theory.*

Proof Let $p \in M(A)$ be a full projection such that $pAp = B$. Let $(A_i)_{i\in I}$ be the collection of all separable C^*-subalgebras of A, ordered by inclusion. For each i, let C_i be the C^*-subalgebra of A generated by pA_ip, pA_i, A_ip and A_ipA_i. Note that the collection $(C_i)_{i\in I}$ is directed by inclusion as the collection (A_i) is. Moreover: each C_i is separable; p acts as a multiplier on each C_i; as p is full in A we have that $\lim_{i\in I} C_i = A$; and as $pAp = B$ we have that $\lim_{i\in I} pC_ip = B$.

To complete the proof, it will thus suffice to show that p is a full projection for each C_i. Indeed, in that case Corollary 2.7.18 implies that the inclusion $pC_ip \to C_i$ induces an isomorphism on K-theory for each i. Thanks to Proposition 2.7.1 we would get therefore that the inclusion

$$\lim_{i\in I} pC_ip \to \lim_{i\in I} C_i$$

induces an isomorphism on K-theory, which is the desired result.

Consider then the C^*-algebra generated by C_ipC_i, which we want to show equals C_i. Using an approximate unit for C_i (which exists by Theorem 1.5.6), however, it is not difficult to see that the C^*-algebra generated by C_ipC_i contains each of the sets pA_ip, pA_i, A_ip and A_ipA_i; as these generate C_i, we are done. □

2.8 Index Elements

In this section we make the index map from Section 2.4 above more explicit and collect some variations that are useful for applications.

Say first that B is a C^*-algebra, and A is an ideal in B. If $u \in B$ is an operator such that the image of u in B/A is invertible, then we get a class $[u] \in K_1(B/A)$, and hence via the index map

$$\text{Ind}\colon K_1(B/A) \to K_0(A)$$

of Definition 2.5.8, a class $\text{Ind}[u] \in K_0(A)$. Looking back at the construction of Ind from Definition 2.4.3 above, we have the following concrete description of the element $\text{Ind}[u]$.

Definition 2.8.1 With notation as above, let $w \in B$ be such that $uw - 1$ and $wu - 1$ are in A. Consider the product

$$v := \begin{pmatrix} 1 & 0 \\ u & 1 \end{pmatrix} \begin{pmatrix} 1 & -w \\ 0 & 1 \end{pmatrix} \begin{pmatrix} 1 & 0 \\ u & 1 \end{pmatrix} = \begin{pmatrix} 1 - wu & -w \\ 2u - uwu & 1 - uw \end{pmatrix} \tag{2.19}$$

in $M_2(B)$, which is an invertible element that agrees with $\begin{pmatrix} 0 & -w \\ u & 0 \end{pmatrix}$ modulo $M_2(A)$ (compare line (2.12) above). Then

$$\mathrm{Ind}[u] := \left[v \begin{pmatrix} 1 & 0 \\ 0 & 0 \end{pmatrix} v^{-1} \right] - \begin{bmatrix} 0 & 0 \\ 0 & 1 \end{bmatrix}.$$

As in Definition 2.5.8 above, the class $\mathrm{Ind}[u]$ does not depend on the choice of w.

Remark 2.8.2 For some applications it is useful that this construction is purely algebraic: it involves no functional calculus, just finite linear combinations of finite products of elements from the set $\{1, u, w\}$. This follows as the definition of v in line (2.19) leads to a purely algebraic formula for v^{-1} in terms of u and w. Spelling this out,

$$v^{-1} = \begin{pmatrix} 1 - wu & w \\ uwu - 2u & 1 - uw \end{pmatrix}$$

and so

$$\mathrm{Ind}[u] = \begin{bmatrix} (1 - wu)^2 & w(1 - uw) \\ u(2 - wu)(1 - wu) & uw(2 - uw) \end{bmatrix} - \begin{bmatrix} 0 & 0 \\ 0 & 1 \end{bmatrix}; \tag{2.20}$$

Thus the formula above gives a representative for the class $\mathrm{Ind}[u]$ where each matrix entry is a sum of products of at most five elements from the set $\{u, w\}$ (and the identity).

The situation is even better if the image of u in B/A is unitary. Indeed, in this case, one can use $w = u^*$, and the resulting formula for $\mathrm{Ind}[u]$ involves only finite linear combinations of finite products of elements from the set $\{1, u, u^*\}$.

Example 2.8.3 The fundamental example occurs when $A = \mathcal{K}(H)$ and $B = \mathcal{B}(H)$ for some Hilbert space H. Let $u \in B$ be invertible in B/A. Then *Atkinson's theorem* (see Exercise 2.11.21) says that u is Fredholm, meaning that it has closed range, and the kernel and cokernel of u are finite-dimensional. Let K be the kernel of u, and let C be the orthogonal complement of its image, which is finite-dimensional. Then u restricts to a bijection $u_0 \colon K^\perp \to C^\perp$, which is invertible by the open mapping theorem. Let $w_0 \colon C^\perp \to K^\perp$ be the inverse of u_0, and let $w \in \mathcal{B}(H)$ be defined to be equal to w_0 on elements of

$C^\perp$ and equal to 0 on elements from C. Then if p_K and p_C are the orthogonal projections onto K and C respectively, we have that

$$uw = 1 - p_C \quad \text{and} \quad wu = 1 - p_K.$$

Substituting these into the formula in line (2.20) above and computing, we get that

$$\mathrm{Ind}[u] = \begin{bmatrix} p_K & 0 \\ 0 & 1 - p_C \end{bmatrix} - \begin{bmatrix} 0 & 0 \\ 0 & 1 \end{bmatrix} = [p_K] - [p_C].$$

Under the isomorphism $K_0(\mathcal{K}) \cong \mathbb{Z}$ of Example 2.7.3, the class $\mathrm{Ind}[u]$ corresponds to

$$\mathrm{rank}(p_K) - \mathrm{rank}(p_C) = \dim(K) - \dim(C),$$

which by definition is the classical Fredholm index of u. This justifies the name 'index map' for $\mathrm{Ind}\colon K_1(B/A) \to K_0(A)$.

There is also an opposite parity case of the index construction: this starts with an element p of B such that the image of p in B/A is a projection and so defines a class $[p] \in K_0(B/A)$.

Definition 2.8.4 With notation as above, say the image of $p \in B$ in B/A is a projection. Then we define

$$\mathrm{Ind}[p] \in K_1(A)$$

to be the image of $[p] \in K_0(B/A)$ under the Bott periodicity map

$$\mathrm{Ind}_2 \circ \beta_{B/A}\colon K_0(B/A) \to K_1(A)$$

of Theorem 2.6.5. Using the description in Remark 2.6.8, we have the following concrete formula

$$\mathrm{Ind}[p] = [e^{-2\pi i p}]$$

for this class.

Note that the formula above is still concrete, but has the disadvantage that it is no longer purely algebraic. For this reason, it is sometimes advantageous to use the identification $K_0(B/A) = K_1(S(B/A))$ so we can apply the formula in Definition 2.8.1 to this case too.

We now give another variant that is useful in geometric applications. Here, it is useful to have a description of the above index elements in the language of graded Hilbert spaces. See Appendix E for conventions on gradings (although we will recall what we need here).

Recall first from Definition E.1.4 that a *grading* on a Hilbert space is a unitary operator $U: H \to H$ such that $U^2 = 1$; this U is also called the *grading operator* for H. A bounded operator F on H is *even* for the grading if $UFU = F$, and is *odd* if $UFU = -F$. If U is a grading on H, then there is a direct sum splitting $H = H_0 \oplus H_1$ into the $+1$ and -1 eigenspaces of U respectively, called the *even* and *odd* parts of H. Writing an operator F on H as a matrix

$$F = \begin{pmatrix} F_{00} & F_{01} \\ F_{10} & F_{11} \end{pmatrix},$$

one computes that F is even if and only if the matrix is diagonal (i.e. $F_{10} = F_{01} = 0$), and is odd if and only if the matrix is off-diagonal (i.e. $F_{00} = F_{11} = 0$).

Say now, in addition, that A is a C^*-algebra, concretely represented on H, and that the grading operator $U: H \to H$ is in the multiplier algebra of A in the sense of Definition 1.7.1. Hence in particular it induces an *inner grading* on A in the sense of Definition E.1.1. Then, writing operators as matrices as above, we have that the projections

$$P_0 := \begin{pmatrix} 1 & 0 \\ 0 & 0 \end{pmatrix} \quad \text{and} \quad P_1 := \begin{pmatrix} 0 & 0 \\ 0 & 1 \end{pmatrix}$$

are equal to $\frac{1}{2}(U+1)$ and $\frac{1}{2}(U-1)$ respectively, and thus also in the multiplier algebra of A. We can therefore think of elements of A as matrices

$$a = \begin{pmatrix} a_{00} & a_{01} \\ a_{10} & a_{11} \end{pmatrix},$$

where $a_{ij} := P_i a P_j \in A$, identified with an operator $H_j \to H_i$ in the natural way. This works in exactly the same way for multipliers of A. In what follows, we will be typically elide the difference between $P_i a P_j$ considered as an operator on H, and considered as an operator $H_j \to H_i$; thus we might say something like '$V: H_0 \to H_1$ is in the multiplier algebra of A' when to be technically correct, we should say '$\left(\begin{smallmatrix} 0 & 0 \\ V & 0 \end{smallmatrix}\right)$ is in the multiplier algebra of A'.

Definition 2.8.5 Let A be a C^*-algebra, concretely represented on a Hilbert space H. Let $F \in \mathcal{B}(H)$ be a bounded operator such that:

(i) F is in the multiplier algebra of A (see Definition 1.7.1 above);
(ii) $F^2 - 1$ is in A.

Then we can define an index class $\text{Ind}[F] \in K_*(A)$ in one of two ways, depending on whether or not H is assumed graded.

(i) Say there is no grading on H. Set $P = \frac{1}{2}(F+1)$ and note that the image of P in $M(A)/A$ is a projection. We define

$$\text{Ind}[F] := \text{Ind}[P] \in K_1(A)$$

using Definition 2.8.4.

(ii) Say now that $H = H_0 \oplus H_1$ is graded in such a way that the grading operator U is in the multiplier algebra of A, and that F is odd with respect to the grading. Writing operators on H as matrices as above, the facts that F is odd and in the multiplier algebra of A imply that F has the form

$$F = \begin{pmatrix} 0 & V \\ W & 0 \end{pmatrix}$$

for some operators $W: H_0 \to H_1$ and $V: H_1 \to H_0$ in the multiplier algebra of A. The condition that $F^2 - 1$ is in A moreover implies that $VW - 1$ and $WV - 1$ are both[16] in A. We then define

$$\text{Ind}[F] := \begin{bmatrix} (1-VW)^2 & V(1-WV) \\ W(2-VW)(1-VW) & WV(2-WV) \end{bmatrix} - \begin{bmatrix} 0 & 0 \\ 0 & 1 \end{bmatrix} \quad (2.21)$$

in $K_0(A)$ quite analogously to line (2.20) above.

See Exercise 2.11.28 for a relationship between the construction of Definition 2.8.5 above and that of Definition 2.8.1 from earlier in this section.

Remark 2.8.6 It is immediate from the construction above that if we perturb F by an element of A, then the class $\text{Ind}[F]$ is unchanged: indeed, the construction only depends on the image of F in $M(A)/A$.

We complete this section with a useful lemma.

Lemma 2.8.7 *If $F^2 = 1$, then either of the constructions from Definition 2.8.5 above give $\text{Ind}[F] = 0$.*

Proof Assume first we are in the ungraded case. Then $P = \frac{1}{2}(F+1)$ is already an idempotent in $M(A)$, and so defines a class in $K_0(M(A))$ that maps to the class $[P]$ on $K_0(M(A)/A)$. Thus by exactness of the K-theory sequence

$$K_0(M(A)) \to K_0(M(A)/A) \to K_1(A),$$

the element $\text{Ind}[F]$ is zero. The graded case is entirely analogous, noting that now $F^2 = 1$ implies that V and W are mutually inverse. □

[16] We are abusing notation again: in $VW - 1$, 1 is the identity operator on H_0, and in $WV - 1$ it is the identity operator on H_1.

2.9 The Spectral Picture of K-theory

Our goal in this section is to introduce the spectral picture of K-theory for a graded C^*-algebra. We will show that if the C^*-algebra is inner graded, spectral K-theory provides a new model for the usual K-theory groups. This is useful as the spectral picture is particularly well suited to discussions of elliptic operators, and of products in K-theory. This material is not used until Part THREE of the book.

In order to define the spectral picture of K-theory, we work in the language of graded C^*-algebras. This is not strictly necessary, but is technically convenient, and is also useful for applications. See Appendix E for a summary of the background we need on gradings. In particular, for the first definition below we need the following notation: $\mathscr{K}$ denotes a standard graded copy of the compact operators on a separable infinite-dimensional Hilbert space as in Example E.1.9; $\text{Cliff}_{\mathbb{C}}(\mathbb{R}^i)$ denotes the Clifford algebra of Example E.1.11; and $\widehat{\otimes}$ denotes the graded spatial tensor product from Definition E.2.9.

Definition 2.9.1 For graded C^*-algebras A and B, let $\{A, B\}$ denote the set of homotopy classes of graded $*$-homomorphisms from A to B; in particular, this means that all homomorphisms in a homotopy should preserve the grading.

Let A be a graded C^*-algebra. Define sets by

$$spK_i(A) := \{\mathscr{S}, A \widehat{\otimes} \text{Cliff}_{\mathbb{C}}(\mathbb{R}^i) \widehat{\otimes} \mathscr{K}\}.$$

Remark 2.9.2 Using the graded $*$-isomorphism

$$\text{Cliff}_{\mathbb{C}}(\mathbb{R}^{d-1}) \widehat{\otimes} \text{Cliff}_{\mathbb{C}}(\mathbb{R}) \cong \text{Cliff}_{\mathbb{C}}(\mathbb{R}^d)$$

of line (E.1) and the graded $*$-isomorphisms

$$\text{Cliff}_{\mathbb{C}}(\mathbb{R}^d) \cong \begin{cases} M_{2^{d/2}}(\mathbb{C}), & \text{d even} \\ M_{2^{(d-1)/2}}(\mathbb{C}) \oplus M_{2^{(d-1)/2}}(\mathbb{C}), & \text{d odd} \end{cases}$$

of line (E.2), we get graded $*$-isomorphisms

$$\text{Cliff}_{\mathbb{C}}(\mathbb{R}^d) \widehat{\otimes} \mathscr{K} \cong \begin{cases} \mathscr{K}, & \text{d even} \\ \text{Cliff}_{\mathbb{C}}(\mathbb{R}) \widehat{\otimes} \mathscr{K}, & \text{d odd} \end{cases}.$$

Moreover, these isomorphisms are unique up to (graded) homotopy equivalence. Thus up to canonical bijection, each set $spK_i(A)$ is the same as one of $spK_0(A)$ or $spK_1(A)$. Thus the sets $spK_i(A)$ satisfy a form of Bott periodicity for purely algebraic reasons. Later we will introduce binary operations on the sets $spK_i(A)$; these operations will be compatible with the periodicity isomorphisms above.

Example 2.9.3 An illustrative (and actually general – see Exercise 2.11.29) example of an element of $spK_0(A)$ arises as follows. Let $\pi : A \to \mathcal{B}(H)$ be a faithful graded representation as in Definition E.1.4. Let D be a (possibly unbounded, essentially) self-adjoint operator on H such that $f(D) \in A$ for all $f \in \mathscr{S}$. Then the functional calculus (see Theorem D.1.7 for the unbounded case) gives a $*$-homomorphism

$$\mathscr{S} \to A, \quad f \mapsto f(D).$$

One can check using Lemma D.1.9 that this $*$-homomorphism is graded if and only if D is odd in the sense of Example E.1.14.

For example, let $\mathscr{S} \to \mathcal{B}(L^2(\mathbb{R}))$ be the usual graded representation as multiplication operators (see Example E.1.10). Then the identity map $\mathscr{S} \to \mathscr{S}$ arises in this way from the unbounded (odd) operator on $L^2(\mathbb{R})$ given by multiplication by the independent variable x, say with domain $C_c(\mathbb{R})$.

The following alternative description of $spK_0(A)$ is often useful, and in some ways simpler. For the statement, recall from Example E.1.8 that a grading ϵ_A on a C^*-algebra A extends uniquely to a grading ϵ_{A^+} on the unitisation A^+.

Lemma 2.9.4 *Let B be a graded C^*-algebra. Then there is a canonical bijection between $\{\mathscr{S}, B\}$ and the set of path components of the following subset of the unitisation B^+*

$$\left\{ u \in B^+ \;\middle|\; \begin{array}{l} u \text{ is unitary, } \epsilon_{B^+}(u) = u^*, \text{ and } u \text{ maps to } 1 \\ \text{under the canonical quotient } B^+ \to \mathbb{C} \end{array} \right\}. \tag{2.22}$$

In particular, for a graded C^*-algebra A, there is a canonical bijection between $spK_0(A)$ and the set defined above in the special case $B = A \widehat{\otimes} \mathscr{K}$.

Proof A graded $*$-homomorphism $\phi \colon \mathscr{S} \to B$ extends uniquely to a $*$-homomorphism between unitisations $\phi^+ \colon \mathscr{S}^+ \to B^+$, and this extension is still graded. Forgetting the grading for a moment, $\mathscr{S}^+$ is isomorphic as a C^*-algebra to $C(\mathbb{R}^+)$, the C^*-algebra of continuous functions on the one point compactification $\mathbb{R}^+$ of $\mathbb{R}$. Now, the Cayley transform

$$c \colon \mathbb{R}^+ \to S^1, \quad x \mapsto \frac{x - i}{x + i}$$

(with $\frac{\infty - i}{\infty + i}$ interpreted as 1) is a homeomorphism identifying $C(\mathbb{R}^+)$ with $C(S^1)$. It follows from the spectral theorem that $C(\mathbb{R}^+)$ is the universal C^*-algebra generated by the unitary c, whence a $*$-homomorphism $C(\mathbb{R}^+) \to B^+$ is uniquely determined by a unitary $u := \phi^+(c)$ in B^+. Moreover, such a $*$-homomorphism comes from the canonical extension of a $*$-homomorphism

$C_0(\mathbb{R}) \to B$ if and only if the corresponding unitary u maps to 1 under the canonical quotient map $B^+ \to \mathbb{C}$.

Reintroducing the gradings, as the grading on $\mathscr{S}^+$ takes c to c^*, the fact that ϕ^+ is graded is equivalent to the unitary $u = \phi^+(c)$ satisfying $\epsilon_{B^+}(u) = u^*$. One can carry homotopies through this whole discussion to get paths of unitaries, which gives the result. □

Definition 2.9.5 For a graded C^*-algebra B, we will call unitaries in the set in line (2.22) *Cayley transforms*. For a graded $*$-homomorphism $\phi \colon \mathscr{S} \to B$, we will denote the element $\phi^+(c) \in B^+$ constructed in the proof of Lemma 2.9.4 u_ϕ, and call it the *Cayley transform* of ϕ.

We now introduce the binary operation on $spK_i(A)$. In fact, we just do this for $spK_0(A)$; as $spK_i(A) := spK_0(A \mathbin{\widehat{\otimes}} \mathrm{Cliff}_{\mathbb{C}}(\mathbb{R}^i))$, this also covers the general case.

Definition 2.9.6 Let $H = H_0 \oplus H_1$ be a choice of Hilbert space underlying $\mathscr{K}$. Choose a unitary isomorphism $U \colon H \oplus H \to H$ that restricts to unitary isomorphisms $H_i \oplus H_i \cong H_i$ for $i \in \{0, 1\}$. For homotopy classes $[\phi]$, $[\psi]$ in $spK_0(A)$ define $[\phi] + [\psi]$ to be the homotopy class represented by the graded $*$-homomorphism

$$\mathscr{S} \to A \mathbin{\widehat{\otimes}} \mathscr{K}, \quad f \mapsto U(\phi(f) \oplus \psi(f))U^*.$$

Remark 2.9.7 Say that $u_\phi, u_\psi \in (A \mathbin{\widehat{\otimes}} \mathscr{K})^+$ are Cayley transforms associated to $\phi, \psi \colon \mathscr{S} \to A \mathbin{\widehat{\otimes}} \mathscr{K}$ respectively. Identify $M_2(\mathscr{K})$ with $\mathscr{K}$ via the same unitary we used to identity $H \oplus H$ with H in Definition 2.9.6. Then the homomorphism underlying $[\phi] + [\psi]$ has Cayley transform $\begin{pmatrix} u_\phi & 0 \\ 0 & u_\psi \end{pmatrix}$.

Lemma 2.9.8 *Let A be a graded C^*-algebra and $i \geqslant 0$. Then the binary operation on $spK_i(A)$ does not depend on the choice of unitary U and makes $spK_i(A)$ into an abelian group.*

Proof It suffices to prove the lemma for $i = 0$. Note first that all unitaries U satisfying the condition in Definition 2.9.6 are homotopic (through unitaries satisfying the same condition), whence the class $[\phi] + [\psi]$ in $\{\mathscr{S}, A \mathbin{\widehat{\otimes}} \mathscr{K}\}$ does not depend on U. This also proves associativity and commutativity of the binary operation. It is moreover clear that the identity element is represented by the zero $*$-homomorphism $\mathscr{S} \to A \mathbin{\widehat{\otimes}} \mathscr{K}$, so $spK_0(A)$ is a commutative monoid. It remains to check that inverses exist.

For this, let $u \in (A \mathbin{\widehat{\otimes}} \mathscr{K})^+$ be a Cayley transform as in Definition 2.9.5, so by Lemma 2.9.4 there is a corresponding class $[u]$ in $spK_0(A)$. We claim that

u^* is a Cayley transform representing the additive inverse of $[u]$; the check that u^* is a Cayley transform is direct, so we leave this to the reader.

To see that $[u^*]$ is the inverse of $[u]$, choose faithful graded representations (H_A, U_A) and $(H_K, U_{\mathcal{K}})$ for A and $\mathcal{K}$ respectively, and let $A \widehat{\otimes} \mathcal{K}$ be equipped with the canonical faithful graded representation on $(H_A \widehat{\otimes} H_{\mathcal{K}}, U)$ of Definition E.2.7 (see also Lemma E.2.8), where $U = U_A \otimes U_{\mathcal{K}}$.

Write now $H_{\mathcal{K}} = H_0 \oplus H_1$ in the usual way and let $s(t) \in \mathcal{B}(H_A \otimes H_{\mathcal{K}})$ be the matrix $\begin{pmatrix} 0 & \sin(t) \\ \sin(t) & 0 \end{pmatrix}$ with respect to the decomposition

$$H_A \otimes H_{\mathcal{K}} = (H_A \otimes H_0) \oplus (H_A \otimes H_1).$$

Consider the homotopy in $M_2((A \widehat{\otimes} \mathcal{K})^+)$ given by

$$\begin{pmatrix} \cos(t)u & s(t) \\ -s(t) & \cos(t)u^* \end{pmatrix}, \quad t \in [0, \pi/2].$$

Then we have that

$$\begin{pmatrix} U & 0 \\ 0 & U \end{pmatrix} \begin{pmatrix} \cos(t)u & s(t) \\ -s(t) & \cos(t)u^* \end{pmatrix} \begin{pmatrix} U & 0 \\ 0 & U \end{pmatrix} = \begin{pmatrix} \cos(t)u^* & -s(t) \\ +s(t) & \cos(t)u \end{pmatrix}$$
$$= \begin{pmatrix} \cos(t)u & s(t) \\ -s(t) & \cos(t)u^* \end{pmatrix}^* .$$

To correct for the fact that this homotopy does not map constantly to one under the canonical quotient map $M_2((A \widehat{\otimes} \mathcal{K})^+) \to M_2(\mathbb{C})$, we replace it with

$$\begin{pmatrix} \cos(t/2)u & -s(t/2) \\ s(t/2) & \cos(t/2) \end{pmatrix} \begin{pmatrix} \cos(t)u & s(t) \\ -s(t) & \cos(t)u^* \end{pmatrix} \begin{pmatrix} \cos(t/2)u & -s(t/2) \\ s(t/2) & \cos(t/2) \end{pmatrix}$$

with $t \in [0, \pi/2]$. This gives a homotopy between $\begin{pmatrix} u & 0 \\ 0 & u^* \end{pmatrix}$ and the identity matrix that passes through Cayley transforms. As the identity matrix is the Cayley transform of the zero $*$-homomorphism $\mathcal{S} \to A \widehat{\otimes} \mathcal{K}$, and as the matrix $\begin{pmatrix} u & 0 \\ 0 & u^* \end{pmatrix}$ is the Cayley transform of the sum $[u] + [u^*]$ (see Remark 2.9.7), this completes the proof. □

Definition 2.9.9 For a graded C^*-algebra A, the abelian groups $spK_i(A)$ are called the *spectral K-theory groups* of A.

Remark 2.9.10 The spectral K-theory groups are covariantly functorial under graded $*$-homomorphisms $\phi \colon A \to B$. Indeed, this follows as one can postcompose a graded $*$-homomorphism

$$\mathcal{S} \to A \widehat{\otimes} \mathrm{Cliff}_{\mathbb{C}}(\mathbb{R}^i) \widehat{\otimes} \mathcal{K}$$

with the induced map

$$\phi \widehat{\otimes} \mathrm{id}_{\mathrm{Cliff}_{\mathbb{C}}(\mathbb{R}^i)} \widehat{\otimes} \mathrm{id}_{\mathscr{K}} : A \widehat{\otimes} \mathrm{Cliff}_{\mathbb{C}}(\mathbb{R}^i) \widehat{\otimes} \mathscr{K} \to B \widehat{\otimes} \mathrm{Cliff}_{\mathbb{C}}(\mathbb{R}^i) \widehat{\otimes} \mathscr{K}.$$

Our remaining task in this section is to relate the spectral K-theory groups to usual K-theory, in the case that the input C^*-algebra is trivially (or more generally, inner) graded. We will actually construct two such isomorphisms; this is useful for explicit computations and applications. Here are the maps underlying the first isomorphism. In the statements, if A is a graded C^*-algebra, then we define the usual K-theory groups $K_i(A)$ by forgetting the grading.

Construction 2.9.11 Assume that A is an inner graded C^*-algebra. Let $\mathscr{K}$ denote a standard graded copy of the compact operators with underlying Hilbert space $H_0 \oplus H_1$, and let $\mathcal{K} = \mathcal{K}(H_0)$. Using Example E.2.13, we have a spatially implemented C^*-algebra isomorphism $A \widehat{\otimes} \mathscr{K} \cong M_2(A \otimes \mathcal{K})$, which is unique on the level of homotopy. Under this isomorphism, the grading on $A \widehat{\otimes} \mathscr{K}$ corresponds to the grading on $M_2(A \otimes \mathcal{K})$ implemented by the unitary multiplier $\begin{pmatrix} 1 & 0 \\ 0 & -1 \end{pmatrix}$ of $M_2(A \otimes \mathcal{K})$.

Now let $\phi \colon \mathscr{S} \to A \widehat{\otimes} \mathscr{K}$ represent a class in $spK_0(A)$, so by the above comments, we may identify ϕ with a graded $*$-homomorphism $\mathscr{S} \to M_2(A \otimes \mathcal{K})$ using the isomorphisms above. Let $u_\phi \in M_2(A \otimes \mathcal{K})^+ \subseteq M_2((A \otimes \mathcal{K})^+)$ denote the Cayley transform of ϕ as in Definition 2.9.5. Define

$$p_\phi := \frac{1}{2}(vu_\phi + 1) \in M_2((A \otimes \mathcal{K})^+),$$

and note that $p_\phi - \frac{1}{2}(v+1)$ is in $M_2(A \otimes \mathcal{K})$, whence the class $[\frac{1}{2}(v+1)] - [p_\phi]$ defines an element of $K_0(A \otimes M_2(\mathcal{K}))$ using Exercise 2.11.6. Define a map $spK_0(A) \to K_0(A)$ to be the composition of the map

$$spK_0(A) \to K_0(M_2(A \otimes \mathcal{K})), \quad [\phi] \mapsto \left[\frac{1}{2}(v+1)\right] - [p_\phi]$$

and the canonical stabilisation isomorphism

$$K_0(M_2(A \otimes \mathcal{K})) \to K_0(A)$$

(see Corollary 2.7.2).

In the K_1 case, let $\phi \colon \mathscr{S} \to A \widehat{\otimes} \mathrm{Cliff}_{\mathbb{C}}(\mathbb{R}) \widehat{\otimes} \mathscr{K}$ represent a class in $spK_1(A)$. Example E.1.11 gives an isomorphism of C^*-algebras $\mathrm{Cliff}_{\mathbb{C}}(\mathbb{R}) \cong \mathbb{C} \oplus \mathbb{C}$ with the grading corresponding to the flip automorphism

$$\mathbb{C} \oplus \mathbb{C} \to \mathbb{C} \oplus \mathbb{C}, \quad (z, w) \mapsto (w, z).$$

Exercise E.3.3 says that graded tensor products are isomorphic (as graded C^*-algebras) to ungraded tensor products if at least one of the gradings is inner, whence we have C^*-algebra isomorphisms

$$A \widehat{\otimes} \mathrm{Cliff}_{\mathbb{C}}(\mathbb{R}) \widehat{\otimes} \mathscr{K} \cong A \otimes \mathrm{Cliff}_{\mathbb{C}}(\mathbb{R}) \otimes \mathscr{K} \cong (A \otimes \mathscr{K}) \oplus (A \otimes \mathscr{K})$$

such that the canonical tensor product grading on the left-hand side corresponds to the grading

$$(a,b) \mapsto \big((\epsilon_A \otimes \epsilon_{\mathscr{K}})(b), (\epsilon_A \otimes \epsilon_{\mathscr{K}})(a)\big)$$

on the right-hand side. Now, let $u_\phi \in (A \otimes \mathscr{K}) \oplus (A \otimes \mathscr{K})$ be the image of the Cayley transform of ϕ under the isomorphism above, so $u_\phi = (u_\phi^{(0)}, u_\phi^{(1)})$ for some unitaries $u, v \in (A \otimes \mathscr{K})^+$. Define a map $spK_1(A) \to K_1(A)$ to be the composition of the map

$$spK_1(A) \to K_1(A \otimes \mathscr{K}), \quad [\phi] \mapsto [u_\phi^{(0)}]$$

and the inverse $K_1(A \otimes \mathscr{K}) \to K_1(A)$ of the canonical stabilisation isomorphism (Corollary 2.7.2).

Proposition 2.9.12 *Let A be an inner graded C^*-algebra. Then the maps of construction 2.9.11 are well-defined isomorphisms. In particular, $spK_i(A) \cong K_i(A)$ for all $i \in \mathbb{N}$, and these isomorphisms are natural for functoriality under $*$-homomorphisms.*

The result fails for general (non-inner) gradings: see Exercise 2.11.31.

Proof Thanks to formal Bott periodicity for spK_i (see Remark 2.9.2) and Bott periodicity for K_i, the statement for all $i \in \mathbb{N}$ follows from the cases $i \in \{0,1\}$, so we just prove those cases.

We first look at $spK_0(A)$. Let $\phi\colon \mathscr{S} \to A \widehat{\otimes} \mathscr{K}$ represent a class in $spK_0(A)$; using the discussion in Construction 2.9.11, such a homomorphism is the same thing as a graded $*$-homomorphism $\phi\colon \mathscr{S} \to M_2(A \otimes \mathcal{K})$, where the latter algebra is graded by the unitary multiplier $v = \begin{pmatrix} 1 & 0 \\ 0 & -1 \end{pmatrix}$. Let $u_\phi \in M_2(A \otimes \mathcal{K})^+ \subseteq M_2((A \otimes \mathcal{K})^+)$ be the Cayley transform of ϕ. Then the correspondence

$$u_\phi = v(2p_\phi - 1) \longleftrightarrow p_\phi = \frac{1}{2}(vu_\phi + 1)$$

sets up a bijection between the collection of Cayley transforms and the collection of projections in $M_2((A \otimes \mathcal{K})^+)$ that are equal to $\frac{1}{2}(v+1) = \begin{pmatrix} 1 & 0 \\ 0 & 0 \end{pmatrix}$ modulo $M_2(A \otimes \mathcal{K})$; moreover, this correspondence preserves homotopy classes. Hence using Lemma 2.9.4, $spK_0(A)$ is naturally in bijection with the collection of path components of the set

$$S := \left\{ p \in M_2((A \otimes \mathcal{K})^+) \;\middle|\; \begin{array}{l} p \text{ a projection that is equal to } \begin{pmatrix} 1 & 0 \\ 0 & 0 \end{pmatrix} \\ \text{modulo } M_2(A \otimes \mathcal{K}) \end{array} \right\}.$$

Now, we have a well-defined map

$$\pi_0(S) \to \text{Kernel}\big(K_0(M_2((A \otimes \mathcal{K})^+)) \to K_0(M_2(\mathbb{C}))\big),$$
$$[p] \mapsto \begin{bmatrix} 1 & 0 \\ 0 & 0 \end{bmatrix} - [p], \quad (2.23)$$

where the map $K_0(M_2((A \otimes \mathcal{K})^+)) \to K_0(M_2(\mathbb{C}))$ is induced by the canonical quotient $*$-homomorphism $(A \otimes \mathcal{K})^+ \to \mathbb{C}$. It is not too difficult to check that this map is moreover a group homomorphism for the group structure on $\pi_0(S)$ inherited from $spK_0(A)$: this essentially follows as one can see the operations as direct sum in both cases.

We claim that the map in line (2.23) is actually an isomorphism. Indeed, for surjectivity, let p, q be projections in some $M_n(M_2(A \otimes \mathcal{K})^+) \subseteq M_{2n}((A \otimes \mathcal{K})^+)$ such that the image of $[p] - [q]$ in $M_{2n}(\mathbb{C})$ under the canonical map is zero. It suffices to show that the class $[p] - [q]$ is in the image of the above map; conjugating by a scalar unitary, we may moreover assume that $p - q = 0$ in $M_{2n}(\mathbb{C})$. Embedding in $M_{4n}((A \otimes \mathcal{K})^+)$ and adding $1 - p$ to both p and q, we may assume that our class is of the form

$$\begin{bmatrix} 1 & 0 \\ 0 & 0 \end{bmatrix} - [p], \quad (2.24)$$

where the matrix entries are in $M_{2n}((A \otimes \mathcal{K})^+)$, and $\left(\begin{smallmatrix} 1 & 0 \\ 0 & 0 \end{smallmatrix}\right) - p$ maps to 0 in $M_{4n}(\mathbb{C})$. It follows from this that both $\left(\begin{smallmatrix} 1 & 0 \\ 0 & 0 \end{smallmatrix}\right)$ and p are in $M_2(M_{2n}(A \otimes \mathcal{K})^+)$; using an isomorphism $M_{2n}(A \otimes \mathcal{K}) \cong A \otimes \mathcal{K}$ induced by a spatially induced isomorphism $M_{2n}(\mathcal{K}) \cong \mathcal{K}$, we may assume that our element is of the form in line (2.24), but now where the matrices are in $M_2((A \otimes \mathcal{K})^+)$. This completes the proof of surjectivity of the map in line (2.23). On the other hand, injectivity of the map in line (2.23) follows on pushing homotopies through all of the above argument.

To summarise in the K_0 case, we now have a chain of isomorphisms

$$\begin{aligned} spK_0(A) &\cong \pi_0(S) \cong \text{Kernel}\big(K_0(M_2((A \otimes \mathcal{K})^+)) \to K_0(M_2(\mathbb{C}))\big) \\ &\cong K_0(A \otimes \mathcal{K}) \cong K_0(A), \end{aligned}$$

the second to last of which is the isomorphism of Exercise 2.11.6, and the last of which is the stabilisation isomorphism of Corollary 2.7.2. This completes the proof in the K_0 case.

We now move on to $spK_1(A)$. Thinking of a graded $*$-homomorphism

$$\mathscr{S} \to A \mathbin{\widehat{\otimes}} \mathrm{Cliff}_{\mathbb{C}}(\mathbb{R}) \mathbin{\widehat{\otimes}} \mathcal{K} \cong (A \otimes \mathscr{K}) \oplus (A \otimes \mathscr{K})$$

as in Construction 2.9.11 via its Cayley transform, such a $*$-homomorphism corresponds to a pair of unitaries

$$(u, v) \in (A \otimes \mathscr{K})^+ \oplus (A \otimes \mathscr{K})^+$$

with $v = (\epsilon_A \otimes \epsilon_{\mathscr{K}})(u)^*$, and such that (u, v) maps to $(1, 1)$ under the canonical quotient $(A \otimes \mathscr{K})^+ \oplus (A \otimes \mathscr{K})^+ \to \mathbb{C} \oplus \mathbb{C}$. This in turn is exactly the same information as just having a single unitary $u \in (A \otimes \mathcal{K})^+$ that maps to one under the canonical quotient map $(A \otimes \mathcal{K})^+ \to \mathbb{C}$. We conclude that $spK_1(A)$ is canonically isomorphic to the set of path components of

$$\{u \in (A \otimes \mathscr{K})^+ \mid uu^* = u^*u = 1 \text{ and } u \mapsto 1 \text{ under } (A \otimes \mathscr{K})^+ \to \mathbb{C}\}.$$

Using the stability isomorphism in K-theory (see Corollary 2.7.2) and arguments much as in the first part, one sees that the set of path components of the right-hand side is canonically isomorphic to $K_1(A)$ (in a way compatible with the operations on spK_1 and K_1); we leave the remaining details to the reader. □

Remark 2.9.13 We have another useful description of $spK_1(A)$ in the trivially (more generally, inner) graded case, for which we need a little notation. For C^*-algebras A and B, let $[A, B]$ denote the set of homotopy classes of $*$-homomorphisms from A to B. Let $\mathcal{K}$ denote a copy of $\mathscr{K}$ where we forget the grading and equip $[A, B \otimes \mathcal{K}]$ with the semigroup structure defined quite analogously to that on spK_0 (but ignoring the irrelevant condition that U be compatible with the grading).

For an inner graded C^*-algebra A there is then a canonical isomorphism (of semigroups, whence of abelian groups)

$$spK_1(A) \cong [C_0(\mathbb{R}), A \otimes \mathcal{K}].$$

Indeed, we have already seen that $spK_1(A)$ is isomorphic to

$$\pi_0\{u \in (A \otimes \mathcal{K})^+ \mid uu^* = u^*u = 1 \text{ and } u \mapsto 1 \text{ under } (A \otimes \mathcal{K})^+ \to \mathbb{C}\}$$

in the proof of Proposition 2.9.12. On the other hand, using the universal property of $C(S^1)$ as in the proof of Lemma 2.9.4, it is not difficult to identify the latter set with $[C_0(\mathbb{R}), A \otimes \mathcal{K}]$; we leave the remaining details to the reader.

If A is an inner graded C^*-algebra, we will sometimes just write $K_i(A)$, or $K_*(A)$ and not be specific about whether we are using the usual picture or the spectral picture of K-theory.

It is possible to give a concrete description of an inverse to the isomorphism in Proposition 2.9.12. This goes as follows.

Lemma 2.9.14 *Let A be a unital and inner graded C^*-algebra, so we have the identification $A \widehat{\otimes} \mathscr{K} \cong M_2(A \otimes \mathcal{K})$ of Example E.2.13. Then the prescription*

$$K_0(A) \to spK_0(A), \quad [p]-[q] \mapsto \left(f \mapsto \begin{pmatrix} f(0)p & 0 \\ 0 & f(0)q \end{pmatrix}\right)$$

is a well-defined inverse to the isomorphism of Construction 2.9.11.

Proof To make sense of the map, note that the top left entry of the matrix in the statement should be an element of $A \otimes \mathcal{K}$; we make sense of $f(0)p$ in here by identifying $M_n(A)$ with a subalgebra of $A \otimes \mathcal{K}$ in the standard way. The bottom right entry is similar. Having explained this, we leave it to the reader to check that the map

$$\mathscr{S} \to M_2(A \otimes \mathcal{K}), \quad f \mapsto \begin{pmatrix} f(0)p & 0 \\ 0 & f(0)q \end{pmatrix}$$

is a well-defined graded $*$-homomorphism, and that the map on K-theory in the statement is well defined.

Now, to check that the above map is an inverse to the one from the proof of Proposition 2.9.12, as we know the latter is an isomorphism it suffices to check that it is a one-sided inverse. Consider then what happens if we apply the map from this lemma, then the one from Proposition 2.9.12, to an element $[p]-[q] \in K_0(A)$.

To compute the composition, note that the map from this lemma extends to the unital map

$$C(S^1) \to (A \otimes \mathcal{K})^+, \quad f \mapsto \begin{pmatrix} f(0)p + f(\infty)(1-p) & 0 \\ 0 & f(0)q + f(\infty)(1-q) \end{pmatrix}$$

(where we identify S^1 with $\mathbb{R} \cup \{\infty\}$ to make sense of this) on the level of unitisations. Applying this to the Cayley transform $f(x) = \frac{x-i}{x+i}$ gives

$$\begin{pmatrix} 1-2p & 0 \\ 0 & 1-2q \end{pmatrix},$$

and then the prescription in the proof of Proposition 2.9.12 takes this to

$$\begin{bmatrix} 1 & 0 \\ 0 & 1 \end{bmatrix} - \left[\frac{1}{2}\left(\begin{pmatrix} 1-2p & 0 \\ 0 & 2q-1 \end{pmatrix} + \begin{pmatrix} 1 & 0 \\ 0 & 1 \end{pmatrix}\right)\right],$$

which simplifies to

$$\begin{bmatrix} 1 & 0 \\ 0 & 0 \end{bmatrix} - \begin{bmatrix} 1-p & 0 \\ 0 & q \end{bmatrix} = [p] - [q],$$

so we are done. □

We are now ready for the second description of the isomorphism $spK_i(A) \to K_i(A)$ for inner graded A. This second description is particularly useful for index theory.

Construction 2.9.15 Let us look at K_0 first. Using Example E.2.13 and that A is inner graded, we have an isomorphism $A \widehat{\otimes} \mathscr{K} \cong A \otimes \mathscr{K}$, where the right-hand algebra has the grading $\mathrm{id} \otimes \epsilon_{\mathscr{K}}$. Represent $A \otimes \mathscr{K}$ in a grading preserving way on a graded Hilbert space H (see Definition E.1.4).

Now, consider a graded homomorphism

$$\phi \colon \mathscr{S} \to A \otimes \mathscr{K} \subseteq \mathcal{B}(H).$$

Let H' be the closed span of $\phi(\mathscr{S})H$, and note that the grading operator preserves H' as ϕ is a graded $*$-homomorphism. Let $C[-\infty, \infty]$ denote the continuous functions on the natural 'two-point' compactification of $\mathbb{R}$, and note that Proposition 1.7.3 gives us an extension

$$\phi \colon C[-\infty, \infty] \to M(\phi(\mathscr{S})) \subseteq \mathcal{B}(H')$$

of our original $*$-homomorphism ϕ; this extension is still graded, as one easily checks.

Let now $f : \mathbb{R} \to [-1, 1]$ be any odd function such that $\lim_{t \to \pm\infty} f(t) = \pm 1$, considered as an element of $C[-\infty, \infty]$, and define $F := \phi(f) \in \mathcal{B}(H)$. Then F is in an odd, self-adjoint, contractive element of the multiplier algebra $M(\phi(\mathscr{S}))$ such that $F^2 - 1 \in \phi(\mathscr{S})$. As the grading on $A \otimes \mathscr{K}$ is inner, the construction of Definition 2.8.5 gives an index class

$$\mathrm{Ind}[F] \in K_0(\phi(\mathscr{S})).$$

Composing with the canonical map on K-theory induced by the inclusion $\phi(\mathscr{S}) \subseteq A \otimes \mathscr{K}$ and the stabilisation isomorphism on K-theory, we get a class

$$\mathrm{Ind}[F] \in K_0(A \otimes \mathscr{K}) \cong K_0(A).$$

The case of K_1 is similar: we use the description $spK_1(A) = [C_0(\mathbb{R}), A \otimes \mathcal{K}]$ from Remark 2.9.13. Choose a function $f \colon \mathbb{R} \to [-1, 1]$ such that $\lim_{t \to \pm\infty} f(t) = \pm 1$ (it no longer matters if f is odd). Representing $A \otimes \mathcal{K}$ on some Hilbert space, just as before, from an element $[\phi] \in spK_1(A)$ we get a homomorphism

$$\phi\colon C[-\infty,\infty] \to A \otimes \mathcal{K} \subseteq \mathcal{B}(H)$$

and therefore a self-adjoint contraction $F := \phi(f)$ in the multiplier algebra $M(\phi(C_0(\mathbb{R})))$ such that $F^2 - 1 \in \phi(C_0(\mathbb{R}))$. The index construction of Definition 2.8.5 gives an element

$$\mathrm{Ind}[F] \in K_1(\phi(C_0(\mathbb{R}))),$$

and postcomposing with the map on K-theory induced by the inclusion $\phi(C_0(\mathbb{R})) \subseteq A \otimes \mathcal{K}$ gives an element of $K_1(A \otimes \mathscr{K}) \cong K_1(A)$.

Theorem 2.9.16 *Construction 2.9.15 induces the same isomorphism $spK_i(A) \to K_i(A)$ as in the proof of Proposition 2.9.12.*

Proof We leave it to the reader to do the direct check that Construction 2.9.15 gives a well-defined map $spK_i(A) \to K_i(A)$ that does not depend on the choice of function $f \in C[-\infty,\infty]$.

We look first at the K_1 case. Given a $*$-homomorphism, Construction 2.9.15 above asks us to choose a function $f\colon \mathbb{R} \to [-1,1]$ with $\lim_{t\to\infty} f(t) = \pm 1$. We may assume that f is strictly increasing. We then form $F = \phi(f)$. Definition 2.8.5 tells us to form $P = \frac{1}{2}(F+1)$ and take its index in the sense of Definition 2.8.4. The element we get is thus

$$e^{-2\pi i P} = e^{-\pi i F} e^{-\pi i} = -e^{\pi i \phi(f)} = \phi(-e^{-\pi i f}) \in (A \otimes \mathcal{K})^+.$$

Now, as t ranges from $-\infty$ to ∞, $f(t)$ increases from -1 to 1 and the function $-e^{-\pi i f}$ goes counterclockwise around the circle once, starting and finishing at one. It is thus homotopic (through functions taking infinity to one) to the Cayley transform $c\colon \mathbb{R}^+ \to S^1$. Hence Construction 2.9.15 gives us the class of the unitary $\phi(c)$ in $K_1(A)$, which is exactly what happens in the proof of Proposition 2.9.12.

Let us now look at the K_0 case. Thanks to Proposition 2.9.12, we know that for A non-unital $spK_i(A) = \mathrm{Ker}(spK_i(A^+) \to spK_i(A))$; as the construction above is compatible with $*$-homomorphisms, it therefore suffices to prove the result in the unital case. In that case, it suffices to prove that Construction 2.9.15 gives a one-sided inverse to the map from Lemma 2.9.14, which we now do.

Let $H_0 \oplus H_1$ be the Hilbert space underlying $\mathscr{K}$, let $\mathcal{K} = \mathcal{K}(H_0) \cong \mathcal{K}(H_1)$ and let $p, q \in A \otimes \mathcal{K}$. Consider the $*$-homomorphism

$$\phi\colon \mathscr{S} \to A \otimes \mathscr{K}, \quad f \mapsto \begin{pmatrix} f(0)p & 0 \\ 0 & f(0)q \end{pmatrix},$$

where the matrix is defined with respect to the grading on the Hilbert space $H_0 \oplus H_1$ underlying $\mathcal{K}$. Represent $A \otimes \mathcal{K}$ faithfully on a Hilbert space H. Let $f \in C[-\infty, \infty]$ be a function with the properties in Construction 2.9.15, so in particular f is odd and so $f(0) = 0$. It follows that $F = \phi(f) = 0$. On the other hand,

$$\overline{\text{span}\phi(\mathcal{S})H} = pH \oplus qH,$$

with the grading given by the direct sum decomposition. As $F = 0$, it follows from the explicit formula from Definition 2.8.5 that we have

$$\text{Ind}[F] = \begin{bmatrix} p & 0 \\ 0 & 0 \end{bmatrix} - \begin{bmatrix} 0 & 0 \\ 0 & q \end{bmatrix} = [p] - [q],$$

which tells us exactly that Construction 2.9.15 inverts the map of Lemma 2.9.14, and we are done. □

2.10 The External Product in K-theory

Our goal in this section is to construct an external product

$$spK_i(A) \otimes spK_j(B) \to spK_{i+j}(A \widehat{\otimes}_{\max} B)$$

on spectral K-theory groups of graded C^*-algebras. We will need to use material from Section 2.9 on spectral K-theory groups and Appendix E on graded C^*-algebras.

To construct the external product on K-theory, we first need to construct the so-called comultiplication on $\mathcal{S}$, which is a $*$-homomorphism

$$\Delta \colon \mathcal{S} \to \mathcal{S} \widehat{\otimes} \mathcal{S}.$$

There are several ways to do this: we proceed in a slightly ad hoc way, partly as it introduces some machinery that we will need again later.

For the statement of the next lemma, see Definition E.1.4 for the definition of a graded Hilbert space and Definition D.1.1 for that of an unbounded operator.

Definition 2.10.1 Let (H_1, U_1) and (H_2, U_2) be graded Hilbert spaces, and let D_1 and D_2 be unbounded operators on H_1 and H_2 with domains S_1 and S_2 respectively, and that are odd[17] for the gradings. Define an unbounded

[17] Compare Example E.1.14: in particular, this assumes that U_1, U_2 preserve the domains S_1, S_2 respectively.

operator on $H_1 \widehat{\otimes} H_2$, with domain the algebraic tensor product $S_1 \odot S_2$ by the formula

$$(D_1 \widehat{\otimes} 1 + 1 \widehat{\otimes} D_2)(u \otimes v) := D_1 u \otimes v + U_1 u \otimes D_2 v. \tag{2.25}$$

The formula in line (2.25) might seem more intuitive when compared with Definition E.2.5 and Remark E.2.6.

For the next lemma, recall from Remark D.1.8 that we may apply the functional calculus of Theorem D.1.7 to essentially self-adjoint operators.

Lemma 2.10.2 *With notation as in Definition 2.10.1, assume moreover that for $j \in \{1,2\}$ the operator D_j is essentially self-adjoint, and that for all $t \in \mathbb{R}$ the operators e^{itD_j} preserve the domain S_j. Then $D_1 \widehat{\otimes} 1 + 1 \widehat{\otimes} D_2$ is odd for the tensor product grading $U_1 \otimes U_2$, and essentially self-adjoint.*

Proof Direct algebraic checks show that D as in the statement is odd and formally self-adjoint; it remains to check essential self-adjointness. The functional calculus (Theorem D.1.7) lets us build bounded operators e^{itD_1} and e^{itD_2} on H_1 and H_2 respectively for $t \in \mathbb{R}$. We may thus form the operators $V_t := e^{itD_1} \widehat{\otimes} e^{itD_2}$ as in Definition E.2.5. Using our conventions on graded tensor products (see Definition E.2.5), if we split e^{itx} into its odd and even parts, then we have

$$V_t = e^{itD_1} \otimes \cos(tD_2) + e^{tD_1} U_1 \otimes i \sin(tD_2).$$

Note that the collection $(V_t)_{t \in \mathbb{R}}$ is a strongly continuous family of unitary operators. Using the part of the Stone–von Neumann theorem in Proposition D.2.1 applied to the self-adjoint closures of D_1 and D_2 (and the usual arguments proving the product rule from calculus), for each $u \in S_1 \odot S_2$ we have that $\lim_{t \to 0} \frac{1}{i} \frac{V_t u - u}{t}$ equals

$$\begin{aligned}
&\frac{1}{i}\Big((ie^{itD_1} D_1 \otimes \cos(tD_2))|_{t=0} + (e^{itD_1} \otimes (-\sin(tD_2)D_2))|_{t=0} \\
&\quad + (ie^{itD_1} D_1 U_1 \otimes i \sin(tD_2))|_{t=0}) + (e^{itD_1} U_1 \otimes i \cos(tD_2)D_2)|_{t=0} \Big) u \\
&= (D_1 \otimes 1 + U_1 \otimes D_2)u.
\end{aligned}$$

Hence from the part of the Stone–von Neumann theorem in Theorem D.2.2, $D_1 \widehat{\otimes} 1 + 1 \widehat{\otimes} D_2$ is essentially self-adjoint on the given domain. □

Let us now identify $\mathscr{S}$ with its image in the natural multiplication representation on $L^2(\mathbb{R})$ with associated grading operator U defined by $(Uu)(x) := u(-x)$ as in Example E.1.10. Let M_x be the unbounded, odd, operator of multiplication by the identity function on $L^2(\mathbb{R})$ with domain $C_c(\mathbb{R})$. It is straightforward to check that this is essentially self-adjoint. Then Lemma 2.10.2 gives us an unbounded, odd, essentially self-adjoint operator

$$C := x \widehat{\otimes} 1 + 1 \widehat{\otimes} x$$

on $L^2(\mathbb{R}) \otimes L^2(\mathbb{R})$ with domain $C_c(\mathbb{R}) \odot C_c(\mathbb{R})$. Hence we may apply the unbounded functional calculus (Theorem D.1.7 and Remark D.1.8) to M_x, getting in particular a $*$-homomorphism

$$\Delta\colon C_0(\mathbb{R}) \to \mathcal{B}(L^2(\mathbb{R}) \otimes L^2(\mathbb{R})), \quad f \mapsto f(M_x). \tag{2.26}$$

Lemma 2.10.3 *The $*$-homomorphism Δ in line* (2.26) *above is graded and takes image in $\mathscr{S} \widehat{\otimes} \mathscr{S}$.*

Proof First one computes that M_x^2 acts as $x^2 \otimes 1 + 1 \otimes x^2$ on $C_c(\mathbb{R}) \odot C_c(\mathbb{R})$, and therefore the $*$-homomorphism in line (2.26) takes the function e^{-x^2} to

$$e^{-M_x^2} = e^{-(x^2 \otimes 1 + 1 \otimes x^2)} = e^{-x^2} \widehat{\otimes} e^{-x^2}$$

(where the right-hand side is interpreted as the operator of multiplication by the given function). Similarly, it takes xe^{-x^2} to

$$M_x e^{-M_x^2} = (x \otimes U + 1 \otimes x)e^{-(x^2 \otimes 1 + 1 \otimes x^2)} = xe^{-x^2} \widehat{\otimes} e^{-x^2} + e^{-x^2} \widehat{\otimes} xe^{-x^2}.$$

In particular, the $*$-homomorphism in line (2.26) takes both of the functions e^{-x^2} and xe^{-x^2} into $\mathscr{S} \widehat{\otimes} \mathscr{S}$; as these functions generate $C_0(\mathbb{R})$ as a C^*-algebra, this shows that the $*$-homomorphism Δ takes image in $\mathscr{S} \widehat{\otimes} \mathscr{S}$. Moreover, from the formulas above, Δ takes e^{-x^2} to an even element; as e^{-x^2} generates the C^*-subalgebra of even elements of $C_0(\mathbb{R})$, this implies that Δ takes even elements to even elements. Finally, it takes xe^{-x^2} to an odd element; as the collection of products of the form $f(x)xe^{-x^2}$ with $f \in \mathscr{S}$ even is dense in the subspace of odd elements in $\mathscr{S}$, Δ takes odd elements to odd elements as well, and thus preserves the grading. □

Definition 2.10.4 The *comultiplication* for $\mathscr{S}$ is the graded $*$-homomorphism

$$\Delta\colon \mathscr{S} \to \mathscr{S} \widehat{\otimes} \mathscr{S}$$

of Lemma 2.10.3.

Remark 2.10.5 Using Corollary E.2.19, we may equally well use the spatial $\mathscr{S} \widehat{\otimes} \mathscr{S}$ or maximal $\mathscr{S} \widehat{\otimes}_{\max} \mathscr{S}$ tensor products in the definition of Δ.

We will need one more fact about Δ, often called *coassociativity*.

Lemma 2.10.6 *The following diagram commutes.*

$$\begin{array}{ccc} \mathcal{S} & \xrightarrow{\Delta} & \mathcal{S} \mathbin{\widehat{\otimes}} \mathcal{S} \\ \downarrow{\scriptstyle \Delta} & & \downarrow{\scriptstyle \Delta \widehat{\otimes} id} \\ \mathcal{S} \mathbin{\widehat{\otimes}} \mathcal{S} & \xrightarrow{id \widehat{\otimes} \Delta} & \mathcal{S} \mathbin{\widehat{\otimes}} \mathcal{S} \mathbin{\widehat{\otimes}} \mathcal{S} \end{array}$$

Proof It suffices to check commutativity on the generators e^{-x^2} and xe^{-x^2} of $\mathcal{S}$. This can be done using the computations from the proof of Lemma 2.10.3: for example, one sees that either composition takes xe^{-x^2} to

$$xe^{-x^2} \mathbin{\widehat{\otimes}} e^{-x^2} \mathbin{\widehat{\otimes}} e^{-x^2} + e^{-x^2} \mathbin{\widehat{\otimes}} xe^{-x^2} \mathbin{\widehat{\otimes}} e^{-x^2} + e^{-x^2} \mathbin{\widehat{\otimes}} e^{-x^2} \mathbin{\widehat{\otimes}} xe^{-x^2}.$$

We leave the explicit computations to the reader. □

We are now in a position to define the external product on spectral K-theory. Let $\mathcal{K}$ be a standard graded copy of the compact operators (see Example E.1.9), and say the underlying Hilbert space is $H = H_0 \oplus H_1$. Choose a unitary isomorphism $U\colon H \otimes H \to H$ that restricts to unitary isomorphisms

$$(H_0 \otimes H_0) \oplus (H_1 \otimes H_1) \to H_0 \quad \text{and} \quad (H_0 \otimes H_1) \oplus (H_0 \otimes H_1) \to H_1$$

(in other words, U preserves the gradings). Then conjugation by U induces a graded $*$-isomorphism $\mathcal{K} \mathbin{\widehat{\otimes}} \mathcal{K} \to \mathcal{K}$; moreover, any two such choices of unitary will be homotopic through unitaries satisfying the same condition.

For the next definition, note that Corollary E.2.19 implies that we may as well use the maximal $\widehat{\otimes}_{\max}$ as spatial tensor $\widehat{\otimes}$ tensor product when considering an element of $spK_0(A)$ as a homomorphism $\mathcal{S} \to A \mathbin{\widehat{\otimes}} \mathcal{K}$.

Definition 2.10.7 Let A and B be graded C^*-algebras. Let $\phi\colon \mathcal{S} \to A \mathbin{\widehat{\otimes}_{\max}} \mathcal{K}$ and $\psi\colon \mathcal{S} \to B \mathbin{\widehat{\otimes}_{\max}} \mathcal{K}$ be $*$-homomorphisms representing classes in $spK_0(A)$ and $spK_0(B)$. Then their *(external) product* is the class in $spK_0(A \mathbin{\widehat{\otimes}_{\max}} B)$ of the composition

$$\mathcal{S} \xrightarrow{\Delta} \mathcal{S} \mathbin{\widehat{\otimes}_{\max}} \mathcal{S} \xrightarrow{\phi \widehat{\otimes} \psi} A \mathbin{\widehat{\otimes}_{\max}} \mathcal{K} \mathbin{\widehat{\otimes}_{\max}} B \mathbin{\widehat{\otimes}_{\max}} \mathcal{K}$$

$$\longrightarrow A \mathbin{\widehat{\otimes}_{\max}} B \mathbin{\widehat{\otimes}_{\max}} \mathcal{K} \mathbin{\widehat{\otimes}_{\max}} \mathcal{K} \xrightarrow{\mathrm{ad}_{1 \otimes U}} A \mathbin{\widehat{\otimes}_{\max}} B \mathbin{\widehat{\otimes}_{\max}} \mathcal{K},$$

where the third map is the canonical 'reordering' isomorphism arising from associativity and commutativity of $\widehat{\otimes}_{\max}$ (Remark E.2.15).

This also gives rise to products

$$spK_i(A) \otimes spK_j(B) \to spK_{i+j}(A \mathbin{\widehat{\otimes}_{\max}} B)$$

defined using the canonical isomorphism $\mathrm{Cliff}_{\mathbb{C}}(\mathbb{R}^i)\widehat{\otimes}_{\max}\mathrm{Cliff}_{\mathbb{C}}(\mathbb{R}^j) \cong \mathrm{Cliff}_{\mathbb{C}}(\mathbb{R}^{i+j})$ arising from the discussion in Example E.2.12 and Corollary E.2.19.

Lemma 2.10.8 *The product above is well defined, does not depend on the choice of U, distributes over the group operations on $spK_n(A)$ and is associative.*

Proof Whether the product is well defined or not comes down to the facts that a tensor product of homotopies is a homotopy, and that the composition of a homotopy with a $*$-homomorphism is a homotopy. It does not depend on the choice of U as any two such unitaries are homotopic through unitaries satisfying the same conditions. Distribution over addition follows directly from the definitions. Associativity follows directly from Lemma 2.10.6. □

2.11 Exercises

2.11.1 Show that if v, w implement a Murray–von Neumann equivalence between idempotents e and f (Definition 2.1.2), then so do $v' := evf$ and $w' = fwe$. Moreover, these new elements satisfy $ev' = v'f = v'$ and $w'e = fw' = w'$.

2.11.2 Let R be a unital ring. Show that the addition operation on $V(R)$ from Definition 2.1.4 is well defined, and makes $V(R)$ into a commutative monoid.

2.11.3 A (right) module P over a ring R is *projective* if for any commutative diagram of module maps

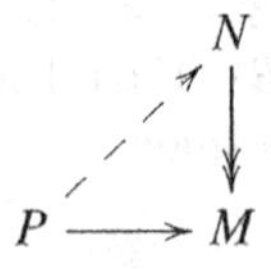

with the vertical arrow surjective, the dashed arrow can be filled in.

(i) Show that a finitely generated right module M over a unital ring R is projective if and only if there is $n \in \mathbb{N}$ and an idempotent $e \in M_n(R)$ with $M \cong e(R^n)$.
Hint: apply the diagram above where $P = M$, the bottom arrow is the identity and the vertical arrow is some choice of quotient map $p\colon R^n \to M$ (which exists as M is finitely generated). This gives a commutative diagram

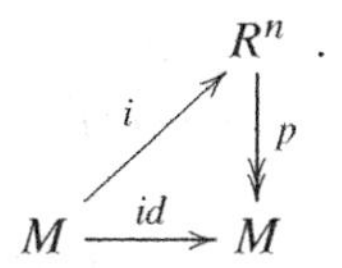

Now define $e = i \circ p \colon R^n \to R^n$ and show that this works, having identified $M_n(R)$ with the collection of R-module maps from R^n to itself in the canonical way.

(ii) Show that two finitely generated projective modules M and N over R are isomorphic as R modules if and only if the following property holds: for any n and any idempotents $e, f \in M_n(R)$ with $eR^n \cong M$ and $fR^n \cong N$, one has that e and f are Murray–von Neumann equivalent.

(iii) Conclude that $V(R)$ can equivalently be described as the collection of all isomorphism classes of finitely generated projective modules over R.

2.11.4 The goal of this exercise is to show that $V(C(X))$ can equivalently be described in terms of isomorphism classes of vector bundles over a compact Hausdorff space X. It requires some background in the basic theory of vector bundles: see Section 2.12, Notes and References, at the end of this chapter. To establish conventions, let us define a vector bundle over a compact Hausdorff space X to be a locally compact, Hausdorff, topological space E satisfying the following conditions:

(a) there is a continuous surjection $\pi : E \to X$;

(b) for each $x \in X$, the *fibre* $E_x := \pi^{-1}(x)$ is equipped with the structure of a finite-dimensional complex vector space;

(c) for each $x \in X$, there is an open set $U \ni x$ and a homeomorphism

$$\phi \colon U \times \mathbb{C}^d \to \pi^{-1}(U)$$

(called a *local trivialisation*) such that $\pi(\phi(x, v)) = x$ for all $(x, v) \in U \times \mathbb{C}^d$, and so that for each $x \in X$, the map

$$\mathbb{C}^d \to E_x, \quad v \mapsto \phi(x, v)$$

is an isomorphism of vector spaces.

(i) Let E be a vector bundle over X. Show that there is N such that E embeds inside the trivial vector bundle $X \times \mathbb{C}^N$.
Hint: choose a finite cover $U_1, \ldots, U_n$ of X such that each U_i is equipped with a homeomorphism $\phi_i : U_i \times \mathbb{C}^{d_i} \to \pi^{-1}(U_i)$ as in the definition of vector bundle. Let $q_i : U_i \times \mathbb{C}^{d_i} \to \mathbb{C}^{d_i}$ be the coordinate

projection. Let $f_1, \ldots, f_n$ *be a partition of unity (see Definition A.1.2) subordinate to this cover, and define*

$$E \to X \times \left(\bigoplus_{i=1}^{n} \mathbb{C}^{d_i} \right), \quad e \mapsto \left(\pi(e), \left(f_i(\pi(e)) q_i(\phi_i^{-1}(e)) \right)_{i=1}^{n} \right).$$

Show this works.

(ii) Show that if E is a vector bundle over X which is embedded inside some trivial bundle $X \times \mathbb{C}^N$, then there is another sub-bundle F of $X \times \mathbb{C}^N$ such that $E \oplus F$ is isomorphic to $X \times \mathbb{C}^N$.
Hint: choose an inner product on $\mathbb{C}^N$*, and define* F *to be*

$$F := \{(x, v) \in X \times \mathbb{C}^N \mid \langle v, w \rangle = 0 \text{ for all } w \in E_x\},$$

i.e. F *is the 'pointwise orthogonal complement of* E*'. Equip* F *with the restriction* $\pi_F \colon F \to X$ *of the coordinate projection* $\pi \colon X \times \mathbb{C}^N \to X$ *and equip each fibre* $F_x := \pi_F^{-1}(x)$ *with the vector space structure it inherits from* $\mathbb{C}^N$*. Finally, show that if* $U \ni x$ *and* $\phi \colon U \times \mathbb{C}^d \to \pi_E^{-1}(U)$ *is a local trivialisation of* E*, we can define a local trivialisation of* F *on a smaller open set* $V \ni x$ *as follows. First, note that for each* $y \in U$*, we have a map*

$$\psi_y \colon \mathbb{C}^d \oplus F_x \to \mathbb{C}^N, \quad (v, w) \mapsto \phi(y, v) + w,$$

thus defining a continuous map

$$\psi \colon U \to Hom(\mathbb{C}^d \oplus F_x, \mathbb{C}^N), \quad y \mapsto \psi_y$$

where the right-hand side has its usual topology (coming, for example, from a choice of identification with $M_N(\mathbb{C})$*). As* ψ_x *is an isomorphism and as the invertible elements in* $Hom(\mathbb{C}^d \oplus F_x, \mathbb{C}^N)$ *are open, there must exist a neighbourhood* $V \ni x$ *such that* ψ_y *is an isomorphism for all* $y \in V$*. Let* $q \colon \mathbb{C}^d \oplus F_x \to F_x$ *be the coordinate projection, and provisionally define a local trivialisation of* F *over* V *to be the inverse of the map*

$$\pi_F^{-1}(V) \to V \times F_x, \quad (y, w) \mapsto q(\psi_y^{-1}(w)).$$

Show that this does indeed define a local trivialisation of F *over* V*.*

(iii) With N as in part (ii), show that there is an idempotent $e_E \in C(X, M_N(\mathbb{C}))$ such that the image of e_E (considered as a fibre-preserving map $X \times \mathbb{C}^N \to X \times \mathbb{C}^N$) is exactly E.
Hint: find F *as in part (ii) above and define* e_E *to be the idempotent with image* E *and kernel* F *in the appropriate sense.*

(iv) We now move in the opposite direction: from idempotents to vector bundles. Let $e \in C(X, M_N(\mathbb{C}))$ be an idempotent for some N, and define

$$E := \{(x, v) \in X \times \mathbb{C}^N \mid v \in \text{range}(e(x))\}.$$

Show that E_e is a well-defined vector bundle over X.
Hint: for local triviality, the key point is that if V, W are finite-dimensional vector spaces, then the invertible elements of $Hom(V, W)$ form an open set.

(v) With notation e_E as in part (iii) and E_e as in part (iv), show that the above processes $e \mapsto E_e$ and $E \mapsto e_E$ give well-defined bijections between the collection $V(X)$ of isomorphism classes of complex vector bundles over X, and the collection $V(C(X))$ from Definition 2.1.4. Show moreover that this bijection is compatible with the natural operations (direct sum of vector bundles on $V(X)$, and the block sum of idempotents from Definition 2.1.4).

2.11.5 Let M be a commutative monoid (or just a semigroup: the presence of the unit makes no real difference). In this exercise, we give two constructions of the universal *Grothendieck group*: an abelian group equipped with a map of monoids $M \to G(M)$ with the universal property that

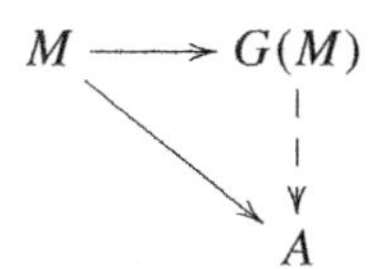

can be filled in for any abelian group A.

(i) Show that $G(M)$ can be constructed as the quotient of the free abelian group with generating set $\{[a] \mid a \in M\}$, subject to the relations

$$[a] +_{G(M)} [b] = [a +_M b].$$

(ii) Show that $G(M)$ can be constructed as the quotient of the direct sum monoid $M \oplus M$ by the relation

$$(a_1, b_1) \sim (a_2, b_2) \quad \text{if } a_1 + b_2 = a_2 + b_1.$$

If $G(M)$ is constructed as in part (i), show moreover that any element of $G(M)$ is equivalent to one of the form $[a] - [b]$ with $a, b \in M$, and that this corresponds to the equivalence class of (a, b) when $G(M)$ is constructed as in part (ii).

2.11.6 In Definition 2.1.13, we defined K_0 of a non-unital $\mathbb{C}$-algebra R to be the kernel of the map

$$\phi_*\colon K_0(R^+) \to K_0(\mathbb{C})$$

where $\phi\colon R^+ \to \mathbb{C}$ is the canonical quotient map from the unitisation of R to $\mathbb{C}$. Show that for any $n \in \mathbb{N}$, $K_0(R)$ also canonically identifies with the kernel of the map

$$K_0(M_n(R^+)) \to K_0(M_n(\mathbb{C}))$$

induced on matrices by the above quotient map.

2.11.7 Let A be a Banach algebra.

(i) Let A^+ be the unitisation of A as in Definition 1.1.9, and equip A^+ with the norm defined by

$$\|(a, \lambda)\| := \|a\|_A + |\lambda|.$$

Show that this is indeed a norm, and that it makes A^+ into a Banach algebra.

(ii) Let $(A^+)^{\oplus n}$ denote the direct sum of n copies of A^+, equipped with the norm

$$\|(a_1, \ldots, a_n)\| := \sum_{i=1}^{n} \|a_i\|_{A^+}.$$

Show that $M_n(A)$ acts on $(A^+)^{\oplus n}$ in the natural way by matrix mutliplication (treating elements of $(A^+)^{\oplus n}$ as column vectors), and that the associated operator norm on $M_n(A)$ makes it into a Banach algebra.

(iii) With respect to the norms introduced above, show that the 'top left corner' inclusion

$$M_n(A) \to M_{n+1}(A), \quad a \mapsto \begin{pmatrix} a & 0 \\ 0 & 0 \end{pmatrix}$$

is an isometry.

2.11.8 Let Q be a unital $\mathbb{C}$-algebra and Q^+ its unitisation. Show that if Q fits into a short exact sequence

$$0 \longrightarrow I \longrightarrow R \longrightarrow Q \longrightarrow 0$$

then the same formula as in Definition 2.4.3 defines a homomorphism

$$\text{Ind}\colon GL_\infty(Q) \to K_0(I)$$

and that the diagram

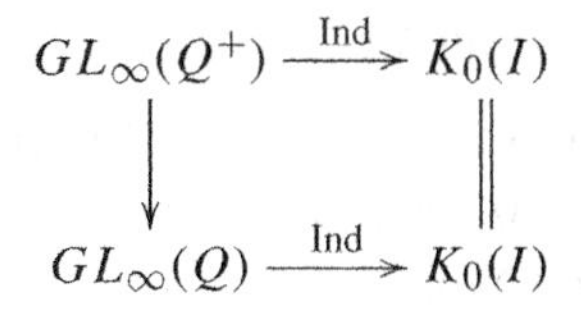

$$\begin{array}{ccc} GL_\infty(Q^+) & \xrightarrow{\text{Ind}} & K_0(I) \\ \downarrow & & \| \\ GL_\infty(Q) & \xrightarrow{\text{Ind}} & K_0(I) \end{array}$$

commutes, where the vertical arrow on the left is induced by $a + \lambda(1_{Q^+} - 1_Q) \mapsto a$. Show the same formulas induce a similar diagram where $GL_\infty(Q^+)$ and $GL_\infty(Q)$ are replaced by $K_1(Q^+)$ and $K_1(Q)$ respectively, and where the vertical map on the left now is now an isomorphism (part of the exercise is to make precise sense of this).

2.11.9 Let A be a Banach algebra. Show that the monoid defined in part (iv) of Proposition 2.2.9 defines the usual monoid $V(A)$ underlying the definition of $K_0(A)$.

2.11.10 With notation as in Proposition 2.3.11, complete the following sketch proof that $I_{\tau,1}$ is an ideal. Let first x be a positive element in $I_{\tau,1}$ and u be a unitary in the unitisation A^+. Use the polarisation identity (line (2.4)) applied to the product $(x^{1/2}u)^*x^{1/2}$ plus the facts that $I_{\tau,1}$ is hereditary and that uxu^* is in $I_{\tau,1}$ to deduce that ux is in $I_{\tau,1}$. As A^+ is spanned by unitaries and $I_{\tau,1}$ is spanned by its positive elements, this suffices.

2.11.11 Prove that under the hypotheses of Proposition 2.3.16, the inclusion $\iota\colon \mathcal{A} \to A$ induces an isomorphism on K_1 groups.

2.11.12 Show that the hypotheses of Proposition 2.3.16 can be weakened to the case that $\mathcal{A}$ is a Fréchet algebra for which the inclusion $\iota\colon \mathcal{A} \to A$ is continuous (and that the conclusion of Exercise 2.11.11 holds in this case too). *Hint: the idea of the proof is similar, but there are some additional subtleties. Assume that A and $\mathcal{A}$ are unital (with the same unit) for notational simplicity.*

First, while it is not true that the invertible elements of a Fréchet algebra always form an open set, show that the invertibles $\mathcal{A}^{-1}$ in $\mathcal{A}$ are open under the assumption that $\mathcal{A}$ is a (dense) inverse closed subalgebra of a C-algebra (or even just a Banach algebra) A. Now appeal to (or prove) the following fact from general topology: if U is an open subset of a complete metric space X then there exists a complete metric on U inducing the subspace topology. In particular, $\mathcal{A}^{-1}$ is a complete, metrisable topological space.*

At this point appeal to (or prove) the following automatic continuity result for $G = \mathcal{A}^{-1}$: if G is a group equipped with a complete metrisable topology for which the multiplication is jointly continuous, then the inverse map $G \to G$

is also continuous. Show that this implies that $\mathcal{A}$ has a holomorphic functional calculus (which agrees with the holomorphic functional calculus on A). The proof can now be completed much as the proof of Proposition 2.3.16.

2.11.13 Use the result of the Exercise 2.11.12 to show that if M is a closed smooth manifold then $K^0(M)$ is generated by classes of the form $[p]$ where $p \in M_n(C(M))$ is a smooth projection.
Hint: show that $\mathcal{A} = C^\infty(M)$ satisfies the assumptions of Exercise 2.11.12.

2.11.14 Show that the trace defined in Example 2.3.4 is indeed an unbounded trace in the sense of Definition 2.3.1, and that it does not depend on any of the choices involved in its construction.

2.11.15 Show that $K_0(\ell^\infty(\mathbb{N}))$ is canonically isomorphic to the collection of all *bounded* maps from $\mathbb{N}$ to $\mathbb{Z}$, and thus that there is no 'naive' isomorphism

$$K_*\left(\prod_{i\in I} A_i\right) \cong \prod_{i\in I} K_*(A_i)$$

when each A_i is $\mathbb{C}$, and I is $\mathbb{N}$.

2.11.16 Let A be a C^*-algebra, $\mathcal{K}$ be the compact operators on an infinite-dimensional Hilbert space and $M(A \otimes \mathcal{K})$ the multiplier algebra of $A \otimes \mathcal{K}$. Generalise the argument of Corollary 2.7.7 to show that $K_*(M(A \otimes \mathcal{K})) = 0$.

2.11.17 With notation as in the previous exercise, let B be the quotient C^*-algebra $M(A \otimes \mathcal{K})/(A \otimes \mathcal{K})$. Show that natural map

$$\pi_0\big(\{p \in B \mid p \text{ a projecton}\}\big) \to K_0(B), \quad [p] \mapsto [p]$$

is a well-defined bijection. Analogously show that

$$\pi_0\big(\{u \in B \mid u \text{ unitary}\}\big) \to K_1(B), \quad [u] \mapsto [u]$$

is a well-defined bijection.
Hint: for the projection case, you might want to start by showing that the class $[1]$ of the identity is zero in K-theory (this follows from Exercise 2.11.17), and that every projection $p \in M_\infty(B)$ is Murray–von Neumann equivalent to a subprojection of 1.

2.11.18 Consider a pushout diagram of C^*-algebras as in Definition 2.7.13. Define

$$C = \{f\colon [0,1] \to A \mid f(0) \in I \text{ and } f(1) \in J\}.$$

Note that the evaluation maps at 0 and 1 gives rise to a short exact sequence

$$0 \longrightarrow SA \longrightarrow C \longrightarrow I \oplus J \longrightarrow 0.$$

Deduce the existence of the pushout Mayer–Vietoris sequence from the six-term exact sequence associated to this short exact sequence.

2.11.19 Consider a pullback diagram of C^*-algebras as in Definition 2.7.13. Consider the C^*-algebra

$$C = \{(f_A, f_B) \in CA \oplus CB \mid \pi_A(f_A(1)) = \pi_B(f_B(1))\}.$$

The evaluation map $(f_A, f_B) \mapsto (f_A(1), f_B(1))$ from C to P fits into a short exact sequence

$$0 \longrightarrow SA \oplus SB \longrightarrow C \longrightarrow P \longrightarrow 0.$$

Deduce the existence of the pushout Mayer–Vietoris sequence from the six-term exact sequence associated to this short exact sequence.

2.11.20 Let $G = \mathbb{Z}/3\mathbb{Z} = \{e, g, g^2\}$, and let $\mathbb{C}[G]$ be its complex group algebra. Note that $\mathbb{C}[G] \cong \mathbb{C}^3$, so $K_0(\mathbb{C}[G]) \cong \mathbb{Z}^3$.

Write elements of $\mathbb{C}[G]$ as formal sums $ae + bg + cg^2$ where $a, b, c \in \mathbb{C}$, and equip $\mathbb{C}[G]$ with the ℓ^1 norm

$$\|ae + bg + cg^2\| := |a| + |b| + |c|$$

and $*$-operation

$$(ae + bg + cg^2)^* := \overline{a}e + \overline{b}g + \overline{c}g^2,$$

so $\mathbb{C}[G]$ is a Banach *-algebra (it is *not* a C^*-algebra!). Now let

$$f = \frac{1}{3}e + \frac{1}{3}\left(-\frac{1}{2} + \frac{\sqrt{3}}{2}\right)g + \frac{1}{3}\left(-\frac{1}{2} - \frac{\sqrt{3}}{2}i\right)g^2,$$

which is an idempotent in $\mathbb{C}[G]$. Show that there is no self-adjoint idempotent $p \in \mathbb{C}[G]$ with $[p] = [f]$. Compare this to the statement of Proposition 2.2.5.

2.11.21 Prove *Atkinson's theorem* as used in Example 2.8.3 above. This says that an operator $T \in \mathcal{B}(H)$ is invertible modulo $\mathcal{K}(H)$ if and only if it has closed range, and finite-dimensional kernel and cokernel.
Hint: the harder direction goes from invertibility modulo $\mathcal{K}(H)$. Start by proving that if T is invertible modulo $\mathcal{K}(H)$, then it is actually invertible modulo finite rank operators. Note that T having closed range actually follows from the fact that cokernel $H/Im(T)$ is finite-dimensional: prove this too!

2.11.22 It follows from the discussion of doubles in Section 2.7 that if I is an ideal in a C^*-algebra A, and e, f are idempotents in A such that $e - f$ is in I, then there is a canonically associated class $[e] - [f]$ in $K_0(I)$. In Definition 2.1.13, we defined $K_0(I)$ in terms of idempotents in matrices over the unitisation I^+; the purpose of this exercise is to give an explicit, and purely algebraic, formula for the class $[e] - [f]$ in terms of matrices over I^+ (all this also works with general rings; the fact that we are working with C^*-algebras is irrelevant).

Passing to the unitisation of A if necessary we may assume that A is unital, and thus identify I^+ with a C^*-subalgebra of A. Define then

$$Z(f) := \begin{pmatrix} f & 0 & 1-f & 0 \\ 1-f & 0 & 0 & f \\ 0 & 0 & f & 1-f \\ 0 & 1 & 0 & 0 \end{pmatrix}.$$

(i) Check that $Z(f)$ is invertible by showing that its inverse is

$$Z(f)^{-1} = \begin{pmatrix} f & 1-f & 0 & 0 \\ 0 & 0 & 0 & 1 \\ 1-f & 0 & f & 0 \\ 0 & f & 1-f & 0 \end{pmatrix}.$$

(ii) Show that the difference below

$$Z(f)^{-1} \begin{pmatrix} e & 0 & 0 & 0 \\ 0 & 1-f & 0 & 0 \\ 0 & 0 & 0 & 0 \\ 0 & 0 & 0 & 0 \end{pmatrix} Z(f) - \begin{pmatrix} 1 & 0 & 0 & 0 \\ 0 & 0 & 0 & 0 \\ 0 & 0 & 0 & 0 \\ 0 & 0 & 0 & 0 \end{pmatrix}$$

is in $M_4(I)$ (and therefore the first matrix is in $M_4(I^+)$), and thus it defines an element of $K_0(I)$.

(iii) Show that this is the same element as the class defined by the image of $[(e, f)] \in K_0(D_A(I))$ under the canonical map $K_0(D_A(I)) \to K_0(I)$.

2.11.23 Let X be a locally compact space, and let U be an open subset. As we commented in Example 2.7.16, there is a 'wrong way map' in K-theory $K^*(U) \to K^*(X)$ induced by the inclusion $C_0(U) \to C_0(X)$. What is the map on the *spatial* (as opposed to C^*-algebraic) level that induces this map on K-theory?

Hint: think about one-point compactifications.

2.11.24 Use the K-theory Mayer–Vietoris sequence(s) to compute the K-theory groups of d-tori, d-spheres and orientable surfaces.

2.11.25 Find explicit formulas for *all* the maps appearing in the two Mayer–Vietoris sequences explicitly (including the vertical maps).

2.11.26 Let $\phi\colon A \to B$ be a $*$-homomorphism between two C^*-algebras. The *mapping cone* of ϕ is the C^*-algebra defined by

$$C(\phi) := \{(a, f) \in A \oplus CB \mid \phi(a) = f(1)\}.$$

Show that the C^*-algebra C appearing in the proof of the 'pullback' Mayer–Vietoris sequence in Exercise 2.11.19 identifies with the mapping cone of the natural inclusion $P \hookrightarrow A \oplus B$.

2.11.27 With notation as in Exercise 2.11.26, show that a $*$-homomorphism $\phi\colon A \to B$ induces an isomorphism on K-theory if and only if $K_*(C(\phi)) = 0$.

2.11.28 Let A be a C^*-algebra. Equip $M_2(A)$ with the standard inner grading induced by the element $\begin{pmatrix} 1 & 0 \\ 0 & -1 \end{pmatrix}$ of its multiplier algebra. Show that the following two classes can be canonically identified.

- Pairs consisting of a graded representation of $M_2(A)$ on a graded Hilbert space H, and an odd operator F in the multiplier algebra of $M_2(A)$ such that $F^2 - 1$ is in $M_2(A)$.
- Pairs consisting of representations of A on a Hilbert space, and pairs of operators $V, W\colon H \to H$ in the multiplier algebra of A such that $VW - 1$ and $WV - 1$ are in A.

2.11.29 Let A be a graded C^*-algebra equipped with a faithful graded representation $A \subseteq \mathcal{B}(H)$ on a graded Hilbert space. Show that if $\phi\colon \mathcal{S} \to A$ is a graded $*$-homomorphism, then there is an odd (essentially) self-adjoint operator D on H such that $\phi(f) = f(D)$ for all $f \in \mathcal{S}$.
Hint: one way to do this is to apply the Stone–von Neumann theorem (Theorem D.2.2) to the Fourier transform of ϕ in an appropriate sense.

2.11.30 In Lemma 2.9.4, if B is a graded C^*-algebra, then we identified the collection $\{\mathcal{S}, B\}$ of homotopy classes of graded $*$-homomorphisms from $\mathcal{S}$ to B with the set of path components of the set

$$\left\{ u \in B^+ \;\middle|\; \begin{array}{l} u \text{ is unitary, } \epsilon_{B^+}(u) = u^*, \text{ and } u \text{ maps to } 1 \\ \text{under the canonical quotient } B^+ \to \mathbb{C} \end{array} \right\}.$$

Now, say B is gradedly represented on a Hilbert space H. Show that $\{\mathcal{S}, B\}$ bijectively identifies with the path components of the set

$$\left\{ F \in \mathcal{B}(H) \;\middle|\; \begin{array}{l} F \text{ is odd, } F = F^*,\ F^2 - 1 \in B \\ \|F\| \leqslant 1, \text{ and } F^3 - F \in B \end{array} \right\}.$$

Hint: the bijection we have in mind is canonical, up to a choice of continuous, increasing, odd bijection $f: \mathbb{R} \to (-1,1)$.

2.11.31 Show that

$$spK_i(\text{Cliff}_{\mathbb{C}}(\mathbb{R})) \cong \begin{cases} 0 & i=0 \\ \mathbb{Z} & i=1 \end{cases}.$$

Note that if we forget the grading, then $Cliff_{\mathbb{C}}(\mathbb{R})$ is isomorphic to $\mathbb{C} \oplus \mathbb{C}$. Hence this shows that the analogue of Proposition 2.9.12 fails if the C^*-algebra is not inner graded.

2.12 Notes and References

There are several good texts on C^*-algebra K-theory. Rørdam, Larsen and Laustsen [221] and Wegge-Olsen [248] are written at an introductory level, and are a good place to start. Blackadar's classic text [33] expects more of the reader, and goes much further. The first two texts work only in the context of C^*-algebras, while the latter sets up basic topological K-theory in the more general context of so-called local Banach algebras; this is very useful for some applications. From an operator algebraic point of view, an important inspiration for K-theory comes from von Neumann's work on continuous geometries (see von Neumann [242]).

Cuntz, Meyer and Rosenberg [71] is a more modern introduction to topological K-theory: it works with the class of so-called bornological algebras, which provide a very satisfactory setting for topological K-theory and related material. The first two chapters are quite similar in spirit to the way we have presented this chapter. Chapter 4 of Higson and Roe [135] also contains a (brief) introduction to C^*-algebra K-theory from a point of view that is well-suited to index-theoretic applications, amongst other things.

In terms of the proofs we skipped, most of the material can be found in all, or almost all, of the references mentioned above. There are two places where we skipped proofs of slightly more non-standard material, however. First, the more algebraic treatments in Sections 2.1 and 2.4 can be found in much more detail in chapter 1 of Cuntz et al. [71]. The result of Brown we quoted as Theorem 2.7.17 is corollary 2.6 in Brown [40].

There are quite a few tools and tricks that we used in this chapter, some non-obvious; unfortunately, we do not know the origins of most of them. Some of the ones we know attributions for (or at least, where we know the references that we learnt the material from) are: our treatment of the algebraic index map, which is based on chapter 3 of Milnor [183]; the formulas in Exercise 2.11.22, which come from section 6 of Kasparov and Yu [153]; and the idea of a quasi-morphism, which is based on work of Cuntz [69].

The material in Section 2.3 is not found in any standard textbook as far as we know, and is based on unpublished lecture notes of John Roe. The material in Lemmas 2.3.14, 2.3.15 and Proposition 2.3.16 can also be found (in the more general version of Exercise 2.11.12) in the paper by Schweitzer [233]. The basic philosophy in that section – that sometimes one needs to pass to a dense subalgebra in order to pair certain objects with K-theory – is fundamental in cyclic (co)homology theory and the so-called theory of *noncommutative differential geometry*; traces are the zero-dimensional part of cyclic cohomology, and thus in some sense the simplest manifestation of the general theory. See for example Connes [58] and chapter 3 of Connes [60] for a look at the ideas involved here.

The spectral picture of K-theory and the associated definition of the external product is also not as standard as the other material, which is why we went over it in more detail. The idea is exposited in the literature by Trout [240] and Lecture 1 of Higson and Guentner [128], and it is closely related to the bivariant E-theory of Connes and Higson [61]. There is also a nice recent exposition in section 1.1 of Zeidler [275] that highlights the connections to index theory, and also in unpublished lecture notes of John Roe. Our exposition is based on all of these sources.

For the reader interested in the purely algebraic theory (which is not directly relevant for this book, but does connect to it in many ways), we recommend the classic introduction to the basics by Milnor [183] and the more modern and comprehensive texts by Rosenberg [224] and Weibel [250]. The survey article by Cortiñas [67] gives a very nice introduction to the relationship between topological and algebraic K-theory.

For the reader interested in the purely topological theory (more or less equivalently, the K-theory of commutative C^*-algebras), we recommend the classic texts of Atiyah [6] and Karoubi [147]. In particular, chapter 1 of Atiyah [6] is an excellent source for the material needed to understand Exercise 2.11.4, amongst other things.

3

Motivation

Positive Scalar Curvature on Tori

In this chapter, we discuss the non-existence of positive scalar curvature metrics on tori. This is a classical problem from Riemannian geometry. We have tried to explain the ideas in a way that one can understand without much background in manifold geometry. We generally do not give complete proofs in this chapter, but most of the material will be justified and expanded on later in the book.

The chapter is structured as follows. Section 3.1 is differential-geometric: we discuss the question of the existence of positive scalar curvature metrics and state a theorem relating this to so-called Dirac operators. In Section 3.2 we introduce analysis: we discuss Hilbert space methods that are used to convert the original geometric question to one about index elements in K-theory groups of associated C^*-algebras. Then, in Section 3.3 we sketch K-theoretic computations that answer this question in the case of tori. The structure of these three sections – starting with a geometric or topological problem, then introducing Hilbert space techniques to frame the problem in terms of operator K-theory and finally solving the K-theoretic problem – mimics that of much research in the area generally.

Section 3.4 then gives some historical discussion, partly to provide more context. Finally, Section 3.5 discussed the content of the rest of the book, using the material in this chapter as a framework.

3.1 Differential Geometry

Let M be a d-dimensional Riemannian manifold: in other words, M is a topological space, locally modelled on $\mathbb{R}^d$ is a smooth way, and there is additional 'Riemannian structure' or 'Riemannian metric' that lets us talk about geometric notions. Intuitively, one can think of M as a subset of some

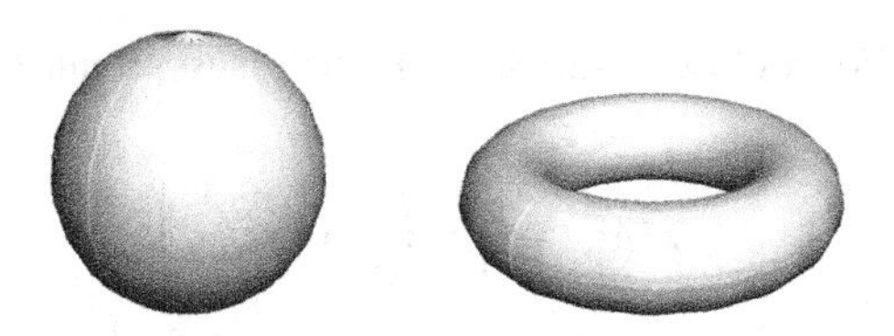

Figure 3.1 Curved sphere and torus.

N-dimensional Euclidean space for $N \geqslant d$, such as the sphere and torus in Figure 3.1 (with $d = 2$ and $N = 3$).

The canonical inner product on $\mathbb{R}^N$ then induces a Riemannian structure on M that induces the geometry 'relevant to life on M'. This means that the induced distance function on M is not just the restriction of the distance function from $\mathbb{R}^N$, but rather that as an ant living on M would see it: the distance between two points on M is the distance you have to walk along the surface of M to get there, not the straight line distance through the ambient Euclidean space.

Now, as well as a distance function on M, the Riemannian structure also induces a measure, so it makes sense to speak of the volume of an r-ball $B(x;r)$ around some $x \in M$. In particular, it makes sense to discuss how this volume differs from the corresponding volume of an r-ball in Euclidean space $\mathbb{R}^d$ of the same dimension as M. This difference is measured quantitatively by the scalar curvature function of M.

The *scalar curvature* of M is the smooth function $\kappa\colon M \to \mathbb{R}$ determined by the following condition: for any $x \in M$ and small $r > 0$ we have

$$\frac{\text{Volume}(B_M(x;r))}{\text{Volume}(B_{\mathbb{R}^d}(x;r))} = 1 - \frac{\kappa(x)}{6(d+2)}r^2 + O(r^4).$$

Thus $\kappa(x)$ measures how much the volume of the ball of radius r about $x \in M$ differs from the volume of a ball of radius r in $\mathbb{R}^d$. If $\kappa(x)$ is positive (for example, if M is the sphere pictured in Figure 3.1), balls in M are smaller than in Euclidean space and if it is negative (for example, if M is a hyperbolic space), they are bigger. In many cases, a Riemannian manifold will have some points where the scalar curvature is negative and others where it is positive: this happens for the torus in Figure 3.1, where the scalar curvature is positive on the bits of the torus looking inwards to the 'doughnut hole' and positive on the 'exterior facing' parts.

The possible values that can be taken by scalar curvature can depend on the qualitative features of the manifold. The fundamental example of this is given by the Gauss–Bonnet theorem.

Example 3.1.1 For a closed[1] surface[2] M, the Gauss–Bonnet theorem is the formula

$$\chi(M) = \frac{1}{4\pi}\int_M \kappa(x)dx$$

for the Euler characteristic $\chi(M)$ of M in terms of the integral[3] of the scalar curvature.[4] This puts restrictions on the possible scalar curvature functions that can be admitted by M in terms of a purely topological invariant. In particular, M admits a metric with positive scalar curvature only if it is a sphere or a projective plane (i.e. if $\chi(M) = 1$ or $\chi(M) = 2$); with a constantly zero scalar curvature function only if M is a torus or a Klein bottle (i.e. $\chi(M) = 0$); and with a negative scalar curvature function only if M is a (possibly non-orientable) higher genus surface (i.e. $\chi(M) < 0$).

In dimensions larger than two, it is thus natural[5] to ask the following question.

Question 3.1.2 For a fixed smooth manifold M, which functions $\kappa\colon M \to \mathbb{R}$ arise as the scalar curvature of some Riemannian metric, and in particular, are there obstructions to the particular types of scalar curvature arising purely from the topology of M?

It seems at first as if this question might not have an interesting answer: indeed, if we increase the dimension beyond that of surfaces, then there are essentially no restrictions on the existence of *negative* scalar curvature metrics thanks to the *Kazhdan–Warner theorem*.

Theorem 3.1.3 (Kazhdan–Warner) *Let M be a smooth manifold of dimension at least 3, and let $\kappa\colon M \to \mathbb{R}$ be a smooth function that takes a negative value at some point of M. Then there is a metric on M for which κ is the scalar curvature function.* □

However, we are still left with the following.

Question 3.1.4 Which smooth manifolds M admit positive scalar curvature, i.e. a scalar curvature function that is everywhere positive?

[1] i.e. compact, with no boundary.
[2] i.e. two-dimensional manifold.
[3] Defined using the measure determined by the Riemannian metric.
[4] The Gauss–Bonnet theorem is more usually stated in terms of the so-called Gaussian curvature K: for a two-dimensional manifold M, the scalar curvature κ is exactly twice Gaussian curvature.
[5] And quite relevant to other areas of mathematics and physics!

Based on the Kazhdan–Warner theorem, it is tempting to guess that there are no obstructions to the existence of a positive scalar curvature metric: perhaps higher-dimensional manifolds are so 'flexible' that they allow metrics of any scalar curvature. It turns out that this is not the case: there are obstructions to the existence of positive scalar curvature metrics for manifolds of any dimension.

One of the most important of these obstructions applies in the special case that M is a *spin* manifold. This is a topological condition on M that is satisfied by many natural examples: for example, all spheres and tori in any dimension are spin. For us, the important property that (Riemannian) spin manifolds have is the existence of a canonically associated differential operator D, the *spinor Dirac operator* that is closely connected to the scalar curvature of M.

In order to precisely state the relevant property of the Dirac operator, we need some notation. Let S be a Hermitian bundle over M: this means that S is a smooth complex vector bundle over M such that for each $x \in M$, the fibre S_x over x is equipped with a non-degenerate Hermitian form $\langle , \rangle_x$ in a way that depends smoothly on x. Let $C_c^\infty(M; S)$ denote the vector space of compactly supported smooth sections of S, and note that $C_c^\infty(M; S)$ is equipped with a positive definite inner product defined by

$$\langle u, v \rangle := \int_M \langle u(x), v(x) \rangle_x dx. \tag{3.1}$$

A linear operator $T\colon C_c^\infty(M; S) \to C_c^\infty(M; S)$ is *adjointable* if there is a linear operator $T^*\colon C_c^\infty(M; S) \to C^\infty(M; S)$ satisfying

$$\langle u, Tv \rangle = \langle T^* u, v \rangle.$$

If T is adjointable, the (necessarily unique) operator T^* satisfying the above condition is called the *adjoint* of T. For example, note that elements of the $*$-algebra $C^\infty(M)$ of smooth functions[6] on M act on $C_c^\infty(M; S)$ by pointwise multiplication, and that all such operators are adjointable: the adjoint of (multiplication by) f is (multiplication by) the complex conjugate of f.

Other important examples of adjointable operators are given by differential operators: these are operators on $C_c^\infty(M; S)$ that can be written in local coordinates in terms of partial derivatives and (fibrewise) endomorphisms of S.

The following theorem is the first of our 'unjustified ingredients'. It is fundamental for the index-theoretic approach to positive scalar curvature.

Theorem 3.1.5 *Let M be a Riemannian spin manifold. Then there is a canonically associated Hermitian* spinor bundle S *and first order differential*

[6] Or elements of $C_c^\infty(M)$, the $*$-algebra of smooth compactly supported functions.

operators ∇ *(the* spinor connection*) and* D *(*the (spinor) Dirac operator*) on* $C_c^\infty(M;S)$ *such that if* $\kappa\colon M \to \mathbb{R}$ *is the scalar curvature of* M*, then*

$$D^2 = \nabla^*\nabla + \frac{\kappa}{4}. \qquad \square$$

Example 3.1.6 Let $M = \mathbb{R}$ be the real line, with its usual metric. The spinor bundle on $\mathbb{R}$ turns out to be the one-dimensional trivial bundle $S = \mathbb{R} \times \mathbb{C}$. The Dirac operator is the differential operator $D = -i\frac{d}{dx}$, and so $D^2 = -\frac{d^2}{dx^2}$ is the negative of the usual Laplacian, i.e. it equals $\nabla^*\nabla$, with $\nabla = \frac{d}{dx}$. In this case, the scalar curvature is zero.

The computation underlying Theorem 3.1.5 is purely differential geometric. In the next section, we discuss how to bring some analysis into play.

3.2 Hilbert Space Techniques

Throughout this section, M is a Riemannian spin manifold, S the spinor bundle over M and D the Dirac operator as in Theorem 3.1.5.

We now bring some Hilbert space techniques into play, and in particular some ideas from unbounded operator theory as briefly sketched in Appendix D. The first step is to complete some of the spaces of sections we were working on in the last section to Hilbert spaces.

Let then $L^2(M;S)$ denote the Hilbert space completion of $C_c^\infty(M;S)$ for the inner product in line (3.1). We consider D as an unbounded operator on $L^2(M;S)$ with domain $C_c^\infty(M;S)$.

We will need to use two results from the theory of analysis on manifolds – Proposition 3.1.5 and Proposition 3.2.4 below – as black boxes. For the first of these, recall the notion of essential self-adjointness from Definition D.1.4.

Proposition 3.2.1 *Assume that the Riemannian metric on* M *is complete.*[7] *Then* D *is essentially self-adjoint.* $\square$

Example 3.2.2 Let $M = \mathbb{R}$ be the real line, S be the one-dimensional trivial bundle over $\mathbb{R}$ and $D = -i\frac{d}{dx}$ be the Dirac operator as in Example 3.1.6. Then $L^2(M;S)$ identifies with $L^2(\mathbb{R})$. The subspace $C_c^\infty(\mathbb{R};S) = C_c^\infty(\mathbb{R})$ of $L^2(\mathbb{R})$ consisting of smooth compactly supported functions identifies under Fourier transform with a subspace of the rapidly decaying functions on $\mathbb{R}$, and the

[7] i.e. Cauchy sequences for the induced distance function converge; for Riemannian manifolds, this is equivalent to the statement that all closed balls are compact by Theorem A.3.6.

operator $-i\frac{d}{dx}$ with the operator of multiplication by x. Exercise 3.6.1 asks you to prove directly that this operator is essentially self-adjoint.

Assume then that M is complete, so D is essentially self-adjoint by Proposition 3.2.1. Then for any continuous bounded function $\phi\colon \mathbb{R} \to \mathbb{C}$, the functional calculus (Theorem D.1.7) lets us define a bounded normal operator $\phi(D)$ on $L^2(M;S)$. We will need that, at least under certain conditions, the operators $\phi(D)$ are closely tied to the geometry of M.

Definition 3.2.3 Let T be a bounded operator on $L^2(M;S)$. The *propagation* of T, denoted $\mathrm{prop}(T)$, is the smallest number r in $[0,\infty]$ with the following property: whenever f_1, f_2 are elements of $C_c^\infty(M;S)$ such that $d(\mathrm{supp}(f_1), \mathrm{supp}(f_2)) > r$, we have that $f_1 T f_2 = 0$.

The operator T is *locally compact* if for any $f \in C_c^\infty(M)$ (thought of as a multiplication operator on $L^2(M;S)$), the operators fT and Tf are compact.

We will need the following result as another black box.

Proposition 3.2.4 *Assume that M is complete.*

(i) If $\phi\colon \mathbb{R} \to \mathbb{C}$ has Fourier transform supported in $[-r,r]$, then $\phi(D)$ has propagation at most r.
(ii) If ϕ is compactly supported, then $\phi(D)$ is locally compact. □

Example 3.2.5 Let $M = \mathbb{R}$, $S = \mathbb{R}\times\mathbb{C}$ and $D = -i\frac{d}{dx}$ be as in Example 3.1.6. Using the Fourier transform, one computes that $\phi(D)$ is (up to a constant, depending on Fourier transform conventions) the operator of convolution by $\widehat{\phi}$. If $\widehat{\phi}$ has support contained in $[-r,r]$ and if $f \in L^2(M;S)$ is supported in some compact set K, we have that $\widehat{\phi} * f$ is supported in

$$N_r(K) := \{x \in \mathbb{R} \mid d(x,K) \leqslant r\}.$$

The fact that $\phi(D)$ has propagation at most r follows from this.

Moreover, if ϕ is compactly supported, then the Fourier transform $\widehat{\phi}$ is Schwartz class. Hence for any $f \in C_c^\infty(M;S)$, the operator $T = \phi(D)f$ is given by

$$(Tu)(x) = \int_{\mathbb{R}} \widehat{\phi}(x-y)f(y)u(y)dy.$$

The kernel $k(x,y) = \widehat{\phi}(x-y)f(y)$ is square-integrable on $\mathbb{R}\times\mathbb{R}$, however, whence Hilbert–Schmidt and in particular compact.

The C^*-algebra in the next definition allows us to bring index theory to bear.

Definition 3.2.6 The *Roe algebra* of $L^2(M;S)$, denoted $C^*(M)$, is the C^*-algebra generated by all the locally compact, finite propagation operators on $L^2(M;S)$.

Note now that $L^2(M;S)$ comes equipped with a grading (see Definition E.1.4) in the case that M is even-dimensional, and that the Dirac operator D is odd for this grading.

Lemma 3.2.7 *Continue to assume that M is complete. Let $\psi: \mathbb{R} \to [-1,1]$ be an odd function such that*

$$\lim_{x\to-\infty} \psi(x) = -1 \quad \textit{and} \quad \lim_{x\to+\infty} \psi(x) = +1.$$

Then the operator $F = \psi(D)$ is a self-adjoint multiplier of $C^*(M)$, which is odd if M is even-dimensional, and such that $1 - F^2$ is an element of $C^*(M)$. In particular it defines an index element

$$\mathit{Index}[F] \in K_i(C^*(M))$$

via Definition 2.8.5, where i agrees with the dimension of M modulo 2. This class does not depend on the choice of χ.

Proof Let $g: \mathbb{R} \to [0,\infty)$ be an even function with compactly supported Fourier transform, and such that $\int_{\mathbb{R}} g(x)dx = 1$. Define g_n by $g_n(x) := ng(nx)$. Then the convolutions $g_n * \psi$ converge in supremum norm to ψ as $n \to \infty$, whence

$$F_n := (g_n * \psi)(D) \to \psi(D) = F$$

in norm as $n \to \infty$. Moreover, as the Fourier transform converts convolution to pointwise multiplication, $g_n * \psi$ has compactly supported Fourier transform. It follows now from Proposition 3.2.4 that F_n has finite propagation.

Now let T be a finite propagation, locally compact operator. To show that F is a multiplier of $C^*(M)$, it suffices to show that $F_n T$ and $T F_n$ are in $C^*(M)$ for all n; we will focus on $F_n T$, the other case being similar.

Say T has propagation s and F_n has propagation r. Let f_1 and f_2 be any elements of $C_c^\infty(M;S)$, and assume that $d(\operatorname{supp}(f_1), \operatorname{supp}(f_2)) > r + s + 1$. Let $f \in C_c^\infty(M)$ be supported in $\{x \in M \mid d(\operatorname{supp}(u), x) \leqslant r + 2/3\}$ and equal to one inside $\{x \in M \mid d(\operatorname{supp}(m), x) \leqslant r + 1/3\}$. As F_n has propagation r, we have that

$$f_1 T F_n f_2 = f_1 T f F_n f_2,$$

which is zero as T has propagation at most s. It follows that $T F_n$ has propagation at most $r + s + 1$.

To see that TF_n is locally compact, let f be any element of $C_c^\infty(M)$. The product fTF_n is then compact as the compact operators are an ideal. On the other hand, it follows just as above from finite propagation of F_n that there exists compactly $f_0 \in C_c^\infty(M)$ with $F_n f = f_0 F_n f$, whence

$$TF_n f = Tf_0 F_n f,$$

which is compact by local compactness of T.

The fact that F is odd when D is follows as ψ is odd. To see that $1 - F^2$ is in $C^*(M)$, note

$$1 - F^2 = (1 - \psi^2)(D).$$

The function $1 - \psi^2$ is in $C_0(\mathbb{R})$, however, so this follows from Proposition 3.2.4. Finally, the fact that $\mathrm{Ind}[F]$ does not depend on the choice of ψ follows as if ψ_1, ψ_2 are any two such functions, then

$$\psi_1(D) - \psi_2(D) = (\psi_1 - \psi_2)(D).$$

As $\psi_1 - \psi_2 \in C_0(\mathbb{R})$, this operator is in $C^*(M)$ and we may apply Remark 2.8.6. □

The following vanishing theorem is the key result needed for applications of operator K-theory to questions on the existence of positive scalar curvature metrics.

Theorem 3.2.8 *Say M is a complete, spin, Riemannian manifold with positive scalar curvature bounded below. Then the element*

$$\mathit{Index}[F] \in K_*(C^*(M))$$

defined in Corollary 3.2.7 is zero.

Proof Theorem 3.1.5, the fact that κ is bounded below and essential self-adjointness of D imply that the spectrum of D does not contain anything in the interval $[-c, c]$ for some $c > 0$. We may choose ψ satisfying the conditions in Lemma 3.2.7 and that satisfies $\psi(x) \in \{+1, -1\}$ for all $x \notin [-c, c]$. It follows from this that $F^2 = 1$, and thus the result now follows from Lemma 2.8.7. □

Here then is the basic strategy: we will show that certain manifolds cannot admit a metric of positive scalar curvature by showing that the class associated to the Dirac operator in $K_*(C^*(M))$ is non-zero.

3.3 K-theory Computations

In this section, let M be the d-dimensional torus, so as a smooth manifold

$$M = \underbrace{S^1 \times \cdots \times S^1}_{d \text{ times}}$$

is the d-fold product of the circle with itself. This is then a spin manifold. We assume that M is equipped with a fixed Riemannian metric, but do *not* assume that M has the standard flat[8] metric coming from the identification $M \cong \mathbb{R}^d/\mathbb{Z}^d$! Let $\widetilde{M}$ denote the universal cover of M. Both the Riemannian metric and spin structure on M can be canonically pulled back to $\widetilde{M}$. Thus $\widetilde{M}$ is diffeomorphic to $\mathbb{R}^d$, but equipped with a possibly non-standard Riemannian metric. Note that the lifted metric must be complete, as completeness is a local property (compare Theorem A.3.6), and $\widetilde{M}$ is locally isometric to the complete (as compact) manifold M.

Let $\widetilde{S}$ and $\widetilde{D}$ be the spinor bundle and Dirac operator on $\widetilde{M}$ respectively. Let $\widetilde{F} = \chi(\widetilde{D})$ denote an operator associated to $\widetilde{D}$ as in Lemma 3.2.7, and let

$$\operatorname{Index}(\widetilde{F}) \in K_*(C^*(\widetilde{M}))$$

be the associated index class.

Now, as well as the Roe algebra $C^*(\widetilde{M})$, we may associate the following C^*-algebra to the universal cover of our n-torus.

Definition 3.3.1 Consider the collection of all functions

$$(T_t)\colon [1,\infty) \to \mathcal{B}(L^2(M;S))$$

with the following properties:

(i) $t \mapsto T_t$ is uniformly continuous and uniformly bounded;
(ii) the propagation $\operatorname{prop}(T_t)$ is finite for all $t \in [1,\infty)$ and tends to zero as t tends to infinity;
(iii) all the operators T_t are locally compact.

These functions form a $*$-algebra. The completion of this $*$-algebra for the norm $\|(T_t)\| := \sup_t \|T_t\|$ is denoted $C_L^*(\widetilde{M})$ and called the *localisation algebra of* $\widetilde{M}$.

Our Dirac operator $\widetilde{D}$ gives rise to a K-theory class for $C_L^*(\widetilde{M})$ via the operator $\widetilde{F}$ and the following construction. The proof is essentially the same as that of Lemma 3.2.7, once we have noted that if ϕ has Fourier transform supported in $[-r,r]$, then $\phi(t^{-1}\cdot)$ has Fourier transform supported in $[-t^{-1}r, t^{-1}r]$.

[8] i.e. locally isometric to Euclidean space.

Lemma 3.3.2 *Let $\chi : \mathbb{R} \to [-1,1]$ be a function with the properties in Lemma 3.2.7, and for $t \in [1,\infty)$ define*

$$\widetilde{F}_t := \chi(t^{-1}\widetilde{D}).$$

Then the family of operators (F_t) is a multiplier of $C_L^*(\widetilde{M})$ such that $(1-\widetilde{F}_t^2)$ is in $C_L^*(\widetilde{M})$, and which is odd if the dimension of M is even.

In particular, there is an index class

$$Index_L(\widetilde{F}) \in K_*(C_L^*(\widetilde{M}))$$

as in Definition 2.8.5. □

Here is the last 'black box' we need. This piece is from the differential-topological aspect of K-theory.

Proposition 3.3.3 *The group $K_i(C_L^*(\widetilde{M}))$ is isomorphic to $\mathbb{Z}$ if i equals the dimension of M modulo 2, and zero otherwise. The class $Index_L(\widetilde{F})$ is a generator.* □

Now, there is clearly an 'evaluation-at-one' $*$-homomorphism

$$\text{ev}\colon C_L^*(\widetilde{M}) \to C^*(\widetilde{M}), \quad (T_t) \mapsto T_1.$$

The following theorem is the main step in the proof that the tori do not admit a metric of positive scalar curvature.

Theorem 3.3.4 *When M is the d-torus equipped with any Riemannian metric, the evaluation-at-one map*

$$ev\colon C_L^*(\widetilde{M}) \to C^*(\widetilde{M})$$

induces an isomorphism on K-theory.

Corollary 3.3.5 *The d-torus does not admit a metric of positive scalar curvature.*

Proof If it did, $\widetilde{M}$ as above would have a metric with uniformly positive scalar curvature. Corollary 3.2.8 forces $\text{Index}(F)$ to be zero, which contradicts the combination of Proposition 3.3.3 and Theorem 3.3.4, and the fact that the evaluation-at-one maps takes the class $\text{Index}_L(\widetilde{F}) \in K_*(C_L^*(\widetilde{M}))$ to $\text{Index}[\widetilde{F}] \in K_*(C^*(\widetilde{M}))$. □

We spend the rest of this section sketching a proof of Theorem 3.3.4: this is not trivial, but the ingredients required all come from elementary operator K-theory. We do not provide full details, as the machinery we develop later will enable us to do so more conveniently.

Let $h\colon \widetilde{M} \to \mathbb{R}^d$ be a diffeomorphic lift of the set-theoretic identity map $M \to \mathbb{T}^d$, where $\mathbb{T}^d = \mathbb{R}^d/\mathbb{Z}^d$ has the usual flat metric. As the derivatives of the identity $M \to \mathbb{T}^d$ must be uniformly bounded by compactness, the same is true for h, which forces h to be bi-Lipschitz (by the mean value theorem). In other words, there is a constant $c > 0$ such that

$$\frac{1}{c} d_{\widetilde{M}}(x,y) \leqslant d_{\mathbb{R}^d}(h(x),h(y)) \leqslant c d_{\widetilde{M}}(x,y).$$

The first thing to note is that the definitions of $C_L^*(\widetilde{M})$ and $C^*(\widetilde{M})$ are not changed under bi-Lipschitz equivalence. It thus suffices to prove the following analogue of Theorem 3.3.4.

Theorem 3.3.6 *The evaluation-at-one map*

$$ev\colon C_L^*(\mathbb{R}^d) \to C^*(\mathbb{R}^d)$$

induces an isomorphism on K-theory.

Sketch proof We proceed by induction on the dimension. In the case $n = 0$, $\mathbb{R}^d$ is a single point. $C^*(\mathbb{R}^0)$ is just a copy of the compact operators, so its K-theory is $\mathbb{Z}$, generated by a rank one projection. One can show directly that $K_*(C_L^*(\mathbb{R}^0))$ is also $\mathbb{Z}$, and is generated by the constant map to any fixed rank one projection; this completes the base case.

For the inductive step, write $\mathbb{R}^d = E \cup F$, where $E = \mathbb{R}^{d-1} \times (-\infty, 0]$ and $F = \mathbb{R}^{d-1} \times [0,\infty)$. One can show by constructing natural pushout diagrams (see Definition 2.7.13) that this decomposition gives rise to a commutative diagram of Mayer–Vietoris sequences connected by the evaluation maps as follows:

$$\begin{array}{ccc}
K_i(C_L^*(E \cap F)) & \xrightarrow{\text{ev}_*} & K_i(C^*(E \cap F)) \\
\downarrow & & \downarrow \\
K_i(C_L^*(E)) \oplus K_i(C_L^*(F)) & \xrightarrow{\text{ev}_*} & K_i(C^*(E)) \oplus K_i(C^*(F)) \\
\downarrow & & \downarrow \\
K_i(C_L^*(\mathbb{R}^d)) & \xrightarrow{\text{ev}_*} & K_i(C^*(\mathbb{R}^d)) \\
\downarrow & & \downarrow \\
K_{i-1}(C_L^*(E \cap F)) & \xrightarrow{\text{ev}_*} & K_{i-1}(C^*(E \cap F)) \\
\downarrow & & \downarrow \\
K_{i-1}(C_L^*(E \cap F)) & \xrightarrow{\text{ev}_*} & K_{i-1}(C^*(E \cap F)).
\end{array}$$

One can show by an Eilenberg swindle 'pushing everything to infinity along $[0,\infty)$' that

$$K_*(C^*(E)) = K_*(C^*(F)) = K_*(C_L^*(E)) = K_*(C_L^*(F)) = 0,$$

whence the second and fifth horizontal maps are isomorphisms. On the other hand, the top and fourth horizontal maps are an isomorphism by the inductive hypothesis, so we are done by the five lemma. □

3.4 Some Historical Comments

Classical index theory proves topological formulas for the index of Fredholm operators as in Example 2.8.3: chapter 5 of Douglas [81] is an excellent reference for background here. As far as we know, the earliest appearance of such a theorem is in a 1920 work by Fritz Noether [192]. From a modern point of view, the most straightforward index theorem is probably that for Toeplitz operators with continuous symbol, which seems to have been rediscovered separately by several different mathematicians: see, for example, Theorem 7.26 of Douglas [81] or section 2.3 of Higson and Roe [135] and surrounding discussion.

Integer Valued Index Theory

To describe the Toeplitz index theorem, and give a flavour of what is meant by a topological formula, let $f: S^1 \to \mathbb{C}$ be a continuous function on the circle. Then f acts on $L^2(S^1)$ by multiplication. The Fourier transform identifies $L^2(S^1)$ and $\ell^2(\mathbb{Z})$, and f acts on this space by a convolution operator C_f. Now, let $V: \ell^2(\mathbb{N}) \to \ell^2(\mathbb{Z})$ be the natural isometric inclusion. The *Toeplitz operator* associated to f is by definition the operator

$$T_f := V^* C_f V$$

on $\ell^2(\mathbb{N})$. It turns out that T_f is Fredholm if and only if $f: S^1 \to \mathbb{C}$ is invertible (i.e. non-zero everywhere). Thus if T_f is Fredholm, the image of f does not contain zero, and it makes sense to talk about the winding number of f around zero.

Theorem 3.4.1 *Let $T_f: \ell^2(\mathbb{N}) \to \ell^2(\mathbb{N})$ be a Toeplitz operator as above with $f: S^1 \to \mathbb{C}$ nowhere vanishing. Then*

$$\textit{Index}(T_f) = -\textit{(winding number)}(f).$$

This is the Toeplitz index theorem. The motivations for these early index theorems come from single operator theory and the so-called Fredholm alternative: this is because vanishing of the index gives a useful criterion for such operators to be invertible.

A little later, motivations for index theory came more from differential topology and geometry. This is the case for the most famous index theorem computing Fredholm indices by a topological formula: the Atiyah–Singer index theorem for elliptic pseudodifferential operators on closed manifolds (see Atiyah and Singer [12]), which dates to about 1963 with the 'canonical' K-theoretic proof appearing in 1968. To give a vague flavour of what this says, we give the cohomological form of its statement from Atiyah and Singer [13].

Theorem 3.4.2 *Let M be a closed, smooth, oriented d-dimensional manifold, and let P be an elliptic pseudodifferential operator on some bundle over M. Then P is a Fredholm on an appropriate bundle of sections, and there are canonical classes $[ch(\sigma_P)]$ and $[Todd(T_{\mathbb{C}}M)]$ in the cohomology ring*[9] *$H^*(M;\mathbb{Q})$ such that if $[M] \in H_d(M;\mathbb{Q})$ is the fundamental class coming from the orientation, then*

$$Index(P) = (-1)^d \langle [ch(\sigma_P)] \cup [Todd(T_{\mathbb{C}}M)], [M] \rangle.$$

This looks (and indeed is!) much more complicated than the Toeplitz index theorem; in fact, the latter is essentially the special case of the Atiyah–Singer theorem for order zero pseudodifferential operators when M is the circle. Having said that, the formula is explicitly computable in many cases, particularly when the operator P is arising in a canonical way from the topology and geometry of M; these are typically the important cases for applications.

Higher Index Theory

The above theorems all describe integer-valued indices of Fredholm operators. However, it was realised fairly early on (at least by the late 1960s) that one also has useful 'indices' defined using traces on von Neumann algebras (see Coburn et al. [56]), and taking values in the representation ring of a compact group (see Atiyah and Segal [11]). These situations come under the following general framework. One has an operator D in some unital algebra B with an ideal I, and so that the image of D is invertible in B/I. The K-theoretic index map

$$K_1(B/I) \to K_0(I)$$

[9] The product is the cup product $\cup$.

then takes the class $[D]$ in $K_1(B/I)$ to a class in $K_0(I)$, and one wants to use this class in $K_0(I)$ to derive information about D (or the underlying geometry and topology). The special case when I is the compact operators $\mathcal{K}$ on some Hilbert space, so $K_0(I) \cong \mathbb{Z}$, corresponds to the classical integer-valued index for Fredholm operators thanks to Example 2.8.3. Indices taking values in the representation ring of a compact group typically correspond to I being the stabilisation $C^*(G) \otimes \mathcal{K}$ of the group C^*-algebra $C^*(G)$, in which case $K_0(I)$ canonically identifies with the representation ring[10] of G. Indices defined using traces of a von Neumann algebra correspond to the case where I is a von Neumann algebra with tracial state τ and where one considers the class $\tau_*(\text{Index}(D)) \in \mathbb{R}$, where $\tau_* : K_0(I) \to \mathbb{R}$ is as in Remark 2.1.16.

The general scheme outlined above, i.e. considering indices taking values in the K-groups $K_*(A)$ of some (C^*-)algebra A, is sometimes called *higher index theory*. Sometimes one pairs $K_*(A)$ with other data to get numerical invariants as in the work of Connes and Moscovici [65];[11] or one might just consider the classes in K-theory in of themselves. Such higher indices may live in the K-theory of any algebras; however, one gets a particularly powerful theory when the algebra is a C^*-algebra, as then important analytic tools coming from positivity and the functional calculus come to bear. Important examples of algebras used here include group C^*-algebras and associated crossed products (for example, Higson [125] and Kasparov [149]), examples related to representation theory of non-compact Lie groups (for example, Atiyah and Schmid [10] and Connes and Moscovici [64]), foliation C^*-algebras (for example, Connes [57, 59], Connes and Skandalis [66], Pflaum, Posthuma and Tang [205] and Zhang [277] for some non-C^*-algebraic approaches to foliated index theory) and Roe-type algebras in coarse geometry (for example, Roe [212, 214] and also Tang, Willett and Yao [239] for a theory combining Roe algebras and foliation C^*-algebras).

Back to Positive Scalar Curvature

Having gone through this very brief survey of index theory and higher index theory, let us get back to the existence of positive scalar curvature metrics. The first index-theoretic approach[12] to the question of existence of positive

[10] In this case, $K_0(I)$ has a ring structure coming from the fact that one can take tensor products of group representations.

[11] Page 346 of this paper seems to be where the term 'higher index' first appears appears

[12] Not counting the Gauss–Bonnet theorem, which can also be viewed as a special case of the Atiyah–Singer theorem.

scalar curvature metrics is due to Lichnerowicz[13] [168] and is based on the Atiyah–Singer index theorem as mentioned above. The Dirac operator on a closed Riemannian spin manifold is Fredholm, meaning that as an operator on smooth sections $C^\infty(M; S)$ of the spinor bundle S it has finite-dimensional kernel and co-kernel, and thus a well-defined integer-valued index

$$\text{Index}(D) := \dim(\text{Kernel}(D)) - \dim(\text{Cokernel}(D)) \in \mathbb{Z}.$$

The Atiyah–Singer theorem in this case specialises to the following result.

Theorem 3.4.3 *Let M be a closed spin Riemannian manifold, and D the associated Dirac operator. Then there is a differential form $\hat{A}(M)$, the $\hat{A}$-form of M, that depends only on the structure of the tangent bundle of M and that satisfies*

$$\textit{Index}(D) = \int_M \hat{A}(M). \qquad \square$$

Now, if M is closed (so in particular, complete) and has positive scalar curvature, it follows from Theorem 3.1.5 and Proposition 3.2.1 that D has index zero. Hence we have the following corollary.

Corollary 3.4.4 *Let M be a closed spin manifold such that $\int_M \hat{A}(M) \neq 0$. Then M does not admit a metric of positive scalar curvature.*

Proof The Atiyah–Singer theorem implies that the index of the Dirac operator is non-zero, and in particular that D is not invertible. Theorem 3.1.5 and Proposition 3.2.1 show that D is invertible in the presence of a positive scalar curvature metric, however. $\square$

This corollary covers many interesting cases: see for example section IV.4 of Lawson and Michelson [164] for examples, including concrete complex algebraic surfaces. However, it is fairly restrictive in some ways: for example, if M has trivialisable tangent bundle then the $\hat{A}$-form is trivial, and Corollary 3.4.4 gives no information. In particular, we get no information for tori,[14] so one cannot hope to apply Corollary 3.4.4 here. The question of whether the d-tours admits a metric of positive scalar curvature was open for many years before being solved for $d \leqslant 7$ by Schoen and Yau [231, 232] using minimal hypersurfaces, and in arbitrary dimensions by Gromov and Lawson

[13] The formula $D^2 = \nabla^*\nabla + \kappa/4$ from Theorem 3.1.5 is often called the *Lichnerowicz formula*.
[14] Tori have trivialisable tangent bundles, as one can see, for example, using that they are Lie groups.

[179] using an index-theoretic proof. See Gromov and Lawson [180] for an inspiring overview of their methods.

The approach of Gromov and Lawson uses integer-valued Fredholm indices; however, there are K-theoretic higher indices lurking in the background, as seems to have first been observed in this context by Rosenberg [222]. The approach discussed in the earlier sections of this chapter is one way of making this higher index machinery precise. It applies in particular to the case of the d-torus, to show that it does not admit a metric of positive scalar curvature. The proof involves ingredients from differential geometry, analysis of differential operators, algebraic topology and operator K-theory; as such it may seem rather intimidating at first!

However, the ingredients that we used as 'black boxes' are all quite general: they would apply to any manifold we were trying to study. We will develop most of these in this book, but they can be regarded as background material. The part of the proof that applied to the d-torus specifically, and that needs to be generalised to apply the theory to other manifolds, is Theorem 3.3.4; the essential ingredients here are operator K-theoretic in this nature, and this will be the main focus of our text.

3.5 Content of This Book

Part TWO of the book is taken up with expounding the purely metric and topological aspects of the theory. One does not need background in manifold topology and geometry to read this material.

- The construction of the Roe algebras and localisation algebras can be carried out in much more generality. For a metric space X (for example, a Riemannian manifold, but discrete spaces are also very interesting here), the basic idea is to consider Hilbert spaces equipped with a representation of $C_0(X)$ (for example, the Hilbert space $L^2(M; S)$ used in Section 3.2 is equipped with a multiplication representation of $C_0(M)$). This allows one to define notions of propagation and local compactness. In Chapter 4 we discuss *geometric modules* over a metric space X, i.e. Hilbert spaces equipped with a representation of $C_0(X)$, and various types of maps between them which model maps between the metric spaces themselves. We also carry out the general constructions in the presence of an action of a discrete group.
- Just before the statement of Theorem 3.4, we used that the Roe algebra $C^*(\widetilde{M})$ of $\widetilde{M}$ is invariant under bi-Lipschitz equivalences. Bi-Lipschitz

equivalences preserve a lot of structure on a space; all that was really relevant here, however, is that they preserve the *large-scale* or *coarse* metric structure. In Chapter 5 we look at the underlying metric space theory and show that the (K-theory of) Roe algebras is a functor on an appropriate category of geometric spaces, where the morphisms preserve only the large-scale structure. We allow also for the presence of a group action.

- On the other hand, the fact that the localisation algebra $C_L^*(\widetilde{M})$ of $\widetilde{M}$ is invariant under bi-Lipschitz equivalence is due to the fact that it preserves the *small-scale* or *topological* structure. In Chapter 6 we develop this much more fully, showing that the K-theory of (a slight variant of) the localisation algebra is a functor on an appropriate category of topological spaces. In fact, it is a model for *K-homology*, the dual generalised homology theory to K-theory. We also establish this in Chapter 6. Again, we allow the presence of a group action.
- The evaluation-at-one map

$$\mathrm{ev}_*\colon K_*(C_L^*(M)) \to K_*(C^*(M))$$

used in Section 3.3 is central to the theory. One can think of ev_* as a 'forget control' map: it forgets the small-scale metric structure of a space in favour of its large-scale structure. It is called the *assembly map*. For reasons such as the application to positive scalar curvature discussed above, one is interested in situations when ev_* is an isomorphism; however, one cannot expect this to happen in general. In Chapter 7 we set up the basic theory of the assembly map and construct a sort of universal assembly map – the *Baum–Connes assembly map* – that one hopes is an isomorphism in general.

In Part THREE of the book, we go back to manifolds.

- In the above study of positive scalar curvature, we used some facts about the analysis of Dirac operators on complete Riemannian spin manifolds as a 'black box'. In Chapter 8 we develop the necessary analysis to make these ideas precise.
- In Chapter 9 we set up some general machinery involving pairings between differential operators and vector bundles and use this to prove the K-theory Poincaré duality isomorphism. This is the key fact that underlies the black box that we used in Proposition 3.3.3 above.
- The Dirac operator is one of several operators with particular geometric importance. In Chapter 10 we sketch the two most important topological and geometric applications of higher index theory. One – to the existence of positive scalar curvature metrics – has been sketched out above and is based

on the Dirac operator. Another – to the topological invariance of higher signature – is based on the *signature operator* and will be discussed more fully there. This material requires more background in differential topology than we have assumed in this book, so the basic ideas will just be sketched. We also discuss some applications to pure (operator) algebra, proving non-existence of idempotents in group C^*-algebras.

Finally, in Part FOUR of the book we look at some results on the universal Baum–Connes assembly map. Thanks to the discussion in Chapter 10, these have important consequences in topology, geometry and C^*-algebra theory.

- In Chapter 11, we discuss what we call almost constant bundles and use these to give an elementary proof that the coarse Baum–Connes assembly map is injective in many cases. This material provides a particularly elementary approach to the non-positively curved situation.
- In Chapter 12 we discuss the coarse Baum–Connes conjecture for spaces that coarsely embed into Hilbert spaces. This is a very general result: despite being around for twenty years at time of writing, it is still more or less the state of the art in terms of checking that a particular group satisfies (for example) either the Novikov or Gromov–Lawson conjectures. The proof takes the whole chapter, uses many of the ideas developed earlier and is the deepest theorem covered by this book.
- In Chapter 13 we discuss counterexamples to the coarse Baum–Connes conjecture arising from expander graphs and sequences of large spheres. Much remains to be understood here in terms of the geometric significance of these examples.

3.6 Exercises

3.6.1 With notation as in Example 3.2.2, use the Fourier transform to show directly that the operator $D = -i\frac{d}{dx}$ on $L^2(\mathbb{R})$ with domain $C_c^\infty(\mathbb{R})$ is essentially self-adjoint.

3.6.2 Justify the computations in Example 3.2.5, including working out the various constants involved using your favourite Fourier transform conventions.

3.6.3 Let A be the C^*-subalgebra of $\ell^2(\mathbb{N})$ generated by all Toeplitz operators T_f as in the discussion around Theorem 3.4.2. One can show that A contains $\mathcal{K}(\ell^2(\mathbb{N}))$ as an ideal, and that there is a short exact sequence

$$0 \longrightarrow \mathcal{K}(\ell^2(\mathbb{N})) \longrightarrow A \longrightarrow C(S^1) \longrightarrow 0$$

with the map $A \to C(S^1)$ determined by the fact it sends to T_f to f. Note that this combined with Atkinson's theorem (see Example 2.8.3) shows that T_f is Fredholm if and only if f is invertible. Use this short exact sequence to prove Theorem 3.4.2.

Hint: by Example 2.8.3, the index map

$$\mathbb{Z} \cong K_1(C(S^1)) \to K_0(\mathcal{K}(\ell^2(\mathbb{N})) \cong \mathbb{Z}$$

takes the class $[f] \in K_1(C(S^1))$ *of an invertible* $f \in C(S^1)$ *to the index of* T_f *in* $\mathbb{Z}$. *It thus suffices to check the formula from Theorem 3.4.2 on a single example. Consider* $f(z) = z$.

PART TWO

Roe Algebras, Localisation Algebras and Assembly

4

Geometric Modules

A geometric module over X is a Hilbert space H_X equipped with a suitable representation of $C_0(X)$. The motivating example is the Hilbert space $L^2(X)$ of square-integrable functions on X with respect to some measure, with $C_0(X)$ acting by multiplication. However, it is convenient to allow more general modules: the idea is to give us a flexible setting in which to do analysis with operators associated to X. Our aim in this chapter is to set up the basic theory of geometric modules, as well as discuss a lot of examples.

The material in this chapter should be regarded as technical background: we recommend readers skim (or just skip) it on a first reading, coming back to it later as necessary.

This chapter is structured as follows. In Section 4.1 we introduce geometric modules, focusing on examples. The examples are important partly for intuition, but also as they will allow us to do explicit computations later in the book. Section 4.2 then discusses covering isometries: these are maps between geometric modules that in some sense model functions between the underlying spaces. Sections 4.3 and 4.4 then specialise the covering isometry machinery to the specific settings relevant to large-scale geometry and small-scale topology respectively. Finally, Section 4.5 discusses how the theory can be adapted in the presence of a group action to take that extra structure into account.

Throughout this chapter, the symbols X and Y will always denote locally compact, second countable, Hausdorff topological spaces. We have collected together the basic facts and definitions in metric space theory and coarse geometry that we will need in Appendix A, but will also repeat any non-standard definitions that come up as we need them.

4.1 Geometric Modules

In this section we introduce geometric modules over topological spaces.

Throughout this section, X and Y are locally compact, second countable, Hausdorff spaces.

For the next definition we will need the notion of a non-degenerate representation of a C^*-algebra from Definition 1.6.4.

Definition 4.1.1 A *(geometric) module over* X, or *(geometric)* X *module* is a separable Hilbert space H_X equipped with a non-degenerate representation $\rho\colon C_0(X) \to \mathcal{B}(H_X)$.

A geometric module H_X is *ample* if no non-zero element of $C_0(X)$ acts as a compact operator, and if H_X is infinite-dimensional.[1]

We will often say something like 'let H_X be a geometric module' without explicitly mentioning the space. Note that if H_X is an ample geometric module, then the associated representation $\rho\colon C_0(X) \to \mathcal{B}(H_X)$ is faithful, but we do not assume faithfulness in general. We will generally abuse notation, omitting ρ unless it seems likely to cause confusion: for example, if $f \in C_0(X)$ and $u \in H_X$, then fu denotes the image of u under $\rho(f)$.

Example 4.1.2 Say μ is a Radon measure on X, and define $H_X := L^2(X, \mu)$ to be the usual Hilbert space of square-integrable functions (modulo those that are zero almost everywhere). Then H_X is a geometric module when equipped with the natural pointwise multiplication action of $C_0(X)$. This is the motivating example, and close to the general case by Exercise 4.6.2. Modules of this form may or may not be ample: see the next two examples.

Example 4.1.3 Let X be a discrete space, and μ be counting measure, so $L^2(X, \mu) = \ell^2(X)$. This X module is never ample. However, if we fix an auxiliary separable infinite-dimensional Hilbert space H and set $H_X := \ell^2(X, H)$ to be the space of square-summable functions from X to H, then we do get an ample X module with the natural multiplication action.

Example 4.1.4 As another special case of Example 4.1.2 above, let X be a Riemannian manifold equipped with the smooth measure μ associated to the metric: for example, X could be $\mathbb{R}^d$ equipped with its usual Euclidean metric and Lebesgue measure. As above, we can build $H_X := L^2(X, \mu)$. This H_X is 'usually'[2] ample.

[1] Infinite dimensionality is automatic if X is non-empty, but it will be technically convenient later that we allow X to be the empty set.

[2] Precisely, it is ample if and only if the dimension of every connected component of X is non-zero.

Example 4.1.5 Let μ be a Radon measure on X as in Example 4.1.2, and let S be a (non-zero) complex vector bundle over X. Assume S is equipped with a Hermitian structure: this means that each fibre S_x is equipped with a Hermitian inner product $\langle , \rangle_x : S_x \times S_x \to \mathbb{C}$ such that for any continuous sections s_1, s_2 of S, the function

$$X \to \mathbb{C}, \quad x \mapsto \langle s_1(x), s_2(x) \rangle_x$$

is continuous. Let $C_c(X; S)$ denote the vector space of continuous, compactly supported sections of S, and define a positive semi-definite inner product[3] on this space by

$$\langle s_1, s_2 \rangle := \int_X \langle s_1(x), s_2(x) \rangle_x d\mu(x).$$

The Cauchy–Schwarz inequality implies that the collection of $s \in C_c(X)$ such that $\langle s, s \rangle = 0$ is a subspace of $C_c(X)$. Taking the vector space quotient by this subspace gives a new vector space on which $\langle , \rangle$ descends to a positive definite inner product, and taking the completion with respect to the associated norm $\|s\| := \sqrt{\langle s, s \rangle}$ defines the Hilbert space $L^2(X; S)$ of square-integrable sections (this process of taking a quotient by vectors of length zero, then completing, is sometimes called taking the *separated completion*). Again, pointwise multiplication makes this into an X module. Note that if X and μ are as in Example 4.1.4, then $L^2(X; S)$ is ample if and only if (every component of) X has positive dimension.

Example 4.1.6 Let H be a separable infinite-dimensional Hilbert space, and $Z \subseteq X$ be a countable dense subset (such a Z exists as X is second countable). Then the Hilbert space $H_X := \ell^2(Z, H)$ is equipped with a natural pointwise multiplication action of $C_0(X)$ by restriction to Z, and thus becomes an X module. As H is infinite-dimensional and Z is dense in X, it is moreover ample. In particular, this example shows that ample (separable) X modules always exist. Note that Example 4.1.3 is the special case of this one where X is discrete.

Now let H_X be an arbitrary X module. Then by Proposition 1.6.11 there is a canonical extension of the representation of $C_0(X)$ on H_X to a unital representation of $B(X)$, the C^*-algebra of bounded Borel functions on X. Moreover, this extension takes pointwise convergent bounded sequences to strongly convergent sequences. If E is a Borel subset of X, we will write χ_E for the characteristic function of E and for the corresponding projection

[3] This means that is satisfies the usual axions of an inner product except that maybe there can be non-zero s with $\langle s, s \rangle = 0$.

operator on H_X. One should think of the subspace $\chi_E H_X$ as the part of H_X supported over E: note that if $X = L^2(X,\mu)$ as in Example 4.1.2 above, then $\chi_E H_X$ identifies with $L^2(E,\mu|_E)$.

The key definition that ties operators on H_X to the structure of X is as follows.

Definition 4.1.7 Let H_X and H_Y be geometric modules, and let $T\colon H_X \to H_Y$ be a bounded operator. The *support* of T, denoted $\text{supp}(T)$, consists of all points $(y,x) \in Y \times X$ such that for all open neighbourhoods U of x and V of y

$$\chi_V T \chi_U \neq 0.$$

Note that the support of an operator is always a closed subset of $Y \times X$.

Definition 4.1.8 Let H_X be a geometric module over a metric space X, and let $T\colon H_X \to H_X$ be a bounded operator. The *propagation* of T is the extended real number

$$\text{prop}(T) := \sup\{d(y,x) \mid (y,x) \in \text{supp}(T)\} \in [0,\infty].$$

We now look at some examples.

Example 4.1.9 Say H_X is an X module and f a bounded Borel function on X. Then the support of the corresponding multiplication operator is contained in the closed subset

$$\overline{\{(x,x) \in X \times X \mid f(x) \neq 0\}}$$

of the diagonal of $X \times X$: see Exercise 4.6.3. If H_X is ample and f is continuous then the support of f is exactly equal to the above set, but not in general. The propagation of such an operator (with respect to any metric) is always zero.

Example 4.1.10 Let μ be a Radon measure on X, and let $H_X = L^2(X,\mu)$ be as in Example 4.1.2. Let T be a bounded operator defined by some continuous kernel function $k\colon X \times X \to \mathbb{C}$, i.e. for u in the dense subset $C_c(X)$ of H_X, Tu is the function defined by

$$(Tu)(x) = \int_X k(x,y)u(y)d\mu(y) \tag{4.1}$$

(note that one needs to put additional conditions on k to ensure boundedness of T: see Exercise 4.6.4 for a useful sufficient condition). Then the support of T is contained in that of k, i.e. the closure of the set

$$\{(x,y) \in X \times X \mid k(x,y) \neq 0\},$$

and is exactly the support of k if H_X is ample (for example, if X is a positive-dimensional Riemannian manifold with the associated measure as in Example 4.1.6): see Exercise 4.6.3. It follows that if X is equipped with a metric d, then

$$\operatorname{prop}(T) \leqslant \sup\{d(x,y) \mid k(x,y) \neq 0\},$$

and this becomes an equality if H_X is ample. More generally, this works for sections of a bundle S over X, in which case k should be a continuous section of the vector bundle over $X \times X$ with fibre over (x, y) given by $\mathcal{B}(S_x, S_y)$, and u in line (4.1) should be taken to be a compactly supported continuous section.

The ideas in this example apply more generally to operators with non-continuous kernels, or even distributional kernels. We will not need it, but for intuition it is worth mentioning that the classical *Schwartz kernel theorem* implies that many natural classes of operators associated to manifolds roughly have this form, where k is now assumed to be a distribution in some appropriate sense.

Example 4.1.11 Let Z be a countable dense subset of X, H a separable infinite-dimensional Hilbert space and $H_X = \ell^2(Z, H)$ be as in Example 4.1.6.

Then any bounded operator $T\colon \ell^2(Z,H) \to \ell^2(Z,H)$ can be represented uniquely as a Z-by-Z matrix $(T_{xy})_{x,y \in Z}$ of bounded operators on H. Indeed, for any $z \in Z$, there is an isometry $V_z\colon H \to \ell^2(Z,H)$ defined by

$$(V_z u)(y) := \begin{cases} u & z = y, \\ 0 & z \neq y, \end{cases}$$

and each matrix entry $T_{xy}\colon H \to H$ is defined by

$$T_{xy} := V_x^* T V_y.$$

The support of T is then the closure of the set

$$\{(x,y) \in Z \times Z \mid T_{xy} \neq 0\}$$

in $X \times X$.

The next lemma records how supports behave under the usual algebraic operations on operators. We need some preliminary definitions. Recall first that if E, F are subsets of $Z \times Y$ and $Y \times X$ respectively, then their *composition* is defined to be

$$E \circ F := \{(z,x) \in Z \times X \mid \text{there exists } y \in Y \text{ such that } (z,y) \in E, (y,x) \in F\} \tag{4.2}$$

and the *inverse* of E is defined by

$$E^{-1} = \{(y,z) \in Y \times Z \mid (z,y) \in E\}.$$

Recall also (see Definition A.1.7) that a map $f: X \to Y$ between topological spaces is *proper* if for all compact $K \subseteq Y$, $f^{-1}(K)$ is also compact.

Definition 4.1.12 A bounded operator $T: H_X \to H_Y$ is *properly supported* if the restrictions of the coordinate projections

$$\pi_X : X \times Y \to X \quad \text{and} \quad \pi_Y : X \times Y \to Y$$

to supp(T) are proper maps.

Lemma 4.1.13 *Let H_X, H_Y, H_Z be modules over X, Y, Z respectively. Let $R, S: H_X \to H_Y$ and $T: H_Y \to H_Z$ be bounded operators. Then:*

(i) $supp(R + S) \subseteq supp(R) \bigcup supp(S)$;
(ii) $supp(T^*) = supp(T)^{-1}$;
(iii) $supp(TS) \subseteq \overline{supp(T) \circ supp(S)}$.

Moreover, if either S or T is properly supported, then condition (iii) *can be replaced with*

(iii′) $supp(TS) \subseteq supp(T) \circ supp(S)$.

Proof The statements on supports of sums and adjoints are immediate from the definition of support. We look first at the condition in line (iii). For technical convenience, we fix metrics on all the spaces involved that induce their topologies.

Let S, T be as given, and $(z, x) \in Z \times X$ be an element of supp(TS). We claim first that for each $n \geqslant 1$, there exists y_n such that

$$\chi_{B(z;1/n)} T \chi_{B(y_n;\epsilon)} S \chi_{B(x;1/n)} \neq 0$$

for all $\epsilon > 0$. Indeed, if not, it would follow that for each $y \in Y$ there exists $\epsilon_y > 0$ such that

$$\chi_{B(z;1/n)} T \chi_{B(y;\epsilon_y)} S \chi_{B(x;1/n)} = 0.$$

As the collection $\{B(y;\epsilon_y)\}_{y \in Y}$ covers Y, Lemma A.1.10 implies that there exists a decomposition $Y = \bigsqcup_{i \in I} E_i$ of Y into countably many disjoint Borel subsets E_i such that

$$\chi_{B(z;1/n)} T \chi_{E_i} S \chi_{B(x;1/n)} = 0$$

for each i. Summing over i, using that the sum $\sum_{i \in I} \chi_{E_i}$ of operators on H_Y converges strongly to the identity (see Proposition 1.6.11), then gives

$$\chi_{B(z;1/n)} T S \chi_{B(x;1/n)} = 0.$$

This contradicts that (z, x) is in supp(TS).

Now, we have that for each n there is $y_n \in Y$ such that for all $\epsilon > 0$

$$\chi_{B(z;1/n)} T \chi_{B(y_n;\epsilon)} \neq 0.$$

Assume that n is large enough so that $\overline{B(z,1/n)}$ is compact (this is possible as X is locally compact). Applying a similar argument to the one that produced y_n, for all $m \geqslant 1$ there exists $z_{nm} \in B(z;1/n)$ such that

$$\chi_{B(z_{nm};1/m)} T \chi_{B(y_n;1/m)} \neq 0.$$

Let $z_n \in \overline{B(z,1/n)}$ be any limit point of the sequence $(z_{nm})_{m=1}^{\infty}$, which exists by compactness. Note then that if $U \ni z_n$ and $V \ni y_n$ are open sets then for suitably large m, $B(z_{nm};1/m) \subseteq U$ and $B(y_n;1/m) \subseteq V$, whence

$$0 \neq \chi_{B(z_{nm};1/m)} T \chi_{B(y_n;1/m)} = \chi_{B(z_{nm};1/m)} \chi_U T \chi_V \chi_{B(y_n;1/m)};$$

this implies that $\chi_U T \chi_V \neq 0$, and thus (z_n, y_n) is in $\text{supp}(T)$. An exactly analogous argument shows that for all n suitably large there exists x_n in $\overline{B(x,1/n)}$ such that (y_n, x_n) is in $\text{supp}(S)$. It follows that for all n suitably large, (z_n, x_n) is in $\text{supp}(T) \circ \text{supp}(S)$; as by construction, we have that $(z_n, x_n) \to (z,x)$ as $n \to \infty$, we have shown that (z,x) is in $\overline{\text{supp}(T) \circ \text{supp}(S)}$.

For the last part of the proof, assume that S and T are properly supported and that (z,x) is an element of $\text{supp}(TS)$. Proceeding exactly as above, for each $n \geqslant 1$ we find elements $x_n \in \overline{B(x,1/n)}$, $y_n \in Y$ and $z_n \in \overline{B(z,1/n)}$ such that (y_n, x_n) is in $\text{supp}(S)$ and $(z_n, x_n) \in \text{supp}(T)$ for all n. Assume that T is properly supported; the case where S is properly supported is similar. Let $\pi_Z \colon \text{supp}(T) \to Z$ be the coordinate projection. Then the sequence $((z_n, y_n))_{n=1}^{\infty}$ in $Z \times Y$ is contained in the compact subset

$$\pi_Z^{-1}(\{z_n \mid n \geqslant 1\} \cup \{z\}) \cap \text{supp}(T)$$

of $Z \times Y$, and thus has a convergent subsequence. In other words, passing to a subsequence, we may assume that (y_n) converges to some $y \in Y$. As $\text{supp}(T)$ and $\text{supp}(S)$ are closed, this gives that $(z,y) \in \text{supp}(T)$, $(y,x) \in \text{supp}(S)$ and thus $(z,x) \in \text{supp}(T) \circ \text{supp}(S)$ as required. □

Corollary 4.1.14 *Let H_X be a module over a metric space X, and let $S,T \colon H_X \to H_X$ be bounded operators. Then:*

(i) prop$(S+T) \leqslant \max\{$prop(S),prop$(T)\}$;
(ii) prop$(T^*) =$ prop(T);
(iii) prop$(TS) \leqslant$ prop$(T) +$ prop(S).

Proof These follow from each of the three points Lemma 4.1.13 in turn. Indeed, the first is obvious, the second follows from symmetry of the metric and the third follows from condition (iii), the triangle inequality and continuity of the metric. □

We finish this section with a technical lemma that will be used many times in the remainder of this chapter. To state it we need a little notation: if F is a subset of $Y \times X$ and K a subset of X, respectively Y, then define[4]

$$F \circ K := \{y \in Y \mid \text{ there is } x \in K \text{ such that } (y,x) \in F\},$$

and

$$K \circ F := \{x \in X \mid \text{ there is } y \in K \text{ such that } (y,x) \in F\}.$$

Lemma 4.1.15 *Let $T : H_X \to H_Y$ be a bounded operator between geometric modules, and $F = supp(T)$. Then for any compact subset K of X, respectively Y, we have*

$$T\chi_K = \chi_{F\circ K}T\chi_K, \quad \chi_K T = \chi_K T \chi_{K\circ F}.$$

Proof We assume $K \subseteq X$; the case $K \subseteq Y$ can be proved similarly. Example 4.1.9 implies that the support of χ_K is a subset of $\{(x,x) \in X \times X \mid x \in K\}$. Hence Lemma 4.1.13 implies that

$$\text{supp}(T\chi_K) \subseteq \overline{\{(y,x) \in Y \times X \mid x \in K, (y,x) \in F\}} \subseteq F \circ K \times K,$$

where the second inclusion uses that K is compact and F closed to deduce that $F \circ K$ is closed. Let now y be an element of $Y \setminus (F \circ K)$. Then for every $x \in K$, $(y,x) \notin \text{supp}(T\chi_K)$, and thus there exist open sets $U_{yx} \ni y$ and $V_{xy} \ni x$ with $\chi_{U_{yx}} T \chi_K \chi_{V_{yx}} = 0$. Take a finite cover $V_{yx_1}, \ldots, V_{yx_n}$ of K from among the open sets V_{yx}. Set $U_y := \bigcap_{i=1}^n U_{yx_i}$. Let $E_1 = K \cap V_{yx_1}$ and for each $i \in \{2, \ldots, n\}$, define

$$E_i := (K \cap V_{yx_i}) \setminus \left(\bigcup_{j=1}^{i-1} V_{yx_j} \right).$$

Then each E_i is Borel, and

$$\chi_{U_y} T \chi_K = \sum_{i=1}^{n} \chi_{U_y} T \chi_K \chi_{E_i} = \sum_{i=1}^{n} \chi_{U_y} \chi_{U_{yx_i}} T \chi_K \chi_{E_i} = 0.$$

On the other hand, applying Lemma A.1.10 to the cover (U_y) of $Y \setminus (F \circ K)$ gives a countable cover $(E_i)_{i\in I}$ of $Y \setminus (F \circ K)$ by disjoint Borel sets such that

[4] One can think of this as essentially the same as the composition defined in line (4.2) above, on identifying K with $\{(x,x) \in X \times X \mid x \in K\}$.

$\chi_{E_i} T \chi_K = 0$ for all n. Using strong convergence of $\sum_{i \in I} \chi_{E_i}$ to $\chi_{Y \setminus (F \circ K)}$, this gives

$$\chi_{Y \setminus (F \circ K)} T \chi_K = \sum_{i \in I} \chi_{E_i} T \chi_K = 0,$$

which in turn is equivalent to the desired formula $\chi_{F \circ K} T \chi_K = T \chi_K$. □

4.2 Covering Isometries

In this section, we discuss isometries between geometric modules that are meant to simulate maps on the spatial level. The machinery we build here underlies the functoriality of both the Roe algebras discussed in Chapter 5 and the localisation algebras discussed in Chapter 6.

Throughout this section, X and Y are locally compact, second countable, Hausdorff spaces.

Definition 4.2.1 Let H_X, H_Y be geometric modules, let $f: X \to Y$ be a function and let $\mathcal{U}$ be an open cover of Y. Then an isometry $V: H_X \to H_Y$ is a $\mathcal{U}$-*cover* of f if

$$\operatorname{supp}(V) \subseteq \bigcup_{U \in \mathcal{U}} \overline{U} \times \overline{f^{-1}(U)}.$$

One should think of the cover $\mathcal{U}$ as governing how good an approximation V is to f: roughly, the set $\bigcup_{U \in \mathcal{U}} \overline{U} \times \overline{f^{-1}(U)}$ can be thought of as a sort of neighbourhood of the graph $\{(f(x), x) \in Y \times X \mid x \in X\}$ of f.

We will show that $\mathcal{U}$-covers always exist later in this section. First, however, we consider some natural examples.

Example 4.2.2 Let X, Y be Riemannian manifolds equipped with geometric modules H_X, H_Y of square-integrable functions as in Example 4.1.2. Let $f: X \to Y$ be a diffeomorphism, and let $J: X \to \mathbb{R}$ be its Jacobian. For each $u \in C_c^\infty(X)$, define $Vu \in C_c^\infty(Y)$ by the formula

$$(Vu)(y) = \frac{u(f^{-1}(y))}{\sqrt{|J(f^{-1}(y))|}}.$$

Then V extends to a unitary operator on H_X that $\mathcal{U}$-covers f for any open cover $\mathcal{U}$ of Y.

This example can be generalised to (some) spaces of square-integrable functions on more general spaces, with the role of the Jacobian being played by the Radon–Nikodym derivative of f.

Example 4.2.3 Let $f: X \to Y$ be an arbitrary function. Let Z_X be a countable dense subset of X, and Z_Y a countable dense subset of Y that contains $f(Z_X)$. Let H be a separable infinite-dimensional Hilbert space, and let $\ell^2(Z_X, H)$ and $\ell^2(Z_Y, H)$ be the ample X and Y modules considered in Example 4.1.6. For each $y \in Y$, choose an isometry

$$V_y: \ell^2(f^{-1}(y), H) \to \ell^2(\{y\}, H)$$

(this is possible as the left-hand side is separable and the right-hand side infinite-dimensional; note that V_y will be zero if $f^{-1}(y) = \varnothing$). Define

$$V := \bigoplus_{y \in Z_Y} V_y : \underbrace{\bigoplus_{y \in Z_Y} \ell^2(f^{-1}(y), H)}_{=H_X} \to \underbrace{\bigoplus_{y \in Z_Y} \ell^2(\{y\}, H)}_{=H_Y}.$$

Then V is a $\mathcal{U}$-cover of f for any open cover $\mathcal{U}$ of Y. This construction motivates our general proof of the existence of $\mathcal{U}$-covers.

Example 4.2.4 Let $X = [0,1]$ and $Y = [0,1] \times [0,1]$. Let $f: X \to Y$ be the natural inclusion as $[0,1] \times \{0\}$, and let H_X, H_Y be the standard Lebesgue L^2-spaces for X and Y. Then for any $\epsilon > 0$, the isometry defined by

$$V_\epsilon : H_X \to H_Y, \quad (V_\epsilon u)(s,t) = \frac{1}{\sqrt{\epsilon}} u(s) \chi_{[0,\epsilon]}(t)$$

has support $\{(s,t) \in [0,1] \times [0,1] \mid t \leqslant \epsilon\}$. Hence for any open cover $\mathcal{U}$ of Y, a compactness argument shows that V_ϵ is a $\mathcal{U}$-cover of f for all suitably small ϵ. However, there is no single $V: H_X \to H_Y$ that $\mathcal{U}$-covers f for all possible $\mathcal{U}$: see Exercise 4.6.7.

Construction 4.2.5 Let $f: X \to Y$ be a Borel function, and let H_X, H_Y be geometric modules, with H_Y ample.

Let $(E_i)_{i \in I}$ be a countable collection of Borel subsets of Y with the following properties:

(i) Y is equal to the disjoint union $\bigsqcup_{i \in I} E_i$ of the sets E_i;
(ii) each E_i has non-empty interior;
(iii) for any compact $K \subseteq Y$, the set

$$\{i \in I \mid E_i \cap K \neq \varnothing\}$$

is finite.

For each i, note that $f^{-1}(E_i)$ is Borel (as f is Borel and E_i is Borel), whence $\chi_{f^{-1}(E_i)} H_X$ makes sense; moreover, the fact that E_i has non-empty interior and ampleness of H_Y implies that $\chi_{E_i} H_Y$ is infinite-dimensional. Hence for each i, we may choose an isometry

$$V_i\colon \chi_{f^{-1}(E_i)} H_X \to \chi_{E_i} H_Y$$

(possibly zero). Set

$$V := \bigoplus_{i \in I} V_i\colon \underbrace{\bigoplus_{i \in I} \chi_{f^{-1}(E_i)} H_X}_{=H_X} \to \underbrace{\bigoplus_{i \in I} \chi_{E_i} H_Y}_{=H_Y},$$

which is an isometry from H_X to H_Y.

Lemma 4.2.6 *Let $f\colon X \to Y$, $(E_i)_{i\in I}$, and $V\colon H_X \to H_Y$ be as in Construction 4.2.5 above. Then*

$$supp(V) \subseteq \bigcup_{i \in I} \overline{E_i} \times \overline{f^{-1}(E_i)}.$$

Proof Say $(y,x) \notin \bigcup_{i \in I} \overline{E_i} \times \overline{f^{-1}(E_i)}$, so for each $i \in I$ there exist open neighbourhoods $W_i \ni y$ and $U_i \ni x$ such that

$$W_i \times U_i \cap E_i \times f^{-1}(E_i) = \varnothing.$$

Let W_0 be a neighbourhood of y with compact closure, so by the properties of the cover (E_i), the set $J = \{i \in I \mid W_0 \cap E_i \neq \varnothing\}$ is finite. Set

$$U = \bigcap_{i \in J} U_i, \quad W = V_0 \cap (\bigcap_{i \in J} V_i),$$

which are open neighbourhoods of y, x respectively. From the choice of W_0 and the U_i, W_i, then

$$\chi_W V \chi_U = \chi_W \Big(\bigoplus_{i \in J} \chi_{E_i} V \chi_{f^{-1}(E_i)} \Big) \chi_U = 0,$$

so (y,x) is not in supp(V). □

Corollary 4.2.7 *Let $f\colon X \to Y$ be a Borel map, and H_X, H_Y be geometric modules with H_Y ample. Then for any open cover $\mathcal{U}$ of Y, there exists an isometry $V\colon H_X \to H_Y$ that $\mathcal{U}$-covers f.*

Proof Using Lemma A.1.10, there exists a cover $(E_i)_{i\in I}$ with the properties in Construction 4.2.5, and moreover so that each E_i is contained in some $U \in \mathcal{U}$. Lemma 4.2.6 tells us that if we apply Construction 4.2.5 starting with this cover (E_i), then the resulting isometry $V\colon H_X \to H_Y$ satisfies

$$\text{supp}(V) \subseteq \bigcup_{i \in I} \overline{E_i} \times \overline{f^{-1}(E_i)}.$$

The right-hand side is contained in $\bigcup_{U \in \mathcal{U}} \overline{U} \times \overline{f^{-1}(U)}$, which is the desired conclusion. □

4.3 Covering Isometries for Coarse Maps

In this section, we produce a specialisation of the above material to the coarse category. The coarse category is looked at in more detail in Section A.3; for the reader's convenience, we repeat the main definitions here.

Definition 4.3.1 For us, metrics are allowed to take the value infinity (but otherwise satisfy all the usual conditions). A metric space X is *proper* if all closed bounded sets are compact.

Let $f: X \to Y$ be any map between (proper) metric spaces. The *expansion function* of f, denoted $\omega_f: [0, \infty) \to [0, \infty]$, is defined by

$$\omega_f(r) := \sup\{d_Y(f(x_1), f(x_2)) \mid d_X(x_1, x_2) \leqslant r\}.$$

The function f is *coarse* if:

(i) $\omega_f(r)$ is finite for all $r \geqslant 0$;

(ii) f is a *proper map*, meaning that for any compact subset K of Y, the pull-back $f^{-1}(K)$ has compact closure.

Two maps $f, g: X \to Y$ are *close* if there exists $c \geqslant 0$ such that for all $x \in X$, $d_Y(f(x), g(x)) \leqslant c$. The *coarse category*, denoted $\mathcal{C}oa$, has objects proper metric spaces, and morphisms closeness classes of coarse maps.

For the rest of the section, X, Y, Z will be objects of the category $\mathcal{C}oa$. We will use usual metric notions: in particular, the ball of radius $r \in (0, \infty)$ around a point $x \in X$ is $B(x; r) := \{y \in X \mid d(x, y) < r\}$.

Remark 4.3.2 Using the assumptions that our metrics are proper, a map $f: X \to Y$ is proper in the sense above if and only if it pulls back bounded sets to finite unions of bounded sets. This is because a compact set is always a finite union of bounded sets. Note, however, that as we allow our metrics to take the value infinity, a finite union of bounded sets need not be bounded: compare Lemma A.3.2.

The definition of covering isometry appropriate to the coarse category is as follows.

Definition 4.3.3 Let H_X, H_Y be geometric modules, and let $f: X \to Y$ be a coarse map. An isometry $V: H_X \to H_Y$ *covers* f, or is a *covering isometry* of f, if there is $t \in (0, \infty)$ such that $d_Y(y, f(x)) < t$ whenever $(y, x) \in \text{supp}(V)$.

Proposition 4.3.4 *Let $f: X \to Y$ be a coarse map, and H_X, H_Y be geometric modules such that H_Y is ample. Then there is a covering isometry $V: H_X \to H_Y$ for f.*

Proof Let $r \in (0, \infty)$, and let $\mathcal{U}$ be the open cover $\{B(y; r) \mid y \in Y\}$ of Y. Note that if $g: X \to Y$ is a coarse map that is close to f, then any covering isometry for g is also a covering isometry for f; using Lemma A.3.12, then, we may assume that f is Borel. Using Corollary 4.2.7 above, there exists a $\mathcal{U}$-cover $V: H_X \to H_Y$ and by definition of a $\mathcal{U}$-cover,

$$\text{supp}(V) \subseteq \bigcup_{U \in \mathcal{U}} \overline{U} \times \overline{f^{-1}(U)}.$$

It thus suffices to find t such that if $(y, x) \in \overline{U} \times \overline{f^{-1}(U)}$ for some $U \in \mathcal{U}$, then $d_Y(y, f(x)) < t$. Indeed, as $x \in \overline{f^{-1}(U)}$, there is $x' \in f^{-1}(U)$ with $d_X(x, x') < r$. Using that $y \in \overline{U}$, there is $y' \in U$ with $d_Y(y, y') < r$, and thus

$$d_Y(y, f(x')) \leqslant d_Y(y, y') + d_Y(y', f(x')) < r + \text{diam}(U) < 3r.$$

Hence

$$d_Y(y, f(x)) \leqslant d_Y(y, f(x')) + d_Y(f(x'), f(x)) < 3r + \omega_f(r),$$

so we may take $t = 3r + \omega_f(r)$. □

A *coarse equivalence* is an isomorphism in the category *Coa* of Definition 4.3.1. In the case that we have a coarse equivalence from X to Y and both modules are ample, we can use some machinery from Section A.3 to do a bit better.

Proposition 4.3.5 *Let $f: X \to Y$ be a coarse equivalence, and H_X, H_Y be ample geometric modules. Then there is a covering isometry $V: H_X \to H_Y$ for f which is also a unitary isomorphism.*

Proof Using Lemma A.3.12, we may assume that f is Borel. Using Exercise A.4.3 and the fact that $f: X \to Y$ is a coarse equivalence there exists $c > 0$ such that for every $y \in Y$ is within c of some point in $f(X)$. Moreover, there exists $s > 0$ such that for all $x \in X$, the diameter of $f(B(x; s))$ is at most s. Let $r = c + s + 1$, and let Z be a $2r$-net in Y as in Definition A.3.10, which exists by Lemma A.3.11. As Y is second countable, Z is countable, so we may enumerate its elements as $z_1, z_2, \ldots$. Iteratively define Borel subsets of Y in the following way. Set

$$E_1 := B(z_1; 3r) \setminus \bigcup_{z \in Z \setminus \{z_1\}} B(z; r),$$

and given $E_1, \ldots, E_{n-1}$, define

$$E_n := B(z_n; 3r) \setminus \left(\bigcup_{z \in Z \setminus \{z_n\}} B(z; r) \cup \bigcup_{i=1}^{n-1} E_i \right).$$

Then $(E_n)_{n \in \mathbb{N}}$ is a Borel cover of Y by disjoint sets, such that each E_n is contained in $B(z_n; 3r)$ (and in particular the family is uniformly bounded) and such that each contains $B(z_n; r)$.

We claim that for each n, each $f^{-1}(E_n)$ contains an open set. Indeed, by choice of c, there is some $x \in X$ with $d(f(x), z_n) < c$. On the other hand, by choice of s, every point in $B(x; 1)$ is mapped into $B(f(x); s)$, and so $B(x; 1)$ is mapped into $B(z_n; c + s)$, which is contained in E_n by choice of r.

As this point ampleness gives that both $\chi_{f^{-1}(E_n)} H_X$ and $\chi_{E_n} H_Y$ are (separable and) infinite-dimensional, and so we may choose a unitary isomorphism $V_n : \chi_{f^{-1}(E_n)} H_X \to \chi_{E_n} H_Y$. Define

$$V := \bigoplus_{n \in \mathbb{N}} V_i : \underbrace{\bigoplus_{n \in \mathbb{N}} \chi_{f^{-1}(E_n)} H_X}_{= H_X} \to \underbrace{\bigoplus_{n \in \mathbb{N}} \chi_{E_i} H_Y}_{= H_Y},$$

which is a unitary isomorphism. Checking that this is a covering isometry for f is quite analogous to the argument in the proof of Proposition 4.3.4, and we leave the remaining details to the reader. □

We finish this section with two technical results that will be useful later. The first uses the notion of properly supported operator from Definition 4.1.12.

Lemma 4.3.6 *Let $f: X \to Y$ be a coarse map, H_X, H_Y be geometric modules and $T: H_X \to H_Y$ a bounded operator such that there exists $t \in [0, \infty)$ with $d(y, f(x)) < t$ for all $(y, x) \in supp(T)$. Then T is properly supported.*

Proof If $K \subseteq Y$, define the t-neighbourhood of K to be $N_t(K) := \bigcup_{y \in K} B(y; t)$, and similarly for $K \subseteq X$. Let $\pi_Y : \text{supp}(T) \to Y$ be the coordinate projection, and let $K \subseteq Y$ be compact. Then

$$\pi_Y^{-1}(K) \subseteq K \times f^{-1}(N_t(K)).$$

Note that as K is compact, it is a finite union of bounded sets (compare Remark 4.3.2), and thus $N_t(K)$ is also a finite union of bounded sets and so has compact closure by properness of Y. As f is proper, the set $f^{-1}(N_t(K))$ thus has compact closure, and thus $\pi_Y^{-1}(K)$ is compact. Similarly, if $\pi_X : \operatorname{supp}(T) \to X$ is the coordinate projection, then

$$\pi_X^{-1}(K) \subseteq N_t(f(K)) \times K.$$

As K is a finite union of bounded sets and $\omega_f(r) < \infty$ for all $r \in [0,\infty)$, $N_t(f(K))$ is also a finite union of bounded sets. Hence $N_t(f(K))$ has compact closure, and thus $\pi_X^{-1}(K)$ is compact. □

Corollary 4.3.7 *Say $V_f : H_X \to H_Y$, $V_g : H_Y \to H_Z$ are covering isometries for coarse maps $f : X \to Y$ and $g : Y \to Z$. Then the composition $V_g \circ V_f$ is a covering isometry of $g \circ f$.*

Proof Let $t \in [0,\infty)$ be as in definition of covering isometry for both f and g. Using Lemma 4.3.6 and Lemma 4.1.13, part (iii′)

$$\operatorname{supp}(V_g V_f) \subseteq \operatorname{supp}(V_g) \circ \operatorname{supp}(V_f),$$

whence if (z,x) is in $\operatorname{supp}(V_g V_f)$ there exists $y \in Y$ such that (z,y), (y,x) are in $\operatorname{supp}(V_g)$ and $\operatorname{supp}(V_f)$ respectively. Hence

$$d(z, g(f(x))) \leqslant d(z, g(y)) + d(g(y), g(f(x))) \leqslant t + \omega_g(t);$$

as $t + \omega_g(t)$ is independent of x and z, this completes the proof. □

4.4 Covering Isometries for Continuous Maps

In this section, we discuss a parametrised version of Construction 4.2.5 that is appropriate for continuous maps between topological spaces.

Throughout this section, X and Y are second countable, locally compact, Hausdorff spaces. Recall that an open cover $\mathcal{U}$ of X *refines* an open cover $\mathcal{V}$ if every $U \in \mathcal{U}$ is contained in some $V \in \mathcal{V}$.

Definition 4.4.1 Let H_X, H_Y be geometric modules, and $f : X \to Y$ a function. Let $(\mathcal{U}_n)$ be a sequence of open covers of Y such that $\mathcal{U}_{n+1}$ refines $\mathcal{U}_n$ for all n.

A family of isometries $(V_t)_{t\in[1,\infty)}$ *covers* f *with respect to the sequence* $(\mathcal{U}_n)$ if:

(i) for all n and all $t \geqslant n$, V_t is a $\mathcal{U}_n$ cover of f;

(ii) the function $t \mapsto V_t$ from $[1,\infty)$ to $\mathcal{B}(H_X, H_Y)$ is uniformly norm continuous.

Example 4.4.2 Let $V\colon H_X \to H_Y$ be a family of isometries that is a $\mathcal{U}$-cover of f for *any* $\mathcal{U}$, as appearing in Examples 4.2.2 and 4.2.3. Then the corresponding constant family defined by setting $V_t = V$ for all t is a cover for f with respect to any sequence $(\mathcal{U}_n)$.

Proposition 4.4.3 *Let $f\colon X \to Y$ be a Borel map, H_X, H_Y be geometric modules with H_Y ample and $(\mathcal{U}_n)$ a sequence of open covers of Y such that $\mathcal{U}_{n+1}$ refines $\mathcal{U}_n$ for all n. Then there exists a cover (V_t) for f with respect to the sequence $(\mathcal{U}_n)$.*

To prove this, we need a general lemma about the existence of paths between isometries.

Lemma 4.4.4 *Let H, H' be separable Hilbert spaces, and let V_0, V_1 be isometries from H into H' such that the subspaces $V_0 H$ and $V_1 H$ have the same (dimension and) codimension. Then there exists a path $(V_t\colon H \to H')_{t\in[0,1]}$ of isometries connecting V_0 and V_1 such that*

$$\|V_t - V_s\| \leqslant 2\pi|s-t|$$

for all $s,t \in [0,1]$.

Moreover, if V_0 and V_1 are unitary, then each V_t may also be chosen to be unitary.

Proof Note that the partial isometry $V_1 V_0^* \in \mathcal{B}(H')$ is a unitary isomorphism from $V_0 H$ to $V_1 H$. Choose an arbitrary partial isometry $W \in \mathcal{B}(H')$ that acts as zero on $V_0 H$ and takes the orthogonal complement of $V_0 H$ onto the orthogonal complement of $V_1 H$ (such exists as $V_0 H$ and $V_1 H$ have the same codimension). Define $U = V_1 V_0^* + W \in \mathcal{B}(H')$, which is a unitary operator. Let $f\colon S^1 \to [0, 2\pi)$ be the (Borel) inverse to the exponential map $t \mapsto e^{it}$, and let $T = f(U) \in \mathcal{B}(H')$ be defined using the Borel functional calculus of Corollary 1.6.12. Then T is a self-adjoint bounded operator on H' of norm at most 2π such that $U = e^{iT}$. The path of isometries

$$\gamma\colon t \mapsto e^{itT} V_0$$

satisfies

$$\|\gamma(t) - \gamma(s)\| = \|e^{itT} - e^{isT}\| \leqslant \|T\| |t-s| \leqslant 2\pi |t-s|$$

for all $s,t \in [0,1]$ by the functional calculus. It moreover satisfies $\gamma(0) = V_0$ and that $\gamma(1) = V_1$, so we are done in the case that V_0 and V_1 are isometries.

The statement about unitaries follows on noting that the above construction automatically gives a path of unitaries if one starts with unitary V_0 and V_1. □

Proof of Proposition 4.4.3 Using Lemma A.1.10, there exists a countable cover $(E_i)_{i\in I}$ of Y by non-empty disjoint Borel sets such that each E_i has compact closure, each is contained in the closure of its interior, each is contained in some element of $\mathcal{U}_1$ and such that only finitely many of the sets E_i intersect any compact subset of Y. Apply Construction 4.2.5 to this cover, with the additional requirement that each isometry

$$V_i \colon \chi_{f^{-1}(E_i)} H_X \to \chi_{E_i} H_Y \tag{4.3}$$

appearing in the construction has range of infinite codimension (we can do this, as the right-hand side above is infinite-dimensional and the left-hand side is separable). Lemma 4.2.6 implies that the corresponding isometry $V \colon H_X \to H_Y$ is a $\mathcal{U}_1$-cover of f; we write V_1 for this isometry and $V_{1,i}$ for each of the isometries as in line (4.3) used to build it.

Now, as each E_i has compact closure and as each E_i is the closure of its interior, we may subdivide E_i into finitely many non-empty Borel pieces $E_i = \bigsqcup_{j=1}^{N_i} E_{ij}$ such that each E_{ij} is contained in the closure of its interior and such that for each i, j, we have that E_{ij} is contained in some element of $\mathcal{U}_2$. For each i, j, there is an isometry

$$V_{2,ij} \colon \chi_{f^{-1}(E_{ij})} H_X \to \chi_{E_{ij}} H_Y$$

with range of infinite codimension. Consider the pair of isometries

$$V_{1,i},\ \oplus_{j=1}^{N_i} V_{2,ij} \ \colon \chi_{f^{-1}(E_i)} H_X \to \chi_{E_i} H_Y.$$

These have range of infinite codimension, whence by Lemma 4.4.4 there is a norm continuous path of isometries $(V_{t,i})_{t\in[1,2]}$ connecting them. Moreover, from the explicit estimate in Lemma 4.4.4, the family of paths

$$(V_{t,i})_{t\in[1,2]}$$

as i ranges over I can be chosen to be Lipschitz, with Lipschitz constant 2π. It follows that the path

$$V_t := \bigoplus_{i\in I} V_{i,t} \colon H_X \to H_Y$$

of isometries is Lipschitz with Lipschitz constant 2π and connects V_1 and $V_2 := \oplus_{i,j} V_{2,ij}$. (The proof of) Corollary 4.2.7 implies that V_t is a $\mathcal{U}_1$-cover of f for all $t \in [1,2]$ and a $\mathcal{U}_2$-cover for $t = 2$.

We may now similarly subdivide each E_{ij} into finitely many non-empty Borel pieces, each of which is contained in the closure of its interior and each of which is contained in some element of $\mathcal{U}_3$; this decomposition can then be used to construct V_3 and a norm continuous path from V_2 to V_3 in much the same way. Connecting our two paths gives a path $(V_t)_{t\in[1,3]}$ such that $t \mapsto V_t$ is Lipschitz with Lipschitz constant 2π, and such that V_t is a $\mathcal{U}_1$-cover of f for all $t \in [1,3]$, a $\mathcal{U}_2$-cover for $t \in [2,3]$ and a $\mathcal{U}_3$-cover for $t = 3$.

Continuing this process gives a norm continuous path $(V_t)_{t\in[1,\infty)}$ with the desired properties. □

Remark 4.4.5 Say $f: X \to Y$ is a homeomorphism, and that H_X, H_Y are geometric modules that are both ample. Then in the above construction we may take all the various isometries appearing to be unitary isomorphisms, and thus can assume that the final result is a continuous family $(V_t: H_X \to H_Y)_{t\in[1,\infty)}$ of unitary isomorphisms with the properties in the statement of Proposition 4.4.3.

The following definition and corollary give a particularly important special case. Recall that Y^+ denotes the one point compactification of Y: see Definition A.1.4.

Definition 4.4.6 Let $f: X \to Y$ be a continuous map, and let H_X, H_Y be geometric modules. Then a *continuous cover of* f is a family of isometries $(V_t: H_X \to H_Y)_{t\in[1,\infty)}$ such that:

(i) the function $t \mapsto V_t$ from $[1,\infty)$ to $\mathcal{B}(H_X, H_Y)$ is uniformly norm continuous;

(ii) for any open subset $U \subseteq Y^+ \times Y^+$ that contains the diagonal, there exists $t_U \geqslant 0$ such that for all $t \geqslant t_U$

$$\operatorname{supp}(V_t) \subseteq \{(y,x) \in Y \times X \mid (y, f(x)) \in U\}.$$

Corollary 4.4.7 *Let $f: X \to Y$ be a continuous map, and let H_X, H_Y be geometric modules with H_Y ample. Then a continuous cover of f exists.*

Proof Fix a metric d on Y^+ that induces the topology. Using compactness of Y^+, it will suffice to find a uniformly continuous family (V_t) such that for all $\epsilon > 0$ we have

$$\operatorname{supp}(V_t) \subseteq \{(y,x) \in Y \times X \mid d(y, f(x)) < \epsilon\}$$

for all suitably large t. Let $\mathcal{U}_n$ be the open cover of Y by balls of radius 2^{-n} for the restricted metric from Y^+, and apply Proposition 4.4.3. Let $\epsilon > 0$, and let n be such that $2^{-n} < \epsilon/2$; we claim $t_\epsilon = n$ works. Indeed, let $t \geqslant n$, and note that V_t is a $\mathcal{U}_n$ cover of f, whence by definition of a $\mathcal{U}_n$-cover (Definition 4.2.1) we have that

$$\mathrm{supp}(V_t) \subseteq \bigcup_{y \in Y} \overline{B(y;2^{-n})} \times \overline{f^{-1}(B(y;2^{-n}))}.$$

Say $(y,x) \in \mathrm{supp}(V)$, whence there is $z \in Y$ with

$$(y,x) \in \overline{B(z;2^{-n})} \times \overline{f^{-1}(B(z;2^{-n}))}.$$

As f is continuous, $x \in \overline{f^{-1}(B(z;2^{-n}))}$ implies that $d(f(x),z) \leqslant 2^{-n}$. Hence

$$d(y,f(x)) \leqslant d(y,z) + d(z,f(x)) \leqslant 2 \cdot 2^{-n} < \epsilon,$$

completing the proof. □

4.5 Equivariant Covering Isometries

In this section, we will discuss geometric modules and covering isometries in the presence of a group action. The basic ideas are largely the same as in the previous sections, but using background from Section A.2 as opposed to Section A.1 for the necessary ingredients. Unfortunately, some proofs end up being technical; for most purposes, the reader would not lose much by treating them as black boxes.

Throughout this section, G denotes a countable discrete group, and X, Y are locally compact, second countable Hausdorff spaces. We always assume that G acts *properly* on all spaces: recall from Definition A.2.2 that this means that for any compact subset K of the relevant space, the set $\{g \in G \mid gK \cap K \neq \varnothing\}$ is finite. As in Proposition A.2.1, there is an induced action of G on $C_0(X)$, which we denote by α; in symbols

$$(\alpha_g f)(x) := f(g^{-1}x) \tag{4.4}$$

for all $g \in G$, $f \in C_0(X)$ and $x \in X$. See Section A.2 for associated definitions and basic facts on group actions.

We will also need to work with unitary representations: a *unitary representation* of a group G is a homomorphism from G to the unitary group $\mathcal{U}(H)$ of some Hilbert space H, and two unitary representations U, V on Hilbert spaces H_U, H_V are *isomorphic* if there is a unitary isomorphism $W\colon H_U \to H_V$ such that $WU_g = V_g W$ for all $g \in G$. See Section C.1 for a summary of definitions and basic facts.

Equivariant Geometric Modules

We now discuss the equivariant geometric modules that we need to set up the theory in the presence of a group action.

Definition 4.5.1 A *(geometric) X-G module*, or *(geometric) module over X-G*, is an X module H_X equipped with a unitary representation $U\colon G \to \mathcal{U}(H_X)$ that spatially implements the action of G of $C_0(X)$ in the sense of Definition C.1.8: precisely,

$$U_g f U_g^* = \alpha_g(f)$$

for all $f \in C_0(X)$ and $g \in G$.

We will almost always denote the G-action on an X-G module H_X by $g \mapsto U_g$ without necessarily explicitly mentioning this.

Say that H_X is an X-G module. Note that the C^*-algebra $B(X)$ of bounded Borel functions on X has a G action defined by the same formula as in line (4.4) above. Moreover, if we extend the representation of $C_0(X)$ on H_X to one of $B(X)$ as in Proposition 1.6.11, checking the construction shows that the unitary representation $U: G \to \mathcal{U}(H_X)$ also spatially implements the G action on $B(X)$. In particular, if F is a subgroup of G and $E \subseteq X$ an F-invariant Borel subset, then for all $g \in F$

$$U_g \chi_E U_g^* = \alpha_g(\chi_E) = \chi_{gE} = \chi_E.$$

In other words, U_g commutes with χ_E for all $g \in F$. It follows that the restriction $U|_F\colon F \to \mathcal{U}(H_X)$ of the unitary representation to F induces a well-defined unitary representation of F on $\chi_E H_X$.

Definition 4.5.2 An X-G module H_X is *locally free* if for any finite subgroup F of G and any F-invariant Borel subset E of X there is a Hilbert space H_E (possibly zero) equipped with the trivial representation of F such that $\chi_E H_X$ and $\ell^2(F) \otimes H_E$ are isomorphic as F representations.

The module H_X is *ample* (as an X-G module) if it is locally free and ample as an X module.

It is possible for an X-G module to be ample as an X module, but not as an X-G module (i.e. not locally free): for example, take X to be a point, G a finite non-trivial group and H any infinite-dimensional Hilbert space equipped with the unital action of $C(X) = \mathbb{C}$, and the trivial action of G. This ambiguity of terminology should not cause any confusion.

Example 4.5.3 Say $X = G$ and $H_X = \ell^2(G, H)$ for some Hilbert space H with $C_0(G)$ acting on H_X by pointwise multiplication. Define a unitary representation of G by $(U_g u)(h) := u(g^{-1}h)$ for all $g, h \in G$ and $u \colon G \to H$. This module is always locally free. Indeed, for an F-invariant (Borel) subset E of G, one can choose a subset S of G such that

$$E = \bigsqcup_{g \in S} Fg.$$

It follows that

$$\chi_E \ell^2(G, H) \cong \ell^2(E, H) \cong \ell^2(F \times S, H) \cong \ell^2(F) \otimes \ell^2(S, H)$$

as F representations, where F acts trivially on S. This representation is ample if and only if H is infinite-dimensional.

Example 4.5.4 Let μ be a Radon measure on X and consider the X module $H_X := L^2(X, \mu)$ from Example 4.1.2. Assume moreover that μ is G-*invariant*, meaning that $\mu(gE) = \mu(E)$ for all Borel subsets $E \subseteq G$. Then the formula

$$(U_g u)(x) := u(g^{-1}x), \quad u \in H_X,\ g \in G,\ x \in X,$$

defines a unitary representation of G on $L^2(X, \mu)$ that makes it into an X-G module.

For example, X could be the real line equipped with Lebesgue measure, and G could be $\mathbb{Z}$ acting by translations in the usual way. More generally, X could be a complete Riemannian manifold with a G action by proper isometries, and μ the measure defined by the Riemannian structure.

We claim that if the G action on X is free, then the module H_X is locally free.[5] Indeed, say F is a finite subgroup of G. One can show (either derive it from Lemma A.2.9 with $G = F$, or do it directly) that there is a Borel subset D of X and a decomposition $X = \bigsqcup_{g \in F} gD$. If then E is an arbitrary F-invariant Borel subset of X, set $H_E := \chi_{E \cap D} H_X$, equipped with the trivial F action and define an operator

$$V \colon \chi_E H_X \to \ell^2(F) \otimes H_E, \quad u \mapsto \sum_{g \in F} \delta_g \otimes U_g^* \chi_{gD} u.$$

We leave it as an exercise for the reader to check that V is an isomorphism of F representations.

[5] This can fail without the freeness assumption: consider again a finite group acting trivially on X.

In particular, if X is a Riemannian manifold of positive dimension with associated measure μ, and G acts freely and properly by isometries on X, then $L^2(X,\mu)$ is naturally an ample X-G module. One could also include a bundle as in Example 4.1.5, as long as it is G-equivariant.

We now go back to generalities.

Lemma 4.5.5 *Ample X-G modules always exist.*

Proof As G is countable and X is second countable, there exists a countable, dense and G-invariant subset Z of X. Let H be an infinite-dimensional separable Hilbert space. Define

$$H_X := \ell^2(Z) \otimes H \otimes \ell^2(G)$$

equipped with the 'diagonal' action of G defined by

$$U_g : \delta_z \otimes u \otimes \delta_h \mapsto \delta_{gz} \otimes u \otimes \delta_{gh}.$$

Define an action of $C_0(X)$ by pointwise multiplication

$$f : \delta_z \otimes u \otimes \delta_h \mapsto f(z)\delta_z \otimes u \otimes \delta_h.$$

Then one computes that

$$\begin{aligned}(U_g f U_g^*)(\delta_z \otimes u \otimes \delta_h) &= (U_g f)(\delta_{g^{-1}z} \otimes u \otimes \delta_{g^{-1}h}) \\ &= U_g(f(g^{-1}z)\delta_{g^{-1}z} \otimes u \otimes \delta_{g^{-1}h}) \\ &= f(g^{-1}z)(\delta_z \otimes u \otimes \delta_h) \\ &= (\alpha_g f)(\delta_z \otimes u \otimes \delta_h),\end{aligned}$$

which proves the covariance relation.

As H is infinite-dimensional and Z is dense in X, H_X is ample as an X module (compare Example 4.1.6), so it remains to check local freeness. Let then F be a finite subgroup of G, and let E be an F-invariant Borel subset of X. We then have that

$$\chi_E H_X = \ell^2(Z \cap E) \otimes H \otimes \ell^2(G).$$

Note that the F action on G identifies with the left multiplication action of F on $F \times (F\backslash G)$ (where the right coset space $F\backslash G$ has trivial F action), whence as F representations

$$\ell^2(G) \cong \ell^2(F) \otimes \ell^2(F\backslash G)$$

and so

$$\chi_E H_X \cong \ell^2(F) \otimes \Big(\ell^2(E \cap Z) \otimes \ell^2(F\backslash G) \otimes H\Big)$$

as F representations. On the other hand, if H_E is defined to be the Hilbert space $\ell^2(E \cap Z) \otimes \ell^2(F \backslash G) \otimes H$ equipped with the trivial F representation, then Fell's trick (Proposition C.2.1) implies that

$$\ell^2(F) \otimes \left(\ell^2(E \cap Z) \otimes \ell^2(F \backslash G) \otimes H\right) \cong \ell^2(F) \otimes H_E$$

as F representations and we are done. □

Equivariant Covering Isometries

In the rest of this section we look at covering isometries: the appropriate notions are similar to those discussed in Sections 4.2, 4.3, and 4.4, but with additional equivariance conditions.

Definition 4.5.6 Let H_X and H_Y be geometric modules, $f : X \to Y$ a function and $\mathcal{U}$ an open cover of Y. Write $G\mathcal{U}$ for the open cover

$$\{gU \mid g \in G, U \in \mathcal{U}\}$$

of Y. An isometry $V : H_X \to H_Y$ is an *equivariant $\mathcal{U}$-cover of f* if it is equivariant for the G representations on H_X and H_Y and if it $G\mathcal{U}$-covers f in the sense of Definition 4.2.1.

The analogue of Construction 4.2.5 in the equivariant case is then as follows.

Construction 4.5.7 Let $f : X \to Y$ be an equivariant Borel function. Let H_X, H_Y be geometric modules, with H_Y ample. Let $(E_i)_{i \in I}$ be a countable collection of Borel subsets of Y with the following properties:

(i) Y is equal to the disjoint union $\bigsqcup_{i \in I} GE_i$ of the sets GE_i;
(ii) each E_i has non-empty interior;
(iii) for any compact $K \subseteq Y$, the set

$$\{i \in I \mid E_i \cap K \neq \varnothing\}$$

is finite;
(iv) for each i, there is a finite subgroup F_i of G such that E_i is F_i-invariant, and such that GE_i is equal to the disjoint union $\bigsqcup_{gF_i \in G/F_i} gE_i$.

Now, by the local freeness condition there is a Hilbert space H_i equipped with a trivial F_i action and an isomorphism

$$\chi_{E_i} H_Y \cong \ell^2(F_i) \otimes H_i$$

of F_i representations. Moreover, as E_i has non-empty interior, ampleness implies that $\chi_{E_i} H_Y$ is infinite-dimensional and thus H_i is too. Note that $f^{-1}(E_i)$ is also F_i invariant, whence $\chi_{f^{-1}(E_i)} H_X$ is equipped with an F_i representation. Corollary C.2.2 thus implies that we may choose an F-equivariant isometry

$$W_i \colon \chi_{f^{-1}(E_i)} H_X \to \chi_{E_i} H_Y. \tag{4.5}$$

Define an isometry $V_i \colon \chi_{f^{-1}(GE_i)} H_X \to \chi_{GE_i} H_Y$ by the formula

$$V_i := \bigoplus_{gF_i \in G/F_i} U_g W_i \chi_{f^{-1}(E_i)} U_g^* \colon \underbrace{\bigoplus_{gF_i \in G/F_i} \chi_{gf^{-1}(E_i)} H_X}_{=\chi_{f^{-1}(GE_i)} H_X} \to \underbrace{\bigoplus_{gF_i \in G/F_i} \chi_{gE_i} H_Y}_{=\chi_{GE_i} H_Y}$$

(convergence in the strong operator topology). This isometry does not depend on the choice of coset representatives from G/F_i by F_i equivariance of W_i and F_i invariance of $\chi_{f^{-1}E_i}$. It is moreover G equivariant as for any $h \in G$

$$\begin{aligned} V_i U_h &= \bigoplus_{gF_i \in G/F_i} U_g W_i \chi_{f^{-1}(E_i)} U_{h^{-1}g}^* \\ &= \bigoplus_{kF_i \in G/F_i} U_{hk} V_{E_i} \chi_{f^{-1}(E_i)} U_k^* = U_h V_i, \end{aligned}$$

where we used the 'change of variables' $k = h^{-1}g$ in the second equality. Define finally

$$V := \bigoplus_{i \in I} V_i \colon \underbrace{\bigoplus_{i \in I} \chi_{f^{-1}(GE_i)} H_X}_{=H_X} \to \underbrace{\bigoplus_{i \in I} \chi_{GE_i} H_Y}_{=H_Y},$$

which is a G-equivariant isometry from H_X to H_Y.

Remark 4.5.8 Assume that H_X is also ample and that we can choose the cover (E_i) above with the additional property that each pullback $f^{-1}(E_i)$ also has non-empty interior. Then we have that both $\chi_{f^{-1}(E_i)} H_X$ and $\chi_{E_i} H_Y$ are isomorphic as F-representations to $\ell^2(F) \otimes H$ for some infinite-dimensional Hilbert space H equipped with the trivial F_i representation. In particular, in line (4.5) above we may choose W_i to be a unitary F-equivariant map. The rest of the construction will then give that V itself is also unitary.

We now have an equivariant version of Lemma 4.2.6.

Lemma 4.5.9 *With notation as in Construction 4.5.7, we have*

$$supp(V) \subseteq \bigcup_{i \in I} \bigcup_{g \in G/F_i} \overline{gE_i} \times \overline{f^{-1}(gE_i)}.$$

Proof This is actually a special case of Lemma 4.2.6 in disguise. With notation as in Construction 4.5.7, Y is covered by the disjoint sets in the collection

$$(gE_i)_{g \in G/F_i, i \in I}.$$

This collection has the properties in the statement of Construction 4.2.5. On the other hand, the isometry V is built as a direct sum of the isometries

$$V_{i,gF_i} := U_g W_i \chi_{f^{-1}(E_i)} U_g^* : \chi_{f^{-1}(gE_i)} H_X \to \chi_{gE_i} H_Y$$

as i ranges over I and gF_i over G/F_i, and these have the properties required in Construction 4.2.5. Lemma 4.2.6 thus applies verbatim. □

Now, let $\mathcal{U}$ be an open cover of Y. Then Lemma A.2.9 implies that we may always find a countable collection (E_i) as in the statement of Construction 4.5.7 such that each E_i is contained in some element of $\mathcal{U}$. The proof of the following result now goes in much the same way as in the non-equivariant case: we leave the details to the reader.

Corollary 4.5.10 *Let $f : X \to Y$ be an equivariant Borel map, and H_X, H_Y be geometric modules with H_Y ample. Then for any open cover $\mathcal{U}$ of Y, there exists an equivariant isometry $V : H_X \to H_Y$ that equivariantly $\mathcal{U}$-covers f.* □

Equivariant Covering Isometries for Continuous Maps

We now turn to the appropriate covering isometries for equivariant continuous maps.

Definition 4.5.11 Let $f : X \to Y$ be an equivariant continuous map, and let H_X, H_Y be geometric modules. Then an *equivariant continuous cover* of f is a family $(V_t : H_X \to H_Y)$ of isometries with the following properties.

(i) the function $t \mapsto V_t$ from $[1, \infty)$ to $\mathcal{B}(H_X, H_Y)$ is uniformly norm continuous;

(ii) for any open subset $U \subseteq Y^+ \times Y^+$ that contains the diagonal, there exists $t_U \geqslant 1$ such that for all $t \geqslant t_U$

$$\text{supp}(V_t) \subseteq \{(y,x) \in Y \times X \mid (y, f(x)) \in U\};$$

(iii) each V_t is G equivariant (in symbols, $U_g V = V U_g$ for all $g \in G$).

Proposition 4.5.12 *Let $f : X \to Y$ be an equivariant continuous map, H_X, H_Y be geometric modules with H_Y ample. Then there exists an equivariant continuous cover (V_t) for f.*

Moreover, if H_X and H_Y are both ample, and f is an equivariant homeomorphism, then there exists an equivariant continuous cover (V_t) for f where each V_t is a unitary isomorphism.

Proof We fix an open set $U \subseteq Y^+ \times Y^+$ containing the diagonal, and first show that we can get a single equivariant isometry $V\colon H_X \to H_Y$ such that $\operatorname{supp}(V) \subseteq \{(y,x) \in Y \times X \mid (y, f(x)) \in U\}$.

Let then $U \subseteq Y^+ \times Y^+$ be given. As $(\infty, \infty) \in U$, there exists a sequence of compact subsets $K_1 \subseteq K_2 \subseteq \cdots$ of Y, each contained in the interior of the next, such that $\overline{Y \setminus K_n} \times \overline{Y \setminus K_n} \subseteq U$ for all n, and such that $Y = \bigcup_{n=1}^{\infty} K_n$. For each, let $S_n = \{g \in G \mid gK_n \cap K_n \neq \varnothing\}$, a finite subset of G by properness. For each n, let $\mathcal{U}_n$ be a finite cover of K_n with the properties that every $W \in \mathcal{U}_n$ is contained in K_{n+1}, and such that for all $g \in S_{n+1}$ we have that

$$\overline{gW} \times \overline{gW} \subseteq U;$$

to see that such a finite cover exists, use finiteness of S_n to get a possibly infinite cover of K satisfying the properties, then use compactness of K_n to get a finite subcover. Set now $\mathcal{U} = \bigcup_{n=1}^{\infty} \mathcal{U}_n$, and let (E_i) be a countable collection of Borel subsets of Y with the properties in Lemma A.2.9 with respect to this cover. Apply Construction 4.5.7 and Lemma 4.5.9 to get an equivariant isometry $V\colon H_X \to H_Y$ such that

$$\operatorname{supp}(V) \subseteq \bigcup_{i \in I} \bigcup_{g \in G/F_i} \overline{gE_i} \times \overline{f^{-1}(gE_i)}.$$

We claim that in fact this V satisfies $\{(y, f(x)) \mid (y,x) \in \operatorname{supp}(V)\} \subseteq U$. Indeed, let (y,x) be an element of $\operatorname{supp}(V)$, so there exist $g \in G$ and $i \in I$ such that $(y,x) \in \overline{gE_i} \times \overline{f^{-1}(gE_i)}$. Then there are $n \in \mathbb{N}$ and $W \in \mathcal{U}_n$ such that $\overline{gE_i} \subseteq \overline{gW}$. If $g \in S_{n+1}$, then we have

$$(y, f(x)) \in \overline{gE_i} \times f(\overline{f^{-1}(gE_i)}) \subseteq \overline{gE_i} \times \overline{gE_i} \subseteq \overline{gW} \times \overline{gW} \subseteq U,$$

where the first set inclusion uses continuity of f, the second that $E_i \subseteq W$ and the third uses the construction of $\mathcal{U}_n$. On the other hand, if $g \notin S_{n+1}$, then $g\overline{W} \cap K_n = \varnothing$ and thus

$$(y, f(x)) \in \overline{gE_i} \times f(\overline{f^{-1}(gE_i)}) \subseteq \overline{gE_i} \times \overline{gE_i} \subseteq \overline{Y \setminus K_n} \times \overline{Y \setminus K_n} \subseteq U.$$

This completes the proof that V has the right properties.

Having explained the above, the construction of a family (V_t) proceeds much as in the proof of Proposition 4.4.3. Indeed, the construction in Lemma 4.4.4 preserves equivariance, so one has a completely analogous equivariant version of that lemma. Having noted this, use metrisability and compactness of Y^+ to construct a decreasing sequence

$$U_1 \supseteq U_2 \supseteq \cdots$$

of (not necessarily G-invariant) open subsets of $Y^+ \times Y^+$ that contain the diagonal and are eventually contained in any open subset $U \subseteq Y^+ \times Y^+$ that contains the diagonal. An iterative construction as in the proof of Proposition 4.4.3 will then complete the construction of a family (V_t) with the right properties: we leave the remaining details to the reader.

For the remaining comment about homeomorphisms, we just note that in that case we may use Remark 4.5.8 and the unitary case of Lemma 4.4.4 to produce unitaries at every stage in the above process. □

Equivariant Covering Isometries for Coarse Maps

In the coarse setting, we will assume that X and Y are proper metric spaces as in Definition 4.3.1, equipped with proper isometric actions of a countable discrete group G. We will be interested in building covering isometries for coarse maps $f: X \to Y$ as in Definition 4.3.1 that are also equivariant.

Definition 4.5.13 Let H_X, H_Y be geometric modules, and let $f: X \to Y$ be an equivariant coarse map between G-spaces as above. An isometry $V: H_X \to H_Y$ is said to *equivariantly cover* f, or to be an *equivariant covering isometry* of f, if it is equivariant and if there is $r \in (0, \infty)$ such that $d_Y(y, f(x)) < r$ whenever $(y, x) \in \operatorname{supp}(V)$.

Proposition 4.5.14 *Let $f: X \to Y$ be a coarse map, and H_X, H_Y be geometric modules such that H_Y is ample. Then there is an equivariant covering isometry $V: H_X \to H_Y$ for f.*

Moreover, any equivariant covering isometry is properly supported, and if $V_f: H_X \to H_Y$ and $V_g: H_Y \to H_Z$ are equivariant covering isometries for $f: X \to Y$ and $g: Y \to Z$ respectively, then $V_g \circ V_f$ is an equivariant covering isometry for $g \circ f$.

Proof Using Lemma A.3.18, we may assume that f is Borel. The proofs of Proposition 4.3.4, Lemma 4.3.6 and Corollary 4.3.7 adapt directly, where we use Corollary 4.5.10 (as opposed to Corollary 4.2.7) as the basic ingredient for existence. □

We would also like an analogue of Proposition 4.3.5 in this context, i.e. a version that shows that any equivariant coarse equivalence can be covered by a unitary isomorphism. This is more technical and requires a somewhat different technique. The key point is the following structure lemma, which says that any ample X-G module is 'locally isomorphic' to $\ell^2(G) \otimes H$ in a controlled way.

Lemma 4.5.15 *Let H_X be an ample X-G module, let $E \subseteq X$ be a G-invariant Borel subset with non-empty interior and let W be an open subset of X such that $GW \supseteq \overline{E}$. Then there is a projection P on H_X with the following properties:*

(i) P has infinite rank;

(ii) $\chi_W P = P\chi_W = P$ (in other words, the image of P is a subspace of the image of χ_W);

(iii) the collection $(U_g P U_g^*)_{g\in G}$ consists of mutually orthogonal projections whose sum converges strongly to χ_E;

(iv) if the tensor product $\ell^2(G) \otimes PH_X$ is equipped with the G representation defined by

$$V_g \colon \delta_h \otimes u \mapsto \delta_{gh} \otimes u,$$

then the formula

$$U \colon \chi_E H_X \to \ell^2(G) \otimes PH_X, \quad u \mapsto \sum_{g\in G} \delta_g \otimes PU_g^* u$$

defines a unitary isomorphism[6] of G-representations.

Proof Note that the closure $\overline{E}$ of E is a proper metric space equipped with a proper isometric G-action. Apply Lemma A.2.9 to the open cover $\mathcal{U} := \{gW \mid g \in G\}$ of $\overline{E}$ to get a countable collection $(E_i)_{i\in I}$ of Borel subsets of $\overline{E}$ with the following properties:

(a) the collection $(GE_i)_{i\in I}$ is a disjoint cover of $\overline{E}$;

(b) for $i \neq j$, $GE_i \cap GE_j = \varnothing$;

(c) each E_i is contained in some set $g_i W$;

(d) each E_i is contained in the closure of its interior (for the induced topology on $\overline{E}$);

(e) for each i there is a finite subgroup $F_i \leqslant G$ such that E_i is invariant under F_i, and such that the function

$$G \times_{F_i} E_i \to GE_i, \quad [g,x] \mapsto gx$$

is an equivariant homeomorphism for the natural G-actions on each side (recall from Example A.2.6 that $G \times_{F_i} E_i$ is the quotient of $G_i \times E_i$ by the F_i action $f \cdot (g,x) := (gf^{-1}, fx)$).

[6] Recall this means that U is a unitary isomorphism such that $V_g U = UU_g$ for all $g \in G$: see Definition C.1.1.

Note that by replacing each E_i with the translate $g_i^{-1}E_i$, we may assume that E_i is contained in W; this does not alter the other properties of the cover.

Now, as H_X is locally free (see Definition 4.5.2) and $E \cap E_i$ is F_i invariant, we get Hilbert spaces H_i (possibly zero) equipped with the trivial F_i representation and unitary isomorphisms

$$\chi_{E\cap E_i} H_X \to \ell^2(F_i) \otimes H_i$$

of F_i representations. Extending these isomorphisms by zero on $\chi_{E\setminus E_i} H_X$ for each i gives a surjective partial isometry

$$V_i : \chi_E H_X \to \ell^2(F_i) \otimes H_i$$

that is equivariant for the F_i actions on both sides. Let now e be the identity element of G, and define

$$P_i := V_i^*(\chi_{\{e\}} \otimes 1_{H_i})V_i,$$

so each P_i is a projection on H_X with image a subspace of $\chi_{E\cap E_i} H_X$. For $i \neq j$ we have $E_i \cap E_j = \varnothing$, and so the projections P_i and P_j are mutually orthogonal. Hence the sum

$$P := \sum_{i\in I} P_i$$

converges strongly to a projection on H_X. We claim that P has the right properties.

Indeed, for property (i), note that E has non-empty interior E°. As the collection $(g(E \cap E_i))_{g\in G, i\in I}$ is a disjoint cover of E, one of the sets $E \cap E_i$ must intersect E°. As E_i is the closure of its interior, this implies that the interior of E_i intersects E°, so in particular that $E_i \cap E^\circ$ contains an open set. Using ampleness, it follows that $\chi_{E\cap E_i}$ has infinite rank, and thus that H_i is infinite-dimensional for this i. Hence P_i is infinite rank, and thus P is infinite rank as P_i is a subprojection of P.

Property (ii) follows as each $E \cap E_i$ is contained in W, whence we have

$$PH_X = \bigoplus_{i\in I} P_i H_X \subseteq \bigoplus_{i\in I} \chi_{E\cap E_i} H_X \subseteq \chi_W H_X.$$

We leave it to the reader to check that property (iv) is a direct consequence of property (iii), so it remains to prove property (iii).

For property (iii), we claim that for each i, the family $(U_g P_i U_g^*)_{g\in G}$ consists of mutually orthogonal projections that sum strongly to the identity on $\chi_{G(E\cap E_i)}$. This will imply property (iii): indeed, as the collection

$(G(E \cap E_i))_{i \in I}$ consists of disjoint sets, this implies that the collection $(U_g P_i U_g^*)_{i \in I, g \in G}$ also consists of orthogonal projections. Moreover, we have

$$\sum_{g \in G} U_g P U_g^* = \sum_{g \in G} \sum_{i \in I} U_g P_i U_g^* = \sum_{i \in I} \sum_{g \in G} U_g P_i U_g^* = \sum_{i \in I} \sum_{g \in G} \chi_{g(E \cap E_i)} = \chi_E,$$

where the last equality uses that the collection $(G(E \cap E_i))_{i \in I}$ is a cover of E.

It now remains to establish the claim. Note first that by F_i-equivariance of V_i, for each $h \in F_i$,

$$U_h P_i U_h^* = V_i^*(\chi_{\{h\}} \otimes 1_{H_i}) V_i.$$

As the projections $(\chi_{\{h\}} \otimes 1_{H_i})_{h \in F_i}$ are mutually orthogonal and sum to the identity on $\ell^2(F_i) \otimes H_i$, and as the restriction of V_i to a map $\chi_{E_i} H_X \to \ell^2(F_i) \otimes H_i$ is a unitary isomorphism, it follows that the projections $(U_h P_i U_h^*)_{h \in F_i}$ are mutually orthogonal and sum to $\chi_{E \cap E_i}$. On the other hand, let $S \subseteq G$ be a set of right coset representatives for G/F_i, i.e. so that $G = \bigsqcup_{g \in S} g F_i$. Using that the map

$$G \times_{F_i} (E \cap E_i) \to G(E \cap E_i), \quad [g,x] \mapsto gx$$

is an equivariant homeomorphism, we have that $G(E \cap E_i) = \bigsqcup_{s \in S} s(E \cap E_i)$, whence there is an orthogonal direct sum decomposition

$$\chi_{G(E \cap E_i)} H_X = \bigoplus_{s \in S} U_s \chi_{E \cap E_i} H_X.$$

Combined with our earlier observations that the collection $(U_h P_i U_h^*)_{h \in F_i}$ consists of mutually orthogonal projections that sum to $\chi_{E \cap E_i}$, this implies that the collection $(U_s U_h P_i U_h^* U_s^*)_{h \in F_i, s \in S}$ consist of mutually orthogonal projections that sum to $\chi_{G(E \cap E_i)}$. As each $g \in G$ can be uniquely represented as sh for $s \in S$ and $h \in F_i$, the collections $(U_s U_h P_i U_h^* U_s^*)_{h \in F_i, s \in S}$ and $(U_g P_i U_g^*)_{g \in G}$ are the same, so we are done. □

Proposition 4.5.16 *Let $f \colon X \to Y$ be an equivariant coarse equivalence, and H_X, H_Y be ample equivariant geometric modules. Then there is an equivariant covering isometry $V \colon H_X \to H_Y$ for f which is also a unitary isomorphism.*

Proof Using Lemma A.3.18, we may assume that f is Borel. Using Exercise A.4.3 and the fact that $f \colon X \to Y$ is a coarse equivalence there exists $c > 0$ such that for every $y \in Y$ is within c of some point in $f(X)$. Moreover, there exists $s > 0$ such that for all $x \in X$, the diameter of $f(B(x;1))$ is at most s. Let $r = c + s + 1$. Using Zorn's lemma, we see that there is a maximal subset Z of Y with the following property: for any distinct $x, y \in Z$, and any $g, h \in G$, $d(gx, hy) \geqslant 3r$. In other words, Z is maximal subject to the condition that the orbits of any two points in Z are $3r$-separated (we do not make any assumptions on how well-separated points are in the same orbit of an element of Z).

As Y is second countable, Z must be countable, so we may enumerate it as $z_1, z_2, \ldots$. Define

$$E_1 := \Big(\bigcup_{g\in G} B(gz_1;4r)\Big) \setminus \Big(\bigcup_{m\neq 1}\bigcup_{g\in G} B(gz_m;r)\Big)$$

and for $n \geqslant 1$, iteratively define

$$E_n := \Big(\bigcup_{g\in G} B(gz_n;4r)\Big) \setminus \Big(\bigcup_{m\neq n}\bigcup_{g\in G} B(gz_m;r) \cup \bigcup_{i=1}^{n-1} E_i\Big).$$

Then the collection $(E_n)_{n\in\mathbb{N}}$ consists of disjoint, Borel, G-invariant sets that cover Y. Moreover, each E_n contains $\bigcup_{g\in G} B(gz_n;r)$ by construction and so in particular has non-empty interior.

For the remainder of the proof, we write U_g for the unitary operators on H_Y inducing the action of G, and U_g^X for those on H_X.

Let now $W_n = B(gz_n;5r)$. Then each W_n is open, and GW_n contains $\overline{E_n}$. Hence using Lemma 4.5.15, there is an infinite rank projection P_n on H_Y such that $P_n = \chi_{W_n} P_n \chi_{W_n}$, so that $\sum_{g\in G} U_g P U_g^* = \chi_{E_n}$ and so that we have an isomorphism

$$U_n\colon \chi_{E_n} H_Y \to \ell^2(G) \otimes PH_Y, \quad u \mapsto \sum_{g\in G} \delta_g \otimes PU_g^* u.$$

For each n, consider now the pullback $f^{-1}(E_n)$, which is a G-invariant Borel subset of X. As any point of Y is within c of a point of $f(X)$, there exists $x \in X$ with $d(f(x), z_n) \leqslant c$. Hence by choice of s and r, we have that $f(B(x;1))$ is completely contained in $B(z_n;r)$, and so in E_n, whence $B(x;1)$ is contained in $f^{-1}(E_n)$. In particular $f^{-1}(E_n)$ has non-empty interior. Moreover, using that f is a coarse equivalence there exists $t \geqslant 0$ and for each n a point $x_n \in X$ such that $\overline{f^{-1}(W_n)} \subseteq B(x_n;t)$. Setting $W_n^X := B(x_n;t)$, we therefore have that W_n^X is an open set such that GW_n^X contains $\overline{f^{-1}(E_n)}$. Hence we may apply Lemma 4.5.15 to get an infinite rank projection P_n^X on H_X such that $P_n^X = \chi_{W_n^X} P_n^X \chi_{W_n^X}$, so that $\sum_{g\in G} U_g^X P_n^X (U_g^X)^* = \chi_{f^{-1}(E_n)}$ and so that we have a unitary isomorphism

$$U_n^X\colon \chi_{E_n} H_X \to \ell^2(G) \otimes P_n^X H_X, \quad u \mapsto \sum_{g\in G} \delta_g \otimes P_n^X (U_g^X)^* u,$$

which is also G-equivariant.

To complete the construction, choose a unitary isomorphism $V_{n,00}\colon P_n^X H_X \to P_n H_Y$ (this is possible as P_n and P_n^X have infinite rank), and define

$$V_{n,0}\colon \ell^2(G)\otimes P_n^X H_X \to \ell^2(G)\otimes P_n H_Y, \quad \delta_g \otimes u \mapsto \delta_g \otimes V_{n,00}u,$$

and

$$V_n\colon \chi_{f^{-1}(E_n)}H_X \to \chi_{E_n}H_Y, \quad V_n := U_n^* V_{n,0} U_n^X.$$

We claim that the unitary isomorphism

$$V := \bigoplus_{n\in\mathbb{N}} V_n\colon \underbrace{\bigoplus_{n\in\mathbb{N}} \chi_{f^{-1}(E_n)}H_Y}_{=H_X} \to \underbrace{\bigoplus_{n\in\mathbb{N}} \chi_{E_n}H_X}_{=H_Y}$$

has the right properties. Indeed, note first that each V_n is G-equivariant, as it is the composition $U_n^* V_{n,0} U_n^X$, and each of these maps is G-equivariant.

It remains to show that V is a covering isometry for f. For this we claim that

$$\operatorname{supp}(V) \subseteq \bigcup_{n\in\mathbb{N}}\bigcup_{g\in G} \overline{gW_n} \times \overline{gW_n^X}. \tag{4.6}$$

The claim will suffice to complete the proof. Indeed, say (y,x) is in $\operatorname{supp}(V)$. Then there is some $g\in G$ and $n\in\mathbb{N}$ such that $(y,x)\in \overline{gW_n}\times\overline{gW_n^X}$. Choose $w\in f^{-1}(W_n)$, so in particular w is contained in W_n^X by choice of W_n^X. We then have that

$$d(y,f(x)) \leqslant d(y,f(w)) + d(f(w),f(x)).$$

As $f(w)$ and y are in $\overline{gW_n} = \overline{B(gz_n;5r)}$, we have that $d(y,f(w))\leqslant 10r$; and as w and x are in $\overline{gW_n^X} = \overline{B(x_n;t)}$, we have that $d(f(w),f(x))\leqslant \omega_f(2t)$, where ω_f is the expansion function of Definition A.3.9. Putting this together, $d(y,f(x))\leqslant 10r+\omega_f(2t)$, so we are done modulo the claim.

To establish the claim in line (4.6), it suffices to show that for some fixed n, $\operatorname{supp}(V_n) \subseteq \bigcup_{g\in G}\overline{gW_n}\times\overline{gW_n^X}$. Assume then that (y,x) is in $\operatorname{supp}(V_n)$, so for arbitrary open neighbourhoods $W_x\ni x$ and $W_y\ni y$ we have that $\chi_{W_y}V_n\chi_{W_x}\neq 0$, or in other words that

$$\chi_{W_y}U_n^* V_{n,0} U_n^X \chi_{W_x} \neq 0.$$

As $\sum_{g\in G} U_g P_n U_g^* = \chi_{E_n}$, and as $\sum_{h\in G} U_h^X P_n^X (U_h^X)^* = \chi_{f^{-1}(E_n)}$, there must exist $g,h\in G$ such that

$$\chi_{W_y} U_g P_n U_g^* U_n^* V_{n,0} U_n^X U_h^X P_n^X (U_h^X)^* \chi_{W_x} \neq 0$$

and so

$$\chi_{W_y} U_n^* U_n U_g P_n U_g^* U_n^* V_{n,0} U_n^X U_h^X P_n^X (U_h^X)^* (U_n^X)^* U_n^X \chi_{W_x} \neq 0.$$

Write $p_g \in \mathcal{B}(\ell^2(G))$ for the rank one projection with range the span of δ_g. Then $U_n U_g P_n U_g^* U_n^* = p_g \otimes \mathrm{id}_{P_n H_Y}$ and $U_n^X U_h^X P_n^X (U_h^X)^* (U_n^X)^* = p_h \otimes \mathrm{id}_{P_n^X H_X}$. Hence from the previous displayed line, we get that

$$\chi_{W_y} U_n^* p_g p_h \otimes V_{n,00} U_n^X \chi_{W_x} \neq 0,$$

which is impossible unless $g = h$, so we now have that

$$\chi_{W_y} U_n^* p_g \otimes V_{n,00} U_n^X \chi_{W_x} \neq 0$$

for some $g \in G$. Going backwards through the same argument but with $g = h$, we get

$$\chi_{W_y} U_g P_n U_g^* U_n^* p_g \otimes V_{n,00} U_n^X U_g^X P_n^X (U_g^X)^* \chi_{W_x} \neq 0,$$

and so

$$\chi_{W_y} U_g P_n U_g^* \neq 0 \quad \text{and} \quad U_g^X P_n^X (U_g^X)^* \chi_{W_x} \neq 0.$$

As $U_g P_n U_g^* = \chi_{gW_n} U_g P_n U_g^*$ and $U_g P_n U_g^* = U_g P_n U_g^* \chi_{gW_n^X}$, this implies that

$$\chi_{W_y} \chi_{gW_n} \neq 0 \quad \text{and} \quad \chi_{gW_n^X} \chi_{W_y} \neq 0.$$

In other words, we have shown that for any open neighbourhoods of $W_x \ni x$ and $W_y \ni y$, there exists $g \in G$ such that the above inequalities hold. Using properness of the action, we may assume that there exists $g \in G$ and $\epsilon > 0$ such that for all open neighbourhoods $W_x \ni x$ and $W_y \ni y$ of diameter at most ϵ, we have

$$\chi_{W_y} \chi_{gW_n} \neq 0 \quad \text{and} \quad \chi_{gW_n^X} \chi_{W_y} \neq 0.$$

This implies that $x \in \overline{gW_n}$ and $y \in \overline{W_n^X}$, completing the claim. □

4.6 Exercises

4.6.1 Let H_X be an X module. The more usual definition of the support of a bounded operator T on H_X in the literature is as follows: the support of T is the complement of all those points (x, y) for which there exist $f, g \in C_0(X)$ for which $f(x) \neq 0$ and $g(y) \neq 0$ and for which $fTg = 0$. Show that this definition is equivalent to Definition 4.1.7 above.

4.6.2 Using the representation theory of commutative C^*-algebras (on separable Hilbert spaces), show that any separable X module identifies (as an X

module: this means there is a unitary isomorphism intertwining the representations) with a direct sum

$$\bigoplus_{n=1}^{\infty} L^2(X, \mu_n)$$

for some collection (μ_n) of Radon measures on X, equipped with the direct sum of the multiplication representations. Find explicit measures that have the above property for Example 4.1.6.

4.6.3 Let H_X be an X module.

(i) Show that if f is a bounded Borel function on X, then the support of the corresponding multiplication operator is contained in

$$\overline{\{(x,x) \in X \times X \mid f(x) \neq 0\}},$$

and that the support is exactly equal to this set if f is continuous and H_X is ample.

(ii) Similarly, if T is a bounded operator associated to a continuous kernel k as in Example 4.1.10, show that the support of T is contained in

$$\overline{\{(x,y) \in X \times X \mid k(x,y) \neq 0\}},$$

and that the support equals this set if H_X is ample.

4.6.4 Let μ be a Radon measure on X, and $H_X = L^2(X,\mu)$. Let $k\colon X \times X \to \mathbb{C}$ be a continuous function, and assume that there is $c > 0$ such that

$$\int_X |k(x,y)| d\mu(x) \leqslant c \quad \text{and} \quad \int_X |k(y,x)| d\mu(x) \leqslant c$$

for all $y \in X$. For $u \in C_c(X)$ define $Tu\colon X \to \mathbb{C}$ by the formula

$$(Tu)(x) := \int_X k(x,y) u(y) d\mu(y).$$

Show that T extends (uniquely) to a bounded operator on H_X.

4.6.5 Let H_X be an X module, and let T be a bounded operator on H_X. Show that $\operatorname{supp}(T)$ is contained in the diagonal $\{(x,x) \in X \times X \mid x \in X\}$ if and only if T commutes with $C_0(X)$.
Hint: the proof of Lemma 6.1.2 might help.

4.6.6 In Lemma 4.1.13, we showed that

$$\operatorname{supp}(TS) \subseteq \overline{\operatorname{supp}(T) \circ \operatorname{supp}(S)} \tag{4.7}$$

for operators T, S on geometric modules such that the above composition is defined, and moreover that taking the closure on the right is not necessary if T and S are properly supported. Show that the closure is necessary in general.

4.6.7 Show that if $X = [0,1]$, $Y = [0,1] \times [0,1]$, $f \colon X \to Y$ is the natural inclusion as $[0,1] \times \{0\}$ and H_X, H_Y are the usual Lebesgue spaces, then there is no isometry $V \colon H_X \to H_Y$ that $\mathcal{U}$-covers f for all open covers $\mathcal{U}$ of Y.

4.6.8 Show that if G acts freely on X, and H_X is an X-G module that is ample as an X module, then it is also ample as an X-G module.

4.6.9 Say G acts by isometries on a Riemannian manifold X of positive dimension with associated measure μ, and that the measure of the set

$$\{x \in X \mid \text{there exists } g \in G \backslash \{e\} \text{ such that } gx = x\}$$

is zero. Show that $L^2(X, \mu)$ is ample as an X-G module. Find 'reasonable' generalisations of this statement to other metric measure spaces.

4.7 Notes and References

The material in this chapter is largely folklore by now, although the details of our approach are somewhat different from those in the existing literature. The idea of using something like X modules as an abstract setting for operators associated to a space X goes back to at least as far as Atiyah's ideas about analytic models for K-homology (see Atiyah [7]). A detailed exposition of a similar idea in the measure-theoretic context can be found in chapter I of Moore and Schochet [186].

Early references for covering isometries are Higson [124] and Higson, Roe and Yu [139]. Chapter 5 of Higson and Roe [135] develops covering isometry ideas for representations of possibly noncommutative C^*-algebras; this requires quite a different approach based around Voiculescu's theorem from general C^*-algebra theory. The idea of using continuous families of covering isometries comes from Yu [270].

5

Roe Algebras

Roe algebras are C^*-algebras associated to metric spaces. The K-theory of Roe algebras provides a natural home for higher indices of elliptic operators, and is a central focus of this book. In this chapter, we introduce Roe algebras and discuss their basic functoriality properties: it turns out that Roe algebras are insensitive to the local topology of a space, seeing only the large-scale, or 'coarse', geometry. We also introduce equivariant Roe algebras, which take into account a group action on the space, and discuss their relationship with group C^*-algebras.

The chapter is structured as follows. In Section 5.1 we introduce Roe algebras and discuss their functoriality properties. In Section 5.2 we introduce group actions into the picture, and sketch the changes this necessitates. Finally, in Section 5.3, we relate equivariant Roe algebras to group C^*-algebras.

Throughout this chapter, X, Y denote proper[1] metric spaces. The metric on X will be denoted d, or d_X if we need to clarify which space it is associated to. See Section A.3 for more discussion and examples.

5.1 Roe Algebras

In this section, we introduce Roe algebras and study their functoriality properties.

Throughout the section, X, Y are proper metric spaces.

Definition 5.1.1 Let H_X be a geometric module (Definition 4.1.1), and let T be a bounded operator on H_X.

[1] Recall that a metric space is *proper* if all closed balls are compact.

(i) T is *locally compact* if for any compact subset K of X, we have that

$$\chi_K T \quad \text{and} \quad T\chi_K$$

are compact operators.

(ii) T has *finite propagation* if the extended real number

$$\text{prop}(T) := \sup\{\rho(y,x) \mid (y,x) \in \text{supp}(T)\} \in [0,\infty]$$

from Definition 4.1.8 is finite.

The following basic result is the special case of Lemma 4.3.6 where $X = Y$ and $f \colon X \to Y$ is the identity map.

Lemma 5.1.2 *A finite propagation operator on H_X is properly supported.* □

Corollary 5.1.3 *With notations as in Definition 5.1.1, the collection of all bounded locally compact operators on H_X is a C^*-algebra. The collection of all bounded finite propagation operators on H_X is a $*$-algebra.*

Proof The locally compact operators are a C^*-algebra as the compact operators are a closed ideal in $\mathcal{B}(H_X)$. The finite propagation operators are a $*$-algebra by Corollary 4.1.14. □

Lemma 5.1.3 implies that the following objects are $*$-algebras.

Definition 5.1.4 Let H_X be a geometric module. The *Roe $*$-algebra* of H_X, denoted $\mathbb{C}[H_X]$, is the $*$-algebra of all finite propagation, locally compact operators on H_X.

The *Roe C^*-algebra*, or just *Roe algebra*, of H_X, denoted $C^*(H_X)$, is the norm closure of $\mathbb{C}[H_X]$ in the bounded operators on H_X.

To get a bit more intuition for Roe algebras, the next examples give more concrete pictures in some motivating special cases.

Example 5.1.5 Let X be a complete Riemannian manifold; note that X is proper as a metric space by Theorem A.3.6. Let μ be the smooth measure defined by the Riemannian structure, and assume that for each $r > 0$ the extended real number

$$\mu(r) := \sup\{\mu(B(x;r)) \mid x \in X\}$$

is finite: this happens, for example, if the Ricci curvature is uniformly bounded below (see the discussion in Example A.3.21). Let $H_X = L^2(X,\mu)$ be the geometric module from Example 4.1.4.

Now, let $k: X \times X \to \mathbb{C}$ be a bounded smooth function such that the 'propagation'

$$\text{prop}(k) := \sup\{d(x, y) \mid k(x, y) \neq 0\}$$

is finite. As in Example 4.1.10, define an integral operator $T: C_c^\infty(X) \to C_c^\infty(X)$ by the formula

$$(Tu)(x) = \int_X k(x, y)u(y)d\mu(y)$$

(the image is in $C_c^\infty(X)$ as k is smooth and has finite propagation, and as X is proper). Our assumptions that $\mu(r)$ is always finite, and that k is bounded and finite propagation imply that T extends (uniquely) to a bounded operator on H_X: indeed, this follows from the result of Exercise 4.6.4. The propagation of T is equal to the propagation of k: compare Exercise 4.6.3. Moreover, if K is a compact subset of X, then the assumptions imply that the operators $\chi_K T$ and $T\chi_K$ are defined by integration against the kernels

$$(x, y) \mapsto \chi_K(x)k(x, y) \quad \text{and} \quad (x, y) \mapsto k(x, y)\chi_K(y)$$

respectively. These functions are bounded and compactly supported, whence L^2-integrable on $X \times X$. Hence the operators $\chi_K T$ and $T\chi_K$ are Hilbert-Schmidt, so in particular compact.

Putting this together, any such kernel operator defines an element of $C^*(H_X)$. With a bit more work, one can show that the collection of all such kernel operators is a dense $*$-subalgebra of $C^*(H_X)$: the reader is asked to do this in Exercise 5.4.3. The construction above can also be adapted to the case of sections of a vector bundle S over X: compare Examples 4.1.5 and 4.1.10.

Example 5.1.6 Let Z be a metric space with *bounded geometry*: as in Definition A.3.19, this means that for every $r \in (0, \infty)$ there exists $n(r) \in \mathbb{N}$ such that all balls of radius r in Z have cardinality bounded above by $n(r)$. Let $H_Z := \ell^2(Z)$ equipped with the multiplication action of $C_0(Z)$ as in Example 4.1.3; this is a Z-module, which is never ample. Then the algebra $\mathbb{C}[H_Z]$ consists precisely of all Z-by-Z indexed matrices $(T_{xy})_{x,y\in Z}$ such that:

(i) each T_{xy} is a complex number;
(ii) there exists $M > 0$ such that $\|T_{xy}\| \leqslant M$ for all $x, y \in X$;
(iii) there exists $r \in [0, \infty)$ such that if $d(x, y) > r$ then $T_{xy} = 0$.

The algebra operations are just the usual matrix operations: see Exercise 5.4.3. Note that without the bounded geometry assumption, it would still be true that operators in $\mathbb{C}[H_Z]$ identify with matrices satisfying conditions (i), (ii) and (iii) above; however, it would not necessarily be true that every such matrix defines an element of $\mathbb{C}[H_Z]$. Indeed determining when such a matrix give rise to an operator in $\mathbb{C}[H_Z]$ comes down to determining whether it defines a bounded operator, which may not be obvious.

In this case, the Roe algebra $C^*(H_Z)$ is usually denoted $C^*_u(Z)$ and called the *uniform Roe algebra of Z*.

Example 5.1.7 Say Z is a metric space with bounded geometry as in Example 5.1.6. Let H be a separable Hilbert space, and $H_Z = \ell^2(Z,H)$ be as in Example 4.1.3, equipped with the pointwise multiplication action of $C_0(Z)$. The algebra $\mathbb{C}[H_Z]$ can be characterised just as in Example 5.1.6 above, except now each matrix entry T_{xy} will be a compact operator on H.

For a general proper metric space X, assume that $Z \subseteq X$.is a net: as in Definition A.3.10 that this means that Z is a discrete subset of X such that for some $r \in (0,\infty)$, $d(z_1,z_2) \geqslant r$ for all $z_1,z_2 \in Z$, and that for all $x \in X$ there is $z \in Z$ with $d(x,z) < r$. Nets always exist, as shown in Lemma A.3.11. Moreover, the restriction of d to such a net Z is a proper metric. Assume moreover that Z has bounded geometry; the existence of such a Z is not automatic, but happens in many interesting cases as discussed in Examples A.3.20 and A.3.21. Then for any ample X module H_X, there is a (spatially implemented) isomorphism of $\mathbb{C}[H_X]$ with $\mathbb{C}[H_Z]$ where $H_Z = \ell^2(Z,H)$ as above for some infinite-dimensional H: see Exercise 5.4.1 (note that this isomorphism does not require bounded geometry).

One takeaway from this discussion is that many Roe $*$-algebras that arise in applications admit a simple 'matricial' description.

The next two lemmas follow directly from the definitions: we leave the proofs to the reader.

Lemma 5.1.8 *If X is bounded, then for any X module H_X, $C^*(H_X) = \mathcal{K}(H_X)$.* □

Lemma 5.1.9 *For any geometric module H_X, the canonical action of the bounded Borel functions $B(X)$ on H_X (see Proposition 1.6.11) makes $\mathbb{C}[H_X]$ and $C^*(H_X)$ into both left and right modules over $B(X)$.* □

In the remainder of this section we will study functoriality properties of Roe C^*-algebras. The Roe C^*-algebras themselves are not precisely functorial for any reasonable class of maps between metric spaces: this is due to the choices involved in their construction. However, this non-canonicality disappears on passage to K-theory: it turns out that the K-theory of Roe algebras is precisely functorial on the *coarse category*.

The coarse category is discussed in Section A.3: for the reader's convenience, we recall the definition here.

Definition 5.1.10 Let $f: X \to Y$ be any map. The *expansion function* of f, denoted $\omega_f: [0,\infty) \to [0,\infty]$, is defined by

$$\omega_f(r) := \sup\{d_Y(f(x_1), f(x_2) \mid d_X(x_1,x_2) \leqslant r\}.$$

The function f is *coarse* if:

(i) $\omega_f(r)$ is finite for all $r \geqslant 0$;
(ii) f is a *proper map*, meaning that for any compact subset K of Y, the pull-back $f^{-1}(K)$ has compact closure.

Two maps $f, g: X \to Y$ are *close* if there exists $c \geqslant 0$ such that for all $x \in X, d_Y(f(x), g(x)) \leqslant c$. Closeness is clearly an equivalence relation on the set of maps from X to Y. The *coarse category*, denoted $\mathcal{C}oa$, has proper metric spaces for objects, and morphisms are closeness classes of coarse maps.

For the reader's convenience, we also repeat Definition 4.3.3.

Definition 5.1.11 Let H_X, H_Y be geometric modules, and let $f: X \to Y$ be a coarse map. An isometry $V: H_X \to H_Y$ is said to *cover* f, or to be a *covering isometry* of f, if there is $t \in (0,\infty)$ such that $d(y, f(x)) < t$ whenever $(y,x) \in \operatorname{supp}(V)$.

Lemma 5.1.12 *Let H_X, H_Y be geometric modules, let $f: X \to Y$ be a coarse map and let $V: H_X \to H_Y$ be a covering isometry for f. The $*$-homomorphism*

$$ad_V : \mathcal{B}(H_X) \to \mathcal{B}(H_Y), \quad T \mapsto VTV^*$$

restricts to $*$-homomorphism from $C^*(H_X)$ to $C^*(H_Y)$ and from $\mathbb{C}[H_X]$ to $\mathbb{C}[H_Y]$.

Moreover, the map induced by ad_V on K-theory depends only on f and not on the choice of V.

Proof We first show that if $T \in \mathcal{B}(H_X)$ has finite propagation and is locally compact, then $\mathrm{ad}_V(T)$ has these properties too.

Assume first that $T \in \mathcal{B}(H_X)$ has finite propagation. Let $t > 0$ be as in Definition 5.1.11 so $d(f(x), y) < t$ whenever $(y,x) \in \operatorname{supp}(V)$. Using Lemma 4.1.13 part (ii), we can also conclude that $d(f(x), y) < t$ when $(x,y) \in \operatorname{supp}(V^*)$. Let (y_1, y_2) be an element of $\operatorname{supp}(VTV^*)$. Combining Lemma 4.3.6 with Lemma 4.1.13, part (iii′), we have that

$$\operatorname{supp}(VTV^*) \subseteq \operatorname{supp}(V) \circ \operatorname{supp}(T) \circ \operatorname{supp}(V^*).$$

Hence there exist $x_1, x_2 \in X$ such that $(y_1, x_1) \in \mathrm{supp}(V)$, $(x_1, x_2) \in \mathrm{supp}(T)$ and $(x_2, y_2) \in \mathrm{supp}(V^*)$. We thus have that $d(y_1, y_2)$ is bounded above by

$$d(y_1, f(x_1)) + d(f(x_1), f(x_2)) + d(f(x_2), y_2) \leqslant 2t + \omega_f(\mathrm{prop}(T)).$$

As this bound is independent of (y_1, y_2), this completes the proof of finite propagation.

Assume now that T is locally compact, let K be a compact subset of Y and write $F = \mathrm{supp}(V)$. Then

$$\chi_K V T V^* = \chi_K V \chi_{K \circ F} T V^*$$

by Lemma 4.1.15. Lemma 4.3.6 tells us that the coordinate projection $\pi_Y : F \to Y$ is a proper map, whence $K \circ F = \pi_X(\pi_Y^{-1}(K))$ is a compact set. Hence by local compactness of T, $\chi_{K \circ F} T$ is a compact operator, and thus $\chi_K V \chi_{K \circ F} T V^*$ is too as the compact operators form an ideal. The case of $V T V^* \chi_K$ is similar.

We have now shown that the $*$-homomorphism $\mathrm{ad}_V : \mathcal{B}(H_X) \to \mathcal{B}(H_Y)$ restricts to a $*$-homomorphism from $\mathbb{C}[H_X]$ to $\mathbb{C}[H_Y]$, whence also from $C^*(H_X)$ to $C^*(H_Y)$.

For the K-theoretic statement, let V_1 and V_2 be isometries that cover $f : X \to Y$. We must show that ad_{V_1} and ad_{V_2} induce the same map on K-theory. It suffices to prove that the $*$-homomorphisms from $C^*(H_X)$ to $M_2(C^*(H_Y))$ defined by

$$\alpha_1 : T \mapsto \begin{pmatrix} V_1 T V_1^* & 0 \\ 0 & 0 \end{pmatrix} \quad \text{and} \quad \alpha_2 : T \mapsto \begin{pmatrix} 0 & 0 \\ 0 & V_2 T V_2^* \end{pmatrix}$$

agree on the level of K-theory. Analogous arguments to the above show that the partial isometries $V_1 V_2^*$ and $V_2 V_1^*$, and projections $V_1 V_1^*$ and $V_2 V_2^*$, are multipliers of $C^*(H_Y)$, whence the operator

$$U = \begin{pmatrix} 1 - V_1 V_1^* & V_1 V_2^* \\ V_2 V_1^* & 1 - V_2 V_2^* \end{pmatrix}$$

is a (unitary) multiplier of $M_2(C^*(H_Y))$. In particular, conjugation by U induces the identity map on $K_*(C^*(H_Y))$ by Proposition 2.7.5. Note, however, that

$$U \alpha_1(T) U^* = \alpha_2(T),$$

which completes the proof. □

Remark 5.1.13 More generally, if $f : X \to Y$ is an isomorphism in *Coa*, and H_X and H_Y are ample modules, then Proposition 4.3.5 gives a unitary isomorphism $V: H_X \to H_Y$ that covers f. Hence ad_V gives $*$-isomorphisms between $\mathbb{C}[H_X]$ and $\mathbb{C}[H_Y]$ and between $C^*(H_X)$ and $C^*(H_Y)$ that 'model' the action of f. Moreover, the map induced on K-theory by ad_V depends only on f.

In particular, we may apply this to the identity map on X, and two ample X modules H_X and H'_X to get $*$-isomorphisms between $\mathbb{C}[H_X]$ any $\mathbb{C}[H'_X]$, and between $C^*(H_X)$ and $C^*(H'_X)$ arising from an isometry that covers the identity map. Thus Roe algebras associated to ample modules are all non-canonically isomorphic. Moreover, any two such isomorphisms induce the same map on K-theory so their K-theory groups are canonically isomorphic.

Proposition 4.3.4 says that covering isometries always exist as long as H_Y is ample. Therefore the following definition makes sense.

Definition 5.1.14 Let H_X, H_Y be geometric modules with H_Y ample and let $f: X \to Y$ be a coarse map. Define

$$f_*: K_*(C^*(H_X)) \to K_*(C^*(H_Y))$$

to be the map on K-theory induced by the $*$-homomorphism

$$\mathrm{ad}_V: C^*(H_X) \to C^*(H_Y)$$

associated to some covering isometry for f as in Lemma 5.1.12 above.

Theorem 5.1.15 *For each X in $\mathcal{Coa}$, choose*[2] *an ample X module H_X. Then the assignments*

$$X \mapsto K_*(C^*(H_X)), \quad f \mapsto f_*$$

give a well-defined functor from $\mathcal{Coa}$ to the category $\mathcal{GA}$ of graded abelian groups.

Moreover, the functor that one gets in this way does not depend on the choice of modules up to canonical equivalence.

Proof Proposition 4.3.4 and Lemma 5.1.12 together imply that the assignments

$$X \mapsto K_*(C^*(H_X)), \quad f \mapsto f_*$$

make sense and are well defined (given the choice of ample X module). To check that this defines a functor from $\mathcal{Coa}$ to $\mathcal{GA}$ it thus suffices to check the following facts: (i) if $f, g: X \to Y$ are close coarse maps, and V covers f,

[2] This is a dubious manoeuvre: the collection of objects of $\mathcal{Coa}$ is not a set! We leave it as an exercise to find a way to do this without getting into set-theoretic difficulties.

then V also covers g; (ii) any isometry covering the identity map induces the identity on K-theory; (iii) if $f\colon X \to Y$ and $g\colon Y \to Z$ are coarse maps, and V_f, V_g cover f, g respectively, then $V_g \circ V_f$ covers $g \circ f$.

Part (i) is straightforward from the definition of covering isometries. Part (ii) follows from Lemma 5.1.12 and the fact that the identity map from H_X to itself is an isometry covering the identity map from X to itself. Part (iii) is immediate from Corollary 4.3.7.

We must now check that the functors defined by any two choices $\{X \mapsto H_X\}$ and $\{X \mapsto H_X'\}$ of assignments of ample geometric modules are naturally equivalent. Indeed, Remark 5.1.13 implies that for each X there exists a unitary isomorphism $V_X\colon H_X \to H_X'$ covering the identity map. Moreover, Lemma 5.1.12 implies that there are maps on K-theory

$$\mathrm{ad}^*_{V_X}\colon K_*(C^*(H_X)) \to K_*(C^*(H_X'))$$

that do not depend on the choice of V_X; note moreover that as V_X is unitary, these maps are isomorphisms. Finally, note that if $f\colon X \to Y$ is a coarse map, covered by $V\colon H_X \to H_Y$ and $V'\colon H_X' \to H_Y'$ then the diagram

$$\begin{array}{ccc} K_*(C^*(H_X)) & \xrightarrow{\mathrm{ad}^*_V = f_*} & K_*(C^*(H_Y)) \\ {\scriptstyle \cong}\big\downarrow{\scriptstyle \mathrm{ad}^*_{V_X}} & & {\scriptstyle \cong}\big\downarrow{\scriptstyle \mathrm{ad}^*_{V_Y}} \\ K_*(C^*(H_X')) & \xrightarrow{\mathrm{ad}^*_{V'} = f_*} & K_*(C^*(H_Y')) \end{array}$$

commutes by the fact that both 'right-down' and 'down-right' compositions are induced by covering isometries for f and point (iii) from the first part of the proof. □

Convention 5.1.16 If H_X is an ample X module, we will usually write $\mathbb{C}[X]$ and $C^*(X)$ for $\mathbb{C}[H_X]$ and $C^*(H_X)$ respectively, and (abusively) refer to these as *the* Roe $*$-/C^*-algebra of X. Justification for this is provided by Remark 5.1.13, which implies that these algebras do not depend on the choice of ample module up to non-canonical isomorphism, and by Theorem 5.1.15, which implies that the K-theory groups of the C^*-algebras do not depend on the choice of modules up to canonical isomorphism.

There is, however, one important way in which the choice of ample module H_X does matter: the $B(X)$ module structure on $C^*(H_X)$ from Lemma 5.1.9 does depend on H_X. It will sometimes be useful to choose H_X so that this module structure has good properties.

We will also often want to consider Roe algebras of a countable discrete group G, where G is considered with a left-invariant bounded geometry metric as in Lemma A.3.13. We use the notations $\mathbb{C}[|G|]$ and $C^*(|G|)$ to denote the Roe $*$-algebras and Roe C^*-algebra associated to G, to avoid confusion with the group algebra $\mathbb{C}[G]$ and group C^*-algebra $C^*(G)$. The notation '$|G|$' refers to G considered as a metric space, without its group structure.

5.2 Equivariant Roe Algebras

In this section, we define equivariant Roe algebras, which also take into account a group action. Equivariant Roe algebras are a setting in which to do higher index theory in the presence of a group action.

For the reader's convenience, let us recall that an action of a discrete group G on a metric space X is *proper* if for any compact subset K of X the set $\{g \in G \mid gK \cap K \neq \varnothing\}$ is finite, and is *isometric* if for any $g \in G$ and $x, y \in X$ we have $d(x, y) = d(gx, gy)$. See Sections A.2 and A.3 for some more conventions and basic facts on group actions.

Throughout this section, X, Y are proper metric spaces as in Definition A.3.3 (recall this means that all closed balls are compact), and G is a countable discrete group acting on X and Y via a proper isometric action. We will assume that all geometric modules appearing in this section are equivariant (see Section 4.5); to avoid too much repetition, we will generally not repeat this.

Definition 5.2.1 Let H_X be a geometric module, and let $\mathbb{C}[H_X]$ be the associated Roe $*$-algebra. The *equivariant Roe $*$-algebra of H_X*, denoted $\mathbb{C}[H_X]^G$, is defined to be the $*$-subalgebra of $\mathbb{C}[H_X]$ consisting of operators T that commute with the group action, i.e. so that $U_g T = T U_g$ for all $g \in G$.

The *equivariant Roe C^*-algebra of H_X*, denoted $C^*(H_X)^G$, is the closure of $\mathbb{C}[H_X]^G$ in the operator norm.

Note that one can equivalently define $\mathbb{C}[H_X]^G$ to be the $*$-algebra of fixed points under the conjugation G action on $\mathbb{C}[H_X]$ defined by

$$T \mapsto U_g T U_g^* \tag{5.1}$$

(we leave it as an exercise to show that this formula does define a G action on $\mathbb{C}[H_X]$: this uses that G acts on X by isometric homeomorphisms). This description inspires the '$\cdot^G$' notation, which often means 'fixed points of $\cdot$'.

Example 5.2.2 The basic, and probably most important, example of an equivariant Roe algebra occurs when $X = G$ itself. Recall that we can equip G with a proper (meaning balls are finite in this case) left-invariant (meaning

$d(gx, gy) = d(x, y)$ for all $g, x, y \in G$) metric. Then G itself is a proper metric space, and the left action of G is by isometries. Let $H_X := \ell^2(G)$ equipped with the left translation of G. Let $\mathbb{C}[G]$ be the group algebra of G, and recall that the right regular representation of G on $\ell^2(G)$ is defined by

$$\rho_g : \delta_h \mapsto \delta_{hg^{-1}}$$

(see Example C.1.3). Then we get a map

$$\mathbb{C}[G] \to \mathbb{C}[H_X]^G, \quad g \mapsto \rho_g;$$

we leave it as an exercise for the reader to check that this is a well-defined $*$-isomorphism. Hence in this case $C^*(H_X)^G$ identifies with the (right) reduced group C^*-algebra $C^*_\rho(G)$ as in Definition C.1.7.

For readers familiar with group von Neumann algebras, it may aid intuition to note that $\mathbb{C}[H_X]^G$ is just the intersection of $\mathbb{C}[H_X]$ with the commutant of the left regular representation, or, in other words, with the von Neumann algebra generated by the right regular representation.

More generally, let H be a separable Hilbert space, and define $H_X := \ell^2(G, H) = \ell^2(G) \otimes H$, equipped with the left translation action of G again. This H_X is ample as an X-G module if H is infinite-dimensional. One checks that

$$\mathbb{C}[G] \odot \mathcal{K}(H) \to \mathbb{C}[H_X]^G, \quad g \otimes T \mapsto \rho_g \otimes T$$

is a $*$-isomorphism, and therefore that $C^*(H_X)^G \cong C^*_\rho(G) \otimes \mathcal{K}(H)$.

Remark 5.2.3 The notation '$C^*(H_X)^G$' suggests that one first completes, then takes the invariant part to define an equivariant Roe algebra; in fact, one first takes the invariant part, and then takes the completion. For many groups, the order 'take invariants, then complete' versus 'complete, then take invariants' does not matter: see Exercise 5.4.13 for an example. It seems plausible, however, that the order does matter in general, even for the basic case of Example 5.2.2 above. See Section 5.5, Notes and References, at the end of this chapter for a little more discussion of this.

Our next task is to discuss functoriality of equivariant Roe algebras. This requires only minor elaborations of our work in Section 5.1 (using material from Section 4.5 rather than Section 4.3 as appropriate), so we will just sketch the arguments.

Here is the category of spaces we will be working with.

Definition 5.2.4 Let G be a countable discrete group. Let $\mathcal{C}oa^G$ be the category with objects given by proper metric spaces, equipped with a proper action of G by isometries. Morphisms in $\mathcal{C}oa^G$ are closeness classes of equivariant coarse maps.

Now, quite analogously to Lemma 5.1.12, one sees that if $V: H_X \to H_Y$ is any isometry equivariantly covering $f: X \to Y$ in the sense of Definition 4.5.13, then ad_V defines $*$-homomorphisms

$$\mathrm{ad}_V: \mathbb{C}[H_X]^G \to \mathbb{C}[H_Y]^G \quad \text{and} \quad \mathrm{ad}_V: C^*(H_X)^G \to C^*(H_Y)^G.$$

Moreover, the map on K-theory induced by $\mathrm{ad}_V: C^*(H_X)^G \to C^*(H_Y)^G$ depends only on f, and not on any of the choices involved in the construction: again, the same arguments used for Lemma 5.1.12 go through. We may thus make the following definition.

Definition 5.2.5 Let $f: X \to Y$ be an equivariant coarse function and let $C^*(H_X)^G$ and $C^*(H_Y)^G$ be equivariant Roe algebras associated to ample G-geometric modules. Define

$$f_*: K_*(C^*(H_X)^G) \to K_*(C^*(H_Y)^G)$$

to be the map on K-theory induced by the $*$-homomorphism

$$\mathrm{ad}_V: C^*(H_X)^G \to C^*(H_Y)^G$$

associated to some equivariant cover for f as in Lemma 5.1.12 above.

The following theorem records the basic functoriality properties of $C^*(H_X)^G$. The proof is essentially the same as that of Theorem 5.1.15 (using the results of Proposition 4.5.14 in place of the non-equivariant versions), and is thus left to the reader.

Theorem 5.2.6 *For each X in $\mathcal{C}oa^G$ choose*[3] *an ample X-G module H_X. Then the assignments*

$$X \mapsto K_*(C^*(H_X)^G), \quad f \mapsto f_*$$

give a well-defined functor from $\mathcal{C}oa^G$ to the category $\mathcal{GA}$ of graded abelian groups.

Moreover, the functor that one gets in this way does not depend on the choice of modules up to canonical equivalence. □

Just as for the non-equivariant Roe algebras, we will often abuse notation and terminology, writing $C^*(X)^G$ for $C^*(H_X)^G$ when H_X is ample and speaking of 'the' equivariant Roe algebra of X.

[3] Again, this can be done in such a way as to avoid set-theoretic difficulties: exercise.

5.3 Relationship to Group C^*-algebras

In this section we will discuss the relationship between equivariant Roe algebras and group C^*-algebras, substantially generalising Example 5.2.2.

Throughout this section, X, Y are proper metric spaces as in Definition A.3.3, and G is a countable discrete group acting on X and Y via a proper isometric action. We will assume that all geometric modules appearing in this section are equivariant (see Section 4.5); to avoid too much repetition, we will generally not repeat this.

We start with a definition.

Definition 5.3.1 The action of G on X is *cobounded* if there is a bounded subset B of X such that $GB = X$.

For the statement of the next theorem, recall from Definition C.1.7 that $C^*_\rho(G)$ denotes the reduced group C^*-algebra of G defined by the right regular representation. We let $\mathcal{K}$ denote an abstract copy of the compact operators.

Theorem 5.3.2 *Say the action of G on X is cobounded. Then there is a canonical family of* $*$*-isomorphisms*

$$C^*_\rho(G) \otimes \mathcal{K} \to C^*(X)^G$$

that all induce the same isomorphism

$$\phi_X : K_*(C^*(X)^G) \to K_*(C^*_\rho(G))$$

on K-theory.

Moreover, if $f: X \to Y$ *is an equivariant coarse equivalence, then the diagram*

$$\begin{array}{ccc} K_*(C^*(X)^G) & \xrightarrow{\phi_X} & K_*(C^*_\rho(G)) \\ \downarrow{\scriptstyle f_*} & & \| \\ K_*(C^*(Y)^G) & \xrightarrow{\phi_Y} & K_*(C^*_\rho(G)) \end{array}$$

commutes.

Proof From the Svarc-Milnor Lemma (Lemma A.3.14), under our coboundedness assumption any orbit inclusion

$$G \to X, \quad g \mapsto gx$$

is an equivariant coarse equivalence. Hence Proposition 4.5.16 gives us an equivariant unitary isomorphism of the modules underlying $C^*(X)^G$ and $C^*(|G|)$ that covers this orbit inclusion, giving a $*$-isomorphism

$$C^*(|G|) \to C^*(X)^G.$$

Moreover, coboundedness implies that any two orbit inclusions $G \to X$ are close, and thus the maps induced on K-theory by any of the $*$-isomorphisms above are the same by Theorem 5.2.6. Using Example 5.2.2, the left-hand side $C^*(|G|)$ identifies canonically with $K_*(C^*_\rho(G) \otimes \mathcal{K})$, which identifies with $K_*(C^*_\rho(G))$ using stability of K-theory, so we are done with the first part.

The second part is an immediate consequence of the description of the isomorphisms given in the first part, and of Theorem 5.2.6. □

For certain applications we need the isomorphism of Theorem 5.3.2 to be more concrete. In the remainder of this section, we thus give a concrete isomorphism that works in a special case. This might also aid intuition a little.

Definition 5.3.3 A *fundamental domain* for X is a Borel subset $D \subseteq X$ such that X is the disjoint union $X = \bigsqcup_{g \in G} gD$.

See Appendix C for the representation-theoretic terminology in the next result.

Proposition 5.3.4 *Assume that there is a bounded fundamental domain D for the action of G on X. Let H_X be an ample X-G module, and let $H_D = \chi_D H_X$. Let the algebraic tensor product $\mathbb{C}[G] \odot \mathcal{K}(H_D)$ be represented on $\ell^2(G) \otimes H_D$ by the tensor product of the right regular and trivial representations. Then the map*

$$U\colon H_X \to \ell^2(G) \otimes H_D, \quad u \mapsto \sum_{g \in G} \delta_g \otimes \chi_D U_g^* u$$

is a well-defined unitary isomorphism, which is equivariant when $\ell^2(G) \otimes H_D$ is equipped with the tensor product of the left regular and trivial representations. Moreover, conjugation by U induces $*$-isomorphisms

$$ad_U\colon \mathbb{C}[H_X]^G \to \mathbb{C}[G] \odot \mathcal{K}(H_D) \quad \text{and}$$

$$ad_U\colon C^*(H_X)^G \to C^*_\rho(G) \otimes \mathcal{K}(H_D).$$

Proof Note that the formula for U at least makes sense when u has bounded support, as then the sum defining Uu has only finitely many non-zero terms. One computes directly that for U as given,

$$\langle Uu, Uu\rangle = \sum_{g \in G} \langle \chi_D U_g^* u, \chi_D U_g^* u\rangle = \left\langle \sum_{g \in G} \chi_{gD} u, u \right\rangle.$$

As $X = \bigsqcup_{g\in G} gD$, the sum $\sum_{g\in D} \chi_{gD}$ converges in the strong operator topology to the identity, and so this equals $\langle u,u\rangle$. Thus the formula for U gives a well-defined isometry on all of H_X. Moreover, one computes that

$$U^*(\delta_g \otimes v) = U_g v,$$

and from this that U is unitary as claimed. The computation

$$\begin{aligned}\lambda_g U u &= \sum_{h\in G} \delta_{gh} \otimes \chi_D U_h^* u = \sum_{k\in G} \delta_k \otimes \chi_D U_{g^{-1}k}^* u = \sum_{k\in G} \delta_k \otimes \chi_D U_k^* U_g u \\ &= U U_g u\end{aligned}$$

(using the change of variables $k = gh$ for the second equality) shows that U is equivariant.

Now, say T is in $\mathbb{C}[X]^G$. Then, using the formula above for U^*, we compute that

$$UTU^* = \sum_{g\in G} \rho_g \otimes \chi_D T U_g \chi_D.$$

For each $g \in G$, we have that

$$\chi_D T U_g \chi_D = \chi_D T \chi_{gD} U_g.$$

This equals zero unless g is in the set $\{g \in G \mid d(D, gD) \leqslant \text{prop}(T)\}$, which is a finite set by properness of the G-action, and properness of the metric on X. Moreover, as T is locally compact, all the terms $\chi_D T U_g \chi_D$ are compact operators. Hence UTU^* is in $\mathbb{C}[G] \odot \mathcal{K}(H_D)$.

Conversely, if $S = \rho_g \otimes T$ is an elementary tensor in $\mathbb{C}[G] \otimes \mathcal{K}(H_D)$ then we compute that

$$U^* S U = \sum_{h\in G} U_h U_g^* T U_h^*.$$

This is clearly G-invariant. To see that it is locally compact, note that if $K \subseteq X$ is compact, then properness of the action combined with the fact that $\chi_D T \chi_D = T$ implies that only finitely many terms in the sums making up $\chi_K USU^*$ and $USU^* \chi_K$ are non-zero; as each term in the sum is compact, this implies that USU^* is compact. Finally, to see finite propagation, say that (x,y) is in the support of USU^*. We may assume using G-invariance of S and the fact that $X = \bigsqcup_{g\in G} gD$ that x is an element of D. Then for any open sets $W_x \ni x$ and $W_y \ni y$, we have $\chi_{W_x} USU^* \chi_{W_y}$ is non-zero, and so

$$0 \neq \chi_{W_x}\Big(\sum_{h\in G} U_h U_g^* T U_h^*\Big)\chi_{W_y} = \sum_{h\in G} U_h U_g^* \chi_{gh^{-1}W_x} T \chi_{h^{-1}W_y} U_h^*.$$

Hence there is some $h \in G$ for which $(gh^{-1}x, h^{-1}y)$ is in the support of T, which implies that $gh^{-1}x$ and $h^{-1}y$ are in the closure of D. As x is in D and G acts properly, there is only a finite subset F of G (independent of x and y, although possibly depending on g) for which $gh^{-1}x$ can be in the closure of D. Hence

$$d(x,y) \leqslant \sup_{h \in F, x \in D} d(x, hgh^{-1}x) + d(hgh^{-1}x, y) \leqslant M + \text{prop}(T)$$

for some absolute bound M, independent of the choice of $x \in D$ and $y \in X$ (although possibly depending on g).

To summarise, we now have that U is an equivariant unitary isomorphism, and that

$$U\mathbb{C}[X]^G U^* \subseteq \mathbb{C}[G] \odot \mathcal{K}(H_D) \quad \text{and} \quad U^*\big(\mathbb{C}[G] \odot \mathcal{K}(H_D)\big)U \subseteq \mathbb{C}[X]^G.$$

This suffices to establish the statement. □

5.4 Exercises

5.4.1 Let X be a proper metric space and Z a net in X (see Definition A.3.10). Let H be a separable infinite-dimensional separable Hilbert space and $H_Z = \ell^2(Z, H)$, equipped with the Z module structure coming from multiplication. Show that if H_X is any ample X module, then there is a unitary isomorphism $U \colon H_Z \to H_X$ such that $U(\mathbb{C}[H_Z])U^* = \mathbb{C}[H_X]$, and similarly on the level of completions.
Note: you can do this by appealing to Proposition 4.3.5, but it is possible to give a slightly simpler proof directly.

5.4.2 Let X be a proper metric space. Show that $C^*(X)$ admits a directed approximate unit of projections (and thus that the K_0 group of $C^*(X)$ can be described as in Corollary 2.7.4).
Hint: use an isomorphism $C^*(X) \cong C^*(Z)$ as in Exercise 5.4.1. Note that $C^*(Z)$ contains $\ell^\infty(Z, \mathcal{K}(H))$ acting on $\ell^2(Z, H)$ in the natural way, and show that the collection of projection-valued functions in $\ell^\infty(Z, \mathcal{K}(H))$ is a (non-sequential) approximate unit for $C^*(Z)$ made up of projections.

5.4.3 With the set-up in Examples 5.1.5 and 5.1.7, show that the operators T satisfying the conditions there are bounded, finite propagation and locally compact (this is significantly easier in the case of Example 5.1.7, but the ideas in both cases are similar).

Show, moreover, that the collection of all kernel operators as in Example 5.1.7 form a dense $*$-subalgebra of $C^*(H_X)$.

5.4.4 Let X be a locally compact, second countable, Hausdorff space, that is equipped with a coarse structure in the sense of Remark A.3.7. For an X module H_X define the associated Roe $*$-algebra $\mathbb{C}[H_X]$ to consist of all locally compact bounded operators T on H_X such that the support of T (see Definition 4.1.7) is a controlled set for the coarse structure. Show that $\mathbb{C}[H_X]$ is a $*$-algebra and that this construction generalises Definition 5.1.4. Also develop functoriality properties for these $*$-algebras (this includes either making up, or looking up, the correct definition of 'coarse map' in this context).

5.4.5 Let X be a proper metric space, and assume that any non-empty open subset of X is infinite. Let $\ell^2(X)$ be the collection of all square-summable functions from X to $\mathbb{C}$. This Hilbert space satisfies all the conditions to be an ample X module, except (if X is uncountable) it is not separable. Show that nonetheless the Roe algebra defined on this space has the same K-theory as 'the' usual version.

5.4.6 Let X be a bounded geometry metric space (see Definition A.3.19). Let $\mathbb{C}_u[X]$ denote the collection of all bounded kernels $k\colon X \times X \to \mathbb{C}$ with support in a set of the form $\{(x, y) \in X \times X \mid d(x, y) < r\}$ for some $r \in (0, \infty)$ (that is allowed to depend on k). Show that $\mathbb{C}_u[X]$ is $*$-algebra for the natural 'matrix operations'. For $k \in \mathbb{C}_u[X]$ provisionally define an operator T_k on $\ell^2(X)$ by

$$T_k\colon \delta_x \mapsto \sum_{y \in X} k(y, x)\delta_y.$$

(i) Show that T_k is a bounded operator, and that $k \mapsto T_k$ defines a faithful $*$-representation of $\mathbb{C}_u[X]$.

The *uniform Roe algebra*, denoted $C_u^*(X)$, is defined to be the completion of $\mathbb{C}_u[X]$ under the norm inherited from this $*$-representation. Let now H be an infinite-dimensional separable Hilbert space, and let H_X be the ample X-module $\ell^2(X) \otimes H$ (equipped with the pointwise multiplication action of $C_0(X)$).

(ii) Show that if $\mathcal{K}(H)$ denotes the compact operators on H, then the natural representation of the spatial tensor product $C_u^*(X) \otimes \mathcal{K}(H)$ on H_X induces an embedding $C_u^*(X) \otimes \mathcal{K}(H) \to C^*(H_X)$, but that this embedding is not surjective if X is infinite.

5.4.7 Lemma 5.1.8 states that if X is a bounded space (so in particular compact) and H_X is ample, then $C^*(H_X) = \mathcal{K}(H_X)$. What happens if X is not assumed bounded, but 'only' compact (recall that our metrics are allowed to take infinite values)?

5.4.8 (i) Show that $C^*(X)$ is not separable if X is not bounded.

(ii) A C^*-algebra is σ-unital if it has a countable approximate unit. Show that $C^*(X)$ is not σ-unital for unbounded X (this is stronger than non-separability).

Hint: with Z as in Exercise 5.4.1, show that a countable approximate unit for $C^*(X)$ would give rise to a countable approximate unit for $\ell^\infty(Z,\mathcal{K})$ by 'restriction to the diagonal'. Use this to derive a contradiction.

(iii) If you know what it means for a C^*-algebra to be exact, show that $C^*(X)$ is also not exact when X is unbounded.

Hint: it suffices to show that $C^*(X)$ contains a non-exact C^*-algebra. With Z as in Exercise 5.4.1, $\ell^\infty(Z,\mathcal{K})$ works.

5.4.9 Let H_X be an ample module for a locally compact, second countable space X. Let $LC^*(X)$ denote the C^*-algebra of all locally compact operators on H_X (see Definition 5.1.1 and Lemma 5.1.3). Show that if X is non-compact, then $K_*(LC^*(X)) = 0$.

Hint: Show that for any non-compact X, $LC^*(X) \cong LC^*(\mathbb{N})$, the latter being defined using the module $\ell^2(\mathbb{N}) \otimes H$ where H is a separable infinite-dimensional Hilbert space. Now do an Eilenberg swindle.

5.4.10 Let X be a proper metric space, and assume that there exists a bounded geometry net $Z \subseteq X$ (see Definitions A.3.10 and A.3.19).

(i) Show that there is a Borel cover $(E_x)_{x\in Z}$ of X and $S > 0$ with the properties in Lemma A.1.10, such that $\mathrm{diam}(E_x) \leqslant S$ for all x and such that each E_x contains x.

(ii) Let $r, S > 0$. Then there exists a constant $c = c(r,S,Z)$ with the following property. For any Borel cover $(E_x)_{x\in Z}$ as above, any X-module H_X and any bounded operator T on H_X, define $T_{xy} = \chi_{E_x} T \chi_{E_y}$. Then if $\mathrm{prop}(T) \leqslant r$, one has

$$\|T\| \leqslant c_r \sup_{x,y\in Z} \|T_{xy}\|.$$

5.4.11 (This requires some background in crossed products). Let G be a discrete group and H be a separable infinite-dimensional Hilbert space. Let H_G be the ample equivariant G-module $\ell^2(G,H)$. Represent the Roe algebra $C^*(|G|)$ on $\ell^2(G,H) \otimes \ell^2(G)$ by the formula

$$T(u \otimes v) = Tu \otimes v$$

(in other words, this is just the amplification of the representation of $C^*(|G|)$ on $\ell^2(G,H)$). By adapting 'Fell's trick' (see Proposition C.2.1) show that there is a unitary U on $\ell^2(G,H) \otimes \ell^2(G)$ such that

$$UC^*(|G|)U^* = \ell^\infty(G,\mathcal{K}(H)) \rtimes_r G.$$

5.4.12 The *Pimsner–Voiculescu theorem* in C^*-algebra K-theory gives an exact sequence

$$\begin{array}{ccccc} K_0(\ell^\infty(\mathbb{Z},\mathcal{K})) & \xrightarrow{1-\alpha} & K_0(\ell^\infty(\mathbb{Z},\mathcal{K})) & \longrightarrow & K_0(\ell^\infty(\mathbb{Z},\mathcal{K}) \rtimes_r \mathbb{Z}), \\ \uparrow & & & & \downarrow \\ K_1(\ell^\infty(\mathbb{Z},\mathcal{K}) \rtimes_r \mathbb{Z}) & \longleftarrow & K_1(\ell^\infty(\mathbb{Z},\mathcal{K})) & \xleftarrow[1-\alpha]{} & K_1(\ell^\infty(\mathbb{Z},\mathcal{K})) \end{array}$$

where $\alpha\colon K_*(\ell^\infty(\mathbb{Z},\mathcal{K})) \to K_*(\ell^\infty(\mathbb{Z},\mathcal{K}))$ is the map on K-theory induced by shifting one unit right on $\mathbb{Z}$. Use this and the result of the previous exercise to compute the groups $K_*(C^*(|\mathbb{Z}|))$ (and thus also to compute $K_*(C^*(|\mathbb{R}|))$). You do not need to understand what a crossed product is to do this exercise. *Hint: show first directly that the K-theory groups of $l^\infty(\mathbb{Z},\mathcal{K})$ are $\mathbb{Z}^{\mathbb{Z}}$ (the abelian group of all maps from $\mathbb{Z}$ to $\mathbb{Z}$) in dimension zero, and 0 in dimension one. Be warned that $l^\infty(\mathbb{Z},\mathcal{K})$ is not isomorphic to $\ell^\infty(\mathbb{Z}) \otimes \mathcal{K}$, and the two do not have the same K-theory; this is related to the discussion of Exercise 5.4.6*

5.4.13 (This requires some background in amenability and its relation to approximation properties of C^*-algebras). Let G be an amenable group. Use Schur multipliers constructed from a Følner sequence to show that the G-invariant part of $C^*(|G|)$ identifies with the completion of $\mathbb{C}[|G|]^G$.

5.4.14 Let G be a finite group, and let $\hat{G}$ be the (finite) set of equivalence classes of irreducible unitary representations of G, and for each $[\pi] \in \hat{G}$, let $\dim(\pi)$ denote its dimension. It follows from elementary representation theory that the reduced group C^*-algebra of G satisfies

$$C^*_\rho(G) \cong \bigoplus_{\pi \in \hat{G}} M_{\dim(\pi)}(\mathbb{C}).$$

Let X be a compact bounded space with a G-action. Show that the K-theory of the equivariant Roe algebra $C^*(X)^G$ is given by $\mathbb{Z}^{\hat{G}}$ in dimension zero and 0 in dimension one.

5.5 Notes and References

A version of the Roe algebra was introduced by Roe [212, 213] in order to do index theory on non-compact manifolds. The theory has since been extensively developed by several authors: some general references include chapter 6 of

Higson and Roe [135], Roe [216] and chapter 4 of Roe [218]. The use of covering isometries to get functoriality goes back at least as far as Higson, Roe and Yu [139]. The material relating equivariant Roe algebras and group C^*-algebras goes back at least as far as the approach to equivariant assembly in chapter 5 of Roe [216].

The technical functional analytic issue mentioned in Remark 5.2.3 is closely related to the *invariant translation approximation property* as discussed in section 11.5.3 of Roe in [218]; in particular, this reference is the source of Exercise 5.4.13. It follows from work of Zacharias [274] that this property is in turn closely related to the AP of Haagerup and Kraus [119], which is known to fail for some groups: for example, Lafforgue and de la Salle [162] showed the AP fails for $SL(3,\mathbb{Z})$. From these results, it seems likely that 'order matters' in the situation of Remark 5.2.3. However, this is not a proof, and it remains an interesting open question whether this is really the case (and even more so whether or not any difference can be detected on the level of K-theory).

The uniform Roe algebra $C_u^*(X)$ of Example 5.1.6 and Exercise 5.4.6 has been extensively studied, partly as some of its C^*-algebraic properties closely mirror the coarse geometric properties of X. See section 5.5 of Brown and Ozawa [44] for a nice discussion of some aspects of this. The algebraic version $\mathbb{C}_u[X]$, sometimes called the *translation algebra* of X, is also quite well-studied: see for example page 262 of Gromov [110] for an early reference, and chapter 4 of Roe [218] for a discussion of some aspects.

Both the Roe algebra and uniform Roe algebra remember a lot of the structure of the underlying metric space, at least in good situations: see for example Braga and Farah [37], Špakula and Willett [244] and White and Willett [256]. This is in some ways quite surprising, as related geometrically or dynamically defined C^*-algebras typically remember very little about the underlying object: see for example Li and Renault [167]. Other results about how coarse geometry of the underlying space is reflected in the structure of the uniform Roe algebra can be found for example in Ara et al. [3], Chen and Wang [53], Chen and Wei [55], Li and Willett [166], Wei [249] and Winter and Zacharias [261].

The result of Exercise 5.4.11 was first observed by Higson and Yu: see for example Lemma 2.4 in Yu [268] or Proposition 5.1.3 in Brown and Ozawa [44]. The Pimsner-Voiculescu theorem of Exercise 5.4.12 was first proved in Pimsner and Voiculescu [207], and is a widely-used tool in C^*-algebra K-theory. Further discussion of the theorem and its generalisations and applications can be found in chapter 10 of Blackadar [33] and chapter 5 of Cuntz, Meyer and Rosenberg [71].

6

Localisation Algebras and K-homology

K-homology for spaces is the dual homology theory to K-theory. The existence of K-homology follows from abstract nonsense. However, for applications, it is useful to have a more concrete construction of the theory. Our goal in this chapter is to construct an analytic model for K-homology. Our approach will be to define the K-homology groups of a space X to be the K-theory groups of an associated C^*-algebra $L^*(X)$, the *localisation algebra* of X.

These localisation algebras might initially look intimidating, and indeed more complicated than the Roe algebras of Chapter 5. This is not the case, however: at least on the level of K-theory, they are much friendlier, more computable objects. Indeed, one has many of the usual tools of algebraic topology, such as Mayer–Vietoris sequences, available to aid in computation.

The approach to K-homology via localisation algebras is originally due to Yu. Our approach here is closely based on Yu's original one, but is not quite the same: this is to ensure that our localisation algebras have better functoriality properties than the original version, and that they are more closely related to elliptic differential operators.

This chapter is structured as follows. In Section 6.1 we give motivation for the definition of the localisation algebra[1] using a precise relationship between small propagation and small commutators: this turns out to be fundamental for many topics discussed later in the book. In Section 6.2 we introduce the localisation algebras themselves based on this motivation, and prove the functoriality properties that we will need to set up K-homology.

[1] The second, very important, motivation for our definition, is that differential operators on manifolds naturally give rise to elements of localisation algebras; we will not discuss that until Chapter 8, however.

The next two sections are the core of the chapter. Section 6.3 proves some of the homological properties of K-homology, like the existence of Mayer–Vietoris sequences, and computes the K-homology groups of the empty set and a point. Section 6.4 then introduces a slightly less concrete version of the localisation algebra in order to prove functoriality in the 'correct' setting, and finish off the establishment of all the homological properties that K-homology 'should' have (compare Section B.2) such as homotopy invariance.

The last three sections are a collection of further topics. Section 6.5 sketches the changes to the definitions one has to make in the presence of a group action, and relates the equivariant theory to the non-equivariant theory of quotient spaces. Section 6.6 gives a variant of the localisation algebra whose K-theory is also a model for K-homology. This is not really used in this chapter, but will be important in Chapter 7 when we come to discuss assembly maps and the Baum–Connes conjecture. Finally, Section 6.7 sketches the relationship to some other analytic models of K-homology in the literature.

Throughout this chapter, X, Y denote locally compact, second countable Hausdorff spaces. Such spaces are metrisable, and we will assume they have a metric when convenient.

6.1 Asymptotically Commuting Families

In this section, we motivate our definition of K-homology. We start with a discussion for a general (unital) C^*-algebra, and then show that the general picture specialises to something geometrically meaningful in the commutative case.

Working in general, let $A \subseteq \mathcal{B}(H)$ be a concrete C^*-algebra, for simplicity containing the unit of $\mathcal{B}(H)$. Let $(P_t)_{t\in[1,\infty)}$ be a family of projections in $\mathcal{B}(H)$ parametrised by $[1,\infty)$ such that:

(i) the map $t \mapsto P_t$ is norm continuous;
(ii) for all t, P_t is a compact operator;
(iii) for each $a \in A$, the commutators $[P_t, a]$ tend to zero as t tends to infinity.

Let $q \in A$ be a projection. Then

$$(P_t q)^2 = P_t q P_t q = P_t^2 q^2 + P_t[q, P_t]q = P_t q + P_t[q, P_t]q,$$

and thus

$$\|(P_t q)^2 - P_t q\| \to 0 \quad \text{as} \quad t \to \infty.$$

Let $\chi : \mathbb{C} \to \mathbb{C}$ be the characteristic function of the subset $\{z \in \mathbb{Z} \mid \mathrm{Re}(z) > 1/2\}$ of $\mathbb{C}$. Then for all suitably large t, χ is holomorphic on an open neighbourhood of the spectrum of the compact operator $P_t q$, and thus by the holomorphic functional calculus (see Theorem 1.4.6) $\chi(P_t q)$ is a well-defined, compact idempotent on H. Moreover, the map $t \mapsto \chi(P_t q)$ is norm continuous by Theorem 1.4.6. Hence the K-theory class

$$[\chi(P_t q)] \in K_0(\mathcal{K}) = \mathbb{Z}$$

is well defined for all large t, and does not depend on the specific choice of t.

One can elaborate on this idea to take into account formal differences of projections in matrix algebras over A, and thus see that (P_t) defines a map

$$P_* : K_0(A) \to \mathbb{Z}, \quad [q_1] - [q_2] \mapsto [\chi(P_t q_1)] - [\chi(P_t q_2)].$$

Thus (P_t) defines a 'functional' on K-theory. These computations suggest trying to build a model for K-homology based on such 'asymptotically commuting' families of operators.

Now, let us specialise to the case $A = C(X)$, where X is a compact metric space. In this case, families of operators that asymptotically commute with A admit a nice geometric description (at least up to an approximation): this is the content of the next result. See Definition 4.1.1 for the definition of an X module and Definition 4.1.8 for the propagation $\text{prop}(T)$ of an operator on an X module.

Proposition 6.1.1 *Let X be a compact metric space, and let H_X be an X module. Let $(T_t)_{t\in[1,\infty)}$ be a norm continuous, uniformly bounded family of operators on H_X. Then the following are equivalent:*

(i) for each $f \in C(X)$, $\lim_{t\to\infty} \|[T_t, f]\| = 0$;

(ii) there exists a norm continuous family $(S_t)_{t\in[1,\infty)}$ of bounded operators on H_X such that $\text{prop}(S_t) \to 0$ as $t \to \infty$ and such that $\|T_t - S_t\| \to 0$ as $t \to \infty$.

For the proof of this, we need a technical lemma. Recall from Definition A.3.9 that if $f : X \to Y$ is a function between metric spaces, then the associated *expansion function* of f is defined by

$$\omega_f : [0,\infty) \to [0,\infty], \quad \omega_f(r) := \sup\{d(f(x), f(y)) \mid d(x,y) \leqslant r\}.$$

Lemma 6.1.2 *Let X be a metric space and let H_X be an X module. Let T be a finite propagation operator on H_X, and let $f : X \to \mathbb{C}$ be a bounded Borel function considered as an operator on H_X via the X module structure. Then*

$$\|fT - Tf\| \leqslant 8\omega_f(\text{prop}(T))\|T\|.$$

Proof Note first that the expansion functions of the real and imaginary parts of f are bounded by the expansion function of f itself. Hence it suffices to prove that

$$\|fT - Tf\| \leqslant 4\omega_f(\mathrm{prop}(T))\|T\|$$

for a real-valued bounded Borel function f on X. As f is bounded, $\epsilon := \omega_f(\mathrm{prop}(T))$ is finite. For each $k \in \mathbb{Z}$, define

$$X_k := f^{-1}[k\epsilon\,, (k+1)\epsilon).$$

Let χ_k be the characteristic function of the Borel set X_k, and define

$$g := \sum_{k\in\mathbb{Z}} k\epsilon\chi_k;$$

as only finitely many of the sets X_k are non-empty, g is a real-valued Borel function. Clearly $\|f - g\| \leqslant \epsilon$, whence

$$\|fT - Tf\| \leqslant \|gT - Tg\| + 2\epsilon\|T\|. \tag{6.1}$$

To estimate the norm of $gT - Tg$, note that if $x \in X_k$, $y \in X_j$ and $|k-j| > 1$, then $|f(x) - f(y)| > \epsilon$ and so $d(x,y) > \mathrm{prop}(T)$ by choice of ϵ. Hence

$$\begin{aligned} gT - Tg &= \sum_{k\in\mathbb{Z}} k\epsilon(\chi_k T - T\chi_k) \\ &= \sum_{k\in\mathbb{Z}} k\epsilon(\chi_k T(\chi_{k-1} + \chi_k + \chi_{k+1}) - (\chi_{k-1} + \chi_k + \chi_{k+1})T\chi_k). \end{aligned}$$

Rearranging the sum and cancelling terms gives

$$gT - Tg = \epsilon\Big(\sum_{k\in\mathbb{Z}} \chi_k T\chi_{k+1} + \sum_{k\in\mathbb{Z}} \chi_{k+1}T\chi_k\Big).$$

As the operators in each sum are mutually orthogonal and have norm at most $\|T\|$, this implies that $\|gT - Tg\| \leqslant 2\epsilon\|T\|$. Combining this with line (6.1) above gives that

$$\|fT - Tf\| \leqslant 4\epsilon\|T\| = 4\omega_f(\mathrm{prop}(T))\|T\|,$$

which completes the proof. □

Proof of Proposition 6.1.1 The fact that the second condition implies the first follows from Lemma 6.1.2 and that a continuous function f on a compact space is uniformly continuous, whence $\lim_{r\to 0} \omega_f(r) = 0$.

For the other implication, let (T_t) be a family of operators that asymptotically commutes with $C(X)$, i.e. that satisfies $[T_t, f] \to 0$ as $t \to \infty$. For each

$n \geqslant 1$, let $(\phi_{i,n})_{i \in I_n}$ be a finite partition of unity subordinate to the cover of X by balls of radius $1/n$ (such exists by Theorem A.1.3). Set $S_{0,t} = T_{0,t}$ and for each n define

$$S_{n,t} := \sum_{i \in I_n} \sqrt{\phi_{i,n}} T_t \sqrt{\phi_{i,n}}.$$

Note that $\text{prop}(S_{n,t}) \leqslant 1/n$ for each n and that

$$S_{n,t} - T_t = \sum_{i \in I_n} \sqrt{\phi_{i,n}} T_t \sqrt{\phi_{i,n}} - \sum_{i \in I_n} \phi_{i,n} T_t = \sum_{i \in I_n} \sqrt{\phi_{i,n}} [T_t, \sqrt{\phi_{i,n}}].$$

As I_n is finite, this tends to zero as t tends to infinity by assumption on (T_t). Hence for each n there exists t_n such that for all $t \geqslant t_n$

$$\|S_{n,t} - T_t\| \leqslant 1/n.$$

We may assume that the sequence (t_n) is strictly increasing and tending to infinity, and that $t_1 > 1$. Set $t_0 = 1$. Let now $(\psi_n)_{n=0}^{\infty}$ be a partition of unity on $[1, \infty)$ such that each ψ_n is supported in $[t_n, t_{n+2}]$. Finally, define

$$S_t = \sum_{n=0}^{\infty} \psi_n(t) S_{n,t}.$$

We leave it as an exercise for the reader to show that this has the right properties. □

To summarise, let X be a compact metric space and let H_X be an X module. Let A be the C^*-subalgebra of the C^*-algebra $C_b([1, \infty), \mathcal{K}(H))$ of bounded continuous functions from $[1, \infty)$ to $\mathcal{K}(H)$ generated by families $(T_t)_{t \in [1,\infty)}$ with $\text{prop}(T_t) \to 0$ as $t \to \infty$. One thinks of elements (T_t) of A as getting more and more 'local' as time t advances. Then our discussion above shows that a projection (P_t) in A defines in a natural way a 'functional'

$$P_* \colon K^0(X) \to \mathbb{Z}.$$

It is not too difficult to extend this idea and show that there is a pairing

$$K_0(A) \otimes K^0(X) \to \mathbb{Z}$$

that extends the above construction of linear functionals. This leads one to guess that $K_0(A)$ might be a good model for the K-homology group $K_0(X)$, and indeed this gives the 'correct' group if H_X is ample (see Definition 4.1.1). Our approach to K-homology will be based on this idea, but the technical details will be a little different in order to facilitate some arguments in both this chapter and Chapter 8.

6.2 Localisation Algebras

In this section, we define the localisation algebras that will form the basis of our treatment for K-homology.

Throughout this section, X, Y denote locally compact, second countable, Hausdorff topological spaces. The one-point compactifications (see Definition A.1.4) of X and Y are denoted X^+ and Y^+ respectively. Such spaces X, Y and their one-point compactifications are metrisable, and we will assume that they are equipped with a metric when convenient. For additional background, see Section A.1 for our conventions on topological spaces and Sections 4.1 and 4.4 for the background on geometric modules and covering isometries that we will need.

Throughout this section, we will be working with functions from $[1,\infty)$ to the C^*-algebra $\mathcal{B}(H)$ of bounded operators on some Hilbert space. We think of such a function as a family of operators $(T_t)_{t\in[1,\infty)}$ parametrised by $t \in [1,\infty)$.

Definition 6.2.1 Let H_X be an X module. Define $\mathbb{L}[H_X]$ to be the collection of all bounded functions (T_t) from $[1,\infty)$ to $\mathcal{B}(H_X)$ such that:

(i) for any compact subset K of X, there exists $t_K \geqslant 0$ such that for all $t \geqslant t_K$, the operators

$$\chi_K T_t \quad \text{and} \quad T_t \chi_K$$

are compact, and the functions

$$t \mapsto \chi_K T_t \quad \text{and} \quad t \mapsto T_t \chi_K$$

are uniformly norm continuous when restricted to $[t_K,\infty)$;

(ii) for any open neighbourhood U of the diagonal in $X^+ \times X^+$, there exists $t_U \geqslant 1$ such that for all $t > t_U$

$$\operatorname{supp}(T_t) \subseteq U.$$

Remark 6.2.2 We stated condition (ii) above in terms of open sets to emphasise its topological nature. In metric terms, if we fix a metric d on X^+ then it is equivalent to the following: for every $\epsilon > 0$ there exists $t_\epsilon \geqslant 1$ such that for all $t > t_\epsilon$, $\operatorname{prop}(T_t) < \epsilon$. The metric formulation is often more convenient to work with in practice, and we will often do so. There are also equivalent conditions in terms of neighbourhoods of the diagonal in $X \times X$, and with respect to metrics on X, although these are only asked to hold 'locally'. The reader is asked to explore all of this in Exercise 6.8.1.

Recall now from Lemma 4.1.13 that supports of operators behave well under sums, adjoints and compositions. Using this and the fact that the compact operators are an ideal in $\mathcal{B}(H_X)$, it follows directly that $\mathbb{L}[H_X]$ is a $*$-algebra.

Definition 6.2.3 Define $L^*(H_X)$ to be the C^*-algebra completion of $\mathbb{L}[H_X]$ for the norm

$$\|(T_t)\| := \sup_t \|T_t\|_{\mathcal{B}(H_X)}.$$

We call $\mathbb{L}[H_X]$ the *localisation $*$-algebra* of H_X and $L^*(H_X)$ the *localisation C^*-algebra*, or just *localisation algebra* of H_X.

Localisation algebras may initially look more complicated than the Roe algebras of Chapter 5. However, from a K-theoretic point of view, they are much simpler: in particular, their K-theory is typically easier to compute.

Remark 6.2.4 Let H_X be an X module, and let f be a bounded Borel function on f. Note that $L^*(H_X)$ is naturally represented on the (non-separable) Hilbert space $\ell^2([1,\infty), H_X)$, and that f defines a bounded operator on this Hilbert space by the formula

$$(f \cdot u)(t) := fu(t)$$

for $u \in \ell^2([1,\infty), H_X)$ and $t \in [1,\infty)$, where the right-hand side uses the action of the bounded Borel functions on the X module H_X (see Proposition 1.6.11). It is not difficult to see that the multiplication operator thus defined is in the multiplier algebra of $L^*(H_X)$ as in Definition 1.7.6. To summarise, bounded Borel functions on X naturally define multipliers of $L^*(H_X)$.

In the remainder of this section, we discuss functoriality of the localisation algebras for proper continuous maps. In Section 6.4 we will bootstrap these results up to functoriality on a larger category, but we deal with the special case now as it is more intuitive (and suffices for many applications). The construction underlying functoriality is based on the machinery of continuous covers from Section 4.4. For the reader's convenience, we repeat Definition 4.4.6.

Definition 6.2.5 Let H_X, H_Y be geometric modules, and $f\colon X \to Y$ a function. A family of isometries $(V_t\colon H_X \to H_Y)_{t\in[1,\infty)}$ is a *continuous cover* of f if:

(i) the function $t \mapsto V_t$ from $[1,\infty)$ to $\mathcal{B}(H_X, H_Y)$ is uniformly norm continuous;

(ii) for any open subset $U \subseteq Y^+ \times Y^+$ that contains the diagonal, there exists $t_U \geqslant 1$ such that for all $t \geqslant t_U$,

$$\operatorname{supp}(V_t) \subseteq \{(y,x) \in Y \times X \mid (y, f(x)) \in U\}.$$

Remark 6.2.6 Analogously to Remark 6.2.2, it is important that we use open neighbourhoods of the diagonal in $Y^+ \times Y^+$, and not in $Y \times Y$, in the above.

If we fix a metric d on Y^+, then we can restate condition (ii) above as follows: for any $\epsilon > 0$ there exists $t_\epsilon \geqslant 1$ such that for all $t \geqslant t_\epsilon$, $\operatorname{supp}(V_t) \subseteq \{(y,x) \in Y \times X \mid d(y, f(x)) < \epsilon\}$; compare Exercise 6.8.1. This is often the more convenient formulation to work with; we stated the 'official' definition in the form above to emphasise its topological nature.

Lemma 6.2.7 *Let H_X, H_Y be geometric modules. Let $f : X \to Y$ be a continuous and proper map, and let (V_t) be a continuous cover for f. Then*

$$(T_t) \mapsto (V_t T_t V_t^*)$$

defines a $*$-homomorphism

$$ad_{(V_t)} : \mathbb{L}[H_X] \to \mathbb{L}[H_Y]$$

that extends to a $*$-homomorphism from $L^*(H_X)$ to $L^*(H_Y)$.

Moreover, the map induced by $ad_{(V_t)}$ on K-theory depends only on f and not on the choice of (V_t).

Proof The formula $\mathrm{ad}_{(V_t)}: (T_t) \mapsto (V_t T_t V_t^*)$ clearly defines a $*$-homomorphism from the C^*-algebra $\ell^\infty([1,\infty), \mathcal{B}(H_X))$ of bounded functions from $[1,\infty)$ to $\mathcal{B}(H_X)$ to $\ell^\infty([1,\infty), \mathcal{B}(H_Y))$. We have to show that it restricts to a map from $\mathbb{L}[H_X]$ to $\mathbb{L}[H_Y]$.

Let us fix a metric d on Y^+ and work with metric language. Let K be any compact subset of Y and for $\epsilon > 0$ let

$$N_\epsilon(K) := \bigcup_{y \in K} B(y;\epsilon)$$

be the ϵ-neighbourhood of K. As Y is locally compact, there is $\epsilon > 0$ such that $\overline{N_\epsilon(K)}$ is a compact subset of Y, whence as f is proper, $f^{-1}(\overline{N_\epsilon(K)})$ is a compact subset of X. Using Lemma 4.1.15 (and with notation as in that lemma),

$$\begin{aligned} K \circ \operatorname{supp}(V_t) &\subseteq \{x \in X \mid \text{there exists } y \in K \text{ with } d(f(x), y) < \epsilon\} \\ &= f^{-1}(N_\epsilon(K)) \end{aligned} \tag{6.2}$$

for all suitably large t. Hence by Lemma 4.1.15,

$$\chi_K V_t T_t V_t^* = \chi_K V_t \chi_{K \circ \mathrm{supp}(V_t)} T_t V_t^*.$$

Line (6.2) and the preceding comments imply that $K \circ \mathrm{supp}(V_t)$ has compact closure, whence condition (i) from Definition 6.2.1 implies that there is t_0 such that for all $t \geqslant t_0$, $\chi_{K \circ \mathrm{supp}(V_t)} T_t$ is a compact operator and the map

$$[t_0, \infty) \to \mathcal{K}(H_Y), \quad t \mapsto \chi_{K \circ \mathrm{supp}(V_t)} T_t$$

is uniformly continuous. The case of $V_t T_t V_t^* \chi_K$ is analogous, using the 'adjoint' of the formula from Lemma 4.1.15. This shows that the family $(V_t T_t V_t^*)$ satisfies condition (i) from Definition 6.2.1.

For condition (ii), let us also fix a metric on X^+. Let t_ϵ be large enough so that for all $t \geqslant t_\epsilon$, if (y, x) is in $\mathrm{supp}(V_t)$ (equivalently by Lemma 4.1.13, if $(x, y) \in \mathrm{supp}(V_t^*)$), then $d(y, f(x)) \leqslant \epsilon$, and moreover such that if $t \geqslant t_\epsilon$, then $\mathrm{prop}(T_t) < \epsilon$. Lemma 4.1.13 gives us that for any t

$$\mathrm{supp}(V_t T_t V_t^*) \subseteq \overline{\mathrm{supp}(V_t) \circ \overline{\mathrm{supp}(T_t) \circ \mathrm{supp}(V_t^*)}}.$$

Let now $t \geqslant t_\epsilon$, and let (y, z) be a point in the support of $V_t T_t V_t^*$. Then the above implies that there are sequences (y_n) and (z_n) in Y and (x_n) in X such that $y_n \to y$ and $z_n \to z$ as $n \to \infty$, and such that $(y_n, x_n) \in \mathrm{supp}(V_t)$ and $(x_n, z_n) \in \overline{\mathrm{supp}(T_t) \circ \mathrm{supp}(V_t^*)}$ for all n. Moreover, for each n there are sequences $(x_{nm})_{m=1}^{\infty}$ and $(v_{nm})_{m=1}^{\infty}$ in X and $(z_{nm})_{m=1}^{\infty}$ in Y such that $x_{nm} \to x_n$ and $z_{nm} \to z_n$ as $m \to \infty$, such that $(x_{nm}, v_{nm}) \in \mathrm{supp}(T_t)$ and such that $(v_{nm}, z_{nm}) \in \mathrm{supp}(V_t^*)$ for all n, m. Now let

$$\omega_f(\epsilon) := \sup\{d(f(x_1), f(x_2)) \mid d(x_1, x_2) \leqslant \epsilon\}.$$

Then, putting the above discussion together, we get

$$\begin{aligned} d(y, z) &= \lim_{n \to \infty} d(y_n, z_n) \\ &\leqslant \limsup_{n \to \infty} d(y_n, f(x_n)) + d(f(x_n), z_n) \\ &\leqslant \epsilon + \limsup_{n \to \infty} \limsup_{m \to \infty} d(f(x_{nm}), f(v_{nm})) + d(f(v_{nm}), z_{nm}) \\ &\leqslant \epsilon + \omega_f(\epsilon) + \epsilon. \end{aligned} \tag{6.3}$$

As $f: X \to Y$ is continuous and proper, it extends (uniquely) to a continuous map $f: X^+ \to Y^+$. As X^+ is compact, f is therefore uniformly continuous for the metrics we are using. Hence the expression $2\epsilon + \omega_f(\epsilon)$ bounding $d(y, z)$ in line (6.3) tends to zero as ϵ tends to zero. We have thus shown that $\mathrm{prop}(V_t T_t V_t^*)$ tends to zero as t tends to infinity, so condition (ii) is satisfied.

We now have that $\mathrm{ad}_{(V_t)}$ defines a $*$-homomorphism from $\mathbb{L}[H_X]$ to $\mathbb{L}[H_Y]$, and therefore also from $L^*(H_X)$ to $L^*(H_Y)$. The statement about K-theory is proved in exactly the same way as the corresponding statement for Roe algebras in Lemma 5.1.12; we leave this to the reader. □

Remark 6.2.8 Using Remark 4.4.5, if $f: X \to X$ is the identity map and H_X, H'_X are both ample X modules, then there is a continuous cover $(V_t: H_X \to H'_X)$ of the identity map that consists of unitary operators. It follows that $\mathbb{L}[H_X]$ and $\mathbb{L}[H'_X]$ are isomorphic, and so are their completions. Thus the localisation $*$-algebras and C^*-algebras do not depend on the modules used to define them up to non-canonical isomorphism, as long as these modules are ample. Moreover, Lemma 6.2.7 above implies that the isomorphism one gets this way is canonical on the level of K-theory.

Corollary 4.4.7 implies that continuous covers exist whenever $f: X \to Y$ is continuous and H_Y is ample. The following definition thus makes sense.

Definition 6.2.9 Let $f: X \to Y$ be a continuous and proper function, and H_X, H_Y be geometric modules with H_Y ample. Let $L^*(H_X)$ and $L^*(H_Y)$ be localisation algebras associated to ample geometric modules. Define

$$f_*: K_*(L^*(H_X)) \to K_*(L^*(H_Y))$$

to be the map on K-theory induced by the $*$-homomorphism

$$\mathrm{ad}_{(V_t)}: L^*(H_X) \to L^*(H_Y)$$

associated to some continuous cover for f as in Lemma 6.2.7 above.

Theorem 6.2.10 *For each second countable, locally compact, Hausdorff space X choose*[2] *an ample X module H_X. Then the assignments*

$$X \mapsto K_*(L^*(H_X)), \quad f \mapsto f_*$$

give a well-defined functor from the category of such spaces and continuous, proper maps to the category $\mathcal{GA}$ of graded abelian groups.

Moreover, the functor that one gets in this way does not depend on the choice of modules up to canonical equivalence.

Proof The proof is very similar to that of Theorem 5.1.15: we leave it to the reader to check the details. □

Analogously to the case of Roe algebras, we make the following convention.

[2] Just as with Theorem 5.1.15, we leave it as an exercise to find a legitimate way to do this.

Convention 6.2.11 If H_X is an ample X module, we will often write $\mathbb{L}[X]$ and $L^*(X)$ for $\mathbb{L}[H_X]$ and $L^*(H_X)$ respectively, and refer to these as *the* localisation $*$-/C^*-algebra of X.

This is justified by Theorem 6.2.10, which in particular implies that at the level of K-theory, $K_*(L^*(X))$ is determined by X up to canonical isomorphism, and by Remark 6.2.8 which implies that $\mathbb{L}[X]$ and $L^*(X)$ are determined by X up to non-canonical isomorphism. As a technical point, note that the multiplier action of the bounded Borel functions $B(X)$ on $L^*(X)$ of Remark 6.2.4 does depend on the choice of X module, however: this sometimes makes particular choices of X module more convenient for certain proofs.

6.3 K-homology

In this section we define the K-homology groups of a space as the K-theory groups of the associated localisation algebra. We then compute the K-homology of the empty set and of a point, and prove that K-homology has Mayer–Vietoris sequences in an appropriate sense.

Throughout this section, X, Y denote second countable, locally compact, Hausdorff topological spaces.

Definition 6.3.1 The *K-homology* groups of X are defined by the formula

$$K_n(X) := K_n(L^*(X)), \quad K_*(X) := K_0(X) \oplus K_1(X).$$

The assignments $X \mapsto K_n(X)$ are well-defined up to canonical equivalence. They are covariant functors from the category of locally compact, second countable, Hausdorff spaces and proper continuous maps to the category of graded abelian groups by Theorem 6.2.10.

Remark 6.3.2 One can also get elements of K-homology from localisation algebras over not-necessarily ample X modules in the following way. Say $H_{X,0}$ is any X module and H_X an ample X module. Then using Corollary 4.4.7 (which only requires ampleness on the target module), there is a continuous cover $(V_t : H_{X,0} \to H_X)$ for the identity map. Using Lemma 6.2.7, this gives us a $*$-homomorphism

$$(\mathrm{ad}_{(V_t)})_* : K_*(L^*(H_{X,0})) \to K_*(L^*(H_X))$$

and thus a map from $K_*(L^*(H_{X,0}))$ to K-homology; moreover, the map induced on K-theory does not depend on the choice of (V_t). Hence elements of $K_*(L^*(H_{X,0}))$ canonically give rise to elements of K-homology.

As the C^*-algebra $L^*(X)$ is fairly large, it is not completely obvious what $K_n(X)$ is, even for very simple X. We start with direct computations for the two simplest cases: the empty set and a single point.

Proposition 6.3.3 *The K-homology of the empty set is given by $K_n(\varnothing) = 0$ for all n.*

If X is a single point space, then

$$K_n(X) = \begin{cases} \mathbb{Z} & n = 0 \mod 2, \\ 0 & n = 1 \mod 2. \end{cases}$$

Moreover, if H_X is an ample X module, then the group $K_0(X) = K_0(L^*(H_X))$ is generated by any constant function from $[1,\infty)$ to a rank one projection in the compact operators on H_X.

Proof First, let X be the empty set $\varnothing$. Then $C(\varnothing) = \{0\}$, and the conditions defining $L^*(X)$ are vacuous. Hence $L^*(X)$ is just the C^*-algebra

$$l^\infty([1,\infty), \mathcal{B}(H))$$

of bounded functions from $[1,\infty)$ to the bounded operators on some separable infinite-dimensional Hilbert space. This has zero K-theory by applying the Eilenberg swindle showing that $\mathcal{B}(H)$ has zero K-theory from Corollary 2.7.7 pointwise in $[1,\infty)$.

Now let X be a single point. Hence $C(X) = \mathbb{C}$ and an ample X module H_X is necessarily a separable infinite-dimensional Hilbert space, equipped with the unital action of $\mathbb{C}$. The conditions defining $\mathbb{L}[X]$ specialise to say that it consists of bounded functions $(T_t)_{t\in[1,\infty)}$ from $[1,\infty)$ to $\mathcal{B}(H_X)$ such that there is some t_0 so that the restriction $(T_t)_{t\in[t_0,\infty)}$ is compact operator valued and uniformly continuous. Define also

$$\mathbb{L}_0[X] := \{(T_t) \in \ell^\infty([1,\infty), \mathcal{B}(H)) \mid \text{ there is } t_0 \text{ such that } T_t = 0 \text{ for } t \geqslant t_0\},$$

which is a $*$-subalgebra of $\mathbb{L}[X]$, and let $L_0^*(X)$ be the closure of $\mathbb{L}_0[X]$, which is an ideal in $L^*(X)$.

Now, write $C_{ub}([1,\infty), \mathcal{K}(H))$ for the C^*-algebra of bounded uniformly continuous functions from $[1,\infty)$ to the compact operators in H. Then the discussion in the previous paragraph shows that the natural inclusion

$$C_{ub}([1,\infty), \mathcal{K}(H)) \to L^*(X) \tag{6.4}$$

induces an isomorphism

$$\frac{C_{ub}([1,\infty), \mathcal{K}(H))}{C_0([1,\infty), \mathcal{K}(H))} \cong \frac{L^*(X)}{L_0^*(X)}.$$

Moreover, both $C_0([1,\infty),\mathcal{K}(H))$ and $L_0^*(X)$ have zero K-theory: the former as it is contractible, and the latter by applying the Eilenberg swindle showing that $\mathcal{B}(H)$ has zero K-theory (see Corollary 2.7.7) pointwise in $[1,\infty)$. Hence the inclusion in line (6.4) induces an isomorphism on K-theory. Thus to complete the proof, it will suffice to show that the evaluation-at-one $*$-homomorphism

$$\mathrm{ev}\colon C_{ub}([1,\infty),\mathcal{K}(H)) \to \mathcal{K}(H)$$

induces an isomorphism on K-theory (note that this also shows that the K-theory is generated by an element of the form claimed). Moreover, using the six-term exact sequence again, it will suffice to show that the kernel I of ev has zero K-theory; we will now proceed to do this using an Eilenberg swindle.

Write

$$H = \bigoplus_{n=0}^{\infty} H_n,$$

where each H_n is infinite-dimensional. For each $n \geqslant 0$ choose a unitary isomorphism $V_n\colon H \to H_n$. Extend each $(T_t) \in I$ to a continuous function on all of $\mathbb{R}$ by defining $T_t = 0$ for $t < 1$ (this is uniformly continuous as $T \in I$ forces $T_1 = 0$). Provisionally define a $*$-homomorphism $\alpha\colon I \to I$ by

$$\alpha((T_t)) = \left(\sum_{n=1}^{\infty} V_n T_{t-(n-1)} V_n^*\right);$$

note that uniform continuity of $\alpha((T_t))$ follows from that of $t \mapsto T_t$, and that for each fixed $t \in [1,\infty)$, $\alpha((T_t))_t$ is in $\mathcal{K}$ as all but finitely many terms in the direct sum are zero; hence α is well-defined.

Now let $\beta\colon I \to I$ be the $*$-homomorphism defined by conjugating by V_0 pointwise in t, which induces an isomorphism on K-theory by Proposition 2.7.5. Then α and β have orthogonal images, whence the map $\gamma^{(0)} := \alpha + \beta$ is a $*$-homomorphism and induces the same map on K-theory as $\alpha_* + \beta_*$ by Lemma 2.7.6. The homotopy defined for $s \in [0,1]$ by

$$\gamma^{(s)}\colon (T_t) \mapsto \left(V_0 T_t V_0^* + \sum_{n=1}^{\infty} V_n T_{t-(n-1)-s} V_n^*\right)$$

between $\gamma^{(0)}$ and a new $*$-homomorphism $\gamma^{(1)}$ shows that these two maps induce the same map on K-theory (here we use uniform continuity to show that this a homotopy). On the other hand, $\gamma^{(1)}$ is conjugate to α via the isometric multiplier of I defined by applying the isometry

$$V := \sum_{n=0}^{\infty} V_{n+1} V_n^*$$

pointwise in t. Hence $\gamma^{(1)}$ and α induce the same map on K-theory by Proposition 2.7.5.

Putting the above together

$$\alpha_* + \beta_* = \gamma_*^{(0)} = \gamma_*^{(1)} = \alpha_*,$$

and cancelling α_* gives that β_* is zero. However, we know that β_* is an isomorphism, so this forces $K_*(I) = 0$. □

Our remaining goal in this section is to prove the existence of Mayer–Vietoris sequences.

Theorem 6.3.4 *Let $X = E \cup F$ be a cover of X by closed subsets. Then there exists an exact Mayer–Vietoris sequence*

$$\begin{array}{ccccc} K_0(E \cap F) & \longrightarrow & K_0(E) \oplus K_0(F) & \longrightarrow & K_0(X) \\ \uparrow & & & & \downarrow \\ K_1(X) & \longleftarrow & K_1(E) \oplus K_1(F) & \longleftarrow & K_1(E \cap F), \end{array}$$

where all the horizontal arrows are those functorially induced by the relevant inclusions.

Moreover, the Mayer–Vietoris sequence is natural in the following sense. Let $W = C \cup D$ be another decomposition into closed subsets, and let $f\colon W \to X$ is a proper continuous map that satisfies $f(C) \subseteq E$ and $f(D) \subseteq F$. Then the diagram in which all diagonal maps are induced by f, commutes.

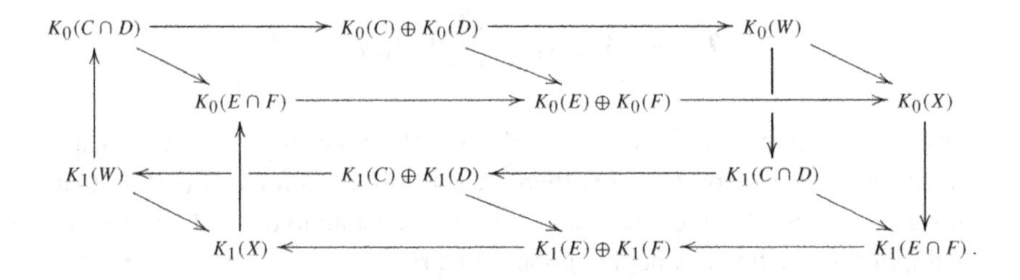

The key technical ingredient in the proof of Theorem 6.3.4 is a construction that models the K-homology of a closed subspace Y of X using the K-theory of an ideal inside $L^*(X)$. This is the content of the next definition and lemma.

Definition 6.3.5 Let Y be a closed subspace of X, and let H_X be an X module. Define $\mathbb{L}_Y[H_X]$ to be the subset of $\mathbb{L}[H_X]$ consisting of (T_t) such that for any open subset U of $X^+ \times X^+$ that contains $Y^+ \times Y^+$, there exists t_U such that for all $t \geqslant t_U$

$$\mathrm{supp}(T_t) \subseteq U.$$

Define $L_Y^*(H_X)$ to be the closure of $\mathbb{L}_Y[H_X]$ inside $L^*(H_X)$.

Note that $\mathbb{L}_Y[H_X]$ is a $*$-ideal in $\mathbb{L}[H_X]$ (we leave this as an exercise for the reader), whence $L_Y^*(H_X)$ is an ideal in $L^*(H_X)$.

Lemma 6.3.6 *Let Y be a closed subspace of X, let H_Y and H_X be ample modules over these spaces and let $(V_t : H_Y \to H_X)_{t \in [1,\infty)}$ be any continuous cover of the inclusion map $Y \to X$. Then the map*

$$ad_{(V_t)} : L^*(H_Y) \to L^*(H_X)$$

of Lemma 6.2.7 takes image in $L_Y^*(H_X)$, and the associated map

$$ad_{(V_t)} : L^*(H_Y) \to L_Y^*(H_X)$$

induces an isomorphism on K-theory.

Proof It is not difficult to show that if $(W_t : H_X \to H_X')$ is any continuous cover of the identity map, then $\mathrm{ad}_{(W_t)}$ takes $L_Y^*(H_X)$ isomorphically onto $L_Y^*(H_X')$. Hence it suffices to prove the lemma for a specific choice of X module, and also of continuous cover (V_t) for the inclusion $Y \to X$. Let Z_X be a countable dense subset of X whose intersection with Y is also dense. Let H be a separable, infinite-dimensional Hilbert space, and define ample X and Y modules respectively by $H_X := \ell^2(Z_X, H)$ and $H_Y := \ell^2(Z_Y, H)$. Then the natural inclusion $H_Y \to H_X$ induced by the inclusion $Z_Y \to Z_X$ gives rise to a constant family $(V_t : H_X \to H_Y)$ of isometries, which is a continuous cover for the inclusion map (compare Example 4.4.2); we will prove the lemma for this particular case.

Let now χ_Y be the characteristic function of Y, which acts as a multiplier of $L^*(H_X)$ (see Remark 6.2.4). Then it is immediate that the corner $\chi_Y L^*(H_X)\chi_Y$ identifies with the image of $L^*(H_Y)$ under $\mathrm{ad}_{(V_t)}$. Clearly $\chi_Y L^*(H_X)\chi_Y$ is a C^*-subalgebra of $L_Y^*(H_X)$, which gives the inclusion. We claim that $L_Y^*(H_X)$ is the ideal in $L^*(H_X)$ generated by χ_Y by $L^*(H_X)\chi_Y L^*(H_X)$ (compare Definition 1.7.8 for notation). This will imply that $\chi_Y L^*(H_X)\chi_Y$ is a full corner in $L_Y^*(H_X)$ in the sense of Definition 1.7.8, which will suffice to complete the proof by Proposition 2.7.19.

Fix a metric d on X^+, and equip X and Y with the induced metrics for the remainder of the proof. First, if (T_t) and (S_t) are elements of $\mathbb{L}[H_X]$ then it is straightforward from Lemma 4.1.15 and the fact that the propagations of T_t, S_t tend to zero that $(S_t \chi_Y T_t)$ is in $\mathbb{L}_Y[H_X]$. This implies that $L_Y^*(H_X)$ contains the ideal generated by χ_Y.

For the converse inclusion, note that $L_Y^*(H_X)$ is a C^*-algebra, and so the collection of products $(S_t) \cdot (T_t)$ of two elements from $\mathbb{L}_Y[H_X]$ is dense in $L_Y^*(H_X)$ (see for example Exercise 1.9.9). It thus suffices to show that any such product is in the ideal generated by χ_Y. Write H as an infinite direct sum

$$H = \bigoplus_{x \in Z_X} H_x$$

parametrised by Z_X, where each H_x is infinite-dimensional; this is possible as Z_X is countable. For each $x \in Z_X$, let $U_x : H \to H_x$ be any unitary isomorphism. Choose a function $f: Z_X \to Z_Y$ such that $d(x, f(x)) \leqslant 2d(x, Z_Y)$ for all $x \in Z_X$; note that f is the identity on Y. For each $x \in Z_X$, let $V_x : H \to H_X$ be the natural isometry with image functions supported at x, and let $W_x : H_X \to H_X$ be the partial isometry defined by $W_x := V_{f(x)} U_x V_x^*$. Note that $W := \bigoplus_{x \in Z_X} W_x : H_X \to H_X$ is then an isometry with range contained in H_Y, where the latter is identified with a subspace of H_X in the natural way. Define new operators

$$S_t' := S_t W^*, \quad T_t' := W T_t.$$

As W is an isometry with range contained in H_Y and $\chi_Y H_Y = H_Y$ we have

$$S_t T_t = S_t W^* W T_t = S_t W^* \chi_Y W T_t = S_t' \chi_{Y^+} T_t'.$$

To complete the proof, it thus suffices to show that (S_t') and (T_t') are in $\mathbb{L}[H_X]$; as (S_t) and (T_t) are arbitrary elements of the $*$-algebra $\mathbb{L}_Y[H_X]$, either case follows from the other on taking adjoints, so we focus on (T_t').

For $x, y \in Z_X$, write $T_{t,xy} := V_y T_t V_x^* : H \to H$ for the '(x, y)th matrix entry' of T_t, and let $\epsilon > 0$. Then for large t, $\operatorname{supp}(T_t)$ is contained in $N_\epsilon(Y) \times N_\epsilon(Y)$ and $\operatorname{prop}(T_t) < \epsilon$. Now, for such t and some $y \in U$, if the matrix entry $T_{t,yz}'$ is not 0, there must exist $x \in Z_X$ with $f(x) = y$ and $T_{t,xz} \neq 0$. Hence $d(x, z) < \epsilon$ and there exists $y' \in Z_Y$ with $d(z, y') < \epsilon$. Hence

$$d(f(x), x) \leqslant 2d(x, Z_Y) \leqslant 2d(x, y') \leqslant 2(d(x, z) + d(z, y')) < 4\epsilon$$

and so

$$d(y,z) = d(f(x),z) \leqslant d(x,f(x)) + d(x,z) < 5\epsilon.$$

This implies that $\mathrm{prop}(T_t')$ tends to zero as t tends to infinity. Now, let $K \subseteq X$ be compact. As $T_t' = WT_t$ it is clear from the corresponding properties for T_t that for suitably large t, the function $t \mapsto T_t'\chi_K$ is uniformly continuous and that the operator $T_t'\chi_K$ is compact. On the other hand, if $\epsilon > 0$ is such that $\overline{N_\epsilon(K)}$ is compact, then from what we already proved about the propagation of T_t', we have

$$\chi_K T_t' = \chi_K T_t' \chi_{\overline{N_\epsilon(K)}}$$

for all suitably large t. Hence $\chi_K T_t'$ is compact and $t \mapsto \chi_K T_t'$ is uniformly continuous for all large t by the discussion of right multiplication by χ_K. □

Proof of Theorem 6.3.4 Fix a metric d on X^+, and equip each of X, E, F and $E \cap F$ with the restriction of this metric. Let $Z_{E\cap F}$ be a countable dense subset of $E \cap F$, and let Z_E, Z_F be countable dense subsets of E, F respectively whose intersection is $Z_{E\cap F}$. Let $Z_X = Z_E \cup Z_F$, which is a countable dense subset of X. Let H be a fixed infinite-dimensional separable Hilbert space, and for each of the four possible choices $Y \in \{X, E, F, E \cap F\}$, define

$$H_Y := \ell^2(Z_Y, H),$$

which is an ample Y module. We will use this H_Y to define $K_*(Y)$ in each case.

We claim first that there is a pushout diagram

$$\begin{array}{ccc} L^*_{E\cap F}(H_X) & \longrightarrow & L^*_E(H_X) \\ \downarrow & & \downarrow \\ L^*_F(H_X) & \longrightarrow & L^*(H_X) \end{array}$$

of C^*-ideals in $L^*(X)$ in the sense of Definition 2.7.13: recall that this means that

$$L^*(H_X) = L^*_E(H_X) + L^*_F(H_X) \quad \text{and} \quad L^*_E(H_X) \cap L^*_F(H_X) = L^*_{E\cap F}(H_X).$$

To see that $L^*(H_X) = L^*_E(H_X) + L^*_F(H_X)$, note that any element (T_t) of $L^*(H_X)$ can be written as a sum $(T_t\chi_E)+(T_t\chi_{X\setminus E}\chi_F)$ of elements of $L^*_E(H_X)$ and $L^*_F(H_X)$ respectively. That $L^*_E(H_X) \cap L^*_F(H_X) \supseteq L^*_{E\cap F}(H_X)$ is clear. The converse inclusion is a consequence of the following fact from metric topology: for a compact metric space Y, closed,[3] subsets A, B of Y and any $\epsilon > 0$ there exists $\delta > 0$ such that $N_\delta(A) \cap N_\delta(B) \subseteq N_\epsilon(A \cap B)$. We leave the proof as an exercise for the reader.

It follows then from Proposition 2.7.15 that there is a Mayer–Vietoris sequence

$$\begin{array}{ccccc} K_0(L^*_{E\cap F}(H_X)) & \longrightarrow & K_0(L^*_E(H_X)) \oplus K_0(L^*_F(H_X)) & \longrightarrow & K_0(L^*(H_X)) \\ \uparrow & & & & \downarrow \\ K_1(L^*(H_X)) & \longleftarrow & K_1(L^*_E(H_X)) \oplus K_1(L^*_F(H_X)) & \longleftarrow & K_1(L^*_{E\cap F}(H_X)). \end{array}$$

On the other hand, there is also a commutative diagram

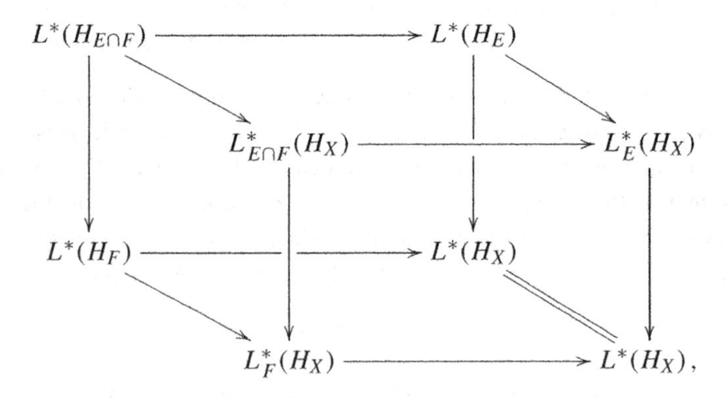

where all the arrows are the natural inclusions. The arrows in the rear square are all defined by families of isometries, constant in t, that continuously cover the respective inclusion maps of spaces, and thus (compare Example 4.4.2) give the maps on K-homology induced by the respective inclusions of spaces. Lemma 6.3.6 implies moreover that the diagonal maps all induce isomorphisms on K-theory. The existence of the Mayer–Vietoris sequence in the statement, and the identification of the horizontal arrows, follows from this.

It remains to prove naturality. Given decompositions $W = C \cup D$ and $X = E \cup F$ satisfying the assumptions of the theorem, and a map between them, one can use the construction above together with that of Example 4.2.3 to build a commutative diagram

[3] This is the only place in the proof that we use the assumption that E and F are closed subsets of X.

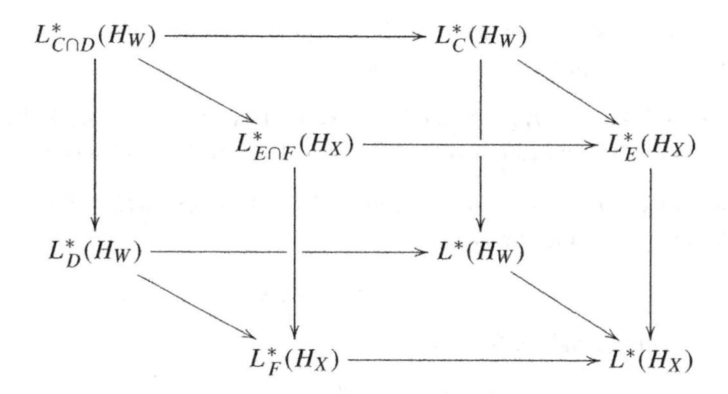

of pushout squares. The result follows from this, the discussion above and naturality of the K-theory Mayer–Vietoris sequence (Proposition 2.7.15). □

6.4 General Functoriality

In this section, we prove functoriality results for a larger category of morphisms. This is important in of itself for applications, and also useful to extend homological properties of K-homology beyond those of the last section. In particular we show that K-homology is a homology theory (see Definition B.2.2) on the category $\mathcal{LC}$ of Definition 6.4.1 below. All of this necessitates working with some variants of the localisation algebra that are a little less concrete but more flexible.

Throughout the section, X and Y denote locally compact, second countable topological spaces. Their one-point compactifications (see Definition A.1.4) are denoted by X^+ and Y^+ respectively, and the point at infinity by ∞.

Definition 6.4.1 The category $\mathcal{LC}$ has as objects second countable, locally compact Hausdorff topological spaces; and morphisms from X to Y are continuous functions $f\colon X^+ \to Y^+$ such that $f(\infty) = \infty$.

See Remark A.1.6 and Proposition A.1.8 for some equivalent descriptions of $\mathcal{LC}$.

Remark 6.4.2 A continuous proper map $f\colon X \to Y$ extends uniquely to a continuous map $f\colon X^+ \to Y^+$ that takes infinity to infinity, and thus a morphism in $\mathcal{LC}$. On the other hand, not every morphism in $\mathcal{LC}$ arises in this way: for example, let $f\colon (\mathbb{R})^+ \to (0,1)^+$ be the map that is the identity on $(0,1)$ and sends all of $(\mathbb{R})^+ \setminus (0,1)$ to $\infty \in (0,1)^+$. In this way, $\mathcal{LC}$ can be regarded as strictly larger than the category with the same objects, and morphisms being proper continuous maps.

Definition 6.4.3 Let H_{X^+} be an X^+ module. Define $\mathbb{L}[H_{X^+};\infty]$ to be the collection of all bounded functions (T_t) from $[1,\infty)$ to $\mathcal{B}(H_{X^+})$ such that:

(i) for any compact subset K of X (*not* of X^+!), there exists $t_K \geqslant 0$ such that for all $t \geqslant t_K$, the operators

$$\chi_K T_t \quad \text{and} \quad T_t \chi_K$$

are compact, and the functions

$$t \mapsto \chi_K T_t \quad \text{and} \quad t \mapsto T_t \chi_K$$

are uniformly norm continuous when restricted to $[t_K,\infty)$;

(ii) for any open neighbourhood U of the diagonal in $X^+ \times X^+$, there exists $t_U \geqslant 1$ such that for all $t > t_U$,

$$\text{supp}(T_t) \subseteq U.$$

Define $L^*(H_{X^+};\infty)$ to be the completion of $\mathbb{L}[H_{X^+};\infty]$ for the norm

$$\|(T_t)\| := \sup_t \|T_t\|_{\mathcal{B}(H_{X^+})}.$$

Similarly to the localisation algebras, one can use that supports of operators behave well under sums, adjoints and compositions (Lemma 4.1.13) and the fact that the compact operators are an ideal in $\mathcal{B}(H_X)$, to see that $\mathbb{L}[H_{X^+};\infty]$ is a $*$-algebra and $L^*(H_{X^+};\infty)$ a C^*-algebra. The notation is meant to suggest that we take the localisation algebra of H_{X^+} 'relative to infinity'.

Now, completely analogously to the corresponding material in Section 6.2 we have the following definitions and results. The proofs are essentially the same as in Section 6.2: we leave it to the reader to check the details.

Lemma 6.4.4 *Let H_{X^+}, H_{Y^+} be geometric modules. Let f be a morphism in $\mathcal{LC}$ from X to Y, and let (V_t) be a continuous cover for the function $f\colon X^+ \to Y^+$. Then*

$$(T_t) \mapsto (V_t T_t V_t^*)$$

defines a $*$-homomorphism

$$ad_{(V_t)}\colon \mathbb{L}[H_{X^+};\infty] \to \mathbb{L}[H_{Y^+};\infty]$$

that extends to a $*$-homomorphism from $L^*(H_{X^+};\infty)$ to $L^*(H_{Y^+};\infty)$.

Moreover, the map induced by $ad_{(V_t)}$ on K-theory depends only on f and not on the choice of (V_t). □

Definition 6.4.5 Let $f: X \to Y$ be a morphism in $\mathcal{LC}$, and H_{X^+}, H_{Y^+} be geometric modules with H_{Y^+} ample. Let $L^*(H_{X^+};\infty)$ and $L^*(H_{Y^+};\infty)$ be localisation algebras associated to ample geometric modules. Define

$$f_*: K_*(L^*(H_{X^+};\infty)) \to K_*(L^*(H_{Y^+};\infty))$$

to be the map on K-theory induced by the $*$-homomorphism

$$\mathrm{ad}_{(V_t)}: L^*(H_{X^+};\infty) \to L^*(H_{Y^+};\infty)$$

associated to some continuous cover of the function $f: X^+ \to Y^+$.

Theorem 6.4.6 *For each X in the second countable, locally compact, Hausdorff space X choose*[4] *an ample X^+ module H_{X^+}. Then the assignments*

$$X \mapsto K_*(L^*(H_{X^+};\infty)), \quad f \mapsto f_*$$

give a well-defined functor from the category $\mathcal{LC}$ to the category $\mathcal{GA}$ of graded abelian groups.

Moreover, the functor that one gets in this way does not depend on the choice of modules up to canonical equivalence. □

Our next goal is to show that this new functor extends the one from Theorem 6.2.10. We first need a canonical way to extend an ample X module H_X to an ample X^+ module. Fix a separable infinite-dimensional Hilbert space H and equip $H_X \oplus H$ with the $C(X^+)$ representation defined on $(u, v) \in H_X \oplus H$ by

$$f \cdot (u, v) := (f|_X u, f(\infty)v).$$

Call this module H_X^+. Note that the natural inclusion $H_X \to H_X^+$ induces a $*$-homomorphism

$$\phi^X: L^*(H_X) \to L^*(H_X^+;\infty). \tag{6.5}$$

Proposition 6.4.7 *For any ample X module H_X the map ϕ^X of line* (6.5) *induces an isomorphism on K-theory. Moreover, if $f: X \to Y$ is a proper continuous map, then the diagram*

$$\begin{array}{ccc} K_*(L^*(H_X)) & \xrightarrow{\phi_*^X} & K_*(L^*(H_X^+;\infty)) \\ \downarrow f_* & & \downarrow f_* \\ K_*(L^*(H_Y)) & \xrightarrow{\phi_*^Y} & K_*(L^*(H_Y^+;\infty)) \end{array}$$

commutes.

[4] Just as with Theorem 5.1.15, we leave it as an exercise to find a legitimate way to do this.

To prove this, we will introduce a technical variation of the localisation algebra that will also be useful for some other purposes.

Definition 6.4.8 Let H_X be a geometric module. Let $\mathbb{L}_0[H_X]$ be the collection of all $(T_t) \in \mathbb{L}[H_X]$ such that for any compact subset K of X there exists $t_K \geqslant 0$ such that for all $t \geqslant t_K$,

$$\chi_K T_t = T_t \chi_K = 0.$$

It is not difficult to see that $\mathbb{L}_0[H_X]$ is a $*$-ideal in $\mathbb{L}[H_X]$. Let $L_0^*(H_X)$ be the closure of $\mathbb{L}_0[H_X]$ inside $L^*(H_X)$, let

$$L_Q^*(H_X) := L^*(H_X)/L_0^*(H_X)$$

be the corresponding quotient C^*-algebra.

Analogously, if H_{X^+} is a geometric module, let $\mathbb{L}_0[H_{X^+};\infty]$ be the collection of all $(T_t) \in \mathbb{L}[H_{X^+};\infty]$ such that for any compact subset K of X there exists $t_K \geqslant 0$ such that for all $t \geqslant t_K$,

$$\chi_K T_t = T_t \chi_K = 0.$$

Again, the closure $L_0^*(H_{X^+};\infty)$ of this in $L^*(H_{X^+};\infty)$ is an ideal, and we let

$$L_Q^*(H_{X^+};\infty) := L^*(H_{X^+};\infty)/L_0^*(H_{X^+};\infty)$$

be the corresponding quotient C^*-algebra.

Remark 6.4.9 We could replace the condition 'for any compact subset K of X there exists $t_K \geqslant 1$ such that for all $t \geqslant t_K$, we have $\chi_K T_t = T_t \chi_K = 0$' appearing in the definition above with 'for any compact subset K of X we have $\lim_{t\to\infty} \chi_K T_t = \lim_{t\to\infty} T_t \chi_K = 0$'. This would make no difference on the level of the C^*-algebraic closure. Indeed, it suffices to show that any (T_t) satisfying the weaker second condition can be approximated arbitrarily well by one satisfying the stronger first condition. This can be done by replacing (T_t) by a suitable compression $(\psi_t T_t \psi_t)$ where (ψ_t) is a $[1,\infty)$-parameterised family of continuous functions from X to $[0,1]$ with the property that for any compact K, $\psi_t|_K$ is one for small t and zero for large t. We leave the details to the reader: see Exercise 6.8.2.

Remark 6.4.10 One can check that the analogues of Lemma 6.2.7 and Theorem 6.2.10 also hold for $L_0^*(H_X)$ and $L_Q^*(H_X)$ as long as we assume the modules are ample; this comes down to the fact that if $(V_t \colon H_X \to H_Y)$ is a continuous cover of a continuous proper map $f \colon X \to Y$, then the family (V_t) conjugates $\mathbb{L}_0(H_X)$ into $\mathbb{L}_0(H_Y)$. Analogously to Convention 6.2.11, we will sometimes streamline the notation and write $L^*_Q(X)$ for $L^*_Q(H_X)$ when the choice of H_X is not important. Similar remarks pertain to $L^*_0(H_{X^+};\infty)$ and $L^*_Q(H_{X^+};\infty)$.

Lemma 6.4.11 *If H_X (respectively, H_{X^+}) is an ample geometric module, then the quotient map $L^*(H_X) \to L^*_Q(H_X)$ (respectively, $L^*(H_{X^+};\infty) \to L^*_Q(H_{X^+};\infty)$) induces an isomorphism on K-theory.*

Proof We will focus on the case of $L^*(H_X)$; the case of $L^*(H_{X^+};\infty)$ is entirely analogous. Using the six-term exact sequence in K-theory it suffices to show that $K_*(L^*_0(H_X)) = 0$. Define H_X^∞ to be the infinite direct sum

$$H_X^\infty = \bigoplus_{n=0}^{\infty} H_X,$$

which is also an ample X module when equipped with the diagonal action of $C_0(X)$. Let $\alpha\colon L^*_0(H_X) \to L^*_0(H_X^\infty)$ be the $*$-homomorphism induced by the inclusion $H_X \to H_X^\infty$ as the first summand. The $*$-homomorphism α is induced by a (constant) family of isometries $(V_t)_{t\in[1,\infty)}$ that continuously covers the identity map, so by Remark 6.4.10 induces an isomorphism on K-theory.

Consider the formula

$$\beta\colon (T_t) \mapsto \left(0 \oplus \bigoplus_{n=1}^{\infty} T_t\right).$$

We claim that this defines a $*$-homomorphism $\mathbb{L}_0(H_X) \to \mathbb{L}_0(H_X^\infty)$: the key point is to show that the image is in $\mathbb{L}_0(H_X^\infty)$, and this follows as for any compact subset K of X we have that

$$\chi_K \phi((T_t)) = \left(0 \oplus \bigoplus_{n=1}^{\infty} \chi_K T_t\right),$$

and for all suitably large t, $\chi_K T_t = 0$. Clearly β extends to a $*$-homomorphism $\beta\colon L^*_0(H_X) \to L^*_0(H_X^\infty)$ on the completions.

Now, α has orthogonal image to β and thus $\alpha + \beta$ is a $*$-homomorphism, and as maps on K-theory $(\alpha+\beta)_* = \alpha_* + \beta_*$ by Lemma 2.7.6. On the other hand, $\alpha + \beta$ is conjugate to β via the isometric multiplier of $L^*(H_X^\infty)$ induced by applying the shift isometry

$$V\colon H_X^\infty \to H_X^\infty, \quad (v_0, v_1, v_2, \ldots) \mapsto (0, v_0, v_1, \ldots)$$

constantly in t, whence by Proposition 2.7.5 $\alpha+\beta$ and β induce the same map on K-theory. Hence

$$\beta_* = (\alpha+\beta)_* = \alpha_* + \beta_*.$$

Cancelling β_*, we conclude that α_* is zero. However, we already noted that it is an an isomorphism, so $K_*(L_0^*(H_X)) = 0$ as claimed. □

Proof of Proposition 6.4.7 Note that ϕ^X takes $L_0^*(H_X)$ into $L_0^*(H_X^+;\infty)$, whence it induces a map on the quotients $\phi_Q^X\colon L_Q^*(H_X) \to L_Q^*(H_X^+;\infty)$. Using Lemma 6.4.10, to show that ϕ^X induces an isomorphism on K-theory, it suffices to show that ϕ_Q^X does. We claim in fact that ϕ_Q^X is actually an isomorphism of C^*-algebras. Indeed, as $L^*(H_X) \cap L_0^*(H_X^+;\infty) = L_0^*(H_X)$, ϕ_Q^X is injective. To see surjectivity, let (T_t) be an element of $L^*(H_X^+;\infty)$, and let $P\colon H_X^+ \to H_X$ be the projection onto the first factor. Then (PT_tP) is in the image of ϕ^X. Moreover, for any compact subset K of X we have that $P\chi_K = \chi_K P = \chi_K$. Hence

$$\chi_K((PT_tP) - (T_t)) = (P\chi_K T_t P - P\chi_K T_t).$$

Fix a metric on X^+. For any $\epsilon > 0$ and all sufficiently large t, we have from Lemma 4.1.15 that $\chi_K T_t = \chi_K T_t N_\epsilon(K)$, where as usual

$$N_\epsilon(K) := \bigcup_{x\in K} B(x;\epsilon)$$

is the ϵ-neighbourhood of K. If ϵ is small enough that $N_\epsilon(K)$ is a subset of X with compact closure, then $\chi_{N_\epsilon(K)}P = P\chi_{N_\epsilon(K)} = \chi_{N_\epsilon(K)}$ and so for all suitably large t

$$P\chi_K T_t P - P\chi_K T_t = P\chi_K T_t \chi_{N_\epsilon(K)} P - P\chi_K T_t \chi_{N_\epsilon(K)} P = 0.$$

This shows that $\chi_K((PT_tP) - (T_t))$ is in $L_0^*(H_X^+;\infty)$, completing the proof that ϕ_Q^X is an isomorphism.

For the naturality statement, let $f\colon X \to Y$ be continuous and proper, and let $(V_t\colon H_X \to H_Y)$ be a continuous cover for f. For each t, define

$$W_t := V_t \oplus \mathrm{Id}\colon H_X \oplus H \to H_Y \oplus H.$$

Then $(W_t\colon H_X^+ \to H_Y^+)$ is a covering isometry for the extended function $f\colon X^+ \to Y^+$. Using Remark 6.4.10, we get induced maps on the quotients

$$\mathrm{ad}_{(V_t)}\colon L_Q^*(H_X) \to L_Q^*(H_Y) \quad \text{and} \quad \mathrm{ad}_{(W_t)}\colon L_Q^*(H_X^+;\infty) \to L_Q^*(H_Y^+;\infty).$$

The diagram

$$\begin{array}{ccc} L_Q^*(H_X) & \xrightarrow{\phi_Q} & L_Q^*(H_X^+;\infty) \\ \Big\downarrow{\scriptstyle \mathrm{ad}_{(V_t)}} & & \Big\downarrow{\scriptstyle \mathrm{ad}_{(W_t)}} \\ L_Q^*(H_Y) & \xrightarrow{\phi_Q} & L^*(H_Y^+;\infty) \end{array}$$

commutes, which suffices to complete the proof. □

To summarise, we now have the following:

Corollary 6.4.12 *The functor defined in Theorem 6.4.6 is an extension of the functor in Theorem 6.2.10, up to canonical equivalence.* □

Convention 6.4.13 From now, on, we will just write $K_*(X)$ and f_* for the results of either of the functors defined in Theorem 6.4.6 or Theorem 6.2.10, and call both K-homology.

There is one situation where it is natural to change the notation for morphisms. Say $i\colon U \to X$ is the inclusion of an open set. Then there is a morphism $c\colon X^+ \to U^+$ in $\mathcal{LC}$ defined by collapsing the complement of U in X^+ to the point at infinity in U^+. Thus we get a 'restriction' map on K-homology $c_*\colon K_*(X) \to K_*(U)$. As we would like to think of this map as induced by the inclusion $i\colon U \to X$, but it goes in the opposite direction, we will denote it by $i^*\colon K_*(X) \to K_*(U)$.

Our first goal using the more flexible model for K-homology is to prove homotopy invariance. The proposition below is the main ingredient.

Proposition 6.4.14 *For any X, the K-homology of $X \times [0, \infty)$ is zero.*

Unfortunately, the proof of this is quite technical. In order to make it more palatable, we split off a K-theoretic lemma.

Lemma 6.4.15 *Let Y be a locally compact space, H_{Y^+} an ample Y^+-module and consider the localisation algebra $L^*(H_{Y^+} \otimes H; \infty)$, where $H_{Y^+} \otimes H$ is equipped with the amplification of the $C(Y^+)$ action on H_{Y^+}. Choose a decomposition $H = \bigoplus_{n \in \mathbb{N}} H_n$ and for each n, let $U_n\colon H \to H$ be an isometry with image H_n, considered as a multiplier of $L^*(H_{Y^+} \otimes H; \infty)$ (constant in t) in the natural way. Finally, assume that there exists a sequence of isometries $(V_n(t)\colon H_{Y^+} \to H_{Y^+})_{n \in \mathbb{N} \cup \{\infty\}, t \in [1,\infty)}$ with the following properties:*

(i) for each $0 \leqslant n \leqslant \infty$, the map

$$(T_t) \to (V_n(t) T_t V_n(t)^*)$$

conjugates $L^*(H_{Y^+}; \infty)$ into itself;

(ii) for all $0 \leqslant n \leqslant \infty$, $(V_n(t) V_{n+1}(t)^*)$ (where $\infty + 1 = \infty$) defines a multiplier of $L^*(H_{Y^+}; \infty)$ and the sums[5]

$$\sum_{1 \leqslant n < \infty} U_n V_n(t) V_{n+1}(t)^* U_n^* \quad \text{and} \quad \sum_{1 \leqslant n < \infty} U_n V_\infty(t) V_\infty(t)^* U_n^*$$

converge strongly for each fixed t, and the functions of t thus defined are multipliers of $L^*(H_{Y^+} \otimes H; \infty)$;

[5] The slightly unusual indexing is to emphasise that $n = \infty$ is *not* included in the sum.

(iii) *for all* $(T_t) \in L^*(H_{Y^+} \otimes H; \infty)$*, the elements*

$$\Big(\sum_{1 \leqslant n < \infty} U_n T_t (V_\infty(t) V_\infty(t)^* - V_{n+1}(t) V_n(t)^*) U_n^* \Big)$$

and

$$\Big(\sum_{1 \leqslant n < \infty} U_n (V_\infty(t) V_\infty(t)^* - V_{n+1}(t) V_n(t)^*) T_t U_n^* \Big)$$

are in $L^*(H_{Y^+} \otimes H; \infty)$*;*

(iv) *for each* $(T_t) \in L^*(H_{Y^+} \otimes H; \infty)$ *and each fixed* t*, the sums*

$$\sum_{1 \leqslant n < \infty} U_n V_n(t) T_t V_n(t)^* U_n^* \quad \text{and} \quad \sum_{1 \leqslant n < \infty} U_n T_t U_n^*$$

converge strongly in $\mathcal{B}(H_{Y^+} \otimes H)$ *and moreover the functions sending* t *to the above sums define multipliers of* $L^*(H_{Y^+} \otimes H; \infty)$*;*

(v) *for any* $(T_t) \in L^*(H_{Y^+} \otimes H; \infty)$*, the difference*

$$\Big(\sum_{1 \leqslant n < \infty} U_n V_n(t) T_t V_n(t)^* U_n \Big) - \Big(\sum_{n=1}^{\infty} U_n V_\infty(t) T_t V_\infty(t)^* U_n \Big)$$

of elements of the multiplier algebra of $L^*(H_{Y^+} \otimes H; \infty)$ *is in* $L^*(H_{Y^+} \otimes H; \infty)$*.*

Then the $*$*-homomorphisms*

$$(T_t) \mapsto V_1(t) T_t V_1(t)^* \quad \text{and} \quad (T_t) \mapsto V_\infty(t) T_t V_\infty(t)^*$$

from $L^*(H_{Y^+}; \infty)$ *to itself induce the same map on* K*-theory.*

Proof To avoid the notation getting too cluttered by indices, write $A = L^*(H_{Y^+} \otimes H; \infty)$, write elements of A as a (rather than (T_t)), and write $u_n = U_n$ and $v_n = V_n(t)$. Let

$$D := \{(a, b) \in M(A) \oplus M(A) \mid a - b \in A\}$$

be the double of $M(A)$ along A as in Definition 2.7.8. Let

$$C = \Big\{ (c, d) \in D \mid d = \sum_{n=0}^{\infty} u_n v_\infty a v_\infty^* u_n^* \text{ for some } a \in A \Big\},$$

which is a C^*-subalgebra of D. Define also

$$w_1 := \sum_{n=0}^{\infty} u_n v_{n+1} v_n^* u_n^*, \quad w_2 := \sum_{n=0}^{\infty} u_n v_\infty v_\infty^* u_n^*$$

(which are elements of $M(A)$ by condition (ii) in the statement) and set $w := (w_1, w_2) \in M(A) \oplus M(A)$. We claim that w is actually in the multiplier algebra of C.

Indeed, if (c,d) is in C, then $dw_2 = w_2 d = d$, so it suffices to show that $cw_1 - d$ and $w_1 c - d$ are in A; we focus on $w_1 c - d$, the other case being similar. We have

$$w_1 c - d = w_1(c - d) + (w_1 d - d),$$

whence as $c - d \in A$ and $w_1 \in M(A)$, it suffices to show that $w_1 d - d$ is in A. There exists $a \in A$ with

$$\begin{aligned} w_1 d - d &= \sum_{n=0}^{\infty} u_n (v_{n+1} v_n^* v_\infty a v_\infty^* - v_\infty a v_\infty^*) u_n^* \\ &= \sum_{n=0}^{\infty} u_n (v_{n+1} v_n^* - v_\infty v_\infty^*) v_\infty a v_\infty^* u_n^*, \end{aligned}$$

and this is in A by condition (iii) of the statement, completing the proof of the claim.

Now, provisionally define $*$-homomorphisms

$$\alpha, \beta \colon A \to C$$

by the formulas

$$\alpha(a) := \left(\sum_{n=0}^{\infty} u_n v_n a v_n^* u_n^* , \sum_{n=0}^{\infty} u_n v_\infty a v_\infty^* u_n^* \right)$$

and

$$\beta(a) := \left(\sum_{n=0}^{\infty} u_n v_{n+1} a v_{n+1}^* u_n^* , \sum_{n=0}^{\infty} u_n v_\infty a v_\infty^* u_n^* \right).$$

It follows from conditions (iv) and (v) that $\alpha \colon A \to C$ is a $*$-homomorphism. That β is a $*$-homomorphism and has image in C follows as w is in the multiplier algebra of C, and as $w\alpha(a)w^* = \beta(a)$ for all $a \in A$. Moreover, a direct computation gives that $\alpha(a)w^*w = \alpha(a)$, whence α and β induce the same map $K_*(A) \to K_*(C)$ by Proposition 2.7.5. Post-composing with the map $K_*(C) \to K_*(D)$ induced by the inclusion of C into D, it follows that α and β induce the same map $K_*(A) \to K_*(D)$.

Let now

$$v = \sum_{1 \leqslant n < \infty} u_{n+1} u_n^*,$$

which converges pointwise (in t) strongly to an isometry in $M(A)$. Then (v, v) is a multiplier of D; conjugating by (v, v) and applying Proposition 2.7.5 shows that β induces the same map $K_*(A) \to K_*(D)$ as the $*$-homomorphism $\gamma : A \to D$ defined by

$$\gamma(a) := \left(\sum_{n=1}^{\infty} u_n v_n a v_n^* u_n^* \, , \, \sum_{n=1}^{\infty} u_n v_\infty a v_\infty^* u_n^* \right).$$

On the other hand, the $*$-homomorphism $\delta : A \to D$ defined by

$$\delta : a \mapsto (u_0 v_\infty a v_\infty^* u_0^* \, , \, u_0 v_\infty a v_\infty^* u_0^*)$$

induces the zero map on K-theory by Lemma 2.7.9 and the fact that $K_*(M(A)) = 0$, which can be proved in exactly the same way as Corollary 2.7.7, using the isometries u_n to perform an Eilenberg swindle. Moreover, δ has orthogonal image to γ. Hence from Lemma 2.7.6 the sum $\epsilon := \gamma + \delta$ is a well-defined $*$-homomorphism that induces the same map on K-theory as β.

Compiling our discussion so far, we have

$$\alpha_* = \beta_* = \gamma_* = \gamma_* + \delta_* = \epsilon_* \tag{6.6}$$

as maps $K_*(A) \to K_*(D)$. Let $\psi_0, \psi_\infty : A \to D$ be the $*$-homomorphisms defined by

$$\psi_0 : a \mapsto (u_0 v_0 a v_0^* u_0^*, 0) \quad \text{and} \quad \psi_\infty : a \mapsto (u_0 v_\infty a v_\infty^* u_0^*, 0),$$

and define $\zeta : A \to D$ by

$$\zeta(a) := \left(\sum_{n=1}^{\infty} u_n v_n a v_n^* u_n^* \, , \, \sum_{n=0}^{\infty} u_n v_\infty a v_\infty^* u_n^* \right).$$

Note that ζ has orthogonal image to ψ_0 and ψ_∞, and that

$$\psi_0 + \zeta = \alpha \quad \text{and} \quad \psi_\infty + \zeta = \epsilon;$$

hence from Lemma 2.7.6 and line (6.6),

$$(\psi_0)_* + \zeta_* = \alpha_* = \epsilon_* = (\psi_\infty)_* + \zeta_*.$$

Cancelling ζ_* thus gives that ψ_0 and ψ_∞ induce the same maps on K-theory.

Finally, note that if $\iota : A \to D$ is the map $\iota(a) = (a, 0)$, then

$$\psi_i(a) = u_0 \iota(\phi_i(a)) u_0^*$$

for all $a \in A$ and $i \in \{0, \infty\}$. This implies the desired result as Proposition 2.7.5 and Lemma 2.7.9 imply respectively that conjugation of D by (u_0, u_0) and $\iota : A \to D$ both induce isomorphisms on K-theory. □

Proof of Proposition 6.4.14 We may replace $X \times [0, \infty)$ by $X \times (0, 1]$. Let H_X be an ample X-module, let $Z = [0, 1] \cap \mathbb{Q}$, and let H be a separable infinite-dimensional Hilbert space. For $f \in C((X \times (0, 1])^+)$ and $z \in Z$, let f_z denote the restriction of f to $X \times \{z\}$, or the scalar $f(\infty)$ if $z = 0$. Let $H_{(X\times(0,1])^+} := H_X \otimes \ell^2(Z) \otimes H$, equipped with the representation of $C((X \times (0, 1])^+)$ defined by

$$f \cdot (u \otimes \delta_z \otimes v) = f_z u \otimes \delta_z \otimes v;$$

it is not difficult to see that $H_{(X\times(0,1])^+}$ is then an ample $(X \times (0, 1])^+$ module.

Choose a decomposition $H = \bigoplus_{z \in Z} H_z$ where each H_z is separable and infinite-dimensional, which is possible as Z is countable, and let $W_z \colon H \to H$ be a choice of isometry with image H_z. For $r \in \mathbb{Q} \cap [0, 1]$, define

$$W(r) \colon H_{(X\times(0,1])^+} \to H_{(X\times(0,1])^+}, \quad u \otimes \delta_z \otimes v \mapsto u \otimes \delta_{(1-r)z} \otimes W_z v.$$

For each $t \in [1, \infty)$ and each $n \in \mathbb{N}_{>0} \cup \{\infty\}$ define $V_n(t) \colon H_{(X\times(0,1])^+} \to H_{(X\times(0,1])^+}$ by the following cases:

(i) if $t < n$, define $V_n(t) := W(0)$;

(ii) if $m \in [n, 2n) \cap \mathbb{N}$, $t \in [m, m + 1)$ and $z \neq 0$, define the action of $V_n(t)$ on $u \otimes \delta_z \otimes v$ to agree with that of

$$\left|\cos\left(\frac{\pi}{2}(t - m)\right)\right| W\left(\frac{m - n}{n}\right)\Bigg| + \left|\sin\left(\frac{\pi}{2}(t - m)\right| W\left(\frac{m + 1 - n}{n}\right);$$

(iii) if $t \in [n, 2n)$, define

$$V_n(t)(u \otimes \delta_0 \otimes v) = u \otimes \delta_0 \otimes W_0 v \ ;$$

(iv) if $t \geqslant 2n$, define $V_n(t) := W(1)$.

The following schematic may help to visualise the operators $V_n(t)$.

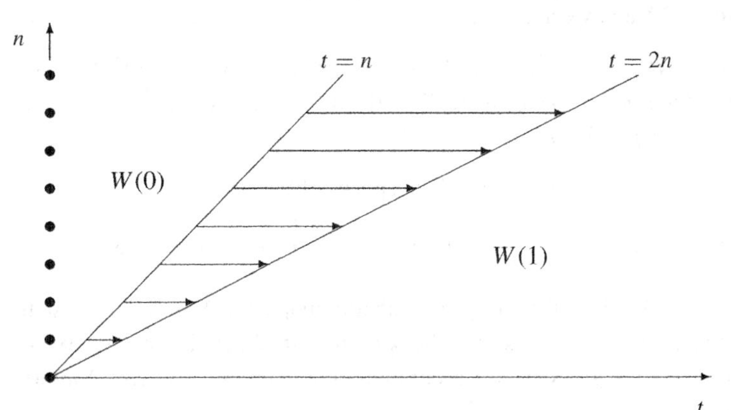

Here $V_n(t)$ is constantly equal to $W(0)$ in the left triangular region, and constantly equal to $W(1)$ in the right triangular region. Along each of the horizontal arrows in the intermediate region, $V_n(t)$ interpolates between $W(0)$ and $W(1)$, taking longer and longer to do so as n increases.

Finally, we also choose a decomposition $H = \bigoplus H_n$, and corresponding isometries $U_n \colon H \to H_n$ as in Lemma 6.4.15. We leave to the reader the tedious, yet essentially elementary, checks that the families of operators $V_n(t)$ and U_n satisfy the hypotheses of Lemma 6.4.15. It follows, then, from the conclusion of that lemma that the $*$-homomorphisms

$$(T_t) \mapsto V_1(t)T_tV_1(t)^* \quad \text{and} \quad (T_t) \mapsto V_\infty(t)T_tV_\infty(t)^* \tag{6.7}$$

from $L^*(H_{(X\times(0,1])^+};\infty)$ to itself induce the same map on K-theory. The conclusion will then follow once we show that the first map in line (6.7) induces the zero map on K-theory, while the second map induces the identity map on K-theory.

Indeed, for the first map this follows as for all $t \geqslant 2$, $V_1(t) = W(1)$, whence the first map in line (6.7) induces the same map on K-theory as conjugation by $W(1)$ by Remark 6.4.10, so it suffices to show that conjugation by $W(1)$ induces the zero map on K-theory. This follows as $W(1)$ conjugates all of $H_{(X\times(0,1])^+}$ into the part of this Hilbert space supported over the point at infinity, on which $L^*(H_{(X\times(0,1])^+};\infty)$ puts no conditions. It follows that we may perform an Eilenberg swindle showing that the map $W(1)$ induces the zero map on K-theory as required; we leave the details to the reader.

For the second map in line (6.7), we have that $V_\infty(t) = W(0)$ for all t, so to complete the proof we must show that $W(0)$ induces the identity map on K-theory. However, $W(0)$ is an isometry in the multiplier algebra of $L^*(H_{(X\times(0,1])^+};\infty)$, whence induces the identity map on K-theory by Proposition 2.7.5 and we are done. □

Theorem 6.4.16 *Let $f^{(0)}, f^{(1)} \colon X \to Y$ be morphisms in $\mathcal{LC}$ that are homotopic through a morphism $h \colon X \times [0,1] \to Y$. Then $f_*^{(0)} = f_*^{(1)}$ as maps from $K_*(X)$ to $K_*(Y)$.*

Proof Consider the following closed subspaces of $\mathbb{R} \times X$:

$$A = (-\infty, 0] \times X, \quad B = [0,\infty) \times X, \quad C = (-\infty, 1] \times X.$$

Then $\mathbb{R} \times X = A \cup B = C \cup B$ and the identity map from X to X gives rise to a map of these decompositions as in the naturality statement in Theorem 6.3.4. Theorem 6.3.4 thus gives rise to a commutative diagram of Mayer–Vietoris sequences

$$
\begin{array}{ccccccccc}
\cdots \longrightarrow & K_i(A \cap B) & \longrightarrow & K_i(A) \oplus K_i(B) & \longrightarrow & K_i(\mathbb{R} \times X) & \longrightarrow \cdots . \\
& \downarrow & & \downarrow & & \| & \\
\cdots \longrightarrow & K_i(C \cap B) & \longrightarrow & K_i(C) \oplus K_i(B) & \longrightarrow & K_i(\mathbb{R} \times X) & \longrightarrow \cdots
\end{array}
$$

Now, all the groups $K_i(A)$, $K_i(B)$ and $K_i(C)$ are zero by Proposition 6.4.14; substituting this information and the identifications $A \cap B \cong X$ and $C \cap B \cong X \times [0,1]$ gives

$$
\begin{array}{ccccccccc}
\cdots \longrightarrow & K_i(X) & \longrightarrow & 0 & \longrightarrow & K_i(\mathbb{R} \times X) & \longrightarrow \cdots . \\
& \downarrow & & \| & & \| & \\
\cdots \longrightarrow & K_i(X \times [0,1]) & \longrightarrow & 0 & \longrightarrow & K_i(\mathbb{R} \times X) & \longrightarrow \cdots
\end{array}
$$

Applying the five lemma, the left-hand vertical arrow, which is induced by the inclusion $i^{(0)}: X \to X \times [0,1]$ defined by $x \mapsto (x,0)$, is an isomorphism.

Now, let $\pi : X \times [0,1] \to X$ be the projection. As $\pi \circ i^{(0)}$ is the identity, it follows that π_* is a one-sided inverse to $i_*^{(0)}$; as the latter is an isomorphism, π_* is itself an isomorphism. Moreover, π_* is therefore also an inverse to $i_*^{(1)}$ (with the obvious notation), as it is a one-sided inverse, and so $i_*^{(0)} = i_*^{(1)}$ as both are inverses to the same map. Finally, note that if $h\colon (X \times [0,1])^+ \to Y^+$ is a homotopy between $f^{(0)}$ and $f^{(1)}$ we get

$$
f_*^{(0)} = h_* \circ i_*^{(0)} = h_* \circ i_*^{(1)} = f_*^{(1)}. \qquad \square
$$

Our next result, the long exact sequence for an open inclusion, now follows directly from the proof of the very general Proposition B.2.3.

Theorem 6.4.17 *Let $i\colon U \to X$ be the inclusion of an open set, $F = X \setminus U$ the complementary closed set, and $j\colon F \to X$ be the inclusion. Then there is a six-term exact sequence*

$$
\begin{array}{ccccc}
K_0(F) & \xrightarrow{j_*} & K_0(X) & \xrightarrow{i^*} & K_0(U) \\
\uparrow & & & & \downarrow \\
K_1(U) & \xleftarrow[i^*]{} & K_1(X) & \xleftarrow[j_*]{} & K_1(F)
\end{array}
$$

that is natural for maps of pairs. $\square$

The final homological property of K-homology that we discuss is that it takes countable disjoint unions to products. First, we have two technical lemmas. To state the first one, recall that by Remark 6.2.4 a bounded Borel function f on X defines a multiplier of $L^*(H_X)$: for example, left multiplication by f is given by the formula $(T_t) \mapsto (fT_t)$. Moreover, this multiplier descends to a well-defined multiplier of $L^*_Q(H_X)$ by Corollary 1.7.4, which implies in particular that the quotient map $L^*(H_X) \to L^*_Q(H_X)$ induces a map on multipliers $M(L^*(H_X)) \to M(L^*_Q(H_X))$ in a natural way.

If our bounded Borel function also happens to be continuous, then we can say more about the multiplier it defines.

Lemma 6.4.18 *Let f be a continuous bounded function on X and H_X a geometric module. Then the multiplier of $L^*_Q(H_X)$ defined by f is central.*

Proof Let (T_t) be an element of $\mathbb{L}(H_X)$. It will suffice to show that the commutator $[f,T_t]$ is in $L^*_0(H_X)$. Let K be a compact subset of X, and fix a metric on X^+. Let $\epsilon > 0$ be such that if $N_\epsilon(K) = \bigcup_{x\in K} B(x;\epsilon)$, then $\overline{N_\epsilon(K)}$ is a compact subset of X. Then the fact that $\text{prop}(T_t)$ tends to zero implies that for all suitably large t we have

$$\chi_K[f,T_t] = \chi_K[f,T_t]\chi_{\overline{N_\epsilon(K)}}$$

(compare Lemma 4.1.15). This tends to zero as t tends to infinity by the first part of Proposition 6.1.1 applied to the compact space $\overline{N_\epsilon(K)}$ and geometric module $\chi_{\overline{N_\epsilon(K)}} \cdot H_X$. □

Lemma 6.4.19 *Let H_{X^+} be an ample X^+ module. Then $L^*_Q(H_{X^+};\infty)$ is quasi-stable, in the sense of Definition 2.7.11.*

Proof Let H_{X^+} be an ample X^+ module, and for each n let $H^{\oplus n}_{X^+}$ be its n-fold direct sum with itself. Note that $L^*_Q(H^{\oplus n}_{X^+};\infty)$ identifies naturally with $M_n(L^*_Q(H_{X^+};\infty))$. Remark 6.2.8 gives a family of unitary isomorphisms $(V_t\colon H^{\oplus n}_X \to H_X)$ covering the identity map on X. Identifying H_X with the first summand $H^{\oplus n}_X$, we can think of (V_t) as an isometry in $\ell^\infty([1,\infty),\mathcal{B}(H_{X^+}))$. As it covers the identity map, it follows that that (V_t) is actually a multiplier of $M_n(L^*_Q(H_{X^+};\infty)) \cong L^*_Q(H^{\oplus n}_{X^+};\infty)$ and is not difficult to check that it has the properties required by quasi-stability. □

Note that as in the proof of Proposition 6.4.7 that $L^*_Q(H_X) \cong L^*_Q(H^+_X;\infty)$, we also get that such f defines a multiplier on the latter algebra.

We are now ready to discuss how K-homology behaves with respect to disjoint unions. Let $(X_n)_{n=0}^\infty$ be a countable collection of (second countable, locally compact, Hausdorff) spaces. Then their *topological disjoint union* is the set-theoretic disjoint union $X := \bigsqcup X_n$ equipped with the topology where $U \subseteq X$ is open if and only if $U \cap X_n$ is open for all n. With this topology, X is itself a locally compact, second countable, Hausdorff space.

In particular, each X_n identifies with an open and closed subset of X. The inclusion maps $i_n\colon X_n \to X$ then induce maps $i_n^*\colon K_*(X) \to K_*(X_n)$ as in Remark 6.4.13. These maps in turn induce

$$\prod i_n^*\colon K_*(X) \to \prod K_*(X_n).$$

Theorem 6.4.20 *With notation as above, the map*

$$\prod i_n^*\colon K_*(X) \to \prod K_*(X_n).$$

is an isomorphism.

Proof Fix a countable dense subset Z_n of each X_n, and set $Z = \bigcup Z_n$, which is a countable dense subset of X. Fix a separable, infinite-dimensional Hilbert space H, and use the ample modules $H_X := \ell^2(Z, H)$ and $H_{X_n} := \ell^2(Z_n, H)$ to define localisation algebras $L^*(H_X)$ and $L^*(H_{X_n})$. Using that any compact subset K of X can only intersect finitely many of the subspaces X_n, one checks that the natural identification $H_X \cong \bigoplus H_{X_n}$ gives rise to an inclusion

$$\iota\colon \prod_n L^*(H_{X_n}) \to L^*(H_X).$$

Using the same fact that compact subsets of X only see finitely many of the X_n, we have that ι takes $\prod_n L_0^*(X_n)$ into $L_0^*(X)$, and so induces a $*$-homomorphism

$$\iota^Q\colon \prod_n L_Q^*(H_{X_n}) \to L_Q^*(H_X).$$

On the other hand, for each n, let $\chi_n\colon X \to \mathbb{C}$ denote the characteristic function of X_n, which is continuous. Define a contractive linear map

$$\kappa\colon L^*(H_X) \to \prod_n L^*(H_{X_n}), \quad (T_t) \mapsto \prod_n \chi_n(T_t)\chi_n,$$

which clearly descends to a well-defined contractive linear map

$$\kappa^Q\colon L_Q^*(H_X) \to \prod_n L_Q^*(H_{X_n}).$$

Note that each χ_n is a continuous bounded function on X, so by Lemma 6.4.18 is a central multiplier of $L_Q^*(H_X)$. It follows this that κ^Q as above is actually a $*$-homomorphism.

Now, it is clear that $\kappa \circ \iota$ is the identity map, whence $\kappa^Q \circ \iota^Q$ is too. On the other hand, let $(T_t) \in L^*(H_X)$ and K be a compact subset of X. Say N is such that $K \cap X_n = \varnothing$ for all $n \geqslant N$. Then

$$\chi_K(\iota \circ \kappa((T_t)) - (T_t)) = \chi_K \sum_{n=0}^{N} \chi_n(T_t)\chi_n - \chi_K(T_t).$$

As the sum is finite, Lemma 6.4.18 implies that this is equal modulo $L_0^*(H_X)$ to

$$\chi_K \sum_{n=0}^{N} (T_t)\chi_n^2 - \chi_K(T_t) = \chi_K(T_t) \sum_{n=0}^{N} \chi_n^2 - \chi_K(T_t) = 0.$$

Hence $\iota^Q \circ \kappa^Q$ is the identity as well.

Hence in particular, κ^Q induces an isomorphism

$$\kappa_*^Q \colon K_*(L_Q^*(H_X)) \to K_*\bigg(\prod_n L_Q^*(H_{X_n})\bigg).$$

Using Lemma 6.4.19 and Proposition 2.7.12, the right-hand side is isomorphic to $\prod_n K_*(L_Q^*(H_{X_n}))$; further, making the identifications in Lemma 6.4.11 gives an isomorphism

$$\kappa_*^Q \colon K_*(X) \to \prod_n K_*(X_n).$$

To complete the proof, we need to show that for each fixed m the composition

$$K_*(X) \xrightarrow{\kappa_*^Q} \prod\nolimits_n K_*(X_n) \longrightarrow K_*(X_m)$$

of κ_*^Q as above and the canonical quotient map $\prod_n K_*(X_n) \to K_*(X_m)$ is the map i_m^* from the statement. As the map i_m^* is an instance of our more general functoriality, we convert our modules H_X and H_{X_m} to X^+ and X_m^+ modules H_X^+ and $H_{X_m}^+$ by adding the point at infinity to Z and Z_m, getting sets Z^+ and Z_m^+ respectively. Define a unitary isomorphism $V \colon \ell^2(Z^+, H) \to \ell^2(Z_m^+, H)$ to be the identity on $\ell^2(Z_m, H)$ and an arbitrary unitary equivalence of the orthogonal complements $\ell^2((Z \setminus Z_m) \cup \{\infty\}, H)$ and $\ell^2(\{\infty\}, H)$. Then the constant family (V_t) with $V_t = V$ for all t is a continuous cover of the 'collapsing' map $c \colon X^+ \to X_m^+$ (compare Remark 6.4.13) that induces i_m^*, and therefore pointwise conjugation ad_V by V induces i_m^*. The proof is completed by noting that the diagram

$$\begin{array}{ccc} L^*(H_{X^+}) & \xrightarrow{\mathrm{ad}_{(V)}} & L^*(H_{X_m^+}) \\ \downarrow & & \downarrow \\ L_Q^*(H_{X^+};\infty) & \xrightarrow{(T_t)\mapsto \chi_m(T_t)\chi_m} & L_Q^*(H_{X_m^+};\infty)\,, \end{array}$$

with vertical arrows given by the canonical quotients, commutes ☐

Note that Proposition 6.3.3 and Theorems 6.3.4, 6.4.6, 6.4.16 and 6.4.20 now combine[6] to show that K-homology is a homology theory in the sense of Definition B.2.2.

6.5 Equivariant K-homology

In this section, we will define equivariant K-homology for spaces equipped with proper actions and prove its basic functoriality properties. We will then prove some 'induction' theorems relating the equivariant K-homology of a space equipped with a group action to the usual K-homology of associated quotient space.

Throughout this section, X and Y denote locally compact second countable Hausdorff spaces equipped with a proper action of a fixed countable discrete group G. Throughout, we will use the machinery of equivariant geometric modules from Section 4.5: see that section for notation and terminology. In particular, we will denote the unitary operators implementing the G action on such a module by U_g, $g \in G$.

Definition 6.5.1 Let H_X be an X-G module. Let $\mathbb{L}[H_X]^G$ be the invariant part of the algebraic localisation algebra under the action defined by

$$(T_t) \mapsto (U_g^* T_t U_g);$$

we call this the *equivariant localisation $*$-algebra* of H_X.

The *equivariant localisation C^*-algebra* or just *equivariant localisation algebra*, denoted $L^*(H_X)^G$, is defined to be the completion of $\mathbb{L}[H_X]^G$ for the norm $\|(T_t)\| := \sup_t \|T_t\|_{\mathcal{B}(H_X)}$.

Remark 6.5.2 Let us say that a subset K of X is *G-compact* if it is G-invariant, and if the associated quotient space K/G is compact. When defining $\mathbb{L}[H_X]^G$, we could replace the condition 'for any compact subset K of X, there exists $t_K \geqslant 0$ such that for all $t \geqslant t_K$, the functions

$$t \mapsto \chi_K T_t \quad \text{and} \quad t \mapsto T_t \chi_K$$

are uniformly norm continuous' in part (i) from Definition 6.2.1 with the condition 'for any G-compact subset K of X, there exists $t_K \geqslant 0$ such that for all $t \geqslant t_K$, the functions

$$t \mapsto \chi_K T_t \quad \text{and} \quad t \mapsto T_t \chi_K$$

[6] Almost: there is a slight gap, in that we did not prove that the naturality statement in Theorem 6.3.4 holds for all functions in the category $\mathcal{LC}$. The reader is asked to bridge this gap in Exercise 6.8.9.

are uniformly norm continuous'. Moreover, we may replace the condition 'for any compact subset K of X, there exists $t_K \geqslant 0$ such that for all $t \geqslant t_K$, the operators $\chi_K T_t$ and $T_t \chi_K$ are compact' in part (i) from Definition 6.2.1 with the condition 'for any G-compact subset K of X, there exists $t_K \geqslant 0$ such that for all $t \geqslant t_K$, the operators $\chi_K T_t$ and $T_t \chi_K$ are locally compact' (see Definition 5.1.1 for local compactness). Thanks to equivariance, it is not too difficult to check that these changes make no difference.

Functoriality works analogously to the non-equivariant case, using Section 4.5 rather than Section 4.4 for the underlying ingredients (compare also Section 5.2 where we discuss this for the equivariant Roe algebra). As a result we just give the definitions and results here, leaving the details to the reader.

The following is Definition 4.5.11, repeated for the reader's convenience.

Definition 6.5.3 Let H_X, H_Y be equivariant geometric modules, and $f\colon X \to Y$ a function. A family of isometries $(V_t\colon H_X \to H_Y)_{t\in[1,\infty)}$ is an *equivariant continuous cover* of f if:

(i) the function $t \mapsto V_t$ from $[1,\infty)$ to $\mathcal{B}(H_X, H_Y)$ is uniformly norm continuous;
(ii) for any open subset $U \subseteq Y^+ \times Y^+$ that contains the diagonal, there exists $t_U \geqslant 1$ such that for all $t \geqslant t_U$,

$$\operatorname{supp}(V_t) \subseteq \{(y,x) \in Y \times X \mid (y, f(x)) \in U\};$$

(iii) each V_t is G equivariant (in symbols, $U_g V_t = V_t U_g$ for all $g \in G$).

The following is a direct consequence of Proposition 4.5.12.

Lemma 6.5.4 *Let H_X, H_Y be equivariant geometric modules with H_Y ample, and $f\colon X \to Y$ an equivariant proper continuous map. Then an equivariant continuous cover for f exists.*

Moreover, if f is an equivariant homeomorphism and H_X is also ample, then there exists an equivariant continuous cover (V_t) for f where each V_t is a unitary isomorphism. □

The next lemma follows from the same argument as given for Lemma 6.2.7: one just needs to check the additional G invariance condition on operators, and this is automatic from equivariance of (V_t).

Lemma 6.5.5 *Let H_X, H_Y be ample equivariant geometric modules, and let $f\colon X \to Y$ be an equivariant, proper, continuous map. Let (V_t) be an equivariant continuous cover for f. Then taking adjoints by (V_t) pointwise in t*

$$(T_t) \mapsto (V_t T_t V_t^*)$$

defines a $*$-homomorphism

$$ad_{(V_t)}\colon \mathbb{L}[H_X]^G \to \mathbb{L}[H_Y]^G$$

that extends to a $*$-homomorphism from $L^*(H_X)^G$ to $L^*(H_Y)^G$.

Moreover, the map induced by $ad_{(V_t)}$ on K-theory depends only on f and not on the choice of (V_t). □

Definition 6.5.6 Let $f\colon X \to Y$ be an equivariant, proper, continuous function and let $L^*(H_X)^G$ and $L^*(H_Y)^G$ be localisation algebras associated to ample geometric modules. Define

$$f_*\colon K_*(L^*(H_X)^G) \to K_*(L^*(H_Y)^G)$$

to be the map on K-theory induced by the $*$-homomorphism

$$\mathrm{ad}_{(V_t)}\colon L^*(H_X)^G \to L^*(H_Y)^G$$

associated to some equivariant continuous cover for f as in Lemma 6.5.5 above.

Theorem 6.5.7 *For each X, choose an ample X-G module H_X. Then the assignments*

$$X \mapsto K_*(L^*(H_X)^G), \quad f \mapsto f_*$$

give a well-defined functor from the category of second countable, locally compact, Hausdorff spaces equipped with proper G actions, and equivariant proper continuous maps to the category $\mathcal{GA}$ of graded abelian groups.

Moreover, the functor that one gets in this way does not depend on the modules chosen up to canonical equivalence. □

Definition 6.5.8 The *equivariant K-homology* of X is defined by

$$K_n^G(X) := K_{-n}(L^*(H_X)^G)$$

for any choice of equivariant ample X-G module H_X.

Remark 6.5.9 Analogously to Section 6.4, one can extend this functor to the category with the same objects, but with morphisms given by continuous equivariant maps $f\colon X^+ \to Y^+$ that take infinity to infinity (here the G actions are extended to the one point compactifications by stipulating that they fix the point at infinity). There is a slight additional subtlety: the extended actions of G are no longer proper. One can get around this by using equivariant modules where the characteristic function of infinity $\chi_{\{\infty\}}$ acts as the projection onto a subspace that identifies as a G representation with $\ell^2(G, H)$. We have no applications of this in this book, so leave the details to interested readers: see Exercise 6.8.13.

In the remainder of this section, we give some applications that relate G equivariant K-homology to F equivariant K-homology, where F is a finite, or even trivial, subgroup of G.

We start with analogues of Definition 6.4.8 and Lemma 6.4.11 that will give useful technical tools.

Definition 6.5.10 Let H_X be an ample equivariant geometric module. Let $\mathbb{L}_0[H_X]^G$ be the collection of all $(T_t) \in \mathbb{L}[H_X]^G$ such that for any G-compact subset K of X there exists $t_K \geqslant 1$ such that for all $t \geqslant t_K$,

$$\chi_K T_t = T_t \chi_K = 0.$$

This is a $*$-ideal in $\mathbb{L}[H_X]^G$. We write $L_0^*(H_X)^G$ for its closure in $L^*(H_X)^G$ and

$$L_Q^*(H_X)^G := L^*(H_X)^G / L_0^*(H_X)^G$$

for the associated quotient.

Remark 6.5.11 Analogously to Remark 6.4.9, we could replace the condition 'for any G-compact subset K of X, there exists $t_K \geqslant 1$ such that for all $t \geqslant t_K$, we have $\chi_K T_t = T_t \chi_K = 0$' appearing above with 'for any G-compact subset K of X, we have $\lim_{t \to \infty} \chi_K T_t = \lim_{t \to \infty} T_t \chi_K = 0$'. This would make no difference on the level of the C^*-algebraic closure.

The proof of the next result is the same as that of Lemma 6.4.11, so it has been omitted here.

Lemma 6.5.12 *Let H_X be an ample geometric module. Then the quotient map $L^*(H_X)^G \to L_Q^*(H_X)^G$ induces an isomorphism on K-theory.* □

For the next result, recall from Example A.2.6 that if F is a subgroup of G, Y is an F space, then the *balanced product* is the quotient space of $Y \times G$ for the F action defined by

$$f \cdot (y, g) := (fy, f^{-1}g).$$

It is equipped with the quotient topology. We write $Y \times_F G$ for the balanced product, and $[y, g]$ for the point corresponding to (y, g).

Proposition 6.5.13 *Let F be a finite subgroup of G, let Y be an F space and let X be the balanced product $Y \times_F G$. Then there is a canonical* induction isomorphism

$$\Psi_Y : K_*^F(Y) \to K_*^G(X).$$

Moreover, this construction is natural in the following sense: if $f : Y_1 \to Y_2$ is an F equivariant continuous proper map, $X_j = Y_j \times_F G$ for $j \in \{1,2\}$, and $\widetilde{f} : X_1 \to X_2$ is defined by $\widetilde{f}([y,g]) = [f(y),g]$, then the diagram

$$\begin{array}{ccc} K_*^F(Y_1) & \xrightarrow{\Psi_{Y_1}} & K_*^G(X_1) \\ \downarrow f_* & & \downarrow \widetilde{f}_* \\ K_*^F(Y_2) & \xrightarrow{\Psi_{Y_2}} & K_*^G(X_2) \end{array}$$

commutes.

Proof Let H_Y be any ample F-Y module. Define

$$H_X := \{u \in \ell^2(G, H_Y) \mid u(gh) = U_h^* u(g) \text{ for all } g \in G,\ h \in F\},$$

which is a closed subspace of $\ell^2(G, H_Y)$ and thus a Hilbert space. Equip H_X with the G-action defined by

$$(V_g u)(k) := u(g^{-1}k), \quad g,k \in G.$$

Identify Y with the image of the subset $\{(y,h) \in Y \times G \mid h \in F, y \in Y\}$ of $G \times Y$ under the quotient map $Y \times G \to Y \times_F G$, and identify H_Y with the collection

$$H_Y = \{u \in H_X \mid u(g) = 0 \text{ for all } g \notin F\} \tag{6.8}$$

(this makes sense as an element of H_X that vanishes off F is uniquely determined by its image at the identity element of G).

Now let α denote the G-action on $C_0(X)$, and define a $C_0(X)$-action ρ on H_X by the formula

$$(\rho(f)u)(g) := \alpha_{g^{-1}}(f|_{gY})u(g);$$

this makes sense, as $\alpha_{g^{-1}}(f|_{gY})$ is supported in Y, so acts on $u(g) \in H_Y$. Computing, for $g,k \in G$, $f \in C_0(X)$ and $u \in H_X$, we get

$$\begin{aligned}(V_g \rho(f) V_g^* u)(k) &= (\rho(f) V_g^* u)(g^{-1}k) = \alpha_{k^{-1}g}(f|_{g^{-1}kY})(V_g^* u)(g^{-1}k) \\ &= \alpha_{k^{-1}}((\alpha_g f)|_{kY}))u(k) = (\rho(\alpha_g f)u)(k).\end{aligned}$$

Hence ρ is compatible with the G action on H_X, which is thus an X-G module. To see that H_X is ample as an X-module, assume that $f \in C_0(X)$ is non-zero. Then $f|_{gY}$ is non-zero for some fixed $g \in G$. Then up to our identification of H_Y with a subspace of H_X as in line (6.8) above, we see that

$V_g^* \chi_{gY} \rho(f) \chi_{gY} V_g$ is non-compact by ampleness of H_Y, and thus $\rho(f)$ is non-compact. We leave the algebraic check that H_X is locally free, and thus that it is ample as an X-G module, to the reader.

Now, provisionally define

$$\Phi \colon L^*(H_X)^G \to L^*(H_Y)^F, \quad (T_t) \mapsto (\chi_Y T_t \chi_Y)$$

and

$$\Psi \colon L^*(H_Y)^F \to L^*(H_X)^G, \quad (T_t) \mapsto \Bigg(\sum_{gF \in G/F} V_g T_t V_g^* \Bigg),$$

where we identify H_Y with a subspace of H_X as in line (6.8) above. We claim that these are well-defined linear maps, and that they descend to well-defined mutually inverse $*$-isomorphisms $L_Q(H_X)^G \cong L_Q(H_Y)^F$. It is straightforward that Φ is a well-defined linear map, using that Y is H-invariant. To see that Ψ is well-defined, note first that F-invariance of $(T_t) \in L^*(H_Y)^F$ implies that $V_g T_t V_g^*$ only depends on the coset gF, so the formula makes sense. The operator $\Psi(T_t)$ is clearly G-invariant, and the properties from Definition 6.2.1 that it needs to satisfy to be in $L^*(H_X)$ can be directly checked once we have observed that any compact subset of $Y \times_F G$ can only intersect finitely many subsets of the G-translates of Y (where Y as before is identified with the image of the subset $\{(y,h) \in Y \times G \mid h \in F, y \in Y\}$ of $Y \times G$ under the quotient map $Y \times G \to Y \times_F G$).

Now, it is clear that Φ sends $L_0^*(H_X)^G$ to $L_0^*(H_Y)^F$, and thus descends to a well-defined map on the quotient. Moreover, the multiplier χ_Y of $L_Q^*(H_X)$ is central by Lemma 6.4.18 from which it follows that Φ defines a $*$-homomorphism on the quotients. That Ψ also induces a well-defined map on the quotients follows from our earlier observation that any compact subset of $Y \times_F G$ can only intersect finitely many G-translates of Y. To complete the proof that Φ and Ψ induce mutually inverse $*$-isomorphisms on the quotients, it remains to prove that $\Psi(\Phi(T_t))$ and $\Phi(\Psi(S_t))$ differ from (T_t) and (S_t) by elements of $L_0^*(H_X)^G$ and $L_0^*(H_Y)^G$ respectively. In fact $\Phi(\Psi(S_t)) = (S_t)$ on the nose, as one checks directly. Computing in the other case,

$$\Psi(\Phi(T_t)) = \Bigg(\sum_{gF \in G/F} V_g \chi_Y T_t \chi_Y V_g^* \Bigg) = \Bigg(\sum_{gF \in G/F} \chi_{gY} T_t \chi_{gY} \Bigg),$$

where the second equality uses G-invariance of $(T_t) \in L^*(H_X)^G$. Passing to the quotient $L_Q^*(H_X)^G$, we may commute χ_{gY} past (T_t) to get that this is equal modulo $L_0^*(H_X)^G$ to

$$\Big(\sum_{gF\in G/F} \chi_{gY} T_t\Big).$$

However, using that

$$X \times_F Y = \bigsqcup_{gF\in G/F} gY,$$

this equals (T_t), and we have completed the proof that Φ and Ψ are mutually inverse $*$-isomorphisms on the level of the quotients. Combining this with Lemma 6.5.12 shows that the induced map

$$\Psi\colon L_Q^*(H_Y)^F \to L_Q^*(H_X)^G$$

induces the required isomorphism $\Psi_Y\colon K_*^F(Y) \to K_*^G(X)$ on K-theory.

Finally, naturality follows as all the constructions in the proof are canonical, and the construction of the module H_X from H_Y can be used to build a covering isometry for $\widetilde{f}$ from one for f. We leave the remaining details to the reader. □

We now turn to the case of a free action of a discrete group on X. In this case, we want to show that $K_*^G(X) \cong K_*(X/G)$. The following construction, which we isolate for later use, is the key idea of the proof.

Construction 6.5.14 Say G acts freely (properly, by homeomorphisms) on X. Let $\pi\colon X \to X/G$ be the quotient map. Let $\mathcal{U}$ be any open cover of X/G such that any compact subset of X/G intersects at most finitely many of the open sets in $\mathcal{U}$, and such that for each $U \in \mathcal{U}$ there is an open set $\widetilde{U} \subseteq X$ (which we fix from now on) such that π restricts to a homeomorphism from $\widetilde{U}$ to U, and such that the map

$$G \times \widetilde{U} \to \pi^{-1}(U), \quad (g,x) \mapsto gx$$

is a homeomorphism. Let $(\phi_U)_{U\in\mathcal{U}}$ be an 'ℓ^2' partition of unity subordinate to $\mathcal{U}$, where the qualifier 'ℓ^2' means that $\sum_{U\in\mathcal{U}} \phi_U(x)^2 = 1$ for all $x \in X/G$.

Now, let H_X be an ample X-G module (with associated unitaries $(U_g)_{g\in G}$ implementing the G action) and $H_{X/G}$ be an ample X/G-module. For each U, choose a continuous cover

$$(V_{U,t}\colon \chi_U H_{X/G} \to \chi_{\widetilde{U}} H_X)_{t\in[1,\infty)}$$

of the homeomorphism $(\pi|_{\widetilde{U}})^{-1}\colon U \to \widetilde{U}$; using Remark 4.4.5 we may assume that each operator $V_{U,t}$ is unitary. Provisionally define maps

$$\Phi\colon L^*(H_{X/G}) \to L^*(H_X)^G, \quad (T_t) \mapsto \Big(\sum_{g\in G, U\in\mathcal{U}} U_g V_{U,t} \phi_U T_t \phi_U V_{\widetilde{U},t}^* U_g^*\Big)$$

and

$$\Psi \colon L^*(H_X)^G \to L^*(H_{X/G}), \quad (T_t) \mapsto \left(\sum_{U \in \mathcal{U}} \phi_U V_{U,t}^* T_t V_{U,t} \phi_U \right).$$

The next theorem is our final goal in this section. To avoid the computations in the proof going on too long, we leave some more details to the reader this time.

Theorem 6.5.15 *Say G acts freely on X. Then the maps $\Phi \colon L^*(H_X)^G \to L^*(H_{X/G})$ and $\Psi \colon L^*(H_{X/G}) \to L^*(H_X)^G$ are well defined and induce mutually inverse $*$-isomorphisms $L_Q^*(H_X)^G \cong L_Q^*(H_{X/G})$ on the quotients. In particular there is an isomorphism*

$$K_*^G(X) \cong K_*(X/G).$$

Proof We leave it to the reader to check that Φ and Ψ descend to well-defined maps on the quotients $L_Q^*(H_X)^G$ and $L_Q^*(H_{X/G})$, and just check that they are $*$-homomorphic, and mutual inverses. To see that Ψ is a $*$-homomorphism, note that it is clearly linear and $*$-preserving. To see multiplicativity, let (T_t) and (S_t) be elements of $L^*(H_X)^G$. Then

$$\Psi(T_t)\Psi(S_t) = \left(\sum_{U,V \in \mathcal{U}} \phi_U V_{U,t}^* T_t V_{U,t} \phi_U \phi_V V_{V,t}^* S_t V_{V,t} \phi_V \right).$$

Using that we are working in the quotient $L_Q^*(H_{X/G})$ and Lemma 6.4.18, we may commute each term $V_{U,t} \phi_U \phi_V V_{V,t}^*$ past (S_t) to get that this equals

$$\left(\sum_{U,V \in \mathcal{U}} \phi_U V_{U,t}^* T_t S_t V_{U,t} \phi_U \phi_V V_{V,t}^* V_{V,t} \phi_V \right).$$

On the other hand, $\phi_V V_{V,t}^* V_{V,t} = \phi_V$, and $\sum_V \phi_V^2$ is the identity, so this equals

$$\left(\sum_{U} \phi_U V_{U,t}^* T_t S_t V_{U,t} \phi_U \right) = \Psi(T_t S_t)$$

and we have multiplicativity. To complete the proof, it thus suffices to prove that both compositions $\Phi \circ \Psi$ and $\Psi \circ \Phi$ are the identity maps on the respective algebras.

Computing, for $(T_t) \in L^*(H_X)^G$

$$\Phi(\Psi(T_t)) = \left(\sum_{U,V \in \mathcal{U}, g \in G} U_g V_{U,t} \phi_U \phi_V V_{V,t}^* T_t V_{V,t} \phi_V \phi_U V_{U,t}^* U_g^* \right).$$

As we are working in the quotient $L_Q^*(H_X)^G$, we may commute the terms $V_{U,t}\phi_U\phi_V V_{V,t}^*$ past (T_t) to get that this equals

$$\Big(\sum_{U,V,g} U_g T_t V_{U,t}\phi_U^2\phi_V^2 V_{U,t} U_g^*\Big).$$

Using that $\sum_V \phi_V^2$ is the identity, and that (T_t) is G-invariant, this equals

$$\Big(\sum_{U,g} T_t U_g V_{U,t}\phi_U^2 V_{U,t}^* U_g^*\Big).$$

Finally, we have that $\sum_{U,g} U_g V_{U,t}\phi_U^2 V_{U,t}^* U_g^*$ is the identity on $H_{X/G}$ by the properties of the original cover $\mathcal{U}$, completing this computation.

On the other hand, for $(T_t) \in L^*(H_{X/G})$,

$$\Psi(\Phi(T_t)) = \Big(\sum_{U,V\in\mathcal{U},g\in G} \phi_V V_{V,t}^* U_g V_{U,t}\phi_U T_t \phi_U V_{U,t}^* U_g^* V_{V,t}\phi_V\Big).$$

Using that we are in the quotient $L_Q^*(H_{X/G})$, we may commute $\phi_V V_{V,t}^* U_g V_{U,t}\phi_U$ with (T_t) to get that this equals

$$\Big(\sum_{U,V,g} T_t\phi_V V_{V,t}^* U_g V_{U,t}\phi_U^2 V_{U,t}^* U_g^* V_{V,t}\phi_V\Big).$$

The properties of the original cover $\mathcal{U}$ and of the partition of unity $\{\phi_U\}$ imply that $\sum_{U,g} U_g V_{U,t}\phi_U^2 V_{U,t}^* U_g^*$ is the identity on H_X, and thus the above equals

$$\sum_{U,V,g} T_t\phi_V V_{V,t}^* V_{V,t}\phi_V.$$

As $\phi_V V_{V,t}^* V_{V,t}\phi_V = \phi_V^2$, and as $\sum_V \phi_V^2$ is the identity on $H_{X/G}$ we are done. □

6.6 The Localised Roe Algebra

In this section, we introduce the localised Roe algebra $C_L^*(X)$. This is a variant of the localisation algebra whose K-theory is also a model for K-homology. This will be important for the next chapter when we discuss assembly maps. The localised Roe algebra $C_L^*(X)$ is more concrete than the localisation algebra $L^*(X)$ and more closely connected to Roe algebras. The reason we do not use $C_L^*(X)$ to define K-homology is that it is only defined for proper metric spaces, it has less good functoriality properties than $L^*(X)$ and it is less closely connected to elliptic differential operators.

Throughout this section, X and Y are proper metric spaces equipped with proper isometric actions of a countable discrete group G as in Section A.3. To avoid too much repetition, all maps, geometric modules, Roe algebras and K-homology groups in this section are considered equivariant; we will not state this explicitly. We allow the case that G is trivial, in which case equivariant modules are the same as usual modules. It might be easier for the reader to simply assume that G is trivial on a first reading, although this does not make a substantial difference.

Definition 6.6.1 Let H_X be a geometric module, and $\mathbb{C}[H_X]^G$ the associated Roe $*$-algebra (see Definition 5.2.1, or Definition 5.1.4 when G is trivial). Define $\mathbb{C}_L[H_X]^G$ to be the $*$-algebra of all uniformly continuous, bounded functions (T_t) from $[1,\infty)$ to $\mathbb{C}[H_X]^G$ such that for any open neighbourhood U of the diagonal in $X^+ \times X^+$, there exists $t_U \geqslant 1$ such that for all $t > t_U$,

$$\mathrm{supp}(T_t) \subseteq U.$$

We let $C_L^*(H_X)^G$ denote the completion of $\mathbb{C}_L[H_X]^G$ for the supremum norm

$$\|(T_t)\| := \sup_t \|T_t\|_{\mathcal{B}(H_X)}.$$

We will call $\mathbb{C}_L[H_X]^G$ the *localised Roe $*$-algebra* associated to H_X and $C_L^*(H_X)^G$ the *localised Roe C^*-algebra*, or just *localised Roe algebra*, associated to H_X.

With $\mathbb{L}[H_X]^G$ as in Definition 6.5.1 (or Definition 6.2.1 when G is trivial), there are inclusions

$$\mathbb{C}_L[H_X]^G \to \mathbb{L}[H_X]^G, \quad C_L^*(H_X)^G \to L^*(H_X)^G$$

of $*$-subalgebras of $\ell^\infty([1,\infty),\mathcal{B}(H_X))$.

Proposition 6.6.2 *Let H_X be an ample geometric module. Then the natural inclusion map*

$$C_L^*(H_X)^G \to L^*(H_X)^G$$

above induces an isomorphism on K-theory.

In particular, $K_*(C_L^*(H_X)^G)$ is canonically isomorphic to the K-homology of X.

Proof Let $\mathbb{L}_0[H_X]^G$ denote the $*$-ideal of elements (T_t) in $\mathbb{L}[H_X]^G$ such that for any G-compact subset K of X, there is a $t_K \geqslant 0$ such that for all $t \geqslant t_K$,

$$\chi_K T_t = T_t \chi_K = 0$$

as in Definition 6.5.10, and $L_0^*(H_X)^G$ denote its closure. Let $C_{L,0}^*(H_X)^G$ denote the intersection

$$C_{L,0}^*(H_X)^G := L_0^*(H_X)^G \cap C_L^*(H_X)^G. \tag{6.9}$$

Lemma 6.5.12 gives that $K_*(L_0^*(H_X)^G) = 0$, and the same proof as Lemma 6.4.11 shows that $K_*(C_{L,0}^*(H_X)^G) = 0$. It will thus suffice to show that the induced inclusion

$$\frac{C_L^*(H_X)^G}{C_{L,0}^*(H_X)^G} \to \frac{L^*(H_X)^G}{L_0^*(H_X)^G} \tag{6.10}$$

induces an isomorphism on the level of K-theory. In fact, we will show that it is an isomorphism of C^*-algebras.

Now, thanks to the definition in line (6.9), the map in line (6.10) above is injective, so it suffices to show that it is surjective, that is, that any element (T_t) in $L^*(H_X)^G$ is equivalent to an element of $C_L^*(H_X)^G$ modulo $L_0^*(H_X)^G$. We will do this in two steps: first we will show that an arbitrary (T_t) in $L^*(H_X)^G$ is equal modulo $L_0^*(H_X)^G$ to an element (S_t) with $\sup_t \operatorname{prop}(S_t) < \infty$. Second we will show that such an (S_t) is equal modulo $L_0^*(H_X)^G$ to an element $(R_t) \in L^*(H_X)^G$ such that $\sup_t \operatorname{prop}(R_t) < \infty$, such that the function

$$[1,\infty) \to \mathcal{B}(H_X), \quad t \mapsto R_t$$

is uniformly continuous and such that each R_t is locally compact in the sense of Definition 5.1.1.

First, we look at replacing (T_t) by a finite propagation family. Fix $r \in (0,\infty)$. The open cover $\{B(x;r) \mid x \in X\}$ of X is equivariant, and thus by Corollary A.2.8 there exists a partition of unity $(\phi_i)_{i\in I}$ on X such that each ϕ_i is supported in some $B(x;r)$, such that any compact subset of X only intersects the support of finitely many of the ϕ_i and such that for each $i \in I$ and $g \in G$, there is $j \in I$ with $\phi_i(gx) = \phi_j(x)$ for all $x \in X$. Provisionally define a map

$$\Phi\colon \mathcal{B}(H_X) \to \mathcal{B}(H_X), \quad T \mapsto \sum_{i\in I} \sqrt{\phi_i}\,T\sqrt{\phi_i}.$$

To see that this makes sense, note that the formula

$$V\colon H_X \to \bigoplus_{i\in I} H_X, \quad u \mapsto \big(\sqrt{\phi_i}\,u\big)_{i\in I}$$

is an isometry and one can compute that $\Phi(T) = VTV^*$ (with convergence of the sum above in the strong operator topology). Hence Φ is a well-defined unital contraction. If we set $S_t := \Phi(T_t)$, it is not too difficult to see that the family (S_t) has uniformly finite propagation. Moreover, for any G-compact subset K of X and any t,

$$\chi_K(S_t - T_t) = \sum_{i \in I} \chi_K \sqrt{\phi_i}([\sqrt{\phi_i}, T_t]).$$

This tends to zero in norm as t tends to infinity: indeed, all the terms are translates under the G action of only finitely many non-zero terms (by local finiteness of the partition of unity), and each of the commutators tends to zero by Lemma 6.1.2. Hence $(S_t - T_t)$ is an element of $L_0^*(H_X)^G$ by Remark 6.5.11 and we are done with the first part.

For the second part, recall from Lemma A.2.5 that the quotient space $Y := X/G$ is a locally compact and Hausdorff space. Let Y^+ be the one-point compactification of Y, and choose a metric on d on Y^+. Let $\pi : X \to Y$ be the quotient map, and consider the function on X defined by

$$\delta : X \to [0, \infty), \quad x \mapsto \frac{1}{d(\pi(x), \infty)}.$$

For each n, let $K_n := \delta^{-1}([0, n])$, so each K_n is G-invariant and G-compact, $K_n \subseteq K_{n+1}$ for each n, and $X = \bigcup_{n=1}^{\infty} K_n$. For each n, let t_n be such that for all $t \geqslant t_n$ we have that $\chi_{K_n} T_t$ and $T_t \chi_{K_n}$ are locally compact (see Definition 5.1.1), and such that the functions $t \mapsto \chi_{K_n} T_t$ and $t \mapsto T_t \chi_{K_n}$ are uniformly continuous; such a t_n exists by Remark 6.5.2. Increasing the t_n if necessary, we may assume that the sequence (t_n) is strictly increasing and that $|t_{n+1} - t_n| > 1$ for all n. Define f_{t_1} to be the zero function from $[0, \infty)$ to $[0, 1]$, and for each $n > 1$, define $f_{t_n} : [0, \infty) \to [0, 1]$ by setting the function to be 1 on $[0, n-1]$, to be 0 on $[n, \infty)$ and to linearly interpolate between 0 and 1 on $[n-1, n]$. Define $f_t : [0, \infty) \to [0, 1]$ by linearly interpolating between f_{t_n} and $f_{t_{n+1}}$ on the interval $[t_n, t_{n+1}]$, and set $f_t = 0$ for $t \leqslant t_1$. Define finally for each $t \in [1, \infty)$,

$$\psi_t : X \to [0, \infty), \quad x \mapsto f_t(\delta(x)),$$

which gives a uniformly continuous function from $[1, \infty)$ to $\mathcal{B}(H_X)$. Now define $R_t := \psi_t S_t \psi_t$. Using the properties of (ψ_t) and of (S_t), it is not too difficult to see that (R_t) is in $C_L^*(H_X)^G$: the key points here are that modifying S_t to R_t can only decrease propagation and that $\chi_{K_n} R_t = 0 = R_t \chi_{K_n}$ whenever $t \leqslant t_n$. Moreover, one checks that $(S_t - R_t)$ is in $L_0^*(H_X)^G$: the key point for this is that ψ_t is constantly equal to one on any G-compact set for all suitably large t. This completes the proof. □

Unfortunately, the constructions of Sections 6.2 and 6.5 do not translate directly to show that $K_*(C_L^*(H_X)^G)$ is functorial under pointed maps $f\colon X \to Y$ (or even proper continuous maps $f\colon X \to Y$); indeed, this is one of the reasons we used $L^*(H_X)^G$ instead of $C_L^*(H_X)^G$ to define K-homology.[7]

However, if $f\colon X \to Y$ is a continuous *and* coarse map and H_X, H_Y are ample geometric modules, then (a slight elaboration on) Proposition 4.5.12 shows that there exists a continuous cover $(V_t\colon H_X \to H_Y)$ of f such that each V_t itself covers f. The proofs of Lemmas 5.1.12 and 6.2.7 translate directly to show that conjugation by (V_t) defines a $*$-homomorphism

$$\mathrm{ad}_{(V_t)}\colon C_L^*(H_X)^G \to C_L^*(H_Y)^G.$$

Theorem 6.6.3 *Let $\mathcal{P}ro^G$ denote the category of proper metric spaces equipped with proper isometric G actions, with morphisms given by continuous equivariant coarse maps. Then choosing an ample module H_X for each object X and continuous cover (V_t) for each morphism f defines a functor*

$$X \mapsto K_*(C_L^*(H_X)^G), \quad f \mapsto (ad_{(V_t)})_*$$

from $\mathcal{P}ro^G$ to $\mathcal{GA}$ that does not depend on any of the choices involved up to canonical equivalence. Moreover, this functor agrees (up to canonical equivalence) with the restriction of 'the' K-homology functor from $\mathcal{LC}^G$ to $\mathcal{P}ro^G$.

Proof The fact that we have a well-defined functor follows from the same arguments as given for Theorems 5.2.6 and 6.5.7. The fact that it agrees with K-homology follows directly from Proposition 6.6.2. □

6.7 Other Pictures of K-homology

This section will not be used in the rest of the text: we include it to help orient readers who are interested in how our picture of K-homology relates to two of the other standard analytic models for K-homology: one based on KK-theory and one based on E-theory.

Throughout this section, X is a locally compact, second countable, Hausdorff topological space. We will write $C_0(X)$ for the continuous functions on

[7] As we mentioned at the start of the chapter, the other reasons are that $C_L^*(H_X)^G$ does not behave very well for spaces that are locally compact but not necessarily proper, and that $L^*(H_X)^G$ is easier to deal with when analysing differential operators.

X vanishing at infinity whether or not X is compact (so if X is compact, $C_0(X)$ identifies canonically with $C(X)$: see Example 1.3.1).

We start with the KK-theoretic picture. The basic building blocks are Fredholm modules as in the following definition. We will deviate slightly from standard conventions by using an ungraded picture of picture of KK-theory as this is more convenient for our purposes.

Definition 6.7.1 An *even Fredholm module* consists of a pair (H_X, F) where

(i) H_X is an X module;
(ii) F is a bounded operator on H_X such that

$$f(1 - FF^*), \quad f(1 - F^*F), \quad [F, f]$$

are compact operators for all $f \in C_0(X)$.

An *odd Fredholm module* over X consists of a pair (H_X, F) where:

(i) H_X is an X module;
(ii) F is a bounded operator on H_X such that

$$f(F - F^*), \quad f(F^2 - 1), \quad [F, f]$$

are compact operators for all $f \in C_0(X)$.

Write $\mathcal{E}^0(X)$, respectively $\mathcal{E}^1(X)$, for the collections of all even, respectively odd, Fredholm modules over X.

Kasparov's K-homology groups can then be defined as follows:

Definition 6.7.2 Let $\sim$ be the equivalence relation on $\mathcal{E}^i(X)$ generated by the following two relations:

(i) (H_X^0, F_0) is *unitarily equivalent* to (H_X^1, F_1) if there is a unitary isomorphism $V : H_X^0 \to H_X^1$ that respects the X module structure and conjugates F_0 to F_1.
(ii) (H_X, F_0) is *operator homotopic* to (H_X, F_1) if there is a family of Fredholm modules of the form $(H_X, F_t)_{t \in [0,1]}$ that agrees with the given modules on the endpoints and is such that the function $t \mapsto F_t$ is norm continuous.

As a set, the *Kasparov K-homology group* is defined to be

$$KK_i(X) := \mathcal{E}^i(X)/\sim$$

for $i = 0, 1$. There is a binary operation defined in both even and odd cases by

$$[H_{X,0}, F_0] + [H_{X,1}, F_1] := [H_{X,0} \oplus H_{X,1}, F_0 \oplus F_1]$$

(and similarly in the odd case by just removing the grading operators) that makes $KK_i(X)$ into an abelian group for $i = 0, 1$.

We now construct a homomorphism $KK_i(X) \to K_i(X)$ from Kasparov's K-homology groups to ours.

Construction 6.7.3 We first deal with the even case. Let (H_X, F) be a Kasparov module. Fix a metric on X and fix n for now. Let $(\phi_i)_{i \in I}$ be a locally finite[8] compactly supported partition of unity on X subordinate to the cover by balls of radius 2^{-n}. Let $V : H_X \to \ell^2(I, H_X)$ be defined by

$$(Vu)(i) := \sqrt{\phi_i}u.$$

Then one checks that V is an isometry with adjoint given by $V^*u = \sum_{i \in I} \sqrt{\phi_i}u$ and that for any $T \in \mathcal{B}(H_X)$ we have

$$VTV^* = \sum_{i \in I} \sqrt{\phi_i}T\sqrt{\phi_i},$$

with convergence in the strong operator topology. Hence in particular,

$$F_n := \sum_{i \in I} \sqrt{\phi_i}F\sqrt{\phi_i}$$

converges in the strong operator topology to an operator of norm at most $\|F\|$. Moreover, using the condition that $[F, \sqrt{\phi_i}]$ is compact for each i and local finiteness of the partition of unity, one checks directly that $f(1 - F_n^*F_n)$ and $f(1 - F_nF_n^*)$ are compact for all $f \in C_c(X)$, (whence also for all $f \in C_0(X)$).

Unfix n, and for $t \in [n, n+1)$ define

$$F_t := (t - n)F_{n+1} + (n + 1 - t)F_n.$$

Then the family $(F_t)_{t \in [1,\infty)}$ defines a multiplier of the localisation algebra $L^*(H_X)$ that is moreover unitary in the quotient $M(L^*(H_X))/L^*(H_X)$. Hence (F_t) defines a class $[(F_t)] \in K_1(M(L^*(H_X))/L^*(H_X)$, and taking its image under the boundary (or index) map

$$\partial : K_1(M(L^*(H_X))/L^*(H_X)) \to K_0(L^*(H_X))$$

appearing in Theorem 2.6.5 gives a class $\partial[(F_t)]$ in $K_0(L^*(H_X))$. Finally, fixing any ample X module H'_X and a continuous cover $(V_t : H_X \to H'_X)$ gives a map

$$(\mathrm{ad}_{(V_t)})_* : K_0(L^*(H_X)) \to K_0(L^*(H'_X))$$

[8] This means that any compact subset of X only intersects the support of finitely many of the ϕ_i.

(compare Remark 6.3.2). Our natural transformation is then defined by

$$KK_0(X) \ni [H_X, F] \mapsto (\mathrm{ad}_{(V_t)})_* \partial[(F_t)] \in K_0(X).$$

The case of KK_1 is similar. This time, one starts with (H_X, F) and sets $P = \frac{1}{2}(F+1)$. Then P satisfies that $f(P^2 - P)$ and $f(P - P^*)$ are compact for all $f \in C_0(X)$. Applying the same construction as above, compressing by partitions of unity gives rise to a multiplier $(P_t)_{t\in[1,\infty)}$ of $L^*(H_X)$ whose image in the quotient $M(L^*(H_X)/L^*(H_X)$ is a projection, and that thus defines a class $[(P_t)] \in K_0(M(L^*(H_x)/L^*(H_X))$. Applying the K-theory boundary map

$$\partial\colon K_0(M(L^*(H_X))/L^*(H_X)) \to K_1(L^*(H_X))$$

defined in Theorem 2.6.5 and using a continuous cover (V_t) to transfer to the localisation algebra over an ample module completes the construction, giving a natural transformation

$$KK_0(X) \ni [H_X, F] \mapsto (\mathrm{ad}_{(V_t)})_* \partial[(P_t)] \in K_1(X).$$

We will not justify this here, but this construction gives the required natural transformation from the KK-theoretic picture of K-homology to ours, and indeed induces an isomorphism

$$KK_i(C_0(X), \mathbb{C}) \to K_i(L^*(X))$$

for all X and $i \in \{0, 1\}$. Once one has shown that it is a natural transformation, proving that it is an isomorphism can be done following a standard pattern for comparing homological functors:

(i) first show that it is an isomorphism for a point (see Exercise 6.8.16);
(ii) then deduce from this that it is an isomorphism for a finite simplicial complex using homotopy invariance, Mayer–Vietoris sequences and the five lemma;
(iii) then use a limiting argument and the fact that every compact metric space is an inverse limit of a sequence of finite simplicial complexes using nerves of increasingly fine finite covers of X;
(iv) finally, deduce the case of general locally compact spaces by considering one point compactifications.

The reader is led through this argument in Exercises B.3.2 and B.3.3.

We now look at E-theory in the sense of Connes and Higson. This time we will construct a natural transformation going from our K-homology groups $K_i(X)$ to the corresponding E-theory group $E_i(X)$.

Definition 6.7.4 Let A and B be (separable) C^*-algebras.

An *asymptotic morphism* from A to B is a $*$-homomorphism

$$\alpha\colon A \to \frac{C_b([1,\infty),B)}{C_0([1,\infty),B)}.$$

Two asymptotic morphisms α_0 and α_1 are *homotopic* if there is an asymptotic morphism α from A to $C([0,1],B)$ that evaluates to α_0 and α_1 at the endpoints.[9] Write $[\![A,B]\!]$ for the collection of homotopy classes of asymptotic morphisms from A to B.

The *E-theory groups* $E(A,B)$ can then be defined by

$$E_i(A,B) := [\![C_0(\mathbb{R}^{1+i})\otimes A, C_0(\mathbb{R})\otimes B\otimes \mathcal{K}]\!]$$

for $i\in\{0,1\}$. The *E-homology groups* of a space X are the special case $E_i(X) := E(C_0(X),\mathbb{C})$.

The basic idea of the natural transformation $K_i(L^*(X)) \to E_i(X)$ that we want to construct goes as follows. Let (P_t) be a projection in $L^*(X)$ representing an element of $K_0(X)$. Consider the map

$$C_c(X) \to \frac{C_b([1,\infty),\mathcal{K})}{C_0([1,\infty),\mathcal{K})}, \quad f \mapsto (t\mapsto P_t f),$$

which makes sense using Lemma 6.1.2. This is norm continuous, so extends to an asymptotic morphism

$$C_0(X) \to \frac{C_b([1,\infty),\mathcal{K})}{C_0([1,\infty),\mathcal{K})}.$$

Thus our projection gives rise to an asymptotic morphism from A to $\mathcal{K}$, and after suspension an element of $E_0(X)$. Unfortunately, it need not be true that $K_i(L^*(X))$ is generated by the classes of projections in (matrices over) $L^*(X)$: one needs to work with formal differences in the unitisation. As such, we give a more sophisticated picture that works in general. For this we need to work with general bivariant E-theory groups; we refer the reader to Section 6.9, Notes and References, at the end of the chapter for more information.

Construction 6.7.5 There is a $*$-homomorphism

$$\alpha\colon C_0(X)\otimes L^*(X) \to \frac{C_b([1,\infty),\mathcal{K})}{C_0([1,\infty),\mathcal{K})}.$$

[9] Warning: this is strictly weaker than α_0 and α_1 being homotopic as $*$-homomorphisms from A to $C_b([1,\infty),B)/C_0([1,\infty),B)$.

defined on elementary tensors $f \otimes (T_t)$ with $f \in C_c(X)$ by $f \otimes (T_t) \mapsto (t \mapsto fT_t)$. For a separable C^*-algebra A and possibly non-separable C^*-algebra B (such as $C_0(X) \otimes L^*(X)$), define

$$E_i^{\text{sep}}(B, A) := \lim_{\leftarrow} E_i(B_0, A), \quad E_i^{\text{sep}}(A, B) := \lim_{\rightarrow} E_i(A, B_0),$$

where the inverse and direct limit are taken over all separable C^*-subalgebras B_0 of B, using the natural functoriality of E-theory (contravariant in the first variable and covariant in the second). Then there is a well-defined product

$$E_i^{\text{sep}}(\mathbb{C}, L^*(X)) \otimes E_j^{\text{sep}}(C_0(X) \otimes L^*(X), \mathbb{C}) \to E_{i+j}(C_0(X), \mathbb{C}).$$

The group $E_i^{\text{sep}}(\mathbb{C}, L^*(X))$ identifies canonically with $K_i(L^*(X))$, while α naturally defines an element of $E_0^{\text{sep}}(C_0(X) \otimes L^*(X), \mathbb{C})$. The map

$$K_i(L^*(X)) \to E_i(X)$$

we want is then defined by taking the product with $\alpha \in E_0^{\text{sep}}(C_0(X) \otimes L^*(X), \mathbb{C})$ in the sense above.

Again, these homomorphisms define a natural transformation from our model of K-homology to E-theory that gives an isomorphism

$$K_i(L^*(X)) \to E_i(X)$$

for all X. Once one has seen that this map is a natural transformation, the argument that it is an isomorphism follows the same pattern sketched above for Kasparov theory (and exposited in general in Exercises B.3.2 and B.3.3).

6.8 Exercises

6.8.1 Let X be a second countable, locally compact, Hausdorff topological space, and H_X a geometric module. Let $(T_t)_{t \in [1,\infty)}$ be a norm continuous family of operators on H_X. Show that the following are equivalent:

(i) for any open neighbourhood U of the diagonal in $X^+ \times X^+$, there exists $t_U \geqslant 1$ such that for all $t > t_U$,

$$\text{supp}(T_t) \subseteq U;$$

(ii) for any open neighbourhood U of the diagonal in $X \times X$ and any $x \in X$ there is an open neighbourhood V of x and $t_{U,V} \geqslant 1$ such that for all $t > t_{U,V}$,

$$\text{supp}(\chi_V T_t) \subseteq U \quad \text{and} \quad \text{supp}(T_t \chi_V) \subseteq U;$$

(iii) for any open neighbourhood U of the diagonal in $X \times X$ and compact subset K of X, there is $t_{U,K} \geqslant 1$ such that for all $t > t_{U,K}$,

$$\text{supp}(\chi_K T_t) \subseteq U \quad \text{and} \quad \text{supp}(T_t \chi_K) \subseteq U.$$

Say X^+ is equipped with a metric inducing the topology. Show that (i) above is equivalent to the analogous condition where 'for any open neighbourhood of U of the diagonal in $X^+ \times X^+$' is replaced with 'for any $\epsilon > 0$', and '$\text{supp}(T_t) \subseteq U$' is replaced by '$\text{prop}(T) < \epsilon$'. Say X is equipped with a metric inducing the topology. Show that (ii) and (iii) are equivalent to the analogous conditions where 'for any open neighbourhood of U of the diagonal in $X \times X$' is replaced with 'for any $\epsilon > 0$', and the conditions of the form '$\text{supp}(\cdot) \subseteq U$' are replaced by '$\text{prop}(\cdot) < \epsilon$'.

6.8.2 Fill in the details in Remark 6.4.9: that is, show that the closures of the sets

$$\{(T_t) \in \mathbb{L}[H_X] \mid \lim_{t\to\infty} \chi_K T_t = \lim_{t\to\infty} T_t \chi_K = 0 \text{ for all compact } K \subseteq X\}$$

and

$$\left\{ (T_t) \in \mathbb{L}[H_X] \;\middle|\; \begin{array}{l} \text{for all compact } K \subseteq X \text{ there is } t_K \geqslant 1 \\ \text{such that } \chi_K T_t = T_t \chi_K = 0 \text{ for all } t \geqslant t_K \end{array} \right\}$$

inside $L^*(H_X)$ are the same.

6.8.3 (i) Show that any element of $L^*(H_X)$ as in Definition 6.2.1 can be represented uniquely as a function (T_t) from $[1, \infty)$ to $\mathcal{B}(H_X)$ (in principle, this is only true for the dense subset $\mathbb{L}[H_X]$).

(ii) Show that a function as in part (i) is in $L_0^*(X)$ (see Definition 6.4.8) if and only if for any compact subset K of X,

$$\lim_{t\to\infty} \chi_K T_t = \lim_{t\to\infty} T_t \chi_K = 0.$$

(iii) Show that $L_Q^*(H_X)$ (Definition 6.4.8 again) is the separated completion of $L^*(H_X)$ for the seminorm

$$\|(T_t)\| = \sup_{K \subseteq X \text{ compact}} \limsup_{t\to\infty} \max\{\|\chi_K T_t\|, \|T_t \chi_K\|\}.$$

6.8.4 Extend the proof of Proposition 6.1.1 to show that the following are equivalent for a continuous family $(T_t)_{t\in[1,\infty)}$ of uniformly bounded operators on an X module H_X:

(i) for all $f \in C_0(X)$, $\lim_{t\to\infty} \|[f, T_t]\| = 0$;

(ii) there exists a continuous family $(S_t)_{t\in[1,\infty)}$ such that $\lim_{t\to\infty}\|S_t - T_t\| = 0$ and such that for any open subset U of $X^+\times X^+$ containing the diagonal, there exists $t_U \geqslant 1$ such that for all $t > t_U$, $\mathrm{supp}(S_t) \subseteq U$.

6.8.5 In Definition 6.4.3, we are careful in condition (i) to say that K is a subset of X, not of X^+. What would we get if we just took K to be a compact subset of X^+ and defined the localisation algebra accordingly (and took its K-theory)? What about if we looked at the compactness and continuity conditions separately? Similarly, what would happen if we replaced $X^+ \times X^+$ with $X \times X$ in part (ii)?

6.8.6 Using the Mayer–Vietoris sequence, homotopy invariance and the computation of the K-homology groups of a point, compute the K-homology groups of spheres, tori and orientable surfaces.

6.8.7 Prove Bott periodicity in K-homology in the form

$$K_i(\mathbb{R}^n \times X) \cong K_{i-n}(X).$$

Hint: Mayer–Vietoris, homotopy invariance and induction on n.

6.8.8 Let Y be a closed subspace of a second countable, locally compact, Hausdorff space X. Prove that for any ample X module H_X there is a projection P_Y in the multiplier algebra of $L^*(H_X)$ such that the corner $P_Y L^*(H_X) P_Y$ is isomorphic to $L^*(Y)$, and such that the ideal generated by

$$L^*(H_X) P_Y L^*(H_X)$$

equals $L^*_Y(H_X)$ as in Definition 6.3.5.
Hint: you can do this directly, or using covering isometries and (the proof of) Lemma 6.3.6.

6.8.9 Prove that the naturality statement of Theorem 6.3.4 extends to an analogous version for general morphisms in the category $\mathcal{LC}$.

6.8.10 (i) Let X be a single point space, and let G be a finite group acting (trivially!) on X. Compute the equivariant K-homology $K_*^G(X)$ in terms of the unitary dual $\hat{G}$ of G as in Exercise 5.4.14: you should get $K_0^G(X) = \mathbb{Z}^{\hat{G}}$ and $K_1^G(X) = 0$.
Hint: mimic the argument of Proposition 6.3.3.

(ii) Show that $K_0^G(G) \cong \mathbb{Z}$ and $K_1^G(G) = 0$, where G acts on itself by translations.
Hint: this can be done from Proposition 6.5.13, but can also be done more directly by mimicking the argument of Proposition 6.3.3.

(iii) Show that the map $K_0^G(G) \to K^G(X)$ induced by the (equivariant) collapse map $G \to X$ is given on K-theory by

$$\mathbb{Z} \to \mathbb{Z}^{\hat{G}}, \quad a \mapsto (a, a, \ldots, a)$$

(compare Exercise 5.4.14).
Hint: one way to do this is to use the argument of Proposition 6.3.3 to show that the map $K_0^G(G) \to K_0^G(X)$ is induced by the canonical inclusion

$$\big(C(G) \otimes \mathcal{K}(\ell^2(G) \otimes H)\big)^G \to \big(\mathcal{K}(\ell^2(G)) \otimes \mathcal{K}(\ell^2(G) \otimes H)\big)^G,$$

where H is an infinite-dimensional Hilbert space with trivial G action, and compute the map induced on K-theory by this.

6.8.11 Let X be a metric space, and let H_X be an ample X module. Show that if X is non-compact, then H_X is also an ample X^+ module. Show that if X is compact, then we may 'extend' H_X to an ample X^+ module by taking $H_{X^+} = H_X \oplus H$ for any separable infinite-dimensional H, and having $f \in C(X^+)$ act on $(v, w) \in H_X \oplus X$ by

$$f \cdot (v, w) = (f|_X v, f(\infty) w).$$

6.8.12 Generalise the Mayer–Vietoris and homotopy invariance results for non-equivariant K-homology to the equivariant case.

6.8.13 Fill in the details in Remark 6.5.9.

6.8.14 Show that the technical issue raised in Exercise 5.4.13 is a non-issue on the level of K-theory for localisation algebras in the following sense: the inclusion of the G-invariant part of $L^*(X)$ into $L^*(X)^G$ induces an isomorphism on K-theory.
Hint: one way to do this is to show that the assignment of X to the K-theory of the G-invariant part of $L^*(X)$ is a homology theory on G-spaces in an appropriate sense.

6.8.15 The original definition of the localisation algebra for a proper metric space is as follows. Let X be a proper metric space, H_X be an ample X module and let $C^*_{L,og}(H_X)$ consist of all bounded uniformly continuous functions from $[1, \infty)$ to the Roe algebra $C^*(H_X)$ (see Definition 5.1.4) such that $\text{prop}(T_t) \to 0$ as $t \to \infty$. Show that with $C^*_L(H_X)$ as in Definition 6.6.1 there is a natural inclusion

$$C^*_{L,og}(H_X) \to C^*_L(H_X)$$

that induces an isomorphism on K-theory.

6.8.16 Let X be a single point space. One can show that a generator of the Kasparov group $KK_0(C(X),\mathbb{C}) \cong \mathbb{Z}$ is represented by the pair $(\mathbb{C},0)$, where $\mathbb{C}$ is made into a $C(X)$ module via the unique unital representation. Show that Construction 6.7.3 takes this generator to a generator of $K_0(L^*(X)) \cong \mathbb{Z}$ (see Proposition 6.3.3)

6.9 Notes and References

The first construction of K-homology comes from homotopy theory, as an application of a general machine using Spanier–Whitehead duality and the K-theory spectrum. See for example [1], particularly III.5 and III.6. This model for K-homology is theoretically very useful, but not so directly connected to index theory. A different topological model for K-homology was given by Baum and Douglas [23] and Baum, Higson and Schick [27]. Cycles for $K_*(X)$ in the Baum–Douglas theory are given by continuous maps from manifolds (with some extra structure) to X, and subject to an equivalence relation that incorporates both bordism and a form of Bott periodicity. The Baum–Douglas theory was directly motivated by, and has interesting applications to, index theory.

The idea of describing K-homology in analytic terms using operators on what we call X modules is due to Atiyah [7], and was also inspired by index theory. Atiyah was able to get a good description of cycles, but not the right equivalence relation. A self-contained analytic model of K-homology was subsequently provided by Kasparov [148], who found the right equivalence relation to impose on Atiyah's cycles. Around the same time, and coming at the issue motivated by single operator theory, Brown, Douglas and Fillmore [42] independently gave an analytic description of K-homology based on extensions of C^*-algebras by the compact operators. Subsequent analytic models for K-homology have been based on Paschke duality (see Paschke [203] and Higson [124]) and on the asymptotic morphisms of Connes and Higson (see Connes and Higson [61] and Guentner, Higson and Trout [112]). All of these models have found applications in various fields, such as index theory, manifold topology, C^*-algebras and single operator theory.

All of these analytic models generalise to give models for K-homology for separable noncommutative C^*-algebras and (to some extent at least) to equivariant and bivariant groups. The different models for K-homology all turn out to be the same for spaces, but there are subtle differences in the noncommutative case. Higson and Roe [135] gives a general introduction to analytic K-homology, including aspects of the Brown–Douglas–Fillmore,

Paschke duality and Kasparov theories, and their applications. Douglas [82] and Kasparov [149] are recommended for further background and breadth: the first for an overview of the Brown–Douglas–Fillmore theory, its applications to single operator theory and a discussion of its relationship to other early models of K-homology and their applications in manifold topology; the second is a much more challenging read that gives an idea of the breadth, depth and power of Kasparov's bivariant, equivariant theory.

The idea to use localisation algebras to describe K-homology comes from Yu [270], inspired by the heat kernel approach to the Atiyah–Singer index theorem (see Roe [217] for example). The original definition of the localisation algebra is the same as the one in Exercise 6.8.15. The arguments for the Mayer–Vietoris sequence and homotopy invariance in Section 6.3 are adapted from that paper. The K-theory of the version of the localisation algebra in Yu [270] was shown to agree with K-homology for simplicial complexes. This was extended to all proper metric spaces by Qiao and Roe [208]. Our changes to the localisation algebra in this text were motivated by the desire to get a homology theory on the 'right' category $\mathcal{LC}$, and by the needs of the analysis of differential operators carried out in later chapters.

Dadarlat, Willett and Yu [75] studies a model for bivariant theories (for possibly noncommutative C^*-algebras) based on a form of localisation algebra. It also proves (disguised versions of) the results discussed in Section 6.7.

The issue raised in Exercise 6.8.14 is intriguing. If one could show that the corresponding inclusion in Exercise 5.4.13 fails to induce an isomorphism on K-theory for some example, the discrepancy could lead to problems with the Baum–Connes type conjectures considered in Chapter 7.

7

Assembly Maps and the Baum–Connes Conjecture

In this chapter we will introduce assembly maps. The basic version of the assembly map is a homomorphism

$$\mu\colon K_*(X) \to K_*(C^*(X)) \tag{7.1}$$

from the K-homology of a space X to the K-theory of its Roe algebra: one can think if it as 'assembling' local topological data into global geometric data.

There are several equivalent ways of constructing the assembly map. One important construction represents cycles for $K_*(X)$ as operators that are invertible modulo $C^*(X)$, and the assembly map takes such an operator to its index class in the sense of Section 2.8. For this reason, the assembly map is also sometimes called the '(higher) index map'.

Our treatment will construct the assembly map as the map induced on K-theory by a directly defined $*$-homomorphism. This approach is quite elementary and intuitive, and is well suited to many applications. However, it has the unfortunate side effect of obscuring the assembly map's index-theoretic connections.

One should not expect the map in line (7.1) to give much information in general: the left-hand side depends only on the small-scale topology of the space, and the right-hand side only on the large-scale geometry. However, one can get more information by allowing the space to vary in an appropriate way. Indeed, if Y is any other space equipped with a coarse equivalence from X, then one also has an assembly map

$$K_*(Y) \to K_*(C^*(X))$$

defined by postcomposing the assembly map $K_*(Y) \to K_*(C^*(Y))$ for Y with the isomorphism $K_*(C^*(Y)) \cong K_*(C^*(X))$ coming from the coarse equivalence. Taking an appropriate limit over all such coarse equivalences

$Y \to X$ gives a sort of 'universal assembly map'; this is the so-called coarse Baum–Connes assembly map

$$\mu\colon KX_*(X) \to K_*(C^*(X)).$$

The coarse Baum–Connes conjecture for the space X predicts that this map is an isomorphism; informally, it says that all elements of $K_*(C^*(X))$, and also all relations between such elements, have a topological origin. There are also equivariant versions of this conjecture that allow for a group action, and we will discuss both in this chapter.

The chapter is structured as follows. We will start the chapter in Section 7.1 by giving our direct construction of the assembly map. We then introduce the Baum–Connes assembly maps as 'universal' assembly maps. The next three sections aim to give more concrete descriptions of this universal assembly map (at least in special cases): Section 7.2 gives a concrete model for the Baum–Connes assembly map in terms of Rips complexes; Section 7.3 discusses uniform contractibility and the particularly nice geometric models one gets in that case; Section 7.4 discusses connections with the classifying spaces of algebraic topology.

Finally, we conclude the chapter with a proof of the coarse Baum–Connes conjecture for Euclidean spaces $\mathbb{R}^d$ in Section 7.5. This provides some of the missing details for the discussion in Section 3.3 that proves tori do not admit metrics of positive scalar curvature.

7.1 Assembly and the Baum–Connes Conjecture

In this section, we introduce assembly maps, prove a basic functoriality result and then use that to show the existence of a universal assembly map, the so-called *Baum–Connes assembly map*.

Throughout this section, X, Y are proper metric spaces equipped with proper isometric actions of a countable discrete group G as in Section A.3. To avoid too much repetition, all maps, geometric modules, (localised) Roe algebras and K-homology groups in this section are equivariant; we will not generally state the equivariance assumption. It might be easier for the reader to assume that G is trivial on a first reading. Given how we have set up the theory in the book so far, this does not make a really substantial difference and still contains all the basic ideas.

Let H_X be an X-G module and let $C^*(H_X)^G$ and $C^*_L(H_X)^G$ be respectively the Roe algebra (Definition 5.1.4) and localised Roe algebra (Definition 6.6.1) of H_X. As $C^*_L(H_X)^G$ consists of functions (T_t) from $[1,\infty)$ to $C^*(H_X)^G$ (satisfying some other properties), there is an evaluation-at-one $*$-homomorphism

$$\mathrm{ev}\colon C^*_L(H_X)^G \to C^*(H_X)^G, \quad (T_t) \mapsto T_1.$$

Assuming H_X is ample and using Proposition 6.6.2 to identify the K-theory of $C^*_L(H_X)^G$ with the K-homology $K^G_*(X)$, passage to K-theory gives rise to a homomorphism

$$\mathrm{ev}_*\colon K^G_*(X) \to K_*(C^*(X)^G)$$

from the K-homology group of X to the K-theory of 'the' Roe algebra of X. We will see shortly that the map ev_* does not depend on the choice of module H_X: this follows from Lemma 7.1.3 below as applied to the identity map.

Definition 7.1.1 The *assembly map*, or *higher index map*, for X is the map

$$\mu_X\colon K^G_*(X) \to K_*(C^*(X)^G)$$

induced by the evaluation-at-one $*$-homomorphism $\mathrm{ev}\colon C^*_L(X)^G \to C^*(X)^G$.

Remark 7.1.2 The terminology 'higher index map' comes about as it is more common to define the left-hand side of the conjecture in terms of some variant of Kasparov's K-homology as in Section 6.7. In that picture, the assembly map can naturally be described using the index constructions of Section 2.8. Our picture loses this direct index-theoretic flavour, but has the advantage that the assembly map is given by a concrete $*$-homomorphism. It also gives a nice intuitive picture of the assembly map as a map that 'forgets local control', or that 'assembles' local data into global data.

The following lemma is key to setting up the universal assembly maps. There are many different models for the assembly map as in Definition 7.1.1 above: an important virtue of our model is that one can treat both sides of assembly maps on a similar footing, and thus lemmas like the one below are straightforward (at least, now we have set up all the machinery).

Lemma 7.1.3 *Let $f\colon X \to Y$ be a continuous, equivariant, coarse map. Then the diagram*

$$\begin{array}{ccc} K^G_*(X) & \xrightarrow{\mu_X} & K_*(C^*(X)^G) \\ \downarrow{\scriptstyle f_*} & & \downarrow{\scriptstyle f_*} \\ K^G_*(Y) & \xrightarrow{\mu_Y} & K_*(C^*(Y)^G) \end{array}$$

commutes.

Proof Recall from Theorem 6.6.3 that the f_* on the left is induced by conjugation by a continuous cover (V_t) for f that has the additional property that V_t is a cover for f for all t. Looking at Definition 5.2.5, we may define the map f_* on the right to be the map induced by conjugation by V_1. With this choice, the diagram commutes on the level of $*$-homomorphisms, so it certainly also does so on the level of K-theory. □

Now, the assembly map

$$\mu_X : K_*^G(X) \to K_*(C^*(X)^G)$$

will not be an isomorphism in general: the group $K_*^G(X)$ sees only the small-scale topological structure of X, and the group $K_*(C^*(X)^G)$ sees only the large-scale geometric structure. For example, if X is a closed manifold and G is trivial, then $K_*^G(X)$ is the K-homology of X and can be quite complicated; however, $C^*(X)^G$ is just a copy of the compact operators, and so $K_*(C^*(X)^G)$ will be a single copy of $\mathbb{Z}$ in dimension zero.

On the other hand, we can also produce elements of $K_*(C^*(X)^G)$ by allowing ourselves more general spaces that are (equivariantly) coarsely equivalent to X. Indeed, say Y is equipped with an equivariant coarse equivalence $p: X \to Y$. Then Theorem 5.2.6 implies that p functorially induces an isomorphism

$$p_* : K_*(C^*(X)^G) \to K_*(C^*(Y)^G).$$

Combining this with the assembly map for Y from Definition 7.1.1 gives a homomorphism

$$\mu_{Y,X} : K_*^G(Y) \xrightarrow{\mu_Y} K_*(C^*(Y)^G) \xrightarrow{(p_*)^{-1}} K_*(C^*(X)^G) . \tag{7.2}$$

Informally, the so-called Baum–Connes conjecture for X predicts that every element of $K_*(C^*(X)^G)$, and every relation between such elements, arises from such generalised assembly maps as Y varies over all spaces that are (equivariantly) coarsely equivalent to X.

We now start work on making this precise. It is convenient to introduce the following category which collects together the spaces Y appearing in the above discussion.

Definition 7.1.4 The *equivariant bounded category over* X, denoted $\mathcal{C}^G(X)$, has as objects pairs (Y, p), where Y is a proper metric space and $p: X \to Y$ is an equivariant coarse equivalence. A morphism in $\mathcal{C}^G(X)$ between (Y, p_Y) and (Z, p_Z) is a continuous equivariant coarse map $f: Y \to Z$ such that the diagram

$$\begin{array}{ccc} Y & \xrightarrow{f} & Z \\ \uparrow{\scriptstyle p_Y} & & \uparrow{\scriptstyle p_Z} \\ X & = & X \end{array}$$

commutes 'up to closeness': $f \circ p_Y$ is close to p_Z in the sense of Definition A.3.9 as functions $X \to Z$.

We will often abuse notation and omit the map p_Y when discussing objects and morphisms in $\mathcal{C}^G(X)$.

Our goal is to find a group $KX_*^G(X)$ that packages all the information contained in the groups $K_*^G(Y)$ as Y ranges over $\mathcal{C}^G(X)$. The Baum–Connes conjecture for the action of G on X will then say that an associated 'universal assembly map'

$$\mu\colon KX_*^G(X) \to K_*(C^*(X)^G)$$

is an isomorphism. Morally, $KX_*^G(X)$ is the 'limit' over the all the groups $K_*^G(Y)$ as Y ranges over the category $\mathcal{C}^G(X)$. To make this precise, we introduce a little category-theoretic terminology.

Definition 7.1.5 Let $\mathcal{C}$ be a category, and $F : \mathcal{C} \to \mathcal{GA}$ a functor from $\mathcal{C}$ to the category of graded abelian groups.

An *F-group* is a graded abelian group A and a collection of homomorphisms $(c_Y : F(Y) \to A)$ parametrised by the objects of $\mathcal{C}$ such that for every morphism $f : Y \to Z$ in $\mathcal{C}$, the diagram

$$\begin{array}{ccc} F(Y) & \xrightarrow{c_Y} & A \\ \downarrow{\scriptstyle F(f)} & & \| \\ F(Z) & \xrightarrow{c_Z} & A \end{array}$$

commutes.

An F-group A with family of homomorphisms $(c_Y : F(Y) \to A)$ is *universal* if for any F-group B with family of homomorphisms $(d_Y : F(Y) \to B)$ there exists a unique homomorphism $\mu : A \to B$ such that for each object Y of $\mathcal{C}$, the diagram

$$\begin{array}{ccc} F(Y) & \xrightarrow{c_Y} & A \\ \| & & \downarrow{\scriptstyle \mu} \\ F(Y) & \xrightarrow{d_Y} & B \end{array}$$

commutes.

The only example we will apply this to is the assignment

$$F : \mathcal{C}^G(X) \to \mathcal{GA}, \quad F(Y) = K_*^G(Y). \tag{7.3}$$

This is a functor by Theorem 6.6.3.

For the next lemma, recall the following terminology from category theory. A category is *small* if the collections of objects and morphisms form a set. A subcategory $\mathcal{C}'$ of a category $\mathcal{C}$ is *full* if whenever A, B are objects of $\mathcal{C}'$, then all morphisms between A and B in $\mathcal{C}$ are actually in $\mathcal{C}'$. A subcategory $\mathcal{C}'$ of a category $\mathcal{C}$ is *skeletal* if every object of $\mathcal{C}$ is isomorphic to a unique object of $\mathcal{C}'$.

Lemma 7.1.6 *With notation as in Definition 7.1.5, assume that $\mathcal{C}$ has a small skeletal subcategory $\mathcal{C}'$. Then a universal F-group exists and is unique up to canonical isomorphism.*

Proof Consider

$$A_0 := \bigoplus_{Y \in \mathcal{C}'} F(Y)$$

and identify each $F(Y)$ with the its image under the natural map to A_0. Let N be the (normal) subgroup of A_0 generated by all elements of the form $x - F(f)(x)$, where x is in some $F(Y)$, and $f : Y \to Z$ is a morphism between elements of $\mathcal{C}'$. Define A to be the quotient A_0/N. For each object Z of $\mathcal{C}$, choose an isomorphism $f : Z \to Z'$ to an object of $\mathcal{C}'$, and define c_Z to be the composition

$$F(Z) \xrightarrow{F(f)} F(Z') \longrightarrow A,$$

where the second map is induced by the canonical inclusion of $F(Z')$ into A_0; the choice of N implies that $c_Z : F(Z) \to A$ does not depend on the choice of Z' or the choice of isomorphism $f : Z \to Z'$. Moreover, it is straightforward to check that the collection of morphisms (c_Z) makes A into an F-group.

To see that this A has the right universal property, say B is any other F-group with family of morphisms $(d_Y : F(Y) \to B)$. Define a map $\mu_0 : A_0 \to B$ by setting the restriction of μ_0 to $F(Y)$ to be equal to $d_Y : F(Y) \to B$. The compatibility properties of the family (d_Y) guarantee that μ_0 contains N in its kernel, and so descends to $\mu : A \to B$. It is straightforward to check that μ has the right properties.

Uniqueness of universal F-groups follows directly from the universal property. □

Having established this abstract machinery, we now apply it to the case of interest. Note that $\mathcal{C}^G(X)$ has a small skeletal category: we leave this to the reader to verify.

Definition 7.1.7 Let F be the functor on $\mathcal{C}^G(X)$ defined by $F(Y) = K_*^G(Y)$ as in line (7.3) above. The *(equivariant) coarse K-homology of X*, denoted $KX_*^G(X)$, is the universal F-group for the category $\mathcal{C}^G(X)$ from Definition 7.1.4 and functor $F(Y) = K_*^G(Y)$ from line (7.3).

Definition 7.1.8 Given an object (Y, p) in $\mathcal{C}^G(X)$, the *X-assembly map for (Y, p)* is the homomorphism $\mu_{Y,X} \colon K_*^G(Y) \to K_*(C^*(X)^G)$ defined as the composition

$$K_*^G(Y) \xrightarrow{\mu_Y} K_*(C^*(Y)^G) \xrightarrow{(p_*)^{-1}} K_*(C^*(X)^G)$$

as in line (7.2); here μ_Y is as Definition 7.1.1 and p_* is as in Theorem 5.2.6.

Lemma 7.1.9 *Equipped with the family of maps $c_Y := \mu_{Y,X}$ from Definition 7.1.8, the group $K_*(C^*(X)^G)$ is an F-group, where F is as in line* (7.3) *above.*

Proof We must show that for any morphism $f \colon Y \to Z$ in $\mathcal{C}^G(X)$, the outer rectangle in the diagram

$$\begin{array}{ccc}
K_*^G(Y) & \xrightarrow{f_*} & K_*^G(Z) \\
\big\downarrow{\scriptstyle \mu_Y} & & \big\downarrow{\scriptstyle \mu_Z} \\
K_*(C^*(Y)^G) & \xrightarrow{f_*} & K_*(C^*(Z)^G) \\
\big\downarrow{\scriptstyle (p_*^Y)^{-1}} & & \big\downarrow{\scriptstyle (p_*^Z)^{-1}} \\
K_*(C^*(X)^G) & = & K_*(C^*(X)^G)
\end{array}$$

commutes. The top square commutes by Lemma 7.1.3, and the bottom square by Theorem 5.2.6. □

Definition 7.1.10 The *Baum–Connes assembly map* is the homomorphism of graded abelian groups

$$\mu \colon KX_*^G(X) \to K_*(C^*(X)^G)$$

coming from the universal property of $KX_*^G(X)$. The *Baum–Connes conjecture* for X asserts that μ is an isomorphism.

There are two special cases of this that are by far the most studied, and that are particularly important for applications (although the general case still seems very interesting). We spell these out separately.

Definition 7.1.11 Say G is the trivial group. Then we usually omit G from the notation and write

$$\mu\colon KX_*(X) \to K_*(C^*(X))$$

for the Baum–Connes assembly map. In this case, μ is called the *coarse Baum–Connes assembly map for* X, and the *coarse Baum–Connes conjecture for* X is the statement that it is an isomorphism.

Say now $X = G$. Then Theorem 5.3.2 canonically identifies $K_*(C^*(X)^G)$ with the K-theory group $K_*(C^*_\rho(G))$ of the reduced group C^*-algebra of G. Thus the Baum–Connes assembly map identifies with a homomorphism

$$\mu\colon KX^G_*(G) \to K_*(C^*_\rho(G)),$$

which is usually just called the *Baum–Connes assembly map for* G. The *Baum–Connes conjecture for* G asserts that this assembly map is an isomorphism.

Remark 7.1.12 It is a remarkable fact that the group $KX^G_*(G)$ can be defined as an F-functor in the same way, but where $\mathcal{C}^G(G)$ is replaced by the subcategory where objects are (G-equivariant) spin^c manifolds,[1] and morphisms are assumed smooth. Unfortunately, a proof of this would take us a little fair afield (see the notes at the end of the chapter for a reference). However, we thought it worth mentioning both as this fact is sometimes useful, and as this picture of $KX^G_*(G)$ is rather closer to the original approach taken by Baum and Connes.

The descriptions of the (coarse) Baum–Connes assembly map in Definition 7.1.10 are theoretically useful, and we hope that they are also quite conceptual. As one might expect, they are not the best descriptions if we actually want to compute the group $KX^G_*(X)$ or prove that the assembly map μ is an isomorphism! In the next three sections we will do some work to make the definitions more concrete.

To finish this section, however, we give an example where we can say something using only the abstract nonsense above. This is the case of the Baum–Connes conjecture for a finite group acting on a compact bounded metric space. Having unpacked the definitions, this is not difficult; nonetheless, it is perhaps instructive to see what happens.

Example 7.1.13 Say X is a compact bounded metric space, equipped with a (proper) isometric action of a finite group G. Let pt be a single-point space equipped with the trivial action and $p\colon X \to pt$ the collapsing map, so the pair (pt, p) is an object of $\mathcal{C}^G(X)$.

[1] A spin^c manifold is one that is 'oriented' in some sense appropriate to K-theory.

The universal F-group for $F(Y) = K_*^G(Y)$ as in line (7.3) is just the group $K_*^G(pt)$ together with the family of maps $c_Y : K_*^G(Y) \to K_*^G(pt)$ defined by collapsing Y to a single point. To see that this data defines an F-group note that the diagram

$$\begin{array}{ccc} F(Y) & \xrightarrow{c_Y} & K_*^G(pt) \\ \downarrow{\scriptstyle F(f)} & & \| \\ F(Z) & \xrightarrow{c_Z} & K_*^G(pt) \end{array}$$

commutes for any morphism $f : Y \to Z$ in $\mathcal{C}^G(X)$ as any space in $\mathcal{C}^G(X)$ admits only one (equivariant, coarse) map to a point. Universality follows as if A is any other F-group with the associated family of morphisms (d_Y) we have a diagram

$$\begin{array}{ccc} F(Y) & \xrightarrow{c_Y} & F(pt) \\ \| & & \downarrow{\scriptstyle d_{pt}} \\ F(Y) & \xrightarrow{d_Y} & A \end{array}$$

that commutes by definition of A being an F-group; hence we can just take the map μ required by universality to be d_{pt}.

In particular, applying this to the F-group $K_*(C^*(X)^G)$ gives that the Baum–Connes assembly map identifies with the X-assembly map for pt

$$\mu_{pt,X} : K_*^G(pt) \to K_*(C^*(X)^G).$$

Moreover, as the collapsing map $p : X \to pt$ is an equivariant coarse equivalence, the map $p_* : K_*(C^*(pt)^G) \to K_*(C^*(X)^G)$ is an isomorphism, so to prove the Baum–Connes conjecture for X, it suffices to show that the assembly map

$$\mu_{pt} : K_*^G(pt) \to K_*(C^*(pt)^G)$$

of Definition 7.1.1 is an isomorphism, i.e. that the evaluation-at-zero map

$$\mathrm{ev} : C_L^*(pt)^G \to C^*(pt)^G$$

induces an isomorphism on K-theory. This follows from an Eilenberg swindle argument just like (but a little simpler than) the one we used in Proposition 6.3.3 to compute the K-homology of a point: see Exercise 7.6.8. In conclusion, the Baum–Connes conjecture holds for actions of finite groups on compact, bounded spaces.

7.2 Rips Complexes

In this section, we introduce a concrete model for the coarse K-homology groups that works in full generality. In later sections, we will use this model as a stepping stone to other, more specialised, situations.

The key tools for doing this are Rips complexes. For technical reasons (essentially those outlined in Exercise 7.6.5 below), the basic building blocks for our Rips complexes will be spherical simplices as in the next definition. To state it, let $\arccos\colon [-1,1] \to [0,\pi]$ be the usual inverse cosine function. Let $S(\mathbb{R}^d)$ be the sphere of radius one in a finite-dimensional Euclidean space, and let the *intrinsic distance* on $S(\mathbb{R}^d)$ be defined by

$$d_{in}(x,y) = \arccos(\langle x,y\rangle).$$

In words, $d_{in}(x,y)$ is the angle between the rays through x and y. Equivalently, it is the length of the shorter arc of a great circle connecting x and y, or is the distance defined by the usual Riemannian metric on $S(\mathbb{R}^d)$.

Definition 7.2.1 Let F be a finite set, and consider the set $\sigma(F)$ of formal sums

$$\sum_{z\in F} t_z z,$$

where $t_z \in [0,1]$ for each $z \in F$, and $\sum_{z\in F} t_z = 1$. Let $S(\mathbb{R}^F)$ be the sphere in the finite-dimensional Euclidean space $\mathbb{R}^F$ spanned by F, and define a bijection

$$f\colon \sigma(F) \to S(\mathbb{R}^F), \quad \sum_{z\in F} t_z z \mapsto \Big(\sum_{z\in F} t_z^2\Big)^{-1/2} \sum_{z\in F} t_z z.$$

The *spherical metric* on $\sigma(F)$ is the metric defined by

$$d(x,y) := \frac{2}{\pi} d_{in}(f(x), f(y)),$$

and $\sigma(F)$ equipped with this metric is called the *spherical simplex* on F.

The factor $2/\pi$ is chosen so that $\sigma(F)$ has diameter one, as follows from the fact that the maximal angle between any two rays defined by points in the image of f is $\pi/2$.

For the next definition, recall that a metric space Z is *locally finite* if any ball contains finitely many points. In particular, a locally finite metric space is proper.

Definition 7.2.2 Let Z be a locally finite metric space and let $r \geqslant 0$. The *spherical Rips complex of Z at scale r*, denoted $S_r(Z)$, consists as a set of all formal sums

$$x = \sum_{z \in Z} t_z z$$

such that each t_z is in $[0,1]$, such that $\sum_{z \in Z} t_z = 1$ and such that the *support* of x defined by $\operatorname{supp}(x) := \{z \in Z \mid t_z \neq 0\}$ has diameter at most r.

Let F be a (finite) subset of Z of diameter at most r. The *simplex* spanned by F is the set $\sigma(F)$ of formal sums $\sum_{z \in Z} t_z z$ that are supported in F. We equip each simplex with the spherical metric.

For points $x, y \in S_r(Z)$, a *simplicial path* γ between them is a finite sequence $x = x_0, \ldots, x_n = y$ of points in $S_r(Z)$ together with a choice of simplices $\sigma_1, \ldots, \sigma_n$ such that each σ_i contains (x_{i-1}, x_i). The *length* of such a path is defined to be

$$\ell(\gamma) := \sum_{i=1}^{n} d_{\sigma_i}(x_{i-1}, x_i).$$

Finally, we define the *spherical distance* between two arbitrary points $x, y \in S_r(Z)$ to be

$$d_{S_r}(x, y) := \inf\{\ell(\gamma) \mid \gamma \text{ a simplicial path between } x \text{ and } y\}$$

(and $d_{S_r}(x, y) = \infty$ if no simplicial path exists).

We leave it as an exercise for the reader to show that d_{S_r} is a well-defined (possibly infinite-valued) metric on $S_r(Z)$.

Lemma 7.2.3 *With notation as in Definition 7.2.2:*

(i) the inclusion $\sigma(F) \to S_r(Z)$ of any simplex is an isometry;
(ii) if $x, y \in S_r(Z)$ are elements of $S_r(Z)$ with disjoint support, then $d_{S_r}(x, y) \geqslant 1$.

Proof Let x, y be two points in $S_r(Z)$ that are a finite distance apart, let $\epsilon > 0$ and let $x = x_0, \ldots, x_n = y$ be a simplicial path from x to y with associated simplices $\sigma_1, \ldots, \sigma_n$, and with the property that

$$\sum_{i=1}^{n} d_{\sigma_i}(x_{i-1}, x_i) \leqslant d_{S_r}(x, y) + \epsilon.$$

Let $F \subseteq Z$ be the union of the supports of all of the x_i, and let $\sigma(F)$ be the spherical simplex on F (which may not be a simplex in $S_r(Z)$, but still makes sense as an abstract spherical simplex). As each σ_i is isometrically included in $\sigma(F)$, the triangle inequality gives that

$$d_{\sigma(F)}(x, y) \leqslant \sum_{i=1}^{n} d_{\sigma_i}(x_{i-1}, x_i),$$

and so by choice of $x_0, \ldots, x_n$,

$$d_{\sigma(F)}(x, y) \leqslant d_{S_r}(x, y) + \epsilon. \tag{7.4}$$

We claim that this inequality proves both parts (i) and (ii).

Indeed, for part (i) note that if x, y are contained in the same simplex σ of $S_r(Z)$, then σ is isometrically included in $\sigma(F)$, and so the inequality in line (7.4) implies that $d_\sigma(x, y) \leqslant d_{S_r}(x, y) + \epsilon$. As the inequality $d_{S_r}(x, y) \leqslant d_\sigma(x, y)$ is clear from the definitions, and as ϵ was arbitrary, this completes the proof of part (i).

For part (ii), note that if x, y have disjoint support, then they are distance one apart in $\sigma(F)$, as the corresponding vectors in $\mathbb{R}^F$ are orthogonal. Hence (ii) follows directly from line (7.4) (and the fact that ϵ was arbitrary). □

As a consequence of part (i) of Lemma 7.2.3, note that if $x_0, \ldots, x_n$ is a simplicial path (with some choice of associated simplices $\sigma_1, \ldots, \sigma_n$), then the length of the path equals

$$\sum_{i=1}^{n} d_{S_r}(x_{i-1}, x_i).$$

We will just use this form for the length of a simplicial path from now on and suppress mention of the simplices involved.

For the next lemma, let us introduce some convenient terminology.

Definition 7.2.4 Let z, w be two points in Z. The *combinatorial distance* between z and w in $S_r(Z)$, denoted $d_c(z, w)$, is the length of a shortest edge path between them, i.e. the smallest n such that there exists a sequence $z = z_0, \ldots, z_n = w$ with each z_i in Z, and such that each consecutive pair z_{i-1}, z_i are in the same simplex in $S_r(Z)$ (and infinity if no such n exists).

Lemma 7.2.5 *Say Z is a locally finite metric space, z is a point of Z and σ is a simplex of $S_r(Z)$. Then for all $y \in \sigma$,*

$$d_{S_r}(z, y) \geqslant \min\{d_c(z, w) \mid w \in Z \text{ a vertex of } \sigma\}.$$

Proof We prove this by induction on the number $n = \min\{d_c(z, w) \mid w \in Z$ a vertex of $\sigma\}$ (the case $n = \infty$ is clear, so we can ignore this). For $n = 0$, there is nothing to prove. Say then we have the result for all values at most $n - 1$ for some $n \geqslant 1$. Let σ be such that $n = \min\{d_c(z, w) \mid w \in Z$ a vertex of $\sigma\}$, and let y be a point of σ.

Let $\epsilon > 0$, and let $z = x_0, \ldots, x_n = y$ be a simplicial path from z to y such that

$$\sum_{i=1}^{n} d_{S_r}(x_{i-1}, x_i) \leqslant d_{S_r}(z, y) + \epsilon. \tag{7.5}$$

Let $k \in \{0, \ldots, n-1\}$ be the largest number such that for all w in the support of x_k, we have that $d_c(z, w) \leqslant n - 1$; certainly x_0 has this property, so k with this property exists. We claim that in fact that for all w in the support of x_k we have $d_c(z, w) = n - 1$. Indeed, if not there is some $w \in \text{support}(x_k)$ with $d_c(z, w) = m < n - 1$. Let τ be a simplex that contains both x_k and x_{k+1} (such a τ exists by definition of a simplicial path). Then w is a vertex of τ, and so all vertices v of τ satisfy $d_c(z, v) \leqslant m + 1 < n$. On the other hand, by choice of k, the fact that x_{k+1} is in τ implies that τ also has a vertex v with $d_c(z, v) = n$, so we have a contradiction.

Now the claim and the inductive hypothesis imply that $d_{S_r}(z, x_k) \geqslant n - 1$. On the other hand, all vertices w in the support of y satisfy $d_c(z, w) \geqslant n$ by assumption, whence x_k and y have disjoint supports, and so $d_{S_r}(x_k, y) \geqslant 1$ by Lemma 7.2.3, part (ii). We now have from these inequalities, line (7.5) and the triangle inequality that

$$d_{S_r}(z, y) \geqslant \sum_{i=1}^{n} d_{S_r}(x_{i-1}, x_i) - \epsilon \geqslant d_{S_r}(z, x_k) + d_{S_r}(x_k, y) - \epsilon \geqslant n - \epsilon.$$

As ϵ was arbitrary, this completes the proof. □

The following corollary is the crucial fact we need about the spherical metric. It is the key reason we are using the spherical metric, as it can fail for the more standard Euclidean metric: see Exercise 7.6.5.

Corollary 7.2.6 *Let Z be a locally finite metric space, $r \geqslant 0$, and z, w be points of Z. Then $d_c(z, w) = d_{S_r}(z, w)$.*

Proof The inequality $d_c(z, w) \geqslant d_{S_r}(z, w)$ is clear, and the opposite inequality follows from Lemma 7.2.5 applied to z and the simplex $\sigma = \{w\}$. □

Remark 7.2.7 At this point, we could use the metric spaces $S_r(Z)$ to produce a concrete model for the coarse Baum–Connes conjecture for Z directly; this is what is usually done in the literature. Indeed, as the canonical inclusions $S_r(Z) \to S_s(Z)$ for $r \leqslant s$ are proper and contractive, we have directed systems

$$K_*(S_1(Z)) \to K_*(S_2(Z)) \to \cdots$$

and

$$K_*(C^*(S_1(Z)) \to K_*(C^*(S_2(Z))) \to \cdots .$$

Moreover, these directed systems are compatible with the evaluation-at-one maps and therefore we get a direct limit map

$$\mathrm{ev}_*\colon \lim_{r\to\infty} K_*(S_r(Z)) \to \lim_{r\to\infty} K_*(C^*(S_r(Z))), \tag{7.6}$$

which, one can show, identifies with the Baum–Connes assembly map for Z (the reader is asked to do this in Exercise 7.6.6). This also works in the presence of a group action on Z. Moreover, one can treat a general proper metric space X (with G-action) along these lines by identifying the Baum–Connes assembly map for X with the Baum–Connes assembly map for a (G-invariant) net Z in X.

Having said this, it is slightly tricky to identify the map in line (7.6) with the Baum–Connes assembly map in our set-up, primarily as the canonical inclusions $Z \to S_r(Z)$ will, in general, not be coarse equivalences (or even coarse maps at all): compare Exercise 7.6.3. For this reason, we now change the metric on $S_r(Z)$ to remedy this defect. Another way around this issue using abstract coarse structures (see Definition A.3.7) is given in Exercise 7.6.4: that method may be more conceptual for readers familiar with that language.

Definition 7.2.8 Let Z be a locally finite metric space, and let $S_r(Z)$ be the associated spherical Rips complex at scale r. A *semi-simplicial path* δ between points x and y in $S_r(Z)$ consists of a sequence of the form

$$x = x_0, y_0, x_1, y_1, x_2, y_2, \ldots, x_n, y_n = y,$$

where each of $x_1, \ldots, x_n$ and each of $y_0, \ldots, y_{n-1}$ are in Z. The *length* of such a path is

$$\ell(\delta) := \sum_{i=0}^{n} d_{S_r}(x_i, y_i) + \sum_{i=0}^{n-1} d_Z(y_i, x_{i+1}).$$

We define the *semi-spherical distance* on $S_r(Z)$ by

$$d_{P_r}(x,y) := \inf\{\ell(\gamma) \mid \gamma \text{ a semi-simplicial path between } x \text{ and } y\}$$

(note that a semi-simplicial path between two points always exists).

The *Rips complex* of Z is defined to be the space $P_r(Z)$ equipped with the metric d_{P_r} above.

Again, we leave it to the reader to check that d_{P_r} is indeed a metric. Note that the Rips complex $P_0(Z)$ identifies isometrically with Z.

The following technical lemma is the key tool to understanding the structure of $P_r(Z)$ as a metric space. To state it, if F is a subset of Z, write $P_r(F)$ for the subset of $P_r(Z)$ consisting of all formal sums $\sum_{z\in Z} t_z z$ supported in F.

Lemma 7.2.9 *For any $z \in Z$ and $s \geqslant 0$, we have the inclusion*

$$B_{P_r}(z;s) \subseteq P_r(B_Z(z;(s+2)(r+1))).$$

Proof Let $y = \sum_{w\in Z} t_w w$ be an element of $B_{P_r}(z;s)$, and let $w \in Z$ be such that $t_w \neq 0$. We must show that $d_Z(z,w) \leqslant (s+2)(r+1)$. By definition of d_{P_r}, there exists a semi-simplicial path

$$z = x_0, y_0, \ldots, x_n, y_n = w$$

with all x_i, y_i in Z, and with

$$\sum_{i=0}^{n} d_{S_r}(x_i, y_i) + \sum_{i=0}^{n-1} d_Z(y_i, x_{i+1}) \leqslant s+2.$$

Write $d_{S_r}(x_i, y_i) = m_i$, and note in particular that

$$\sum_{i=0}^{n} m_i \leqslant s+2. \tag{7.7}$$

Then by Corollary 7.2.6, each m_i is an integer, and for each i there is a sequence $x_i = z_0^{(i)}, \ldots, z_{m_i}^{(i)} = y_i$ in Z with each $z_j^{(i)}$ an element of Z and each consecutive pair $(z_j^{(i)}, z_{j+1}^{(i)})$ in the same simplex. By definition of the spherical Rips complex we then have that

$$d_Z(x_i, y_i) \leqslant m_i r. \tag{7.8}$$

At this point we have from the triangle inequality in Z and lines (7.7) and (7.8) that

$$\begin{aligned} d_Z(z,w) &\leqslant \sum_{i=0}^{n} d_Z(x_i, y_i) + \sum_{i=0}^{n-1} d_Z(y_i, x_{i+1}) \leqslant \sum_{i=0}^{n} m_i r + (s+2) \\ &\leqslant (s+2)(r+1) \end{aligned}$$

as claimed. □

The next lemma gives a simple description of the topology on $P_r(Z)$.

Lemma 7.2.10 *Let F be a finite subset of Z. Then a sequence*

$$\left(\sum_{z\in Z} t_z^{(n)} z\right)_{n\in\mathbb{N}}$$

in $P_r(F)$ *converges to a point* $\sum_{z\in Z} t_z z$ *in* $P_r(F)$ *if and only if the sequence* $(t_z^{(n)})_{n\in\mathbb{N}}$ *converges to* t_z *in* $[0,1]$ *for all* $z \in F$.

Proof Note first that we may assume d_{P_r} is finite on $P_r(F)$ (otherwise just work one component at a time for the equivalence relation defined by $x \sim y$ if $d_{P_r}(x,y) < \infty$).

Let $\sigma(F)$ be the spherical simplex on F, and consider $P_r(F)$ as a subset of $\sigma(F)$. We claim that there is $c > 0$ such that for all $x, y \in P_r(F)$,

$$c^{-1} d_{\sigma(F)}(x,y) \leqslant d_{P_r}(x,y) \leqslant c d_{\sigma(F)}(x,y). \tag{7.9}$$

As the conclusion of the lemma clearly holds inside $\sigma(F)$, this will suffice to complete the proof.

First, let $\epsilon > 0$ and let $x = x_0, y_0, \ldots, x_n, y_n = y$ be a semi-simplicial path between x and y with

$$d_{P_r}(x,y) + \epsilon \geqslant \sum_{i=0}^{n} d_{S_r}(x_i, y_i) + \sum_{i=0}^{n-1} d_Z(y_i, x_{i+1}). \tag{7.10}$$

Note that $d_{S_r}(x_i, y_i) = d_{\sigma(F)}(x_i, y_i)$ as x_i and y_i are in the same simplex (see Lemma 7.2.3). On the other hand, if $a := \min\{d_Z(z,w) \mid z, w \in F, z \neq w\}$ (which is positive as F is finite), then $d_Z(y_i, x_{i+1}) \geqslant a = a d_{\sigma(F)}(y_i, x_{i+1})$, as each of $y_0, \ldots, y_{n-1}$ and $x_1, \ldots, x_n$ are in F, whence $d_{\sigma(F)}(y_i, x_{i+1}) = 1$. Putting this discussion together with line (7.10) gives that

$$\begin{aligned} d_{P_r}(x,y) + \epsilon &\geqslant \sum_{i=0}^{n} d_{\sigma(F)}(x_i, y_i) + \sum_{i=0}^{n-1} a d_{\sigma(F)}(y_i, x_{i+1}) \\ &\geqslant \min 1, a\Big(d_{\sigma(F)}(x_i, y_i) + d_{\sigma(F)}(y_i, x_{i+1})\Big) \\ &\geqslant \min\{1, a\} d_{\sigma(F)}(x,y), \end{aligned}$$

where the last inequality is the triangle inequality. As ϵ was arbitrary, this gives one of the inequalities in line (7.9).

For the other inequality in line (7.9), say $x = \sum_{z\in F} t_z^x z$ and $y = \sum_{z\in F} t_z^y z$. Let $F_x, F_y \subseteq F$ be the supports of x and y respectively. We split the proof into the cases where $F_x \cap F_y$ is empty or non-empty. If this intersection is empty, then $d_\sigma(F)(x,y) = 1$, and so $d_{P_r}(x,y) \leqslant \operatorname{diam}_{P_r}(P_r(F)) d_{\sigma(F)}(x,y)$. On the other hand, if the intersection is non-empty, then define

$$w_x := \sum_{z\in F_x\cap F_y} t_z^x z \quad \text{and} \quad w_y := \sum_{z\in F_x\cap F_y} t_z^y z.$$

Then

$$d_{P_r}(x,y) \leqslant d_{S_r}(x,y) \leqslant d_{S_r}(x,w_x) + d_{S_r}(w_x,w_y) + d_{S_r}(w_y,y)$$
$$= d_{\sigma(F)}(x,w_x) + d_{\sigma(F)}(w_x,w_y) + d_{\sigma(F)}(w_y,y),$$

where the last equality uses that each of the pairs appearing is in the same simplex of $S_r(F)$, and Lemma 7.2.3. On the other hand, it is straightforward to check that from the definition of $d_{\sigma(F)}$ that each of $d_{\sigma(F)}(x,w_x)$, $d_{\sigma(F)}(w_x,w_y)$, and $d_{\sigma(F)}(w_y,y)$ is bounded above by $d_{\sigma(F)}(x,y)$, whence $d_{P_r}(x,y) \leqslant 3d_{\sigma(F)}(x,y)$. Combining the two cases, we get

$$d_{P_r(x,y)} \leqslant \max\{3, \operatorname{diam}(P_r(F))\} d_{\sigma(F)}(x,y)$$

and are done. □

Proposition 7.2.11 *Let Z be a countable, locally finite metric space.*

(i) The Rips complex $P_r(Z)$ is a proper, second countable metric space.
(ii) For each $s \geqslant r \geqslant 0$ the canonical inclusion $i_{sr}\colon P_r(Z) \to P_s(Z)$ is a homeomorphism onto its image, and a coarse equivalence.

Proof Using Lemma 7.2.9, each finite radius ball in $P_r(Z)$ is contained in $P_r(F)$ for some finite subset F of Z. It follows directly from this and Lemma 7.2.10 that any bounded sequence in $P_r(Z)$ has a convergent subsequence, whence $P_r(Z)$ is proper. Moreover, it is separable as Lemma 7.2.10 implies that the countable set

$$\left\{ \sum_{z \in Z} t_z z \in P_r(Z) \mid t_z \in \mathbb{Q} \text{ for all } z \in Z \right\}$$

is dense, and a separable metric space is always second countable. The fact that the canonical inclusions $i_{sr}\colon P_r(Z) \to P_s(Z)$ are all homeomorphisms onto their images also follows directly from Lemma 7.2.10 and Lemma 7.2.9.

To see that these inclusions i_{sr} are all coarse equivalences, it suffices by Exercise A.4.3 to show that for any points x, y in $P_r(Z)$ then: (a) for all t there exists t' such that if $d_{P_r}(x,y) \leqslant t$ then $d_{P_s}(x,y) \leqslant t'$; and (b) for all t there exists t' such that if $d_{P_s}(x,y) \leqslant t$ then $d_{P_r}(x,y) \leqslant t'$. Part (a) is clear as the map i_{sr} is contractive by definition, so it remains to prove part (b).

Assume then that $d_{P_s}(x,y) \leqslant t$ for some t. Let $z_x, z_y \in Z$ be vertices in the same simplices in $P_r(Z)$ as x and y respectively. Then $d_{P_s}(z_x, x_y) \leqslant t+2$. Using Lemma 7.2.9, we have that $d_Z(z_x,z_y) \leqslant (t+4)(s+1)$, and so as the canonical map $Z \to P_r(Z)$ is clearly contractive, $d_{P_r}(z_x,z_y) \leqslant (t+4)(s+1)$. Finally this implies that $d_{P_r}(x,y) \leqslant (t+4)(r+1)+2$, which completes the proof. □

In the presence of a group action, we also get corresponding structure on the Rips complex.

Lemma 7.2.12 *Let G be a countable discrete group acting properly by isometries on a locally finite metric space Z, and let $r \geqslant 0$. Then for $g \in G$, the formula*

$$g: \sum_{z \in Z} t_z z \mapsto \sum_{z \in Z} t_z (gz)$$

defines a proper isometric action on $P_r(Z)$. Moreover, the canonical inclusions $P_r(Z) \to P_s(Z)$ for $0 \leqslant r \leqslant s$ are equivariant for this action.

Proof That the action is isometric follows as the action on Z is isometric, and it is clear that the inclusions $P_r(Z) \to P_s(Z)$ are equivariant. Properness follows from properness of the action on Z and Lemma 7.2.9. □

For the next lemma, recall that two maps $f_0, f_1 : X \to Y$ with codomain a metric space are *close* if there is $c > 0$ such that for all $x \in X, d_Y(f_0(x), f_1(x)) \leqslant c$.

Lemma 7.2.13 *Say Y is a proper metric space, equipped with a proper isometric action of a countable group G. Let $f_0, f_1 : Y \to P_r(Z)$ are continuous, proper, equivariant maps that are also close. Then there is $s \geqslant r$ such that the compositions*

$$Y \xrightarrow{f_0} P_r(Z) \xrightarrow{i_{rs}} P_s(Z) \quad \text{and} \quad Y \xrightarrow{f_1} P_r(Z) \xrightarrow{i_{rs}} P_s(Z)$$

are equivariantly properly homotopic.

Proof Let $f_0, f_1 : Y \to P_r(Z)$ be close, continuous and equivariant, so in particular there is $c \geqslant 0$ such that $d_{P_r}(f_0(y), f_1(y)) \leqslant c$ for all $y \in Y$. Let z_0 and z_1 be any points of Z that are in the same simplex as $f_0(y)$ and $f_1(y)$ respectively. In particular, $d_{P_r}(z_0, z_1) \leqslant c + 2$, and so by Lemma 7.2.9, $d_Z(z_0, z_1) \leqslant (c+4)(r+1)$. Let $s := (c+4)(r+1)$. Then by the choice of s, for any $y \in Y$ all points in the support of $f_0(y)$ and $f_1(y)$ are in the same simplex of $P_s(Z)$. Thinking of points of $P_s(Z)$ as formal sums of the form $\sum_{z \in Z} t_z z$ in the usual way, it follows that

$$h: Y \times [0,1] \to (1-t) f_0(y) + t f_1(y)$$

is a well-defined equivariant map from Y to $P_s(Z)$, which is moreover continuous by Lemma 7.2.10. Properness of h follows straightforwardly from properness of f_0 and f_1 and the fact that any compact subset of $P_s(Z)$ is contained in a set of the form $P_s(F)$ for some finite $F \subseteq Z$ by Lemma 7.2.9. □

Lemma 7.2.14 *Let Z be a countable, locally finite metric space equipped with a proper isometric action of a countable group G. Let Y be a proper metric space, and $p\colon Z \to Y$ a coarse equivalence. Then there are $r \geqslant 0$ and a continuous equivariant coarse equivalence $f\colon Y \to P_r(Z)$ such that the diagram*

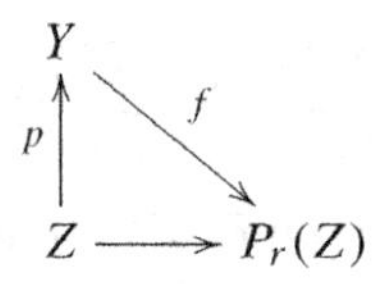

commutes 'up to closeness', i.e. $f \circ p\colon Z \to P_r(Z)$ is close to the canonical inclusion $Z \to P_r(Z)$.

Proof Write $W = p(Z) \subseteq Y$. We first note that as p is a coarse equivalence, there is some absolute bound r_0 on the diameter of all the sets $p^{-1}(w)$ as w ranges over W. For each $w \in W$, let $|p^{-1}(w)|$ denote the cardinality of this finite set, and define a function $f_0\colon W \to P_{r_0}(Z)$ by

$$f_0(w) = \frac{1}{|p^{-1}(w)|} \sum_{z \in p^{-1}(w)} z.$$

Note that f is equivariant. Moreover, it is continuous as W is discrete set: indeed, W is locally finite as Z is locally finite, and p is a coarse equivalence. Further, for any $z \in Z$, $f_0(p(z))$ is in the same simplex as z, whence $d_{P_r}(z, f_0(p(z))) \leqslant 1$, and so in particular $f_0 \circ p\colon Z \to P_{r_0}(Z)$ is close to the canonical inclusion $Z \to P_{r_0}(Z)$. It thus suffices to show that there is $r \geqslant r_0$ and a continuous equivariant function $f\colon Y \to P_r(Z)$ whose restriction to W is close to the composition of f_0 and the canonical inclusion $P_{r_0}(Z) \to P_r(Z)$.

As p is a coarse equivalence, there exists $r_1 \geqslant 0$ such that $\bigcup_{w \in W} B_Y(w; r_1)$ covers Y. As the action on Y is proper Corollary A.2.8 implies that there is an equivariant subordinate partition of unity, say $(\phi_w)_{w \in W}$. Let r be such that if $z, z' \in W$ satisfy $d_Y(f(z), f(z')) \leqslant 2r_1$, then $d(z, z') \leqslant r$. Provisionally define now

$$f\colon Y \to P_r(Z), \quad f(y) = \sum_{w \in W} \phi_w(y) f_0(w).$$

To see that f does indeed take image in $P_r(Z)$, it suffices to check that for any $y \in Y$, and any $z, z' \in Z$ with $\phi_{f(z)}(y) \neq 0$ and $\phi_{f(z')}(y) \neq 0$, we have that $d_Z(z, z') \leqslant r$. Indeed, note that if $\phi_{f(z)}(y)$ and $\phi_{f(z')}(y)$ are both non-zero, then y is in both $B(f(z); r_1)$ and $B(f(z'); r_1)$, whence $d_Y(f(z), f(z')) \leqslant 2r_1$. Hence $d_Z(z, z') \leqslant r$ by choice of r.

We check the claimed properties of f. First note that f is continuous: this follows from Lemma 7.2.10, and the fact that f_0 and the functions in the partition of unity are continuous. Finally, to complete the proof, it suffices to check that the diagram

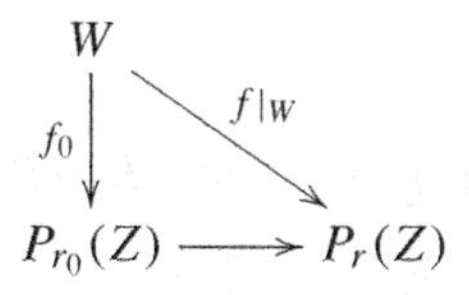

commutes up to closeness. We leave the remaining details to the reader: we think it's the sort of proof that it's better to think through than to read. □

Now, let X be a proper metric space equipped with an isometric action of G. Choose a G-invariant, coarsely dense, locally finite subset Z of X: such exists by a slight variation on Lemma A.3.11 that we leave to the reader. Applying Lemma 7.2.14 to the inclusion $Z \to X$ gives $r \geqslant 0$ and a continuous, equivariant map $p_r : X \to P_r(Z)$ whose restriction to Z is close to the inclusion $Z \to P_r(Z)$. Hence p_r is a coarse equivalence. Moreover, for each $s \geqslant r$, the composition

$$p_s := i_{sr} \circ p_r : X \to P_r(Z) \to P_s(Z) \tag{7.11}$$

is also a coarse equivalence. Fix these maps from now on and use them to think of the pairs $(P_s(Z), p_s)$ as elements of $\mathcal{C}^G(X)$ for all suitably large s.

Consider the group

$$\lim_{r\to\infty} K_*^G(P_r(Z)),$$

where the limit is defined with respect to the inclusion maps $i_{sr} : P_r(Z) \to P_s(Z)$ from part (ii) of Lemma 7.2.11 above. For each object Y of $\mathcal{C}^G(X)$, choose $r \geqslant 0$ and a continuous equivariant coarse equivalence $f : Y \to P_r(Z)$ with the properties given in Lemma 7.2.14. Define

$$c_Y : K_*^G(Y) \to \lim_{r\to\infty} K_*^G(P_r(Z)) \tag{7.12}$$

to be the homomorphism induced by $f_* : K_*^G(Y) \to K_*^G(P_r(Z))$, and note that Lemma 7.2.13 implies that c_Y does not depend on the choices of f and r.

Lemma 7.2.15 *The maps in line* (7.12) *above make* $\lim_{r\to\infty} K_*^G(P_r(Z))$ *into a universal* F*-group for the functor* $F(Y) = K_*^G(Y)$.

Proof To see that the maps in line (7.12) make $\lim_{r\to\infty} K_*^G(P_r(Z))$ into an F-group, we must check that any diagram of the form

$$\begin{array}{ccc} K_*^G(Y_1) & \xrightarrow{c_{Y_1}} & \lim_{r\to\infty} K_*^G(P_r(Z)) \\ \downarrow f_* & & \| \\ K_*^G(Y_2) & \xrightarrow{c_{Y_2}} & \lim_{r\to\infty} K_*^G(P_r(Z)) \end{array}$$

commutes. This is clear from Lemma 7.2.13, and invariance of K-homology under (equivariant) proper homotopies. To see universality, let A be another F-group with associated morphisms $(d_Y\colon K_*^G(Y) \to A)$. Note first that for any $s \geqslant r \geqslant 0$ we have a commutative diagram

$$\begin{array}{ccc} K_*^G(P_r(Z)) & \xrightarrow{d_{P_r(Z)}} & A \\ \downarrow (i_{sr})_* & & \| \\ K_*^G(P_s(Z)) & \xrightarrow{d_{P_s(Z)}} & A\,, \end{array}$$

which commutes by definition of A being an F-group. Hence taking the direct limit of the maps $(d_{P_r(Z)})_{r\geqslant 0}$ gives a well-defined homomorphism

$$d_\infty\colon \lim_{r\to\infty} K_*^G(P_r(Z)) \to A.$$

On the other hand, for any Y and all suitably large r there is a diagram

$$\begin{array}{ccc} K_*^G(Y) & \xrightarrow{f_*} & K_*^G(P_r(Z)) \\ \| & & \downarrow d_{P_r(Z)} \\ K_*^G(Y) & \xrightarrow{d_Y} & A \end{array}$$

(where $f\colon Y \to P_r(Z)$ is our fixed choice defining c_Y coming from Lemma 7.2.14), which commutes by definition of A being an F-group. Taking the limit over r then gives

$$\begin{array}{ccc} K_*^G(Y) & \xrightarrow{c_Y} & \lim_{r\to\infty} K_*^G(P_r(Z)) \\ \| & & \downarrow d_\infty \\ K_*^G(Y) & \xrightarrow{d_Y} & A\,, \end{array}$$

which is the diagram whose existence is required by universality. □

Finally, putting everything together, we get the main result of this section, which is immediate from our work above.

Theorem 7.2.16 *Let X be a proper metric space equipped with an isometric action of a countable group G, and let Z be a locally finite G-invariant net in X. Then the coarse K-homology group of X identifies with $\lim_{r\to\infty} K_*^G(P_r(Z))$, and the Baum–Connes assembly map with the direct limit of the X-assembly maps*

$$\lim_{r\to\infty} \mu_{P_r(Z),X} : \lim_{r\to\infty} K_*^G(P_r(Z)) \to K_*(C^*(X)^G).$$

□

7.3 Uniformly Contractible Spaces

In this section we give a particularly concrete formulation of the coarse Baum–Connes conjecture that holds in many cases of interest. Roughly, this says that if a metric space X is uniformly contractible in the sense of the next definition, then the coarse Baum–Connes conjecture for X identifies with the assembly map for X itself.

Definition 7.3.1 A proper metric space X is *uniformly contractible* if for all $r \geqslant 0$ there exists $s \geqslant r$ such that for all $x \in X$ the inclusion $B(x;r) \to B(x;s)$ is homotopic to a constant map.

Example 7.3.2 Euclidean space $\mathbb{R}^d$ with its usual metric is uniformly contractible: indeed, one can just take $r = s$ in the definition. From Exercise 7.6.12, it follows that any metric on $\mathbb{R}^d$ which is coarsely equivalent to the original metric is also uniformly contractible. Much more generally, any space of 'non-positive curvature' in a suitable sense has a similar property: we will discuss such spaces in Chapter 11. It is, however, certainly possible to put metrics on $\mathbb{R}^d$ that are *not* uniformly contractible: see Exercise 7.6.13.

Example 7.3.3 Say X is a contractible metric space that admits a cocompact isometric group action. Then (Exercise 7.6.11) X is uniformly contractible. A particularly nice class of examples, that is also very important for applications, comes when M is a closed Riemannian manifold with contractible universal cover. Then the universal cover of M equipped with the lifted Riemannian metric is uniformly contractible, as the action of the covering group is isometric and cocompact.

We will also need a technical condition on simplicial complexes.

Definition 7.3.4 Let X be a proper metric space. A simplicial complex structure on X is *good* if it is finite-dimensional, if the vertex set has bounded geometry (see Definition A.3.19) and if the inclusion of the vertex set is a coarse equivalence.

Examples 7.3.5 The Rips complex of a discrete bounded geometry metric space is a good simplicial complex.

Say M is a closed Riemannian manifold, and assume that M has also been given a finite simplicial complex structure. If the universal cover of M is given the lifted simplicial complex structure and lifted Riemannian metric, then it is a good simplicial complex.

Here is the main result of this section.

Theorem 7.3.6 *Let X be a uniformly contractible, good simplicial complex. Then the coarse Baum–Connes conjecture for X identifies with the assembly map*

$$\mu_X : K_*(X) \to K_*(C^*(X))$$

for X itself.

The key step in the proof is the following lemma.

Lemma 7.3.7 *Let Y be a uniformly contractible metric space. Let X be a good simplicial complex, and let $f : X \to Y$ be a coarse map. Then there is a continuous coarse map $g : X \to Y$ that is close to f. Moreover, if f is already continuous on a subcomplex X' of X, then we may assume that f and g are the same on X'.*

Proof Let X' be either as in the statement, or empty if no X' is given. Define $g = f$ on the union of X' and the vertex set, which is continuous. We will extend g one dimension at a time. Indeed, let X_k be the k-skeleton of X, and assume that we have already defined a continuous map $g : X' \cup X_k \to Y$ that is close to f. Consider any $k+1$ simplex Δ. As g is a coarse map on $X' \cup X_k$, there is $r \geqslant 0$ (independent of the particular Δ) such that g takes the boundary $\partial\Delta$ into some ball $B(y;r)$ of radius r. As Y is uniformly contractible, there is $s \geqslant r$ (depending only on r) such that the inclusion $B(y;r) \to B(y;s)$ is nullhomotopic. Hence we may extend g to a continuous map on $X' \cup X_k \cup \Delta$ in such a way that the image $g(\Delta)$ in in $B(y;s)$: precisely, if $x_0 \in \Delta$ is the centre of this simplex, and $(h_t : B(y;r) \to B(y;s))_{t\in[0,1]}$ is a null-homotopy with h_1 the identity and h_0 a constant map, then every point in Δ can be written uniquely as $tz + (1-t)x_0$ for some z in the boundary $\partial\Delta$ and $t \in [0,1]$ and we can define g on Δ by

$$g(tz + (1 - t)x_0) := h_t(z).$$

The result will still be close to f, and in particular will be a coarse map. Doing this for every $k + 1$-simplex and using uniformity of the constants r and s gives our extension of g to $X' \cup X_{k+1}$. Finite dimensionality of X shows that the process terminates, so we are done. □

Proof of Theorem 7.3.6 Let Z be the vertex set of X. According to Lemma 7.2.14 there is $r \geqslant 0$ and a continuous coarse equivalence $f: X \to P_r(Z)$ that restricts to the identity on Z. We may use the coarse equivalence f to define the X assembly map of line (7.2) for $P_r(Z)$, and so get a commutative diagram

$$\begin{array}{ccc} K_*(X) & \xrightarrow{\mu_X} & K_*(C^*(X)) \\ \downarrow{\scriptstyle f_*} & & \| \\ K_*(P_r(Z)) & \xrightarrow{\mu_{P_r(Z),X}} & K_*(C^*(X)). \end{array}$$

Now let $i_{ts}: P_s(Z) \to P_t(Z)$ be the canonical inclusion of Lemma 7.2.11, part (ii), and for each $s \geqslant r$, let $f_s: X \to P_s(Z)$ be the composition $i_{sr} \circ f$. Taking the direct limit as the Rips parameter tends to infinity, we get a commutative diagram

$$\begin{array}{ccc} K_*(X) & \xrightarrow{\mu_X} & K_*(C^*(X)) \\ \downarrow{\scriptstyle f_\infty} & & \| \\ \lim\limits_{s\to\infty} K_*(P_s(Z)) & \xrightarrow{\mu} & K_*(C^*(X)), \end{array}$$

where f_∞ is the direct limit of the maps $(f_s)_*: K_*(X) \to K_*(P_s(Z))$. Theorem 7.2.16 identifies the bottom horizontal arrow with the coarse Baum–Connes assembly map for X, so to complete the proof, it suffices to show that f_∞ is an isomorphism.

To see this, note that if we start with the inclusion map $Z \to X$, then Lemma 7.3.7 allows us to inductively construct continuous coarse equivalences $g_s: P_s(Z) \to X$ for each $s \in \mathbb{N}$ such that for $t \geqslant s$, $g_t \circ i_{ts} = g_s$, and that all restrict to the identity on Z. We thus get a map

$$g_\infty: \lim_{s\to\infty} K_*(P_s(Z)) \to K_*(X)$$

defined as the direct limit of the maps

$$(g_s)_*: K_*(P_s(Z)) \to K_*(X).$$

We claim g_∞ is the inverse to f_∞. Note first that for any $s \geqslant r$, g_s and f_s are mutually inverse coarse equivalences, as both restrict to the identity on Z.

Consider first the composition $g_s \circ f_s \colon X \to X$. Equip $[0,1] \times X$ with a reasonable product metric and simplicial structure so that is a good simplicial complex, and such that the natural maps $X \to [0,1] \times X$, $x \mapsto (t,x)$ are all coarse equivalences (we leave it as an exercise to find such structures). Let $h \colon \{0,1\} \times X \to X$ be the map that is the identity on $\{0\} \times X$ and equal to $g_s \circ f_s$ on $\{1\} \times X$. Lemma 7.3.7 then implies that h extends to a continuous coarse equivalence $\widetilde{h} \colon [0,1] \times X \to X$, and this map is necessarily a proper homotopy between $g_s \circ f_s$ and the identity, so the two induce the same map on $K_*(X)$.

On the other hand, consider the composition $f_s \circ g_s \colon P_s(Z) \to P_s(Z)$. This is again close to the identity, so Lemma 7.2.13 gives $t \geqslant s$ such that the composition $i_{ts} \circ f_s \circ g_s \colon P_s(Z) \to P_t(Z)$ is properly homotopic to the identity. This is enough to complete the proof. □

7.4 Classifying Spaces

In this section we use classifying spaces to give concrete models for the Baum–Connes conjecture for a group G. We give two variations: one in terms of the classical classifying space BG which is perhaps the most concrete, but does not always work, and one in terms of the classifying space for proper actions $\underline{E}G$, which does always work. This section assumes a little more topology than we normally would: this is all to do with covering space theory and fundamental groups, which we hope is still quite accessible.

Definition 7.4.1 Let G be a countable discrete group. A *classifying space*[2] for G is a connected CW complex BG with fundamental group G, and contractible universal cover.

Examples 7.4.2 The following are basic examples of classifying spaces.

(i) If $G = \mathbb{Z}$, then a classifying space for G is the circle S^1 with universal cover $\mathbb{R}$.

(ii) Generalising (i), if $G = F_n$ is a free group on n generators, then a classifying space for G is the wedge product $\bigvee^n S^1$ of n circles, with universal cover the tree in which every vertex has $2n$ edges coming out of it.

(iii) Generalising (i) in another direction, if $G = \mathbb{Z}^d$ is the free abelian group on d generators, then a classifying space for G is the d-torus $(S^1)^d$ with universal cover $\mathbb{R}^d$.

[2] Also called a $K(G,1)$ space.

(iv) Say G is the fundamental group of a closed orientable surface Σ_g of genus $g > 1$. Then a classifying space for G is Σ_g itself, with universal cover the hyperbolic plane.

(v) Say G is the integral Heisenberg group

$$G := \left\{ \begin{pmatrix} 1 & x & z \\ 0 & 1 & y \\ 0 & 0 & 1 \end{pmatrix} \in M_3(\mathbb{R}) \;\middle|\; x, y, z \in \mathbb{Z} \right\}.$$

Let H be the real Heisenberg group, which is defined in the same way but with x, y and z allowed to be any real numbers. Then the closed 3-manifold H/G is a classifying space for G, with universal cover H.

The free group and free abelian group examples can be built from the example for $G = \mathbb{Z}$ using general facts about free products and direct products: see Exercise 7.6.10.

The following theorem from topology summarises the key properties of classifying spaces. We will not prove it here: see the notes and references at the end of the chapter.

Theorem 7.4.3 *Let G be a countable discrete group. Then a classifying space for G exists.*

Moreover, if BG is a classifying space with basepoint y_0 and X is a CW complex with basepoint x_0 then for any map $\pi_1(X, x_0) \to \pi_1(BG, y_0)$ there is a continuous map $f: X \to BG$ taking x_0 and y_0, and that is unique up to homotopy (through maps taking x_0 to y_0). □

Our first goal in this section is the following theorem.

Theorem 7.4.4 *Assume G is finitely generated and torsion free, and admits a classifying space BG which is a finite CW complex.*[3] *Then the Baum–Connes assembly map for G (acting on itself) identifies with the composition*

$$K_*(BG) \cong K_*^G(EG) \xrightarrow{\mu_{EG}} K_*(C^*(EG)^G) \cong K_*(C_\rho^*(G)),$$

where the first isomorphism is that of Theorem 6.5.15, the middle map is the assembly map for EG of Definition 7.1.1 and the last isomorphism comes from Theorem 5.3.2.

Proof Lemma 7.2.14 implies that there is $r \geqslant 0$ and a continuous, equivariant map $f: EG \to P_r(G)$ that restricts to the identity on G (considered as included in EG by an orbit map $g \mapsto gx$ for some fixed $x \in EG$). Note

[3] This actually implies that G is finitely generated and torsion free, so these assumptions are redundant.

that as the action of G on EG is cocompact, f is also a coarse equivalence by the Svarc–Milnor lemma (Lemma A.3.14). This gives rise to a commutative diagram

$$\begin{array}{ccc} K_*^G(EG) & \xrightarrow{\mu_{EG,G}} & K_*(C^*(G)^G) \\ \downarrow f_* & & \| \\ K_*^G(P_r(G)) & \xrightarrow{\mu_{P_r(G),G}} & K_*(C^*(G)^G). \end{array}$$

Let $i_{ts}\colon P_s(G) \to P_t(G)$ be the canonical inclusion, and for each $s \geqslant r$, let $f_s\colon EG \to P_s(G)$ be the composition $i_{sr} \circ f$. Taking the direct limit as the Rips parameter tends to infinity, we get a commutative diagram

$$\begin{array}{ccc} K_*^G(EG) & \xrightarrow{\mu_X} & K_*(C^*(G)^G) \\ \downarrow f_\infty & & \| \\ \lim\limits_{s\to\infty} K_*^G(P_s(G)) & \xrightarrow{\mu} & K_*(C^*(G)^G), \end{array}$$

where f_∞ is the direct limit of the maps $(f_s)_*\colon K_*^G(EG) \to K_*^G(P_s(G))$. To complete the proof, it suffices to show that f_∞ is an isomorphism.

To see this, we will construct an inverse. Note now that by increasing r as above if necessary, finite generation of G implies that $P_r(G)$ is connected. For now fix $s \geqslant r$, and consider the space $X_s := P_s(G)/G$ defined as the quotient of $P_s(G)$ by the canonical G action. Equip X_s with the CW complex structure arising from the simplicial structure of $P_s(G)$. As G is torsion free the action of G on $P_s(G)$ is free (see Exercise 7.6.1), and it is always proper (see Lemma 7.2.12). Hence $P_s(G)$ is a covering space of X_s with deck transformation group G. It follows that there is an associated map of fundamental groups $\pi_1(X_s) \to \pi_1(BG)$ (here we use the image of $e \in G$ as the basepoint of X_s, and fix any basepoint of BG to make sense of this). Theorem 7.4.4 gives a continuous map $g_{s,0}\colon X_s \to BG$ that is unique up to homotopy equivalence. Moreover, this map lifts to a continuous equivariant map $g_s\colon P_s(G) \to EG$ on covers. If we perform the same construction for some $t \geqslant s$, then $X_s \subseteq X_t$ and the maps $g_{s,0}, g_{t,0}|_{X_s}\colon X_s \to BG$ that we get are homotopic. This homotopy lifts to a proper (by compactness of BG) homotopy of the maps $g_s, g_t|_{P_s(G)}\colon P_s(G) \to EG$ on covers. It follows that for any $t \geqslant s$ the following diagram

$$\begin{array}{ccc} K_*^G(P_s(G)) & \xrightarrow{(g_s)_*} & K_*^G(EG) \\ \downarrow (i_{ts})_* & & \| \\ K_*^G(P_t(G)) & \xrightarrow{(g_t)_*} & K_*^G(EG) \end{array}$$

commutes. Hence we get a map on the direct limit

$$g_\infty\colon \lim_{s\to\infty} K_*^G(P_s(G)) \to K_*^G(EG)$$

defined as the direct limit of the maps $(g_s)_*\colon K_*^G(P_s(G)) \to K_*^G(EG)$. We claim g_∞ is the inverse to f_∞. Note first that for any $s \geqslant r$, g_s and f_s are (equivariant, continuous) coarse equivalences that are mutually inverse up to closeness, as both restrict to the identity on G.

To show that g_∞ is the inverse to f_∞, consider first the composition $g_s \circ f_s\colon EG \to EG$. This is an equivariant, continuous map. As $g_s \circ f_s$ is equivariant, it induces a map $BG \to BG$ on quotients, which is homotopic to the identity using Theorem 7.4.3. Lifting this homotopy shows that $g_s \circ f_s$ is itself homotopic to the identity. On the other hand, consider the composition $f_s \circ g_s\colon P_s(G) \to P_s(G)$. This is close to the identity, from which it follows that for some suitable large $t \geqslant s$, the composition $i_{ts} \circ f_s \circ g_s\colon P_s(G) \to P_t(G)$ is equivariantly properly homotopic to the identity via a straight-line homotopy: see Lemma 7.2.13. □

Examples 7.4.5 In the following examples, the left-hand side of the Baum–Connes conjecture for the given group G can be computed directly using standard Mayer–Vietoris arguments (see Exercise 7.6.9). Many other naturally occurring examples can be handled similarly, assuming a little more topology, manifold theory or geometric group theory.

(i) Say $G = \mathbb{Z}^n$. Then the torus $\mathbb{T}^n$ has fundamental group G, and its universal cover is $\mathbb{R}^n$, which is contractible. Hence we may take $BG = \mathbb{T}^n$. It follows that

$$KX_*^G(G) \cong K_*(\mathbb{T}^n) \cong \underbrace{\mathbb{Z}^{2^n-1}}_{K_0} \oplus \underbrace{\mathbb{Z}^{2^n-1}}_{K_1}.$$

(ii) Say G is a free group on n generators, and let $X_n := \bigvee^n S^1$ be a wedge product of n circles. Then $\pi_1(X_n) = G$, and the universal cover of X_n is a tree where every vertex has degree $2n$, which is contractible. Hence

$$KX_*^G(G) \cong K_*(X_n) \cong \underbrace{\mathbb{Z}}_{K_0} \oplus \underbrace{\mathbb{Z}^n}_{K_1}.$$

(iii) Say G is the fundamental group of an oriented closed surface Σ_g of genus g. Then by the uniformisation theorem, the universal cover of Σ_g identifies with the hyperbolic plane, which is contractible. Hence

$$KX_*^G(G) \cong K_*(\Sigma_g) \cong \underbrace{\mathbb{Z}^2}_{K_0} \oplus \underbrace{\mathbb{Z}^{2g}}_{K_1}.$$

As the examples above illustrate, the left-hand side of the Baum–Connes conjecture can be computed very explicitly when there is a good model for the classifying space BG, say one that is a finite simplicial complex. However, the existence of such a good model is a fairly restrictive condition. We now move on to a classifying space construction that works in general.

For the statement of the next definition, if X is a simplicial complex with vertex set X_0, then an action of G on X is *simplicial* if the restricted action permutes X_0 and if the G action takes the simplex with vertices $x_0, \ldots, x_d$ to the simplex with vertices $gx_0, \ldots, gx_d$ by affine extension of the map on the vertices. The most important example for us will occur when G acts isometrically on a locally finite metric space Z, and $X = P_r(X)$ is the Rips complex.

Definition 7.4.6 Let G be a countable group. A *classifying space for proper actions for* G is a topological space $\underline{E}G$ equipped with a proper action of G that has the following universal property: for any proper G simplicial complex[4] X there is a continuous, proper equivariant map $f : X \to \underline{E}G$ that is unique up to proper homotopy equivariant.

Theorem 7.4.7 *Let G be a countable group. Then there exists a locally compact, second countable classifying space for proper actions for G.*

Proof Define

$$Z := \{\mu \in \ell^1(G) \mid \mu \geqslant 0 \text{ and } 1/2 < \|\mu\| \leqslant 1\},$$

equipped with the weak-$*$ topology inherited from the natural identification $C_0(G)^* \cong \ell^1(G)$. We claim Z works as a model for $\underline{E}G$. First, note that

$$Z = \{\mu \in \ell^1(G) \mid \mu \geqslant 0 \text{ and } \|\mu\| \leqslant 1\} \setminus \\ \{\mu \in \ell^1(G) \mid \mu \geqslant 0 \text{ and } 1/2 < \|\mu\|\};$$

as the first set on the right is compact (by Banach–Alaoglu) and the second is open, Z is locally compact. Second countability follows from separability of $C_0(G)$.

To see that the action of G on Z is proper, let $K \subseteq Z$ be compact. For each non-negative-valued, compactly supported function f on G with supremum norm 1, let

$$U_f := \{\mu \in Z \mid \mu(f) > 1/2\}.$$

[4] One can alter the definition by altering the class of proper G spaces X the universal property applies to; our definition with proper G simplicial complexes is fairly restrictive. The literature is not completely consistent on this issue; however, from a practical point of view, it does not usually matter exactly what choice one makes here.

These sets are weak-$*$ open and it is straightforward to check that they cover Z. Hence our compact set K is covered by finitely many of these sets, say $U_{f_1}, \ldots, U_{f_n}$. Let F be the union of the supports of the f_i, a finite subset of G, and note that for any $g \in G$ outside the finite set $\{g \in G \mid gF \cap F \neq \varnothing\}$, we have that

$$g \cdot \left(\bigcup_{i=1}^{n} U_{f_i} \right) \cap \left(\bigcup_{i=1}^{n} U_{f_i} \right) = \varnothing$$

(the bounds on the mass of our measures are crucial here) and therefore that $gK \cap K = \varnothing$.

Finally, let us check the universal property. Let X be as in the statement. Let X_0 be the vertex set of X. Then each G orbit in X_0 identifies as a G-space with G/F for a finite subgroup F of G by properness. Consider the point $\mu := \frac{1}{|F|} \sum_{g \in F} \delta_g$ in Z. This is fixed by F, and hence the orbit inclusion

$$G \to Z, \quad g \mapsto g\mu$$

descends to a well-defined map $G/F \to Z$. Working on each orbit separately in this way, we get a (continuous equivariant) map $f_0 \colon X_0 \to Z$. Now extend f_0 to all of X by taking convex combinations. It is not too difficult to see that this has the right properties. □

The following theorem can now be proved in much the same way as Theorem 7.4.4 above.

Theorem 7.4.8 *Let G be a countable discrete group, and $\underline{E}G$ be a classifying space for proper actions for G. Then the Baum–Connes assembly map for G (acting on itself) identifies with the map*

$$\lim_{Y \subseteq \underline{E}G} K_*^G(\underline{E}G) \xrightarrow{\mu_{\underline{E}G}} K_*(C^*(G)^G),$$

which is the direct limit over all assembly maps $K_*^G(Y) \xrightarrow{\mu_{Y,G}} K_*(C^*(G)^G)$, where $Y \subseteq \underline{E}G$ is a proper cocompact subset.

Proof As $\underline{E}G$ is second countable and locally compact, there exists a countable nested collection $Y_1 \subseteq Y_2 \subseteq \cdots$ of cocompact, equivariant subsets of $\underline{E}G$ such that for any cocompact subset of $\underline{E}G$ is eventually contained in one of the Y_n. Fix an orbit inclusion $G \ni g \mapsto gy \in Y$. Using Lemma 7.2.14, there is $r_1 \geqslant 0$ and a continuous equivariant map $f_1 \colon Y_1 \to P_{r_1}(G)$ that extends our fixed orbit inclusion. Similarly, there is $r_{2,0} \geqslant 0$ such that there is a continuous equivariant map $f_{2,0} \colon Y_2 \to P_{r_{2,0}}(G)$ that extends the same orbit inclusion. As f_1 and $f_{2,0}$ both extend the orbit inclusion, there is $r_2 \geqslant \max\{r_1, r_{2,0}\}$ such that these two maps become close, and therefore properly equivariantly homotopic when seen as maps into $P_{r_2}(G)$. Continuing in this way, we get a sequence $r_1 \leqslant r_2 \leqslant \cdots$ of non-negative real numbers and for each n a continuous equivariant coarse equivalence $f_n\colon Y_n \to P_{r_n}(G)$ such that the diagram

$$\begin{array}{ccccccc}
K_*^G(Y_1) & \longrightarrow & K_*^G(Y_2) & \longrightarrow & K_*^G(Y_3) & \longrightarrow & \cdots \\
\downarrow {\scriptstyle (f_1)_*} & & \downarrow {\scriptstyle (f_2)_*} & & \downarrow {\scriptstyle (f_3)_*} & & \\
K_*^G(P_{r_1}(G)) & \longrightarrow & K_*^G(P_{r_2}(G)) & \longrightarrow & K_*^G(P_{r_2}(G)) & \longrightarrow & \cdots
\end{array}$$

commutes. Taking the limits along both lines gives a map

$$f_\infty\colon \lim_{Y \subseteq \underline{E}G} K_*^G(\underline{E}G) \to \lim_{r \to \infty} K_*^G(P_r(G)).$$

Noting that the diagram

$$\begin{array}{ccc}
\lim\limits_{Y \subseteq \underline{E}G} K_*^G(\underline{E}G) & \xrightarrow{\ \mu_{\underline{E}G}\ } & K_*(C^*(G)^G) \\
\downarrow {\scriptstyle f_\infty} & & \| \\
\lim\limits_{r \to \infty} K_*^G(P_r(G)) & \xrightarrow{\ \mu\ } & K_*(C^*(G)^G)
\end{array}$$

commutes, it is enough to show that f_∞ is an isomorphism.

On the other hand, using the universal property of $\underline{E}G$, for each there is a continuous equivariant map $g_1\colon P_{r_1}(G) \to \underline{E}G$. As the action of G on $P_{r_k}(G)$ is cocompact, there is some Y_{n_1} such that the image of g_1 is actually contained in Y_{n_1}. Similarly, let $n_{2,0}$ be such that there is a continuous equivariant map $g_2\colon P_{r_2}(G) \to Y_{n_{2,0}}$, and let n_2 be such that g_1 and the restriction of $g_{2,0}$ to $P_{r_1}(G)$ are equivariantly properly homotopic (such exists by the universal property of $\underline{E}G$), and consider g_2 as having image in Y_{n_2}. Continuing in this way gives a commutative diagram

$$\begin{array}{ccccccc}
K_*^G(Y_{n_1}) & \longrightarrow & K_*^G(Y_{n_2}) & \longrightarrow & K_*^G(Y_{n_3}) & \longrightarrow & \cdots . \\
{\scriptstyle (g_1)_*} \uparrow & & {\scriptstyle (g_2)_*} \uparrow & & {\scriptstyle (g_3)_*} \uparrow & & \\
K_*^G(P_{r_1}(G)) & \longrightarrow & K_*^G(P_{r_2}(G)) & \longrightarrow & K_*^G(P_{r_2}(G)) & \longrightarrow & \cdots
\end{array}$$

Taking limits then gives a map

$$g_\infty\colon \lim_{r \to \infty} K_*^G(P_r(G)) \to \lim_{Y \subseteq \underline{E}G} K_*^G(Y).$$

Now, note that any composition $g_k \circ f_n\colon Y_n \to Y_{n_k}$ (where defined) becomes properly equivariantly homotopic to the identity on increasing n_k

by the universal property of $\underline{E}G$, and that any (defined) composition $f_n \circ g_k \colon P_{r_k}(G) \to P_{r_n}(G)$ is again is close to the identity, and therefore becomes equivariantly properly homotopic to the identity (by a straight-line homotopy) on increasing the Rips parameter. This shows that f_∞ and g_∞ are mutually inverse, and is enough to complete the proof. □

Although Theorem 7.4.7 gives a reasonably explicit model for a space $\underline{E}G$ that works in general, it is often possible to give much more geometrically natural examples that even allow $\lim_{Y \subseteq \underline{E}G} K_*^G(Y)$ to be computed. We give some below: we will not justify this here, but it can be done in each case below by some argument based on non-positive curvature.

Examples 7.4.9 (i) Let D_∞ be the infinite dihedral group: this is the group of isometries of $\mathbb{R}$ generated by integer translations and reflection about the origin. Then $\mathbb{R}$ is an $\underline{E}G$ for this action.

(ii) Let $G = PSL(2,\mathbb{Z})$ be the group of 2×2 integer matrices with determinant one, modulo the central subgroup $\{\pm 1\}$, where 1 is the identity matrix. This group acts on the upper half plane $\mathbb{H} := \{z \in \mathbb{C} \mid \mathrm{Im}(z) > 0\}$ via Möbius transformations:

$$\begin{pmatrix} a & b \\ c & d \end{pmatrix} : z \mapsto \frac{az+b}{cz+d}.$$

Moreover, this action is by isometries if one gives $\mathbb{H}$ the standard hyperbolic metric. Then $\mathbb{H}$ is an example of an $\underline{E}G$.

(iii) Let $G = PSL(2,\mathbb{Z})$ again. Using *Bass–Serre* theory and the free product decomposition $PSL(2,\mathbb{Z}) = (\mathbb{Z}/2\mathbb{Z}) * (\mathbb{Z}/3\mathbb{Z})$, one can build an action of $PSL(2,\mathbb{Z})$ on a tree T. This tree is also an example of an $\underline{E}G$. It is cocompact, which makes it a little simpler than $\mathbb{H}$ to work with in some ways: in particular, one has that $\lim_{Y \subseteq T} K_*^G(Y) \cong K_*^G(T)$.

Note that

$$\lim_{Y \subseteq \mathbb{H}} K_*^G(Y) \cong K_*^G(T)$$

as both groups identify with $KX_*^G(G)$.

Example 7.4.10 To see how such computations are at least sometimes possible, let $G = D_\infty$ and let us compute $K_*^G(\mathbb{R})$. This computes the domain of the assembly map ('the left-hand side of the Baum–Connes conjecture') in this case.

Let τ be the element of G which translates to the left by one, and σ be the element that reflects around the origin, so G is generated by σ and τ, subject to the relations $\sigma^1 = e$ (where e is the identity) and $\sigma\tau\sigma = \tau^{-1}$. Write $\mathbb{R} = E \cup F$, where

$$E = \bigsqcup_{n \in \mathbb{Z}} [n + 1/3, n + 2/3] \quad \text{and} \quad F = \bigsqcup_{n \in \mathbb{Z}} [n - 1/3, n + 1/3].$$

There is then a Mayer–Vietoris sequence in equivariant K-homology (see Exercise 6.8.12), which looks like

$$\begin{array}{ccccc} K_0^G(E \cap F) & \longrightarrow & K_0^G(E) \oplus K_0^G(F) & \longrightarrow & K_0^G(\mathbb{R}) \\ \uparrow & & & & \downarrow \\ K_1^G(\mathbb{R}) & \longleftarrow & K_1^G(E) \oplus K_1^G(F) & \longleftarrow & K_1^G(E \cap F) \,. \end{array} \tag{7.13}$$

Note first that $E \cap F$ is the set $\{n + 1/3, 1/3 - n \mid n \in \mathbb{Z}\}$, and that this space is isomorphic to G as a G-space. Hence by Proposition 6.5.13,

$$K_i^G(E \cap F) \cong K_i^G(G) \cong K_i(\text{point}) \cong \begin{cases} \mathbb{Z}, & i = 0, \\ 0, & i = 1. \end{cases}$$

To compute $K_i^G(E)$, note first that the linear homotopy contracting each interval $[n + 1/3, n + 2/3]$ to the singleton $n + 1/2$ is equivariant. Write $E_0 = \{n + 1/2 \mid n \in \mathbb{Z}\}$. Note that the point $1/2$ is fixed by $\tau\sigma$, which generates a subgroup H of G of order two. One computes from this that E_0 is isomorphic to the balanced product $G \times_H \{1/2\}$ as a G-space (with the trivial action of H on $\{1/2\}$: see the discussion before Proposition 6.5.13 for notation). Hence

$$K_i^G(E) \cong K_i^G(E_0) \cong K_i^H(\text{point}) \cong \begin{cases} \mathbb{Z} \oplus \mathbb{Z}, & i = 0, \\ 0, & i = 1, \end{cases}$$

where the second isomorphism is Proposition 6.5.13, and the third comes from Exercise 6.8.10 (or if you prefer, from Example 7.1.13). Similarly, we have that F is equivariantly homotopy equivalent to $F_0 := \mathbb{Z}$. Using that the point 0 is stabilised by the order two subgroup K of G generated by σ, one checks that F_0 is isomorphic to $G \times_K \{0\}$ as a G-space. We thus have

$$K_i^G(F) \cong K_i^G(F_0) \cong K_i^K(\text{point}) \cong \begin{cases} \mathbb{Z} \oplus \mathbb{Z}, & i = 0, \\ 0, & i = 1, \end{cases}$$

completely analogously to the case of E.

Now, putting all of this information into the diagram in line (7.13) gives an exact sequence

$$\begin{array}{ccccc} \mathbb{Z} & \xrightarrow{\alpha} & \mathbb{Z}^4 & \longrightarrow & K_0^G(\mathbb{R}) \\ \uparrow & & & & \downarrow \\ K_1^G(\mathbb{R}) & \longleftarrow & 0 & \longleftarrow & 0. \end{array}$$

It remains to compute the map labelled α above, which is induced by the inclusions $E \cap F \to E$ and $E \cap F \to F$. Consider first the map $K_0^G(E \cap F) \to K_0^G(F)$. We may identify $E \cap F$ with $G \times_K \{\pm 1/2\}$ and F with $G \times_K \{0\}$; the map $E \cap F \to F$ is then induced by the (K-equivariant) collapse map $\pm 1/2 \mapsto 0$. Using naturality of the isomorphism of Proposition 6.5.13, the map $K_0^G(E \cap F) \to K_0^G(F)$ that we are interested in is equivalent to the map

$$K_0^K(\{-1/2, 1/2\}) \to K_0^K(\{0\}).$$

Using Exercise 6.8.10, we see that this is the map $\mathbb{Z} \to \mathbb{Z} \oplus \mathbb{Z}$, $a \mapsto (a,a)$ on K-theory. Completely analogously, the map $K_0^G(E \cap F) \to K_0^G(E)$ is again $a \mapsto (a,a)$. Putting this together, the map α is injective, and the map

$$\mathbb{Z}^4 \to \mathbb{Z}^3, \quad (a,b,c,d) \mapsto (a-b, b-c, c-d)$$

is surjective with kernel exactly the image of α. We conclude that the left-hand side of Baum–Connes is

$$K_i^G(\mathbb{R}) \cong \begin{cases} \mathbb{Z}^3 & i = 0, \\ 0 & i = 1. \end{cases}$$

The computation for $PSL(2,\mathbb{Z})$ acting on T can be handled quite similarly. Indeed, the above computation for D_∞ acting on $\mathbb{R}$ can be thought of as an action of the free product $D_\infty \cong (\mathbb{Z}/2) * (\mathbb{Z}/2)$ acting on its Bass–Serre tree, and the case of $PSL(2,\mathbb{Z}) \cong (\mathbb{Z}/2) * (\mathbb{Z}/3)$ can be handled very similarly: the result of the computation is

$$K_i^{PSL(2,\mathbb{Z})}(T) \cong \begin{cases} \mathbb{Z}^4 & i = 0, \\ 0 & i = 1. \end{cases}$$

These computations generalise to any free product of finite groups (assuming one knows their representation theory). We leave the computations for the interested reader who knows enough about Bass–Serre theory. It is also an interesting exercise to compare what happens for the action of $PSL(2,\mathbb{Z})$ on the hyperbolic plane $\mathbb{H}$ for readers who know enough about that action.

We conclude this section by mentioning the *descent principle*. We will not prove it here as it is well covered elsewhere in the literature and the methods we are using here would not lead to any substantial difference with previous treatments.

Theorem 7.4.11 *Say G is a discrete group, which admits a finite CW-complex as a model for BG. Then if the coarse assembly map*

$$K_*(EG) \to K_*(C^*(|G|))$$

is an isomorphism, the Baum–Connes assembly map

$$K_*(BG) \to K_*(C_r^*(G))$$

is injective. □

This is important for applications, partly as the injectivity statement is more closely connected to Novikov-type statements (see Section 10.3).

7.5 The Coarse Baum–Connes Conjecture for Euclidean Space

In this section, we move back from generalities and prove the coarse Baum–Connes conjecture for the metric space $\mathbb{R}^d$. This case already has quite non-trivial consequences: for example, as discussed in Section 3.3 it implies that the d-torus does not admit a metric of positive scalar curvature.

Theorem 7.5.1 *The coarse Baum–Connes conjecture holds for $\mathbb{R}^d$.*

The proof uses ideas that in the main we have already developed, as we explain in the rest of this section.

Using Theorem 7.3.6 and the fact that $\mathbb{R}^d$ is uniformly contractible (see Example 7.3.2), it suffices to prove that if $X = \mathbb{R}^d$ then the evaluation-at-one map

$$\mathrm{ev}\colon C_L^*(X) \to C^*(X)$$

induces an isomorphism on K-theory. We will do this by induction on d using a Mayer–Vietoris argument.

The base case $d = 0$ is the coarse Baum–Connes conjecture for a point, which is true by Example 7.1.13. The inductive step will follow from a Mayer–Vietoris argument applied to the decomposition

$$\mathbb{R}^d = \big(\mathbb{R}^{d-1} \times (-\infty, 0]\big) \cup \big(\mathbb{R}^{d-1} \times [0, \infty)\big), \tag{7.14}$$

and the fact that the assembly map is an isomorphism for any metric space of the form $X \times [0,\infty)$ where X is a proper metric space and $X \times [0,\infty)$ has the metric

$$d_{X\times[0,\infty)}((x_1,t_1),(x_2,t_2)) := \sqrt{d_X(x_1,x_2)^2 + |t_1 - t_2|^2}.$$

Proposition 7.5.2 *Let X be a proper metric space, and equip $X \times [0,\infty)$ with the metric above, we have*

$$K_*(C_L^*(X \times [0,\infty))) = K_*(C^*(X \times [0,\infty))) = 0.$$

In particular, the assembly map

$$\mu\colon K_*(X \times [0,\infty)) \to K_*(C^*(X \times [0,\infty))$$

is an isomorphism for any metric space of this form.

Proof The group $K_*(C_L^*(X \times [0,\infty)))$ is the same as the K-homology group $K_*(X \times [0,\infty))$ and is thus zero by Proposition 6.4.14. We will show that $C^*(X \times [0,\infty))$ has zero K-theory by an Eilenberg swindle.

Assume that $C^*(X \times [0,\infty))$ is defined using an ample geometric module $H_{X\times[0,\infty)}$ of the form $H_X \otimes L^2[0,\infty)$ for some ample X module H_X. Let $H^\infty_{X\times[0,\infty)}$ be the infinite direct sum

$$H^\infty_{X\times[0,\infty)} := \bigoplus_{n=1}^{\infty} H_{X\times[0,\infty)},$$

which is also an ample X module with the structure inherited from $H_{X\times[0,\infty)}$. For each n, let $V_n\colon L^2[0,\infty) \to L^2[0,\infty)$ be the isometry defined for $u \in L^2[0,\infty)$ by

$$(V_n u)(t) = \begin{cases} u(t-1) & t \geqslant 1, \\ 0 & t < 1. \end{cases}$$

Let $W_n\colon H_{X\times[0,\infty)} \to H^\infty_{X\times[0,\infty)}$ be the isometry including $H_{X\times[0,\infty)}$ as the nth summand. Define a $*$-homomorphism $\phi\colon C^*(H_{X\times[0,\infty)}) \to C^*(H^\infty_{X\times[0,\infty)})$ by the formula

$$T \mapsto \bigoplus_{n=1}^{\infty} W_n(1 \otimes V_n)T(1 \otimes V_n^*)W_n^*;$$

the image of this lies in $C^*(H^\infty_{X\times[0,\infty)})$ as for any compact $K \subseteq X \times [0,\infty)$ only finitely many summands of $\chi_K\phi(T)$ and $\phi(T)\chi_K$ are non-zero. Let $\psi\colon C^*(H_{X\times[0,\infty)}) \to C^*(H^\infty_{X\times[0,\infty)})$ be the $*$-homomorphism induced by the inclusion of $H_{X\times[0,\infty)}$ onto the first summand of $H^\infty_{X\times[0,\infty)}$.

Now, for any $T \in C^*(H_{X\times[0,\infty)})$, the elements

$$\begin{pmatrix} \phi(T) & 0 \\ 0 & \psi(T) \end{pmatrix}, \quad \begin{pmatrix} \phi(T) & 0 \\ 0 & 0 \end{pmatrix}$$

of $M_2(C^*(H^{\infty}_{X\times[0,\infty)}))$ are conjugate to each other by an isometry in the multiplier algebra of $M_2(C^*(H^{\infty}_{X\times[0,\infty)}))$, whence

$$\phi_* + \psi_* = \phi_*$$

as maps on K-theory by Proposition 2.7.5, and so $\psi_* = 0$. However, ψ_* covers the identity map, so is an isomorphism by Theorem 5.1.15; the only way this is possible is if $K_*(C^*(X \times [0,\infty)))$ is zero, so we are done. □

To complete the proof we will use a Mayer–Vietoris sequence for the K-theory of the Roe algebra. One needs an excision condition appropriate to coarse geometry to make this work.

Definition 7.5.3 Let $X = E \cup F$ be a cover of X by closed subsets. The cover is said to be *coarsely excisive* if for all $r > 0$, there exists $s > 0$ such that

$$N_r(E) \cap N_r(F) \subseteq N_s(E \cap F).$$

Example 7.5.4 The cover $\mathbb{R} = (-\infty, 0] \cup [0, \infty)$ is coarsely excisive, as is the decomposition in line (7.14) above. Any decomposition with empty intersection is not coarsely excisive. For a less trivial example, let X be the pictured subset of $\mathbb{R}^2$ with the restricted metric, and let E and F be the closed top and bottom halves respectively (so they intersect at the midpoint of the vertical segment).

The resulting decomposition is not coarsely excisive.

Here, then, is the Mayer–Vietoris sequence we want. The proof is closely related to, but more straightforward than, that of Theorem 6.3.4, so we will not be as detailed here.

Theorem 7.5.5 *Let $X = E \cup F$ be a coarsely excisive decomposition of a proper metric space. Then there is a six-term exact sequence*

$$\begin{array}{ccccc} K_0(C^*(F \cap F)) & \longrightarrow & K_0(C^*(E)) \oplus K_0(C^*(F)) & \longrightarrow & K_0(C^*(X)) \\ \uparrow & & & & \downarrow \\ K_1(C^*(E)) & \longleftarrow & K_1(C^*(E)) \oplus K_1(C^*(F)) & \longleftarrow & K_1(C^*(E \cap F)) \end{array}$$

and similarly for the localised Roe algebras.

Moreover, the evaluation-at-one maps naturally map one of these exact sequences to the other in the sense that the diagram

$$\begin{array}{ccccccc}
\longrightarrow K_i(C_L^*(E\cap F)) & \longrightarrow & K_i(C_L^*(E))\oplus K_i(C_L^*(F)) & \longrightarrow & K_i(C_L^*(X)) \longrightarrow \\
\downarrow ev_* & & \downarrow ev_*\oplus ev_* & & \downarrow ev_* \\
\longrightarrow K_i(C^*(E\cap F)) & \longrightarrow & K_i(C^*(E))\oplus K_i(C^*(F)) & \longrightarrow & K_i(C^*(X)) \longrightarrow
\end{array}$$

commutes.

Proof It will suffice to prove the theorem for a given choice of ample modules. Let then Z_E, Z_F be countable dense subsets of E, F respectively such that $Z_E \cap Z_F$ is a countable dense subset of $E \cap F$, and let $Z = Z_E \cup Z_F$, a countable dense subset of X. Let H be a separable infinite-dimensional Hilbert space, and let $H_{E\cap F}$, H_E, H_F, H_X denote the highly ample modules $\ell^2(Z_E \cap Z_F, H)$, $\ell^2(Z_E, H)$, $\ell^2(Z_F, H)$, $\ell^2(Z, H)$ over $E \cap F$, E, F, X respectively; we will use these modules to define all the Roe algebras and localised Roe algebras involved in the statement.

Let χ_E denote the characteristic function of E, which is a multiplier of all the C^*-algebras involved, and let $C_X^*(E)$, $C_{L,X}^*(E)$ the C^*-subalgebras of $C^*(X)$ and $C_L^*(X)$ generated by products of the forms

$$S\chi_E T, \quad (S_t)\chi_E(T_t)$$

(where S, T are in $C^*(X)$ and (S_t), (T_t) are in $C_L^*(X)$) respectively; note that $C_X^*(E)$ and $C_{L,X}^*(E)$ are ideals in $C^*(X)$ and $C_L^*(X)$ respectively. We define $C_X^*(F)$, $C_{L,X}^*(F)$, $C_X^*(E \cap F)$ and $C_{L,X}(E \cap F)$ similarly. Then there is a commutative diagram of pushout diagrams

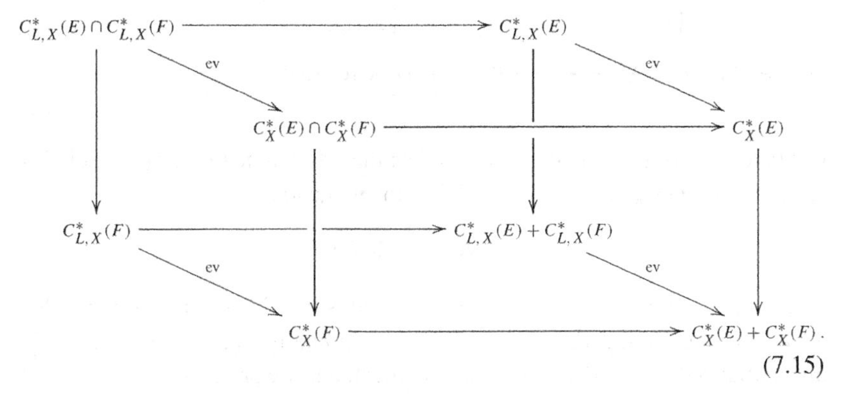

(7.15)

We claim that: (i) $C_X^*(E) \cap C_X^*(F) = C_X^*(E \cap F)$; (ii) $C_X^*(E) + C_X^*(F) = C^*(X)$, and similarly for the localised Roe algebras. For point (i), the inclusion

$$C_X^*(E \cap F) \subseteq C_X^*(E) \cap C_X^*(F)$$

is clear, while the converse inclusion follows from the definition of coarsely excisive pair: up to passing to dense subalgebras, the left-hand side consists of operators supported in some neighbourhood of $E \cap F$, while the right-hand side consists of operators supported in the intersection of a neighbourhood of E and one of F. The corresponding statement for the localised versions is similar, but simpler, using only the closedness assumption. Point (ii) follows from the decomposition

$$T = \chi_E T + (1 - \chi_E)T$$

for any $T \in C^*(X)$ and similarly for the localised case.

The commutative cube in line (7.15) above thus simplifies to

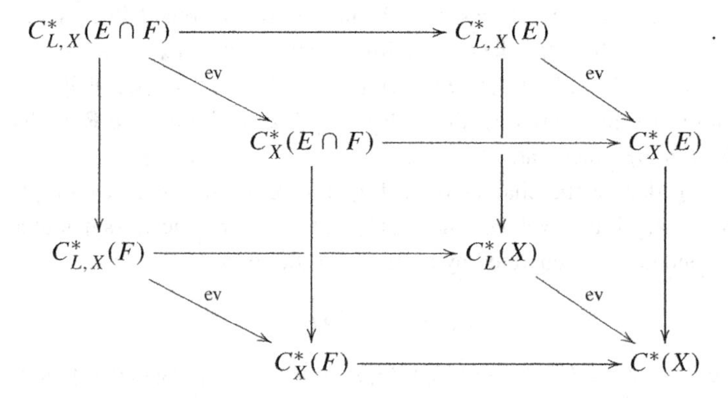

and we may apply Proposition 2.7.15 to get a commutative diagram

$$\begin{array}{ccccccc}
\longrightarrow K_i(C^*_{L,X}(E \cap F)) & \longrightarrow & K_i(C^*_{L,X}(E)) \oplus K_i(C^*_{L,X}(F)) & \longrightarrow & K_i(C^*_L(X)) \longrightarrow \\
\downarrow \mathrm{ev}_* & & \downarrow \mathrm{ev}_* \oplus \mathrm{ev}_* & & \downarrow \mathrm{ev}_* \\
\longrightarrow K_i(C^*_X(E \cap F)) & \longrightarrow & K_i(C^*_X(E)) \oplus K_i(C^*_X(F)) & \longrightarrow & K_i(C^*(X)) \longrightarrow
\end{array} \tag{7.16}$$

of Mayer–Vietoris sequences. To complete the proof, note that $C^*(E)$ includes as the full corner $\chi_E C^*_X(E) \chi_E$, and thus the inclusion

$$C^*(E) \to C^*_X(E)$$

induces an isomorphism on K-theory by Proposition 2.7.19, and similarly for F, $E \cap F$ and the localised versions. The commutative diagram in line (7.16) above thus reduces to the one in the statement, and we are done. □

The proof of the inductive step now follows from Proposition 7.5.2, Example 7.5.4, Theorem 7.5.5 and the five lemma; thus the proof of Theorem

7.5.1 is complete. As discussed in Section 3.3, this proves in particular that the d-torus does not admit a Riemannian metric with positive scalar curvature.

7.6 Exercises

7.6.1 Let G be a countable discrete group, equipped with a bounded geometry left invariant metric. Show that the action of G on all of its Rips complexes $P_r(G)$ is free if and only if G is torsion free.

7.6.2 Prove that if X and Y are coarsely equivalent metric spaces, then the coarse Baum–Connes assembly map is an isomorphism for X if and only if it is for Y.

7.6.3 Let Z be a locally finite metric space. With notation as in Definition 7.2.2, characterise when the inclusion $Z \to S_r(Z)$ of the vertex set is a coarse equivalence for some r.

7.6.4 With notation as in Definition 7.2.2, show that the topology on $S_r(Z)$ is the same as the one it inherits by considering the collections of formal sums $\sum_{z\in Z} t_z z$ defining its points as a subset of the unit ball of $\ell^1(Z)$ (equipped with its usual norm topology).

7.6.5 Let Z be a locally finite metric space, and let $E_r(Z)$ be defined analogously to $S_r(Z)$ as in Definition 7.2.2, but using the metric on each simplex $\sigma(F)$ that it inherits from the natural identification with the usual simplex

$$\left\{\sum_{z\in F} t_z z \in \mathbb{R}^F \mid t_z \in [0,1],\ \sum t_z = 1\right\}$$

in the Euclidean space $\mathbb{R}^F$ (equipped with its usual Euclidean metric). Note that the set-theoretic identity map defines a natural map between $S_r(Z)$ and $E_r(Z)$.

(i) Show that Corollary 7.2.6 fails in general for $E_r(Z)$.
(ii) Show that the set-theoretic identity map defines a coarse equivalence between $E_r(Z)$ and $S_r(Z)$ when Z has bounded geometry (see Definition A.3.19), but not in general.
(iii) Relatedly to part (ii) above, show that if we defined $P_r(Z)$ analogously to Definition 7.2.8, but starting with $E_r(Z)$ rather than $S_r(Z)$, then the natural inclusion $Z = P_0(Z) \to P_r(Z)$ need not be a coarse equivalence (contra part (ii) of Proposition 7.2.11 for the usual definition of $P_r(Z)$).

7.6.6 As in Remark 7.2.7, show that the direct limit of assembly maps

$$\lim_{r\to\infty} K_*(S_r(Z)) \to \lim_{r\to\infty} K_*(C^*(S_r(Z)))$$

identifies with the coarse Baum–Connes assembly map.
Hint: the topologies on $S_r(Z)$ *and* $P_r(Z)$ *are the same, so the left-hand side equals* $\lim_{r\to\infty} K_*(P_r(Z))$*, which is the left-hand side of the coarse Baum–Connes assembly map as in Theorem 7.2.16. For the right-hand side, show that one can choose appropriate geometric modules so that* $\lim_{r\to\infty} C^*(S_r(Z))$ *identifies canonically with* $C^*(X)$*.*

7.6.7 Let $S_r(Z)$ be as in Definition 7.2.2, equipped with the topology defined by its metric (or equivalently, with the topology from Exercise 7.6.4). Say a subset E of $S_r(Z) \times S_r(Z)$ is *controlled* if the collection of numbers

$$\{d_Z(z,w) \mid \text{there is } (x,y) \in S_r(Z) \times S_r(Z) \text{ with } z \in \text{supp}(x) \text{ and } w \in \text{supp}(y)\}$$

is bounded. Show that this collection defines an abstract coarse structure in the sense of Remark A.3.7. Show moreover that this coarse structure is such that the canonical inclusion $i : Z \to S_r(Z)$ is a coarse equivalence: this means that a subset E of $Z \times Z$ is controlled for the metric on Z if and only if $(i \times i)(E)$ is controlled for the coarse structure on $S_r(Z) \times S_r(Z)$, and that there exists a controlled set F such that for every $y \in S_r(Z)$ there is $z \in Z$ with $(y, i(z)) \in F$.

7.6.8 Show that if X is a single-point space and G a finite group then the assembly map

$$\mu\colon K_*^G(X) \to K_*(C^*(X)^G)$$

of Definition 7.1.1 is an isomorphism.
Hint: use an Eilenberg swindle as in the proof of Proposition 6.3.3 to show that the kernel of the evaulation-at-one map

$$ev\colon C_L^*(X)^G \to C^*(X)^G$$

has trivial K*-theory.*

7.6.9 Use Mayer–Vietoris sequences (compare Theorem 6.3.4) to compute the K-homology groups in Examples 7.4.5.

7.6.10 Let G, H be countable discrete groups.

(i) Show that a CW complex BG is a classifying space for G in the sense of Definition 7.4.1 if and only if it has fundamental group G, and all of its higher homotopy groups vanish.

Hint: you will need to know how higher homotopy groups behave on taking covering spaces, and also Whitehead's theorem on weak homotopy equivalences between CW complexes being homotopy equivalences.

(ii) Show that if G, H have classifying spaces BG, BH respectively, then a classifying space for the free product $G * H$ is given by the wedge $BG \vee BH$.
Hint: part (i) might also help, as well as knowing how homotopy groups interact with wedge products.

(iii) Show that if G, H have classifying spaces BG, BH respectively, then a classifying space for the direct product is given by the product $BG \times BH$.
Hint: again, part (i) might help, as well as knowing how higher homotopy groups interact with products.

7.6.11 Prove that a contractible proper metric space that admits a cocompact isometric action of some group is uniformly contractible.

7.6.12 Let X be a uniformly contractible proper metric space. Show that if $f: X \to Y$ is simultaneously a homeomorphism and a coarse equivalence, then Y is also uniformly contractible. Deduce that if d' is a new metric on X inducing the same topology, and coarsely equivalent to the original metric, then (X, d') is uniformly contractible.

7.6.13 Find a metric on $\mathbb{R}$ that induces the usual topology, and is *not* uniformly contractible. Is this possible if the metric is a *Riemannian* metric? Find a Riemannian metric on $\mathbb{R}^2$ that is not uniformly contractible (or at least draw some pictures to convince yourself that one exists).

7.6.14 Elaborate on the argument in Theorem 7.5.5 to prove that if $X = Y \cup Z$ is a coarsely excisive decomposition, then there is a commutative diagram of Mayer–Vietoris sequences

$$\begin{array}{ccccccc}
\longrightarrow KX_i(Y \cap Z) & \longrightarrow & KX_i(Y) \oplus KX_i(Z) & \longrightarrow & KX_i(X) \longrightarrow \\
\downarrow & & \downarrow & & \downarrow \\
\longrightarrow K_i(C^*(Y \cap Z)) & \longrightarrow & K_i(C^*(Y)) \oplus K_i(C^*(Z)) & \longrightarrow & K_i(C^*(X)) \longrightarrow
\end{array}$$

where the vertical maps are coarse Baum–Connes assembly maps.

7.7 Notes and References

Variations on the assembly map appear in the work of several authors starting in the mid-1970s and early 1908s: perhaps the earliest places are the works of Miščenko [185], Kasparov [150] and Baum and Connes [21] (although all of these look quite different to the route we have taken). The name 'assembly map' is by analogy with the assembly map of surgery theory (see Wall [246]), a part of manifold topology. The connection to surgery was originally, and continues to be, a major motivation for the development of the study of assembly maps.

Our model of the assembly map has the virtue of simplicity, being induced by a $*$-homomorphism. There are several other models of the (Baum–Connes) assembly map, and all have technical benefits in various situations: the most widely used are probably those based on Paschke duality (see chapter 12 of Paschke [135]), on descent in KK-theory (see Baum, Connes and Higson [22]), and on descent in E-theory (see Guentner, Higson and Trout [112]). There are still quite a few others: based on noncommutative simplicial complexes (see Cuntz [70]), on localisation of triangulated categories (see Meyer and Nest [177]), on Toeplitz-type algebras (see chapter 5 of Cuntz, Meyer and Rosenberg [71]), and also an approach that subsumes both the Baum–Connes assembly map and its cousins in algebraic topology as special cases (see Davis and Lück [77]).

The Baum–Connes conjecture for groups was developed by Baum and Connes [21], with the definitive version due to Baum, Connes and Higson [22]. The description of the left-hand side in Remark 7.1.12 is closer to the approach of [21], and its identification with our version can be justified using the results of Baum, Higson and Schick [28].

An early version of the coarse Baum–Connes conjecture appears in section 6 of Roe [214], while the modern version was developed by Higson and Roe [133] and Yu [269]. Our approach to both conjectures via localised Roe algebras is based on the approach used by Yu [270]. We will have more to say about situations where the coarse Baum–Connes conjectures do and do not hold, and applications of the conjectures, in later chapters.

Rips complexes are an import to higher index theory from geometric group theory: the first major applications are perhaps in the theory of hyperbolic groups as discussed for example in chapter 4 of Ghys and de la Harpe [105]. Our treatment of the spherical metric is partially inspired by the corresponding material from Wright's thesis (see section 5.1 of Wright [262]). The treatment of the coarse Baum–Connes conjecture using the spherical metric directly as in Remark 7.2.7 and Exercise 7.6.6 is the more standard one in the literature; we differ from this as it is technically convenient to alter the spherical metric on the Rips complex in order to guarantee that the canonical inclusions $Z \to P_r(Z)$ are coarse equivalences. Another closely related construction is in terms of so-called anti-Čech sequences: see sections 3 and 6 of Higson and Roe [133] or section 5.2 of Wright [262].

The definition of uniform contractibility is due to Weinberger. Most of the related material in Section 7.3 comes from section 3 of Higson and Roe [133]. The bounded geometry assumption in Theorem 7.3.6 is necessary: see Dranishnikov, Ferry and Weinberger [83], which gives a particularly exotic metric on (high-dimensional) Euclidean space for which the theorem fails. The Cartan–Hadamard theorem mentioned in Example 7.3.7 can be found in many texts on Riemannian geometry, for example section 19 of Milnor [182] (see also section II.4 of Bridson and Haefliger [38] for a generalisation outside the world of smooth manifolds).

Classifying spaces for groups are a classical part of algebraic and geometric topology, and are intimately tied up with group (co)homology: see for example Brown [39] for background on this. An elementary treatment of the classifying space BG (there called $K(G,1)$) can be found for example in section 1.B of Hatcher [123]; in particular the results we stated without proof as Theorem 7.4.3 follow from example 1B.7 and proposition 1B.9 of Hatcher [123]. Section 1.3 of Hatcher [123] contains an exposition of the material we used in the proof of Theorem 7.4.4 about covering spaces, such as results about lifting maps and homotopies to covering spaces.

On the other hand, the classifying space $\underline{E}G$ for proper actions was introduced by Baum and Connes [20]; the particularly nice model in Theorem 7.4.7 is due to Kasparov and Skandalis (see section 4 of [152]). In algebraic topology, what we have called '$\underline{E}G$' is often written '$E_{\mathcal{FIN}}G$' and called 'the classifying space for actions with finite stabilisers': both it and EG are part of a family of classifying spaces for different families of subgroups as explained for example in section 2 of Lück and Reich [173].

The descent principle of Theorem 7.4.11 appears in several guises in the literature: see chapter 8 of Roe [216] or section 12.6 of Higson and Roe [135] for expositions and proofs of the version we stated. A more powerful (but less widely applicable) version due to Higson works well in some analytic contexts, and can be found in [127].

The argument of Section 7.5 is adapted from work of Higson, Roe and Yu [139].

PART THREE

Differential Operators

8

Elliptic Operators and K-homology

Our goal in this chapter is to show that elliptic differential operators on a manifold M give rise to K-homology classes. This is one of the main motivations for the development of analytic models for K-homology: the groups $K_*(M)$ organise elliptic operators in a way particularly well suited to studying the interactions of these operators with the topology of the manifold.

The chapter is structured as follows. In Section 8.1 we discuss self-adjointness for differential operators. This is crucial: our main tool will be the functional calculus of Theorem D.1.7 and self-adjointness is needed to get this off the ground. In Section 8.2 we then move on to discussing propagation estimates. The main tool here is the wave equation associated to the differential operators we are studying. Using this, we are able to construct a multiplier (F_t) of the localisation algebra $L^*(M)$ out of a differential operator.

In order to make further progress, we discuss ellipticity in Section 8.3. This is the key additional assumption needed to build classes in $K_*(M)$ out of the multipliers (F_t). The idea is to do a concrete analysis in the case of tori using Fourier theory, and then to transplant these results from tori to general manifolds. At this point, we have achieved our main goals.

Finally, in Section 8.4, we perform a more careful analysis to relate the index classes we have constructed to Schatten ideals. This more delicate theory is important mainly for applications that go beyond the scope of this text, although we do have one concrete application: we need the Schatten class theory for our study of the Kadison–Kaplansky conjecture in Section 10.1.

Throughout this chapter, M will denote a smooth (i.e. infinitely differentiable) manifold and S a (finite-dimensional) smooth complex vector bundle over M. If U is an open subset of M, S_U denotes the restriction of S to U, and for $x \in M$ we write S_x for the fibre of S over x. The endomorphism bundle of S, with fibre over $x \in M$ that is equal to the algebra $\mathrm{End}(S_x)$ of linear maps from S_x to itself, is denoted $\mathrm{End}(S)$. Throughout, $C_c^\infty(M; S)$ (respectively,

$C_c^\infty(U; S)$) denotes the vector space of smooth, compactly supported sections of S (respectively, of S_U), and $C_c^\infty(M)$ will denote the space of smooth, compactly supported complex valued functions on M.

8.1 Differential Operators and Self-Adjointness

Our main objects of study in this chapter are differential operators as in Definition 8.1.1 below. Throughout, notation and conventions are as in the introduction to this chapter, so in particular M is a smooth manifold and S is a smooth complex vector bundle over M.

Definition 8.1.1 A (first order, linear) *differential operator* on S is a linear operator $D\colon C_c^\infty(M; S) \to C_c^\infty(M; S)$ with the following properties:

(i) if $f \in C_c^\infty(M; S)$ is supported in some open subset U of M, then Df is also supported in U;

(ii) if $U \subseteq M$ is a coordinate patch with local coordinates $(x_1, \ldots, x_d)$, then there exist smooth sections $a_1, \ldots, a_d, b$ of $\mathrm{End}(S_U)$ such that for all $u \in C_c^\infty(M; S)$ supported in U and all $x \in U$, we have

$$(Du)(x) = \sum_{i=1}^{d} a_i(x) \frac{\partial u}{\partial x_i}(x) + b(x)u(x). \tag{8.1}$$

The *support* of D is the smallest closed subset F of M such that for all $u \in C_c^\infty(M \backslash F; S)$, we have that $Du = 0$. A (first order, linear) differential operator as above is *zeroth order* if all the a_i terms in any local representation as in line (8.1) above are zero.

Examples 8.1.2 (i) Perhaps the most basic (but also very important) example occurs when $M = \mathbb{R}$ is the real line, S the trivial line bundle (so $C_c^\infty(M; S) = C_c^\infty(\mathbb{R})$) and D is the usual differentiation operator $D = \frac{d}{dx}$.

(ii) Let M be any manifold, and Ω be the bundle of differential forms on M. Set $S = \Omega \otimes \mathbb{C}$ to be the associated complexified bundle, so the fibre S_x of S at $x \in M$ is the complexified exterior algebra $\Lambda^* T_x^* M \otimes \mathbb{C}$ of the cotangent bundle to M at x. Set $D = d \otimes 1_{\mathbb{C}}$ to be the complexificiation of the usual de Rham exterior derivative operator on differential forms. Then D has the locality property (i) from Definition 8.1.1: this follows for example from the Leibniz rule

$$d(f\omega) = df \wedge \omega + f d\omega$$

for $f \in C_c^\infty(M)$ and ω a section of S. Moreover, in local coordinates $(x_1, \ldots, x_d)$, D is given by

$$D = \sum_{i=1}^{d} (dx_i \wedge \cdot) \frac{\partial}{\partial x_i},$$

where '$(dx_i \wedge \cdot)$' is the element of $\mathrm{End}(\Omega \otimes \mathbb{C})$ given (locally) by 'take exterior product with dx_i', and so D has the right form.

We will only work with linear, first order operators in what follows. As such, 'differential operator' will always be shorthand for 'first order linear differential operator' unless explicitly stated otherwise.

A key tool in our analysis will be the symbol of a differential operator, which we now introduce. Let D be a differential operator, given in some local coordinate patch U by the formula

$$(Du)(x) = \sum_{i=1}^{d} a_i(x) \frac{\partial u}{\partial x_i}(x) + b(x)u(x).$$

Let $g \in C_c^\infty(U)$, and consider g as a multiplication operator on $C_c^\infty(M; S)$. Then the commutator $[D, g]$ acts on $u \in C_c^\infty(U; S)$ by the formula

$$([D, g]u)(x) = \sum_{i=1}^{d} a_i(x) \frac{\partial g}{\partial x_i}(x)u(x),$$

i.e. $[D, g]$ is acting via the element

$$x \mapsto \sum_{i=1}^{d} a_i(x) \frac{\partial g}{\partial x_i}(x) \tag{8.2}$$

of $C_c^\infty(U; \mathrm{End}(S))$. Moreover, this section depends only on the exterior derivative dg, not on g itself (as follows from the formula).

Now let $\pi : T^*M \to M$ be the cotangent bundle of M, and $\pi^*\mathrm{End}(S)$ be the pullback of $\mathrm{End}(S)$ to T^*M.

Definition 8.1.3 Let D be a differential operator on S. Let (x, ξ) be a point of T^*M, and write $\xi = dg|_x$ for some $g \in C_c^\infty(M)$. Let v be an element of S_x and write $v = u(x)$ for some $u \in C_c^\infty(M; S)$. Then the *symbol* of D is the smooth section

$$\sigma_D \in C^\infty(T^*M; \pi^*\mathrm{End}(S))$$

of the bundle $\pi^*\mathrm{End}(S)$ defined by

$$\sigma_D(x, \xi)v = ([D, g]u)(x).$$

To see that this is well defined, let $(x_1, \ldots, x_d)$ be local coordinates near x and write $\xi = \xi_1 dx_1 + \cdots + \xi_d dx_d$. The computation leading to line (8.2) shows that

$$\sigma_D(x,\xi) = \sum_{i=1}^{d} \xi_i a_i(x),$$

and thus that $\sigma_D(x,\xi)$ does not depend on the choices of g or u.

Examples 8.1.4 Looking back at the examples from 8.1.2, we have the following.

(i) For $M = \mathbb{R}$ and S the trivial bundle, $C^\infty(T^*M;\pi^*\mathrm{End}(S))$ identifies with $C^\infty(TM) \cong C^\infty(\mathbb{R}^2)$. If $D = \frac{d}{dx}$, then $\sigma_D(x,\xi) = \xi$.

(ii) For any M, if $S = \Omega \otimes \mathbb{C}$ is the complexified bundle of differential forms, we have that the fibre of $\pi^*\mathrm{End}(\Omega \otimes \mathbb{C})$ over a point (x,ξ) is the space of endomorphisms of $\Lambda^* T_x^* M \otimes \mathbb{C}$, the complexified exterior algebra of the cotangent space to M at x. If D is the (complexified) de Rham operator on S, then the symbol $\sigma_D(x,\xi)$ is the operator of exterior multiplication by ξ.

We now introduce Hilbert space techniques. For this, we need to assume some extra structure on M and S. From now on, then, assume that M is Riemannian and that S is equipped with a smooth Hermitian structure, i.e. each fibre S_x is equipped with a Hermitian inner product $\langle , \rangle_x$ such that if s_1, s_2 are smooth sections of S, then the function

$$M \to \mathbb{C}, \quad x \mapsto \langle s_1(x), s_2(x) \rangle_x$$

is a smooth function.

We will need to do integration of sections of S. For this purpose, write $\int_M f(x)dx$ for the integral of a suitable function $f\colon M \to \mathbb{C}$ with respect to the measure induced by the Riemannian structure. Define a positive definite inner product on $C_c^\infty(M;S)$ by

$$\langle f,g \rangle := \int_M \langle f(x), g(x) \rangle_x dx.$$

The Hilbert space $L^2(M;S)$ of L^2-sections of S is defined to be the completion of $C_c^\infty(M;S)$ for this inner product. Note that $L^2(M;S)$ is a geometric module over M in the sense of Definition 4.1.1. A differential operator D on S defines a potentially unbounded operator on $L^2(M;S)$ with domain $C_c^\infty(M;S)$ in the sense of Definition D.1.1. We also denote this unbounded operator by D.

Now, we will want to apply the spectral theorem (Theorem D.1.7) to D, and in order to do this we need some sort of self-adjointness assumption. Recall the following definition from the general theory of unbounded operators (compare Definition D.1.3).

Definition 8.1.5 A differential operator D on S is *formally self-adjoint* if for all $u, v \in C_c^\infty(M; S)$,

$$\langle Du, v \rangle = \langle u, Dv \rangle.$$

Formally self-adjoint operators arise naturally from geometric constructions, as we will see in Chapters 9 and 10. In order to apply the spectral theorem for unbounded operators, formal self-adjointness is not enough: we need essential self-adjointness as in the next definition.

Definition 8.1.6 A formally self-adjoint differential operator D on S is *essentially self-adjoint* if, whenever $v, w \in L^2(M; S)$ are such that

$$\langle Du, v \rangle = \langle u, w \rangle \quad \text{for all} \quad u \in C_c^\infty(M; S), \tag{8.3}$$

there is a sequence (v_n) in $C_c^\infty(M; S)$ such that $v_n \to v$ and $Dv_n \to w$ in L^2-norm.

To explain this definition a little, note that if v were in $C_c^\infty(M; S)$ then formal self-adjointness implies that the condition in line (8.3) is equivalent to saying that

$$\langle u, Dv \rangle = \langle u, w \rangle$$

and thus that $Dv = w$ by density of $C_c^\infty(M; S)$. Thus the condition in line (8.3) says that the pair (v, w) 'weakly satisfies the equation $w = Dv$' in some sense; essential self-adjointness says that any such pair is a norm limit of pairs (v_n, Dv_n) that honestly satisfy this equation.

Unfortunately, it is not automatic that a formally self-adjoint differential operator will be essentially self-adjoint: see Exercise 8.5.2. Our goal for the rest of this section will be to develop a sufficient condition for a formally self-adjoint operator to be essentially self-adjoint.

The key tool, which will also be useful later when analysing the connection of differential operators to localisation algebras, is as follows.

Definition 8.1.7 Let D be a differential operator on S, and σ_D its symbol. The *propagation speed* of D at a point x is defined to be

$$c_D(x) := \sup_{\xi \in T_x^*M, \ \|\xi\|=1} \|\sigma_D(x, \xi)\|.$$

The *propagation speed* of D is

$$c_D = \sup_{x \in M} c(x)$$

(possibly infinite).

Here is our main result on essential self-adjointness.

Proposition 8.1.8 *Let D be a formally self-adjoint differential operator. Assume moreover that either the support of D is compact, or that M is complete and the propagation speed is finite. Then D is essentially self-adjoint.*

In order to prove this, we need the existence of Friedrich's mollifiers.

Definition 8.1.9 Let K be a compact subset of M. A family of *Friedrich's mollifiers* is a sequence of bounded operators $(F_n : L^2(K; S) \to L^2(M; S))_{n=1}^{\infty}$ with the following properties:

(i) each F_n is a contraction;
(ii) the image of each F_n is contained in $C_c^\infty(M; S)$;
(iii) for all $v \in L^2(M; S)$, $F_n v \to v$ and $F_n^* v \to v$ in norm as $n \to \infty$;
(iv) for any differential operator D, the sequences $[D, F_n]$ and $[D, F_n^*]$ of operators on $C_c^\infty(K; S)$ are uniformly bounded in operator norm.

Lemma 8.1.10 *Let K be a compact subset of M. Then a family of Friedrich's mollifiers exists.*

Proof Let us first assume that M is $\mathbb{R}^d$, and that S is a trivial bundle. Let K be a compact subset of $\mathbb{R}^d$. Let $h : \mathbb{R}^d \to \mathbb{R}$ be any smooth, non-negative compactly supported function that satisfies $\int_{\mathbb{R}^d} h(x)dx = 1$. For each $n \geqslant 1$, define $h_n(x) = n^d h(nx)$, and let $F_n : L^2(K; S) \to L^2(M; S)$ be the associated convolution operator, i.e. F_n is defined by

$$(F_n u)(x) = \int_{\mathbb{R}^d} h_n(x - y)u(y)dy.$$

Here we use that S is trivial to make sense of this. Looking at the properties in the statement, (i) now follows as the norm of a convolution operator is bounded by the L^1-norm of the corresponding function, which is one in this case; property (ii) follows as h is smooth and compactly supported; property (iii) follows from standard estimates. To see property (iv), note that if D is given in coordinates by

$$D = \sum_{i=1}^{d} a_i \frac{\partial}{\partial x_i} + b,$$

then the derivatives actually commute with F_n and one computes using integration by parts that $[D, F_n]$ is the operator given by

$$([D, F_n]u)(x) = \int_{\mathbb{R}^d} \sum_{i=1}^{d} \frac{\partial}{\partial y_i}\big(h_n(x-y)(a_i(y)-a_i(x)\big)u(y)dy$$

$$+ \int_{\mathbb{R}^d} h_n(x-y)(b(x)-b(y))u(y)dy.$$

It is not too difficult to see that this is bounded independently of n (using that we are working on a compact set).

To complete the proof for general n, one can cover $K \subseteq M$ by finitely many coordinate charts over which S can be trivialised, and use a finite smooth partition of unity consisting of compactly supported functions to patch together mollifiers constructed as above. Note that the compactness assumptions guarantees that the process of transferring from $\mathbb{R}^d$ to the manifold in each coordinate patch only distorts norms a bounded amount. □

Proof of Proposition 8.1.8 Assume that $v, w \in L^2(M; S)$ are such that for all $u \in C_c^\infty(M; S)$,

$$\langle Du, v\rangle = \langle u, w\rangle.$$

We must show that there exists a sequence (v_n) in $C_c^\infty(M; S)$ converging to v in $L^2(M; S)$, and such that (Dv_n) converges to w in $L^2(M; S)$. As in Definition D.1.4, we will say that v is in the *maximal domain* of D if there is a w satisfying the first condition, and that it is in the *minimal domain* if there is a w satisfying the second condition.

Assume first that v is supported in some compact subset K of M. Let (F_n) be a family of Friedrich's mollifiers for K as in Definition 8.1.9 and consider the sequence $(F_n v)$, which is in $C_c^\infty(M; S)$ and converges in norm to v. For any $u \in C_c^\infty(M; S)$,

$$\begin{aligned}\langle DF_n v, u\rangle &= \langle v, F_n^* Du\rangle = \langle v, [F_n^*, D]u\rangle + \langle v, DF_n^* u\rangle \\ &= \langle v, [F_n^*, D]u\rangle + \langle w, F_n^* u\rangle.\end{aligned}$$

Hence in particular, $\|DF_n v\| \leqslant \|[F_n^*, D]\| + \|w\|$ for all n, so the sequence $(DF_n v)$ is uniformly bounded. Replacing it with a subsequence, we may assume that $(DF_n v)$ is weakly convergent. The limit must be w: indeed, for any $u \in C_c^\infty(M; S)$,

$$\langle u, DF_n v\rangle = \langle Du, F_n v\rangle \to \langle Du, v\rangle = \langle u, w\rangle.$$

The Hahn–Banach theorem then implies that there is a sequence (v_n) in $C_c^\infty(M; S)$ consisting of convex combinations of the sequence $(F_n v)$, and such that $v_n \to v$ and $Dv_n \to w$ in norm as $n \to \infty$. We are done in the case that v is compactly supported.

Assume next that D is supported in some compact set K, and let v be an arbitrary element of the maximal domain of D. Let U be an open neighbourhood of K with compact closure. Using a smooth partition of unity, write $v = v_0 + v_1$, where v_0 is supported in U and v_1 in $M \setminus K$. Writing v_1 as a norm limit of elements of $C_c^\infty(M \setminus K)$ shows that v_1 is in the minimal domain of D, while v_0 is in the minimal domain of D by the argument above; this completes the proof when D has compact support.

Assume now that M is complete, and D has finite propagation speed c_D. Working one connected component at a time if necessary, we may assume that M is connected, so in particular the distance function is valued in $[0,\infty)$. Let v be an arbitrary element in the maximal domain of D. Let $(f_n : \mathbb{R} \to [0,1])$ be a sequence of smooth, compactly supported functions such that $f(t) = 1$ for all $t \leqslant n$, and such that the sequence $(\sup_{t\in\mathbb{R}} |f_n'(t)|)$ of real numbers tends to zero as n tends to infinity. Fix $x_0 \in M$, and define $g_n : M \to [0,\infty)$ by $g_n(x) = f_n(d(x,x_0))$. As M is complete, it is proper by the Hopf–Rinow theorem (see Theorem A.3.6). Hence the assumptions on (f_n) imply that each g_n is smooth and compactly supported, that the sequence $(\sup_{x\in M} \|dg_n(x)\|)_{n=0}^\infty$ tends to zero as n tends to infinity and that (g_n) tends to one uniformly on compact subsets of M.

For each n, let $v_n := g_n v$. Then v_n is in the maximal domain of D for all n as

$$\langle Du, v_n\rangle = \langle [g_n, D]u, v\rangle + \langle g_n Du, w\rangle = \langle u, \sigma_D(dg_n)v + g_n w\rangle$$

for all $u \in C_c^\infty(M;S)$. Moreover, v_n is compactly supported by compact support of g_n, and thus v_n is in the minimal domain of D by the first part of the proof. Note also that the sequence (v_n) converges to v in $L^2(M;S)$, as g_n converges to 1 uniformly on compact sets. Replacing D by its closure (see Definition D.1.2) so that Dv_n makes sense, for any $u \in C_c^\infty(M;S)$ we have

$$\langle Dv_n, u\rangle = \langle v, g_n Du\rangle = \langle v, [g_n, D]u\rangle + \langle v, Dg_n u\rangle = \langle v, [g_n, D]u\rangle + \langle w, g_n u\rangle,$$

and

$$|\langle v, [g_n, D]u\rangle| \leqslant \|v\| c_D \|dg_n\| \|u\| \to 0$$

as $n \to \infty$, whence

$$\lim_{n\to\infty} \langle Dv_n, u\rangle = \lim_{n\to\infty} \langle w, g_n u\rangle = \langle w, u\rangle.$$

Hence (Dv_n) converges to w, showing that v is in the minimal domain of D as required. $\square$

8.2 Wave Operators and Multipliers of $L^*(M)$

Our goal in this section is to show that a formally self-adjoint differential operator D on a smooth Hermitian bundle S over a Riemannian manifold M defines a family of multipliers of the localisation algebra $L^*(L^2(M;S))$.

The key point is to control the propagation of associated operators that we build out of D. To do this, we will consider solutions of the wave equation on M associated to D. Say $u\colon \mathbb{R} \times M \to \mathbb{C}$ is a function $u(t,x)$ of 'time' $t \in \mathbb{R}$ and 'space' $x \in M$. The *wave equation* associated to D is

$$\frac{\partial^2 u}{\partial t^2} + D^2 u = 0.$$

This equation governs the development of waves on M: if initial conditions $u(x)$ are given, then the resulting wave u is given by

$$u(t,x) = (e^{itD}u)(x),$$

as long as the *wave operators* e^{itD} makes sense. For example, this will be the case if D is essentially self-adjoint. The speed at which the wave u propagates turns out to be governed by the propagation speed of D (hence the name!). Establishing this is the key technical tool: more general functions $f(D)$ of D can then be treated using Fourier theory.

There are additional technicalities if D is not essentially self-adjoint: the same basic idea works, but we can only make sense of e^{itD} 'locally' and the construction gets more technical.

Here is the fundamental result about wave operators. For the statement, recall that if K is a subset of a metric space X and $r > 0$, the $N_r(K)$ denotes the r-neighbourhood $\{x \in X \mid d(x,K) < r\}$ of K.

Proposition 8.2.1 *Let D be an essentially self-adjoint differential operator on the bundle S over M. Let $u \in C_c^\infty(M;S)$, and let $c > 0$ be such that the propagation speed of D satisfies $c_D(x) \leqslant c$ for all $x \in supp(u)$. Then for any $t \in \mathbb{R}$ we have*

$$supp(e^{itD}u) \subseteq N_{c|t|}(supp(u)).$$

Proof Replacing D with $-D$ if necessary, we may assume that t is positive. Let $\epsilon > 0$ and write $K = \mathrm{supp}(u)$. Let $\delta > 0$ and let $g \in C_c^\infty(M)$ have the following properties:

(i) $g(x) = 0$ if $d(x,K) \geqslant \delta$ and $g(x) = 1$ for $x \in K$;
(ii) $\sup_{\{x \in M \mid d(x,K) \leqslant \delta\}} \left| g(x) - \left(1 - \frac{1}{\delta} d(x,K)\right)\right| < \epsilon$;
(iii) $\|dg\| \leqslant \frac{1}{\delta} + \epsilon$.

It is not too difficult to see that such a function exists. Let $f : \mathbb{R} \to [0,1]$ be any smooth non-decreasing function such that $f(t) = 1$ if and only if $t \geqslant 1$. Define $\gamma := \sup_{x \in N_\delta(K)} c_D(x)(1 + \delta\epsilon)$ and define a function

$$M \times (0,\infty) \to [0,1], \quad (x,t) \mapsto h_t(x)$$

by

$$h_t(x) = f\left(g(x) + \frac{\gamma}{\delta}t\right).$$

Note that h is smooth in both x and t, and that the conditions that $f(t) \geqslant 1$ only when $t \geqslant 1$, and that $g(x)$ is within ϵ of $1 - \frac{1}{\delta}d(x,K)$ on the relevant regions imply that

$$\{x \in M \mid h_t(x) = 1\} \subseteq N_{\gamma t + \epsilon}(K). \tag{8.4}$$

Now, write $\dot{h}_t$ as shorthand for $\partial h / \partial t$. Computing, we see that

$$\dot{h}_t(x) = \frac{\gamma}{\delta} f'\left(g(x) + \frac{\gamma}{\delta}t\right) \tag{8.5}$$

and if d denotes the exterior derivative in the M direction, then

$$dh_t(x) = f'(g(x) + \frac{\gamma}{\delta}t)dg(x) = \frac{\delta}{\gamma}\dot{h}_t(x)dg(x).$$

Hence the section $[D, h_t]$ of End(S) is given by

$$[D, h_t](x) = \sigma_D(dh_t)(x) = \frac{\delta}{\gamma}\dot{h}_t(x)\sigma_D(dg)(x).$$

It follows that for each s, the self-adjoint operator $\dot{h}_t - i[D, h_t]$ on $L^2(M;S)$ acts as an End(S)-valued function, whose value at $x \in M$ is given by

$$(\dot{h}_t - i[D, h_t])(x) = \dot{h}_t(x)\left(1 - \frac{\delta}{\gamma}i\sigma_D(dg)(x)\right). \tag{8.6}$$

On the other hand, the fact that $c_D(x) \leqslant \gamma/(1+\epsilon)$ for all x in the support of g implies that

$$\|\sigma_D(dg)(x)\| \leqslant \frac{\gamma}{1 + \delta\epsilon}\|dg\| \leqslant \frac{\gamma}{1 + \delta\epsilon}(1/\delta + \epsilon) = \frac{\gamma}{\delta},$$

where we have used assumption (iii) above on g for the second inequality. Hence line (8.6) and non-negativity of $\dot{h}_t$ (which follows from line (8.5), and the fact that f is non-decreasing) implies that the operator $\dot{h}_t - i[D, h_t]$ is positive.

To complete the argument, write $u_t = e^{itD}u$ (so $\dot{u_t} = iDu_t$ by Proposition D.2.1) and consider

$$\frac{\partial}{\partial t}\langle h_t u_t, u_t\rangle = \langle \dot{h_t} u_t, u_t\rangle + \langle h_t i D u_t, u_t\rangle + \langle h_t u_t, i D u_t\rangle$$
$$= \langle (\dot{h_t} - i[D,h_t])u_t, u_t\rangle \geqslant 0.$$

It follows that

$$\langle h_t u_t, u_t\rangle \geqslant \langle h_0 u_0, u_0\rangle$$

for all t. Moreover, the facts that $h_0(x) = 1$ for all $x \in \text{supp}(u)$ and that e^{itD} is unitary then imply that

$$\langle h_t u_t, u_t\rangle \geqslant \langle h_0 u_0, u_0\rangle = \langle u_0, u_0\rangle = \langle u_t, u_t\rangle$$

for all t. The equality case of Cauchy–Schwarz combined with the fact that h_t is a norm one operator now forces $h_t u_t = u_t$, and thus (using line (8.4) for the second inclusion)

$$\text{supp}(u_t) \subseteq \{x \in M \mid h_t(x) = 1\} \subseteq N_{\gamma t+\epsilon}(K).$$

Let $c_\delta := \sup_{x\in N_\delta(K)} c_D(x)$. Then letting ϵ tend to zero gives that $\text{supp}(u_t) \subseteq N_{c_\delta t}(K)$. Letting δ tend to zero then completes the proof. □

Using Fourier analysis, we can generalise the previous result (which is the case $f(x) = e^{itx}$, with Fourier transform the point mass at t) to the following. For the statement, recall that if $f : \mathbb{R} \to \mathbb{C}$ is a bounded Borel function, then its *distributional Fourier transform* is the distribution (i.e. functional on $C_c^\infty(\mathbb{R})$) defined by

$$\widehat{f} : u \mapsto \int_{\mathbb{R}} f(x)\widehat{u}(x)dx.$$

The *support* of this distribution is the complement of the largest open set U such that $\widehat{f}(u) = 0$ for all u supported in U.

Corollary 8.2.2 *Let D be an essentially self-adjoint differential operator on S. Let $f : \mathbb{R} \to \mathbb{C}$ be any bounded Borel function such that the (distributional) Fourier transform $\widehat{f}$ of f is supported in $[-r,r]$ for some $r \geqslant 0$. Let $u \in C_c^\infty(M;S)$, and let $c > 0$ be such that $c_D(x) \leqslant c$ for all $x \in supp(u)$. Then*

$$supp(f(D)u) \subseteq N_{cr}(supp(u)).$$

Proof Let $u,v \in C_c^\infty(M;S)$ be such that $d(\text{supp}(u), \text{supp}(v)) > cr$. It will suffice to show that

$$\langle f(D)u, v\rangle = 0.$$

Proposition D.2.3 implies that the inner product on the left is equal to the pairing of the distribution $\widehat{f}$ with the function

$$g\colon \mathbb{R} \to \mathbb{R}, \quad t \mapsto \frac{1}{2\pi}\langle e^{itD}u, v\rangle.$$

However, Proposition 8.2.1 implies that $g(t) = 0$ for all t with $|t| \leqslant cr$, whence it pairs with $\hat{f}$ to zero. □

The following corollary is a direct consequence of Corollary 8.2.2 in the special case that the global propagation speed c_D is finite.

Corollary 8.2.3 *Let D be an essentially self-adjoint differential operator on S. Assume moreover that c_D is finite. Then if $f\colon \mathbb{R} \to \mathbb{C}$ is a bounded Borel function such that the distributional Fourier transform is supported in $[-r,r]$, we have that prop$(f(D)) \leqslant rc_D$.* □

We now want to work towards building multipliers of the localisation algebra $L^*(M)$. This requires study of families of operators (F_t) parametrised by $t \in [1,\infty)$, for which the following definition gives a useful technical tool.

Definition 8.2.4 A differentiable function $f\colon \mathbb{R} \to \mathbb{C}$ has *slow oscillation at infinity* if

$$\sup_{x\in\mathbb{R}} |xf'(x)| < \infty.$$

Write $C_{so}(\mathbb{R})$ for the C^*-subalgebra of $C_b(\mathbb{R})$ generated by bounded functions with slow oscillation at infinity.

The proof of the following lemma is the reason for considering this class of functions.

Lemma 8.2.5 *Assume that D is an essentially self-adjoint differential operator on S. Assume moreover that c_D is finite and that M is complete. Let $f\colon \mathbb{R} \to \mathbb{C}$ be a bounded Borel function and consider the function*

$$[1,\infty) \to \mathcal{B}(L^2(M;S)), \quad t \mapsto F_t := f(t^{-1}D).$$

If f has slow oscillation at infinity, then the function $t \mapsto F_t$ is Lipschitz. Moreover, if f is in $C_{so}(\mathbb{R})$, then the function $t \mapsto F_t$ is uniformly continuous.

Proof The spectral theorem implies that for any s,t in $[1,\infty)$ with $s \leqslant t$ we have

$$\|F_t - F_s\| = \|f(t^{-1}D) - f(s^{-1}D)\| \leqslant \sup_{x\in\mathbb{R}} |f(t^{-1}x) - f(s^{-1}x)|. \quad (8.7)$$

The mean value theorem implies that

$$\begin{aligned}|f(t^{-1}x)-f(s^{-1}x)| &\leqslant \sup_{c\in[t^{-1}x,s^{-1}x]} |f'(c)||t^{-1}x-s^{-1}x| \\ &\leqslant \sup_{c\in[t^{-1}x,s^{-1}x]} |f'(c)t^{-1}x|\frac{|t-s|}{s} \\ &\leqslant \sup_{c\in[t^{-1}x,s^{-1}x]} |f'(c)c||t-s|.\end{aligned}$$

The slow oscillation condition implies that $\sup_{c\in\mathbb{R}}|f'(c)c| \leqslant C$ for some $C>0$, so combining this with line (8.7) gives that

$$\|F_t - F_s\| \leqslant C|t-s|,$$

and thus (F_t) is Lipschitz. The case of general f in $C_{so}(\mathbb{R})$ follows as a uniform limit of Lipschitz functions is uniformly continuous. □

Theorem 8.2.6 *Assume that D is a formally self-adjoint differential operator on S with c_D finite, and that M is complete. Let $f:\mathbb{R}\to\mathbb{C}$ be an element of $C_{so}(\mathbb{R})$. Define a function*

$$[1,\infty)\to\mathcal{B}(L^2(M;S)), \quad t\mapsto F_t := f(t^{-1}D).$$

Then the family (F_t) defines a multiplier of $L^*(L^2(M;S))$.

Proof To show that (F_t) is a multiplier of $L^*(L^2(M;S))$ it will suffice to show that the family (F_t) is uniformly bounded, uniformly continuous in t and that the propagation of (F_t) tends to zero as t tends to infinity (note that the propagation condition and Lemma 6.1.2 imply that $[F_t,h]$ converges to zero for any $h\in C_c(M)$).

Proposition 8.1.8 implies that D is essentially self-adjoint, whence the family (F_t) is well-defined and uniformly bounded. Up to an approximation, we may assume that f has slow oscillation at infinity. Moreover, it is not difficult to see that the convolution of f and a Schwartz-class function still has slow oscillation at infinity; up to another approximation, then, we may replace f with a convolution by a function with smooth and compactly supported Fourier transform, and thus assume that (the distribution) $\widehat{f}$ is supported in $[-r,r]$ for some r.

Lemma 8.2.5 then implies that the function $t\mapsto F_t$ is uniformly continuous. Note that the Fourier transform of the function $x\mapsto f(t^{-1}x)$ is supported in $[t^{-1}r,t^{-1}r]$ for all $t\in[1,\infty)$. It follows from Corollary 8.2.2 that

$$\langle F_t u, v\rangle = 0$$

whenever $u, v \in C_c^\infty(M;S)$ are such that $d(\mathrm{supp}(u), \mathrm{supp}(v)) > c_D t^{-1} r$, and thus that $\mathrm{prop}(F_t) \leqslant c_D t^{-1} r$. □

In the remainder of this section, we consider the general case when D is not necessarily essentially self-adjoint: we aim to build multipliers of $L^*(M)$ from formally self-adjoint differential operators. We need two technical lemmas.

Lemma 8.2.7 *Let D_1 and D_2 be essentially self-adjoint operators on S, and let K be a compact subset of M such that $D_1 = D_2$ on some open set U containing K. Then there exists $R > 0$ such that for all bounded Borel functions $f: \mathbb{R} \to \mathbb{C}$ such that the (distributional) Fourier transform $\widehat{f}$ of f is contained in $[-R, R]$ we have equalities of operators*

$$f(D_1)g = f(D_2)g \quad \text{and} \quad gf(D_1) = gf(D_2)$$

for all bounded functions $g: M \to \mathbb{C}$ supported in K.

Proof We first look at the special case $f(x) = e^{itx}$. Up to an approximation in the strong operator topology, we may assume that g is smooth. Choose $r > 0$ and a compact subset K' of M such that $N_r(K) \subseteq K' \subseteq U$ (this is possible by local compactness of M). Let c be a bound for the propagation speed of $D_1 = D_2$ on K. Let u be any element of $C_c^\infty(M)$ and write

$$u_{1,t} = e^{itD_1} gu, \quad u_{2,t} = e^{itD_2} gu,$$

which are smooth families of elements of $L^2(M;S)$ such that $u_{1,0} = u_{2,0} = gu$. For $|t| < r/c$, Proposition 8.2.1 implies that $u_{1,t}$ and $u_{2,t}$ are supported in K', whence

$$\dot{u_{1,t}} = iD_1 u_{1,t}, \quad \text{and} \quad \dot{u_{2,t}} = iD_2 u_{2,t} = iD_1 u_{2,t}.$$

Hence

$$\begin{aligned}\frac{d}{dt}\|u_{1,t} - u_{2,t}\|^2 &= \langle iD_1(u_{1,t} - u_{2,t}), u_{1,t} - u_{2,t}\rangle \\ &\quad + \langle u_{1,t} - u_{2,t}, iD_1(u_{1,t} - u_{2,t})\rangle,\end{aligned}$$

which is zero as iD_1 is 'formally skew-adjoint' in the natural sense. Hence we have $u_{1,t} = u_{2,t}$ for all suitably small t, and as u was arbitrary, this gives the desired result. The case '$ge^{itD_1} = ge^{itD_2}$' follows on taking adjoints.

The general case now follows from the formula

$$f(D_1)g = \frac{1}{2\pi}\int_{\mathbb{R}} \widehat{f}(t) e^{itD_1} g\,dt$$

from Proposition D.2.3 (see also the statement of that lemma for the exact interpretation of the right-hand side). □

Lemma 8.2.8 *Let D be a formally self-adjoint differential operator on S. Then there exists a family $(g_t)_{t\in[1,\infty)}$ of functions on M with the following properties:*

(i) each g_t is a smooth function from M to $[0,1]$;
(ii) the function

$$[1,\infty) \to C_b(M), \quad t \mapsto g_t$$

is norm continuous;
(iii) for any compact subset K of M there exists t_K such that g_t is identically equal to 1 on K for all $t \geqslant t_K$;
(iv) each operator $g_t D g_t$ (with domain $C_c^\infty(M;S)$) is essentially self-adjoint.

Note that if M is complete and D has finite propagation speed, then we may take g_t to be the constant function with value 1 for all t.

Proof Let M^+ be the one-point compactification of M, and write ∞ for the point at infinity. Fix any (non-Riemannian!) metric d on M^+ that induces the original topology on M, and consider the continuous function

$$h_0 \colon M \to [0,\infty), \quad x \mapsto \frac{1}{d(x,\infty)}.$$

Let $h\colon M \to [0,\infty)$ be any smooth function such that the supremum norm $\|h-h_0\|$ is at most 1, and note that h is proper as h_0 is. Now, let $(f_t)_{t\in[1,\infty)}$ be any norm-continuous family of smooth functions in $C_0[0,\infty)$ such that each f_t is supported in $[0,t+1]$, and identically one on $[0,t]$. Set $g_t = f_t \circ h$. Properties (i), (ii), and (iii) follow directly from the construction, while property (iv) follows from Proposition 8.1.8 and the fact that each $g_t D g_t$ has compact support. □

Definition 8.2.9 Let f be an element of $C_{so}(\mathbb{R})$ as in Definition 8.2.4, and let (g_t) be as in the statement of Lemma 8.2.8. We call the pair $(f,(g_t))$ *multiplier data* for D.

Here, then, is our general construction of multipliers of $L^*(L^2(M;S))$.

Theorem 8.2.10 *Assume that D is a formally self-adjoint differential operator on M and that $(f,(g_t))$ is multiplier data for D. Define a function*

$$[1,\infty) \to \mathcal{B}(L^2(M;S)), \quad t \mapsto F_t := f(t^{-1} g_t D g_t).$$

Then the family (F_t) defines a multiplier of $L^*(L^2(M;S))$.

Proof First note that as $g_t D g_t$ is essentially self-adjoint for all t, so the functional calculus may be applied, and the operator $f(t^{-1} g_t D g_t)$ makes sense, and is uniformly bounded in t. Just as in the proof of Theorem 8.2.6, up to an approximation, we may assume that the distributional Fourier transform $\widehat{f}$ has support in some interval $[-r,r]$ and that f has slow oscillation at infinity. It will suffice to prove the following two properties of (F_t).

(i) for any $h \in C_c(M)$, the functions $t \mapsto F_t h$, $t \mapsto hF_t$ are uniformly continuous for all t suitably large;

(ii) for any $h \in C_c(M)$, the commutator $[F_t, h]$ tends to zero in norm as t tends to infinity (compare Exercise 6.8.4).

For point (i), let K be a compact subset of M containing some neighbourhood of supp(h). Let T be so large that $g_t(x) = 1$ for all $x \in K$, and all $t \geqslant T$. It follows from Corollary 8.2.7 that by increasing T if necessary (whence decreasing the support of the Fourier transform of $x \mapsto f(t^{-1}x)$), we may assume that

$$hf(t^{-1}g_t Dg_t) = hf(t^{-1}g_s Dg_s), \quad f(t^{-1}g_t Dg_t)h = f(t^{-1}g_s Dg_s)h$$

for all $s,t \geqslant T$. From the spectral theorem, we then have that for any $t \geqslant s \geqslant T$

$$\begin{aligned} \|hF_t - hF_s\| &= \|hf(t^{-1}g_t Dg_t) - hf(s^{-1}g_s Dg_s)\| \\ &= \|hf(t^{-1}g_t Dg_t) - hf(s^{-1}g_t Dg_t)\| \\ &\leqslant \sup_{x \in \mathbb{R}} |f(t^{-1}x) - f(s^{-1}x)|. \end{aligned}$$

The rest of the argument for point (i) can be completed just as in the proof of Theorem 8.2.6.

For point (ii), we may again appeal to Corollary 8.2.7 to conclude that there exists a T such that

$$hf(t^{-1}g_t Dg_t) = hf(t^{-1}g_T Dg_T), \quad f(t^{-1}g_t Dg_t)h = f(t^{-1}g_T Dg_T)h$$

for all $t \geqslant T$. Let c be a bound for $c_D(x)$ on supp(g_T). Assume moreover that T is so large that

$$K := \overline{N_{crT^{-1}}(\text{supp}(h))}$$

is compact. It follows then from Proposition 8.2.2 that for all $t \geqslant T$,

$$[h, F_t] = [h, \chi_K f(t^{-1}g_T Dg_T)\chi_K],$$

and that prop$(\chi_K f(t^{-1}g_T Dg_T)\chi_K) \leqslant ct^{-1}r$. The result follows from Lemma 6.1.2. □

8.3 Ellipticity and K-homology

In the previous section, we showed how to use a formally self-adjoint differential operator D together with a choice of multiplier data $(f,(g_t))$ (see Definition 8.2.9) to construct a multiplier (F_t) of $L^*(M)$. Our main goal in this section is to adapt this construction to produce K-theory elements of $L^*(M)$ under the additional assumption that D is elliptic as in the following definition.

Definition 8.3.1 Let D be a differential operator with symbol σ_D, and U be an open subset of M. We say that D is *elliptic* over U if for all $(x,\xi) \in T^*U$ with $\xi \neq 0$ we have that $\sigma_D(x,\xi) \in \mathrm{End}(S_x)$ is invertible.

Example 8.3.2 Looking back at Examples 8.1.4, the operator $\frac{d}{dx}$ on the trivial bundle over $\mathbb{R}$ is elliptic. The exterior differentiation operator is not elliptic, however: for example, $\sigma_D(x,\xi)$ will contain any top-dimensional exterior form in its kernel, whatever ξ is.

The key technical result, which will take most of the section to prove, is the next theorem.

Theorem 8.3.3 *Assume that D is a formally self-adjoint elliptic differential operator on M, let $(f,(g_t))$ be a collection of multiplier data for D and let (F_t) be the multiplier of $L^*(L^2(M;S))$ constructed in Theorem 8.2.10. Then if f is in $C_0(\mathbb{R})$, the family (F_t) is an element of $L^*(M)$.*

The first step in the proof of this theorem is to consider the case of the d-torus $\mathbb{T}^d$; tori are particularly amenable to analysis, as one can use the Fourier transform to change questions about constant coefficient differential operators on $\mathbb{T}^d$ to questions about multiplication operators on $\mathbb{Z}^d$. We will then 'transfer' local results on the d-torus to local results on other d-manifolds, and patch these together to get global results.

In order to carry out the details for this, identify the d-torus $\mathbb{T}^d$ with $\mathbb{R}^d/\mathbb{Z}^d$. Write $(x_1,\ldots,x_d)$ for the local coordinates on $\mathbb{T}^d$ induced from those on $\mathbb{R}^d$, and $\frac{\partial}{\partial x_i}$ for the associated partial derivatives (which are globally well-defined, even though $x_1,\ldots,x_d$ are not). Let S be a trivial rank r Hermitian bundle on $\mathbb{T}^d$; we identify sections of S with functions $u\colon \mathbb{T}^d \to \mathbb{C}^r$. Define a norm on $C^\infty(\mathbb{T}^d;S)$ by

$$\|u\|_{1,2}^2 := \int_{\mathbb{T}^d} \|u(x)\|_{\mathbb{C}^r}^2\,dx + \sum_{j=1}^{d} \int_{\mathbb{T}^d} \left\| \frac{\partial u(x)}{\partial x_i} \right\|_{\mathbb{C}^r}^2 dx.$$

Definition 8.3.4 With notation as above, the *Sobolev space* of $\mathbb{T}^d$, denoted $H^1(\mathbb{T}^d;S)$, is the completion of $C^\infty(\mathbb{T}^d;S)$ for the norm above.

Clearly $H^1(\mathbb{T}^d; S)$ identifies with a dense subspace of $L^2(\mathbb{T}^d; S)$, and we can thus think of elements of this space as functions from $\mathbb{T}^d$ to $\mathbb{C}^r$ and speak of notions like support in the usual way. In some arguments below, we will need to use both the $L^2(\mathbb{T}^d; S)$ and $H^1(\mathbb{T}^d; S)$ norms of some function $u\colon \mathbb{T}^d \to \mathbb{C}^r$; to avoid confusion, we will write $\|u\|_{L^2}$ for the former and $\|u\|_{H^1}$ for the latter.

It will also be convenient to work in the Fourier transform of this picture. For $m = (m_1, \ldots, m_d) \in \mathbb{Z}^d$ and $x = (x_1, \ldots, x_d) \in \mathbb{R}^d$, let

$$m \cdot x := \sum_{j=1}^{d} m_j x_j$$

denote the standard inner product. For $x \in \mathbb{T}^d$, we also write $e^{-2\pi i m \cdot x}$ where we use any lift of $x \in \mathbb{T}^d$ to $\mathbb{R}^d$ to define the inner product (thanks to periodicity of the exponential, the choice of lift does not matter). We define the *Fourier transform* of $u \in C_c^\infty(\mathbb{T}^d; S)$ to be the function $\widehat{u}\colon \mathbb{Z}^d \to \mathbb{C}^r$ defined by

$$\widehat{u}(m) = \int_{\mathbb{T}^d} u(x) e^{-2\pi i m \cdot x} dx.$$

The Fourier transform extends to a unitary isomorphism

$$F\colon L^2(\mathbb{T}^d; S) \to \ell^2(\mathbb{Z}^d, \mathbb{C}^r), \quad u \mapsto \widehat{u}.$$

Write $|m| := \sqrt{m_1^2 + \cdots + m_d^2}$ for the usual Euclidean norm on $\mathbb{R}^d$ restricted to $\mathbb{Z}^d$. Then under the Fourier isomorphism above, the subspace $C^\infty(\mathbb{T}^d; S)$ is taken to the space $C^\infty(\mathbb{Z}^d, \mathbb{C}^r)$ of all *rapidly decaying* functions $\widehat{u}\colon \mathbb{Z}^d \to \mathbb{C}^r$, i.e. those functions such that for each $k \in \mathbb{N}$ there is a constant $c = c(k) > 0$ such that

$$\|\widehat{u}(m)\|_{\mathbb{C}^r} \leqslant (1 + |m|)^{-k}.$$

Moreover, for $j \in \{1, \ldots, d\}$, let M_j denote the multiplication operator

$$(M_j \widehat{u})(m) = m_j \widehat{u}(m), \quad m = (m_1, \ldots, m_d), \tag{8.8}$$

considered as an unbounded operator on $\ell^2(\mathbb{Z}^d, \mathbb{C}^r)$ with domain $C^\infty(\mathbb{Z}^d, \mathbb{C}^r)$. Then the Fourier transform conjugates the partial derivative $\frac{\partial}{\partial x_j}$ (considered as an unbounded operator on $L^2(\mathbb{T}^d; S)$ with domain $C^\infty(\mathbb{T}^d; S)$) to the operator $2\pi i M_j$.

It follows from this discussion that $H^1(\mathbb{T}^d; S)$ corresponds under the Fourier transform to the collection $h^1(\mathbb{Z}^d, \mathbb{C}^r)$ of all functions $\widehat{u}\colon \mathbb{Z}^d \to S$ such that the associated norm

$$\|\widehat{u}\|_{h^1}^2 := (2\pi)^2 \sum_{m \in \mathbb{Z}^d} (1 + |m|^2) \|\widehat{u}(m)\|_{\mathbb{C}^r}^2$$

is finite. We will use the notation $\|\cdot\|_{\ell^2}$ and $\|\cdot\|_{h^1}$ for the norms on $\ell^2(\mathbb{Z}^d,\mathbb{C}^r)$ and $h^1(\mathbb{Z}^d,\mathbb{C}^r)$ when we need to use both at once.

We start with a simple version of the *Rellich lemma.*

Lemma 8.3.5 *The inclusion* $I\colon H^1(\mathbb{T}^d;S) \to L^2(\mathbb{T}^d;S)$ *is a compact operator.*

Proof We apply the Fourier transform, and consider instead the corresponding inclusion

$$h^1(\mathbb{Z}^d,\mathbb{C}^r) \to \ell^2(\mathbb{Z}^d,\mathbb{C}^r).$$

Let

$$I_N\colon h^1(\mathbb{Z}^d,\mathbb{C}^r) \to \ell^2(\mathbb{Z}^d,\mathbb{C}^r), \quad \widehat{u} \mapsto \widehat{u}|_{\{m\in\mathbb{Z}^d \mid |m|\leqslant N\}}$$

be the operator sending an element of $h^1(\mathbb{Z}^d,\mathbb{C}^r)$ to its restriction to the ball of radius N, but now considered as an element of $\ell^2(\mathbb{Z}^d,\mathbb{C}^r)$. Then we have

$$\|I_N\widehat{u} - I\widehat{u}\|^2_{\ell^2} = \sum_{m\in\mathbb{Z}^d,\,|m|\geqslant N} |\widehat{u}(m)|^2 \leqslant (1+N)^{-2}\|u\|^2_{h^1}, \tag{8.9}$$

which tends to zero as N tends to infinity. As I_N has finite rank, this implies that I is compact. □

In fact, we get something a little more precise, which will be useful in the next section.

Remark 8.3.6 Let $T\colon H_1 \to H_2$ be an operator between two Hilbert spaces and $n \in \mathbb{N}$. Then the nth *singular value* of T, is the number

$$s_n(T) := \inf\{\|T - S\| \mid \mathrm{rank}(S) < n\}.$$

For $p \in [1,\infty)$, a bounded operator $T\colon H_1 \to H_2$ between Hilbert spaces is *Schatten p-class* if the associated *Schatten p-norm* defined by

$$\|T\| := \left(\sum_{n=1}^{\infty} s_n(T)^p\right)^{1/p}$$

is finite. Looking at the proof of Lemma 8.3.5, we see that the rank of the operator I_N is roughly N^d: more precisely, there is a constant $c > 0$ (depending on the geometry of balls in $\mathbb{Z}^d$ and on the rank of S) such that

$$c^{-1}N^d \leqslant \mathrm{rank}(I_N) \leqslant cN^d. \tag{8.10}$$

The estimate in line (8.9) shows that

$$\|I_N - I\|_{\mathcal{B}(h^1(\mathbb{Z}^d;\mathbb{C}^r),\ell^2(\mathbb{Z}^d;\mathbb{C}^r))} \leqslant (1+N)^{-1}.$$

It follows from this and line (8.10) that I is a Schatten p-class operator for any $p > d$. We will come back to this later.

The next lemma we need is a version of *Gårding's inequality*. In order to state it, we first note that if D is a first order formally self-adjoint operator on the trivial bundle S over $\mathbb{T}^n$, then the closure (see Definition D.1.2) of D clearly restricts to a bounded operator from $H^1(\mathbb{T}^n; S)$ to $L^2(\mathbb{T}^n; S)$. As usual, we will elide the distinction between D and its closure in what follows.

Lemma 8.3.7 *Say D is a formally self-adjoint operator on a trivial rank r Hermitian bundle S over $\mathbb{T}^d$. Assume that D is elliptic over some open subset U of $\mathbb{T}^d$, and that K is a compact subset of U. Then there is a constant $c > 0$ such that for all $u \in H^1(\mathbb{T}^d; S)$ with support in K we have*

$$\|u\|_{H^1} \leqslant c(\|u\|_{L^2} + \|Du\|_{L^2}).$$

Proof We first consider the case that D has constant coefficients: more specifically, D is of the form

$$D = \sum_{j=1}^{d} a_j \frac{\partial}{\partial x_j} + b,$$

where the a_j and b are constant $r \times r$ matrices, and U is all of $\mathbb{T}^d$. In this case, ellipticity means that the symbol

$$\sum_{j=1}^{d} \xi_j a_j$$

is invertible in $M_r(\mathbb{C})$ for all non-zero $(\xi_1, \ldots, \xi_d) \in \mathbb{R}^d$. In particular there is a constant $c_0 > 0$ such that

$$\left\| \sum_{j=1}^{d} \xi_j a_j v \right\|_{\mathbb{C}^r} \geqslant c_0 \|v\|_{\mathbb{C}^r},$$

for all $\xi = (\xi_1, \ldots, \xi_d) \in \mathbb{R}^d$ of norm one, and all $v \in \mathbb{C}^r$. Now, under Fourier transform D corresponds to the operator

$$M := 2\pi i \sum_{j=1}^{d} a_j M_j + b,$$

where M_j is the multiplication operator from line (8.8) above. It follows that for $u \in H^1(\mathbb{T}^d; S)$

$$\begin{aligned} \|Du\|_{L^2}^2 &= \|M\widehat{u}\|_{\ell^2}^2 \\ &\geqslant (2\pi)^2 \sum_{m \in \mathbb{Z}^d} \|a_j m_j \widehat{u}(m)\|_{\mathbb{C}^r}^2 - \|b\widehat{u}\|_{\ell^2} \\ &\geqslant (2\pi)^2 c_0^2 \sum_{m \in \mathbb{Z}^d} |m|^2 \|\widehat{u}(m)\|_{\mathbb{C}^r}^2 - \|b\| \|\widehat{u}\|_{\ell^2}^2 \\ &= c_0^2 \|\widehat{u}\|_{h^1}^2 - (\|b\| + 1)\|\widehat{u}\|_{\ell^2}^2. \end{aligned}$$

Rearranging this and reversing the Fourier transforms, we have

$$\|u\|_{H^1}^2 \leqslant \frac{1}{c_0^2}(\|Du\|_{L^2}^2 + (1 + \|b\|)\|u\|_{L^2}^2) \leqslant \frac{1 + \|b\|}{c_0^2}(\|Du\|_{L^2}^2 + \|u\|_{L^2}^2).$$

Gårding's inequality in the constant coefficient case follows from this and the inequality $\sqrt{x^2 + y^2} \leqslant x + y$ for x, y non-negative real numbers.

We now look at the general case, so D is of the form

$$D = \sum_{j=1}^{d} a_j \frac{\partial}{\partial x_j} + b,$$

where a_j and b are smooth functions from $\mathbb{T}^d$ to $r \times r$ matrices. For $x \in \mathbb{T}^d$, let D_x be the constant coefficient operator obtained by 'freezing coefficients' at the point x, i.e. D_x is the constant coefficient operator

$$D_x := \sum_{j=1}^{d} a_j(x) \frac{\partial}{\partial x_j} + b(x).$$

Then it is not too difficult to see that for a fixed $\epsilon > 0$ and any $x \in \mathbb{T}^d$, there is a neighbourhood $V_{x,\epsilon}$ of x such that

$$\|Du - D_x u\|_{L^2} \leqslant \epsilon \|u\|_{H^1}$$

for all $u \in H^1(\mathbb{T}^d; S)$. Hence applying the first part of the proof and choosing ϵ suitably small, for each $x \in U$ we may find a constant $c_x > 0$ and an open set $V_x \ni x$ such that for all $u \in H^1(\mathbb{T}^d; S)$ with support in V_x we have that

$$\|u\|_{H^1} \leqslant c_x(\|u\|_{L^2} + \|Du\|_{L^2}). \tag{8.11}$$

Now, cover the compact set K by finitely many of these sets V_x, say $V_1, \ldots, V_N$, and let c_0 be the largest of the associated constants. Let $\phi_1, \ldots, \phi_N$ be a smooth partition of unity on K, with each ϕ_i supported in V_i.

Let $u \in H^1(\mathbb{T}^d; S)$ with support in K be given, and define $u_i := \phi_i u$. As each ϕ_i is smooth one sees that multiplication by ϕ_i defines a bounded operator on $H^1(\mathbb{T}^n)$ (see Exercise 8.5.4), whence there is a constant $c_1 > 0$ such that

$$\|u\|_{H^1} \leqslant c_1 \sum_{i=1}^{N} \|u_i\|_{H^1}.$$

On the other hand, we can bound this using our 'local Gårding's inequalities' as in line (8.11) for the sets V_i to get

$$\begin{aligned}\|u\|_{H^1} &\leqslant c_0 c_1 \sum_{i=1}^{N} (\|u_i\|_{L^2} + \|Du_i\|_{L^2}) \\ &\leqslant c_0 c_1 \sum_{i=1}^{N} (\|\phi_i u\|_{L^2} + \|\sigma_D(d\phi_i)u\|_{L^2} + \|\phi_i Du\|_{L^2}).\end{aligned}$$

Using that each ϕ_i has norm at most one as a multiplication operator on $L^2(\mathbb{T}^d; S)$ and that each $\sigma_D(d\phi_i)$ is bounded by some constant c_2, this implies that

$$\begin{aligned}\|u\|_{H^1} &\leqslant c_0 c_1 \sum_{i=1}^{N} ((1 + c_2)\|u\|_{L^2} + \|Du\|_{L^2}) \\ &\leqslant c_0 c_1 (1 + c_2) N (\|u\|_{L^2} + \|Du\|_{L^2}),\end{aligned}$$

completing the proof. □

Proposition 8.3.8 *Let D be an essentially self-adjoint differential operator on a trivial bundle over $\mathbb{T}^d$, and assume that D is elliptic over some open subset U of $\mathbb{T}^d$. Let ϕ be a smooth function on $\mathbb{T}^d$ which is supported in U. Then the operators*

$$(D \pm i)^{-1}\phi \quad \text{and} \quad \phi(D \pm i)^{-1}$$

are compact.

Proof We will look only at the case of $(D + i)^{-1}\phi$; the case of $(D - i)^{-1}\phi$ is similar to this, and the other cases follow on taking adjoints. We claim that $(D + i)^{-1}\phi$ defines a bounded operator from $L^2(\mathbb{T}^d; S)$ to the Sobolev space $H^1(\mathbb{T}^d; S)$. The result follows from this and the Rellich Lemma (Lemma 8.3.5), as then as an operator from $L^2(\mathbb{T}^d; S)$ to $L^2(\mathbb{T}^d; S)$, $(D+i)^{-1}\phi$ is equal to a composition of a bounded operator and the compact inclusion operator $I \colon H^1(\mathbb{T}^d; S) \to L^2(\mathbb{T}^d; S)$ from that lemma.

To see the claim, note that Gårding's inequality (Lemma 8.3.7) gives us a constant c such that for any $u \in L^2(\mathbb{T}^d; S)$

$$\begin{aligned}\|(D+i)^{-1}\phi u\|_{H^1}^2 &\leqslant c(\|(D+i)^{-1}\phi u\|_{L^2}^2 + \|D(D+i)^{-1}\phi u\|_{L^2}^2) \\ &= c\|(D+i)(D+i)^{-1}\phi u\|_{L^2}^2 \\ &\leqslant c'\|u\|_{L^2}^2,\end{aligned}$$

where the equality uses that D is essentially self-adjoint. □

Having thus analysed the case of a trivial bundle on $\mathbb{T}^d$ in detail, we can now deduce Theorem 8.3.3 by patching together results on coordinate patches.

Corollary 8.3.9 *Let D be an essentially self-adjoint differential operator on a Hermitian bundle S over a Riemannian manifold M. Assume that D is supported on some closed set K which is diffeomorphic to a closed Euclidean ball, and elliptic over some open set $U \subseteq K$. Then for any $\phi \in C_c^\infty(M)$ with support in U, we have that*

$$(D \pm i)^{-1}\phi \quad \textit{and} \quad \phi(D \pm i)^{-1}$$

are compact operators on $L^2(M; S)$.

Proof For notational convenience, focus on the case of $(D+i)^{-1}$; the case of $(D-i)^{-1}$ is similar. We also focus on the case of $(D+i)^{-1}\phi$; the case of $\phi(D-i)^{-1}$ follows on taking adjoints. Let d be the dimension of M. The assumptions on K imply that there exists an open subset V of $\mathbb{T}^d$, diffeomorphic to a ball in $\mathbb{R}^d$, and a diffeomorphism $F\colon K \to \overline{V}$. Note that both F and F^{-1} have bounded derivatives by compactness. Let $W = F^{-1}(V)$. As W is contractible, we may assume that the restriction S_W of S to W is a trivial bundle, i.e. that $S_W = W \times \mathbb{C}^r$ for some r, in a way compatible with the Hermitian structure. The function F then defines a linear operator

$$T_F\colon C_c^\infty(V; \mathbb{C}^r) \to C_c^\infty(U; \mathbb{C}^r), \quad u \mapsto u \circ F$$

that extends to a bounded invertible linear operator

$$T_F\colon L^2(U; S) \to L^2(V; S).$$

Now, the differential operator

$$T_F D T_F^{-1}\colon C_c^\infty(V; \mathbb{C}^r) \to C_c^\infty(V; \mathbb{C}^r)$$

is elliptic on $F(U)$. We may consider this as an operator on all of $\mathbb{T}^d$ by extending by zero outside of V; the result is still a differential operator, as we are assuming that D is supported in K. Hence by Proposition 8.3.8, the operators

$$(T_F^{-1} D T_F + i)^{-1} T_F^{-1} \phi T_F$$

are compact for any ϕ with support in U. However, these operators are equal to

$$T_F^{-1}(D+i)^{-1} \phi T_F$$

and the result follows from boundedness of T_F and Proposition 8.3.8. □

Theorem 8.3.10 *Let D be an essentially self-adjoint elliptic operator on S, and let U be an open subset of M over which D is elliptic. Let g be an element of $C_c(U)$. Then the operators*

$$g(D \pm i)^{-1}, \quad (D \pm i)^{-1} g$$

on $L^2(M; S)$ are compact.

Proof Let K be the support of g. Let $U_1, \ldots, U_N$ be a finite cover of K by open subsets of U such that the closure of each U_i is diffeomorphic to a Euclidean ball. Let $V_1, \ldots, V_N$ be a cover of K by open sets V_i such that $\overline{V_i} \subseteq U_i$. Let $(\phi_i : M \to [0,1])_{i=1}^N$ be smooth functions such that each ϕ_i is supported in V_i, and such that

$$\sum_{i=1}^{N} \phi_i(x)^2 = 1$$

for all $x \in K$. For each i, let $\psi_i : M \to [0,1]$ be a smooth function, supported in U_i, and equal to one on V_i. Note that Corollary 8.3.9 implies that each of the operators

$$\phi_i(\psi_i D \psi_i \pm i)^{-1} \phi_i : L^2(M; S) \to L^2(M; S)$$

is compact. We will show that the operator $(D+i)^{-1} g$ is compact; the other cases are similar.

Note then that

$$\begin{aligned}(D+i)^{-1} g &- \sum_{i=1}^{N} \phi_i(\psi_i D \psi_i + i)^{-1} \phi_i g \\ &= \left(\sum_{i=1}^{N} (D+i)^{-1} \phi_i^2 - \phi_i(\psi_i D \psi_i + i)^{-1} \phi_i \right) g;\end{aligned}$$

it suffices to prove that each of the summands is compact. Looking at one summand, then, and removing subscripts to simplify notation, we have that

$$(D+i)^{-1}\phi^2 - \phi(\psi D\psi + i)^{-1}\phi$$
$$= (D+i)^{-1}\Big(\phi(\psi D\psi + i) - (D+i)\phi\Big)(\psi D\psi + 1)^{-1}\psi. \qquad (8.12)$$

Using that ψ is equal to one on the support of ϕ, the central term in the larger parentheses is equal to

$$\phi\psi D\psi - D\phi = \phi\psi D\psi - \psi D\psi\phi,$$

whence equal to

$$\big[\phi, \psi D\psi\big] = -\sigma_{\psi D\psi}(d\phi);$$

as everything is compactly supported, this is a bounded section of $\mathrm{End}(S)$ with support contained in the support of ϕ. Going back to line (8.12), the right-hand side is equal to

$$(D+i)^{-1}\sigma_{\psi D\psi}(d\phi)(\psi D\psi + 1)^{-1}\phi.$$

The term $(D+i)^{-1}$ is bounded as the function $x \mapsto (x+i)^{-1}$ is bounded, and the term $\sigma_{\psi D\psi}(d\phi)(\psi D\psi + 1)^{-1}\phi$ is compact by Corollary 8.3.9, so we are done. □

Theorem 8.3.3 now follows.

Proof of Theorem 8.3.3 Given the results of Theorem 8.2.10, we must only show that for any $g \in C_c^\infty(M)$ and all t suitably large, the operators

$$F_t g, \quad gF_t$$

are compact. Note that as g is compactly supported, for all suitably large t, the operators $g_t D g_t$ are (essentially self-adjoint and) elliptic over a neighbourhood of the support of g, whence Theorem 8.3.10 implies that

$$g(g_t D g_t \pm i)^{-1} \quad \text{and} \quad (g_t D g_t \pm i)^{-1} g$$

are compact. The case of a general f follows as the functions $x \mapsto (x \pm i)^{-1}$ generate $C_0(\mathbb{R})$ as a C^*-algebra. □

Here then is the construction of K-theory classes. The odd case is slightly simpler, so we address this first. We will use the spectral picture of K-theory from Section 2.9 as this is technically a very clean approach.

Construction 8.3.11 Let D be a differential operator on a manifold M, and let $(g_t)_{t\in[1,\infty)}$ be a family of functions as in the definition of multiplier data for M (Definition 8.2.9). Theorem 8.3.3 (together with the functional calculus) implies that the assignment

$$C_0(\mathbb{R}) \mapsto (f(g_t D g_t))_{t\in[1,\infty)}$$

defines a $*$-homomorphism $C_0(\mathbb{R}) \to L^*(L^2(M;S))$. Tensoring with a rank one projection in an abstract copy $\mathcal{K}$ of the compact operators on a separable infinite-dimensional Hilbert space gives a $*$-homomorphism

$$C_0(\mathbb{R}) \to L^*(L^2(M;S)) \otimes \mathcal{K},$$

and thus an element of $K_1(L^*(L^2(M;S)))$ by Remark 2.9.13. We define $[D] \in K_1(L^*(L^2(M;S)))$ to be this class. Note that as long as M has positive[1] dimension, $L^2(M;S)$ is an ample M module (see Example 4.1.5), so $K_*(L^*(L^2(M;S)))$ identifies canonically with $K_*(M)$, and so we have defined a class $[D] \in K_1(M)$.

Construction 8.3.12 Let D be a differential operator on a manifold M acting on some bundle S, and assume moreover that S is equipped with a splitting $S = S_- \oplus S_+$ such that D interchanges $C_c^\infty(M;S_+)$ and $C_c^\infty(M;S_-)$. Note that the decomposition $S = S_- \oplus S_+$ gives rise to a decomposition

$$L^2(M;S) = L^2(M;S_-) \oplus L^2(M;S_+);$$

we let U be the grading operator (see Definition E.1.4) on this Hilbert space defined by multiplication by ± 1 on $L^2(M;S_\pm)$. Conjugation by this unitary preserves $L^*(L^2(M;S))$ making this into a graded C^*-algebra; as U is a multiplier of $L^*(L^2(M;S))$, this grading is inner. Note also that U preserves the domain of D, and that $UDU = -D$, so D is odd for U.

Let (g_t) be as in the definition of multiplier data for M (Definition 8.2.9), and note that each g_t acts as an even operator on $L^2(M;S)$. As in Construction 8.3.11, Theorem 8.3.3 gives a $*$-homomorphism

$$C_0(\mathbb{R}) \to L^*(L^2(M;S)), \quad f \mapsto (f(g_t D f_t))_{t\in[1,\infty)};$$

in this case, the homomorphism is also graded where the domain is taken to be $\mathscr{S}$ as in Example E.1.10. Let $\mathscr{K}$ be a standard graded copy of the compact operators as in Example E.1.9, and choose an even rank one projection $p \in \mathscr{K}$. Tensoring by p, we get a graded $*$-homomorphism

$$\mathscr{S} \to L^*(L^2(M;S)) \widehat{\otimes} \mathscr{K}$$

[1] One can slightly modify the construction so that it also works when M is zero-dimensional: this is an exercise for the reader.

and so an element of $spK_0(L^*(L^2(M;S)))$, where the latter is considered as a graded group. However, the grading on $L^*(L^2(M;S))$ is inner, so this is the same as the usual K-theory group $K_0(L^*(L^2(M;S)))$ by Proposition 2.9.12. Finally, as $L^2(M;S)$ is ample (see Example 4.1.5 – we leave the trivial case where M is zero-dimensional to the reader), $K_0(L^*(M;S))$ is the same as the K-homology group $K_0(M)$ of M, so we have defined a class $[D] \in K_0(M)$.

To conclude this section, we have the following result saying that the class $[D]$ is well-defined.

Proposition 8.3.13 *The classes $[D] \in K_i(M)$ defined in Constructions 8.3.11 and 8.3.12 above do not depend on any of the choices involved.*

Proof The two cases are essentially the same, so we treat both at once. If (g_t^1) and (g_t^2) are different families of functions satisfying the relevant conditions in Definition 8.2.9, then for any compact subset K of M we have

$$\chi_K f(t^{-1} g_t^1 D g_t^1) = \chi_K f(t^{-1} g_t^2 D g_t^2)$$

as soon as g_t^1 and g_t^2 are equal to one on K, essentially by Lemma 8.2.7 (compare the proof of Theorem 8.2.10). Hence the $*$-homomorphisms built above differ by something with values in the ideal $L_0^*(M)$ of Definition 6.4.8, and thus define the same K-theory class by Lemma 6.4.11 (consider them as taking values in the quotient $L_Q^*(M)$, where they are the same).

The only other thing to check is that the class $[D]$ does not depend on the choice of rank one (even) projection in $\mathcal{K}$ (in $\mathscr{K}$). Any two such projections are homotopic through projections of the same type, however, so we are done. □

Remark 8.3.14 In the presence of a grading, we may construct a class $[D]$ either in $K_0(M)$, or in $K_1(M)$ by forgetting the grading. It is a fact, however, that the class in $K_1(M)$ is zero: see Exercise 8.5.5.

Remark 8.3.15 We can also define the class $[D]$ in the following more traditionally 'index-theoretic' way. For the K_0 case, choose an odd function $f\colon \mathbb{R} \to [-1,1]$ in the class $C_{so}(\mathbb{R})$ of Definition 8.2.4 and with the property that $\lim_{t\to\pm\infty} f(t) = \pm 1$. Then we can use Theorem 8.2.10 applied to multiplier data $(f,(g_t))$ for some choice of (g_t) to build a multiplier (F_t) of $L^2(M;S)$ associated to an odd elliptic operator D. Moreover, each (F_t) is odd as f and D are, and the choice of f implies that $f^2 - 1 \in C_0(\mathbb{R})$, whence by Theorem 8.3.3, $F^2 - 1 \in L^*(L^2(M;S))$. Hence Definition 2.8.5 gives us an index class

$$\mathrm{Ind}[F] \in K_0(L^*(L^2(M;S))).$$

Thanks to Theorem 2.9.16, this is the same class $[D]$ as we defined earlier using the spectral picture of K-theory. The case of K_1 can be handled very much analogously: just forget the grading (and use a not-necessarily odd function $f\colon \mathbb{R} \to [-1,1]$ with the property that $\lim_{t\to\pm\infty} f(t) = \pm 1$).

8.4 Schatten Classes

In this section, we look at some more refined theory coming from the theory of trace class operators. We need this for an application to the so-called covering index theorem in Section 10.1. First, we recall some definitions already more or less given in Remark 8.3.6.

Definition 8.4.1 Let $T\colon H_1 \to H_2$ be a bounded operator between Hilbert spaces. The *nth singular value* of T is the number

$$s_n(T) := \inf\{\|T - S\| \mid \operatorname{rank}(S) < n\}.$$

For a number $p \in [1,\infty)$, define the *Schatten p-norm* of T to be

$$\|T\|_p := \left(\sum_{n=1}^{\infty} s_n(T)^p\right)^{1/p}.$$

Let $S_p(H_1, H_2)$ denote the collection of all bounded operators $T\colon H_1 \to H_2$ for which the above norm is finite, which are called *Schatten p-class operators*; we shorten this to $S_p(H)$ when $H = H_1 = H_2$, or just S_p when the Hilbert spaces are clear from context. Operators in S_1 are also called *trace class*.

We will need the following basic facts about Schatten-class operators. We will not prove these here: see Section 8.6, Notes and References, at the end of the chapter.

Theorem 8.4.2

(i) If T is a Schatten p-class operator and S is any bounded operator then ST and TS are Schatten p-class whenever the compositions make sense.

(ii) If $p,q,r \in [1,\infty)$, $r^{-1} = p^{-1} + q^{-1}$, T is Schatten p-class, S is Schatten q-class and ST makes sense, then ST is Schatten r-class. □

The following key result connects differential operators to Schatten-class operators.

Proposition 8.4.3 *Let M be a d-dimensional Riemannian manifold. Let D be an essentially self-adjoint differential operator on a Hermitian bundle S*

over M, considered as an unbounded operator on the Hilbert space $L^2(M;S)$. Let U be an open subset of M on which D is elliptic. Let $f \in C_0(\mathbb{R})$ be a continuous function such that for some $s > d+1$, the function

$$x \mapsto (1+x^2)^s f(x)$$

is bounded. Let g be an element of $C_c(U)$. Then the operators

$$gf(D), \quad f(D)g$$

on $L^2(M;S)$ are trace class.

The estimate on s in the theorem is not optimal, but all we need for our applications is that there is some s that works.

Proof Write $f(x) = (1+x^2)^{-s}h(x)$, where $h\colon \mathbb{R} \to \mathbb{R}$ is a bounded continuous function. It follows from the functional calculus that

$$f(D) = (1+D^2)^{-s}h(D).$$

Using that $h(D)$ is bounded, and that $(1+D^2)^{-s}$ and $h(D)$ commute, it suffices to show that $(1+D^2)^{-s}g$ is trace class (that $g(1+D^2)^{-s}$ is trace class follows on taking adjoints). For this, it suffices to show that $(1+D^2)^{-(d+1)}g$ is trace class; we will in fact show by induction on k that for $k \in \{1,\dots,n+1\}$, $(1+D^2)^{-k}g$ is Schatten $\frac{n+1}{k}$-class.

To see this, first note that Remark 8.3.6 tells us that the canonical Sobolev space inclusion

$$I\colon H^1(\mathbb{T}^n;S) \to L^2(\mathbb{T}^n;S)$$

studied there is Schatten $(d+1)$-class. Tracing the proofs from there of the various ingredients leading up to Theorem 8.3.10 and using the ‘ideal property’ (i) from Theorem 8.4.2 above, we see that

$$(D \pm i)^{-1}\theta \in S_{d+1}(L^2(M;S)) \quad \text{for any} \quad \theta \in C_c^\infty(U;S). \tag{8.13}$$

The base case $k=1$ of the induction follows from this and the ‘ideal property’ (i) from Theorem 8.4.2 as

$$(1+D^2)^{-1}g = (D+i)^{-1}(D-i)^{-1}\theta g,$$

where $\theta \in C_c^\infty(U;S)$ is constantly one on the support of g, and using that $(D+i)^{-1}$ is bounded.

For the inductive step, fix $k \in \{1,\ldots,n\}$ and let $\phi \in C_c^\infty(U;S)$ be constantly equal to one on the support of g. Then

$$\begin{aligned}(1+D^2)^{-(k+1)}g &= (D+i)^{-1}(1+D^2)^{-k}(D-i)^{-1}\phi g \\ &= (D+i)^{-1}(1+D^2)^{-k}[(D-i)^{-1},\phi]g \\ &\quad + (D+i)^{-1}(1+D^2)^{-k}\phi(D-i)^{-1}g. \end{aligned} \tag{8.14}$$

We have that

$$\begin{aligned}[(D-i)^{-1},\phi] &= (D-i)^{-1}[D-i,\phi](D-i)^{-1} \\ &= -(D-i)^{-1}\sigma_D(d\phi)(D-i)^{-1} \\ &= -(D-i)^{-1}\psi\sigma_D(d\phi)(D-i)^{-1},\end{aligned}$$

where $\psi \in C_c^\infty(U;S)$ is constantly equal to one on the support of ϕ. Substituting this into line (8.14) above gives

$$\begin{aligned}(1+D^2)^{-(k+1)}g = &-(D+i)^{-1}(D-i)^{-1}(1+D^2)^{-k}\psi\sigma_D(d\phi)(D-i)^{-1}\phi g \\ &+ (D+i)^{-1}(1+D^2)^{-k}\phi(D-i)^{-1}\phi g.\end{aligned} \tag{8.15}$$

Combining this with the inductive hypothesis, the 'multiplicative property' (ii) from Theorem 8.4.2 and line (8.13) above completes the proof. □

The following corollary, which says that K-homology classes associated to differential operators can be represented by cycles with particularly nice properties, will be useful later.

Corollary 8.4.4 *Let M be a complete Riemannian manifold and D be an odd elliptic operator on a graded Hermitian bundle S over M with finite propagation speed. Then there is an odd, self-adjoint, contractive element (F_t) of the multiplier algebra of the localisation algebra $L^*(L^2(M;S))$ whose index class in $K_0(L^*(M))$ represents $[D]$ and with the following properties:*

(i) $\mathrm{prop}(F_t) \to 0$ as $t \to \infty$;
(ii) for any $g \in C_c(M)$ and all suitably large t the operators $g(1-F_t^2)$ and $g(1-F_t^2)$ are trace class.

Proof Using Remark 8.3.15, it suffices to show that we can find an odd function $f\colon \mathbb{R} \to [-1,1]$ in the class $C_{so}(\mathbb{R})$ such that $\lim_{t\to\pm\infty} f(t) = \pm 1$, so that for some $s \geqslant \dim(M)+1$ we have that $(1+t^2)^s(f(t)^2-1)$ is bounded, and so that the distributional Fourier transform $\widehat{f}$ is compactly supported. Indeed, in that case Proposition 8.4.3 combined with (the proof of) Theorem 8.2.6 will give the desired result. To build such an f, take

$$f_0(t) := \begin{cases} -1 & t \leqslant -1, \\ t & -1 < t < 1, \\ 1 & t \geqslant 1 \end{cases}$$

and define f to be the convolution $g * f_0$ where $g\colon \mathbb{R} \to [0, \infty)$ is an even Schwartz-class function with compactly supported Fourier transform and total integral one. We leave it to the reader to check that this works. □

8.5 Exercises

8.5.1 In the text, we look only at formally self-adjoint operators. In practice, this is not that much of a restriction as a general operator is the 'same thing as a formally self-adjoint operator, plus a grading'. Indeed, if D is an arbitrary differential operator on S, show using a computation in local coordinates that there is a 'formal adjoint' $D^\dagger$ such that

$$\langle Du, v\rangle = \langle u, D^\dagger v\rangle$$

for all $u, v \in C_c^\infty(M; S)$. Show that the operator

$$\begin{pmatrix} 0 & D^\dagger \\ D & 0 \end{pmatrix}$$

on the naturally graded bundle $C_c^\infty(M; S \oplus S)$ is then formally self-adjoint and odd.

8.5.2 Show that the operator $i\frac{d}{dx}$ acting on $L^2(0,1)$ with domain $C_c^\infty(0,1)$ is formally self-adjoint, but not essentially self-adjoint.
Hint: show that the constant function with value one is in the maximal domain, but note the minimal one.

8.5.3 (For readers who know the terminology). Assume that $S = \mathbb{C}$ is the trivial bundle on the d-torus $\mathbb{T}^d$, and let $H^1(\mathbb{T}^d)$ be the associated Sobolev space as in Definition 8.3.4. Show that $H^1(\mathbb{T}^d)$ can be described as the space of all functions in $L^2(\mathbb{T}^d)$ whose distributional derivatives with respect to each of the coordinates $x_1, \ldots, x_d$ are also in $L^2(\mathbb{T}^d)$ (equipped with the same norm).

8.5.4 Show that multiplication by a smooth function defines a bounded operator on the Sobolev space $H^1(\mathbb{T}^d; S)$ of Definition 8.3.4.

8.5.5 Show that if D is acting on a graded bundle, and we use the method of Construction 8.3.11 to construct $[D] \in K_1(M)$, then $[D] = 0$.

8.5.6 Show that any paracompact Riemannian d-manifold M can can be covered by d open sets $U_1, \ldots, U_d$ such that each U_i is a disjoint union of disjoint sets of uniformly bounded diameter, and all diffeomorphic to the standard ball in Euclidean space.

8.5.7 Show that it $T : H_1 \to H_2$ is a compact operator between Hilbert spaces, then the singular values $(s_n(T))_{n=1}^{\infty}$ of Definition 8.4.1 are the same thing as the eigenvalues of $(T^*T)^{1/2}$ (or of $(TT^*)^{1/2}$).

8.5.8 Prove part (i) of Theorem 8.4.2 by showing first that $s_n(ST) \leqslant \|S\| s_n(T)$ (and similarly for TS).

8.5.9 Prove part (ii) of Theorem 8.4.2 in the special case that S and T commute (you can also try the general case of course, but this is harder).

8.6 Notes and References

The analysis in this section is based heavily on that given by Higson and Roe [135]: most of our arguments are adapted from theirs. We thought it was worth the duplication as we need somewhat different results, and in order to keep this text self-contained. The exception is the material in Section 8.4, which is inspired by section 4 of Roe [214].

The idea of using propagation speed in this context is due to Roe [212]. For an introduction to unbounded operators (and the many possible pitfalls that can arise), chapter VIII of Reed and Simon [211] is a nice reference. The background we need about Schatten-class operators in Section 8.4 (and much more) can be found in the first two chapters of Simon [236].

9

Products and Poincaré Duality

Our goal in this chapter is to define products and pairings between K-theory and K-homology groups, and use these to prove the K-theory Poincaré duality theorem. A basic version of K-theory Poincaré duality says that if M is a smooth closed manifold that is 'oriented' in an appropriate sense, then there is a canonical isomorphism

$$K^*(M) \to K_*(M)$$

between the K-theory and K-homology of M.

The mere existence of such an isomorphism is interesting in itself, but for our applications the specific form of this isomorphism is also important. Indeed, it turns out that orientability in K-theory is closely tied up with a special class of differential operators called Dirac operators, and the Poincaré duality isomorphism is induced by a product with a Dirac operator. This has quite strong structural consequences tying K-homology to analysis and geometry: for example, it implies the existence of particularly nice representatives for classes in $K_*(M)$. The consequences of the specific form of the Poincaré duality isomorphism are crucial for the applications of the assembly map that we will discuss in Chapter 10.

To carry out the details of the above discussion, we need to construct various pairings between K-theory and K-homology. The basic point underlying all of these pairings is that the localisation algebra $L_Q^*(M)$ commutes with $C_0(M)$ (see Lemma 6.4.18). Combined with the particularly nice description of the external product on K-theory available in the spectral picture (Section 2.9), we can give fairly concrete and specific forms of these pairings.

This chapter is structured as follows. First, in Section 9.1 we give a concrete form of the basic pairing

$$K_0(X) \otimes K^0(X) \to \mathbb{Z}.$$

This will be used to study the assembly map in Chapter 11. However, it is not so well suited to generalisations – it is even difficult to describe the pairing $K_1(X) \otimes K^1(X) \to \mathbb{Z}$ in an analogous way – so we move on to another picture.

In Section 9.2 we use the external product

$$K_i(A) \otimes K_j(B) \to K_{i+j}(A \widehat{\otimes}_{\max} B)$$

of Section 2.10 to construct the duality pairing between K-theory and K-homology, a partial pairing (a sort of slant product) and an external product on K-homology. We also prove some compatibilities between these. Having set up this basic machinery, we are prepared to prove a version of the Bott periodicity theorem in Section 9.3. In our current context, Bott periodicity should be regarded as the special case of the Poincaré duality theorem when the manifold is a Euclidean space $\mathbb{R}^d$; as such, it is a stepping stone to the general Poincaré duality theorem rather than a goal in its own right.

In order to define the pairings underlying Poincaré duality for general[1] non-compact manifolds, our work in Section 9.2 is not enough. We need a variant of K-homology called representable K-homology, which is introduced in Section 9.4. For readers who know the terminology, we remark that representable K-homology is a compactly supported theory analogous to classical homology, while our usual K-homology groups are a locally finite theory, analogous to locally finite homology in the classical case.

Having introduced representable K-homology, we are ready to introduce the last pairing, the cap product, in Section 9.5; this uses both K-homology and representable K-homology in its definition. It also admits a nice geometric interpretation in the case of differential operators on manifolds, which we explain in this section. Finally, in Section 9.6 we use the cap product to set up and prove the Poincaré duality theorem in a fairly general form for non-compact manifolds and deduce some consequences. These consequences will be needed for the applications studied in Chapter 10.

9.1 A Concrete Pairing between K-homology and K-theory

In this section, we give a relatively concrete picture of the pairing between the zeroth K-theory and K-homology groups that uses the usual description of K_0 in terms of projections. Later in the chapter, we will switch to the spectral picture of K-theory from Section 2.9.

Throughout this section, X is a proper metric space and H_X is a fixed ample X module (see Definition 4.1.1). We will use the K-theory of $C_L^*(H_X)$ as in

[1] We can get away with a more naive argument for $\mathbb{R}^d$ in Section 9.3 because $\mathbb{R}^d$ is contractible.

Definition 6.6.1 as a model for the K-homology of X; this is legitimate by Proposition 6.6.2. We will write $C_L^*(X)$ for $C_L^*(H_X)$.

For each n, represent $M_n(\mathbb{C})$ on $\ell^2(\mathbb{N})$ by having it act in the usual way on $\ell^2(\{1,\ldots,n\})$ and by zero on $\ell^2(\{m \in \mathbb{N} \mid m > n\})$. Represent $M_n(C_0(X)) = C_0(X) \otimes M_n(\mathbb{C})$ on $H_X \otimes \ell^2(\mathbb{N})$ via the tensor product representation. Represent $C_L^*(X)$ on $H_X \otimes \ell^2(\mathbb{N})$ via the amplification (see Remark 1.8.7) of its defining representation on H_X.

Let now $\mathcal{K} := \mathcal{K}(\ell^2(\mathbb{N}))$ and $\mathcal{B} := \mathcal{B}(\ell^2(\mathbb{N}))$, and consider the C^*-algebra double $D := D_{\mathcal{B}}(\mathcal{K})$, which we recall from Definition 2.7.8 is defined by

$$D_{\mathcal{B}}(\mathcal{K}) := \{(S,T) \in \mathcal{B} \oplus \mathcal{B} \mid S - T \in \mathcal{K}\}.$$

Let $C_{ub}([1,\infty),D)$ denote the C^*-algebra of uniformly continuous, bounded functions from $[1,\infty)$ to D, and let D_∞ denote the quotient C^*-algebra

$$D_\infty := \frac{C_{ub}([1,\infty),D)}{C_0([1,\infty),D)}.$$

Lemma 9.1.1 *With notation as above, let $p,q \in M_n(C_0(X)^+)$ be projections for some n such that $p - q \in M_n(C_0(X))$. Then there is a well-defined $*$-homomorphism*

$$\phi^{p,q} \colon C_L^*(X) \to D_\infty, \quad (T_t) \mapsto (pT_t, qT_t).$$

Moreover, the map induced on K-theory by this $*$-homomorphism depends only on the class $[p]-[q] \in K_0(X)$.

Proof It follows from the definition of $C_L^*(X)$ that the map $\phi^{p,q}$ takes image in D_∞. A slight variant on Lemma 6.1.2 shows that $[p,T_t]$ and $[q,T_t]$ tend to zero in norm as t tends to infinity. It follows from this and the fact that p and q are projections that $\phi^{[p,q]}$ is a $*$-homomorphism.

To see that the map $\phi_*^{p,q}$ induced on K-theory only depends on $[p]-[q]$ it suffices to show that it takes homotopies of pairs satisfying $p-q \in M_n(C_0(X))$ to homotopies of $*$-homomorphisms, and that if r is a third projection, then $\phi_*^{p,q} = \phi_*^{p\oplus r,q\oplus r}$. The fact that homotopies go to homotopies is clear. For the remaining fact, it follows from Lemma 2.7.6 that $\phi_*^{p\oplus r,q\oplus r} = \phi_*^{p,q} + \phi_*^{r,r}$, so it suffices to show that $\phi_*^{r,r} = 0$.

Include $\mathcal{B}$ in $D_{\mathcal{B}}(\mathcal{K})$ via the diagonal inclusion $b \mapsto (b,b)$, and use this to define an inclusion

$$\frac{C_{ub}([1,\infty),\mathcal{B})}{C_0([1,\infty),\mathcal{B})} \to D_\infty.$$

Then $\phi^{r,r}$ factors through this inclusion. However, the left-hand side has zero K-theory by the same Eilenberg swindle showing that $\mathcal{B}$ has zero K-theory (see Lemma 2.7.7), so $\phi_*^{r,r}$ is zero as required. □

Lemma 9.1.2 *There is a natural isomorphism*

$$\psi\colon K_0(D_\infty) \overset{\cong}{\to} \mathbb{Z}.$$

Proof As $C_0([1,\infty), D)$ is contractible, the quotient map $C_{ub}([1,\infty), D) \to D_\infty$ is an isomorphism on K_0-groups, so it suffices to prove the result for $C_{ub}([1,\infty), D)$. Let $\text{ev}\colon C_{ub}([1,\infty), D) \to D$ be the evaluation-at-one map. Then an Eilenberg swindle very similar to that used in the proof of Proposition 6.3.3 shows that the kernel of this has trivial K-theory, and thus that ev induces an isomorphism on K-theory. Lemma 2.7.9 gives a canonical isomorphism $K_0(D) \cong K_0(\mathcal{K}) \oplus K_0(\mathcal{B})$, and finally we use the canonical identifications $K_0(\mathcal{B}) = 0$ and $K_0(\mathcal{K}) = \mathbb{Z}$. □

Definition 9.1.3 Let α be a class in $K_0(X) = K_0(C_L^*(X))$, and $\beta = [p] - [q]$ be a class in $K^0(X) = K_0(C_0(X))$. Then the *pairing of α and β* is defined by

$$\langle \alpha, \beta \rangle := \psi(\phi_*^{p,q}(\alpha)).$$

Remark 9.1.4 If $\alpha = [(P_t)] \in K_0(X)$ is represented by a single projection $(P_t) \in C_L^*(X)$, and $\beta = [p] \in K^0(X)$ is represented by a single projection in $M_n(C_0(X))$, then tracing through the various identifications involved, we see that the pairing of α and β can be defined as follows. As $p^2 = p$ and as $[p, P_t] \to 0$, for all suitably large t, $q := pP_tp$ will be a compact operator such that $\|q^2 - q\| < 1/4$. It follows that the characteristic function $\chi_{(1/2,\infty)}$ of $(1/2,\infty)$ is continuous on the spectrum of q, and so $\chi_{(1/2,\infty)}(q)$ is a compact projection for all suitably large t. We then have that for all suitably large t

$$\langle \alpha, \beta \rangle = \text{rank}\big(\chi_{(1/2,\infty)}(pP_tp)\big).$$

This formula can be adapted to give something that works well in general, and that is more concrete that the version in Definition 9.1.3. However, it is easier to prove that the version in Definition 9.1.3 is well defined and has good formal properties.

9.2 General Pairings and Products

Our goal in this section is to use the external product on K-theory that we introduced in Section 2.10 to construct the duality pairing

$$K_i(X) \otimes K^i(X) \to \mathbb{Z},$$

partial duality pairing

$$K_i(X \times Y) \otimes K^j(Y) \to K_{i+j}(X),$$

and external product

$$K_i(X) \otimes K_j(Y) \to K_{i+j}(X \times Y).$$

We will need notation from Appendix E: in particular, $\mathscr{S}$ will denote $C_0(\mathbb{R})$ with the grading given by the usual notions of even and odd functions as in Example E.1.10, $\widehat{\otimes}$ will denote the graded spatial tensor product of Definition E.2.9, and $\widehat{\otimes}_{\max}$ will denote the maximal graded tensor product of Definition E.2.14.

Throughout this section, X, Y denote locally compact, second countable, Hausdorff topological spaces. We will use notation for localisation algebras as in Chapter 6: in particular the algebras $L^*(X)$ of Convention 6.2.11 (see also Definition 6.2.3) and $L^*_Q(X)$ of Remark 6.4.10 (see also Definition 6.4.8). For our purposes, it will be convenient to represent elements of the K-homology of a space X as elements of $spK_*(L^*_Q(X))$, i.e. as $*$-homomorphisms $\mathscr{S} \to L^*_Q(X)$. It is legitimate to use $L^*_Q(X)$ rather than $L^*(X)$ in this context by Lemma 6.4.11.

We will also allow $L^*(X)$, and therefore also $L^*_Q(X)$, to be graded by an inner automorphism coming from a grading operator on H_X for which the action of $C_0(X)$ is even; a good example to bear in mind is that used in Construction 8.3.12 to build K-homology classes out of odd operators. The reader can safely ignore this for now, but it will be useful to have the extra generality when discussing differential operators as in Construction 8.3.12.

The key point in the construction of duality and partial duality is the following lemma.

Lemma 9.2.1 *The formula*

$$(T_t) \otimes f \mapsto (T_t f)$$

on elementary tensors induces a well-defined $*$-homomorphism

$$\pi^{X,Y} : L^*_Q(X \times Y) \otimes C_0(Y) \to L^*_Q(X).$$

Proof Let H_X and H_Y be ample X and Y modules respectively, and use $H_X \otimes H_Y$ to define $L^*_Q(X \times Y)$. We can also think of $H_X \otimes H_Y$ as an ample X module via the amplification of the $C_0(X)$ representation on H_X, and use it to define $L^*_Q(X)$. Let now $f \in C_0(Y)$ and $(T_t) \in L^*_Q(X \times Y)$. Using Lemma 6.4.18 we have that $(T_t f) = (f T_t)$ in $L^*_Q(X \times Y)$. Hence Lemma 1.8.13 gives us a $*$-homomorphism

$$L^*_Q(X \times Y) \otimes C_0(Y) \to L^*_Q(X \times Y)$$

defined on elementary tensors by $(T_t) \otimes f \mapsto (T_t f)$. Moreover, if f is in $C_c(Y)$, one checks directly that the product $(T_t f)$ is actually in $L^*_Q(X)$: this boils down to the fact that if K is a compact subset of X, then $K \times \mathrm{supp}(f)$ is a compact subset of $X \times Y$. Hence by an approximation argument the image of the above $*$-homomorphism is always in $L^*_Q(X)$, which gives the result. □

Note that the external product defines a map

$$spK_i(L^*_Q(X \times Y)) \otimes spK_j(C_0(Y)) \to spK_{i+j}(L^*_Q(X \times Y) \,\widehat{\otimes}_{\max}\, C_0(Y)).$$

However, Corollary E.2.19 lets us replace $\widehat{\otimes}_{\max}$ with $\widehat{\otimes}$ on the left-hand side, and Exercise E.3.3, plus the fact that $C_0(Y)$ is trivially graded, lets us replace $\widehat{\otimes}$ with $\otimes$.

Definition 9.2.2 The *partial pairing* between K-theory and K-homology

$$K_i(X \times Y) \otimes K^j(Y) \to K_{i+j}(X), \quad \alpha \otimes \beta \mapsto \alpha/\beta$$

is defined to be the composition

$$\begin{array}{ccc} spK_i(L^*_Q(X \times Y)) \otimes spK_j(C_0(Y)) & \xrightarrow{\times} & spK_{i+j}(L^*_Q(X \times Y) \otimes C_0(Y)) \\ & & \Big\downarrow \pi_*^{X,Y} \\ & & spK_{i+j}(L^*_Q(X)) \end{array}$$

of the external product from Definition 2.10.7, and the map on K-theory induced by the $*$-homomorphism from Lemma 9.2.1.

The notation above is inspired by the slant product from classical (co)homology theory, of which the partial pairing above is a natural analogue.

Definition 9.2.3 The *pairing* between K-theory and K-homology

$$K_i(Y) \otimes K^j(Y) \to K_{i+j}(\mathrm{pt}), \quad \alpha \otimes \beta \mapsto \langle \alpha, \beta \rangle$$

is the specialisation of the partial pairing to the case that X is a point (and thus that $K_{i+j}(X) = \mathbb{Z}$ when $i + j = 0 \bmod 2$, and is zero otherwise).

Remark 9.2.4 In the case $i = j = 0$, this agrees with the pairing of Section 9.1: see Exercise 9.7.1 below.

We next construct the external product, which starts with an analogue of Lemma 9.2.1. To state it, let U_X be the unitary multiplier inducing the grading on $L^*_Q(X)$ (possibly just the identity).

Lemma 9.2.5 *The formula*

$$(S_t)\,\widehat{\otimes}\,(T_t) \mapsto (S_t U_X \otimes T_t)$$

on elementary tensors of homogeneous elements induces a well-defined $*$-homomorphism

$$\sigma^{X,Y}: L_Q^*(X)\,\widehat{\otimes}_{\max}\,L_Q^*(Y) \to L_Q^*(X \times Y).$$

Proof Choose ample modules H_X and H_Y for X and Y respectively, so $H_X \otimes H_Y$ is an ample module for $X \times Y$; if gradings are present, we may assume that these are also spatially induced here. Use H_X, H_Y, and $H_X \,\widehat{\otimes}\, H_Y$ to build $L^*(X)$, $L^*(Y)$ and $L^*(X \times Y)$ respectively. Then direct checks show that the map defined on elementary tensors by

$$(S_t) \otimes (T_t) \mapsto (S_t U_X \otimes T_t)$$

gives rise to a $*$-homomorphism $L_Q^*(X)\,\widehat{\odot}\,L_Q^*(Y) \to L_Q^*(X \times Y)$ on the level of the algebraic tensor product. Using the universal property of $\widehat{\otimes}_{\max}$ (Remark E.2.17), this extends to $L_Q^*(X)\,\widehat{\otimes}_{\max}\,L_Q^*(Y)$, so we are done. □

Definition 9.2.6 The *external product* on K-homology

$$K_i(X) \otimes K_j(Y) \to K_{i+j}(X \times Y), \quad \alpha \otimes \beta \mapsto \alpha \times \beta$$

is defined to be the composition

$$\begin{array}{ccc} spK_i(L_Q^*(X)) \otimes spK_j(L_Q^*(Y)) & \xrightarrow{\times} & spK_{i+j}(L_Q^*(X)\,\widehat{\otimes}_{\max}\,L_Q^*(Y)) \\ & & \Big\downarrow \sigma_*^{X,Y} \\ & & spK_{i+j}(L_Q^*(X \times Y)) \end{array}$$

of the external product from Definition 2.10.7, and the map on K-theory induced by the $*$-homomorphism from Lemma 9.2.5.

We will need the following technical lemma about compatibility of these products.

Lemma 9.2.7 *Let $\alpha \in K_i(X)$, $\beta \in K_j(Y)$, $\gamma \in K^i(X)$ and $\delta \in K^j(Y)$. Then*

$$(\alpha \times \beta)/\delta = \alpha \times (\langle \beta, \delta \rangle) \in K_i(X)$$

and

$$\langle \alpha \times \beta \rangle \langle \gamma \times \delta \rangle = \langle \alpha, \gamma \rangle \langle \beta, \delta \rangle \in \mathbb{Z}.$$

Proof We will just prove the first identity in detail; the second is similar. For notational simplicity, let us ignore the copies of $\mathcal{K}$ and the Clifford algebras in the definitions of the spectral K-theory groups. Let us also abuse notation by eliding the difference between a homomorphism and the class it defines in K-theory. So, we assume we are working with graded $*$-homomorphisms

$$\alpha\colon \mathcal{S} \to L_Q^*(X), \quad \beta\colon \mathcal{S} \to L_Q^*(Y), \quad \text{and} \quad \delta\colon \mathcal{S} \to C_0(Y).$$

Then $\alpha \times \beta$ is represented by the graded $*$-homomorphism

$$\mathcal{S} \xrightarrow{\Delta} \mathcal{S} \mathbin{\widehat{\otimes}} \mathcal{S} \xrightarrow{\alpha \mathbin{\widehat{\otimes}} \beta} L_Q^*(X) \mathbin{\widehat{\otimes}}_{\max} L_Q^*(Y) \xrightarrow{\sigma^{X,Y}} L_Q^*(X \times Y)\ ;$$

we will abuse notation slightly and also write $\alpha \times \beta$ for this $*$-homomorphism. Hence the left-hand side $(\alpha \times \beta)/\delta$ of the equation we are trying to establish is represented by the graded $*$-homomorphism

$$\mathcal{S} \xrightarrow{\Delta} \mathcal{S} \mathbin{\widehat{\otimes}} \mathcal{S} \xrightarrow{(\alpha\times\beta)\mathbin{\widehat{\otimes}}\delta} L_Q^*(X \times Y) \mathbin{\widehat{\otimes}} C_0(Y) \xrightarrow{\pi^{X,Y}} L_Q^*(X)\ .$$

On the other hand, $\langle \beta, \delta \rangle$ is represented by the composition

$$\mathcal{S} \xrightarrow{\Delta} \mathcal{S} \mathbin{\widehat{\otimes}} \mathcal{S} \xrightarrow{\beta \mathbin{\widehat{\otimes}} \delta} L_Q^*(Y) \mathbin{\widehat{\otimes}} C_0(Y) \xrightarrow{\pi^{pt,Y}} L_Q^*(pt)\ ;$$

we will again abuse notation and also write $\langle \beta, \delta \rangle$ for this homomorphism. Hence the right-hand side $\alpha \times \langle \beta, \delta \rangle$ of the equation we are trying to establish is given by the graded $*$-homomorphism

$$\mathcal{S} \xrightarrow{\Delta} \mathcal{S} \mathbin{\widehat{\otimes}} \mathcal{S} \xrightarrow{\alpha \mathbin{\widehat{\otimes}} \langle \beta, \delta \rangle} L_Q^*(X) \mathbin{\widehat{\otimes}} L_Q^*(pt) \xrightarrow{\sigma^{X,pt}} L_Q^*(X \times pt) = L_Q^*(X).$$

Putting these descriptions together (and using the fact from Corollary E.2.19 that $\widehat{\otimes}_{\max}$ and $\widehat{\otimes}$ agree when one of the arguments is commutative), we get the diagram

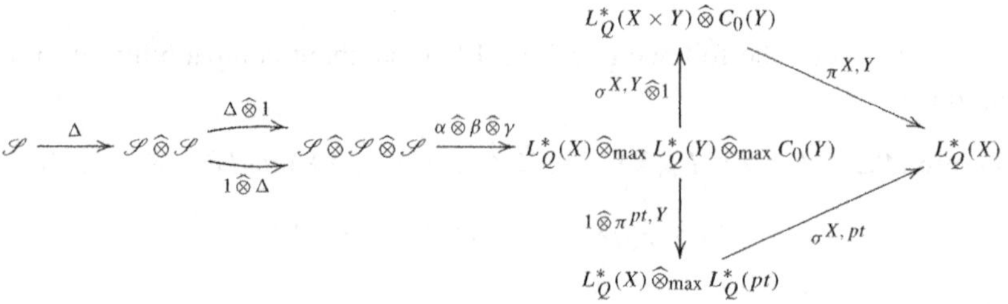

$$\begin{array}{ccccccc}
 & & & & L_Q^*(X \times Y) \mathbin{\widehat{\otimes}} C_0(Y) & & \\
 & & & & \uparrow{\scriptstyle \sigma^{X,Y} \widehat{\otimes} 1} & \searrow{\scriptstyle \pi^{X,Y}} & \\
\mathcal{S} \xrightarrow{\Delta} \mathcal{S} \mathbin{\widehat{\otimes}} \mathcal{S} & \underset{1 \widehat{\otimes} \Delta}{\overset{\Delta \widehat{\otimes} 1}{\rightrightarrows}} & \mathcal{S} \mathbin{\widehat{\otimes}} \mathcal{S} \mathbin{\widehat{\otimes}} \mathcal{S} & \xrightarrow{\alpha \widehat{\otimes} \beta \widehat{\otimes} \gamma} & L_Q^*(X) \mathbin{\widehat{\otimes}}_{\max} L_Q^*(Y) \mathbin{\widehat{\otimes}}_{\max} C_0(Y) & & L_Q^*(X) \\
 & & & & \downarrow{\scriptstyle 1 \widehat{\otimes} \pi^{pt,Y}} & \nearrow{\scriptstyle \sigma^{X,pt}} & \\
 & & & & L_Q^*(X) \mathbin{\widehat{\otimes}}_{\max} L_Q^*(pt) & &
\end{array}$$

with the upper composition corresponding to $(\alpha \times \beta)/\delta$ and the lower composition corresponding to $\alpha \times \langle \beta, \delta \rangle$. The upper and lower paths in the first part of the diagram

$$\mathscr{S} \xrightarrow{\Delta} \mathscr{S} \widehat{\otimes} \mathscr{S} \underset{1 \widehat{\otimes} \Delta}{\overset{\Delta \widehat{\otimes} 1}{\rightrightarrows}} \mathscr{S} \widehat{\otimes} \mathscr{S} \widehat{\otimes} \mathscr{S}$$

are the same by coassociativity of Δ (Lemma 2.10.6). The part

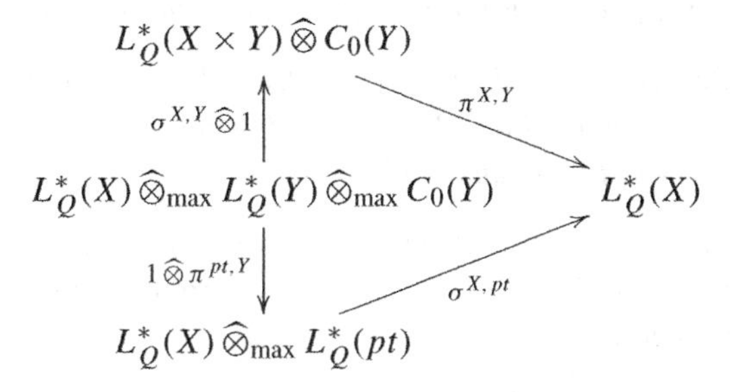

commutes by direct checks that we leave to the reader. These two commutativity statements complete the proof. □

We conclude this section with a statement of the *universal coefficient theorem*. We will not prove it here as it would take us a little too far afield and there are good expositions available. However, it is often a useful tool for computations (indeed, we will use it a little below), so it is worth mentioning. To state it, for an abelian group G let $\mathrm{Ext}(G, \mathbb{Z})$ denote the ext functor from homological algebra: one way to describe this is to say that if

$$0 \longrightarrow K \longrightarrow F \longrightarrow G \longrightarrow 0$$

is an extension of abelian groups with F free abelian, then there is an induced exact sequence

$$0 \longrightarrow \mathrm{Hom}(G, \mathbb{Z}) \longrightarrow \mathrm{Hom}(F, \mathbb{Z}) \longrightarrow \mathrm{Hom}(K, \mathbb{Z}),$$

which need *not* be exact at the right-hand point, i.e. the map $\mathrm{Hom}(F, \mathbb{Z}) \to \mathrm{Hom}(K, \mathbb{Z})$ need not be surjective.[2] The group $\mathrm{Ext}(G, \mathbb{Z})$ has the property that it fits into an exact sequence

$$0 \longrightarrow \mathrm{Hom}(G, \mathbb{Z}) \longrightarrow \mathrm{Hom}(F, \mathbb{Z}) \longrightarrow \mathrm{Hom}(K, \mathbb{Z}) \longrightarrow \mathrm{Ext}(G, \mathbb{Z}) \longrightarrow 0$$

and thus in some sense measures the failure of the functor $G \mapsto \mathrm{Hom}(G, \mathbb{Z})$ to be exact. As the notation suggests, $\mathrm{Ext}(G, \mathbb{Z})$ does not depend on the choice of quotient map $F \to G$ up to canonical isomorphism, as long as F is free

[2] This happens, for example, for $G = \mathbb{Z}/2\mathbb{Z}$, $F = \mathbb{Z}$, and $F \to G$ is the canonical quotient map, as the reader can check.

abelian. We leave this as an exercise for the reader: the point is to use freeness to construct appropriate commutative diagrams.

Theorem 9.2.8 *For each i there is a natural short exact sequence*

$$0 \to Ext(K^i(X), \mathbb{Z}) \to K_{i+1}(X) \to Hom(K^{i+1}(X), \mathbb{Z}) \to 0,$$

where the map $K_{i+1}(X) \to Hom(K^{i+1}(X), \mathbb{Z})$ is induced by the pairing. □

Corollary 9.2.9 *If $K^i(X)$ is free, then $K_{i+1}(X) \cong Hom(K^{i+1}(X), \mathbb{Z})$ (canonically) and if, in addition, $K^{i+1}(X)$ is free and finitely generated, then $K_{i+1}(X) \cong K^{i+1}(X)$ (non-canonically).*

Proof From the description given above of $\mathrm{Ext}(G, \mathbb{Z})$, it clearly vanishes if G is free, as we may take $F = G$. The result follows from this and the short exact sequence from Theorem 9.2.8. □

9.3 The Dirac Operator on $\mathbb{R}^d$ and Bott Periodicity

In this section, we prove a version of Bott periodicity. More or less equivalently, this is the Poincaré duality theorem for $\mathbb{R}^d$. Our eventual goal in this chapter is to bootstrap this to prove a version of Poincaré duality for general (appropriately oriented) manifolds.

Definition 9.3.1 Let $d \in \mathbb{N}$. Let $\mathrm{Cliff}_{\mathbb{C}}(\mathbb{R}^d)$ denote the Clifford algebra over $\mathbb{R}^d$ as in Example E.1.11. Fix an orthonormal basis for $\{e_1, \ldots, e_d\}$ for $\mathbb{R}^d$, and let $x_1, \ldots, x_d$ be the corresponding coordinates. For $v \in \mathbb{R}^d$, write $\widehat{v}$ for the operator on $\mathrm{Cliff}_{\mathbb{C}}(\mathbb{R}^d)$ defined on a homogeneous element w by the formula

$$\widehat{v}\colon w \mapsto (-1)^{\partial w} wv.$$

Thus $\widehat{v}$ is the operator of right multiplication by v, twisted by the grading. Let H_d be the same underlying vector space as $\mathrm{Cliff}_{\mathbb{C}}(\mathbb{R}^d)$, but equipped with the Hilbert space structure discussed in Example E.2.12. Let $L^2(\mathbb{R}^d; H_d)$ denote the Hilbert space of square-summable functions from $\mathbb{R}^d$ to H_d, and let $\mathcal{S}(\mathbb{R}^d; H_d)$ be the dense subspace of $L^2(\mathbb{R}^d; H_d)$ consisting of Schwartz-class functions from $\mathbb{R}^d$ to H_d. Define the *Bott* and *Dirac* operators on $L^2(\mathbb{R}^d; H_d)$ to be the unbounded operators with domain $\mathcal{S}(\mathbb{R}^d; H_d)$ satisfying

$$(Cu)(x) := \sum_{i=1}^{n} e_i x_i u(x), \quad (Du)(x) := \sum_{i=1}^{n} \widehat{e_i} \frac{\partial u}{\partial x_i}(x)$$

for all $u \in L^2(\mathbb{R}^d; H_d)$ and all $x \in \mathbb{R}^d$.

Remark 9.3.2 From the discussion in Example E.2.12, one has an identification

$$L^2(\mathbb{R}^d; H_d) \cong \underbrace{L^2(\mathbb{R}; H_1) \widehat{\otimes} \cdots \widehat{\otimes} L^2(\mathbb{R}; H_1)}_{d \text{ times}}$$

of graded Hilbert spaces (see Definition E.2.5 for the graded tensor product of Hilbert spaces, and also for the definitions of graded tensor products of unbounded operators). With respect to this decomposition, we have

$$C = \sum_{i=1}^{d} 1 \widehat{\otimes} \cdots \widehat{\otimes} 1 \widehat{\otimes} \underbrace{C_1}_{i\text{th place}} \widehat{\otimes} 1 \widehat{\otimes} \cdots \widehat{\otimes} 1,$$

where C_1 is the one-dimensional version of C, and where both operators in the line above are considered as unbounded operators with domain

$$\underbrace{\mathcal{S}(\mathbb{R}; H_1) \odot \cdots \odot \mathcal{S}(\mathbb{R}; H_1)}_{d \text{ times}}$$

(note that C is still essentially self-adjoint for this domain). An analogous formula holds for D.

Analogously (and more or less equivalently), thinking of the decomposition $\mathbb{R}^d = \mathbb{R}^{d-1} \times \mathbb{R}$, we have decompositions of the form

$$C = C_{d-1} \widehat{\otimes} 1 + 1 \widehat{\otimes} C_1 \quad \text{and} \quad D = D_{d-1} \widehat{\otimes} 1 + 1 \widehat{\otimes} D_1. \tag{9.1}$$

The Bott operator can be seen to be essentially self-adjoint by direct checks. On the other hand, the Dirac operator can be seen to be essentially self-adjoint by considering its Fourier transform. Thus we may apply the functional calculus for unbounded operators (Theorem D.1.7). In the case of the Bott operator, if $f \in C_0(\mathbb{R})$ it is not difficult to check that $f(C)$ is in $C_0(\mathbb{R}^d, \mathrm{Cliff}_{\mathbb{C}}(\mathbb{R}^d))$, so we get a functional calculus graded $*$-homomorphism

$$\mathscr{S} \to C_0(\mathbb{R}^d, \mathrm{Cliff}_{\mathbb{C}}(\mathbb{R}^d)), \quad f \mapsto f(C).$$

On the other hand, the symbol of the Dirac operator satisfies $\sigma(x,\xi)^2 = \|\xi\|^2$, whence it is elliptic, and thus we get a $*$-homomorphism

$$\mathscr{S} \to L^*(L^2(\mathbb{R}^d, H_d)), \quad f \mapsto f(t^{-1}D)$$

by Theorem 8.3.3, where the expression '$f(t^{-1}D)$' is slightly sloppy shorthand for the function $t \mapsto f(t^{-1}D)$. Note further that

$$C_0(\mathbb{R}^d, \mathrm{Cliff}_{\mathbb{C}}(\mathbb{R}^d)) \cong C_0(\mathbb{R}^d) \widehat{\otimes} \mathrm{Cliff}_{\mathbb{C}}(\mathbb{R}^d)$$

and using that $\mathcal{B}(H_d) \cong \mathrm{Cliff}_{\mathbb{C}}(\mathbb{R}^d)) \widehat{\otimes} \mathrm{Cliff}_{\mathbb{C}}(\mathbb{R}^d)$ (where we consider the first copy of $\mathrm{Cliff}_{\mathbb{C}}(\mathbb{R}^d)$ as acting on H_d on the left and the second on the right), we get

$$L^*(L^2(\mathbb{R}^d; H_d)) \cong L^*(L^2(\mathbb{R}^d, \mathrm{Cliff}_{\mathbb{C}}(\mathbb{R}^d))) \widehat{\otimes} \mathrm{Cliff}_{\mathbb{C}}(\mathbb{R}^d).$$

Tensoring by a rank one even projection in $\mathcal{K}$, these $*$-homomorphisms give rise to *Bott* and *Dirac* classes

$$[C] \in spK_d(C_0(\mathbb{R}^d)) \quad \text{and} \quad [D] \in spK_d(L^*(\mathbb{R}^d)).$$

The next general lemma takes a while to state, but less time to prove.

Lemma 9.3.3 *Let A and B be graded C^*-algebras represented faithfully in a grading preserving way on graded Hilbert spaces H_A and H_B respectively as in Lemma E.1.6. Let (S_A, D_A) and (S_B, D_B) be odd, possibly unbounded, essentially self-adjoint operators on H_A and H_B (see Example E.1.14). Assume that $f(D_A)$ is in A and $f(D_B)$ is in B for all $f \in \mathcal{S}$. Hence we have a graded $*$-homomorphism*

$$\phi_A : \mathcal{S} \to A, \quad f \mapsto f(D_A)$$

and similarly for B, giving K-theory classes $[\phi_A] \in K_0(A)$ and $[\phi_B] \in K_0(B)$.

Assume moreover that $D := D_A \widehat{\otimes} 1 + 1 \widehat{\otimes} D_B$ (see Definition E.2.5) makes sense as an odd self-adjoint operator on some dense domain in $H_A \widehat{\otimes} H_B$ that contains $S_A \odot S_B$. Assume finally that there is a graded C^*-algebra $C \subseteq \mathcal{B}(H_A \widehat{\otimes} H_B)$ such that the natural tensor product representation

$$\psi : A \widehat{\otimes}_{\max} B \to \mathcal{B}(H_A \widehat{\otimes} H_B)$$

defined analogously to Definition E.2.7 takes image in C, and such that $f(D)$ is in C for all $f \in \mathcal{S}$.

Then the associated functional calculus $*$-homomorphism

$$\phi : \mathcal{S} \to C, \quad f \mapsto f(D)$$

satisfies

$$[\phi] = \psi_*([\phi_A] \times [\phi_B])$$

in $K_0(C)$.

Proof The class $\psi_*([\phi_A] \times [\phi_B])$ is represented by the $*$-homomorphism

$$\mathcal{S} \xrightarrow{\Delta} \mathcal{S} \widehat{\otimes} \mathcal{S} \xrightarrow{\phi_A \widehat{\otimes} \phi_B} A \widehat{\otimes}_{\max} B \xrightarrow{\psi} C . \tag{9.2}$$

We claim that this homomorphism agrees with ϕ, which will certainly suffice to complete the proof. It suffices to check on the generators e^{-x^2} and xe^{-x^2} of $\mathcal{S}$. As in the proof of Lemma 2.10.3, we have the formulas $\Delta(e^{-x^2}) = e^{-x^2} \widehat{\otimes} e^{-x^2}$ and $\Delta(xe^{-x^2}) = xe^{-x^2} \widehat{\otimes} e^{-x^2} + e^{-x^2} \widehat{\otimes} xe^{-x^2}$. Hence the composition in line (9.2) acts as follows on the generators

$$e^{-x^2} \mapsto e^{-D_A^2} \widehat{\otimes} e^{-D_B^2}, \quad xe^{-x^2} \mapsto D_A e^{-D_A^2} \widehat{\otimes} e^{-D_B^2} + D_B e^{-D_A^2} \widehat{\otimes} e^{-D_B^2}.$$

We leave it to the reader to check that ϕ is represented by the same homomorphism: compare the proof of Lemma 2.10.3. □

Lemma 9.3.4 follows directly from the above Lemma 9.3.3, induction on d and the formulas in line (9.1).

Lemma 9.3.4 *Letting $[C_d] \in spK_d(C_0(\mathbb{R}^d))$ and $[D_d] \in K_d(L^*(\mathbb{R}^d))$ denote the Bott and Dirac classes respectively, we have that*

$$[C_d] = \underbrace{[C_1] \times \cdots \times [C_1]}_{d \text{ times}} \quad \text{and} \quad [D_d] = \underbrace{[D_1] \times \cdots \times [D_1]}_{d \text{ times}},$$

where the products are the exterior products in K-theory (Definition 2.10.7) and K-homology respectively (Definition 9.2.6). □

The following theorem is a version of the fundamental Bott periodicity theorem.

Theorem 9.3.5 *The image of $[D] \otimes [C]$ under the pairing $K_d(\mathbb{R}^d) \otimes K^d(\mathbb{R}^d) \to \mathbb{Z}$ of Definition 9.2.3 is one.*

Proof Using Lemma 9.3.4 and Lemma 9.2.7 repeatedly, it suffices to prove this when $d = 1$, which we now do. Choose a norm one vector $e \in \mathbb{R}$, so $\{1, e\}$ is an orthonormal basis for H_1 (considered as a Hilbert space). Let B be the unbounded operator on $L^2(\mathbb{R}; H_1)$ with domain the Schwartz-class functions defined by $B = D + C$. As our choice of (ordered) basis identifies $\text{Cliff}_{\mathbb{C}}(\mathbb{R})$ with $\mathbb{C} \oplus \mathbb{C}$ (as a Hilbert space), we may write B as a 2×2 matrix with entries unbounded operators on $\mathbb{R}$. One checks that

$$B = \begin{pmatrix} 0 & x - \frac{d}{dx} \\ x + \frac{d}{dx} & 0 \end{pmatrix},$$

with respect to this basis. Hence

$$B^2 = \begin{pmatrix} x^2 - \frac{d^2}{dx^2} - 1 & 0 \\ 0 & x^2 - \frac{d^2}{dx^2} + 1 \end{pmatrix} = \begin{pmatrix} H & 0 \\ 0 & H + 2 \end{pmatrix}, \tag{9.3}$$

where $H := x^2 - \frac{d^2}{dx^2} - 1$ is the harmonic oscillator of Definition D.3.1, an unbounded operator on $L^2(\mathbb{R})$ with domain the Schwartz-class functions.

Now, ignoring the copy of $\mathcal{K}$ in the definition of $spK_0(C(pt))$ for notational simplicity, by definition the image of $[D] \otimes [C]$ under the pairing in the statement is given by the $*$-homomorphism

$$\phi \colon \mathcal{S} \to L_Q^*(pt),$$

where we use the ample pt module $L^2(\mathbb{R}; H_1)$ to define the localisation algebra $L_Q^*(pt)$. Let us compute the image of the element e^{-x^2} of $\mathcal{S}$ under this $*$-homomorphism. We have $\Delta(e^{-x^2}) = e^{-x^2} \widehat{\otimes} e^{-x^2}$, whence

$$\phi(e^{-x^2}) = e^{-t^{-1}C^2} e^{-t^{-1}D^2}.$$

Using Lemma 6.4.18 on commutativity of elements of L_Q^* with multiplication operators, we have that as elements of $L_Q^*(pt)$,

$$e^{-t^{-1}C^2} e^{-t^{-1}D^2} = e^{-\frac{1}{2}t^{-1}C^2} e^{-t^{-1}D^2} e^{-\frac{1}{2}t^{-1}C^2}.$$

Moreover, if

$$\alpha(t) := \frac{\cosh(2t) - 1}{2\sinh(2t)} \quad \text{and} \quad \beta(t) := \frac{\sinh(2t)}{2}$$

are as in Corollary D.3.7, then we have the power series expansions

$$\alpha(t) = \frac{t}{2} + O(t^3) \quad \text{and} \quad \beta(t) = t + O(t^3)$$

from which it follows that if we set $\alpha = \alpha(t^{-1})$ and $\beta = \beta(t^{-1})$ then

$$e^{-\frac{1}{2}t^{-1}C^2} e^{-t^{-1}D^2} e^{-\frac{1}{2}t^{-1}C^2} = e^{-\alpha C^2} e^{-\beta D^2} e^{-\alpha C^2}$$

as elements of $L_Q^*(pt)$. Using the formulas

$$D^2 = \begin{pmatrix} -\frac{d}{dx^2} & 0 \\ 0 & -\frac{d}{dx^2} \end{pmatrix} \quad \text{and} \quad C^2 = \begin{pmatrix} x^2 & 0 \\ 0 & x^2 \end{pmatrix}$$

we have that

$$e^{-\alpha C^2} e^{-\beta D^2} e^{-\alpha C^2} = \begin{pmatrix} e^{-\alpha x^2} e^{-\beta \frac{d^2}{dx^2}} e^{-\alpha x^2} & 0 \\ 0 & e^{-\alpha x^2} e^{-\beta \frac{d^2}{dx^2}} e^{-\alpha x^2} \end{pmatrix}.$$

Using Mehler's formula as in Corollary D.3.7, this equals

$$\begin{pmatrix} e^{-\alpha x^2} e^{-\beta \frac{d^2}{dx^2}} e^{-\alpha x^2} & 0 \\ 0 & e^{-\alpha x^2} e^{-\beta \frac{d^2}{dx^2}} e^{-\alpha x^2} \end{pmatrix} = \begin{pmatrix} e^{-t^{-1}} e^{-t^{-1} H} & 0 \\ 0 & e^{-t^{-1}} e^{-t^{-1} H} \end{pmatrix}.$$

Moreover, by line (9.3) we get that

$$\begin{pmatrix} e^{-t^{-1}} e^{-t^{-1} H} & 0 \\ 0 & e^{-t^{-1}} e^{-t^{-1} H} \end{pmatrix} = \begin{pmatrix} e^{-t^{-1}} & 0 \\ 0 & e^{t^{-1}} \end{pmatrix} e^{-t^{-1} B^2}.$$

As an element of $L_Q^*(pt)$, this just equals $e^{-t^{-1}B^2}$, so we conclude that

$$\phi(e^{-x^2}) = e^{-t^{-1}B^2}.$$

An analogous computation starting with the formula

$$\Delta(xe^{-x^2}) = xe^{-x^2} \hat{\otimes} e^{-x^2} + e^{-x^2} \hat{\otimes} xe^{-x^2}$$

gives that

$$\phi(xe^{-x^2}) = t^{-1} B e^{-t^{-1}B^2}.$$

As e^{-x^2} and xe^{-x^2} generate $\mathscr{S}$ as a C^*-algebra, we thus have that

$$\phi(f) = f(t^{-1}B)$$

for all $f \in \mathscr{S}$, where as usual the expression '$f(t^{-1}B)$' is shorthand for the image of the function $t \mapsto f(t^{-1}B)$ in $L_Q^*(pt)$.

Now, it suffices to show that this function represents the generator of $K_0(L_Q(pt))$. For $s \in (0, 1]$ consider the functions

$$\psi_s \colon \mathscr{S} \to L_Q^*(pt), \quad f \mapsto f(s^{-1}t^{-1}B).$$

Moreover, let p be the projection onto the one-dimensional kernel of the harmonic oscillator (see Proposition D.3.3) and define

$$\psi_0 \colon \mathscr{S} \to L_Q^*(pt), \quad f \mapsto f(0) \begin{pmatrix} p & 0 \\ 0 & 0 \end{pmatrix}.$$

Using the eigenvector decomposition of the harmonic oscillator (Proposition D.3.3), we see that $(\psi_s)_{s \in [0,1]}$ is a continuous path of $*$-homomorphisms connecting $\psi = \psi_1$ to the $*$-homomorphism ψ_0. Combining Lemma 2.9.14 with Proposition 6.3.3 and Lemma 6.4.11, one sees that the map ψ_0 indeed represents a generator of $K_0(L_Q^*(pt)) \cong \mathbb{Z}$, so we are done. □

9.4 Representable K-homology

Our goal in this section is to introduce representable K-homology: as well as being interesting in its own right, representable K-homology will be used in the construction of the cap product and thus plays an important role in our eventual statement and proof of the Poincaré duality theorem.

Throughout this section, X, Y denote locally compact, second countable, Hausdorff topological spaces. We use notation for localisation algebras as in Chapter 6.

To explain the basic idea of representable K-homology, we review some classical algebraic topology. The singular homology groups $H_*(X)$ have a natural locally finite[3] analogue $H_*^{lf}(X)$. The difference between these two versions of homology can be described as follows. The groups $H_n(X)$ and $H_n^{lf}(X)$ are both constructed from formal linear combinations of continuous maps $\Delta_n \to X$ from the standard n-dimensional simplex to X. The linear combinations used to build $H_n(X)$ are finite, while those used for $H_n^{lf}(X)$ are 'locally finite': this means that for any compact $K \subseteq X$, there can only be finitely many maps $\Delta_n \to X$ appearing in the linear combination with image intersecting K. The two theories agree for compact spaces, but not in general: for example $H_1(\mathbb{R})$ is zero, but $H_1^{lf}(\mathbb{R}) \cong \mathbb{Z}$. Moreover, they have different functoriality properties: $H_n(X)$ is functorial for all continuous maps, but $H_n^{lf}(X)$ only for those continuous maps which are also proper.

In terms of their functoriality properties (among other things), the K-homology groups $K_*(\cdot)$ defined in Chapter 6 are analogues of the locally finite homology groups $H_*^{lf}(\cdot)$ discussed above. There is also, however, an analogue $RK_*(\cdot)$ of the classical homology groups $H_*(\cdot)$, and our goal in this section is to define and study these groups. Our main purpose is to get to the results on cap products and Poincaré duality discussed later in the chapter: if we are using non-compact manifolds, then the statements of these require the RK_* groups.

Definition 9.4.1 Let H_X be an X module. Define $R\mathbb{L}[H_X]$ to be the collection of all elements (T_t) of $\mathbb{L}[H_X]$ such that there exists a compact subset K of X and $t_K \geqslant 1$ such that

$$T_t = \chi_K T_t \chi_K$$

for all $t \geqslant t_K$. Define $RL^*(H_X)$ to be the completion of $R\mathbb{L}[H_X]$ for the norm

$$\|(T_t)\| := \sup_t \|T_t\|_{\mathcal{B}(H_X)}.$$

[3] Also called Borel–Moore homology.

Functoriality of $RL^*(H_X)$ works much as it does for the localisation algebras, although for a different class of maps. The basic point is the following analogue of Lemma 6.2.7. See Definition 4.4.6 for the definition of continuous cover. The proof is essentially the same as that of Lemma 6.2.7, so we leave it to the reader.

Lemma 9.4.2 *Let H_X, H_Y be geometric modules. Let $f: X \to Y$ be a continuous map, and assume there exists a continuous cover (V_t) for f. Then*

$$(T_t) \mapsto (V_t T_t V_t^*)$$

defines a $*$-homomorphism

$$ad_{(V_t)}: R\mathbb{L}[H_X] \to R\mathbb{L}[H_Y]$$

that extends to a $*$-homomorphism from $RL^*(H_X)$ to $RL^*(H_Y)$. Moreover, the map on K-theory induced by $ad_{(V_t)}$ depends only on f and not on the choice of (V_t). □

Now, given a continuous map $f: X \to Y$ and geometric modules H_X, H_Y with H_Y ample, Corollary 4.4.7 implies that there exists a continuous cover (V_t) for f. The following analogue of Definition 6.2.9 therefore makes sense.

Definition 9.4.3 Let $f: X \to Y$ be a continuous function and let $RL^*(H_X)$ and $RL^*(H_Y)$ be associated to ample geometric modules. Define

$$f_*: K_*(RL^*(H_X)) \to K_*(RL^*(H_Y))$$

to be the map on K-theory induced by the $*$-homomorphism

$$\mathrm{ad}_{(V_t)}: RL^*(H_X) \to RL^*(H_Y)$$

associated to some continuous cover for f as in Lemma 9.4.2 above.

Collecting the above together, we get the following analogue of Theorem 6.2.10. To state it, let $\mathcal{C}ont$ be the category of locally compact second countable Hausdorff spaces and continuous maps, and $\mathcal{GA}$ the category of $\mathbb{Z}/2\mathbb{Z}$-graded abelian groups and graded group homomorphisms.

Theorem 9.4.4 *For each X in $\mathcal{C}ont$ choose an ample X module H_X. Then the assignments*

$$X \mapsto K_*(RL^*(H_X)), \quad f \mapsto f_*$$

give a well-defined functor from $\mathcal{C}ont$ to $\mathcal{GA}$.

Moreover, the functor that one gets in this way does not depend on the choice of modules up to canonical equivalence. □

Definition 9.4.5 The *representable*[4] *K-homology* of X is defined to be the K-theory group

$$RK_*(X) := K_*(RL^*(H_X))$$

for any choice of ample X module H_X.

Just as in Convention 6.2.11, if H_X is ample we will often write $RL^*(X)$ for $RL^*(H_X)$.

Remark 9.4.6 With notation as in Definition 6.4.8, define

$$R\mathbb{L}_0[H_X] := \mathbb{L}_0[H_X] \cap R\mathbb{L}[H_X].$$

More concretely, $R\mathbb{L}_0[H_X]$ consists of elements (T_t) of $R\mathbb{L}[H_X]$ such that there exists $t_0 \geqslant 1$ with the property that for all $t \geqslant t_0$, $T_t = 0$. Let $RL_0^*(H_X)$ to be the closure of this collection inside $RL^*(H_X)$, a C^*-ideal of $RL^*(H_X)$. If H_X is ample, essentially the same proof as in Lemma 6.4.11 shows that $RL_0^*(H_X)$ has zero K-theory, and thus the K-theory of $RL_Q^*(H_X) := RL^*(H_X)/RL_0^*(H_X)$ is another model for representable K-homology. Analogously to Lemma 6.4.18, elements of the C^*-algebra $C_b(X)$ of bounded continuous functions on X act as central multipliers on $RL_Q^*(H_X)$.

These facts can be established in much the same way as the corresponding facts for localisation algebras: we leave the details to the reader.

Our next goal is to discuss the relationship of RK_* to K_*. The key fact is that the representable K-homology of a space X is determined by the compact subsets of X in the sense of the next result.

Proposition 9.4.7 *For any X, let $(K_i)_{i\in I}$ be the net of compact subsets of X, ordered by inclusion. Then the maps $RK_*(K_i) \to RK_*(X)$ functorially induced by the inclusions $K_i \to X$ induce a natural isomorphism*

$$\lim_{i\in I} RK_*(K_i) \cong RK_*(X).$$

Proof As X is second countable and locally compact, there exists a sequence $(K_n)_{n=1}^{\infty}$ of compact subsets of X such that each K_n is the closure of its interior, such that $K_n \subseteq K_{n+1}$ and such that any compact subset of X is contained in all the K_n for n suitably large: indeed, we may take a countable basis $(U_m)_{m=1}^{\infty}$ of X such that each U_m has compact closure, and then set $K_n = \bigcup_{m=1}^{n} \overline{U_m}$. It suffices to show that the inclusions $K_n \to X$ induce an isomorphism

[4] The name is based on the fact that $RK_*(\cdot)$ is a representable functor on an appropriate category, but this will not be important for us.

$$\lim_{n} RK_*(K_n) \cong RK_*(X).$$

To see that this is true, let H_X be any ample X module, and χ_n be the characteristic function of K_n. As K_n is the closure of its interior, $H_{K_n} := \chi_{K_n} H_X$ is an ample K_n module. The inclusion isometry $V_n \colon H_{K_n} \to H_X$ defines a constant family $(V_n)_{t \in [1,\infty)}$ of isometries covering the identity inclusion $K_n \to X$, and thus the associated map $RK_*(K_n) \to RK_*(X)$ is induced by the map

$$\mathrm{ad}_{(V_n)} \colon RL^*(H_{K_n}) \to RL^*(H_X).$$

This descends to a map on the quotients $\mathrm{ad}_{(V_n)} \colon RL^*_Q(H_{K_n}) \to RL^*_Q(H_X)$ as in Remark 9.4.6, which also induces the required map on K-theory.

Now, recall the fact that if an element (T_t) of $\mathbb{L}[H_Y]$ is in $R\mathbb{L}[H_Y]$ for some space Y, then there exists a compact subset K of X and $t_K \geqslant 1$ such that

$$T_t = \chi_K T_t \chi_K$$

for all $t \geqslant t_K$. From this it follows that

$$\bigcup_{n=1}^{\infty} \mathrm{ad}_{(V_n)}\big(R_Q L^*(H_{K_n})\big)$$

is dense in $R_Q L^*(H_X)$. The result now follows from continuity of K-theory (Proposition 2.7.1). □

Recall that K-homology K_* is a functor on the category $\mathcal{LC}$ with locally compact, second countable, Hausdorff spaces as objects, and where a morphism from X to Y is given by a choice of an open subset U of X and a continuous, proper map from U to Y. Seen in this way, both $\mathcal{LC}$ and $\mathcal{C}ont$ naturally identify with sub-categories of the category with objects the second countable, locally compact, Hausdorff spaces, and where the morphisms from X to Y are given by a choice of an open subset U of X and a continuous map from U to Y. Thinking like this, it makes sense to talk about the intersection $\mathcal{LC} \cap \mathcal{C}ont$: this again has second countable, locally compact, Hausdorff spaces as objects, and morphisms are continuous proper maps.

After this discussion, the following result is just a check of the definitions.

Proposition 9.4.8 *For each locally compact, second countable, Hausdorff space X, choose an ample module H_X. Then $RL^*(H_X)$ is a C^*-subalgebra of $L^*(H_X)$. The collection of maps on K-theory induced by the inclusions*

$$RL^*(H_X) \to L^*(H_X), \quad X \in \mathcal{LC} \cap \mathcal{C}ont$$

define a natural transformation between the restrictions of the functors RK_* and K_* to $\mathcal{LC} \cap Cont$.

Moreover, if X is a compact space, then the associated map $RK_*(X) \to K_*(X)$ as above is an isomorphism. □

On the other hand, K-homology and representable K-homology differ even for the simplest non-compact spaces, as the following example shows.

Example 9.4.9 The (graded) group $RK_*([0,\infty))$ identifies with the limit $\lim_n K_*([0,n])$; homotopy invariance and the computation of the K-homology of a point in Theorem 6.4.16 and Proposition 6.3.3 show that this limit is $\mathbb{Z}$ in dimension zero and zero in dimension 1. On the other hand, the K-homology of $[0,\infty)$ is zero in both degrees by Proposition 6.4.14.

The following technical corollary is almost immediate: we again leave the details to the reader. We will use it to deduce homological properties of RK_* from those of K_*.

Corollary 9.4.10 *For any locally compact, second countable, Hausdorff space X, let $(K_i)_{i\in I}$ be the net of compact subsets of X, ordered by inclusion. This defines a directed system $(K_*(K_i))_{i\in I}$ of K-homology groups, and $RK_*(X)$ canonically identifies with the direct limit $\lim_{i\in I} K_*(K_i)$.*

Moreover, this identification is functorial in the following sense. If $f\colon X \to Y$ is continuous and $(K_i)_{i\in I}$, $(K_j)_{j\in J}$ are the nets of compact subsets of X and Y respectively, then for each compact $K \subseteq X$ we have a map

$$(f|_{K_i})_*\colon K_*(K_i) \to K_*(f(K_i)) \to \lim_{j\in J} K_*(K_j),$$

where the map on the right exists by definition of the direct limit. These maps are compatible with the inclusions defining the nets (K_i) and (K_j), and the associated diagram commutes.

$$\begin{array}{ccc} \lim\limits_{i\in I} K_*(K_i) & \xrightarrow{\lim_{i\in I}(f|_{K_i})_*} & \lim\limits_{j\in J} K_*(K_j) \\ \downarrow\cong & & \downarrow\cong \\ RK_*(X) & \xrightarrow{f_*} & RK_*(Y). \end{array}$$

□

We conclude this section with some homological properties of representable K-homology.

Corollary 9.4.11 *If $h\colon X \times [0,1] \to Y$ is a continuous homotopy between $f,g\colon X \to Y$, then $f_* = g_*$ as maps $RK_*(X) \to RK_*(Y)$.*

Proof Let (K_n) be an increasing sequence of compact subsets of X whose union is all of X. Let (K'_n) be an increasing sequence of compact subsets of Y whose union is all of Y and such that $h([0,1] \times K_n) \subseteq K'_n$ for each n. Using homotopy invariance of K-homology (Theorem 6.4.16), the restrictions of f and g to each K_n induce the same map

$$K_*(K_n) \to K_*(K'_n).$$

The result follows on passing to a direct limit. □

Corollary 9.4.12 *Let X be a disjoint union of countably many closed and open subsets $X = \bigsqcup_{n\in\mathbb{N}} X_n$. Then the maps $RK_*(X_n) \to RK_*(X)$ induced by the inclusions $X_n \to X$ induce an isomorphism*

$$\bigoplus_{n\in\mathbb{N}} RK_*(X_n) \to RK_*(X).$$

Proof Let K be a compact subset of X, and write $K_n = K \cap X_n$ so each K_n is a compact subset of X_n and only finitely many K_n are non-empty. A special case of Theorem 6.4.20 using that all but finitely many of the groups $K_*(K_n)$ are zero then gives that

$$\bigoplus_{n\in\mathbb{N}} K_*(K_n) \cong K_*(K).$$

Taking the direct limit over all compact subsets of X gives the result. □

Finally, we finish with a Mayer–Vietoris sequence. This can be derived from the Mayer–Vietoris sequence for K-homology (Proposition 6.3.4) and Corollary 9.4.10, but it will be convenient for later results to give a direct proof.

Proposition 9.4.13 *Let $X = U \cup V$ be a union of two open sets. Then there is a six-term Mayer–Vietoris sequence*

$$\begin{array}{ccccc} RK_0(U\cap V) & \longrightarrow & RK_0(U)\oplus RK_0(V) & \longrightarrow & RK_0(X) \\ \uparrow & & & & \downarrow \\ RK_1(X) & \longleftarrow & RK_1(U)\oplus RK_1(V) & \longleftarrow & RK_1(U\cap V), \end{array}$$

where the horizontal arrows are induced by the inclusions, and which is natural for such decompositions.

Proof Fix an ample X-module. For $Y \in \{U, V, U \cap V\}$, set $R\mathbb{L}_Y[H_X]$ to be the collection of (T_t) in $R\mathbb{L}[H_X]$ such that there is a compact subset K of Y and $t_K \geqslant 1$ such that

$$T_t = \chi_K T_t \chi_K$$

for all $t \geqslant t_K$. Let $RL_Y^*(H_X)$ denote the closure of $R\mathbb{L}_Y[H_X]$ inside $RL^*(H_X)$. Note that if K is a compact subset of an open set Y in a locally compact space Z then there is an open set W such that $K \subseteq W$ and $\overline{W} \subseteq Y$, and also so that $\overline{W}$ is compact; using this and Lemma 4.1.15, one checks that $RL_Y^*(H_X)$ is an ideal in $RL^*(H_X)$.

Let now (ϕ_U, ϕ_V) be a partition of unity on X subordinate to the cover (U, V). Then for any $(T_t) \in R\mathbb{L}[H_X]$ one can check using Lemma 4.1.15 again that $(\phi_U T_t) \in R\mathbb{L}_U[H_X]$, and $(\phi_V T_t) \in R\mathbb{L}_V[H_X]$. As $T_t = \phi_U T_t + \phi_V T_t$ for each t, this implies that $RL^*(H_X) = RL_U^*(H_X) + RL_V^*(H_X)$. On the other hand, it is immediate from the definitions that

$$R\mathbb{L}_U[H_X] \cap R\mathbb{L}_V[H_X] = R\mathbb{L}_{U\cap V}[H_X].$$

This implies that $RL_U^*(H_X) \cap RL_V^*(H_X) \supseteq RL_{U\cap V}^*(H_X)$. For the opposite inclusion, let (T_t) be an element of $RL_U^*(H_X) \cap RL_V^*(H_X)$. Let $\epsilon > 0$ and let (T_t^U) and (T_t^V) be in $R\mathbb{L}_U[H_X]$ and $R\mathbb{L}_V[H_X]$ respectively, and such that both are within ϵ of (T_t). Let $K \subseteq U$ be a compact set and $t_K \geqslant 1$ be such that $T_t^U = \chi_K T_t^U \chi_K$ for all $t \geqslant t_K$. Let $\phi\colon X \to [0,1]$ be a compactly supported function that is equal to one on K and supported in U, and define

$$\phi_t := \begin{cases} 1 & t \leqslant t_K, \\ (t+1-t_K) + (t-t_K)\phi & t \in (t_K, t_K+1], \\ \phi & t \geqslant t_K + 1. \end{cases}$$

Then $\phi_t T_t^U = T_t^U$ for all t, whence $(\phi_t T_t^V)$ is within 2ϵ of (T_t^U). However, $(\phi_t T_t^V)$ is in $R\mathbb{L}_U[H_X] \cap R\mathbb{L}_V[H_X]$ whence (T_t) is within 2ϵ of $R\mathbb{L}_U[H_X] \cap R\mathbb{L}_V[H_X]$ and so (T_t) is in $RL_U^*(H_X) \cap RL_V^*(H_X)$.

Putting all this together, we have a pushout diagram

$$\begin{array}{ccc} RL_{U\cap V}^*(H_X) & \longrightarrow & RL_U^*(H_X) \\ \downarrow & & \downarrow \\ RL_V^*(H_X) & \longrightarrow & RL^*(H_X)\,. \end{array} \tag{9.4}$$

Now, for $Y \in \{U, V, U\cap V\}$ write $H_Y = \chi_Y H_X$, which by openness of Y is an ample Y module. Passing to the quotient by $RL_0^*(H_X)$, we have that the natural inclusions $RL^*(H_Y) \to RL_Y^*(H_X)$ give rise to identifications

$$RL_Q^*(H_Y) \cong \frac{RL_Y^*(H_X)}{RL_Y^*(H_X) \cap RL_0^*(H_X)}\,;$$

the point, which is immediate from the definitions, is that for any element (T_t) of $R\mathbb{L}_Y[H_X]$ there is an element (S_t) of $R\mathbb{L}[H_Y]$ and $t_0 \geqslant 1$ such that $S_t = T_t$ for all $t \geqslant t_0$. The pushout diagram in line (9.4) above thus gives rise to a pushout diagram

$$\begin{array}{ccc} RL^*_Q(H_{U\cap V}) & \longrightarrow & RL^*_Q(H_U) \\ \downarrow & & \downarrow \\ RL^*_Q(H_V) & \longrightarrow & RL^*_Q(H_X)\,. \end{array}$$

From the construction, one checks that this diagram is natural for maps between decompositions of X into open sets. It gives rise to the desired six-term exact Mayer–Vietoris sequence by Proposition 2.7.15. □

9.5 The Cap Product

Our goal in this section is to construct and study a general form of the cap product pairing: this will be a map

$$K_*(X) \otimes K^*(U) \to RK_*(U)$$

defined for an open subset $U \subseteq X$.

Throughout this section X and Y will denote second countable locally compact spaces. We start with the following analogue of Lemma 9.2.1; the proof is essentially the same, so it is omitted here.

Lemma 9.5.1 *For any open subset U of X, there is a $*$-homomorphism*

$$\pi^U : L^*_Q(X) \otimes C_0(U) \to RL^*_Q(U)$$

that satisfies $(T_t) \otimes f \mapsto (T_t f)$ on elementary tensors. □

Definition 9.5.2 For an open subset U of X, the *cap product* is defined to be the map

$$K_i(X) \otimes K^j(U) \to RK_{i+j}(U), \quad \alpha \otimes \beta \mapsto \alpha \cap \beta$$

induced by the composition

$$\begin{array}{ccc} spK_i(L^*_Q(X)) \otimes spK_j(C_0(U)) & \xrightarrow{\times} & spK_{i+j}(L^*_Q(X) \otimes C_0(U)) \\ & & \downarrow{\scriptstyle \pi^U_*} \\ & & spK_{i+j}(RL^*_Q(U)) \end{array}$$

of the external product from Definition 2.10.7, and the map on K-theory induced by the $*$-homomorphism from Lemma 9.5.1 (we also use that $\widehat{\otimes}_{\max}$ and $\widehat{\otimes}$ agree on commutative algebras by Corollary E.2.19, and that $C_0(U)$ is trivially graded to replace $\widehat{\otimes}$ with $\otimes$).

Remark 9.5.3 If $U = X$, the cap product becomes a map

$$K_i(X) \otimes K^j(X) \to RK_{i+j}(X).$$

Composing this with the map $p_*\colon RK_{i+j}(X) \to RK_{i+j}(pt)$ functorially induced from the collapse map $p\colon X \to pt$ from X to a single point, we recover the pairing

$$K_i(X) \otimes K^j(X) \to K_{i+j}(pt)$$

of Definition 9.2.3. We leave the check of this as an exercise for the reader: see Exercise 9.7.4 below.

We now discuss how these products behave with respect to some of the functoriality and homological properties of K-theory and K-homology. To get a suitably general version of this, we have to be a little careful and proceed as follows. Let $f\colon X \to Y$ be a morphism in the category $\mathcal{LC}$, so we may think of f as a continuous proper map $f\colon U \to Y$, for some open subset U of X (compare Proposition A.1.8). In this way, f induces a map $f_*\colon K_*(X) \to K_*(Y)$ on K-homology, a map $f^*\colon K^*(Y) \to K^*(X)$ on K-theory, another map $f^*\colon K^*(Y) \to K^*(U)$ on K-theory (through which the first factors) and a map $f_*\colon RK_*(U) \to RK_*(Y)$ on representable K-homology.

Proposition 9.5.4 *Let $f\colon X \to Y$ be a morphism in the category $\mathcal{LC}$, and let α, β be classes in $K_*(X)$ and $K^*(Y)$ respectively. Then with notation as above*

$$f_*(\alpha) \cap \beta = f_*(\alpha \cap f^*(\beta)) \quad \textit{and} \quad \langle f_*(\alpha), \beta\rangle = \langle \alpha, f^*(\beta)\rangle.$$

Proof The first of these identities implies the second on applying p_*, where $p\colon Y \to pt$ is the collapse function to a single-point space (compare Remark 9.5.3). Hence it suffices to prove the first identity.

For this, it will be technically convenient to choose specific modules. Let Z_X be a countable dense subset of X^+ that contains the point at infinity. Let H be a separable infinite-dimensional Hilbert space. Let $H_{X^+} = \ell^2(Z_X, H)$, $H_X := \ell^2(Z_X \cap X, H)$ and $H_U := \ell^2(Z_X \cap U, H)$, which are ample modules for X^+, X and U respectively. Let Z_Y be a countable dense subset of Y^+ that contains ∞ and $f(Z_X \cap U)$, and define $H_{Y^+} := \ell^2(Z_Y, H)$. Choose a decomposition $H = \bigoplus_{z\in Z_X} H_z$, where each H_z is infinite-dimensional, and isometries $W_z\colon H \to H_z$. Define

$$V\colon H_X \to H_Y, \quad (Vu)(z) = \bigoplus_{x\in f^{-1}(z)} W_x u(x) W_x^*.$$

Setting $V_t = V$ for each t defines a continuous cover (V_t) of f considered as a map from X^+ to Y^+. On noting that the same formula for the pairing works if we start with an element of $L^*(H_{X^+};\infty)$ as in Definition 6.4.3, the desired identity $f_*(\alpha) \cap \beta = f_*(\alpha \cap f^*(\beta))$ now follows from direct checks, which we leave to the reader. □

We isolate below a particularly important instance of functoriality; it is essentially a special case of the results in Proposition 9.5.4. To state it, recall that if $i \colon U \to X$ is an inclusion of an open set, then we get a map $c \colon X^+ \to U^+$ from the one-point compactification of X to that of U that takes infinity to infinity and collapses $X \setminus U$ to the point at infinity in U^+. Taking the maps on K-theory and K-homology induced by c gives us maps

$$i_* \colon K^*(U) \to K^*(X) \quad \text{and} \quad i^* \colon K_*(X) \to K_*(U)$$

(compare Example 6.4.13).

Corollary 9.5.5 *Let $i \colon U \to X$ be an inclusion of an open set, and let $\alpha \in K_*(X)$ be a K-homology class. Then the following diagram commutes.*

$$\begin{array}{ccc} K^*(U) & \xrightarrow{i_*} & K^*(X) \\ \downarrow{\scriptstyle i^*(\alpha)\cap\cdot} & & \downarrow{\scriptstyle \alpha\cap\cdot} \\ RK_*(U) & \xrightarrow{i_*} & RK_*(X)\,. \end{array}$$

Proof The corollary says that for any $\beta \in K^*(U)$ we have

$$i_*(\alpha \cap c^*(\beta)) = c_*(\alpha) \cap \beta,$$

where $c \colon X^+ \to U^+$ is the collapse map discussed just before the statement of the lemma. Checking our notational conventions, this is a special case of Lemma 9.5.4. □

For the next lemma, recall that if $X = U \cup V$ is a decomposition into open sets, then there are Mayer–Vietoris sequences in both K-theory and representable K-homology as Example 2.7.16 and Proposition 9.4.13.

Lemma 9.5.6 *Let X be a union of two open sets $X = U \cup V$, and let α be class in $K_j(X)$. For each $Y \in \{U, V, U \cap V\}$, let $i_Y \colon Y \to X$ be the inclusion, and let $\alpha_Y := i_Y^*(\alpha)$, where i_Y^* is as in Convention 6.4.13. Then the diagram of Mayer–Vietoris sequences and cap products*

$$\begin{array}{ccccccccc} \longrightarrow & K^i(U\cap V) & \longrightarrow & K^i(U)\oplus K^i(V) & \longrightarrow & K^i(X) & \longrightarrow \\ & \downarrow{\scriptstyle \alpha_{(U\cap V)}\cap\cdot} & & \downarrow{\scriptstyle (\alpha_U\cap\cdot)\oplus(\alpha_V\cap\cdot)} & & \downarrow{\scriptstyle \alpha\cap\cdot} & \\ \longrightarrow & RK_{i+j}(U\cap V) & \longrightarrow & RK_{i+j}(U)\oplus RK_{i+j}(V) & \longrightarrow & RK_{i+j}(X) & \longrightarrow \end{array}$$

commutes.

Proof We have that $C_0(X)$ is a sum of ideals $C_0(U)$ and $C_0(V)$ with intersection $C_0(U \cap V)$. On the other hand, the proof of Proposition 9.4.13 writes $RL^*_Q(X)$ as a direct sum of ideals that naturally identify with $RL^*_Q(U)$ and $RL^*_Q(V)$, and with intersection $RL^*_Q(U \cap V)$. The Mayer–Vietoris sequences above both arise from these decompositions into ideals and Proposition 2.7.15. Checking the definitions involved (for this it is convenient to work with modules of the form $\ell^2(Z, H)$ as in the proof of Proposition 9.5.4), one sees that the various maps induced between these algebras by the $*$-homomorphisms appearing in Lemma 9.5.1 are compatible; we leave the details to the reader. □

Our last goal in this section is to give a geometric interpretation of the cap product in the special case that we are working with differential operators and smooth vector bundles on a closed manifold. It says essentially that if $[p] \in K^0(M)$ and $[D] \in K_i(M)$ are classes represented by a smooth projection and an elliptic operator respectively, then the class $[D] \cap [p]$ is represented by a new operator which is essentially 'D with coefficients in the vector bundle underlying p'.

Lemma 9.5.7 *Let M be a closed smooth manifold. Let D be an elliptic operator acting on a smooth vector bundle S over M, and let $p \in M_n(C(M))$ be a smooth projection with corresponding smooth vector bundle E over M. Let $S \otimes \mathbb{C}^n$ be the tensor product of S and the trivial rank n vector bundle, and let $D \otimes id_{\mathbb{C}^n}$ and $id_S \otimes p$ both act on $L^2(M; S \otimes \mathbb{C}^n)$ in the natural way. Then the operator*

$$pDp := (id_S \otimes p)(D \otimes 1_n)(id_S \otimes p) \tag{9.5}$$

thought of as acting on $L^2(M; S \otimes E)$ with domain the smooth sections of $S \otimes E$ is an elliptic differential operator (see Definition 8.3.1), so defines a class $[pDp]$ in $K_*(M)$ (which equals $RK_*(M)$ as M is closed). Moreover, we have that

$$[D] \cap [p] = [pDp].$$

The notation is line (9.5) is potentially a little misleading: this operator is not the same thing as $D \otimes p$ acting on $L^2(M; S) \otimes L^2(M; \mathbb{C}^n)$! Indeed, $L^2(M; S \otimes \mathbb{C}^n)$ is not the same thing as $L^2(M; S) \otimes L^2(M; \mathbb{C}^n)$.

Proof If D is given in local coordinates on a section u of S by

$$(Du)(x) = \sum_{i=1}^{d} a_i(x) \frac{\partial u}{\partial x_i}(x) + b(x)u(x)$$

as in Definition 8.1.1, then one computes that pDp is given in local coordinates on a section u of $S \otimes \mathbb{C}^n$ by

$$((pDp)u)(x) = \sum_{i=1}^{d} a_i(x) \otimes p(x) \left(\frac{\partial u}{\partial x_i}(x) + \mathrm{id}_S \otimes \frac{\partial p}{\partial x_i}(x)u(x) \right)$$
$$+ (b(x) \otimes p(x))u(x).$$

It follows from this that the symbol of pDp in the sense of Definition 8.1.3 is given in local coordinates by

$$\sigma_{pDp}(x,\xi) = \sum_{i=1}^{d} \xi_i a_i(x) \otimes p(x),$$

or in other words,

$$\sigma_{pDp}(x,\xi) = \sigma_D(x,\xi) \otimes \mathrm{id}_E$$

as an endomorphism of $S \otimes E$. It follows from this that ellipticity of D in the sense of Definition 8.3.1 implies ellipticity of pDp. We thus get a class $[pDp] \in K_*(M)$ by either of Constructions 8.3.11 or 8.3.12 depending on whether or not D is assumed odd for some grading (note that if D is odd for some grading on S, then pDp is also odd for the induced grading on $S \otimes E$).

Now, the class $[pDp]$ is by definition the class associated to the $*$-homomorphism

$$\mathscr{S} \to L^*(L^2(M; S \otimes E)), \quad f \mapsto f(t^{-1}pDp)$$

(either graded or ignoring gradings, depending on whether D was odd for some grading to begin with). Let $H = H_0 \oplus H_1$ be an auxiliary graded Hilbert space with H_0 and H_1 separable and infinite-dimensional. Include E in $M \times \mathbb{C}^n$ as the image of p, and include $\mathbb{C}^n$ into H_0. We thus get an inclusion

$$L^2(M; S \otimes E) \subseteq L^2(M; S) \otimes H.$$

This inclusion is an isometry covering the identity map, so the constant family consisting of just this isometry gives rise to an inclusion

$$L^*(L^2(M; S \otimes E)) \to L^*(L^2(M; S) \otimes H),$$

which induces an isomorphism on K-theory by Lemma 6.2.7. In summary, of we write $L^*(M) := L^*(L^2(M; S) \otimes H)$, then we have that the class $[pDp]$ is represented by the $*$-homomorphism

$$\mathscr{S} \to L^*(M), \quad f \mapsto f(t^{-1}pDp).$$

Moreover, Lemma 6.4.11 says that the canonical quotient map $L^*(M) \to L^*_Q(M)$ induces an isomorphism on K-theory, so we equally well replace the codomain above with $L^*_Q(M)$.

Let us now compute a representative for the class of $[D] \cap [p]$. Write $\mathcal{K}$ for the compact operators on H_0, and grade $M_2(C(M) \otimes \mathcal{K})$ by the unitary multiplier $\left(\begin{smallmatrix} 1 & 0 \\ 0 & -1 \end{smallmatrix}\right)$. Include $\mathbb{C}^n$ into H, and use this to identify $M_n(C(X))$ with a subalgebra of $C(M) \otimes \mathcal{K}$. Using Lemma 2.9.14, the class $[p] \in K^0(M)$ is represented as an element of spectral K-theory by the graded $*$-homomorphism

$$\mathscr{S} \to M_2(C(M) \otimes \mathcal{K}), \quad f \mapsto \begin{pmatrix} f(0)p & 0 \\ 0 & 0 \end{pmatrix}.$$

From now on, we will just write $f(0)p$ for the element $\left(\begin{smallmatrix} f(0)p & 0 \\ 0 & 0 \end{smallmatrix}\right)$ of $M_2(C(M) \otimes \mathcal{K})$; this should not cause any confusion. We claim that the cap product $[D] \cap [p]$ is represented by the class of the $*$-homomorphism

$$\mathscr{S} \to L^*(L^2(M; S) \otimes H), \quad f \mapsto pf(t^{-1}D)p.$$

Write $L^*(M) := L^*(L^2(M; S) \otimes H)$. Computing on generators e^{-x^2} and xe^{-x^2}, one checks that the product map

$$\mathscr{S} \to \mathscr{S} \widehat{\otimes} \mathscr{S} \to L^*_Q(M) \otimes M_2(C(M) \otimes \mathcal{K})$$

takes a function f to

$$f(t^{-1}D) \otimes p. \tag{9.6}$$

Indeed, checking for xe^{-x^2}, we see that

$$\begin{aligned} xe^{-x^2} &\overset{\Delta}{\mapsto} xe^{-x^2} \widehat{\otimes} e^{-x^2} + e^{-x^2} \widehat{\otimes} xe^{-x^2} \\ &\mapsto t^{-1}De^{-t^{-2}D^2} \otimes e^{-0^2} p + e^{-t^{-2}D^2} \otimes 0 \\ &= t^{-1}De^{-t^{-2}D^2} \otimes p, \end{aligned}$$

and the case of e^{-x^2} is similar (and simpler). As $p^2 = p$, and as we have

$$pf(t^{-1}D) = f(t^{-1}D)p$$

in $L^*_Q(M)$ by Lemma 6.4.18, we may conclude that the class $[D] \cap [p]$ is represented by the $*$-homomorphism

$$\mathscr{S} \to L_Q^*(M), \quad f \mapsto pf(t^{-1}D)p$$

as claimed.

Now, summarising the discussions in the last two paragraphs, we have that $[pDp]$ and $[D] \cap [p]$ are represented by the $*$-homomorphisms

$$\mathscr{S} \to L_Q^*(M), \quad f \mapsto f(t^{-1}pDp) \quad \text{and} \quad f \mapsto pf(t^{-1}D)p$$

respectively. We want to show that these represent the same class in K-theory; we will actually show that they are the same $*$-homomorphism. As the functions $x \mapsto (x \pm i)^{-1}$ generate $\mathscr{S}$ as a C^*-algebra, it will suffice to show that the elements

$$(t^{-1}pDp \pm ip)^{-1} \quad \text{and} \quad p(t^{-1}D \pm i)^{-1}p$$

of $L_Q^*(M)$ are the same (here and throughout we will identify D and p with the operators $D \otimes \mathrm{id}_{\mathbb{C}^n}$ and $\mathrm{id}_S \otimes p$ from the statement; moreover to make sense of $(t^{-1}pDp \pm ip)^{-1}$ we are taking the inverse as an operator on $L^2(M; S \otimes E)$, then including this as a subspace of $L^2(M; S) \otimes H$). We will just do this computation for the case of $(x+i)^{-1}$; the case of $(x-i)^{-1}$ is similar.

Let us then look at the difference

$$(t^{-1}pDp + ip)^{-1} - p(t^{-1}D + i)^{-1}p;$$

we claim that this is zero in $L_Q^*(M)$. Now, using the so-called resolvent identity $a^{-1} - b^{-1} = a^{-1}(b-a)b^{-1}$, the above equals

$$(t^{-1}pDp + ip)^{-1}\big((t^{-1}D + i) - (t^{-1}pDp + ip)\big)(t^{-1}D + i)^{-1}p.$$

As $(t^{-1}pDp + ip)^{-1} = (t^{-1}pDp + ip)^{-1}p$, this equals

$$\begin{aligned}&(t^{-1}pDp + ip)^{-1}p\big(t^{-1}D + i - t^{-1}pDp - ip\big)(t^{-1}D + i)^{-1}p \\ &= (t^{-1}pDp + ip)^{-1}p[p, t^{-1}D](t^{-1}D + i)^{-1}p.\end{aligned}$$

Using Lemma 6.4.18 (and the fact that the commutator $[p, t^{-1}D]$ is a bounded operator), we have that $(t^{-1}D + i)^{-1}p = p(t^{-1}D + i)^{-1}p$ in $L_Q^*(M)$, and so the above equals

$$(t^{-1}pDp + ip)^{-1}\big(p[p, t^{-1}D]p\big)(t^{-1}D + i)^{-1}p.$$

However, $p[p, t^{-1}D]p = pt^{-1}Dp - pt^{-1}Dp$ is zero, so we are done. □

9.6 The Dirac Operator on a Spin^c Manifold and Poincaré Duality

Our goal in this section is to formulate and prove a version of Poincaré duality for K-theory and deduce some consequences. This is the main theorem of this chapter.

Recall that the usual version of Poincaré duality for a (boundaryless, but not necessarily compact) oriented d-manifold M states that there is a fundamental class $[M]$ in the locally finite[5] homology $H_d^{lf}(M)$ such that taking the cap product with $[M]$ induces an isomorphism

$$H^*(M) \to H_*(M), \quad \alpha \mapsto [M] \cap \alpha.$$

Our goal is to get a similar isomorphism $K^*(M) \to RK_*(M)$ using the cap product with an appropriate fundamental class. Following a standard classical proof, our goal will be to prove this first for $\mathbb{R}^d$ and then deduce the general case by a Mayer–Vietoris argument.

Now, the Bott periodicity theorem (Theorem 9.3.5) suggests that the Dirac operator on $\mathbb{R}^d$ is the right thing to use for a a fundamental class in that case, so we need some sort of global analogue of this on a general manifold. However, a general manifold will not admit an appropriate globally defined Dirac operator. Indeed, by analogy with the classical case one should expect the need for some sort of orientation condition. The correct orientation condition to use is the *spin^c* condition, and we start by describing this.

First we give an auxiliary definition that makes sense for any Riemannian manifold.

Definition 9.6.1 Let M be a Riemannian manifold with tangent bundle TM. As the Clifford algebra construction is a continuous[6] functor, there is an associated (smooth) *Clifford bundle* $\mathrm{Cliff}_{\mathbb{C}}(TM)$ of Clifford algebras over TM, where each fibre $\mathrm{Cliff}_{\mathbb{C}}(TM)_x$ for $x \in M$ identifies with the Clifford algebra $\mathrm{Cliff}_{\mathbb{C}}(T_xM)$ of the tangent space T_xM at x.

For the next definition, let S be a (smooth) Hermitian bundle over M, and assume S is also equipped with a grading i.e. a bundle isomorphism (so in particular, a homeomorphism) $U : S \to S$ such that the induced map $U_x : S_x \to S_x$ on each fibre is a self-adjoint unitary (compare Definition E.1.4). Say A is a bundle of $*$-algebras over M, graded by a bundle isomorphism $\epsilon : A \to A$, meaning that the restriction $\epsilon_x : A_x \to A_x$ of ϵ to each fibre is

[5] Also called Borel–Moore.

[6] i.e. takes continuous morphisms of finite-dimensional vector spaces to continuous morphisms of finite-dimensional algebras; see Section 9.8, Notes and References, at the end of the chapter.

an order two $*$-isomorphism. Then a *left action* of A on S consists of a bundle map $m\colon A \times S \to S$ such that the induced map

$$m_x\colon A_x \times S_x \to S_x, \quad (a,s) \mapsto ax$$

on each fibre satisfies the usual associativity and bilinearity rules for module multiplication fibrewise as well as compatibility with the adjoint, meaning

$$\langle as_1, s_2 \rangle_x = \langle s_1, a^* s_2 \rangle_x,$$

and compatibility with the grading, meaning that

$$\epsilon_x(a)s = U_x a U_x^* s.$$

A right action of A on S is defined similarly.

Definition 9.6.2 Let M be a Riemannian manifold of dimension d with associated Clifford bundle $\mathrm{Cliff}_{\mathbb{C}}(TM)$ as above. Then a *spin^c structure* on M consists of a graded, complex, Hermitian bundle S over M equipped with:

(i) a right action of $\mathrm{Cliff}_{\mathbb{C}}(TM)$, and
(ii) a left action of the trivial bundle of graded $*$-algebras $M \times \mathrm{Cliff}_{\mathbb{C}}(\mathbb{R}^d)$

such that for each $x \in X$ there is an open set and a section $s\colon U \to S|_U$ such that for each $y \in U$ the left action

$$\mathrm{Cliff}_{\mathbb{C}}(\mathbb{R}^d) \to S_y, \quad c \mapsto cs(y)$$

defines an isomorphism of graded Hermitian spaces.

Remark 9.6.3 If we required that both the left and right actions on S were by $\mathrm{Cliff}_{\mathbb{C}}(TM)$, then the existence of S satisfying these conditions would be trivial: just take $S = \mathrm{Cliff}_{\mathbb{C}}(TM)$, acting on itself by left and right multiplication. Similarly, the existence of a bundle as above with both left and right actions of the trivial bundle $M \times \mathrm{Cliff}_{\mathbb{C}}(\mathbb{R}^d)$ is trivial: take $S = M \times \mathrm{Cliff}_{\mathbb{C}}(\mathbb{R}^d)$. However, we need to require different actions; this has to do with the need for the Bott element to be globally defined, while the Dirac operator will be local.

The existence of such an S turns out to be a non-trivial condition. Indeed, it is governed by the first two *Steifel–Whitney classes* $w_1(M) \in H^1(M;\mathbb{Z}/2\mathbb{Z})$ and $w_2(M) \in H^2(M;\mathbb{Z}/2\mathbb{Z})$. Existence of such an S is equivalent to both of the following holding:

(i) $w_1(M)$ must vanish;
(ii) $w_2(M)$ must be in the image of the canonical map

$$H^2(M;\mathbb{Z}) \to H^2(M;\mathbb{Z}/2\mathbb{Z})$$

induced by the canonical quotient $\mathbb{Z} \to \mathbb{Z}/2\mathbb{Z}$.

Now, vanishing of $w_1(M)$ is equivalent to orientability of M. The condition on $w_2(M)$ can be seen as 'just a little more than orientability'. It is satisfied, for example, by all orientable surfaces, all symplectic manifolds, all orientable manifolds with $H^2(M;\mathbb{Z}/2\mathbb{Z}) = 0$ and many others; indeed, it is quite difficult to find an example of an orientable manifold that does not also admit a spin^c structure (such manifolds do exist, however: see Section 9.8, Notes and References, at the end of the chapter).

Remark 9.6.4 Associated to a real vector space V there is also a real Clifford algebra Cliff(V) defined precisely analogously to $\mathrm{Cliff}_{\mathbb{C}}(V)$ but as a real, rather than complex, algebra. One can then ask for a real bundle S satisfying the 'real analogues' of the conditions in Definition 9.6.2; we will call such a bundle a *spin structure* on M. If M has a spin structure, then there is a canonically associated spin^c structure: just tensor everything by $\mathbb{C}$. There is no way to go from a spin^c structure to a spin structure in general: in fact, a manifold admits a spin structure if and only if $w_1(M) = w_2(M) = 0$, and this is a strictly stronger condition that M being spin^c. For example, the complex projective plane $\mathbb{C}P^2$ is a spin^c, but not spin, four-manifold.

Definition 9.6.5 Let S be a spin^c structure over M, let $C^\infty(M;S)$ denote the smooth sections of S, and $C^\infty(M;TM)$ denote the smooth sections of the tangent bundle. A *Dirac connection* on M is a linear map

$$\nabla\colon C^\infty(M;TM) \odot C^\infty(M;S) \to C^\infty(M;S), \quad (X,s) \mapsto \nabla_X s$$

that is a connection in the usual sense[7] satisfying in addition:

(i) for all vector fields X on M and all sections s,t of S

$$X\langle s,t\rangle = \langle \nabla_X s, t\rangle(x) + \langle s, \nabla_X t\rangle;$$

(ii) for all vector fields X, Y and sections s of S,

$$\nabla_X(s \cdot Y) = (\nabla_X s)\cdot Y + x \cdot (\nabla_X^{LC} Y),$$

where '$\cdot$' denotes multiplication in the Clifford algebra and ∇^{LC} is the Levi-Civita connection associated to the Riemannian metric.

[7] This means that for all $f, g \in C^\infty(M)$, $s \in C^\infty(M;S)$ and $X, Y \in C^\infty(M;TM)$, we have the rules $\nabla_{(fX+gY)}s = f\nabla_X s + g\nabla_X s$ and $\nabla_X(fu) = f\nabla_X u + X(f)u$.

If ∇ is a Dirac connection on S, then the associated *Dirac operator* is the composition

$$C^\infty(M;S) \xrightarrow{\nabla} C^\infty(M;S) \otimes C^\infty(M;\Omega M) \xrightarrow{\langle,\rangle} C^\infty(M;S) \otimes C^\infty(M;TM),$$
$$\downarrow c$$
$$C^\infty(M;S)$$

where: the first map is the connection considered (via the duality between vectors and covectors) as a map from sections of S to sections of S tensored by 1-forms; the second map is the identity on sections of S tensored by the isomorphism between vector fields and 1-forms induced by the metric; and the third map is Clifford multiplication.

Example 9.6.6 Let $M = \mathbb{R}^d$ with its usual metric. As $\mathbb{R}^d$ has trivial(isable) tangent bundle, we may identify the Clifford bundle $\text{Cliff}_\mathbb{C}(T\mathbb{R}^d)$ with $\mathbb{R}^d \times \text{Cliff}_\mathbb{C}(\mathbb{R}^d)$. We may thus take a spin^c structure to be equal as a graded bundle to the trivial bundle $\mathbb{R}^d \times \text{Cliff}_\mathbb{C}(\mathbb{R}^d)$, with the left action of $\mathbb{R}^d \times \text{Cliff}_\mathbb{C}(\mathbb{R}^d)$ defined by left multiplication, and the right action of $\text{Cliff}_\mathbb{C}(T\mathbb{R}^d) \cong \mathbb{R}^d \times \text{Cliff}_\mathbb{C}(\mathbb{R}^d)$ defined by right multiplication, but 'twisted' by the grading: precisely, for homogeneous elements $s \in S$ and $v \in \mathbb{R}^d \times \text{Cliff}_\mathbb{C}(\mathbb{R}^d)$ we define

$$s \cdot v := (-1)^{\partial s \partial v} sv.$$

A vector field on $\mathbb{R}^d$ is then given by an expression

$$X = \sum_{i=1}^{d} f_i \frac{\partial}{\partial x_i}$$

for some smooth functions $f_i : \mathbb{R}^d \to \mathbb{R}$, and we define a connection by

$$\nabla_X s := \sum_{i=1}^{d} f_i \frac{\partial s}{\partial x_i}.$$

One can then (and the reader should!) check that this defines a Dirac connection, and that the associated Dirac operator is the same as the one defined directly in Definition 9.3.1.

Several aspects of the following definitions should be viewed as exercises: the reader should check everything can be made sense of. As a preliminary remark that will be helpful in making sense of some of what follows, note that for finite-dimensional real vector spaces V and W, $\text{Cliff}_\mathbb{C}(V \oplus W)$ is canonically isomorphic to $\text{Cliff}_\mathbb{C}(V) \widehat{\otimes} \text{Cliff}_\mathbb{C}(W)$: see Exercise E.3.4. Moreover, for manifolds M and N, using that $T(M \times N)$ canonically identifies with the exterior direct sum bundle $TM \oplus TN$ whose fibre at $(x, y) \in M \times N$ is $T_x M \oplus T_y N$, this induces a canonical isomorphism of Clifford bundles

$$\mathrm{Cliff}_{\mathbb{C}}(TM) \widehat{\otimes} \mathrm{Cliff}_{\mathbb{C}}(TN) \cong \mathrm{Cliff}_{\mathbb{C}}(T(M \times N));$$

the tensor product here is the exterior graded tensor product of graded algebra bundles, so the fibre at a point $(x, y) \in M \times N$ of $\mathrm{Cliff}_{\mathbb{C}}(TM) \widehat{\otimes} \mathrm{Cliff}_{\mathbb{C}}(TN)$ is by definition equal to the graded tensor product of fibers

$$\mathrm{Cliff}_{\mathbb{C}}(TM)_x \widehat{\otimes} \mathrm{Cliff}_{\mathbb{C}}(TN)_y.$$

Definition 9.6.7 Let M and N be Riemannian manifolds of dimensions m and n, and with spin^c structures S_M and S_N respectively. Then the *product spin^c structure* is given by the exterior tensor product bundle $S_M \otimes S_N$ over $M \times N$ with the tensor product grading, and tensor product left and right actions of $\mathrm{Cliff}_{\mathbb{C}}(\mathbb{R}^{m+n}) \cong \mathrm{Cliff}_{\mathbb{C}}(\mathbb{R}^m) \widehat{\otimes} \mathrm{Cliff}_{\mathbb{C}}(\mathbb{R}^n)$ and $\mathrm{Cliff}_{\mathbb{C}}(T(M \times N)) \cong \mathrm{Cliff}_{\mathbb{C}}(TM) \widehat{\otimes} \mathrm{Cliff}_{\mathbb{C}}(TN)$ respectively.

Let D_M and D_N be spin^c Dirac operators on M and N respectively, built using the connections ∇_M and ∇_N respectively. Then we set $\nabla := \nabla_M \widehat{\otimes} 1 + 1 \widehat{\otimes} \nabla_N$, which is a Dirac connection on the product spin^c bundle. The associated Dirac operator with respect to the connection ∇_M then satisfies

$$D_{M \times N} = D_M \widehat{\otimes} 1 + 1 \widehat{\otimes} D_N.$$

Lemma 9.6.8 *Let M and N be spin^c Riemannian manifolds with associated Dirac classes $[D_M] \in K_*(M)$ and $[D_N] \in K_*(N)$, and also $[D_{N \times M}] \in K_*(N \times M)$ for the associated product spin^c structure. Then*

$$[D_{N \times M}] = [D_N] \times [D_M],$$

where the right-hand side uses the external product in K-homology of Definition 9.2.6.

Proof This is a direct check using Definition 9.6.7 and Lemma 9.3.3. □

Lemma 9.6.9 *Let M be a spin^c Riemannian manifold, let $i: U \to M$ be the inclusion of an open set, and let D be the spin^c Dirac operator on M. Then $i^*[D_M] = [D_U]$.*

Proof The map i^* is induced functorially by the 'collapse' map $c: M^+ \to U^+$ that is the identity on U and sends $M^+ \setminus U$ to the point at infinity in U^+. Let S be the spin^c bundle over M, and let H be an auxiliary separable infinite-dimensional Hilbert space. Let $H_{M^+} := L^2(M; S) \oplus H$, equipped with the $C(M^+)$ action defined on $(u, v) \in L^2(M; S) \oplus H$ by

$$f \cdot (u, v) := (f|_M u, f(\infty) v).$$

Then H_{M^+} is an ample M^+ module (other than in the trivial case where M is zero-dimensional, which we can safely leave to the reader to treat in an ad hoc way of her choosing). Let H_{U^+} be the same Hilbert space as H_{M^+}, but equipped with the action of $C(U^+)$ defined by

$$f \cdot u := fu + \chi_{M^+ \setminus U} f(\infty)u;$$

this gives a well-defined representation of $C(U^+)$, which is ample (other than in the trivial case $U = M$, which again we can safely leave to the reader). The identity map on H_{M^+} considered as a unitary isomorphism $V := H_{M^+} \to H_{U^+}$ is then such that the constant family (V_t) with $V_t = V$ for all $t \in [1, \infty)$ is a continuous cover (see Definition 6.2.5) for $c\colon M^+ \to U^+$.

Now, we can realise the classes $[D_M]$ and $[D_U]$ in natural ways on these modules. Let (g_t^M) and (g_t^U) be as in the definition of multiplier data (Definition 8.2.9) for D_M and D_U. Let $f \in \mathscr{S}$ and let $F_t^M := f^M(t^{-1} g_t D g_t)$ and $F_t^U := f^U(t^{-1} g_t D g_t)$ define the associated multipliers (F_t^M) and (F_t^U) of $L^*(L^2(M; S))$ and $L^*(L^2(U; S))$ respectively (compare Theorem 8.2.10). As in Proposition 6.4.7, the inclusions

$$L^*(L^2(M; S)) \to L^*(H_M^+; \infty) \quad \text{and} \quad L^*(L^2(U; S)) \to L^*(H_U^+; \infty)$$

induce isomorphisms on K-theory. Thanks to the definition of functoriality of K-homology (Definition 6.4.5) and the construction of the classes $[D_M]$ and $[D_U]$ (Constructions 8.3.11 and 8.3.12), it will thus suffice to show that $(V F_t^M V^*)$ and (F_t^U) differ by an element of $L_0^*(H_U^+; \infty)$.

This now follows from essentially the same arguments used to show that the classes involved do not depend on the choice of the associated multiplier data as in Proposition 8.3.13; we leave the details to the reader. □

Lemma 9.6.10 *Let M be a spin^c manifold, let g_0 and g_1 be two Riemannian metrics on M and let D_0 and D_1 be the associated spin^c Dirac operators. Then the classes $[D_0]$ and $[D_1]$ in $K_*(M)$ are the same.*

Proof Consider the product manifold $M \times \mathbb{R}$ equipped with a metric that agrees with $g_0 + dt^2$ on $(-\infty, 0) \times M$ and with $g_1 + dt^2$ on $(1, \infty)$ (and interpolates, say linearly, between these metrics on the remaining part). Let $[D_{M \times \mathbb{R}}] \in K_i(M \times \mathbb{R})$ be the class of the Dirac operator for the product of the spin structures. Let $U_0 = M \times (-1, 0)$ and $U_1 = M \times (1, 2)$. Using Lemma 9.6.9, the images of $[D]$ under the restriction maps $K_*(M \times \mathbb{R}) \to K_*(U_i)$ agree with the classes in $K_*(U_i)$ of the associated Dirac operators, or in other words, thanks to Lemma 9.6.8, with $[D_0] \times [D_{(-1,0)}]$ and $[D_1] \times [D_{(1,2)}]$ where the Dirac operators on the open intervals are defined using the spin^c structure on $\mathbb{R}$ from Example 9.6.6. Using Lemma 9.6.9 again, however, these classes are the restrictions of the Dirac operators $[D_i] \times [D_{\mathbb{R}}]$ on $M \times \mathbb{R}$, where these classes are defined using the product metrics $g_0 + dt^2$ and $g_1 \times dt^2$ respectively. As the restriction maps are all proper homotopy equivalences, at this point we have that

$$[D_{M\times\mathbb{R}}] = [D_1] \times [D_{\mathbb{R}}] = [D_0] \times [D_{\mathbb{R}}].$$

Hence to complete the proof, it suffices to prove that the map

$$K_*(M) \to K_*(M \times \mathbb{R}), \quad \alpha \mapsto \alpha \times [D_{\mathbb{R}}]$$

is injective. This follows as the partial pairing (Definition 9.2.2) with the Bott class $[C] \in K^1(\mathbb{R})$ defines an inverse using Bott periodicity (Theorem 9.3.5) and Lemma 9.2.7. □

Here is the Poincaré duality theorem.

Theorem 9.6.11 *Let D be the Dirac operator on a spin^c manifold M. Then cap product with D induces an isomorphism*

$$K^*(M) \to RK_*(M), \quad \alpha \mapsto [D] \cap \alpha.$$

Proof We will proceed in stages. First assume that M is a single open ball, diffeomorphic to Euclidean space. Thanks to Lemma 9.6.10, we may assume that the metric on this ball is just the usual Euclidean metric. Thanks to Remark 9.5.3, the result in this case is Bott periodicity as in Theorem 9.3.5.

Now assume M is a countable disjoint union of open balls $M = \bigsqcup_{n\in\mathbb{N}} U_n$, where each U_n is diffeomorphic to Euclidean space. Note that both K-theory and representable K-homology are covariantly functorial under inclusions of open sets, and that this is compatible with the cap products: see Corollary 9.5.5. Hence if

$$i_n : U_n \to M$$

denotes the inclusion, then using Lemma 9.6.9 for each n we have a commutative diagram

$$\begin{array}{ccc} K^*(M) & \xrightarrow{[D_M]\cap} & RK_*(M). \\ \uparrow{\scriptstyle (i_n)_*} & & \uparrow{\scriptstyle (i_n)_*} \\ K^*(U_n) & \xrightarrow{[D_{U_n}]\cap} & RK_*(U_n) \end{array}$$

Moreover, both K-theory and representable K-homology are additive for disjoint unions (see Corollary 9.4.12 for representable K-homology), so we get a commutative diagram

$$\begin{array}{ccc} K^*(M) & \xrightarrow{[D_M]\cap} & RK_*(M). \\ \Big\uparrow {\scriptstyle \oplus (i_n)_*} \, \cong & & \Big\uparrow {\scriptstyle \oplus (i_n)_*} \, \cong \\ \oplus_n K^*(U_n) & \xrightarrow{\oplus [D_{U_n}]\cap} & \oplus_n RK_*(U_n) \end{array}$$

Each map $[D_{U_n}]\cap$ is an isomorphism as already noted by Bott periodicity, so we are done in this case too.

Finally, we consider the general case. Note that if M can be written as a union $M = U \cup V$ of two open subsets, then there are Mayer–Vietoris sequences in both K-theory and representable K-homology that are compatible with cap products: see Lemma 9.5.6. Combining this with Lemma 9.6.9 then gives a commutative diagram

$$\begin{array}{ccccccc} \longrightarrow & K^*(U \cap V) & \longrightarrow & K^*(U) \oplus K^*(V) & \longrightarrow & K^*(M) & \longrightarrow \\ & \Big\downarrow {\scriptstyle [D_{U\cap V}]\cap \cdot} & & \Big\downarrow {\scriptstyle ([D_U]\cap \cdot) \oplus ([D_V]\cap \cdot)} & & \Big\downarrow {\scriptstyle [D_M]\cap \cdot} & \\ \longrightarrow & RK_*(U \cap V) & \longrightarrow & RK_*(U) \oplus K^*(V) & \longrightarrow & RK_*(M) & \longrightarrow \end{array} .$$

On the other hand, using the finite covering dimension of M, M can be covered by finitely many open sets $U^1, \ldots, U^N$ such that each U^i is a disjoint union $U^i = \bigsqcup_n U^i_n$, where each U^i_n is an open set diffeomorphic to a Euclidean ball, and moreover, so that any finite intersection of the open sets U^i_n is either diffeomorphic to a Euclidean ball, or empty: see Exercise 9.7.7. The result follows from the previous special case and induction. □

The following corollaries are needed for some of the applications in Chapter 10.

Corollary 9.6.12 *Let M be a contractible manifold. Then*

$$RK_*(M) \cong K^*(M) \cong K_*(M) \cong \mathbb{Z}.$$

Moreover, $K_*(M)$ is generated by the class of the Dirac operator associated to any spin^c structure[8] *and Riemannian metric on M.*

Proof That $RK_*(M)$ is isomorphic to $\mathbb{Z}$ follows from contractibility and the fact that representable K-homology is homotopy invariant (Corollary 9.4.11). The case of $K^*(M)$ then follows from Poincaré Duality (Theorem 9.6.11). The case of $K_*(M)$ follows from this and the universal coefficient theorem (Theorem 9.2.8).

[8] As M is contractible, it has trivialisable tangent bundle, so it is spin^c.

To see that the Dirac operator generates $K_*(M)$ note that Theorem 9.6.11 implies that the cap product induces a homomorphism

$$K_*(M) \to \mathrm{Hom}(K^*(M), RK_*(M))$$

sending the class of the Dirac operator to an isomorphism; as all the groups involved are $\mathbb{Z}$, this is impossible unless the Dirac operator generates $K_*(M)$. □

Corollary 9.6.13 *Let M be a closed even-dimensional spin^c Riemannian manifold. Then $K_0(M)$ is generated as an abelian group by the index classes of odd, self-adjoint, contractive multipliers (F_t) such that prop$(F_t) \to 0$ as $t \to \infty$, and such that $1 - F_t^2$ is trace class for all t.*

Proof Note first that as M is compact, $RK_0(M) = K_0(M)$, so we can just work with K_0 throughout. Theorem 9.6.11 gives that the cup product with the Dirac class $[D]$ is an isomorphism $K^*(M) \to K_*(M)$, so it suffices to show that every class of the form $[D] \cap \alpha$, $\alpha \in K^*(M)$ is represented in the claimed form. As $C(M)$ is unital, we may assume that $\alpha = [p]$ is the class of some single projection $p \in M_n(C(M))$ for some n. Moreover, by Exercise 2.11.13, we may assume that $p \in M_n(C(M))$ is smooth.

Lemma 9.5.7 gives that $[D] \cap [p]$ is the same as the class $[pDp]$ defined there. The class $[pDp]$ can be represented by an index class with the claimed properties by Corollary 8.4.4, so we are done. □

9.7 Exercises

9.7.1 Show that the basic pairing from Definition 9.1.3 above defines a special case of the general pairing $K_*(X) \otimes K^*(X) \to \mathbb{Z}$ from Definition 9.2.3.
Hint: the isomorphism between the usual and spectral pictures of K-theory given in Lemma 2.9.14 should help.

9.7.2 Find an analogous formula to the one in Remark 9.1.4 that works for general classes $\alpha \in K_0(X)$ and $\beta \in K^0(X)$ represented by formal difference of projections.

9.7.3 Instead of the proof given in the text, derive Proposition 9.4.13 from Proposition 6.3.4 and Corollary 9.4.10. Use your proof to generalise the results to the case where U and V are any locally compact subsets of X.

9.7.4 Check the details of the discussion in Remark 9.5.3.

9.7.5 Formulate and prove functoriality results for the partial pairing and external product (Definitions 9.2.2 and 9.2.6) as proved for the cap product and pairing in Proposition 9.5.4.

9.7.6 Formulate and prove a version of Lemma 9.5.7 for non-compact manifolds.

9.7.7 For a Riemannian d-manifold M, recall that any point $x \in M$ there is $r > 0$ such that the ball around x of radius r_x is *geodesically convex*: there is a unique geodesic between any two points in the ball. Choose such an r_x for each $x \in M$, and let $\mathcal{U} = (B(x; r_x))_{x \in M}$ be the corresponding open cover. On the other hand, as M has covering dimension equal to d, there is a subcover $\mathcal{V}$ of $\mathcal{U}$ such that any intersection of $d + 2$ distinct elements of $\mathcal{V}$ is empty. Use these facts to show that a cover of M with the properties used at the end of the proof of Theorem 9.6.11 exists.
Hint: recursively define U^i to be the union of a maximal disjoint collection of elements of $\mathcal{V}$ such that no element used already in $U^1, \ldots, U^{i-1}$ appears.

9.7.8 Define equivariant representable K-homology, and formulate and prove analogues of the results of Section 6.5 for your theory.

9.8 Notes and References

The idea of combining localisation algebras with the spectral picture of K-theory first appears in Zeidler [275]. It is particularly well suited to the discussion of products between K-theory and K-homology groups.

The approach to Bott periodicity that we give here is based on section 1.13 of Guentner and Higson [128]; although our formalism is somewhat different, the underlying details of the computation are more-or-less the same as in this reference. A similar approach (again set in quite a different framework) that makes the connections to quantum mechanics and the idea of passage to the classical limit explicit, can be found in Elliott, Natsume and Nest [88].

The general version of the universal coefficient theorem is due to Rosenberg and Schochet [226]. The version we stated is due to Brown [41], where it first appears in the context of Brown–Douglas–Fillmore theory; a textbook exposition of the theorem that adapts directly to our setting can be found in section 7.6 of Higson and Roe [135].

Representable K-homology was introduced by Kasparov (see [149] for the definitive version). The various products we have discussed here are all special cases of the so-called *Kasparov product* in KK-theory from that paper.

Another exposition of KK-theory (although not touching on representable K-theory) can be found in the later sections of Blackadar [33].

In Definition 9.6.1, we cited without proof the fact that a continuous functor on vector spaces gives rise to an associated functor on vector bundles in order to construct the Clifford bundle. A proof of this, and justification of the other operations that we need to perform on vector bundles, can be found in section 1.2 of Atiyah [6]. A somewhat different approach to the Clifford bundle in terms of principal bundles can be found in definition II.3.4 of Lawson and Michelsohn [164].

The realisation that the notion of a spin^c structure provides the 'right' notion of orientation for (complex topological) K-theory, and the particular presentation we have given is essentially due to Baum; see for example Baum and Douglas [24]. The original connection of spin algebra to K-theory was made by Atiyah, Bott and Shapiro [9]. A discussion of spin^c structures and of the characterisation of spin^c manifolds in terms of Stiefel–Whitney classes from Remark 9.6.3 can be found in Appendix D of Lawson and Michelson [164]. This reference also gives various examples, including of orientable manifolds that are not spin^c: the 5-manifold $SU(3)/SO(3)$ is perhaps the simplest example. Another treatment of spin^c structures, closer in spirit to ours but giving more information, can be found in section 11.2 of Higson and Roe [135].

The notion of a spin structure that we briefly mentioned in Remark 9.6.4 is a huge topic. A wide-ranging discussion on spin structures and the associated differential operators, and geometry and topology, can be found in chapter II of Lawson and Michelsohn [164].

For a classical treatment of Poincaré duality on manifolds, similar in spirit to the one we give here, see section 5 of Bott and Tu [36]. A very general version of K-theory Poincaré duality expressed in terms of Kasparov's bivariant K-theory can be found in section 4 of Kasparov [149]. There are interesting analogues of Poincaré duality for general C^*-algebras: see section VI.4.β of Connes [60] for background, and Emerson [89] for an attractive example.

10

Applications to Algebra, Geometry and Topology

In this chapter, we look at some of the applications of the Baum–Connes conjectures: to the Kadison–Kaplansky conjecture in operator algebras; to the existence of positive scalar curvature metrics in differential geometry; and to the Novikov conjecture in manifold topology. These three topics are covered in Sections 10.1, 10.2 and 10.3 respectively. Section 10.1 on the Kadison–Kaplansky conjecture does not really use any material that we have not covered in this book: while we do not get to the most general possible results, we are still able to prove some interesting and non-trivial theorems and give an idea of what is involved in more general cases. Section 10.2 on positive scalar curvature is sketchier, as we have to use some facts from differential and spin geometry as a black box. Section 10.3 is sketchier still: the machinery needed to set up the operator algebraic approach to the Novikov conjecture in detail requires more manifold topology than we are prepared to assume in this text, so we just aim to provide a brief introduction to some of the ideas involved.

Section 10.5, Notes and References, is fairly long: as well as giving background for the techniques explicitly discussed in this chapter, we attempt to give a brief survey of the literature as related to assembly maps of the sort we study in this book, and their purely algebraic cousins.

10.1 The Kadison–Kaplansky Conjecture

In this section, we will look at an application of the Baum–Connes conjecture to purely (C^*-)algebraic questions. Here is the motivating problem, which is usually called the *Kadison–Kaplansky conjecture*.

Conjecture 10.1.1 *Let G be a torsion-free discrete group. The the reduced group C^*-algebra $C^*_\rho(G)$ contains no idempotents other than zero and one.*

There is also a closely related purely algebraic conjecture, called *Kaplansky's conjecture.*

Conjecture 10.1.2 *Say G is a torsion-free group, and K is a field. Then the group ring $K[G]$ contains no idempotents other than zero or one.*

Clearly the Kadison–Kaplansky conjecture for a group G implies that Kaplansky's conjecture holds for $K[G]$ where K is any subfield of $\mathbb{C}$. Moreover, this fact bootstraps up to show that the Kadison–Kaplansky conjecture implies Kaplansky's conjecture for $K[G]$ where K is any characteristic zero field: indeed, if $e \in K[G]$ is a non-trivial idempotent, then e lives in the subring $K'[G]$ where $K' \subseteq K$ is the extension of $\mathbb{Q}$ generated by the finitely many non-zero coefficients of e; however, any finitely generated extension of $\mathbb{Q}$ is isomorphic to a subfield of $\mathbb{C}$.

Remark 10.1.3 The assumption that G is torsion free is necessary for $K[G]$ to have no non-trivial (i.e. not zero or one) idempotents in the characteristic zero[1] case, and therefore also necessary for $C^*_\rho(G)$ to have no non-trivial idempotents. Indeed, if $g \in G$ has order $n > 1$, then

$$p = \frac{1}{n}\sum_{k=1}^{n} g^k$$

is a non-trivial idempotent in the group ring $\mathbb{Q}[G]$ with rational coefficients.

In this section, our goal is to show that the Baum–Connes conjecture for a torsion-free group G implies the Kadison–Kaplansky conjecture, at least in a special case that suggests how the general argument should go. To begin, we need some preliminaries about traces on group C^*-algebras and equivariant Roe algebras.

Definition 10.1.4 Let G be a countable discrete group. The *canonical trace* on the group C^*-algebra $C^*_\rho(G)$ is the positive linear functional

$$\tau\colon C^*_\rho(G) \to \mathbb{C}, \quad a \mapsto \langle \delta_e, a\delta_e \rangle$$

corresponding to the Dirac mass at the identity in $\ell^2(G)$.

For the next lemma, recall that a positive linear functional $\phi\colon A \to \mathbb{C}$ on a C^*-algebra is *faithful* if whenever $a \in A$ is positive and non-zero we must have $\phi(a) > 0$, and *tracial* if $\phi(ab) = \phi(ba)$ for all $a, b \in A$.

[1] Not in general: for example, if K is the field with two elements and G the group with two elements, then $K[G]$ contains no non-trivial idempotents as one can directly check.

Lemma 10.1.5 *The canonical trace on $C^*_\rho(G)$ is faithful and tracial.*

Proof To see that τ is tracial, let a,b be elements of $\mathbb{C}[G]$ and write $a = \sum_{g\in G} a_g g$, $b = \sum_{g\in G} b_g g$. Computing, we have

$$\tau(ab) = \langle \delta_e, ab\delta_e\rangle = \sum_{g,h\in G} a_h b_g \langle \delta_e, \delta_{h^{-1}g^{-1}}\rangle = \sum_{g\in G} a_g b_{g^{-1}}.$$

This formula is symmetric in a and b, so τ restricts to a trace on $\mathbb{C}[G]$, and is thus a trace on $C^*_\rho(G)$ by continuity.[2]

To see that τ is faithful, let $a \in C^*_\rho(G)$ be positive. Then with λ denoting the left regular representation (Example C.1.3) we have that λ_g commutes with a for any $g \in G$ and thus

$$\tau(a) = \langle \delta_e, a\delta_e\rangle = \langle \lambda_g^*\lambda_g\delta_e, a\delta_e\rangle = \langle \delta_g, \lambda_g a\delta_e\rangle = \langle \delta_g, a\delta_g\rangle.$$

Hence if $\tau(a) = 0$, then $\langle \delta_g, a\delta_g\rangle = 0$ for all $g \in G$. As a is positive, we may take its square root, and the above gives that

$$0 = \langle \delta_g, a\delta_g\rangle = \langle a^{1/2}\delta_g, a^{1/2}\delta_g\rangle = \|a^{1/2}\delta_g\|^2$$

for all $g \in G$. As $(\delta_g)_{g\in G}$ spans a dense subset of $\ell^2(G)$, this forces $a^{1/2} = 0$, and thus $a = (a^{1/2})^2 = 0$ as required. □

As well as the trace appearing above, we will also need unbounded traces as in Section 2.3: see that section for basic facts and background.

Definition 10.1.6 Let G be a countable torsion-free group, and let X be a proper metric space equipped with a proper, co-compact isometric action of G. Let $D \subseteq X$ be a fundamental domain for the action as in Definition 5.3.3, i.e. D is a Borel subset of X with compact closure such that X equals the disjoint union $X = \bigsqcup_{g\in G} gD$ of the translates of D.

Let H_X be an ample X-G module, which we use to define the equivariant Roe algebra $C^*(X)^G$ (Definition 5.2.1). From Proposition 5.3.4, we have an isomorphism

$$C^*(X)^G \to C^*_\rho(G) \otimes \mathcal{K}(\chi_D H_X).$$

Let

$$\tau^D : C^*(X)^G_+ \to [0,\infty]$$

[2] The same formal computation also works for $a,b \in C^*_\rho(G)$ directly, as elements of this C^*-algebra can be represented uniquely by infinite linear combinations of elements in G: we leave it as an exercise to make this precise.

be the unbounded trace on $C^*(X)^G$ defined by taking the composition of the $*$-isomorphism above with the tensor product of the canonical traces on $C^*_\rho(G)$ and $\mathcal{K}(\chi_D H_X)$ as in Example 2.3.4.

We will not need this, but it is worth noting that the trace above does not depend on the choice of fundamental domain D: this is because for any two choices of D, the resulting isomorphisms only differ by conjugation by unitary multipliers of the algebras involved, and the traces we are using are insensitive to such differences.

The key technical result we need is as follows.

Proposition 10.1.7 *Let G be a countable torsion-free group, and let X be a proper metric space equipped with a proper, co-compact, isometric action of G. Assume, moreover, that the Hilbert space H_X underlying $C^*(X)^G$ is graded by a unitary operator in the multiplier algebra of $C^*(X)^G$ (see Definition E.1.4). Then there is $\epsilon > 0$ with the following property. With notation as in Definition 10.1.6, let F be an odd, self-adjoint, contractive element of the multiplier algebra of $C^*(X)^G$ considered as a subalgebra of $\mathcal{B}(H_X)$ and such that:*

(i) the propagation of F is at most ϵ;
(ii) for any $g \in C_c(X)$, the products $g(1-F^2)$ and $(1-F^2)g$ are trace class.

Then if $\alpha \in K_0(C^*(X)^G)$ is the index class of F (see Definition 2.8.5), we have that $\tau^D_*(\alpha)$ is an integer.

Unfortunately, the algebra involved in the proof is easy to get lost in, but the idea is not so complicated: we want to use the smallness of the propagation of F to show that F also descends to X/G. Moreover, the usual trace of the index class of F in $C^*(X/G) = \mathcal{K}$ agrees with the τ^D trace of the index class of F in $C^*(X)^G$; the former is integer-valued, however, so this will complete the proof.

Proof With notation as in Definition 10.1.6, set $H_D := \chi_D H_X$. Let us first give a concrete formula for $\tau^D(T)$ when $T \in C^*(X)^G_+$ to get a sense of what we are trying to prove. Using Proposition 5.3.4, the isomorphism $\phi\colon C^*(X)^G \to C^*_\rho(G) \otimes \mathcal{K}(H_D)$ determined by D is given by

$$\phi(T) = \sum_{g \in G} \rho_g \otimes \chi_D T U_g \chi_D,$$

and thus if $\mathrm{Tr}\colon \mathcal{K}(H_D)_+ \to [0, \infty]$ denotes the canonical densely defined trace on the compact operators of Example 2.3.3 we see that

$$\tau^D(T) = \mathrm{Tr}(\chi_D T \chi_D).$$

Now, the index class of F is represented by a formal difference $[p]-[q]$ of idempotents in the unitisation of $C^*(X)^G$ as in the formula from Definition 2.8.5. Using Remark 2.3.19 (and abusing notation slightly, writing Tr for the canonical trace on $M_n(\mathcal{K}(H_D)) \cong \mathcal{K}(H_D^{\oplus n})$), we get that

$$\tau_*^D([p]-[q]) = \mathrm{Tr}(\chi_D p \chi_D - \chi_D q \chi_D).$$

If $\chi_D p \chi_D$ and $\chi_D q \chi_D$ were idempotents, we would be done. To see this, note that $K_0(\mathcal{K}) \cong \mathbb{Z}$ is generated by finite rank projections, whence $\mathrm{Tr}_*: K_0(\mathcal{K}) \to \mathbb{R}$ takes image in the integers as it just takes a projection to its rank. On the other hand, say $e = \chi_D p \chi_D$ and $f = \chi_D q \chi_D$ were idempotents. Then, as in Remark 2.3.19, the integer $\mathrm{Tr}_*([e]-[f]) \in \mathbb{Z}$ satisfies the formula

$$\mathrm{Tr}_*([e]-[f]) = \mathrm{Tr}(e-f),$$

so $\mathrm{Tr}(e-f)$ is an integer. Our goal for the rest of the proof is to find idempotents $[p_D], [q_D]$ in the unitisation of $M_2(\mathcal{K}(H_D))$ such that the difference is trace class, and so that

$$\mathrm{Tr}(\chi_D p \chi_D - \chi_D q \chi_D) = \mathrm{Tr}(p_D - q_D);$$

by the above discussion, this will complete the proof.

Let T be a finite propagation operator on H_X, and define

$$T_D := \sum_{g \in G} \chi_D T U_g \chi_D;$$

this makes sense as the facts that D has compact closure, T has finite propagation, X is proper and the action is proper and by isometries together imply that only finitely many of the terms are non-zero. Moreover, if T and S both have finite propagation and are G-invariant then

$$T_D S_D = \sum_{g,h \in G} \chi_D T U_g \chi_D \chi_D S U_h \chi_D = \sum_{g,h \in G} \chi_D T \chi_{gD} S U_{gh} \chi_D,$$

and making the change of variables $k = gh$ and using that $X = \bigsqcup_{g \in G} gD$ this equals

$$\sum_{k \in G} \chi_D T S U_k \chi_D = (TS)_D.$$

In other words, the process $T \mapsto T_D$ is multiplicative on the collection of finite propagation, G-invariant operators.

Now, as the action is proper, free (this follows from properness, and as G is torsion-free) and co-compact, and by isometries there is $r > 0$ such that the quotient map $\pi: X \to X/G$ restricts to a homeomorphism on balls of radius r.

Note that this implies that if $g \in G$ is not the identity, then for any $x \in X$, $d(gx,x) \geqslant r$. Set $\epsilon = r/7$, and let F have the properties in the statement. The computation

$$(1 - F_D^2) = \chi_D(1 - F^2) \sum_{g \in G} U_g \chi_D$$

shows that the operator $1 - F_D^2$ on H_D is trace class, and similarly for $1 - F_D^2$. From the explicit formula for the index class in Remark 2.8.2, it thus follows that the index class of F_D is a represented by a formal difference of idempotents $[p_D] - [q_D]$ such that each of p_D and q_D has propagation at most 5ϵ. To complete the proof, it will suffice to show that if $[p] - [q]$ represents the index class of the original F, then

$$\mathrm{Tr}(\chi_D p \chi_D - \chi_D q \chi_D) = \mathrm{Tr}(p_D - q_D).$$

Using the multiplicative property of the process $T \mapsto T_D$ and the formula for the index class from Definition 2.8.5, it will suffice to show that

$$\mathrm{Tr}(\chi_D T \chi_D) = \mathrm{Tr}(T_D)$$

whenever T is a G-invariant operator on H_X such that hT and Th are trace class for $h \in C_c(X)$, and with propagation at most 5ϵ.

Choose an orthonormal basis (v_n) for H_D such that the diameter of the support of each v_n is at most ϵ (we leave it as an exercise for the reader that choosing a 'localised' basis is always possible). Then

$$\mathrm{Tr}(T_D) = \sum_{n=1}^{\infty} \left\langle v_n, \chi_D T \left(\sum_{g \in G} U_g \right) \chi_D v_n \right\rangle \tag{10.1}$$

and

$$\mathrm{Tr}(\chi_D T \chi_D) = \sum_{n=1}^{\infty} \langle v_n \chi_D T \chi_D v_n \rangle.$$

It thus suffices to show that all the terms in the sum in line (10.1) where g is not the identity element are zero. Indeed, say g is not the identity, and for contradiction that $\langle v_n, T U_g v_n \rangle$ is non-zero for some n. Say x is in the support of $T U_g v_n$. Then as the propagation of T is at most 5ϵ, there is y in the support of v_n such that $d(x, gy) \leqslant 5\epsilon$. On the other hand, $d(y, gy) \geqslant r$ by assumption on r. Hence for any z in the support of v_n we have

$$d(z,x) \geqslant d(y,gy) - d(x,gy) - d(y,z) \geqslant r - 5\epsilon - \epsilon = \epsilon.$$

This implies that the supports of v_n and $T U_g v_n$ can have no points in common, so the given inner product is zero and we are done. □

Using this, we get the following result.

Theorem 10.1.8 *Assume that G is the fundamental group of a closed, aspherical, smooth, spin^c manifold M, and that G is countable and torsion-free.*[3] *Assume moreover that the Baum–Connes conjecture holds for G. Then the Kadison–Kaplansky conjecture holds for G.*

Proof Say for contradiction that e is a non-trivial idempotent in $C^*_\rho(G)$. As in Proposition 2.2.5, e is equivalent to some projection p, which must also be non-trivial. As $\tau\colon C^*_\rho(G) \to \mathbb{C}$ is a faithful trace with $\tau(1) = 1$, we must then have that $0 < \tau(p) < 1$. Let $\widetilde{M}$ be the universal cover of M. Fix a Riemannian metric on M, and lift this to $\widetilde{M}$, so that the deck transformation action of G on $\widetilde{M}$ is by isometries. As M is aspherical, it is a model for BG. Hence using Theorem 7.4.4, the Baum–Connes assembly map for G identifies with the assembly map

$$\mu\colon K^G_*(\widetilde{M}) \to K_*(C^*(\widetilde{M})^G).$$

Using Proposition 10.1.7, it will suffice to show that for any $\epsilon > 0$, any element of $K_0(C^*(\widetilde{M})^G)$ can be represented as the index class of some element F in the multiplier algebra of $C^*(\widetilde{M})^G$ with the following properties: F is odd; F has propagation at most ϵ, and for any $g \in C_c(\widetilde{M})$, the operators $g(1 - F^2)$ and $(1 - F^2)g$ are trace class.

To see this, note first that Corollary 9.6.13 implies that any element of $K_0(M)$ can be represented by the index class of some multiplier (F_t) such that $\text{prop}(F_t) \to 0$ as $t \to \infty$, and such that for any $g \in C_c(M)$, $g(1 - F_t^2)$ and $(1 - F_t^2)g$ are trace class. With notation as in Construction 6.5.14, let

$$\widetilde{F}_t := \Phi(F_t),$$

and note that $(\widetilde{F}_t)$ defines a multiplier of $L^*(\widetilde{M})^G$ with propagation tending to zero, and such that $g(1 - \widetilde{F}_t^2)$ and $(1 - \widetilde{F}_t^2)g$ are trace class for all t and all $g \in C_c(\widetilde{M})$. Taking $F = \widetilde{F}_t$ for some suitably large t gives an operator with the properties we want. □

Using the idea of the proof of Theorem 10.1.8 and the description of the assembly map in Remark 7.1.12, one can deduce the following much more general result. We cannot give a complete proof here as justifying Remark 7.1.12 would require more topology than we have developed, but we give the general statement for reference.

[3] These last two assumptions are redundant: the fundamental group of any closed aspherical manifold will always satisfy them.

Theorem 10.1.9 *Say G is a countable, torsion-free group, and that the Baum–Connes conjecture holds for G. Then the Kadison–Kaplansky conjecture holds for G.* □

As a final remark, the key point of the proof above is to use index theory to show that for $\tau\colon C^*_\rho(G) \to \mathbb{C}$ the canonical trace, the induced map

$$\tau_*\colon K_0(C^*_\rho(G)) \to \mathbb{R}$$

is integer valued. Now, if G has elements of finite order $n > 1$, then the range of τ must at least contain $1/n$ by Example 10.1.3. One could speculate that there should not be much other than such rational numbers in the range of τ. It turns out that something like this is actually predicted by the Baum–Connes conjecture, but it requires a more detailed analysis. Moreover, the situation is more subtle than one might think: the first reasonable conjecture one could make is that the range of τ is contained in the subgroup of $\mathbb{Q}$ generated by

$$\{1/n \mid G \text{ has a subgroup of order } n\}$$

and this turns out to be wrong. See Section 10.5, Notes and References, at the end of this chapter for more discussion on this.

10.2 Positive Scalar Curvature and Secondary Invariants

In this section, we briefly introduce the so-called analytic surgery sequence, and an application to the theory of positive scalar curvature.

Let M be a smooth, closed manifold, which we assume equipped with a Riemannian metric. Let G be the fundamental group of M, acting on the universal cover $\widetilde{M}$ of M by deck transformations. Equip $\widetilde{M}$ with the Riemannian metric lifted from M, so the action of G is by isometries. Recall from Section 7.1 that the assembly map for M is the map on K-theory induced by the evaluation-at-one map

$$\mathrm{ev}\colon C^*_L(\widetilde{M})^G \to C^*(\widetilde{M})^G$$

from the equivariant localised Roe algebra of $\widetilde{M}$ to the equivariant Roe algebra of $\widetilde{M}$. This map fits into a short exact sequence

$$0 \longrightarrow C^*_{L,0}(\widetilde{M})^G \longrightarrow C^*_L(\widetilde{M})^G \xrightarrow{\mathrm{ev}} C^*(\widetilde{M})^G \longrightarrow 0,$$

where $C^*_{L,0}(\widetilde{M})^G$ is the kernel of ev; more concretely, $C^*_{L,0}(\widetilde{M})^G$ consists of those (T_t) in $C^*_L(\widetilde{M})^G$ with $T_1 = 0$. This short exact sequence gives rise to a long exact sequence in K-theory in the usual way

$$\cdots \longrightarrow K_i(C^*_{L,0}(\widetilde{M})^G) \longrightarrow K_i(C^*_L(\widetilde{M})^G) \xrightarrow{\mu} K_i(C^*(\widetilde{M})^G) \longrightarrow \cdots,$$

where μ is the assembly map. Moreover, we have canonical isomorphisms

$$K_*(C^*(\widetilde{M})^G) \cong K_*(C^*_\rho(G)) \quad \text{and} \quad K_*(C^*_L(\widetilde{M})^G) \cong K_*(M),$$

using Proposition 5.3.4 for the first of these, and Proposition 6.6.2 and Theorem 6.5.15 for the second. Thus our long exact sequence becomes

$$\cdots \longrightarrow K_i(C^*_{L,0}(\widetilde{M})^G) \longrightarrow K_i(M) \xrightarrow{\mu} K_i(C^*_\rho(G)) \longrightarrow \cdots.$$

Note that although we used the Riemannian structure on M to make sense of the various algebras above, none of the K-theory groups involved end up depending on the choice: this follows as for any two Riemannian metrics on M, the identity map on $\widetilde{M}$ is an equivariant bi-Lipschitz isomorphism between the two metric spaces defined by the lifted metrics, as compactness of M forces its gradient to be uniformly bounded.

Definition 10.2.1 Let M be a closed smooth manifold. The *analytic structure group*, denoted $\mathcal{S}_i^{\mathrm{an}}(M)$, is defined to be $K_i(C^*_{L,0}(\widetilde{M})^G)$. The *analytic surgery exact sequence* is the long exact sequence

$$\cdots \longrightarrow K_{i+1}(C^*_\rho(G)) \longrightarrow \mathcal{S}_i^{\mathrm{an}}(M) \longrightarrow K_i(M) \xrightarrow{\mu} K_i(C^*_\rho(G)) \longrightarrow \cdots$$

arising from the above discussion.

The analytic structure group is closely connected to the Baum–Connes conjecture (see Definition 7.1.11) thanks to the following lemma.

Lemma 10.2.2 *Assume that M is a closed, smooth, aspherical manifold. Then the Baum–Connes conjecture holds for the fundamental group of M if and only if the analytic structure group $\mathcal{S}_i^{an}(M)$ is zero.*

Proof If M is aspherical, then it is a model for the classifying space BG, and Theorem 7.4.4 identifies the Baum–Connes assembly map with the assembly map for G acting on $\widetilde{M}$. □

The analytic structure exact sequence is also interesting when M is not necessarily aspherical. We will spend the rest of this section discussing one way in which elements of $\mathcal{S}_i^{\mathrm{an}}(M)$ arise, and an application to the theory of positive scalar curvature metrics. In the next section, we will briefly discuss another such application, although to topology rather than to geometry.

We start with some differential geometry, which we will use as a black box. We need to work with spin manifolds: this is a topological assumption that is satisfied for a large class of examples that we introduced in Remark 9.6.4.

Theorem 10.2.3 *Let M be a spin Riemannian d-manifold, and let $\kappa : M \to \mathbb{R}$ be the scalar curvature function of M as discussed in Section 3.1. Then there is a canonically*[4] *associated* spinor Dirac operator *D on M with the following properties:*

(i) D is an elliptic first order differential operator on a canonically associated bundle S over M.
(ii) D is odd with respect to a canonically associated grading operator if M is even-dimensional.
(iii) D has globally finite propagation speed.
(iv) The class of D in $K_d(M)$ is non-zero.
(v) $D^2 = \Delta + \frac{\kappa}{4}$ for some self-adjoint unbounded Laplacian-type operator Δ with non-negative spectrum. □

A spin structure on a Riemannian manifold induces a canonical spin^c structure, and the Dirac operator *is* 'the' spin^c Dirac operator associated to this spin^c structure. Hence most of the above follows from the work we did in Section 9.6, with part (iv) in particular following from Poincaré duality (Theorem 9.6.11). The exception is part (v): this is special to the spin, as opposed to spin^c case, and requires some local computations with the Riemannian curvature tensor that we will not get into here (see Section 10.5, Notes and References, at the end of the chapter).

Now, let M be a closed spin manifold, and lift the Dirac operator D as above to an operator $\widetilde{D}$ on the universal cover $\widetilde{M}$ of M. The universal cover is also a spin Riemannian manifold with the lifted structures, and $\widetilde{D}$ turns out to be its Dirac operator, so $\widetilde{D}$ satisfies the assumptions of Theorem 10.2.3. Theorem 8.2.6 then gives an element (F_t) of the multiplier algebra of $L^*(L^2(\widetilde{M};\tilde{S}))^G$, where $F_t = f(t^{-1}\widetilde{D})$ for some appropriate function $f : \mathbb{R} \to [-1,1]$. Using the index construction as in Remark 8.3.15, we get a class $[\widetilde{D}]$ of $K_*(C_L^*(\widetilde{M})^G)$.

Now, assume that M has strictly positive scalar curvature, so by compactness, there is $c > 0$ such that $\kappa(x) \geqslant c$ for all $x \in M$. This is also true on $\widetilde{M}$, as it has the lifted Riemannian metric. Part (v) of Theorem 10.2.3 then implies that the spectrum of $\widetilde{D}$ does not contain any points in $[-d,d]$ for some $d > 0$. We may choose f so that it is constantly equal to -1 on $(-\infty,-d]$, and to 1 on

[4] Unlike the Dirac operator associated to a spin^c structure, one can build D with no choices involved.

$[d,\infty)$. Checking the details of Remark 8.3.15, we see that the index element $[\widetilde{D}] \in K_*(C_L^*(\widetilde{M})^G)$ is actually coming via the inclusion

$$C_{L,0}^*(\widetilde{M})^G \to C_L^*(\widetilde{M})^G$$

from a canonically defined element of $K_*(C_{L,0}^*(\widetilde{M})^G)$: the point is that the choice of f means that F_1 satisfies $F_1^2 = 1$ precisely, and therefore Lemma 2.8.7 implies that the corresponding index class is zero at $t = 1$.

Definition 10.2.4 Let M be a spin Riemannian d-manifold with positive scalar curvature metric g. The *higher ρ-invariant*, denoted $\rho(g)$, is the class in $\mathcal{S}_d^{\mathrm{an}}(M) = K_d(C_{L,0}^*(\widetilde{M})^G)$ constructed above.

Lemma 10.2.5 *Say M is a Riemannian spin d-manifold with a positive scalar metric. Then if $[\widetilde{D}] \in K_d^G(\widetilde{M})$ is the class of the Dirac operator, we have that $\mu[\widetilde{D}] = 0$ in $K_d(C_\rho^*(G))$.*

Proof This follows from exactness of the analytic surgery sequence and the fact that $\rho(g)$ maps to $[\widetilde{D}]$. □

The following important theorem connects the Baum–Connes conjecture to a conjecture of Gromov and Lawson.

Theorem 10.2.6 *If M is a closed, spin, aspherical d-manifold, and if the Baum–Connes assembly map for the fundamental group of M (see Conjecture 7.1.11) is injective, then M cannot admit a positive scalar curvature metric.*

Proof The class $[\widetilde{D}]$ of the lifted Dirac operator in $K_d(\widetilde{M})$ is non-zero by Theorem 10.2.3, part (iv). Hence if $G = \pi_1(M)$, the class $[\widetilde{D}]$ of $\widetilde{D}$ is also non-zero in $K_i^G(\widetilde{M})$ as the two classes are compatible with the forgetful map $K_i^G(\widetilde{M}) \to K_i(\widetilde{M})$ induced by the inclusion $L^*(\widetilde{M})^G \to L^*(\widetilde{M})$. As the Baum–Connes assembly map is injective, the class $\mu[\widetilde{D}]$ is therefore non-zero in $K_i(C_\rho^*(G))$. This, however, is false in the presence of positive scalar curvature by Lemma 10.2.5. □

A similar result holds if we assume that the coarse assembly map for $\widetilde{M}$ is injective. This latter result does not require that M be spin: see Exercise 10.4.1.

We have just scratched the surface of the theory here: in particular, we did not really make substantial use of the higher ρ-invariant from Definition 10.2.4 other than to deduce Lemma 10.2.5. This is losing a lot of information: the higher ρ-invariant should be regarded as a 'reason' why $[D] = 0$, and there may be more than one such reason. Indeed, different positive scalar metrics $g_0 \neq g_1$ on M can give rise to different higher ρ-invariants $\rho(g_0) \neq \rho(g_1)$ in $\mathcal{S}_*^{\mathrm{an}}(M)$. On the other hand, if the metrics g_0, g_1 were in the same path-component of the space of positive scalar curvature metrics, then the associated higher ρ invariants would be the same. In this way, information about $\mathcal{S}_*^{\mathrm{an}}(M)$ can be used to deduce information about the topology of the space of all positive scalar metrics on M. This is currently a very active area of research: see Section 10.5, Notes and References, at the end of this section.

10.3 The Novikov Conjecture

The Novikov conjecture is a fundamental conjecture in the topology of high-dimensional manifolds. For simplicity, let us assume that all manifolds appearing in this section are smooth (much of what we say can be made to work more generally). Let us start by motivating (a special case of) the Novikov conjecture using the so-called Borel conjecture together with an important theorem of Novikov.

For our discussion of the Borel conjecture, let us say that a manifold M is *rigid* if, whenever N is another manifold and $f: M \to N$ is a homotopy equivalence, we have that f is homotopic to a homeomorphism. For example, the classical Poincaré conjecture says that the three-sphere S^3 is rigid.

For non-rigid examples, consider the following quotients of S^3: let n be an integer, let a,b be integers relatively prime to n and let $\mathbb{Z}/n\mathbb{Z}$ act on $\mathbb{C}^2$ by stipulating that the usual generator $1 \in \mathbb{Z}/n\mathbb{Z}$ acts as

$$(z,w) \mapsto (e^{2\pi ia/n}z, e^{2\pi ib/n}w).$$

This action is isometric, so it restricts to an action on the unit sphere $S^3 \subseteq \mathbb{C}^2$. As the action is free, the resulting quotient $S^3/(\mathbb{Z}/n\mathbb{Z})$ is a manifold, called a *lens space* and denoted $L(n;a,b)$. Now, it is known that $L(n;a,b)$ and $L(n;c,d)$ are homotopy equivalent if and only if $ab = cd \bmod n$; on the other hand, they are known to be homeomorphic if and only if (a,b) is the same as (c,d) up to change of order and change of sign (in either, or both variable(s)). The proofs of these facts go beyond the scope of this text, but let us at least note that one method uses the algebraic K-theory of the group ring of $\mathbb{Z}/n\mathbb{Z}$. Thus, for example, the lens spaces $L(5;2,2)$ and $L(5;-1,1)$ are homotopy equivalent, but not homeomorphic, and in particular the three-manifold $L(5;1,-1)$ is not rigid.

Now, recall that a manifold (or more generally, a CW complex) M is said to be aspherical if its universal cover $\widetilde{M}$ is contractible. Examples include tori and fundamental groups of surfaces of non-positive genus. On the other hand, lens spaces as introduced above are not aspherical: their universal cover is S^3. Here is the *Borel conjecture*.

Conjecture 10.3.1 *Closed aspherical manifolds*[5] *are rigid.*

On the face of it, the Borel conjecture is about the global topology of M; the other half of our motivation for the Novikov conjecture comes from the infinitesimal topology of M. To describe it, let us recall that to any real vector bundle E over M and a natural number k, there is an associated *Pontrjagin class* $p_k(E) \in H^{4k}(M;\mathbb{Z})$. Losing some information, one can also consider the *rational Pontrjagin classes* $p_k(E;\mathbb{Q})$, which are the images of the $p_k(E)$ under the natural change of coefficients map $H^{4k}(M;\mathbb{Z}) \to H^{4k}(M;\mathbb{Q})$ induced by the inclusion $\mathbb{Z} \to \mathbb{Q}$. The *rational Pontrjagin classes of M* itself are then defined to be the classes $p_k(M;\mathbb{Q}) := p_k(TM;\mathbb{Q}) \in H^{4k}(M;\mathbb{Q})$ associated to the tangent bundle of M. A priori, these depend on the smooth structure of M, but in fact one has the following remarkable result, often called *Novikov's theorem*.

Theorem 10.3.2 *Let $f\colon M \to N$ be an orientation-preserving homeomorphism between closed oriented manifolds. Then $f^*(p_k(N;\mathbb{Q})) = p_k(M;\mathbb{Q})$ for all $k \in \mathbb{N}$.* □

We should remark here that, starting in dimension four, there is a very rich theory of pairs of manifolds M and N that are homeomorphic but not diffeomorphic. Milnor gave the first examples, showing that there are manifolds that are homeomorphic, but not diffeomorphic, to the seven-sphere S^7. Milnor and Kervaire later showed that there are exactly twenty-eight manifolds that are diffeomorphic, but not homeomorphic, to S^7 (taking orientation into account). Moreover, the invariants used to tell these *exotic spheres* apart come from the Pontrjagin classes of a tangent bundle.[6] Thus Novikov's theorem is quite surprising, as these results on non-unique differentiable structures suggest that the rational Pontrjagin classes should only be diffeomorphism, and not homeomorphism, invariants. Note moreover that aspherical examples with these sort of properties show that one cannot replace 'homeomorphic' by 'diffeomorphic' in the statement of the Borel conjecture.

Now, combining Conjecture 10.3.1 and Theorem 10.3.2, one is led to the following conjecture, which is the special case of the Novikov conjecture when the underlying manifold happens to be aspherical.

[5] The conjecture is usually stated for topological manifolds, rather than smooth as in our standing convention.

[6] Crucially, the tangent bundle of a different manifold that is used in the construction! Note that there are no interesting Pontrjagin classes for a 7-sphere as its $4k$-dimensional cohomology is trivial.

Conjecture 10.3.3 *Let $f\colon M \to N$ be an orientation-preserving homotopy equivalence between closed aspherical oriented manifolds. Then $f^*(p_k(N;\mathbb{Q})) = p_k(M;\mathbb{Q})$ for all $k \in \mathbb{N}$.*

Thanks to Theorem 10.3.2, Conjecture 10.3.3 is implied by the Borel conjecture. Indeed, one can think of the (rational) Pontrjagin numbers $p_k(M;\mathbb{Q})$ as 'infinitesimal' or 'linearised' invariants of the manifold, as they are invariants of the tangent bundle, which is itself an infinitesimal or linearised version of the manifold itself; the Novikov conjecture can thus in some sense be thought of as an infinitesimal version of the Borel conjecture, which is a statement of a more global nature.

To move towards explaining what this has to do with index theory, let us give a more standard, and general, form of the Novikov conjecture. Let $H^*(M;\mathbb{Q})$ be the usual rational cohomology ring of M, where the multiplication is given by the cup product $\cup$. Let $L(M) \in H^*(M;\mathbb{Q})$ be the *Hirzebruch L-class* of M, which is a cohomology class defined as a certain universal polynomial in the Pontrjagin classes. The class $L(M)$ is thus zero in dimensions not a multiple of four, and one has explicit formulas for the component $L_k(M) \in H^{4k}(M;\mathbb{Q})$ in terms of the rational Pontrjagin classes. For example

$$L_1(M) = \frac{1}{3}p_1(M;\mathbb{Q}),$$

$$L_2(M) = \frac{1}{45}(7p_2(M;\mathbb{Q}) - p_1(M;\mathbb{Q}) \cup p_1(M;\mathbb{Q})), \quad \ldots.$$

The L-class is thus a priori an invariant of the differentiable structure on M: one expects that homeomorphic manifolds that are not diffeomorphic would have different L-classes. However, at least some information contained in $L(M)$ is even a homotopy invariant. Indeed, the *signature* of M, denoted $\mathrm{sign}(M)$, is defined to be the signature in the usual sense of algebra of the non-degenerate symmetric form defined by

$$H^{2k}(M;\mathbb{R}) \times H^{2k}(M;\mathbb{R}) \to \mathbb{R}, \quad (x,y) \mapsto \langle x \cup y, [M]\rangle.$$

This signature is defined purely in terms of the cohomology ring $H^*(M;\mathbb{R})$, the pairing between homology and cohomology and the fundamental class $[M]$, and is thus invariant under orientation-preserving homotopy equivalences. One has the *Hirzebruch signature theorem*:

Theorem 10.3.4 *Let M be a closed oriented manifold of dimension $4k$, and let $L_k(M) \in H^{4k}(M;\mathbb{Q})$ be the top-dimensional component of the L-class. Then*

$$\langle L_k(M), [M]\rangle = sign(M). \qquad \square$$

In particular, the component of $L_k(M)$ in $H_{4k}(M;\mathbb{Q})$ is a homotopy invariant. Now, it was shown by Browder and Novikov that $L_k(M) \in H^{4k}(M;\mathbb{Q})$, with M $4k$-dimensional, is the only universal polynomial in the Pontrjagin classes that is invariant under orientation-preserving homotopy equivalences within the class of simply connected manifolds; thus in the simply connected case, the signature theorem delineates the only homotopy-invariant information one gets from the Pontrjagin classes.

In the non-simply connected case, one can hope to get more homotopy invariant information from the Pontrjagin classes. Indeed, let $G = \pi_1(M)$, and assume for simplicity that BG is a finite CW complex. Let $c\colon M \to BG$ be a continuous map that induces an isomorphism on fundamental groups; as BG is aspherical, such a map exists (and is determined up to homotopy equivalence by the map it induces between the fundamental groups). For each $\alpha \in H^*(M;\mathbb{Q})$ define the *higher signature* to be the class

$$\operatorname{sign}(M,\alpha,c) := \langle L(M) \cup c^*(\alpha), [M] \rangle.$$

Here is a more standard statement of the Novikov conjecture.

Conjecture 10.3.5 *Let M and N be closed oriented manifolds, and let $f\colon N \to M$ be an orientation-preserving homotopy equivalence. Then for any $c\colon M \to BG$ and $\alpha \in H^*(BG;\mathbb{Q})$ as above, we have $sign(M,\alpha,c) = sign(N,\alpha,c \circ f)$.*

This conjecture is often summarised by saying *higher signatures are homotopy invariant*. For aspherical manifolds, Conjecture 10.3.5 is equivalent to Conjecture 10.3.3, as one can just take $BG = M$, and it suffices to consider $c\colon M \to BG$ the identity map. In this case, one can recover all the Pontrjagin classes of M from the higher signatures using non-degeneracy of the pairing between (rational) homology and cohomology.

Finally, we are ready to explain what this has to do with index theory. On an oriented smooth Riemannian manifold M, there is an elliptic differential operator D_M called the *signature operator*, giving rise to a class $[D_M] \in K_*(M)$. Given a map $c\colon M \to BG$ as in the definition of higher signatures (and continuing to assume that BG is a finite CW complex) we thus get a class $c_*[D_M] \in K_*(BG)$.

The following result was perhaps first seen by Lusztig and Miščenko, and has been reproved a number of times since.

Theorem 10.3.6 *Let M and N be closed oriented manifolds, and let $f\colon N \to M$ be an orientation-preserving homotopy equivalence. Let $G = \pi_1(M)$. Then for any $c\colon M \to BG$ as above we have that*

$$\mu(c \circ f)_*[D_N] = \mu c_*[D_M],$$

where $\mu\colon K_*(BG) \to K_*(C^*_\rho(G))$ *is the Baum–Connes assembly map.* □

The key C^*-algebraic input to the proof is that one can use the spectral theorem to diagonalise quadratic forms over a C^*-algebra. This is not always possible over a general ring, and it is not at all clear when it is possible over a more general Banach algebra such as $\ell^1(G)$.

To see how this relates to the Novikov conjecture, note that there is a homology Chern character

$$\mathrm{ch}_*\colon K_*(M) \to H_*(M;\mathbb{Q}).$$

This is determined by the more usual K-theory Chern character $\mathrm{ch}^*\colon K^*(M) \to H^*(M;\mathbb{Q})$ and the relationship

$$\langle \mathrm{ch}_*(x), \mathrm{ch}^*(y)\rangle = \langle x, y\rangle$$

for all $x \in K_*(M)$, $y \in K^*(M)$, where the pairings are the usual ones between K-homology and K-theory, and between homology and cohomology; this is well-defined by rational non-degeneracy of these pairings (which in turn follows from the UCT in each case: see Theorem 9.2.8).

Applying this, we see that if μ is injective then from Theorem 10.3.6 we have that $(c \circ f)_*[D_N] = c_*[D_M]$, and so

$$\mathrm{ch}_*(c \circ f)_*[D_N] = \mathrm{ch}_* c_*[D_M] \in H_*(BG;\mathbb{Q}).$$

Equivalently, by rational non-degeneracy of the pairing between homology and cohomology and naturality of the Chern character, we have that

$$\langle (c \circ f)_* \mathrm{ch}_*[D_N], \alpha\rangle = \langle c_* \mathrm{ch}_*[D_M], \alpha\rangle$$

for all $\alpha \in H^*(BG;\mathbb{Q})$. The Atiyah–Singer index theorem implies that the K-homology Chern character takes $[D_M]$ to $c_d(L(M) \cap [M])$, where $c_d \in \mathbb{Q}$ is some non-zero number depending only on the dimension d of M. Hence the above implies that

$$\langle L(N) \cap [N], (c \circ f)^*\alpha\rangle = \langle L(M) \cap [M], c^*\alpha\rangle$$

for all $\alpha \in H^*(M;\mathbb{Q})$. Using that cap product is the adjoint of cup product, we may move the '$L(M)\cap$' on the left to '$L(M)\cup$' on the right, this is exactly the statement of the Novikov conjecture.

Summarising the outcome of the argument above, we get the following.

Proposition 10.3.7 *Say that* G *is a group such that* BG *admits a finite CW complex model, and such that* $\mu\colon K_*(BG) \to K_*(C^*_\rho(G))$ *is injective.*

Then the Novikov conjecture holds for all closed oriented manifolds with fundamental group G. □

For this reason, injectivity of the Baum–Connes assembly map[7] for G is often called the *strong Novikov conjecture*.

Now, in the spirit of the analytic surgery exact sequence, one can do somewhat better than the above classical argument. Indeed, consider the analytic surgery exact sequence that we already discussed above in Section 10.2 (see Definition 10.2.1) for the space BG

$$\cdots \longrightarrow K_{i+1}(C^*_\rho(G)) \longrightarrow S^{\mathrm{an}}_i(BG) \longrightarrow K_i(BG) \stackrel{\mu}{\longrightarrow} K_i(C^*_\rho(G)) \longrightarrow \cdots$$

(for simplicity, we continue to assume that BG is finite). Given an orientation-preserving homotopy equivalence $f\colon N \to M$ between closed manifolds and a map $c\colon M \to BG$ inducing an isomorphism on fundamental groups, one can construct a class

$$\sigma(N,M,f,c) \in S^{\mathrm{an}}_*(BG)$$

that maps to

$$c_*[D_M] - (c \circ f)_*[D_N] \in K_*(BG)$$

under the map $S^{\mathrm{an}}_*(BG) \to K_*(BG)$ appearing in the analytic surgery exact sequence. The existence of such a class is non-trivial: indeed, its 'mere existence' immediately implies Theorem 10.3.6, much as the mere existence of the higher rho class of Definition 10.2.4 implies the vanishing result of Lemma 10.2.5.

Similarly to the positive scalar curvature higher rho invariant, however, it does more than this: it provides an explicit 'reason' for the vanishing of $c_*[D_M] - (c \circ f)_*[D_N]$, and different 'reasons' might exist. For example, if we then fix M up to homeomorphism, there might be many different manifolds N that are homotopy equivalent to M, and these different N could give rise to different classes $\sigma(N,M,f,c)$.

Looking more broadly, this construction lies at the heart of the *mapping surgery to analysis* program. This aims to connect the *surgery exact sequence* from algebraic topology and the analytic surgery exact sequence discussed briefly above. We discuss this much more in Section 10.5, Notes and References, at the end of the chapter.

[7] Or injectivity of one of several variants: for example, rational injectivity.

10.4 Exercises

10.4.1 Show that if M is a closed manifold (spin or not) with positive scalar curvature, and if the coarse Baum–Connes assembly map is injective for the universal cover $\widetilde{M}$, then M does not have a positive scalar curvature metric. *Hint: $\widetilde{M}$ is contractible, so spin, and has an associated Dirac operator satisfying the condition of Theorem 10.2.3 part* (v). *Adapt the argument for Theorem 10.2.6.*

10.5 Notes and References

This section will be rather longer than most of our usual Notes and References sections, as we will attempt to give brief summaries of some parts[8] of the literature.

The Covering Index Theorem

Atiyah's covering index theorem comes from Atiyah [8]; see also Wang and Wang [247] for an interesting recent generalisation. Connes generalised this to a 'measured' index theorem for foliations (see Moore and Schochet [186] for a book-length exposition of these ideas), and this in turn was a key motivation for the development of Connes' index-theoretic study of foliations [57] using C^*-algebraic and K-theoretic methods. The connection between the Baum–Connes conjecture, the covering index theorem and the Kadison–Kaplansky conjecture was already observed in section 7 of the original paper of Baum and Connes on their conjecture (see Baum and Connes [21]. With a much more careful study, one can get more information on the predictions of the Baum–Connes conjecture for the range of the canonical trace: see Lück [172] for the definitive results in this direction.

Positive Scalar Curvature

For detailed background on spin geometry and topology, see Lawson and Michelsohn [164]. Parts of Roe [217] also give a nice introduction. In particular, both of these references contain proofs of the formula in part (v)

[8] We make no claim to be definitive: the discussion omits a great deal, either due to ignorance or carelessness. Our apologies, and of course we would be very grateful for suggestions for improvements.

of Theorem 10.2.3. The index-theoretic approach to studying the existence of positive scalar curvature metrics on spin manifolds was initiated by Lichnerowicz [168], based on the (re)discovery of the spinor Dirac operator by Atiyah and Singer [13] in the lead-up to their proof of the index theorem [12]. A particularly striking early application of these ideas comes from work of Hitchin [141]: this shows in particular that there are nine-dimensional manifolds that are homeomorphic to the nine-dimensional sphere, but do not admit a metric of positive scalar curvature. In the simply connected setting, definitive results on the existence of positive scalar curvature metrics have been obtained by Stolz [237]: these use Dirac operator techniques based on those of Hitchin [141], and deep machinery from algebraic topology.

In the presence of a non-trivial fundamental group, one has many other tools to attack the (non-) existence of positive scalar curvature metrics. Methods combining Dirac operator techniques with coarse geometry and the fundamental group were pioneered by Gromov and Lawson: their 1983 paper on the subject [180] is still highly recommended reading. In particular, they first solved the problem of the existence of a positive scalar curvature metric on the d-torus for general d (Schoen and Yau [231, 232] solved this by different methods for $d \leqslant 7$, and seem to have recently pushed their techniques to work for general d). Soon afterwards, Rosenberg [222] connected the techniques of Gromov and Lawson to the assembly maps of Kasparov and Baum–Connes, leading to the introduction of operator algebraic, and K-theoretic, methods. See Rosenberg [225] for a survey of of this work (and some more recent material), and Schick [230] for an interesting counterexample to the so-called (unstable) Gromov–Lawson–Rosenberg conjecture about the relationship of positive scalar curvature and index-theoretic invariants.

More recently, there is continuing interest in index-theoretic approaches to the problem of the existence of positive scalar curvature metrics: see for example Chang, Weinberger and Yu [51] and Hanke and Schick [121]. There has also been a lot of activity centred on higher rho invariants and the moduli space of positive scalar curvature metrics: in particular, there is a 'mapping surgery to analysis' program (see for example Piazza and Schick [206] and Xie and Yu [265]) relating the positive scalar curvature exact sequence of Stolz (see for example Rosenberg and Stolz [227]) to the analytic surgery exact sequence discussed above. This theory can be used, amongst other things, to detect the size of the space of positive scalar curvature metrics in some sense: see for example Xie and Yu [266].

The Novikov Conjecture

Novikov's theorem on homeomorphism invariance of the rational Pontrjagin numbers comes from section 11 of Novikov [193]. The original statement of Novikov's conjecture comes from [194]. For a survey of the history and ideas around the Novikov conjecture (going up to the early 1990s), we recommend Ferry, Ranicki and Rosenberg [99]. A survey more specifically on coarse geometric approaches to the Novikov conjecture (covering the same period) can be found in Ferry and Weinberger [101]. For general ideas around surgery theory, which we just hinted at above, we recommend Part I of Weinberger [252] for an inspiring high-level overview, while Wall [246] is the canonical classical reference. See also Weinberger [253] for an inspiring overview of many ideas around the Borel conjecture.

The index-theoretic approach to the Novikov conjecture was pioneered by Lusztig [174] and Miščenko [185] (although hints in that direction already appear in the writings of Novikov himself). In particular, Theorem 10.3.6 comes from their work; see also Kaminker and Miller [146] for a more recent proof, and Hilsum and Skandalis [140] for a more general result related to foliations. This line of thinking was pushed forward a great deal by Kasparov [150], with his powerful equivariant KK-theory machine reaching a very sophisticated form by the late 1980s (see Kasparov [149]). The index-theoretic approach to the Novikov conjecture has now touched on many other interesting parts of mathematics such as cyclic homology (Connes and Moscovici [65]), amenability (Higson and Kasparov [129]) and Banach space geometry (Kasparov and Yu [154] and Yu [272]).

Recently, the mapping-surgery-to-analysis paradigm has also taken off. The ideas perhaps first appear in Roe's lectures (see [216]), and were fully developed in the series of papers by Higson and Roe [136–138]. A different model for this based on Baum's geometric model for K-homology has since been given by Deeley and Goffeng [79], and another model based on tangent groupoids was recently given by Zenobi [276]; this latter also extends the maps to the setting of topological (as opposed to smooth) manifolds.

These ideas have been used to get interesting information on several rigidity type problems: for example, Weinberger and Yu [255] study the so-called finite part of the K-theory of the group C^*-algebra and use this to get quantitative results on the size of the structure set. Another recent result along these lines comes from Weinberger, Xie and Yu [254]: here the authors show that the map between structure sets is a homomorphism (as opposed to just a set map in the original formulation) when one uses the topological (as opposed to smooth) surgery exact sequence; this allows one to prove some quantitative non-rigidity results.

Finally, we mention that more algebraic and differential-geometric approaches to related material connecting surgery (theoretic ideas) to analysis can be found, for example, in Leichtnam and Piazza [165], Lott [170], Moriyoshi and Piazza [187], Wahl [245] and Xie and Yu [267] and the references these papers contain. There is a huge theory here, and this is a very active area right now; these references are just to give a small sense or some directions, and are by no means meant to be complete (or even completely representative).

A Very Brief Guide to the Literature on Assembly Maps

We give here a by no means exhaustive collection of references for other methods for studying results on injectivity and surjectivity of assembly type maps, both in the C^*-algebraic and algebraic settings.

In the C^*-algebraic setting, one of the most important ideas is the so-called Dirac-dual-Dirac method. This approach is due to Kasparov and relies on bivariant K-theory for its precise formulation: Kasparov [150] is an early survey, and Kasparov [149] gives the definitive version of Kasparov's machinery; these results rely on the geometry of non-positive curvature. These ideas were pushed further by Kasparov and Skandalis [151, 152] to a remarkably general class of spaces called bolic spaces that exhibit some of the characteristics of non-positive curvature. In another, related, direction, Higson and Kasparov used the Dirac-dual-Dirac method to get results using infinite-dimensional flat geometry (Higson and Kasparov [129]); this proves the Baum–Connes conjecture for the large class of a-T-menable groups. Yu [272] subsequently used these infinite-dimensional techniques to prove the coarse Baum–Connes conjecture for the class of spaces that coarsely embed into Hilbert space; we give a new proof of this last result in Chapter 12.

Higson and Kasparov's approach to the Baum–Connes conjecture is obstructed, however, much beyond the case of a-T-menable groups, thanks to a rigidity property of groups called property (T). The attendant difficulties were circumvented in many cases by Lafforgue [159] using a version of (bivariant) K-theory for Banach algebras rather than just C^*-algebras. Lafforgue's work was used by Mineyev and Yu [184] to prove the Baum–Connes conjecture for the large and interesting class of hyperbolic groups; Lafforgue [161] subsequently gave a different proof of this that also covers the case with coefficients. Unfortunately, Lafforgue [160] was also able to show that many important groups (for example, lattices in higher rank Lie groups such as $SL(3,\mathbb{Z})$) enjoy very strong forms of property (T) that obstruct both the traditional Dirac-dual-Dirac method, and also his methods; the Baum–Connes conjecture remains open in all these cases.

There has also been a great deal of work on permanence properties of the Baum–Connes conjecture: see for example Chabert and Echterhoff [48] and Meyer and Nest [177] (the former is relatively concrete; the latter works in a general abstract setting based on the machinery of triangulated categories). These techniques are very much tied up with the version of the Baum–Connes conjecture with coefficients, and as such have not been touched on at all in this text. These (and other) permanence properties combine with some of the above work using the Dirac-dual-Dirac method and the structure theory of almost connected groups to give a proof of the Baum–Connes conjecture for almost connected groups (see Chabert, Echterhoff and Nest [49]).

Going back to the coarse Baum–Connes conjecture, a separate proof paradigm comes from the work of Yu in the case of spaces of finite asymptotic dimension (see Yu [271]). Here the idea is to use an approximate version of K-theory as exposited in Oyono-Oyono and Yu [198] and concrete (approximate) Mayer–Vietoris sequences to prove the coarse Baum–Connes conjecture. These ideas have recently been pushed into the setting of group actions (Guentner, Willett and Yu [117, 118]). This sort of approach to computing K-theory has also recently been used by Oyono-Oyono and Yu to get results of purely C^*-algebraic interest on the so-called Künneth formula (Oyono-Oyono and Yu [199]); we expect that much more remains to be said here. See also Chabert, Echterhoff and Oyono-Oyono [50] for a completely different take on connections between the world of the Baum–Connes conjecture and the Künneth formula.

In the C^*-algebraic setting, a final broad theme that we will mention builds around ideas based on the so-called Higson corona (see sections 6 and 7 of Roe [214]). This is an interesting compactification of a space that was observed by Higson to be close to K-homology; if one can show that the Higson corona is topologically trivial enough, then results on injectivity of the assembly map follow. These ideas were studied in Dranishnikov, Keesling and Uspenskii [85] and Keesling [155] from the point of view of general topology, while related ideas in algebraic K-theory were studied by Carlsson and Pedersen [46]. Subsequently, Emerson and Meyer [90, 91] realised that a 'stable' version of the Higson corona has better formal properties and is quite intrinsically bound up with the Dirac-dual-Dirac method. Recent work of Wulff [264] proves some conjectures of Roe [215] relating the (stable) Higson corona and the assembly map in a foliated setting; it seems that much remains to be understood here.

There is a parallel theory in algebraic topology centred around the Farrell–Jones conjecture that is more powerful in some ways, less so in others: see for example Davis and Lück [77] for the general machinery setting up the

Farrell–Jones and related conjectures. One great advantage is that it is more closely connected to topology and therefore allows for progress on (for example) the Borel conjecture, while the analytic theory does not; on the other hand, the analytic theory allows for applications to problems in the existence of positive scalar curvature metrics, and the algebraic theory seems to have no direct connections to this. Higson, Pederson and Roe [132], Rosenberg [223] and Weinberger [251] survey some interactions between the analytical and topological theories.

Loday [169] seems to be the first appearance of the algebraic K-theory assembly map (the L-theory assembly map appears earlier in the surgery exact sequence in Wall [246]). Work in the 1980s (and before) towards applications to algebraic topology was pioneered by Farrell, Hsiang and Jones: this approach maybe starts with work of Farrell and Hsiang [93] on non-positively curved manifolds. See also Hsiang's ICM talk (Hsiang [143]) for an interesting survey from around this time. See for example some of the work of Farrell and Jones [94, 95, 97, 98] and particularly [96] for a survey of some of the ideas and history. Ferry and Weinberger [100] could also be seen as fitting into the broad scheme discussed here. Some of this work predates what is now called the Farrell–Jones conjecture, and in some sense aims more directly at the Borel conjecture itself; however, from a modern point of view much of it can be seen as proving special cases of that conjecture.

Finite asymptotic dimension, and its weaker relative finite decomposition complexity, have also been used in the purely algebraic setting to great effect using a variety of more or less direct techniques: see for example Bartels [14], Dranishnikov, Ferry and Weinberger [84], Guentner, Tessera and Yu [116] and Ramras, Tessera and Yu [210].

More recently, the Farrell–Jones conjecture has been fitted into an abstract and more algebraic machine: see for example the work of Davis and Lück [77] on general assembly maps, and the paper of Bartels, Farrell, Jones and Reich [17]. A recent sequence of papers (Bartels [15], Bartels and Bestvina [16], Bartels and Lück [18] and Bartels, Lück and Reich [19], amongst others) use this more algebraic approach (although still with significant geometric machinery) to make very impressive progress on the Farrell–Jones conjecture. This includes a proof of the Borel conjecture for groups that act properly co-compactly and isometrically on CAT(0) spaces (Bartels and Lück [18]); a proof of the Baum–Connes conjecture under similar hypotheses seems completely out of reach with current ideas, so the fact that the Borel conjecture is known in that setting is particularly intriguing. We should also remark that this spectacular recent progress proves the Borel conjecture for all known examples of aspherical manifolds.

We should also remark that in the purely algebraic theory there are some results that are true in a really striking degree of generality: Bökstedt, Hsiang and Madsen [35] proves injectivity of the algebraic K-theory assembly map under very general finiteness conditions, while Yu [273] proves injectivity of the algebraic K-theory map for completely general groups (!), although with analytically flavored coefficients (see also the later, more algebraic approach to this result in Cortiñas and Tartaglia [68]). Unfortunately, these results seem to have no direct applications to manifold topology; nonetheless, they provide some very interesting inspiration as to what one might hope for in the case of those assembly maps that are more directly connected to geometric topology.

We finally remark that one important and mysterious distinction between the algebraic and C^*-algebraic settings concerns various functoriality properties: for example, the right-hand side of the Baum–Connes assembly map is not known to be a functor of the input group (the left-hand side is). Other problems in the analytic theory arise as certain functors can fail to be exact (Higson, Lafforgue and Skandalis [131]); the analogous exactness properties are a non-issue in the algebraic case. Recent attempts to get around some such functoriality problems in the C^*-algebraic case can be found in Antonini, Azzali and Skandalis [2], Baum, Guentner and Willett [25] and Buss, Echterhoff and Willett [45], but much remains to be understood here.

PART FOUR

Higher Index Theory and Assembly

11

Almost Constant Bundles

In this chapter we study what we call almost constant bundles. This is one of the simplest cases where one can get non-trivial information about the assembly map: indeed, our machinery is set up partly to make arguments like this relatively straightforward.

11.1 Pairings

Throughout this section, X is a proper metric space.

Definition 11.1.1 Let $\mathcal{K} = \mathcal{K}(\ell^2(\mathbb{N}))$, and for each k, let $1_k \colon X^+ \to \mathcal{K}$ denote the constant map with image the projection onto the first k basis elements. An *almost constant* sequence for X is a sequence of pairs $(p_n, 1_{k_n})$ of functions from X^+ to $\mathcal{K}$ with the following properties:

(i) for each n, p_n is a projection;
(ii) for each n, the difference $p_n - 1_{k_n}$ is in $C_c(X, \mathcal{K})$;
(iii) for each $r > 0$,

$$\lim_{n\to\infty} \sup\{\|p_n(x) - p_n(y)\| \mid x, y \in X,\ d(x,y) \leqslant r\} = 0.$$

A class α in $K^0(X)$ is *almost constant* if there exists an almost constant sequence $(p_n, 1_{k_n})$ such that $\alpha = [p_n] - [1_{k_n}]$ for all n.

Example 11.1.2 Say X is a bounded metric space. Assume, moreover, that $\alpha \in K^0(X)$ is represented by an almost constant sequence of pairs $(p_n, 1_{k_n})$ such that each p_n and 1_{k_n} actually take image in some fixed corner

$$\mathcal{B}(\ell^2\{1, \ldots, N\}) = 1_N \mathcal{K}(\ell^2(\mathbb{N})) 1_N \subseteq \mathcal{K}(\ell^2(\mathbb{N})).$$

Then α is a multiple of the class of the identity: see Exercise 11.3.1. This is false either if one drops the assumption that X is bounded, or drops the assumption that the maps p_n and 1_{k_n} all take image in the same finite-dimensional corner of $\mathcal{K}$.

Example 11.1.3 For any d, any class α in $K^0(\mathbb{R}^d)$ is almost constant. Indeed, write $\alpha = [p] - [1_k]$ for some $p \colon X^+ \to \mathcal{K}$, which is equal to 1_k in a neighbourhood of ∞. For each $n \in \mathbb{N}$, define

$$p_n(x) := p(x/(n+1))$$

(with the convention that $\infty/(n+1) = \infty$). As each of the maps

$$\mathbb{R}^d \to \mathbb{R}^d, \quad x \mapsto x/(n+1)$$

is proper and properly homotopic to the identity, each of the classes $[p_n] - [1_k]$ is equal to our original α. On the other hand, as p is equal to 1_k in a neighbourhood of infinity, p is uniformly continuous for the standard metric on $\mathbb{R}^d$, and thus $(p_n, 1_k)$ satisfies condition (iii) in Definition 11.1.1.

We will explore the geometry underlying the above example and generalise it in the next section.

The next result explains the usefulness of almost constant classes for studying assembly maps.

Proposition 11.1.4 *Let $C^*(X)$ be the Roe algebra of X and let $(p_n, 1_{k_n})$ be an almost constant sequence. Then there is a well-defined homomorphism*

$$\phi \colon K_0(C^*(X)) \to \frac{\prod_n \mathbb{Z}}{\oplus_n \mathbb{Z}}$$

such that if

$$\mu \colon K_*(X) \to K_*(C^*(X))$$

is the assembly map from Definition 7.1.1 then

$$\phi(\mu(\beta)) = [\, \langle \beta, [p_0] - [1_{k_0}] \rangle, \langle \beta, [p_1] - [1_{k_1}] \rangle, \ldots] \tag{11.1}$$

for all $\beta \in K_0(X)$; here $\langle \beta, [p_n] - [1_{k_n}] \rangle$ denotes the result of the usual pairing between K-theory and K-homology (see Definition 9.2.3), and the right-hand side in the formula above denotes the class of the sequence of integers

$$(\langle \beta, [p_0] - [1_{k_0}] \rangle, \langle \beta, [p_1] - [1_{k_1}] \rangle, \ldots)$$

in $\prod_n \mathbb{Z} / \oplus_n \mathbb{Z}$.

Proof Let H_X be an ample X module, and let H be an auxiliary separable Hilbert space. The natural amplified representations of $C(X^+)$ and $\mathcal{K}(H)$ on $H_X \otimes H$ commute, giving rise to a representation of $C(X^+, \mathcal{K}(H))$ on this Hilbert space. Assume that $C^*(X)$ is defined using the X module H_X and consider $C^*(X)$ as acting on $H_X \otimes H$ via the amplified representation.

Write now $\mathcal{B}$ for $\mathcal{B}(H_X \otimes H)$, and $\mathcal{K}$ for $\mathcal{K}(H_X \otimes H)$. Let

$$D_{\mathcal{B}}(\mathcal{K}) := \{(S,T) \in \mathcal{B} \oplus \mathcal{B} \mid S - T \in \mathcal{K}\}$$

denote the double of $\mathcal{B}$ along $\mathcal{K}$ as in Definition 2.7.8. Consider the C^*-algebra

$$D_\infty := \frac{\prod_{n \in \mathbb{N}} D_{\mathcal{B}}(\mathcal{K})}{\bigoplus_{n \in \mathbb{N}} D_{\mathcal{B}}(\mathcal{K})}.$$

It follows from (a slight elaboration of) Lemma 6.1.2 that for any $T \in C^*(X)$, the commutator $[p_n, T]$ tends to zero as n tends to infinity, whence we have a well-defined $*$-homomorphism

$$\psi \colon C^*(X) \to D_\infty, \quad T \mapsto [(p_1 T, 1_{k_1} T), (p_2 T, 1_{k_2} T), \ldots].$$

Using Lemma 2.7.9, and the fact that $K_*(\mathcal{B}) = 0$ (see Corollary 2.7.7), there is a natural isomorphism $K_0(D_{\mathcal{B}}(\mathcal{K})) \cong \mathbb{Z}$. Piecing these isomorphisms together gives us a map

$$K_0(D_\infty) \to \frac{\prod_{n \in \mathbb{N}} \mathbb{Z}}{\bigoplus_{n \in \mathbb{N}} \mathbb{Z}}$$

and the map ϕ from the statement is by definition the composition of this and the map $\psi_* \colon K_0(C^*(X)) \to K_0(D_\infty)$ induced by ψ.

The compatibility of this with the pairing with K-homology is clear from the 'naive' description of the pairing from Definition 9.1.3. □

The point of Proposition 11.1.4 is that it gives us a simple way of showing that certain elements of K-homology are not sent to zero by the assembly map.

Corollary 11.1.5 *Say $\beta \in K_0(X)$ is a K-homology class such that there exists an almost constant K-theory class $\alpha \in K^0(X)$ with $\langle \beta, \alpha \rangle \neq 0$. Then $\mu(\beta) \neq 0$.*

Proof Let $(p_n, 1_{k_n})$ be any almost constant sequence such that $[p_n] - [1_{k_n}] = \alpha$ for all n. Let

$$\phi \colon K_0(C^*(X)) \to \frac{\prod_n \mathbb{Z}}{\bigoplus_n \mathbb{Z}}$$

be the homomorphism from Proposition 11.1.4. Then Proposition 11.1.4 implies that $\phi(\mu(\beta))$ is the image of the constant sequence with value $\langle \beta, \alpha \rangle$ in $\prod \mathbb{Z} / \oplus \mathbb{Z}$, and this is non-zero. □

This corollary is relatively simple, but already quite powerful. As an example of how it can be used, here is another proof that the d-torus does not admit a metric of positive scalar curvature, at least for d even.

Theorem 11.1.6 *The d-torus does not admit a metric of positive scalar curvature for d even.*

Proof Using Exercise 10.4.1, it suffices to prove that the assembly map

$$\mu\colon K_*(\mathbb{R}^d) \to K_*(C^*(\mathbb{R}^d))$$

is injective. However, we know from Bott periodicity as in Theorem 9.3.5 and Exercise 6.8.7 that $K_0(\mathbb{R}^d) \cong \mathbb{Z}$, and that the class of the Bott bundle in $K^0(\mathbb{R}^d)$ pairs non-trivially with the canonical generator given by the Dirac operator. Example 11.1.3 and Corollary 11.1.5 complete the proof. □

In the rest of this section, we quickly sketch how to extend the above to get similar results in K_1.

Definition 11.1.7 Let $i \geqslant 0$. An *almost constant sequence* of projections for $\mathbb{R}^i \times X$ is a sequence of pairs $(p_n, 1_{k_n})$ of functions from $(\mathbb{R}^i \times X)^+$ to $\mathcal{K}$ with the following properties:

(i) for each n, p_n is a projection;
(ii) for each n, the difference $p_n - 1_{k_n}$ is in $C_c(\mathbb{R}^i \times X, \mathcal{K})$;
(iii) for each $r > 0$,

$$\limsup_{n\to\infty} \sup_{t\in\mathbb{R}^i} \{|p_n(t,x) - p_n(t,y)| \mid x,y \in X,\ d(x,y) \leqslant r\} = 0.$$

A class x in $K^{-i}(X) = K^0(\mathbb{R}^i \times X)$ is *almost constant* if there exists an almost constant sequence (p_n, q_n) of projections for $\mathbb{R}^i \times X$ such that $x = [p_n] - [1_{k_n}]$ for all n.

Proposition 11.1.8 *Let $i \geqslant 0$. Let $C^*(X)$ be the Roe algebra of X and let $(p_n, 1_{k_n})$ be an almost constant sequence for $\mathbb{R}^i \times X$. Then there is a well-defined homomorphism*

$$\phi\colon K_{-i}(C^*(X)) \to \frac{\prod_n \mathbb{Z}}{\bigoplus_n \mathbb{Z}}$$

such that if

$$\mu\colon K_*(X) \to K_*(C^*(X))$$

is the assembly map from Definition 7.1.1, then

$$\phi(\mu(y)) = [\langle \beta, [p_0] - [q_0]\rangle, \langle \beta, [p_1] - [q_1]\rangle, \ldots] \tag{11.2}$$

for all β in $K_i(X)$.

Proof The argument from Proposition 11.1.4 applied pointwise in the $\mathbb{R}^i$ variable gives a homomorphism

$$\phi\colon K_{-i}(C^*(X)) = K_0(C_0(\mathbb{R}^i, C^*(X))) \to \frac{\prod_n K^0(\mathbb{R}^{2i})}{\bigoplus_n K^0(\mathbb{R}^{2i})}.$$

Now apply Bott periodicity to get the image in $\prod_n \mathbb{Z}/\bigoplus_n \mathbb{Z}$. □

The following corollary is proved in exactly the same way as Corollary 11.1.9.

Corollary 11.1.9 *Say $\beta \in K_1(X)$ is a K-homology class such that there exists an almost constant K-theory class $\alpha \in K^1(X)$ with $\langle \beta, \alpha \rangle \neq 0$. Then $\mu(\beta) \neq 0$.* □

Using an obvious analogue of the argument in Example 11.1.3, we can show that any class in $K^{-i}(\mathbb{R}^k)$ is almost constant. Thus we may remove the dimension restriction in Corollary 11.1.6.

Theorem 11.1.10 *The d-torus does not admit a metric of positive scalar curvature.* □

In the next section, we will push this idea quite a bit further, using the geometry of non-positive and negative curvature to give injectivity results for assembly maps (with trivial group action).

11.2 Non-positive Curvature

In this section we look at some examples where one can produce almost flat sequences. Throughout this section, X is a proper geodesic metric space: i.e. X is a metric space, all closed balls in X are compact, for any $x, y \in X$ there is a continuous function $\gamma\colon [0, d(x,y)] \to X$ such that $\gamma(0) = x$, $\gamma(1) = y$, and such that the length of γ equals the distance from x to y (see Definition A.3.4 for more on this). Such a function γ is called a *geodesic*. Note that we do not demand any uniqueness for geodesics between two points (although this will be true for our main examples).

Definition 11.2.1 Let a, b, c be any three points in X. Let A, B, C be any three points in the Euclidean plane such that we have equality of pairwise distances:

$$d_{\mathbb{R}^2}(A,B) = d_X(a,b), \quad d_{\mathbb{R}^2}(B,C) = d_X(b,c), \quad d_{\mathbb{R}^2}(A,C) = d_X(a,c)$$

(it is easy to see that such A, B, C always exist and are unique up to an isometry of the ambient plane). Let γ_{AB} and γ_{AC} be the geodesics from A to B and from B to C respectively in $\mathbb{R}^2$.

The triple (a,b,c) satisfies the *CAT(0)-inequality*[1] if, for any geodesics γ_{ab} and γ_{ac} from a to b and a to c respectively, and for any $t_b \in [0, d(a,b)]$ and any $t_c \in [0, d(a,c)]$, we have that

$$d_X(\gamma_{ab}(t_b), \gamma_{ac}(t_c)) \leqslant d_{\mathbb{R}^2}(\gamma_{AB}(t_b), \gamma_{AC}(t_c)).$$

The metric space X is *CAT(0)* if any triple (a,b,c) satisfies the CAT(0) inequality.

Intuitively, the above definition says that triangles in X are 'no fatter' than triangles in the Euclidean plane.

Examples 11.2.2 (i) Euclidean spaces are (tautologically) CAT(0).

(ii) Let T be a tree (i.e. a connected, undirected graph with no loops, at most one edge between any two points and no non-trivial circuits), and metrise the underlying topological space of T by stipulating that each edge is isometric to $[0,1]$ and the distance between two points is the length of the shortest path between them. Then T is CAT(0): the key point is that triangles in T have 'zero width' in an appropriate sense, and therefore are certainly not fatter than their Euclidean counterparts.

(iii) The hyperbolic plane, and more generally hyperbolic n-space, is CAT(0).

We now collect some basic facts about CAT(0) spaces: our first goal is Proposition 11.2.7, which gives us information about the assembly map for CAT(0) spaces. We will then combine that with the Cartan–Hadamard theorem (Theorem 11.2.10) and Exercise 10.4.1 to deduce that certain manifolds cannot carry metrics of positive scalar curvature.

Lemma 11.2.3 *Say X is a CAT(0) space, and let $\gamma_1 \colon [0,b_1] \to X$ and $\gamma_2 \colon [0,b_2] \to X$ be any two geodesics starting from the same point, i.e. such that $\gamma_1(0) = \gamma_2(0)$. Then for any $t \in [0,1]$,*

$$d(\gamma_1(tb_1), \gamma_2(tb_2)) \leqslant t d(\gamma_1(b_1), \gamma_2(b_2)).$$

Proof This follows directly from the CAT(0) inequality for the triple

$$(\gamma_1(0), \gamma_1(b_1), \gamma_2(b_2))$$

and the corresponding fact for the Euclidean plane. □

The following corollary is immediate.

Corollary 11.2.4 *Let X be a CAT(0) space and x, y be points in X. Then there is a unique geodesic from x to y.* □

[1] Note that this does not depend on the choice of (A,B,C).

Corollary 11.2.5 *CAT(0) spaces are contractible.*

Proof Let X be CAT(0) and fix a basepoint $x_0 \in X$. For each $x \in X$ let $\gamma_{x_0 x}$ be the unique geodesic from x_0 to x as in Corollary 11.2.4. A contracting homotopy is then defined by

$$H\colon [0,1] \times X \to X, \quad H(x,t) = \gamma_{x_0 x}(td(x_0,x))$$

(note that continuity of H follows from Lemma 11.2.3). □

Corollary 11.2.6 *Let X be a (proper) CAT(0) space. Fix a basepoint $x_0 \in X$, and for each $x \in X$ let $\gamma_{x_0 x}$ be the unique geodesic from x_0 to x as in Corollary 11.2.4. Define a function*

$$s\colon X \to X, \quad x \mapsto \gamma_{x_0 x}\left(\frac{d(x_0,x)}{2}\right).$$

Then s is continuous, proper and properly homotopic to the identity. Moreover,

$$d(s(x),s(y)) \leqslant \frac{1}{2}d(x,y) \tag{11.3}$$

for all $x, y \in X$.

Proof Continuity and the inequality in line (11.3) are immediate from Lemma 11.2.3. Properness of s follows as any compact subset K of X is contained in some ball $B(x_0;r)$, and so $s^{-1}(K)$ is contained in $B(x_0;2r)$ and thus has compact closure by properness of X. A homotopy between s and the identity is given by

$$H\colon [0,1] \times X \to X, \quad H(t,x) = \gamma_{x_0 x}\left(\frac{d(x_0,x)}{2-t}\right);$$

properness of H follows by a similar argument to properness of s, and continuity of H again follows from Lemma 11.2.3. □

Proposition 11.2.7 *Say X is a CAT(0) space. Then every class in $K^*(X)$ is almost constant.*

Proof Let $\alpha = [p] - [1_k]$ be a class in $K^{-i}(X)$, where p is a projection in $C((\mathbb{R}^i \times X)^+, \mathcal{K})$ and $p - 1_k$ is in $C_c(\mathbb{R}^i \times X, M_k(\mathbb{C}))$; any class can be represented in this form. Let s be as in Lemma 11.3 and define

$$\tilde{s}\colon \mathbb{R}^i \times X \to \mathbb{R}^i \times X, \quad (t,x) \mapsto (t,s(x)).$$

Set $p_n = p \circ \tilde{s}^n$. Corollary 11.2.6 combined with uniform continuity of p shows that the sequence $(p_n, 1_k)$ has the properties required by Definition 11.1.7 to show that $[p] - [1_k]$ is almost constant. □

Corollary 11.2.8 *Say X is a CAT(0) space such that $K^*(X)$ is finitely generated. Then the assembly map*

$$\mu\colon K_*(X) \to K_*(C^*(X))$$

as in Definition 7.1.1 is rationally injective.

Proof This follows immediately from Proposition 11.2.7, Corollary 11.1.9 and the universal coefficient theorem (Theorem 9.2.8). □

Definition 11.2.9 The space X has *non-positive curvature* if for any $x \in M$ there exists $r > 0$ such that for any $y, z \in B(x; r)$, the triple (x, y, z) satisfies the CAT(0) inequality from Definition 11.2.1.

Thus a space X has non-positive scalar curvature if it satisfies the CAT(0) inequality on 'small scales'. For example, this is true if X is the quotient of a CAT(0) space by a group acting freely, properly co-compactly (see Definition A.2.2) by isometries. This applies to many examples of classical interest, such as when X is a closed surface of genus at least one, and higher-dimensional analogues of such spaces.

We now turn our attention to manifolds. The following is a version of the *Cartan–Hadamard theorem*. We will not prove this here: see Section 11.4, Notes and References, at the end of the chapter.

Theorem 11.2.10 *Say M is a complete Riemannian manifold with non-positive curvature. Then the universal cover $\widetilde{M}$ with the lifted metric is a CAT(0) space.* □

Theorem 11.2.11 *Let M be a closed Riemannian manifold, and assume that the metric has non-positive curvature. Then M cannot admit a (different) metric of positive scalar curvature.*

Proof The Cartan–Hadamard Theorem 11.2.10 implies that the universal cover $\widetilde{M}$ is CAT(0) and thus contractible by Corollary 11.2.5. Hence by Corollary 9.6.12 $K_i(\widetilde{M})$ is isomorphic to $\mathbb{Z}$ if $i = \dim(X) \pmod 2$ and is zero otherwise, and the non-zero group is generated by the class of the Dirac operator (note that $\widetilde{M}$ is spin^c as contractible). Corollary 11.2.8 implies that the assembly map

$$\mu\colon K_*(\widetilde{M}) \to K_*(C^*(\widetilde{M}))$$

is rationally injective; given that the left-hand side is just a copy of $\mathbb{Z}$, however, this is the same as injectivity. The result follows from Exercise 10.4.1. □

11.3 Exercises

11.3.1 Prove the claim in Remark 11.1.2: if X is bounded, any class in $K^0(X)$ that is represented by an almost constant sequence of projections with image in a fixed corner of $\mathcal{K}$ is a multiple of the class of the identity in K-theory.
Hint: the space of projections in $M_N(\mathbb{C})$ is locally contractible, so any projection-valued map $p\colon X \to M_N(\mathbb{C})$ whose image has small enough diameter is homotopic to a constant map.

11.3.2 Show that a CAT(0) space is uniformly contractible (see Definition 7.3.1).
It follows from this and Theorem 7.3.6 that if we can give such an X the structure of a good simplicial complex, then the coarse Baum–Connes assembly map identifies with the assembly map for X. Corollary 11.2.8 combined with the UCT then implies that the coarse Baum–Connes assembly map is rationally injective.

11.4 Notes and References

Bridson and Haefliger [38] contains a great deal of information about the geometry of non-positive curvature. In particular, it includes a proof (chapter II.1 and appendix of Bridson and Haefliger [38]) that in the case of a manifold, the definition of non-positive curvature that we used (Definition 11.2.9 above) is equivalent to the standard definition of non-positive sectional curvature, defined using the Riemannian curvature tensor. It also proves a more general version of the Cartan–Hadamard theorem: see chapter II.4 of Bridson and Haefliger [38]. The class of CAT(0) spaces is large and interesting: one particularly beautiful class of examples comes from the work of Davis [78], who shows that there are closed smooth manifolds with universal covers admitting a CAT(0) metric (so in particular, the original manifold is aspherical), but where the cover is not homeomorphic to Euclidean space (such a manifold cannot admit a metric of non-positive Riemannian curavture).

Our Corollary 11.2.8 proves rational injectivity of the coarse Baum–Connes assembly map for certain CAT(0) spaces. In fact, the coarse assembly map is know to be an isomorphism in this case, as one can see using the ideas in Higson and Roe [133] (and by other methods). One also knows injectivity of the coarse Baum-Connes assembly map for arbitrary subspaces of non-positively curved manifolds (Shan and Wang [235]). This, however, is open for arbitrary subspaces of CAT(0) spaces; moreover, surjectivity of the coarse assembly map can fail for (bounded geometry) subspaces of CAT(0) spaces, as one can see by combining results from Kondo [157] and Willett and Yu [258].

The result of Theorem 11.2.11 that a closed manifold of non-positive sectional curvature does not admit a metric of positive scalar curvature is due to Gromov and Lawson: for an example of some of their work, see [180]. The technique they use is essentially the same as ours, but looks much more differential-geometric than our soft-and-topological proof.

One can get a more general notion than our almost constant classes by considering 'almost flat classes', also due to Gromov and Lawson. See for example Carrión and Dadarlat [47] or Hunger [144] for foundational material. Subsequent work of Connes, Gromov and Moscovici [62, 63] pushed related ideas much further. More recent applications of almost flat ideas include: proofs of the Novikov conjecture for 'low-degree' classes by Hanke and Schick [122] and Mathai [175]; recent work of Kubota [158] on relative index theory; and connections to quasi-diagonality in abstract C^*-algebra theory due to Dadarlat [73].

12

Higher Index Theory for Coarsely Embeddable Spaces

Our goal in this chapter is to prove the coarse Baum–Connes conjecture for bounded geometry metric spaces that coarsely embed into Hilbert space.

Definition 12.0.1 Let X and Y be metric spaces. A map $f\colon X \to Y$ is a *coarse embedding* if there exist non-decreasing functions $\rho_-, \rho_+\colon [0,\infty) \to [0,\infty)$ such that

$$\rho_-(d_X(x_1,x_2)) \leqslant d_Y(f(x_1), f(x_2)) \leqslant \rho_+(d_X(x_1,x_2))$$

for all $x_1, x_2 \in X$, and such that $\rho_-(t) \to \infty$ as $t \to \infty$.

The space X is said to *coarsely embed* into the space Y if a coarse embedding from X to Y exists.

A coarse embedding between two metric spaces may or may not exist. For example, if Y is bounded, then there can be no coarse embedding of an unbounded metric space X into Y. On the other hand, *any* separable metric space X coarsely embeds into $\ell^\infty(\mathbb{N})$: see Exercise 12.7.1. Here is the main theorem we aim to prove in this chapter.

Theorem 12.0.2 *Let X be a bounded geometry metric space that coarsely embeds into a Hilbert space. Then the coarse assembly map*

$$\mu\colon KX_*(X) \to K_*(C^*(X))$$

is an isomorphism.

Much of the power of Theorem 12.0.2 comes as the existence of a coarse embedding into an infinite-dimensional Hilbert space is a fairly weak condition, satisfied by many spaces (and groups) of classical interest.

On a vague level, the idea of this theorem is that the existence of a coarse embedding of X into a Hilbert space H says that one can draw a 'good' picture of X inside H. The very well-behaved geometry of Hilbert spaces then allows us to prove Theorem 12.0.2 via an index-theoretic localisation technique. The key technical ingredient is a careful analysis of the so-called Bott–Dirac operator of Section 9.3 on finite-dimensional Hilbert spaces, and the closely related classical harmonic oscillator of mathematical physics.

All of this chapter is devoted to the proof of Theorem 12.0.2, which is the deepest result in this book. The chapter is structured as follows.

Sections 12.1 and 12.2 provide background. Section 12.1 studies some properties of the Bott–Dirac operator on finite-dimensional Hilbert spaces. Section 12.2 establishes notation and proves some combinatorial and analytic facts coming from bounded geometry. Both Sections 12.1 and 12.2 are somewhat technical if read in isolation; we recommend that the reader just skim them (or not read them at all) on the first reading and refers back as needed.

Sections 12.3 and 12.4 contain the proof of Theorem 12.0.2 in the case of a coarse embedding into a finite-dimensional Hilbert space E; this is the heart of the proof and contains all the key ideas. Section 12.3 uses the (higher) index theory of the Bott–Dirac operator to replace the Roe algebra $C^*(X)$ with another C^*-algebra $A(X;E)$ without losing any K-theoretic information. The C^*-algebra $A(X;E)$ is 'local' in some sense, which allows one to prove the analogue of the coarse Baum–Connes conjecture for it by Mayer–Vietoris arguments; this step is carried out in Section 12.4. Together, these two sections establish Theorem 12.0.2 in the case that the Hilbert space is finite-dimensional.

The remaining two sections 12.5 and 12.6 give the proof of Theorem 12.0.2 in the case of a coarse embedding into an infinite-dimensional Hilbert space. Section 12.5 uses Mayer–Vietoris arguments to reduce to a statement about a sort of 'uniform' version of the coarse Baum–Connes conjecture for a sequence (X_n) of metric spaces that coarsely embed into a sequence (E_n) of finite-dimensional Euclidean spaces in an appropriate sense. Finally, Section 12.6 completes the proof of Theorem 12.0.2 by explaining how to adapt the finite-dimensional proof of Sections 12.3 and 12.4 to the uniform statement arrived at in Section 12.5.

12.1 The Bott–Dirac Operator

In this section, we introduce the Bott–Dirac operator and study some of its key properties. See Section 12.8, Notes and References, at the end of this chapter for a concrete discussion of the Bott–Dirac operator in the one-dimensional case.

To fix terminology, let us say that by a *Euclidean space* we mean a Hilbert space over the real numbers. It is straightforward to check that a metric space X

coarsely embeds into a real Hilbert space if and only if it coarsely embeds into a complex Hilbert space (see Exercise 12.7.2), so we do not lose any generality by restricting to the real case.

Let then E be a finite-dimensional, even-dimensional Euclidean space. We write $|v|$ for the norm of an element v of E. As in Example E.1.11, the *(complex) Clifford algebra* of E, denoted $\mathrm{Cliff}_{\mathbb{C}}(E)$, is the universal unital complex algebra containing E as a (real) subspace and subject to the multiplicative relations

$$xx = |x|^2$$

for all $x \in E$ (on the left-hand side we write "xx" for the multiplication in $\mathrm{Cliff}_{\mathbb{C}}(E)$, and on the right-hand side $|x|^2$ means the scalar $|x|^2$ multiplied by the identity of $\mathrm{Cliff}_{\mathbb{C}}(E)$). Just as in the discussion in Example E.2.12, we can treat $\mathrm{Cliff}_{\mathbb{C}}(E)$ as either a graded C^*-algebra or a graded Hilbert space; considered in the latter way, we write it H_E.

Let $\mathcal{L}^2_E$ denote the Hilbert space of square integrable functions from E to H_E and let $\mathcal{S}_E$ denote the subspace of Schwartz-class functions from E to H_E. Note that both $\mathcal{L}^2_E$ and $\mathcal{S}_E$ inherit a grading from the grading on $\mathrm{Cliff}_{\mathbb{C}}(E)$. Fix for now an orthonormal basis $\{e_1, \ldots, e_d\}$ of E, and let $x_1, \ldots, x_d \colon E \to \mathbb{R}$ be the corresponding coordinates. The *Clifford* and *Dirac* operators, which we think of as unbounded operators on $\mathcal{L}^2_E$ with domain $\mathcal{S}_E$, are defined by the formulas

$$(Cu)(v) = \sum_{i=1}^{d} (x_i e_i) \cdot u(x) \quad \text{and} \quad Du = \sum_{i=1}^{d} \hat{e}_i \frac{\partial u}{\partial x_i} \tag{12.1}$$

just as in Definition 9.3.1. One can check that C and D do not depend on the choice of orthonormal basis of E.

Definition 12.1.1 The *Bott–Dirac operator* is the unbounded operator

$$B = D + C$$

on $\mathcal{L}^2_E$ with domain $\mathcal{S}$.

Note that B is an odd operator. Moreover, it maps its domain into itself and thus powers of B make sense. The following result records the basic properties of B^2 that we will need.

Proposition 12.1.2 *The eigenvalues of B^2 are exactly the non-negative even integers, with the kernel being one-dimensional and spanned by the function*

$$E \to \mathbb{R}, \quad x \mapsto e^{-\frac{1}{2}|x|^2}.$$

Moreover, if H_{2n} is the eigenspace corresponding to the eigenvalue $2n$, then H_{2n} is a finite-dimensional subspace of $\mathcal{S}$, and there is an orthogonal direct sum decomposition

$$\mathcal{L}_E^2 = \bigoplus_{n=0}^{\infty} H_{2n}.$$

In particular, the Bott–Dirac operator B is essentially self-adjoint, and has compact resolvent.

Finally, we have the formula

$$B^2 = D^2 + C^2 + N,$$

where the so-called number operator $N := CD + DC$ *extends to a bounded self-adjoint operator on $\mathcal{L}_E^2$ with norm equal to the dimension d of E.*

Proof The discussion in Remark 9.3.2 and the choice of orthonormal basis gives rise to a tensorial decomposition

$$\mathcal{L}_E^2 \cong \underbrace{L^2(\mathbb{R}, H_{\mathbb{R}}) \widehat{\otimes} \cdots \widehat{\otimes} L^2(\mathbb{R}, H_{\mathbb{R}})}_{d \text{ copies}},$$

where the ith copy of $\mathbb{R}$ is identified with $\text{span}\{e_i\}$. Analogously to Remark 9.3.2, we have that if B_1 is the one-dimensional Bott–Dirac operator, then

$$B = \sum_{i=1}^{d} 1 \widehat{\otimes} \cdots \widehat{\otimes} 1 \widehat{\otimes} \underbrace{B_1}_{i\text{th place}} \widehat{\otimes} 1 \widehat{\otimes} \cdots \widehat{\otimes} 1.$$

Choosing an orthonormal basis $\{1, e\}$ for H_1 where $e \in \mathbb{R}$ is a norm one vector, we have

$$B_1 = \begin{pmatrix} 0 & x - \frac{d}{dx} \\ x + \frac{d}{dx} & 0 \end{pmatrix},$$

just as in the proof of Theorem 9.3.5. It follows from the last two displayed lines that

$$B^2 = \sum_{i=1}^{d} 1 \widehat{\otimes} \cdots \widehat{\otimes} 1 \widehat{\otimes} \underbrace{B_1^2}_{i\text{th place}} \widehat{\otimes} 1 \widehat{\otimes} \cdots \widehat{\otimes} 1$$

where

$$B_1^2 = \begin{pmatrix} H & 0 \\ 0 & H+2 \end{pmatrix},$$

with H with harmonic oscillator of Definition D.3.1. The result follows from this, the eigenspace decomposition for the harmonic oscillator from Proposition D.3.3 and direct computations that we leave to the reader. □

Now, the origin $0 \in E$ plays the role of a basepoint for B: for example, the kernel of B is spanned by the Gaussian $e^{-\frac{1}{2}|x|^2}$ centred at zero. We will need to consider other basepoints for B, and also to introduce another parameter that will govern a deformation.

For $x \in E$, let c_x be the bounded operator on $\mathcal{L}^2_E$ defined by (left) Clifford multiplication by the fixed vector x. Note that c_x is self-adjoint and $c_x^2 = |x|^2$, whence the spectrum of c_x consists precisely of $\pm|x|$ and the norm of c_x is $|x|$.

Definition 12.1.3 Let $s \in [1,\infty)$ and $x \in E$. The *Bott–Dirac operator associated to* (s,x) is the unbounded operator

$$B_{s,x} = s^{-1}D + C - c_x$$

on $\mathcal{L}^2_E$ with domain $\mathcal{S}$.

For $x \in E$ define $V_x : \mathcal{L}^2_E \to \mathcal{L}^2_E$ to be the unitary translation given by

$$(V_x u)(y) = u(y - x) \tag{12.2}$$

and for each $s \in [1,\infty)$ define a unitary shrinking operator $S_s : \mathcal{L}^2_E \to \mathcal{L}^2_E$ by the formula

$$(S_s u)(x) := s^{-\dim(E)/2} u(sx). \tag{12.3}$$

Then V_x and S_s both preserve the space $\mathcal{S}_E$ of Schwartz-class functions, and so the operator $V_x S_s B S_s^* V_x^*$ makes sense as an unbounded operator on $\mathcal{L}^2_E$ with domain $\mathcal{S}$. Moreover, it is straightforward to check that

$$B_{s,x} = s^{-1/2} V_x S_{\sqrt{s}} B S^*_{\sqrt{s}} V_x^* \tag{12.4}$$

as operators on $\mathcal{S}$. The following corollary is immediate from this formula and Proposition 12.1.2.

Corollary 12.1.4 *Each $B^2_{s,x}$ is an essentially self-adjoint, odd operator of compact resolvent. It has eigenvalues equal to $s^{-1/2}$ times the non-negative even integers, finite-dimensional eigenspaces and one-dimensional kernel spanned by the function*

$$E \to \mathbb{C} \subseteq \mathit{Cliff}_{\mathbb{C}}(E), \quad y \mapsto e^{-\frac{s}{2}|y-x|^2}.$$

Moreover, we have the formula

$$B_{s,x}^2 = s^{-2}D^2 + (C - c_x)^2 + s^{-1}N,$$

where N is the same number operator as in Proposition 12.1.2. □

In our set-up, it is convenient to work with bounded versions of the Bott–Dirac operator. We build these using the functional calculus for unbounded operators (see Theorem D.1.7).

Definition 12.1.5 Let $x \in E$ and $s \in [1, \infty)$, and let $B_{s,x} = s^{-1}D + C - c_x$ be as in Definition 12.1.3 above. Define

$$F_{s,x} := B_{s,x}(1 + B_{s,x}^2)^{-1/2}.$$

Our aim in the remainder of this section is to prove some useful properties of the operators $F_{s,x}$: see Proposition 12.1.10 below.

We will need the following integral representation of $F_{s,x}$. For the statement (and at several other points below), recall that a net (T_i) of bounded operators on a Hilbert space H converges *strongly* to an operator T if for all $u \in H$, $T_i u \to u$ as $i \to \infty$. Similarly, (T_i) converges *strong-$*$* to T if (T_i) converges strongly to T, and (T_i^*) converges strongly to T^*. Recall also that c_x denotes the operator on $\mathcal{L}_E^2$ of Clifford multiplication by the fixed vector $x \in E$

Lemma 12.1.6 *For all $s \in [1, \infty)$ and all $x \in E$, we have that*

$$F_{s,x} = \frac{2}{\pi} \int_0^\infty B_{s,x}(1 + \lambda^2 + B_{s,x}^2)^{-1} d\lambda,$$

with the integral on the right converging in the strong-$*$ operator topology.

Moreover, for any $s \in [1, \infty)$ and $x, y \in E$ we have

$$\begin{aligned} F_{s,x} - F_{s,y} &= c_{x-y}(1 + B_{s,x}^2)^{-1/2} \\ &\quad + \frac{2}{\pi} \int_0^\infty B_{s,y}(1 + \lambda^2 + B_{s,y}^2)^{-1}\Big(B_{s,y}c_{x-y} + c_{x-y}B_{s,x}\Big) \\ &\quad \times (1 + \lambda^2 + B_{s,x}^2)^{-1}\Big)d\lambda, \end{aligned}$$

where convergence again takes place in the strong-$*$ topology.

Proof The integral formula follows from the formula

$$\frac{x}{\sqrt{1 + x^2}} = \frac{2}{\pi} \int_0^\infty \frac{x}{1 + \lambda^2 + x^2} d\lambda$$

and the functional calculus: indeed, strong-$*$ convergence of the integral is straightforward to check given the description of the eigenspace decomposition of $B_{s,v}$ in Corollary 12.1.4 (one only gets strong-$*$ convergence rather than norm convergence as the integral does not converge uniformly for all x, only uniformly on compact subsets of $x \in \mathbb{R}$).

For the second formula, we have

$$F_{s,x} - F_{s,y} = (B_{s,x} - B_{s,y})(1 + B_{s,x}^2)^{-1/2} - B_{s,y}\big((1 + B_{s,y}^2)^{-1/2} - (1 + B_{s,x}^2)^{-1/2}\big).$$

Using (a slight variation on) the integral formula and the computation $B_{s,x} - B_{s,y} = c_{x-y}$, this equals

$$c_{x-y}(1 + B_{s,x}^2)^{-1/2} - \frac{2}{\pi}\int_0^\infty B_{s,y}(1 + \lambda^2 + B_{s,y}^2)^{-1} d\lambda + \frac{2}{\pi}\int_0^\infty B_{s,y}(1 + \lambda^2 + B_{s,x}^2)^{-1} d\lambda.$$

Combining the integrals and using the formulas

$$(1 + \lambda^2 + B_{s,y}^2)^{-1} - (1 + \lambda^2 + B_{s,x}^2)^{-1} = (1 + \lambda^2 + B_{s,y}^2)^{-1}(B_{s,y}^2 - B_{s,x}^2)(1 + \lambda^2 + B_{s,x}^2)^{-1}$$

and

$$B_{s,y}^2 - B_{s,x}^2 = B_{s,y}(B_{s,y} - B_{s,x}) + (B_{s,y} - B_{s,x})B_{s,x} = B_{s,y}c_{y-x} + c_{y-x}B_{s,x}$$

to manipulate the integrand gives the result. □

We record the following corollary for later use.

Corollary 12.1.7 *For all s and all $x, y \in E$,*

$$\|F_{s,x} - F_{s,y}\| \leqslant 3|x - y|.$$

Proof We estimate using the formula for $F_{s,x} - F_{s,y}$ from Lemma 12.1.6. First, note that the functional calculus gives

$$\|c_{x-y}(1 + B_{s,x}^2)^{-1/2}\| \leqslant \|c_{x-y}\|\|(1 + B_{s,x}^2)^{-1/2}\| \leqslant |x - y|.$$

Second, note that by the functional calculus $\|B_{s,y}^2(1 + \lambda^2 + B_{s,y}^2)^{-1}\| \leqslant 1$ and $\|(1 + \lambda^2 + B_{s,x}^2)^{-1}\| \leqslant (1 + \lambda^2)^{-1}$, whence

$$\left\|\frac{2}{\pi}\int_0^\infty B_{s,y}^2(1 + \lambda^2 + B_{s,y}^2)^{-1}c_{x-y}(1 + \lambda^2 + B_{s,x}^2)^{-1} d\lambda\right\| \leqslant \frac{2}{\pi}\int_0^\infty \frac{|x - y|}{1 + \lambda^2} d\lambda = |x - y|.$$

Finally, the elementary estimate

$$\sup_{x\in[0,\infty)} \frac{x}{1+\lambda^2+x^2} \leqslant \frac{1}{2\sqrt{1+\lambda^2}}.$$

and the functional calculus imply that

$$\left\| \frac{2}{\pi}\int_0^\infty B_{s,y}(1+\lambda^2+B_{s,y}^2)^{-1}c_{x-y}B_{s,y}(1+\lambda^2+B_{s,y}^2)^{-1}d\lambda \right\|$$
$$\leqslant \frac{2}{\pi}\int_0^\infty \frac{|x-y|}{4(1+\lambda^2)} \leqslant |x-y|.$$

Combining these three estimates with the formula for $F_{s,x}-F_{s,y}$ from Lemma 12.1.6 completes the proof. □

To state the next lemma, for each $x \in E$ and $R \geqslant 0$, let $\chi_{x,R}$ denote the characteristic function of the ball in E, centred at x and of radius R.

Lemma 12.1.8 *Let d be the dimension of E. Then for any $R \geqslant 0$, $\lambda \in [0,\infty)$, $x \in E$ and $s \in [2d,\infty)$, we have that*

$$\|(1+\lambda^2+B_{s,x}^2)^{-1/2}(1-\chi_{R,x})\| \leqslant \left(\frac{1}{2}+\lambda^2+R^2\right)^{-1/4}.$$

Proof For notational simplicity, let us assume that $x=0$, and write $\chi_R = \chi_{R,0}$; this makes no real difference to the proof. As in Corollary 12.1.4, we have

$$B_{s,0}^2 = s^{-2}D^2 + s^{-1}N + C^2,$$

where N is the number operator, a self-adjoint operator of norm d. Hence

$$(1+\lambda^2+B_{s,0}^2)^{-1/2} = (1+\lambda^2+s^{-2}D^2+s^{-1}N+C^2)^{-1/2}.$$

Now, let $u \in \mathcal{L}_E^2$ be Schwartz-class and have norm one. Define

$$v := (1+\lambda^2+B_{s,0}^2)^{-1}u = (1+\lambda^2+s^{-2}D^2+s^{-1}N+C^2)^{-1}u,$$

which is also Schwartz-class and has norm at most one. The operators D^2 and $\frac{1}{2}-s^{-1}N$ are positive (the latter follows as s is in $[2d,\infty)$ and as N is self-adjoint with norm d) so we have that

$$1 \geqslant |\langle u,v\rangle| = \langle (1+\lambda^2+s^{-2}D^2+s^{-1}N+C^2)v,v\rangle \geqslant \left\langle \left(\frac{1}{2}+\lambda^2+C^2\right)v,v\right\rangle.$$

Hence

$$\left\|\left(\frac{1}{2}+\lambda^2+C^2\right)^{1/2}v\right\|^2 = \left|\left\langle\left(\frac{1}{2}+\lambda^2+C^2\right)v,v\right\rangle\right| \leqslant 1. \tag{12.5}$$

On the other hand, for any Schwartz-class $w \in \mathcal{L}^2_E$ and any $x \in E$, we have that $\|(Cw)(x)\|_{\mathrm{Cliff}_{\mathbb{C}}(E)} \geqslant |x| \|w(x)\|_{\mathrm{Cliff}_{\mathbb{C}}(E)}$. Combining this with line (12.5) above gives that

$$\begin{aligned}\|(1-\chi_R)v\| &\leqslant \frac{1}{\sqrt{\frac{1}{2}+\lambda^2+R^2}} \left\| \left(\frac{1}{2}+\lambda^2+C^2\right)^{1/2} (1-\chi_R)v \right\| \\ &\leqslant \frac{1}{\sqrt{\frac{1}{2}+\lambda^2+R^2}} \left\| \left(\frac{1}{2}+\lambda^2+C^2\right)^{1/2} v \right\| \\ &\leqslant \frac{1}{\sqrt{\frac{1}{2}+\lambda^2+R^2}}.\end{aligned}$$

This implies that for any Schwartz-class function u of norm one, we get that

$$\|(1-\chi_R)(1+\lambda^2+B_{s,0}^2)^{-1}v\| \leqslant \frac{1}{\sqrt{\frac{1}{2}+\lambda^2+R^2}}.$$

As the Schwartz-class functions are dense in $\mathcal{L}^2$, this gives that

$$\|(1-\chi_R)(1+\lambda^2+B_{s,0}^2)^{-1}\| \leqslant \frac{1}{\sqrt{\frac{1}{2}+\lambda^2+R^2}}.$$

To complete the argument, note that by the C^*-identity, this implies that

$$\begin{aligned}\|(1+\lambda^2+B_{s,0}^2)^{-1/2}(1-\chi_R)\|^2 &= \|(1-\chi_R)(1+\lambda^2+B_{s,0}^2)^{-1}(1-\chi_R)\| \\ &\leqslant \|(1-\chi_R)(1+\lambda^2+B_{s,0}^2)^{-1}\| \\ &\leqslant \left(\frac{1}{2}+\lambda^2+R^2\right)^{-1/2}.\end{aligned}$$

Taking square roots gives the estimate we want. □

The following result is probably the most technically difficult of this section.

Proposition 12.1.9 *For any $r, \epsilon > 0$ there exists $R_0 > 0$ such that for all $R \geqslant R_0$, all $s \in [2d, \infty)$ and all $x, y \in E$ with $|x-y| \leqslant r$, we have*

$$\|(F_{s,x} - F_{s,y})(1-\chi_{x,R})\| < \epsilon.$$

Proof It will suffice to consider the case where $x = 0$: the general case differs from this by conjugation by the unitary V_x from line (12.2). For simplicity, write then $\chi_R := \chi_{0,R}$. Consider the formula from Lemma 12.1.6

$$F_{s,0} - F_{s,y} = \underbrace{c_{-y}(1+\lambda^2+B_{s,0}^2)^{-1/2}}_{\alpha}$$
$$\underbrace{+\frac{2}{\pi}\int_0^\infty B_{s,y}(1+\lambda^2+B_{s,y}^2)^{-1}\big(B_{s,y}c_{-y}+c_{-y}B_{s,0}\big)(1+\lambda^2+B_{s,0}^2)^{-1}d\lambda\,.}_{\beta} \tag{12.6}$$

It will suffice to show the corresponding estimate for each of the terms labelled α and β separately.

First, let us look at the term labelled α in line (12.6). Lemma 12.1.8 gives that for any $R > 0$

$$\|c_{-y}(1+B_{s,0}^2)^{-1/2}(1-\chi_R)\| \leqslant \frac{|y|}{\sqrt{R}},$$

which implies the desired statement for the term labelled α.

On the other hand, the expression β in line (12.6) above splits into a sum of two terms:

$$\frac{2}{\pi}\int_0^\infty B_{s,y}^2(1+\lambda^2+B_{s,y}^2)^{-1}c_{-y}(1+\lambda^2+B_{s,0}^2)^{-1}d\lambda \tag{12.7}$$

and

$$\frac{2}{\pi}\int_0^\infty B_{s,y}(1+\lambda^2+B_{s,y}^2)^{-1}c_{-y}B_{s,0}(1+\lambda^2+B_{s,0}^2)^{-1}d\lambda. \tag{12.8}$$

We will look at each separately.

First, look at the term in line (12.7). The functional calculus gives that

$$\|B_{s,y}^2(1+\lambda^2+B_{s,y}^2)^{-1}\| \leqslant 1.$$

Hence

$$\left\|\frac{2}{\pi}\int_0^\infty B_{s,y}^2(1+\lambda^2+B_{s,y}^2)^{-1}c_{-y}(1+\lambda^2+B_{s,0}^2)^{-1}d\lambda(1-\chi_R)\right\|$$
$$\leqslant \frac{2|y|}{\pi}\int_0^\infty \|(1+\lambda^2+B_{s,0})^{-1/2}\|\|(1+\lambda^2+B_{s,0})^{-1/2}(1-\chi_R)\|d\lambda.$$

The functional calculus gives that

$$\|(1+\lambda^2+B_{s,0})^{-1/2}\| \leqslant \frac{1}{\sqrt{1+\lambda^2}}$$

and Lemma 12.1.8 gives that

$$\|(1+\lambda^2+B_{s,0})^{-1/2}(1-\chi_R)\| \leqslant \left(\frac{1}{2}+\lambda^2+R^2\right)^{-1/4}.$$

Hence the term in line (12.7) multiplied by $1-\chi_R$ has norm bounded above by

$$\frac{2r}{\pi}\int_0^\infty \frac{1}{\sqrt{1+\lambda^2}\sqrt[4]{\frac{1}{2}+\lambda^2+R^2}}d\lambda.$$

This tends to zero as R tends to infinity by the dominated convergence theorem (at a rate depending on r, but not on $s\in[2d,\infty)$), so we are done with the term in line (12.7).

It remains to deal with the term in line (12.8). The functional calculus gives that

$$\|B_{s,y}(1+\lambda^2+B_{s,y}^2)^{-1}\|\leqslant \sup_{x\in[0,\infty)}\frac{x}{1+\lambda^2+x^2}.$$

Elementary calculus gives that for fixed λ, the function $f(x)=\frac{x}{1+\lambda^2+x^2}$ attains a maximum of $\frac{1}{2\sqrt{1+\lambda^2}}$ on the interval $[0,\infty)$, and so we get

$$\|B_{s,y}(1+\lambda^2+B_{s,y}^2)^{-1}\|\leqslant\frac{1}{2\sqrt{1+\lambda^2}}.$$

Hence

$$\left\|\frac{2}{\pi}\int_0^\infty B_{s,y}(1+\lambda^2+B_{s,y}^2)^{-1}c_{-y}B_{s,0}(1+\lambda^2+B_{s,0}^2)^{-1}d\lambda(1-\chi_R)\right\|$$

$$\leqslant\frac{2|y|}{\pi}\int_0^\infty\frac{1}{2\sqrt{1+\lambda^2}}\|B_{s,0}(1+\lambda^2+B_{s,0})^{-1/2}\|$$

$$\|(1+\lambda^2+B_{s,0})^{-1/2}(1-\chi_R)\|d\lambda.$$

The functional calculus again gives that

$$\|B_{s,0}(1+\lambda^2+B_{s,0})^{-1/2}\|\leqslant 1$$

and Lemma 12.1.8 again gives that

$$\|(1+\lambda^2+B_{s,0})^{-1/2}(1-\chi_R)\|\leqslant\frac{1}{\sqrt[4]{\frac{1}{2}+\lambda^2+R^2}}.$$

Hence the term in line (12.8) multiplied by $1-\chi_R$ has norm bounded above by

$$\frac{2r}{\pi}\int_0^\infty\frac{1}{2\sqrt{1+\lambda^2}}\frac{1}{\sqrt[4]{\frac{1}{2}+\lambda^2+R^2}}d\lambda.$$

Again this tends to zero as R tends to infinity by the dominated convergence theorem, and we are done. □

Having got through the above technical proof, the next result summarises the facts we will need about the Bott–Dirac operator. Fortunately, we have done most of the work already. For the statement, if $x \in E$ and $R \geqslant 0$, write again $\chi_{x,R}$ for the characteristic function of the ball in E centred at x and radius R.

Proposition 12.1.10 *For each $\epsilon > 0$ there exists an odd function $\Psi\colon \mathbb{R} \to [-1,1]$ such that $\Psi(t) \to \pm 1$ as $t \to \pm\infty$, and with the following properties.*

(i) *For all s and all x, $\|F_{s,x} - \Psi(B_{s,x})\| < \epsilon$.*

(ii) *There exists $R_0 > 0$ such that for all $s \in [1,\infty)$ and all $v \in E$, $\mathrm{prop}(\Psi(B_{s,x})) \leqslant s^{-1}R_0$.*

(iii) *For all $s \in [1,\infty)$ and all $x \in E$, the operator $\Psi(B_{s,x})^2 - 1$ is compact.*

(iv) *For all $s \in [1,\infty)$ and all $x, y \in E$, the operator $\Psi(B_{s,x}) - \Psi(B_{s,y})$ is compact.*

(v) *There exists $c > 0$ such that for all $s \in [1,\infty)$ and all $x, y \in E$,*

$$\|\Psi(B_{s,x}) - \Psi(B_{s,y})\| \leqslant c|x - y|.$$

(vi) *For all $x \in E$, the function*

$$[1,\infty) \to \mathcal{B}(\mathcal{L}^2_E) \quad s \mapsto \Psi(B_{s,x})$$

is strong-$*$ continuous.

(vii) *The family of functions*

$$[1,\infty) \to \mathcal{B}(\mathcal{L}^2_E) \quad s \mapsto \Psi(B_{s,x})^2 - 1$$

is norm equicontinuous as x varies over E and s varies over any fixed compact subset of $[1,\infty)$.

(viii) *For any $r \geqslant 0$, the family of functions*

$$[1,\infty) \to \mathcal{B}(\mathcal{L}^2_E) \quad s \mapsto \Psi(B_{s,x}) - \Psi(B_{s,y})$$

is norm equicontinuous as (x, y) varies over the elements of $E \times E$ with $|x - y| \leqslant r$, and s varies over any fixed compact subset of $[1,\infty)$.

(ix) *There exists $R_1 > 0$ such that for all $R \geqslant R_1$, all $s \in [1,\infty)$ and all $x \in E$ we have that*

$$\|(\Psi(B_{s,x})^2 - 1)(1 - \chi_{x,R})\| < 3\epsilon.$$

(x) *For any $r > 0$ there exists $R_2 > 0$ such that for all $R \geqslant R_2$ and all $s \in [2d,\infty)$ and all $x, y \in E$ with $|x - y| \leqslant r$ we have that*

$$\|(\Psi(B_{s,x}) - \Psi(B_{s,y}))(1 - \chi_{x,R})\| < 3\epsilon.$$

Remark 12.1.11 It will be important when we move to the infinite-dimensional case that the various constants appearing in almost all of the above are independent of the dimension of E. The only exceptions are that the various continuity statements in the parameter s do not give equicontinuity as the dimension varies, and the explicit appearance of the dimension in part (x).

Proof of Proposition 12.1.10 We claim first that there is an odd function $\Psi\colon \mathbb{R} \to [-1,1]$ such that $\Psi(t) \to \pm 1$ as $t \to \pm\infty$, such that $\sup_{t\in\mathbb{R}} |\Psi(t) - x(1+x^2)^{-1/2}| < \epsilon$, and such that the (distributional) Fourier transform of Ψ is compactly supported. To see this, let $g\colon \mathbb{R} \to [0,\infty)$ be a smooth even function of integral one, and with compactly supported Fourier transform. Define Ψ to the convolution

$$\Psi(x) := \int_{\mathbb{R}} y(1+y^2)^{-1/2}\delta g(\delta^{-1}(x-y))dy$$

for some suitably small $\delta > 0$. It is not too difficult to check that this works: for the statement on support of the Fourier transform, use that the Fourier transform converts convolution to multiplication.

Part (i) is now immediate from the functional calculus.

For part (ii), let R_0 be such that the support of the Fourier transform of Ψ is contained in $[-R_0, R_0]$. Then Corollary 8.2.3 combined with the fact that the propagation speed (see Definition 8.1.7) of $B_{s,x}$ is s^{-1} gives that $\text{prop}(\Psi(B_{s,x})) \leqslant s^{-1}R_0$.

Part (iii) follows as $\Psi^2 - 1$ is in $C_0(\mathbb{R})$, and from the eigenspace decomposition of $B_{s,x}$ as in Corollary 12.1.4.

For part (iv), let $\delta > 0$, define $\Psi_\delta(t) := \Psi(\delta t)$. Note that

$$B_{s,x} - B_{s,y} = c_{x-y}$$

extends to a bounded operator on $\mathcal{L}^2_E$ with norm $|x-y|$, whence Lemma D.2.4 and the conditions on Ψ imply there exists $c > 0$ (depending on Ψ) such that

$$\|\Psi_\delta(B_{s,x}) - \Psi_\delta(B_{s,y})\| \leqslant c\|\delta(B_{s,x} - B_{s,y})\| = c\delta|x-y|. \tag{12.9}$$

In particular, then, the difference

$$\begin{aligned}(\Psi - \Psi_\delta)(B_{s,x}) &- (\Psi - \Psi_\delta)(B_{s,y}) \\ &= \Psi_\delta(B_{s,x}) - \Psi_\delta(B_{s,y}) + \Psi(B_{s,y}) - \Psi(B_{s,x}) \end{aligned} \tag{12.10}$$

converges to $\Psi(B_{s,x}) - \Psi(B_{s,y})$ as $\delta \to 0$. Note further that for any δ the function $\Psi - \Psi_\delta$ is in $C_0(\mathbb{R})$, whence by the eigenspace decompositions from

Corollary 12.1.4 we have that $(\Psi - \Psi_\delta)(B_{s,x})$ and $(\Psi - \Psi_\delta)(B_{s,y})$ are both compact, which gives the result.

Part (v) is immediate from line (12.9) (with $\delta = 1$).

Part (vi) is straightforward from the eigenspace decomposition of Corollary 12.1.4, strong-$*$ continuity of the map $s \mapsto S_{\sqrt{s}}$ where $S_{\sqrt{s}}$ is the shrinking operator from line (12.3) and the formula in line (12.4).

For part (vii), norm continuity of the map $s \mapsto \Psi(B_{s,0})^2 - 1$ follows again from the eigenspace decomposition of Corollary 12.1.4, strong-$*$ continuity of the map $s \mapsto S_{\sqrt{s}}$ where $S_{\sqrt{s}}$ is the shrinking operator from line (12.3), the formula in line (12.4) and the fact that the function $\Psi^2 - 1$ is in $C_0(\mathbb{R})$. Equicontinuity of the family follows as all the functions involved are conjugates of this one by the unitaries V_x from line (12.2).

For part (viii) with notation as in the proof of part (iv) above, consider the function

$$s \mapsto (\Psi - \Psi_\delta)(B_{s,x})$$

for fixed $\delta > 0$. This is norm continuous by the eigenspace decomposition of $B_{s,v}$ from Corollary 12.1.4 and strong-$*$ continuity of the map $s \mapsto S_{\sqrt{s}}$ where $S_{\sqrt{s}}$ is the shrinking operator from line (12.3), and the formula in line (12.4). Moreover, it is equicontinuous as x varies over E and s over compact subsets of $[1, \infty)$ as all the functions involved are unitary conjugates of each other by the operators V_x of line (12.2). To summarise, putting this together with the estimate in line (12.9) and the formula in line (12.10), we have shown that for (x, y) varying across the subset of $E \times E$ consisting of elements with $|x - y| \leqslant r$, the family of functions $s \mapsto \Psi(B_{s,x}) - \Psi(B_{s,y})$ can be approximated uniformly by an equicontinuous family, which is enough to complete the proof.

For part (ix), note that it suffices by part (i) to prove that for any $\epsilon > 0$ there exists $R_1 > 0$ such that for all $R \geqslant R_1$ and all $s \in [1, \infty)$ and all $x \in E$, we have $\|(F_{s,x}^2 - 1)(\chi_{x,R} - 1)\| < \epsilon$. We have, however, from Lemma 12.1.8 that

$$\begin{aligned} \|(F_{s,x}^2 - 1)(1 - \chi_{x,R})\| &= \|(1 + B_{s,x}^2)^{-1}(1 - \chi_{x,R})\| \\ &\leqslant \|(1 + B_{s,x}^2)^{-1/2}\| \|(1 + B_{s,x}^2)^{-1/2}(1 - \chi_{x,R})\| \\ &\leqslant \left(\frac{1}{2} + R^2\right)^{-1/4}, \end{aligned}$$

which implies the desired result.

Finally, part (x) is straightforward from Proposition 12.1.9. □

12.2 Bounded Geometry Spaces

This section establishes notational conventions, and proves a few combinatorial and analytic facts, about Roe algebras associated to (Rips complexes of) bounded geometry metric spaces as in Definition A.3.19.

Throughout this section, we let X be a bounded geometry metric space equipped with a coarse embedding $f: X \to E$ to some finite-dimensional, even-dimensional Euclidean space.

For each $r \geqslant 0$, let $P_r := P_r(X)$ be the Rips complex of X at scale r as in Definition 7.2.8.

Definition 12.2.1 A *good covering system* for X consists of a collection $(B_{r,x})_{r \geqslant 0, x \in X}$ of with the following properties:

(i) for each r, the collection $(B_{r,x})_{x \in X}$ is a cover of P_r by disjoint Borel sets;
(ii) for each r, x, $B_{r,x}$ contains x;
(iii) for each r, x, $B_{r,x}$ is contained in the union of the simplices that contain x;
(iv) for each $r \leqslant s$, $B_{r,x} \subseteq B_{s,x}$.

We leave it as an exercise for the reader to show that a good covering system exists: see Exercise 12.7.4 below.

Definition 12.2.2 For each $r \geqslant 0$, extend f to a Borel map $f_r: P_r \to E$ by stipulating that f_r takes all points in $B_{r,x}$ to $f(x)$.

Note that the maps f_r are compatible with the inclusions $P_r \to P_s$ for $r \leqslant s$, meaning that f_s restricts to f_r on P_r (and all restrict to $f = f_0$ on X itself). We will generally therefore abuse notation, and write f for all of them.

For each $r \geqslant 0$, let Z_r be the collection of all points in P_r such that all the coefficients t_x as in Definition 7.2.8 take rational values; thus Z_r is a countable dense subset of P_r. Let H be a separable, infinite-dimensional Hilbert space, and define $H_{P_r} := \ell^2(Z_r) \otimes H$ and $H_{P_r,E} := \ell^2(Z_r) \otimes H \otimes \mathcal{L}^2_E$. Then H_{P_r} is an ample P_r module (see Definition 4.1.1), and $H_{P_r,E}$ is both an ample P_r module and an ample E module. For a bounded operator T on $H_{P_r,E}$, write $\text{prop}_P(T)$ and $\text{prop}_E(T)$ for the propagation of T (see Definition 4.1.8) considered with respect to the P_r module structure and the E module structure respectively.

Let $C^*(H_{P_r})$ and $C^*(H_{P_r,E})$ denote the Roe algebras of P_r constructed using the P_r modules H_{P_r} and $H_{P_r,E}$ respectively (see Definition 5.1.4). It will be useful to keep these algebras distinct, so contrary to our usual conventions we do not write $C^*(P_r)$ for either of them. We will consider $C^*(H_{P_r})$ as represented on $H_{P_r,E} = H_{P_r} \otimes \mathcal{L}^2_E$ via the amplified representation $T \mapsto T \otimes 1$; note that in this way $C^*(H_{P_r})$ becomes a subalgebra of the multiplier algebra of $C^*(H_{P_r,E})$. We will also let $C^*_L(H_{P_r})$ and $C^*_L(H_{P_r,E})$ denote the localised Roe algebras (Definition 6.6.1) of P_r associated to these P_r modules.

It will be convenient to represent operators on H_{P_r} and $H_{P_r,E}$ as X-by-X matrices. For each $x \in X$, define $H_{x,P_r} := (\chi_{B_{r,x}} \otimes 1_H)H_{P_r}$. We may think of a bounded operator on H_{P_r} (respectively, $H_{P_r,E}$) as an X-by-X matrix $(T_{xy})_{x,y \in X}$ where each T_{xy} is a bounded operator $H_{y,P_r} \to H_{x,P_r}$ (respectively, $H_{y,P_r} \otimes \mathcal{L}^2_E \to H_{x,P_r} \otimes \mathcal{L}^2_E$). Note that according to the definition of the metric on P_r from Definition 7.2.8 and the fact that the canonical inclusion $X \to P_r$ is a coarse equivalence (Part (ii) of Proposition 7.2.11), there exists a proper non-decreasing function $f_r : [0, \infty) \to [0, \infty)$ such that

$$\mathrm{prop}_P(T) - 2 \leqslant \sup\{d_X(x,y) \mid T_{xy} \neq 0\} \leqslant f_r(\mathrm{prop}_P(T)). \tag{12.11}$$

For later use, note that for $r \leqslant s$ there are canonical inclusions $Z_r \to Z_s$, whence canonical isometric inclusions

$$H_{P_r} \to H_{P_s} \quad \text{and} \quad H_{P_r,E} \to H_{P_s,E}, \tag{12.12}$$

which in turn give rise to canonical inclusions of C^*-algebras

$$C^*(H_{P_r}) \to C^*(H_{P_s}) \quad \text{and} \quad C^*(H_{P_r,E}) \to C^*(H_{P_s,E}). \tag{12.13}$$

These inclusions are compatible with the matrix representations $(T_{xy})_{x,y \in X}$ defined above in the sense that increasing the Rips parameter does not alter the matrix representation: this follows from part ((iv)) of Definition 12.2.1

We finish this section with some facts about bounded geometry spaces and matrices indexed by them. These will be used several times later in the chapter.

Lemma 12.2.3 *Say X is a bounded geometry metric space. Then for each $s \geqslant 0$ there exists $N \in \mathbb{N}$ such that:*

(i) there is a decomposition

$$\{(x,y) \in X \times X \mid d(x,y) \leqslant s\} = \bigsqcup_{n=1}^{N} F_n$$

such that for each $x \in X$ there is at most one element of the form (x,y) or (y,x) in F_n;

(ii) there is a decomposition

$$X = \bigsqcup_{n=1}^{N} X_n$$

such that for each $x, y \in X_n$ with $x \neq y$, $d(x,y) > s$.

Proof For part (i), let F_1 be a maximal subset of

$$\{(x,y) \in X \times X \mid d(x,y) \leqslant s\}$$

such that for each $x \in X$ there is at most one element of the form (x,y) or (y,x) in F_1 (such a maximal subset exists by Zorn's lemma). Having defined $F_1,\ldots,F_n$, define F_{n+1} to be a maximal subset of

$$\{(x,y) \in X \times X \mid d(x,y) \leqslant s\} \setminus (F_1 \cup \cdots \cup F_n)$$

such that for each $x \in X$ there is at most one element of the form (x,y) or (y,x) in F_n. We claim that F_n is empty for all suitably large n, which will suffice to complete the proof. Indeed, if not, then for any n there is $(x,y) \in X \times X$ with $d(x,y) \leqslant s$, and $(x,y) \notin (F_1 \cup \cdots \cup F_n)$. Hence by maximality of each F_i, there are either at least $\lfloor n/2 \rfloor$ distinct points y within distance s of x, or at least $\lfloor n/2 \rfloor$ distinct points x within distance s of y; in either case, this contradicts bounded geometry for n suitably large.

Part (ii) is similar: one defines X_1 to be a maximal subset of X with the property that $d(x,y) > s$ for all distinct points $x,y \in X_1$, defines $X_2, X_3, \ldots$ iteratively and shows that X_n must be empty for suitably large n, otherwise bounded geometry is contradicted. □

Lemma 12.2.4 *For any $s,r \geqslant 0$, there exists $N \in \mathbb{N}$ such that whenever $T = (T_{xy})_{x,y \in X}$ is a bounded operator on $H_{P_r,E}$ such that $\mathrm{prop}_P(T) \leqslant s$, then*

$$\|T\| \leqslant N \sup_{x,y \in X} \|T_{xy}\|.$$

Proof Let $f_r \colon [0,\infty) \to [0,\infty)$ be as in line (12.11), and let $F_1,\ldots,F_N$ be as in part (i) of Lemma 12.2.3 for the parameter $f_r(s)$. For each $n \in \{1,\ldots,N\}$ define $T^{(n)}$ to be the operator with matrix entries

$$T^{(n)}_{xy} = \begin{cases} T_{xy} & (x,y) \in F_n, \\ 0 & \text{otherwise.} \end{cases}$$

Using the fact that for each $x \in X$ there is at most one element of the form (x,y) or (y,x) in F_n, one sees that

$$\|T^{(n)}\| \leqslant \sup_{(x,y) \in F_n} \|T_{xy}\|,$$

whence each $T^{(n)}$ is indeed a well-defined bounded operator. Moreover, as $s \geqslant \mathrm{prop}_P(T)$, we have by line (12.11) above that whenever $T_{xy} \neq 0$, $d_X(x,y) \leqslant f_r(s)$. It follows that $T = \sum_{n=1}^{N} T^{(n)}$ whence

$$\|T\| \leqslant \sum_{n=1}^{N} \|T^{(n)}\| \leqslant N \sup_{x,y \in X} \|T_{xy}\|$$

as required. □

The following corollary, which says one can detect continuity of a one-parameter family of operators from the associated matrix entries, is immediate.

Corollary 12.2.5 *Let*

$$[a,b] \to \mathcal{B}(H_{P_r,E}), \quad s \mapsto T_s$$

be a bounded map so that $\sup_s prop_P(T_s) < \infty$, *and write* $T_s = (T_{s,xy})_{x,y \in X}$. *Then* (T_s) *is norm continuous if and only if the family of maps*

$$(s \mapsto T_{s,xy})_{x,y \in X}$$

is norm equicontinuous. □

12.3 Index Maps

This section contains the index-theoretic part of the proof of Theorem 12.0.2 in the finite-dimensional case.

Throughout then, X is a bounded geometry metric space, and $f : X \to E$ is a coarse embedding into a finite-dimensional and even-dimensional Euclidean space. We will use the notation introduced in Section 12.2 for Rips complexes P_r, dense subset $Z_r \subseteq P_r$, good covering systems $(B_{r,x})_{r \geqslant 0, x \in X}$ (Definition 12.2.1) and associated Hilbert spaces and Roe algebras. We will work at a fixed 'Rips scale' $r \geqslant 0$ throughout the whole section; as such, we will generally drop r from the notation and write $P = P_r(X)$ for the Rips complex, $B_x = B_{r,x}$ for the Borel sets in the fixed good covering system and $Z = Z_r$ for the dense set of rational points in P. We also write $f : P \to E$ for the fixed extension of the coarse embedding from a function on X to a function on P given in Definition 12.2.2, and also $H_P := \ell^2(Z) \otimes H$ and $H_{P,E} := H_P \otimes \mathcal{L}^2_E$ for the associated Hilbert spaces. Finally, we have for each $x \in X$ an associated subspace $H_{x,P} := (\chi_{B_x} \otimes \mathrm{id}_H) H_P$ of H_P. We will generally think of a bounded operator T on $H_{P,E}$ as a matrix $(T_{xy})_{x,y \in X}$, where each T_{xy} is a bounded operator

$$T_{xy} : H_{y,P} \otimes \mathcal{L}^2_E \to H_{x,P} \otimes \mathcal{L}^2_E.$$

The following notation will be convenient shorthand for several of our constructions. Recall from line (12.2) that for $x \in E$, $V_x : \mathcal{L}^2_E \to \mathcal{L}^2_E$ denotes the unitary translation operator by x.

Definition 12.3.1 Let T be a bounded operator on $\mathcal{L}^2_E$. Define a bounded operator T^V on $H_{P,E} = \ell^2(Z) \otimes H \otimes \mathcal{L}^2_E$ by the formula

$$T^V : \delta_x \otimes \xi \otimes u \mapsto \delta_x \otimes \xi \otimes V_{f(x)} T V^*_{f(x)} u.$$

In other words, with respect to our usual matrix conventions, T^V is the operator with matrix entries

$$T^V_{xy} = \begin{cases} 1_H \otimes V_{f(x)} T V^*_{f(x)} & x = y, \\ 0 & \text{otherwise.} \end{cases}$$

We will, in particular, want to apply the above definition to the operator χ_R of multiplication by the characteristic function of the ball in E centred at the origin and of radius R. In this case, note that $V_{f(x)} \chi_R V^*_{f(x)}$ equals the operator $\chi_{f(x),R}$ of multiplication by the ball centred at $f(x)$ and of radius R.

Here are the algebras that we will use.

Definition 12.3.2 Let $C_b([1,\infty), C^*(H_{P,E}))$ denote the C^*-algebra of all bounded continuous functions from $[1,\infty)$ to $C^*(H_{P,E})$. Write elements of this $*$-algebra as parametrised matrices $(T_s)_{s\in[1,\infty)} = (T_{s,xy})_{s\in[1,\infty),x,y\in X}$, and equip it with the norm

$$\|(T_s)\| := \sup_s \|T_s\|_{\mathcal{B}(H_{P,E})}. \tag{12.14}$$

Let $\mathbb{A}(X;E)$ denote the $*$-subalgebra of $C_b([1,\infty), C^*(H_{P,E}))$ consisting of elements satisfying the following conditions.

(i) $\sup_{s\in[1,\infty)} \text{prop}_P(T_s) < \infty$.

(ii) $\lim_{s\to\infty} \text{prop}_E(T_s) = 0$.

(iii)

$$\lim_{R\to\infty} \sup_{s\in[1,\infty)} \|\chi^V_R T_s - T_s\| = \lim_{R\to\infty} \sup_{s\in[1,\infty)} \|T_s \chi^V_R - T_s\| = 0.$$

(iv) If (p_i) is the net of finite rank projections on $\mathcal{L}^2_E$, then for each $s \in [1,\infty)$

$$\lim_{i\to\infty} \|p^V_i T_s - T_s\| = \lim_{i\to\infty} \|T_s p^V_i - T_s\| = 0.$$

Define $A(X;E)$ to the closure of $\mathbb{A}(X;E)$ inside $C_b([1,\infty), C^*(H_{P,E}))$.

Note the different role of s in conditions (iii) and (iv): the former holds uniformly in s, while the latter only holds in each s separately.

Note that $C_b([1,\infty), C^*(H_{P,E}))$, and therefore also $A(X;E)$ is represented on $L^2([1,\infty), H_{P,E})$ in a natural way. The Hilbert space $L^2([1,\infty), H_{P,E})$ is equipped with the grading (in the sense of Definition E.1.4) induced from the grading on $\mathcal{L}^2_E$.

Definition 12.3.3 Define $\mathbb{A}_L(X;E)$ to be the collection of uniformly continuous bounded functions (T_t) from $[1,\infty)$ to $\mathbb{A}(X;E)$ such that the P-propagation of (T_t) tends to zero as t tends to infinity. More precisely, an element $(T_t) = (T_{t,s})$ of $\mathbb{A}_L(X;E)$ is an element of $C_b([1,\infty), A(X;E))$ that satisfies the following conditions.

(i) $\lim_{t\to\infty} \sup_{s\in[1,\infty)} \mathrm{prop}_P(T_{t,s}) = 0$.

(ii) for each $t \in [1,\infty)$, $\lim_{s\to\infty} \mathrm{prop}_E(T_{t,s}) = 0$.

(iii) for each $t \in [1,\infty)$,

$$\lim_{R\to\infty} \sup_{s\in[1,\infty)} \|\chi_R^V T_{t,s} - T_{t,s}\| = \lim_{R\to\infty} \sup_{s\in[1,\infty)} \|T_{t,s}\chi_R^V - T_{t,s}\| = 0.$$

(iv) If (p_i) is the net of finite rank projections on $\mathcal{L}^2_E$, then for each $s,t \in [1,\infty)$

$$\lim_{i\to\infty} \|p_i^V T_{s,t} - T_{s,t}\| = \lim_{i\to\infty} \|T_{s,t} p_i^V - T_{s,t}\| = 0.$$

Define $A_L(X;E)$ to the completion of $\mathbb{A}_L(X;E)$ for the norm $\|(T_t)\| := \sup_t \|T_t\|_{A(X;E)}$.

Our main goal in this section is to construct index maps

$$K_*(C^*(H_P)) \to K_*(A(X;E)) \quad \text{and} \quad K_*(C_L^*(H_P)) \to K_*(A_L(X;E))$$

and (partially) compute the effect of these maps on K-theory. The key ingredient for this is the following family of operators on $H_{P,E}$, built out of the Bott–Dirac operators of Section 12.1. For the next definition, let d denote the dimension of the ambient Euclidean space E.

Definition 12.3.4 For each $s \in [1,\infty)$, let $F_{s,0}\colon \mathcal{L}^2_E \to \mathcal{L}^2_E$ be the operator from Definition 12.1.5 above. Let F_s be the bounded operator on $H_{P,E}$ defined by

$$F_s := F^V_{s+2d,0}.$$

Let F be the operator on $L^2([1,\infty), H_{P,E})$ defined by $(Fu)(s) := F_s u(s)$.

Analogously, if Ψ has the properties in Proposition 12.1.10 (for some $\epsilon > 0$), we write F_s^Ψ and F^Ψ for the operators built in the same way as F_s and F, but starting with $\Psi(B_{s,0})$ in place of $F_{s,0}$.

In order to construct index maps out of F, we need some lemmas. For the first of these, recall that a net (T_i) of bounded operators converges to a bounded operator T in the *strong-$*$* topology if for all v in the underlying Hilbert space, $T_i v \to v$ and $T_i^* v \to T^* v$ in norm. Equivalently, for any finite rank projection P, $T_i P \to TP$ and $PT_i \to PT$ in norm.

Lemma 12.3.5 *Let $\mathcal{S}$ and $\mathcal{T}$ be norm bounded sets of operators on a Hilbert space H, such that $\mathcal{T}$ consists only of compact operators. Equip $\mathcal{S}$ with the strong-$*$ topology and $\mathcal{T}$ with the norm topology. Then if $\mathcal{K}$ denotes the compact operators on H, the product maps*

$$\mathcal{S} \times \mathcal{T} \to \mathcal{K} \quad \text{and} \quad \mathcal{T} \times \mathcal{S} \to \mathcal{K}$$

are jointly continuous.

Proof Let (S_i) be a net in $\mathcal{S}$ converging strong-$*$ to some $S \in \mathcal{S}$, and let (T_j) be a net of compact operators in $\mathcal{T}$ converging to some T in norm. We will show that the net $(S_i T_j)$ converges to ST (the case with the products reversed is essentially the same). Let M be a norm bound for all operators in $\mathcal{S}$ and $\mathcal{T}$, and let $\epsilon > 0$. As (T_j) is a norm convergent net of compact operators, there exists a finite rank projection P and j_0 such that for all $j \geqslant j_0$, $\|(1-P)T_j\| < \epsilon$. Then for any i and any $j \geqslant j_0$,

$$\begin{aligned}\|S_i T_j - ST\| &\leqslant \|(S_i - S)PT_j\| + \|(S_i - S)(1-P)T_j\| + \|S(T_j - T)\| \\ &\leqslant \|(S_i - S)P\|M + 2M\epsilon + M\|T_j - T\|.\end{aligned}$$

Taking the limsup over both i and j, we get

$$\limsup_{i,j} \|S_i T_j - ST\| \leqslant 2M\epsilon$$

for any ϵ, and letting ϵ tends to zero gives the result. □

Lemma 12.3.6 *The operator F is a self-adjoint, norm one, odd operator in the multiplier algebra of $A(X; E)$.*

Proof The operator F is self-adjoint, norm one and odd as each $F_{s,0}$ has these properties. Let $\epsilon > 0$ and let Ψ be as in Proposition 12.1.10 for this ϵ. Then, if F^Ψ is as in Definition 12.3.4, we have that

$$\|F - F^\Psi\| \leqslant \sup_{x,s} \|F_{s,f(x)} - \Psi(B_{s,f(x)})\| \leqslant \epsilon$$

by part (i) of Proposition 12.1.10 and Lemma 12.2.4. As the collection of multipliers of any concrete C^*-algebra is closed, it will suffice to check that any such F^Ψ is a multiplier of $A(X;E)$. Moreover, as $A(X;E)$ is generated as a C^*-algebra by elements satisfying the properties in Definition 12.3.2, it will suffice to show that if (T_s) satisfies those properties, then $(T_s F_s^\Psi)$ does too (as F^Ψ is self-adjoint, we do not need to check the other product $(F_s^\Psi T_s)$ separately).

We first claim that the map

$$s \mapsto T_s F_s^\Psi$$

is norm continuous. For this, it suffices to show that it is continuous when restricted to any compact subset $[1,b]$ of $[1,\infty)$. Now, by compactness of $[1,b]$ and norm continuity of the map $s \mapsto T_s$, part (iv) of Definition 12.3.2 gives us that for any $\epsilon > 0$ there is a finite rank projection p on $\mathcal{L}_E^2$ such that

$$\|T_s F_s^\Psi - T_s p^V F_s^\Psi\| \leqslant \|T_s - T_s p^V\| \|F_s^\Psi\| < \epsilon.$$

Hence it suffices to show that the map $s \mapsto T_s p^V F_s^\Psi$ is norm continuous. Part (vi) of Proposition 12.1.10 together with Lemma 12.3.5 gives that the map

$$[1,b] \to \mathcal{B}(\mathcal{L}_E^2), \quad s \mapsto p\Psi(B_{s,0})$$

is norm continuous, whence the family

$$[1,b] \to \mathcal{B}(\mathcal{L}_E^2), \quad s \mapsto V_{f(y)} p\Psi(B_{s,0}) V_{f(y)}^*$$

is equicontinuous as y ranges over X. Moreover, the collection $s \mapsto T_{s,xy}$ as x, y range over X is equicontinuous on $[1,b]$ by Corollary 12.2.5, and so the collection

$$s \mapsto T_{s,xy} V_{f(y)} p\Psi(B_{s,0}) V_{f(y)}^*$$

is equicontinuous as x, y range over X. This collection is precisely the collection of matrix entries of the functions $s \mapsto T_s p^V F_s^\Psi$, so this function is also continuous by Corollary 12.2.5.

For part (i) of Definition 12.3.2, note that as each F_s^Ψ clearly has P-propagation zero, Corollary 4.1.14 gives that the P-propagation of $T_s F_s^\Psi$ is bounded by that of T_s. For part (ii), note that the E-propagation of F_s^Ψ tends to zero as s tends to infinity by part (ii) of Proposition 12.1.10, whence Corollary 4.1.14 again gives that the E-propagation of $T_s F_s^\Psi$ also tends to zero. Part (iii) of Definition 12.3.2 follows from the fact that the E-propagation of F_s^Ψ is uniformly bounded in s.

Finally, for part (iv) let (p_i) be the net of finite rank projections on $\mathcal{L}^2_E$. Then for any s,

$$\|p_i^V T_s F_s^\Psi - T_s F_s^\Psi\| \leqslant \|p_i^V T_s - T_s\|,$$

which tends to zero as i tends to infinity. On the other hand, for any $\epsilon > 0$ and i, we may choose a finite rank projection $q_i \geqslant p_i$ such that

$$\|q_i \Psi(B_{s,0}) p_i - \Psi(B_{s,0}) p_i\| < \epsilon,$$

from which it follows that

$$\|q_i^V \Psi(B_{s,0})^V p_i^V - \Psi(B_{s,0})^V p_i^V\| < \epsilon.$$

Hence

$$\begin{aligned}\|T_s F_s^\Psi - T_s F_s^\Psi p_i^V\| &\leqslant \|T_s(q_i^V F_s^\Psi p_i^V - F_s^\Psi p_i^V)\| + \|(T_s - T_s q_i^V) F_s^\Psi p_i^V\| \\ &\leqslant \|T_s\|\epsilon + \|T_s - T_s q_i^V\|.\end{aligned}$$

Taking the lim sup over i gives that

$$\limsup_i \|T_s F_s^\Psi - T_s F_s^\Psi p_i^V\| < \|T_s\|\epsilon.$$

Hence as ϵ was arbitrary

$$\lim_i \|T_s F_s^\Psi - T_s F_s^\Psi p_i^V\| = 0$$

as required. □

Lemma 12.3.7 *Let $\mathcal{T}$ be a norm compact subset of $\mathcal{K}(\mathcal{L}^2_E)$, for each $x \in E$ let V_x be the translation operator as in line* (12.2) *above, and let $r \geqslant 0$. Then the sets*

$$\{V_x A \mid |x| \leqslant r \text{ and } A \in \mathcal{T}\} \quad \text{and} \quad \{A V_x \mid |x| \leqslant r \text{ and } A \in \mathcal{T}\}$$

are norm compact.

Proof The map

$$E \to \mathcal{B}(\mathcal{L}^2_E), \quad x \mapsto V_x$$

is strong-$*$ continuous: one way to see this is to note first that if $u \in \mathcal{L}^2_E$ is continuous and compactly supported, then

$$x \mapsto V_x u$$

is continuous, and then approximate a general $u \in \mathcal{L}^2_E$ by continuous elements of compact support. Hence the set $\{V_x \mid |x| \leqslant r\}$ is strong-$*$ compact; it is moreover norm bounded as it consists entirely of unitary operators. Lemma 12.3.5 now shows that the sets in the statement are the image of a compact set under a continuous map, so compact. □

Lemma 12.3.8 *Considered as represented on $L^2[1,\infty) \otimes H_{P,E}$ via the amplification of the identity representation on H_P, $C^*(H_P)$ is a subalgebra of the multiplier algebra of $A(X;E)$.*

Proof It will suffice to show that if S is in $\mathbb{C}[H_P]$ and if (T_s) is in $\mathbb{A}(X;E)$, then (ST_s) is in $A(X;E)$. It is clear that the function $s \mapsto ST_s$ is bounded and norm continuous. The fact that (ST_s) satisfies conditions (i) and (ii) from Definition 12.3.2 follows from Corollary 4.1.14 and the fact that S has finite P-propagation and has E-propagation zero.

Let us now looks at condition (iii) from Definition 12.3.2. The condition

$$\lim_{R\to\infty} \sup_{s\in[1,\infty)} \|ST_s\chi_R^V - ST_s\| = 0$$

is clear, so we need to check

$$\lim_{R\to\infty} \sup_{s\in[1,\infty)} \|\chi_R^V ST_s - ST_s\| = 0.$$

For this, assume that

$$R \geqslant \sup_{d(x,y)\leqslant \mathrm{prop}(S)} |f(x) - f(y)|$$

and define

$$R_1 := R - \sup_{d(x,y)\leqslant \mathrm{prop}(S)} |f(x) - f(y)|.$$

Then for any $x, y \in X$ and any $s \in [1,\infty)$,

$$\begin{aligned}
(S\chi_{R_1}^V T_s)_{xy} &= \sum_{\{z\in X \mid d(x,z)\leqslant \mathrm{prop}(S)\}} (S_{xz} \otimes 1_{\mathcal{L}_E^2})(1_H \otimes \chi_{f(z),R_1})T_{s,zy} \\
&= \sum_{\{z\in X \mid d(x,z)\leqslant \mathrm{prop}(S)\}} (S_{xz} \otimes \chi_{f(x),R}\chi_{f(z),R_1})T_{s,zy} \\
&= \chi_{f(x),R} \sum_{\{z\in X \mid d(x,z)\leqslant \mathrm{prop}(S)\}} (S_{xz} \otimes \chi_{f(z),R_1})T_{s,zy} \\
&= (\chi_R^V S\chi_{R_1}^V T_s)_{xy}.
\end{aligned}$$

Hence for any s,

$$S\chi_{R_1}^V T_s = \chi_R^V S\chi_{R_1}^V T_s,$$

and so

$$\|\chi_R^V ST_s - ST_s\| \leqslant \|\chi_R^V S(\chi_{R_1}^V T_s - T_s)\| + \|S(\chi_{R_1}^V T_s - T_s)\|.$$

This tends to zero as R tends to infinity (uniformly in s) by assumption.

It remains to check part (iv) from Definition 12.3.2. The only thing we need to check is that if (p_i) is the net of finite rank projections, then

$$\lim_{i\to\infty} \|p_i^V ST_s - ST_s\| = 0.$$

For this, it suffices to show that for any $\epsilon > 0$ we can find a finite rank projection p on $\mathcal{L}_E^2$ such that $\|p^V ST_s - ST_s\| < \epsilon$. Let q be any finite rank projection on $\mathcal{L}_E^2$ such that

$$\|q^V T_s - T_s\| < \frac{\epsilon}{3\|S\|}. \tag{12.15}$$

Computing matrix coefficients for any s and $x, y \in X$

$$\begin{aligned}(Sq^V T_s)_{xy} &= \sum_{\{z\in X \mid d(x,z)\leqslant \mathrm{prop}(S)\}} (S_{xz} \otimes 1_{\mathcal{L}_E^2})(1_H \otimes V_{f(z)} q V_{f(z)}^*) T_{s,zy} \\ &= \sum_{\{z\in X \mid d(x,z)\leqslant \mathrm{prop}(S)\}} (S_{xz} \otimes V_{f(z)} q V_{f(z)}^*) T_{s,zy}. \end{aligned} \tag{12.16}$$

Using Lemma 12.3.7, the set

$$A := \left\{ V_u q V_u^* \mid |u| \leqslant \sup_{d(x,y)\leqslant \mathrm{prop}(S)} |f(x) - f(y)| \right\}$$

is a norm compact set of compact operators. Hence there is a finite rank projection p on $\mathcal{L}_E^2$, which we may assume dominates q, that satisfies

$$\|(p-1)a\| < \frac{\epsilon}{3\|S\|\|T\|NM} \tag{12.17}$$

for all $a \in A$, where N is as in Lemma 12.2.4 for the parameter prop(S) and where M is an absolute bound on the number of points in a ball of radius prop(S) in X. Now, for any $x, z \in X$ such that $S_{xz} \neq 0$, we have that

$$(V_{f(x)} p V_{f(x)}^* - 1) V_{f(z)} q V_{f(z)}^* = V_{f(x)} (p-1) V_{f(z)-f(x)} q V_{f(z)-f(x)}^* V_{f(x)}^*,$$

whence by line (12.17)

$$\|(V_{f(x)} p V_{f(x)}^* - 1) V_{f(z)} q V_{f(z)}^*\| < \frac{\epsilon}{3\|S\|\|T\|NM}.$$

It follows from this and the computation in line (12.16) above that

$$\begin{aligned}
&\|(Sq^V T_s)_{xy} - (p^V Sq^V T_s)_{xy}\| \\
&= \left\| \sum_{\{z\in X \mid d(x,z)\leqslant \mathrm{prop}(S)\}} (S_{xz} \otimes (1 - V_{f(x)} p V_{f(x)}^*) V_{f(z)} q V_{f(z)}^*) T_{s,zy} \right\| \\
&\leqslant M \sup_{\{z\in X \mid d(x,z)\leqslant \mathrm{prop}(S)\}} \|(V_{f(x)} p V_{f(x)}^* - 1) V_{f(z)} q V_{f(z)}^*\| \|S\| \|T\| \\
&< \frac{\epsilon}{3N}.
\end{aligned}$$

Hence by Lemma 12.2.4, for each s,

$$\|Sq^V T_s - p^V Sq^V T_s\| < \frac{\epsilon}{3}.$$

Combining this with line (12.15), we thus have that

$$\begin{aligned}
\|p^V ST_s - ST_s\| &\leqslant \|p^V S(T_s - q^V T_s)\| \\
&\quad + \|Sq^V T_s - p^V Sq^V T_s\| + \|S(q^V T_s - T_s)\| \\
&< \|S\| \frac{\epsilon}{3\|S\|} + \frac{\epsilon}{3} + \|S\| \frac{\epsilon}{3\|S\|}
\end{aligned}$$

and we are done. □

Lemma 12.3.9 *For any $T \in C^*(H_P)$ and $s \in [1,\infty)$, the function $s \mapsto [T, F_s]$ is in $A(X; E)$.*

Proof We may assume that $T \in \mathbb{C}[H_P]$. Let $\epsilon > 0$, and let Ψ be as in Proposition 12.1.10. Then

$$\|[T, F_s] - [T, F_s^\Psi]\| = \|[T, F_s - F_s^\Psi]\| < 2\epsilon \|T\|.$$

It will thus suffice to prove that any function of the form $s \mapsto [T, F_s^\Psi]$ with Ψ as in Proposition 12.1.10 is in $\mathbb{A}[X; E]$. First note that the function $s \mapsto [T, F_s^\Psi]$ is bounded, while it is continuous by part (viii) of Proposition 12.1.10, finite propagation of T and Lemma 12.2.4.

It remains to check conditions (i) through (iv) from Definition 12.3.2 for the function $s \mapsto [T, F_s^\Psi]$. First note that using that F_s^Ψ has P-propagation zero, part (i) of Definition 12.3.2 is clear. Part (ii) follows as T has E-propagation zero, using part (ii) of Proposition 12.1.10.

For part (iii), we first compute matrix coefficients

$$[T, F_s^\Psi]_{xy} = T_{xy} \otimes (F_{s,f(y)}^\Psi - F_{s,f(x)}^\Psi).$$

Hence

$$(\chi_R^V [T, F_s^\Psi])_{xy} = T_{xy} \otimes \chi_{f(x),R}(F_{s,f(y)}^\Psi - F_{s,f(x)}^\Psi).$$

It follows from this and part (x) of Proposition 12.1.10 that for any $\epsilon > 0$ there exists $R_2 \geqslant 0$ such that for all x, y with $T_{xy} \neq 0$ and all s, we have that

$$\|(\chi_R^V[T, F_s^\Psi])_{xy} - ([T, F_s^\Psi])_{xy}\| < \|T_{xy}\|\|(1 - \chi_{f(x),R})(F_{s,f(y)}^\Psi - F_{s,f(x)}^\Psi)\| < 3\epsilon\|T\|.$$

Hence from Lemma 12.2.4 we have that

$$\lim_{R\to\infty} \sup_{s\in[1,\infty)} \|[T, F_s^\Psi] - \chi_R^V[T, F_s^\Psi]\| = 0.$$

The case of $[T, F_s^\Psi]\chi_R^V$ is essentially the same.

Finally, it remains to check condition (iv) from Definition 12.3.2. For this we compute that for any s and any finite rank projection p on $\mathcal{L}_E^2$,

$$\begin{aligned}(p^V[T, F_s^\Psi])_{xy} &= T_{xy} \otimes V_{f(x)} p V_{f(x)}^* (F_{s,f(y)}^\Psi - F_{s,f(x)}^\Psi) \\ &= T_{xy} \otimes V_{f(x)} p (F_{s,f(y)-f(x)}^\Psi - F_{s,0}^\Psi) V_{f(x)}^*.\end{aligned}$$

Now, the map

$$E \to \mathcal{K}(\mathcal{L}_E^2), \quad x \mapsto F_{s,x}^\Psi - F_{s,0}^\Psi$$

is norm continuous by part (v) of Proposition 12.1.10, whence the collection

$$\{F_{s,f(y)-f(x)}^\Psi - F_{s,0}^\Psi \in \mathcal{K}(\mathcal{L}_E^2) \mid d(x,y) \leqslant \mathrm{prop}(T)\}$$

has compact closure. It follows that for any $\epsilon > 0$ there exists a finite rank projection p on $\mathcal{L}_E^2$ such that whenever $d(x, y) \leqslant \mathrm{prop}(T)$, we have that

$$\|(1 - p)(F_{s,f(y)-f(x)}^\Psi - F_{s,0}^\Psi)\| < \epsilon.$$

For this p, we therefore get that for any x, y,

$$\|(1 - p^V)[T, F_s^\Psi])_{xy}\| < \|T\|\epsilon.$$

Hence with N as in Lemma 12.2.4 for the parameter $\mathrm{prop}(T)$, we get that

$$\|(1 - p^V)[T, F_s^\Psi]\| < \|T\|N\epsilon,$$

which gives that if (p_i) is the net of finite rank projections on $\mathcal{L}_E^2$, then

$$\lim_{i\to\infty} \|(1 - p_i^V)[T, F_s^\Psi]\| = 0.$$

The case of $\lim_{i\to\infty} \|[T, F_s^\Psi](1 - p_i^V)\|$ is similar, so we are done. □

Lemma 12.3.10 *Let $p \in C^*(H_P)$ be a projection. Then the function*

$$s \mapsto (pF_s p)^2 - p$$

is in the corner $pA(X; E)p$.

Proof Using Lemma 12.3.9, that p is a projection and that F and p are multipliers of $A(X;E)$ (see Lemmas 12.3.6 and 12.3.8, respectively), it suffices to show that the function $s \mapsto F_s^2 p - p$ is in $A(X;E)$. For this, it suffices to show that if q is a finite propagation approximant to p (which need no longer be a projection), then $s \mapsto F_s^2 q - q$ is in $A(X;E)$. Note then that the (x,y)th matrix entry of $F_s^2 q - q$ is

$$q_{xy} \otimes (F_{s,f(x)}^2 - 1) = q_{xy} \otimes V_{f(x)}(F_{s,0}^2 - 1)V_{f(x)-f(y)}^* V_{f(y)}^*.$$

The fact that $B_{s,0}$ has compact resolvent shows that $F_{s,0}^2 - 1$ is compact. Corollary 12.3.7 then shows that the collection

$$\{(F_{s,0}^2 - 1)V_{f(x)-f(y)}^* \mid d(x,y) \leqslant \operatorname{prop}_P(q)\}$$

is compact. The result follows from this and computations very similar to (and easier than) those in the last few lemmas: we leave the details to the reader. □

It follows that for a projection $p \in C^*(H_P)$, the odd self-adjoint operator $(pF_s p)_{s\in[1,\infty)}$ on the graded Hilbert space $p\big(L^2[1,\infty) \otimes H_{P,E}\big)$ can be used to build an index class in $K_0(pA(X;E)p)$ as in Definition 2.8.5. Being a little more specific about this, the fact that $pF_s p$ is odd and self-adjoint means it has the form

$$pF_s p = \begin{pmatrix} 0 & u^* \\ u & 0 \end{pmatrix}$$

when decomposed as a matrix with respect to the grading on $p\big(L^2[1,\infty) \otimes H_{P,E}\big)$, while the fact that p is even means it has the form

$$p = \begin{pmatrix} p_0 & 0 \\ 0 & p_1 \end{pmatrix}$$

when decomposed as a matrix with respect to the grading. The fact that

$$(pF_s p)^2 - p \in pA(X;E)p$$

from Lemma 12.3.10 translates to saying that the matrix

$$\begin{pmatrix} u^*u - p_0 & 0 \\ 0 & uu^* - p_1 \end{pmatrix}$$

(again, decomposed with respect to the grading) is in $pA(X;E)p$. We may now form the explicit formal difference

$$\operatorname{Ind}(pF_s p) = \begin{pmatrix} (p_0 - uu^*)^2 & u^*(p_1 - uu^*) \\ u(2 - u^*u)(1 - u^*u) & uu^*(p_1 - uu^*)^2 \end{pmatrix} - \begin{pmatrix} 0 & 0 \\ 0 & p_1 \end{pmatrix} \tag{12.18}$$

(compare line (2.21)) of idempotents. The idempotents are in the multiplier algebra of $pA(X;E)p$ and their difference is in $pA(X;E)p$ so we get a class

$$\begin{bmatrix} (p_0 - uu^*)^2 & u^*(p_1 - uu^*) \\ u(2 - u^*u)(1 - u^*u) & uu^*(p_1 - uu^*)^2 \end{bmatrix} - \begin{bmatrix} 0 & 0 \\ 0 & p_1 \end{bmatrix} \in K_0(pA(X;E)p). \tag{12.19}$$

Now, composing with the map on K_0 groups induced by the inclusion $pA(X;E)p \to A(X;E)$, we get an element of $K_0(A(X;E))$ that we call $\mathrm{Ind}_F[p]$.

Lemma 12.3.11 *For each $s \in [1,\infty)$, the process above gives a well-defined homomorphism*

$$Ind_F \colon K_0(C^*(H_P)) \to K_0(A(X;E)).$$

Proof Note first that using the result of Exercise 5.4.2, $K_0(C^*(H_P))$ is generated by classes of projections in matrix algebras over $C^*(H_P)$, despite this algebra being non-unital.[1] Replacing H with $H^{\oplus n}$ in the definitions of H_P and $H_{P,E}$, it is straightforward to check that the above construction makes good sense for projections in $M_n(C^*(H_P))$ for some n. To see that we get a well-defined homomorphism on K_0, it suffices to show that if $p,q \in M_n(C^*(H_P))$ are homotopic, then $\mathrm{Ind}_F[p] = \mathrm{Ind}_F[q]$, and that if $p,q \in M_n(C^*(H_P))$ are orthogonal, then $\mathrm{Ind}_F[p+q] = \mathrm{Ind}_F[p] + \mathrm{Ind}_F[q]$. Indeed, using the concrete formula for the index in line (12.19) above, homotopies carry through to homotopies, while orthogonality is preserved as the corners $pA(X;E)p$ and $qA(X;E)q$ are themselves orthogonal. □

Passing to suspensions and applying the above construction pointwise, we similarly get a map

$$\mathrm{Ind}_F \colon K_0(C_0(\mathbb{R}, C^*(H_P))) \to K_0(C_0(\mathbb{R}, A(X;E))),$$

i.e. up to the usual canonical identifications, an index map on the level of K_1. Similarly, we also get an index map on the localised level, for both K_0 and K_1, by applying the above constructions pointwise in t.

Definition 12.3.12 For $s \in [1,\infty)$, the *index maps* associated to F_s are the homomorphisms

$$\mathrm{Ind}_F \colon K_*(C^*(H_P)) \to K_*(A(X;E))$$

[1] Actually, one does not even need matrix algebras, but we do not need to use this.

and

$$\mathrm{Ind}_{F_L}\colon K_*(C_L^*(H_P)) \to K_*(A_L(X;E))$$

constructed above.

For any $s \in [1,\infty)$, let now $\iota^s\colon A(X;E) \to C^*(H_{P,E})$ be the map defined by evaluation at the parameter s (in symbols, $(T_s) \mapsto T_s$). The following is the most important result of this section.

Proposition 12.3.13 *With notation as above, for any $s \in [1,\infty)$ the composition*

$$K_*(C^*(H_P)) \xrightarrow{Ind_F} K_*(A(X;E)) \xrightarrow{\iota^s_*} K_*(C^*(H_{P,E}))$$

is an isomorphism. The analogous statement holds for the localised algebras.

The point of the proposition is not the fact that the groups $K_*(C^*(H_P))$ and $K_*(C^*(H_{P,E}))$ are isomorphic! Indeed, the groups $K_*(C^*(H_P))$ and $K_*(C^*(H_{P,E}))$ are isomorphic for the much more elementary reason that the underlying C^*-algebras are themselves isomorphic (see Remark 5.1.13). The point is that there is an isomorphism between them that factors through $K_*(A(X;E))$.

Proof We will focus on the case directly in the statement; the localised case follows from the same argument applied pointwise.

Define a map $\kappa\colon E \to E$ by the formula

$$\kappa(x) = \begin{cases} \frac{x}{|x|}(|x|-1) & |x| \geqslant 1, \\ 0 & |x| < 1 \end{cases}$$

(thus κ 'moves each element of E one unit closer to the origin'). For simplicity of notation, fix s and write F_x for $F_{s,x}$. For each $n \in \mathbb{N} \cup \{0\}$, define

$$F^{(n)}\colon H_{P,E} \to H_{P,E}, \quad \delta_x \otimes \xi \otimes u \mapsto \delta_x \otimes \xi \otimes F_{\kappa^n(f(x))}u$$

(so $F^{(0)} = F$ as in Definition 12.3.4) and analogously define $F^{(\infty)}$ by the formula

$$F^{(\infty)}\colon H_{P,E} \to H_{P,E}, \quad \delta_x \otimes \xi \otimes u \mapsto \delta_x \otimes \xi \otimes F_0 u.$$

Note that all the operators $F^{(n)}$ are odd multipliers of $C^*(H_{P,E})$. Let $p \in M_m(C^*(H_P))$ be a projection representing a class $[p]$ in $K_0(C^*(H_P))$. For notational simplicity, let us assume that p is actually in $C^*(H_P)$; the matricial case amounts to replacing H with $H^{\oplus m}$ in the definition of $H_{P,E}$ and is analogous (or can be avoided using an argument based on quasi-stability).

Note that p is a multiplier of $C^*(H_{P,E})$, whence the compression $pF^{(n)}p$ is too for each $n \in \mathbb{N} \cup \{0, \infty\}$. An argument analogous to (but simpler than) the proof of Lemma 12.3.10 shows that the difference $(pF^{(n)}p)^2 - p$ is in the corner $pC^*(H_{P,E})p$, so we get an index class in $K_0(pC^*(H_{P,E})p)$ via the construction of Definition 2.8.5, and hence via the inclusion $pC^*(H_{P,E})p \to C^*(H_{P,E})$ an element $\mathrm{Ind}_{F^{(n)}}[p] \in K_0(C^*(H_{P,E}))$. Applying this process pointwise to the suspended algebras, we get a similar construction on the level of K_1. Quite analogously to (but again, with a simpler proof than) Lemma 12.3.11, we thus get a homomorphism

$$\mathrm{Ind}_{F^{(n)}} : K_*(C^*(H_P)) \to K_*(C^*(H_{P,E})) \tag{12.20}$$

for each $n \in \mathbb{N} \cup \{0, \infty\}$. Clearly $\mathrm{Ind}_{F^{(0)}}$ is the same map as in the statement. We will complete the proof by showing first that $\mathrm{Ind}_{F^{(0)}} = \mathrm{Ind}_{F^{(\infty)}}$, and then that $\mathrm{Ind}_{F^{(\infty)}}$ is an isomorphism.

Now let $(\mathcal{L}^2_E)^{\oplus\infty}$ be the direct sum of infinitely many copies of $\mathcal{L}^2_E$. Define

$$H_{P,E,\infty} := \ell^2(Z) \otimes H \otimes (\mathcal{L}^2_E)^{\oplus\infty}$$

and let $C^*(H_{P,E,\infty})$ be the corresponding Roe algebra. The corresponding 'top left corner inclusion'

$$C^*(H_{P,E}) \to C^*(H_{P,E,\infty}), \quad T \mapsto \begin{pmatrix} T & 0 & \dots \\ 0 & 0 & \\ \vdots & & \ddots \end{pmatrix} \tag{12.21}$$

is then induced by a covering isometry for the identity map, so induces an isomorphism on K-theory by Theorem 5.1.15. To show that $\mathrm{Ind}_{F^{(0)}} = \mathrm{Ind}_{F^{(\infty)}}$, it thus suffices to show that their compositions with the map on K-theory induced by this top corner inclusion is the same.

We will focus on K_0; the case of K_1 can be handled similarly using a suspension argument. Let $p \in M_n(C^*(H_P))$ be a projection. As the classes of such projections generate $K_0(C^*(H_P))$ (see Exercise 5.4.2), it suffices to show that $\mathrm{Ind}_{F^{(0)}}[p] = \mathrm{Ind}_{F^{(\infty)}}[p]$. For notational simplicity, let us assume that p is actually in $C^*(H_P)$; the case when p is in some matrix algebra $M_m(C^*(H_P))$ over $C^*(H_P)$ amounts to replacing H with $H^{\oplus m}$ in the definition of $H_{P,E}$ and is quite analogous.

Analogously to Remark 2.8.2, for each n, the class $\mathrm{Ind}_{F^{(n)}}[p]$ is represented by a concrete difference of projections, say

$$[p^{(n)}] - [q]$$

(the second such projection does not depend on $F^{(n)}$) in $M_2(pC^*(H_{P,E})p)$. Consider now the projections

$$\begin{pmatrix} p^{(0)} & 0 & 0 & \cdots \\ 0 & p^{(1)} & 0 & \\ 0 & 0 & p^{(2)} & \\ \vdots & & & \ddots \end{pmatrix}, \quad \begin{pmatrix} p^{(\infty)} & 0 & 0 & \cdots \\ 0 & p^{(\infty)} & 0 & \\ 0 & 0 & p^{(\infty)} & \\ \vdots & & & \ddots \end{pmatrix}$$

in the multiplier algebra of $C^*(H_{P,E,\infty})$. Using the fact that for any $x \in P$, $F_{\kappa^n(f(x))} = F_0$ for all n suitably large (depending on x), it is not too difficult to see that the difference of these projections is in $C^*(H_{P,E,\infty})$ whence the formal difference

$$a := \begin{bmatrix} p^{(0)} & 0 & 0 & \cdots \\ 0 & p^{(1)} & 0 & \\ 0 & 0 & p^{(2)} & \\ \vdots & & & \ddots \end{bmatrix} - \begin{bmatrix} p^{(\infty)} & 0 & 0 & \cdots \\ 0 & p^{(\infty)} & 0 & \\ 0 & 0 & p^{(\infty)} & \\ \vdots & & & \ddots \end{bmatrix}$$

defines a class in $K_0(C^*(H_{P,E,\infty}))$. On the other hand, there is also a class $b \in K_0(C^*(H_{P,E,\infty}))$ defined by the difference

$$b := \begin{bmatrix} p^{(0)} & 0 & 0 & \cdots \\ 0 & 0 & 0 & \\ 0 & 0 & 0 & \\ \vdots & & & \ddots \end{bmatrix} - \begin{bmatrix} p^{(\infty)} & 0 & 0 & \cdots \\ 0 & 0 & 0 & \\ 0 & 0 & 0 & \\ \vdots & & & \ddots \end{bmatrix}.$$

We claim next that $a + b = a$, whence $b = 0$; this will complete the proof that $\mathrm{Ind}_{F^{(0)}} = \mathrm{Ind}_{F^{(\infty)}}$.

Indeed, consider the path $(F^{(n),r})_{r \in [0,1]}$ of operators on $H_{P,E}$ defined by

$$F^{(n),r} : \delta_x \otimes \xi \otimes u \mapsto \delta_x \otimes \xi \otimes F_{s,(1-r)\kappa^n(f(x))+r\kappa^{n+1}(f(x))}u,$$

which by Corollary 12.1.7 satisfies

$$\|F^{(n),r} - F^{(n),r'}\| \leqslant 3|r - r'|$$

for each n. It follows that a is homotopic to the element

$$\begin{bmatrix} p^{(1)} & 0 & 0 & \cdots \\ 0 & p^{(2)} & 0 & \\ 0 & 0 & p^{(3)} & \\ \vdots & & & \ddots \end{bmatrix} - \begin{bmatrix} p^{(\infty)} & 0 & 0 & \cdots \\ 0 & p^{(\infty)} & 0 & \\ 0 & 0 & p^{(\infty)} & \\ \vdots & & & \ddots \end{bmatrix}$$

of $K_*(C^*(H_{P,E,\infty}))$, and thus by a rotation homotopy 'moving everything up a step' that $a + b = a$ as claimed.

To complete the proof, we need to show that $\mathrm{Ind}_{F(\infty)}$ is an isomorphism. Let p_0 be the projection onto the one-dimensional kernel of F_0, spanned by the unit vector v_0, say. Recall that $F_0 = f(s^{-1}D + C)$, where $f(x) = x(1 + x^2)^{-1/2}$. Now, introduce another parameter $r \in [1, \infty]$, and consider the path F_0^r defined by

$$F_0^r := f(r(s^{-1}D + C)).$$

This defines a norm continuous homotopy between $F_0 = F_0^1$ and the operator F_0^∞, which decomposes with respect to the grading as

$$F_0^\infty = \begin{pmatrix} 0 & 1 \\ 1 - p_0 & 0 \end{pmatrix}.$$

It follows from this that $\mathrm{Ind}_{F(\infty)}$ equals $\mathrm{Ind}_{F_0^\infty}$. However, using the fact that F_0^r commutes with all operators in $C^*(H_P)$ for all r, a computation quite analogous to (and easier than) that of Example 2.8.3 shows that for any $p \in C^*(H_P)$ we have

$$\mathrm{Ind}_{F_0^\infty}[p] = [p \otimes p_0]. \tag{12.22}$$

It follows that $\mathrm{Ind}_{F(\infty)}$ is the same as the map induced on K-theory by the inclusion

$$\ell^2(Z) \otimes H \to \ell^2(Z) \otimes H \otimes \mathcal{L}_E^2, \quad v \mapsto v \otimes v_0.$$

As this is an isometry covering the identity map, it induces an isomorphism on K-theory by Theorem 5.1.15, and we are done. □

12.4 The Local Isomorphism

We use the notation introduced in Sections 12.2 and 12.3 for the space X, and associated Rips complexes, C^*-algebras and so on. We make one notational change as follows: in Definitions 12.3.2 of $A(X; E)$ and 12.3.3 of $A_L(X; E)$ above, there is an implicit 'Rips parameter' $r \geqslant 0$, which was kept fixed throughout Section 12.3. In this section, we will need to 'unfix' this parameter. As such, we introduce it into the notation, writing $A^r(X; E)$ and $A_L^r(X; E)$ for the C^*-algebras of Definitions 12.3.2 and 12.3.3.

For each r, there is an analogue of the usual evaluation-at-one homomorphism

$$\mathrm{ev}\colon A_L^r(X; E) \to A^r(X; E), \quad (T_t) \mapsto T_1. \tag{12.23}$$

Recall that for $r \leqslant s$ we have inclusions of Hilbert spaces $H_{P_r,E} \to H_{P_s,E}$ as in line (12.12) above. These give rise to a commutative diagram

$$\begin{array}{ccc} A_L^r(X;E) & \xrightarrow{\text{ev}} & A^r(X;E) \\ \downarrow & & \downarrow \\ A_L^s(X;E) & \xrightarrow{\text{ev}} & A^s(X;E)) \end{array}$$

for each $r \leqslant s$. In particular, it makes sense to take the direct limit of the maps in line (12.23) as the Rips parameter r tends to infinity. Our goal in this section is to prove the following result.

Proposition 12.4.1 *The map*

$$ev_*\colon \lim_{r\to\infty} K_*(A_L^r(X;E)) \to \lim_{r\to\infty} K_*(A^r(X;E))$$

induced by the direct limit of the maps in line (12.23) *is an isomorphism.*

Before embarking on the proof of this, let us see how it implies our main result in the finite-dimensional case.

Theorem 12.4.2 *Let X be a bounded geometry metric space, and say there exists a coarse embedding of X into a finite-dimensional real Hilbert space. Then the coarse Baum–Connes conjecture holds for X.*

Proof We may assume the Hilbert space is even-dimensional (if not, just take the direct sum with $\mathbb{R}$). For each $r \geqslant 0$, we consider the diagram

$$\begin{array}{ccc} K_*(C_L^*(H_{P_r})) & \longrightarrow & K_*(C^*(H_{P_r})) \\ \downarrow{\scriptstyle \mathrm{Ind}_{F_L}} & & \downarrow{\scriptstyle \mathrm{Ind}_F} \\ K_*(A_L^r(X;E)) & \longrightarrow & K_*(A^r(X;E)) \\ \downarrow & & \downarrow \\ K_*(C_L^*(H_{P_r,E})) & \longrightarrow & K_*(C^*(H_{P_r,E})), \end{array}$$

where: all the horizontal arrows are evaluation-at-one maps in the 'X-localisation' ('t') variable; the first pair of vertical maps are the index maps of Definition 12.3.12; and the second pair of vertical maps are induced by evaluation-at-one in the 'E-localisation' ('s') variable. It is immediate from the definitions that this commutes. The compositions of the two vertical maps on either side are isomorphisms for all r by Proposition 12.3.13.

For $r \leqslant s$, the inclusions $H_{P_r,E} \to H_{P_s,E}$ of line (12.12) induce inclusions on all the algebras appearing in the diagram above, and the whole diagram commutes with these inclusions. Thus we may take the limit as r tends to infinity getting the following commutative diagram

$$\begin{array}{ccc}
\lim_{r\to\infty} K_*(C_L^*(H_{P_r})) & \longrightarrow & \lim_{r\to\infty} K_*(C^*(H_{P_r})) \\
\downarrow \mathrm{Ind}_{F_L} & & \downarrow \mathrm{Ind}_F \\
\lim_{r\to\infty} K_*(A_L^r(X;E)) & \longrightarrow & \lim_{r\to\infty} K_*(A^r(X;E)) \\
\downarrow & & \downarrow \\
\lim_{r\to\infty} K_*(C_L^*(H_{P_r,E})) & \longrightarrow & \lim_{r\to\infty} K_*(C^*(H_{P_r,E})).
\end{array}$$

The vertical compositions are direct limits of isomorphisms, so isomorphisms. The middle horizontal map is an isomorphism by Proposition 12.4.1. It now follows from a diagram chase that the top horizontal map is injective and the bottom horizontal map is surjective. However, both the top and bottom horizontal maps identify with the coarse Baum–Connes assembly map for X (see Theorem 7.2.16), so we are done. □

We now turn back to the proof of Proposition 12.4.1. Let F be a closed subset of E, and let

$$A^r(X;F) := (1 \otimes \chi_F)A^r(X;E)(1 \otimes \chi_F)$$

be the corner of $A^r(X;E)$ defined by the idempotent $1 \otimes \chi_F$ in its multiplier algebra. Similarly, $1 \otimes \chi_F$ canonically defines a multiplier of $A_L^r(X;E)$ and we let $A_L^r(X;F)$ be the associated corner

$$A_L^r(X;F) := (1 \otimes \chi_F)A_L^r(X;E)(1 \otimes \chi_F).$$

Note that the evaluation-at-one map restricts to a $*$-homomorphism

$$\mathrm{ev}\colon A_L^r(X;F) \to A^r(X;F)$$

for any closed subset F of E. The rough idea of the proof of Proposition 12.4.1 is to show that the assignments

$$F \mapsto \lim_{r\to\infty} K_*(A_L^r(X;F)), \quad F \mapsto \lim_{r\to\infty} K_*(A^r(X;F)) \tag{12.24}$$

are 'homology theories on the collection of closed subsets of E' in an appropriate sense, and that ev_* is a natural transformation between them. This reduces the check that the map in Proposition 12.4.1 is an isomorphism to proving that the map

$$\mathrm{ev}\colon \lim_{r\to\infty} K_*(A_L^r(X;F)) \to \lim_{r\to\infty} K_*(A^r(X;F))$$

is an isomorphism when F is a disjoint union of uniformly bounded closed subsets of E, which can be done directly (see Lemma 12.4.4 below).

As a preliminary ingredient, we need an abstract K-theoretic lemma. See Definition 2.7.11 for the notion of quasi-stability.

Lemma 12.4.3 *Let A be a quasi-stable C^*-algebra, and let $C_{ub}([1,\infty),A)$ denote the C^*-algebra of uniformly continuous bounded functions from $[1,\infty)$ to A. Then the natural evaluation-at-one map*

$$C_{ub}([1,\infty),A) \to A$$

induces an isomorphism on K-theory.

See Exercise 12.7.8 for some related results.

Proof Using the six-term exact sequence, it suffices to show that the kernel $B = \{f \in C_{ub}([1,\infty),A) \mid f(1) = 0\}$ of the evaluation-at-one map has zero K-theory. Let

$$X := \bigsqcup_{n\geqslant 2,\ n \text{ even}} [n,n+1] \quad \text{and} \quad Y := \bigsqcup_{n\geqslant 3,\ n \text{ odd}} [n,n+1].$$

Note that B fits into a pullback diagram (see Definition 2.7.14)

$$\begin{array}{ccc} B & \longrightarrow & C_{ub}(X,A) \\ \downarrow & & \downarrow \\ C_0((1,2],A) \oplus C_{ub}(Y,A)) & \longrightarrow & \prod_{n\geqslant 1} A. \end{array}$$

We thus get an associated Mayer–Vietoris sequence by Proposition 2.7.15, which (on homotoping away the cone $C_0((1,2],A)$) looks like

$$\longrightarrow K_i(B) \longrightarrow K_i(C_{ub}(X,A)) \oplus K_i(C_{ub}(Y,A)) \longrightarrow \prod_{n\geqslant 2} K_i(A) \longrightarrow .$$

Homotopy invariance of K-theory (combined with uniform continuity) gives us that the natural inclusion

$$\prod_{n\geqslant 2,\ n \text{ even}} A \to C_{ub}(X,A),$$

where the nth copy of A is included as constant functions on the interval $[n,n+1]$ induces an isomorphism on K-theory. Moreover, we can use the quasi-stability of A and Proposition 2.7.12 to commute the products and K-functors, so that we have a natural isomorphism

$$K_i\Big(\prod_{n\geqslant 2,\, n \text{ even}} A\Big) \cong \prod_{n\geqslant 2,\, n \text{ even}} K_i(A)$$

and similarly for Y. Our Mayer–Vietoris sequence thus reduces to

$$\longrightarrow K_i(B) \longrightarrow \prod_{n\geqslant 2,\, n \text{ even}} K_i(A) \oplus \prod_{n\geqslant 3,\, n \text{ odd}} K_i(A) \longrightarrow \prod_{n\geqslant 2} K_i(A) \longrightarrow .$$

Using the description in Proposition 2.7.15, we can also compute that the map

$$\prod_{n\geqslant 2 \text{ even}} K_i(A) \oplus \prod_{n\geqslant 3 \text{ odd}} K_i(A) \longrightarrow \prod_{n\geqslant 2} K_i(A)$$

takes a pair of sequences $(\alpha_n)_{n\geqslant 2 \text{ even}}$, $(\beta_n)_{n\geqslant 3 \text{ odd}}$ to the sequence whose first few terms are $\alpha_2, \alpha_2 - \beta_3, \alpha_4 - \beta_3, \alpha_4 - \beta_5, \alpha_6 - \beta_5$ and so on. This map is both injective and surjective, whence $K_i(B) = 0$ as required. □

Lemma 12.4.4 *Let F be a closed subset of E which splits as a disjoint union $F = \bigsqcup_{n=1}^{\infty} F_n$ of closed subsets such that there exists $R > 0$ such that for each n there exists $x_n \in X$ such that*

$$F_n \subseteq B(f(x_n); R).$$

Then the evaluation-at-one map

$$ev_* : \lim_{r\to\infty} K_*(A_L^r(X;F)) \to \lim_{r\to\infty} K_*(A^r(X;F))$$

is an isomorphism.

Proof Consider the C^*-algebraic product $\prod_n A^r(X;F_n)$ as represented on the Hilbert space $L^2([1,\infty)) \otimes H_{P_r,E}$, and define the 'restricted product'

$$\prod_n^{res} A^r(X;F_n) := \Big(\prod_n A^r(X;F_n)\Big) \cap A^r(X;F)$$

and similarly for $\prod_n^{res} A_L^r(X;F_n)$. There is then a commutative diagram

$$\begin{array}{ccc} A_L^r(X;F) & \xrightarrow{\text{ev}} & A^r(X;F) \\ \uparrow & & \uparrow \\ \prod_n^{res} A_L^r(X;F_n) & \xrightarrow{\prod \text{ev}} & \prod_n^{res} A^r(X;F_n) \end{array}$$

in which the vertical maps are the tautologous inclusions of subalgebras. Moreover, quite analogously to the proof of Theorem 6.4.20, the vertical maps induce isomorphisms on K-theory for each r. It thus suffices to prove that the canonical evaluation-at-one map

$$\prod_n^{res} A_L^r(X;F_n) \to \prod_n^{res} A^r(X;F_n)$$

induces an isomorphism on K-theory on taking the limit as r tends to infinity.

Now, for a closed subset Y of P_r and closed subset G of E, define

$$A^r(Y;G) = (\chi_Y \otimes 1)A^r(X;G)(\chi_Y \otimes 1)$$

to be the associated corner, and similarly for $A_L^r(Y;G)$. Condition (iii) from Definition 12.3.2 implies that for each $r > 0$ and each n, we have that

$$A^r(X;F_n) = \lim_{m\to\infty} A^r(B_X(x_n;m);F_n),$$

and moreover that this limit is 'uniform' in the sense that

$$\prod_n^{res} A^r(X;F_n) = \lim_{m\to\infty} \prod_n^{res} A^r(B_X(x_n;m);F_n).$$

On the other hand, it is clear from the definitions that

$$\lim_{r\to\infty}\lim_{m\to\infty} \prod_n^{res} A^r(B_X(x_n;m);F_n) = \lim_{m\to\infty}\lim_{r\to\infty} \prod_n^{res} A^r(B_X(x_n;m);F_n)$$

(note that both limits are just increasing unions of subalgebras). This all works similarly for the localised versions, whence it suffices to prove that

$$\mathrm{ev}_*\colon \lim_{r\to\infty} K_*(\prod_n^{res} A_L^r(B_X(x_n;m);F_n)) \to \lim_{r\to\infty} K_*(\prod_n^{res} A^r(B_X(x_n;m);F_n))$$

is an isomorphism for each fixed m. Note, however, that this limit stabilises: for all $r \geqslant 2m$, $P_r(B(x_n;m))$ is just equal to the full simplex on $B_X(x_n;m)$, which we denote Δ_n. It thus suffices to prove that

$$\mathrm{ev}_*\colon K_*(\prod_n^{res} A_L^r(\Delta_n;F_n)) \to K_*(\prod_n^{res} A^r(\Delta_n;F_n))$$

is an isomorphism. Moreover, the inclusion of the single point $x_n \to \Delta_n$ induces a commutative diagram

$$\begin{array}{ccc} K_*(\prod_n^{res} A_L^r(\Delta_n;F_n)) & \longrightarrow & K_*(\prod_n^{res} A^r(\Delta_n;F_n)) \\ \uparrow & & \uparrow \\ K_*(\prod_n^{res} A_L^r(\{x_n\};F_n)) & \longrightarrow & K_*(\prod_n^{res} A^r(\{x_n\};F_n)), \end{array}$$

where the vertical maps are isomorphisms by analogues of Theorem 6.4.16 for the left-hand side and Theorem 5.1.15 for the right-hand side. Hence it suffices to prove that the evaluation-at-zero map induces an isomorphism

$$\mathrm{ev}_*\colon K_*\left(\prod_n^{res} A_L^r(\{x_n\}; F_n)\right) \to K_*\left(\prod_n^{res} A^r(\{x_n\}; F_n)\right).$$

At this point, however, the condition that $\mathrm{prop}_{P_r}(T) \to 0$ defining the 'L' version on the left is vacuous, whence the left-hand side $\prod_n^{res} A_L^r(\{x_n\}; F_n)$ is simply the C^*-algebra of uniformly bounded continuous functions to the right-hand side $\prod_n^{res} A^r(\{x_n\}; F_n)$. The result follows from Lemma 12.4.3. □

Lemma 12.4.5 *Let s be a positive real number. Then there exists $M \in \mathbb{N}$ (depending on s, X, and the coarse embedding $f\colon X \to E$) and a decomposition*

$$X = X_1 \sqcup \cdots \sqcup X_M$$

such that for each i and all $x \neq y$ with $x, y \in X_i$, we have $\overline{B(f(x); s)} \cap \overline{B(f(y); s)} = \varnothing$.

Proof This follows from the fact that f is a coarse embedding and part (ii) of Lemma 12.2.3. □

Proof of Proposition 12.4.1 Fix $s > 0$ for the moment, and let

$$W_s = \overline{N_s(f(X))}. \tag{12.25}$$

Using Lemma 12.4.5 there exists a decomposition

$$W_s = \bigcup_{i=1}^{M} \underbrace{\overline{\bigsqcup_{x \in X_i} B(f(x); s)}}_{=: W_s^i}.$$

Lemma 12.4.4 implies that each evaluation-at-one map

$$\mathrm{ev}_*\colon \lim_{r\to\infty} K_*(A_L^r(X; W_i^s)) \to \lim_{r\to\infty} K_*(K_*(A^r(X; W_i^s)))$$

is an isomorphism. It therefore follows from a Mayer–Vietoris argument using an analogue of Theorem 6.3.4 (and Lemma 12.4.4 again to deal with intersections) that

$$\mathrm{ev}_*\colon \lim_{r\to\infty} K_*(A_L^r(X; W^s)) \to \lim_{r\to\infty} K_*(A^r(X; W^s))$$

is an isomorphism. Finally, note that Definition 12.3.2 implies that for any r we have that

$$\lim_{s\to\infty} A^r(X; W^s) = A^r(X; E),$$

and similarly for the localised versions. The proposition follows. □

12.5 Reduction to Coarse Disjoint Unions

We will now work towards the proof of Theorem 12.0.2 – that a bounded geometry metric space that coarsely embeds into Hilbert space satisfies the coarse Baum–Connes conjecture – in the case that the underlying Hilbert space is infinite-dimensional.

Our goal in this section is to reduce the proof to a 'uniform' statement for a sequence of finite metric spaces. The advantage of this is that if a finite metric space coarsely embeds into a Hilbert space then, it coarsely embeds into a finite-dimensional Hilbert space, and so we may use our earlier work in the finite-dimensional case.

We now start working towards precise statements.

Definition 12.5.1 Let $(X_n)_{n=1}^{\infty}$ be a sequence of finite metric spaces (with finite-valued distance functions). The sequence (X_n) has *bounded geometry* if for all $r \geqslant 0$ there exists $N \in \mathbb{N}$ such that for all n, all r balls in X_n have cardinality at most N.

A metric space (X, d) is a *coarse union* of the sequence (X_n) if it is equal as a set to the disjoint union $\bigsqcup_{n=1}^{\infty} X_n$, and if the metric d satisfies the following conditions:

(i) d is finite-valued;
(ii) d restricts to the original metric on each X_n;
(iii) $d(X_n, X\backslash X_n) \to \infty$ as $n \to \infty$.

The *separated coarse union* of (X_n) is the metric space X, which is again equal to $\bigsqcup X_n$ as a set, and is equipped with the metric

$$d(x,y) = \begin{cases} d_{X_n}(x,y) & \text{there exists } n \text{ with } x, y \in X_n, \\ \infty & \text{otherwise.} \end{cases}$$

The sequence (X_n) *uniformly coarsely embeds into Hilbert space* if there exists a sequence of (real) finite-dimensional Hilbert spaces (E_n), a sequence of maps $f_n \colon X_n \to E_n$ and non-decreasing maps $\rho_{\pm} \colon [0,\infty) \to [0,\infty)$ such that for all n and all $x, y \in X_n$

$$\rho_-(d_{X_n}(x,y)) \leqslant \|f(x) - f(y)\|_{E_n} \leqslant \rho_+(d_{X_n}(x,y))$$

and $\rho_-(t) \to \infty$ as $t \to \infty$.

Remark 12.5.2 The conditions above on the metric on a coarse union $X = \bigsqcup X_n$ do not determine the metric uniquely in terms of the metrics on the spaces X_n. However, if d, d' are two metrics satisfying these conditions, then the set-theoretic identity map $(X,d) \to (X,d')$ is a coarse equivalence (see Definition A.3.9 – we leave this as an exercise). Hence the choice of such a metric on $\bigsqcup X_n$ does not matter up to coarse equivalence; in particular, we will sometimes abuse terminology and speak of 'the' coarse union of a sequence (X_n).

We now give two lemmas that reduce the proof of Theorem 12.0.2 to the proof of the coarse Baum–Connes conjecture for bounded geometry separated coarse unions of sequences that uniformly coarsely embed into Hilbert space.

Lemma 12.5.3 *To prove the coarse Baum–Connes conjecture for all bounded geometry metric spaces that coarsely embed into Hilbert space, it suffices to prove the coarse Baum–Connes conjecture for any bounded geometry coarse union that coarsely embeds into Hilbert space.*

Proof Assume we know the coarse Baum–Connes conjecture for all bounded geometry coarse unions that coarsely embed into Hilbert space. We must show that if X is an arbitrary bounded geometry that coarsely embeds into Hilbert space, then X satisfies the coarse Baum–Connes conjecture. Fix a basepoint $x_0 \in X$ and for each $n \geqslant 0$ let

$$X_n := \{x \in X \mid n^3 - n \leqslant d(x,x_0) \leqslant (n+1)^3 + (n+1)\}.$$

Define $Y = \bigsqcup_{n \text{ even}} X_n$ and $Z = \bigsqcup_{n \text{ odd}} X_n$ (both metrised as subspaces as X). Note that as X coarsely embeds into Hilbert space, the metric on X must be finite-valued, whence $X = Y \cup Z$. Moreover, note that Y, Z and $Y \cap Z$ are coarse unions of the sequences $(X_n)_{n \text{ even}}$, $(X_n)_{n \text{ odd}}$ and $(X_n \cap Z)_{n \text{ even}}$ respectively, and that they have bounded geometry and coarsely embed into Hilbert space, as they are subspaces of X. Hence the coarse Baum–Connes conjecture is true for Y, Z and $Y \cap Z$ by assumption.

We claim now that the cover $X = Y \cup Z$ of X is coarsely excisive in the sense of Definition 7.5.3, meaning that if for a subset W of X, we write

$$N_r(W) := \{x \in X \mid d(x,w) \leqslant r \text{ for some } w \in W\}$$

for the r-neighbourhood of W, then for each $r > 0$ there is $s > 0$ such that

$$N_r(Y) \cap N_r(Z) \subseteq N_s(Y \cap Z).$$

Indeed, let $n > r$, and choose $s = (n+1)^3$; we claim this works. Let then $x \in N_r(Y) \cap N_r(Z)$. If $d(x, x_0) \leqslant s$, then $x \in N_s(Y \cap Z)$, as $x_0 \in Y \cap Z$. Otherwise there must exist

$$y \in Y \setminus B(x_0; n^3) \quad \text{and} \quad z \in Z \setminus B(x_0; n^3)$$

such that $d(x,y) \leqslant r$ and $d(x,z) \leqslant r$. Hence in particular, $d(y,z) \leqslant 2r$. We claim that either $y \in Y \cap Z$, or $z \in Y \cap Z$. Indeed, if not then

$$y \in \bigsqcup_{m \geqslant n,\ m \text{ even}} \{x \in X \mid n^3 + n \leqslant d(x,x_0) \leqslant (n+1)^3 - (n+1)\}$$

and

$$z \in \bigsqcup_{m \geqslant n,\ m \text{ odd}} \{x \in X \mid n^3 + n \leqslant d(x,x_0) \leqslant (n+1)^3 - (n+1)\}.$$

It follows that $d(y,z) \geqslant 2n \geqslant 2r$, which is a contradiction.

Having observed this coarse excisiveness, the elaboration of Theorem 7.5.5 from Exercise 7.6.14 gives a commutative diagram of long-exact sequences

$$\begin{array}{ccccccc} \longrightarrow KX_i(Y \cap Z) & \longrightarrow & KX_i(Y) \oplus KX_i(Z) & \longrightarrow & KX_i(X) \longrightarrow \\ \downarrow & & \downarrow & & \downarrow \\ \longrightarrow K_i(C^*(Y \cap Z)) & \longrightarrow & K_i(C^*(Y)) \oplus K_i(C^*(Z)) & \longrightarrow & K_i(C^*(X)) \longrightarrow, \end{array}$$

where the vertical maps are coarse Baum–Connes assembly maps. The coarse Baum–Connes conjecture for X now follows from the conjecture for Y, Z and $Y \cap Z$ together with the five lemma. □

Lemma 12.5.4 *To prove the coarse Baum–Connes conjecture for any bounded geometry metric space that coarsely embeds into Hilbert space, it suffices to prove the coarse Baum–Connes conjecture for any separated coarse union of a bounded geometry sequence that uniformly coarsely embeds into Hilbert space.*

Proof Assume we know the coarse Baum–Connes conjecture to hold for any separated coarse union of a bounded geometry sequence that uniformly coarsely embeds into Hilbert space. It suffices by Lemma 12.5.3 to prove the coarse Baum–Connes conjecture for a bounded geometry coarse union that coarsely embeds into Hilbert space.

Let then (X_n) be a sequence of finite metric spaces, and X a corresponding coarse union, which we assume has bounded geometry and coarsely embeds into Hilbert space. As the coarse Baum–Connes conjecture is insensitive to coarse equivalences (see Exercise 7.6.2), and using Remark 12.5.2, we may as well assume that the metric d on X satisfies

$$\text{if } n > m \text{ then } \quad d(X_n, X_m) > \operatorname{diam}(X_0 \cup \cdots \cup X_n). \tag{12.26}$$

Note that the sequence (X_n) uniformly coarsely embeds into Hilbert space: to see this, just restrict the coarse embedding from X to each X_n. Let Y be the separated coarse union of the sequence (X_n); by our assumption, the coarse Baum–Connes conjecture is true for Y.

For each $r \geqslant 0$, it follows directly from the definition of the assembly map (Definition 7.1.1) that there is a commutative diagram

(12.27)

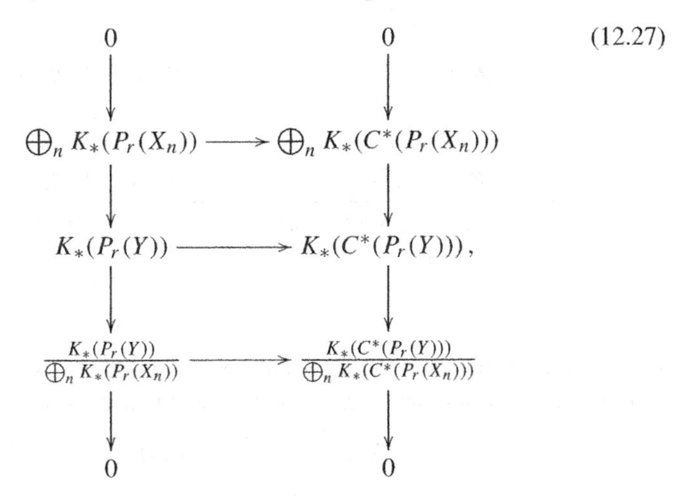

where the horizontal maps are induced by assembly and the vertical maps are induced by the natural inclusions. These diagrams are compatible with increasing the Rips parameter r, so we may take the limit as r tends to infinity. In the limit, the central horizontal map is an isomorphism as we are assuming the coarse Baum–Connes conjecture for Y. Moreover, the topmost horizontal map is an isomorphism in the limit as $r \to \infty$: indeed, using homotopy invariance of K-homology and coarse invariance of K-theory of Roe algebras, each summand

$$K_*(P_r(X_n)) \to K_*(C^*(P_r(X_n)))$$

identifies with the assembly map for a point as soon as $r \geqslant \operatorname{diameter}(X)$, which is an isomorphism (see Example 7.1.13). Hence by the five lemma, the bottom-most horizontal map in diagram (12.27) is an isomorphism.

Looking now at X, if we fix $r \geqslant 0$ then we may write

$$P_r(X) = \Delta_r \sqcup \bigsqcup_{n \geqslant N_r} P_r(X_n)$$

for some $N_r \in \mathbb{N}$ such that $\Delta_r = P_r(X_0 \sqcup \cdots \sqcup X_{N_r-1})$ is a single simplex (such a decomposition exists by the assumption in line (12.26) above). Then we have a similar commutative diagram,

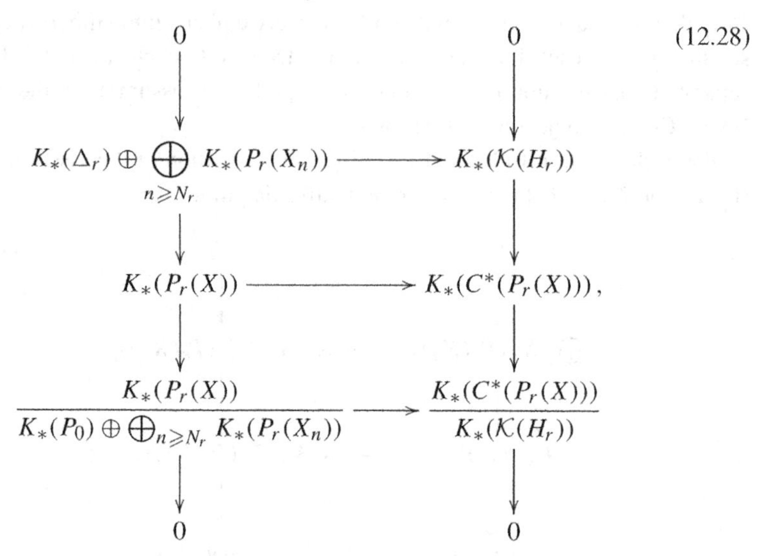

$$\begin{array}{ccc} 0 & & 0 \\ \downarrow & & \downarrow \\ K_*(\Delta_r) \oplus \bigoplus_{n \geqslant N_r} K_*(P_r(X_n)) & \longrightarrow & K_*(\mathcal{K}(H_r)) \\ \downarrow & & \downarrow \\ K_*(P_r(X)) & \longrightarrow & K_*(C^*(P_r(X))), \\ \downarrow & & \downarrow \\ \dfrac{K_*(P_r(X))}{K_*(P_0) \oplus \bigoplus_{n \geqslant N_r} K_*(P_r(X_n))} & \longrightarrow & \dfrac{K_*(C^*(P_r(X)))}{K_*(\mathcal{K}(H_r))} \\ \downarrow & & \downarrow \\ 0 & & 0 \end{array} \tag{12.28}$$

where the horizontal maps are all induced by assembly. Note that $K_*(\Delta_r)$ identifies with the K-homology of a point, and the restriction

$$K_*(\Delta_r) \to K_*(\mathcal{K}(H_r)) \tag{12.29}$$

of the topmost horizontal map to the summand $K_*(\Delta_r)$ is an isomorphism. The diagram is again compatible with increasing the Rips parameter r, so we again may take the limit as r tends to infinity. In the limit as r tends to infinity, the topmost horizontal map identifies with any of the maps in line (12.29) above, and is thus an isomorphism. On the other hand, for each $r \geqslant 0$, $P_r(Y)$ is naturally a subspace of $P_r(X)$, whence we may use the same Hilbert space, say H_r, to define $C^*(P_r(X))$ and $C^*(P_r(Y))$. Using this we see that for any $r \geqslant 0$,

$$C^*(P_r(X)) = \mathcal{K}(H_r) + C^*(P_r(X))$$

and

$$C^*(P_r(Y)) \cap \mathcal{K}(H_r) = \bigoplus_n C^*(P_r(X_n)),$$

whence elementary algebra gives a canonical isomorphism

$$\frac{C^*(P_r(X))}{\mathcal{K}(H_r)} \cong \frac{C^*(P_r(Y))}{\bigoplus_n C^*(P_r(X_n))}.$$

It follows from this and a similar argument on the K-homology level that the bottom-most horizontal map in diagram (12.28) identifies with the bottom-most horizontal map in diagram (12.27) (even before taking the limit as $r \to \infty$), whence the right-hand vertical map in diagram (12.28) is an isomorphism after taking the limit in r. The five lemma now gives that the central vertical map in diagram (12.28) is an isomorphism in the limit as $r \to \infty$, and we are done. □

12.6 The Case of Coarse Disjoint Unions

Using Lemma 12.5.4, the next proposition is enough to complete the proof of Theorem 12.0.2, and thus to complete our work in this chapter.

Proposition 12.6.1 *Let X be a separated coarse union of a bounded geometry sequence (X_n) that uniformly coarsely embeds into Hilbert space. Then the coarse Baum–Connes conjecture holds for X.*

The idea of the proof is to carry out a 'uniform version' of the proof of Theorem 12.4.2. Rather than repeat the whole argument for that theorem, we just sketch out the necessary changes, and where it is important that we have uniformity in certain arguments.

First, notation. Let $r \geqslant 0$. Let $P_{r,n}$ be the Rips complex of X_n at scale r and P_r the Rips complex of X at scale r. Note that as $d(X_n, X_m) = \infty$ for $n \neq m$, $P_r = \bigsqcup_{n=1}^{\infty} P_{r,n}$. For each n, let $f_n : X_n \to E_n$ be a map as in the definition of uniform coarse embeddability (Definition 12.5.1 above). We may assume that each E_n is even-dimensional on replacing it with $E_n \oplus \mathbb{R}$ if necessary. Throughout this section, we use X as a shorthand for the sequence (X_n) and (E_n) as a shorthand for the sequence (E_n).

Let $Z_{r,n} \subseteq P_{r,n}$ consist of all the rational points in $P_{r,n}$, i.e. those points such that all the coefficients t_x as in Definition 7.2.2 take rational values. Let

$$H_{r,n} := \ell^2(Z_{r,n}) \otimes H \quad \text{and} \quad H_{r,n,E} := \ell^2(Z_{r,n}) \otimes H \otimes \mathcal{L}^2_{E_n}.$$

We use these modules to build Roe algebras $C^*(H_{r,n})$ and $C^*(H_{r,n,E})$ of $P_{r,n}$ as in Section 12.2. Define also

$$H_r := \bigoplus_n H_{r,n} \quad \text{and} \quad H_{r,E} = \bigoplus_n H_{r,n,E}$$

and again use these to define Roe algebras $C^*(H_r)$ and $C^*(H_{r,E})$ of P_r. As before, we consider $C^*(H_{r,n})$ and $C^*(H_r)$ in their amplified representations on $H_{r,n,E}$ and $H_{r,E}$ respectively, and thus as subalgebras of the multiplier algebra of $C^*(H_{r,n,E})$ and $C^*(H_{r,E})$ respectively. Note that because $d(X_n, X_m) = \infty$ for $n \neq m$ there is a canonical inclusion

$$C^*(H_r) \subseteq \prod_n C^*(H_{r,n})$$

and thus an element T of $C^*(H_r)$ can be written as a sequence $T = (T_n)_{n=1}^{\infty}$ with each $T_n \in C^*(H_{r,n})$; all this works analogously on replacing $H_{r,n}$ and H_r with $H_{r,n,E}$ and $H_{r,E}$. We fix a good covering system $\{B_{x,r,n}\}_{x\in X,\ n\in\mathbb{N},\ r\geqslant 0}$ for each X_n as in Definition 12.2.1, and use it to consider each element of a sequence (T_n) as above (for either H_r or $H_{r,E}$) as a matrix $T_n = (T_{n,xy})_{x,y\in X_n}$ with respect to the corresponding decomposition

$$H_{r,n,E} = \bigoplus_{x\in X_n} H_{x,r,n,E}, \quad \text{with} \quad H_{x,r,n,E} = (\chi_{B_{x,r,n}} \otimes 1)H_{r,n,E}.$$

This whole discussion works similarly for the purely algebraic versions of the Roe algebras: for example

$$\mathbb{C}[H_r] \subseteq \prod_n \mathbb{C}[H_{r,n}],$$

which will also be used below.

We now define analogues of the Roe algebras with coefficients of Definition 12.3.2; the basic idea is the same, but we must enforce some more uniformity over n. We use the notation χ_R^V and p_i^V analogously to Definition 12.3.2 (see also Definition 12.3.1), but for each X_n separately.

Definition 12.6.2 Let $\prod_{n\in\mathbb{N}} C_b([1,\infty), C^*(H_{r,n,E}))$ denote the product C^*-algebra of the C^*-algebras of all bounded continuous functions from $[1,\infty)$ to $C^*(H_{r,n,E})$. Write elements of this $*$-algebra as collections $(T_{n,s})_{n,s\in[1,\infty)}$. We may further consider each $T_{n,s}$ as a matrix $(T_{n,s,xy})_{s\in[1,\infty),n\in\mathbb{N},x,y\in X_n}$. Let $\mathbb{A}^r(X;E)$ denote the $*$-subalgebra of $\prod_{n\in\mathbb{N}} C_b([1,\infty), C^*(H_{r,n,E}))$ consisting of elements satisfying the following conditions.

(i) $\displaystyle\sup_{s\in[1,\infty),n\in\mathbb{N}} \operatorname{prop}_P(T_{n,s}) < \infty.$

(ii) $\displaystyle\lim_{s\to\infty} \sup_n \operatorname{prop}_E(T_{n,s}) = 0.$

(iii) $\displaystyle\lim_{R\to\infty} \sup_{s\in[1,\infty),n\in\mathbb{N}} \|\chi_R^V T_{n,s} - T_{n,s}\| = \lim_{R\to\infty} \sup_{s\in[1,\infty)} \|T_{n,s}\chi_R^V - T_{n,s}\| = 0.$

(iv) If (p_i) is the net of finite rank projections on $\mathcal{L}^2_{E_n}$, then for each $s \in [1,\infty)$ and each $n \in \mathbb{N}$

$$\lim_{i\to\infty} \|p_i^V T_{n,s} - T_{n,s}\| = \lim_{i\to\infty} \|T_{n,s}p_i^V - T_{n,s}\| = 0.$$

Define $A^r(X;E)$ to the closure of $\mathbb{A}^r(X;E)$ inside $\prod_{n\in\mathbb{N}} C_b([1,\infty), C^*(H_{r,n,E}))$.

Analogously to Definition 12.3.3, define $\mathbb{A}_L^r(X;E)$ to be the collection of uniformly continuous bounded functions (T_t) from $[1,\infty)$ to $\mathbb{A}(X;E)$ such that the P-propagation of (T_t) tends to zero as t tends to infinity. Define $A_L^r(X;E)$ to the completion of $\mathbb{A}_L^r(X;E)$ for the norm $\|(T_t)\| := \sup_t \|T_t\|_{A^r(X;E)}$.

Now, let $F_{n,s}$ be as in Definition 12.1.5 for E_n. Let d_n be the dimension of E_n, and use the collection $F = (F_s)$ where $F_s := (F_{n,s+2d_n})_{n=1}^{\infty}$ to define an index map

$$\mathrm{Ind}_F : K_*(C^*(H_r)) \to K_*(A^r(X;E))$$

just as in the process that led up to Lemma 12.3.11. We then have the following analogue of Proposition 12.3.13.

Proposition 12.6.3 *Let $\iota^s : A^r(X;E) \to C^*(H_{r,E})$ be the map induced by evaluation at any $s \in [1,\infty)$. Then for each $s \in [1,\infty)$, the composition*

$$K_*(C^*(H_r)) \xrightarrow{Ind_F} K_*(A^r(X;E)) \xrightarrow{\iota_*^s} K_*(C^*(H_{r,E}))$$

is an isomorphism. The analogous statement holds for for the localised algebras.

Sketch proof The proof of Proposition 12.6.3 proceeds by applying the argument for Proposition 12.3.13 in all the factors in a sequence (T_n) simultaneously. To make the homotopies used to show that $\mathrm{Ind}_{F(0)} = \mathrm{Ind}_{F(\infty)}$ continuous over the whole sequence, one needs to use that the estimate in part Lemma 12.1.7 is independent of the dimension of the space involved. We then see that $\mathrm{Ind}_F = \mathrm{Ind}_{F(\infty)}$ just as in the proof of Proposition 12.3.13.

The last step in the argument is the homotopy in the variable r from the proof of Proposition 12.3.13 that is used to show that

$$\mathrm{Ind}_{F(\infty)}[p] = [p \otimes p_0]$$

as in line (12.22) above. There is a problem here: this homotopy is not equicontinuous as the dimension of the Hilbert space increases (the problem is that the parameter s is replaced by $s + 2d_n$, with d_n the dimension of E_n, and the larger s is, the worse the modulus of continuity).

To get around this problem, we use a 'stacking argument': the idea is that one can exchange 'space for speed'. As the details are notationally messy, we explain the idea here, leaving a careful write-up in the specific case at hand to the diligent reader. Say then that $(p_t)_{t\in[0,1]}$ is a homotopy of projections in a quasi-stable C^*-algebra A. We want to show that p_0 and p_1 define the same class in K-theory using only homotopies with uniformly bounded (independent of (p_t), or even of the ambient algebra A) modulus of continuity. Fix some large N, and consider the formal difference

$$\begin{bmatrix} p_0 & 0 & \cdots & 0 \\ 0 & p_{1/N} & \cdots & 0 \\ \vdots & \vdots & \ddots & \vdots \\ 0 & 0 & \cdots & p_1 \end{bmatrix} - \begin{bmatrix} 0 & 0 & \cdots & 0 \\ 0 & p_{1/N} & \cdots & 0 \\ \vdots & \vdots & \ddots & \vdots \\ 0 & 0 & \cdots & p_1 \end{bmatrix}, \tag{12.30}$$

where we use quasi-stability of A to make sense of the matrix representations. Clearly this formal difference represents $[p_0]$ in K-theory. On the other hand, a 'short' homotopy from $p_{n/N}$ to $p_{(n-1)/N}$ in each entry of the second matrix, followed by a rotation homotopy between

$$\begin{pmatrix} 0 & 0 & 0 & \cdots & 0 \\ 0 & p_0 & 0 & \cdots & 0 \\ 0 & 0 & p_{1/N} & \cdots & 0 \\ \vdots & \vdots & \vdots & \ddots & \vdots \\ 0 & 0 & 0 & \cdots & p_{(n-1)/N} \end{pmatrix} \quad \text{and} \quad \begin{pmatrix} p_0 & 0 & \cdots & 0 & 0 \\ 0 & p_{1/N} & \cdots & 0 & 0 \\ \vdots & \vdots & \ddots & \vdots & \vdots \\ 0 & 0 & \cdots & p_{(n-1)/N} & 0 \\ 0 & 0 & \cdots & 0 & 0 \end{pmatrix}$$

shows that the formal difference in line (12.30) above defines the same class in K-theory as

$$\begin{bmatrix} p_0 & 0 & \cdots & 0 \\ 0 & p_{1/N} & \cdots & 0 \\ \vdots & \vdots & \ddots & \vdots \\ 0 & 0 & \cdots & p_1 \end{bmatrix} - \begin{bmatrix} p_0 & 0 & \cdots & 0 \\ 0 & p_{1/N} & \cdots & 0 \\ \vdots & \vdots & \ddots & \vdots \\ 0 & 0 & \cdots & 0 \end{bmatrix},$$

which represents $[p_1]$. We can make the modulus of continuity of the homotopies between each $p_{n/N}$ and $p_{(n-1)/N}$ as small as we like by increasing N enough, and the rotation homotopy used has an absolutely bounded modulus of continuity. This completes the argument. □

We now have the following analogue of Proposition 12.4.1.

Proposition 12.6.4 *The map*

$$ev_*\colon \lim_{r\to\infty} K_*(A_L^r(X;E))) \to \lim_{r\to\infty} K_*(A^r(X;E))$$

induced by the direct limit of the evaluation-at-zero maps on K-theory is an isomorphism.

Sketch proof One can carry out the proof of Lemma 12.4.4 across all n at once, as long as the number R in the statement of the lemma is assumed to be the same across all X_n. The very last step in the proof of Lemma 12.4.4 that invokes Lemma 12.4.3 can be redone using quasi-stability again to commute the product over n and the K-functors (Proposition 2.7.12); thus we only have to use continuity in s one n at a time to invoke Lemma 12.4.3, and the lack of uniformity of the continuity in s as n varies is irrelevant.

Analogously with the rest of the proof of Proposition 12.4.1, we use the sets W_s defined in line (12.25) for each X_n in place of X. We apply a Mayer–Vietoris sequence simultaneously across all n to cut the sets W_s into pieces where we can apply Lemma 12.4.4. It is important for this that we have the uniform estimates in part (iii) of Definition 12.3.2, as this means we only need a number of cutting-and-pasting steps that is independent of the (dimension of the) ambient Euclidean space E_n for each of the sets W_s. □

The proof of Proposition 12.6.1 now proceeds quite analogously to the finite-dimensional case. This is enough to complete the proofs.

12.7 Exercises

12.7.1 Let X be a separable metric space. Let $(x_n)_{n=0}^{\infty}$ be a dense sequence in X, and for each $x \in X$, define a map $f_x : \mathbb{N} \to \mathbb{R}$ by

$$f_x(n) = d(x, x_n) - d(x_0, x_n),$$

and show that each f_x is bounded. Show that

$$f : X \to \ell^{\infty}(\mathbb{N}), \quad x \mapsto f_x$$

is an isometric (so in particular, coarse) embedding.

12.7.2 Show that a metric space X coarsely embeds into a real Hilbert space if and only if it coarsely embeds into a complex Hilbert space.

12.7.3 Let T be the vertex set underlying a (connected, undirected) tree, equipped with the *edge metric* defined by setting $d(x, y)$ to be the smallest number of edges in a path connecting x and y. Write E for the edge set of the tree. Observe that for any $x, y \in T$, there is a unique minimal set $\gamma_{xy} \subseteq E$ of edges that is contained in any path from x to y (and moreover, that forms such a path itself). Fix a basepoint $x_0 \in T$, and define

$$f : T \to \ell^2(E), \quad x \mapsto \sum_{e \in \gamma_{x_0 x}} \delta_e.$$

Show that f satisfies

$$\|f(x) - f(y)\|^2 = d(x, y)$$

for all $x, y \in T$, and thus in particular that f is a coarse embedding of T into a Hilbert space.

12.7.4 Show that a good Borel covering system as in Definition 12.2.1 exists. *Hint: one way to do this is to write $X = (x_n)$ as a sequence (we can do this, as it is a bounded geometry metric space with finite-valued metric, so countable); now iteratively choose B_{r,x_n} to consist of all points where the coordinate t_{x_n} of Definition 7.2.2 is non-zero, and that have not appeared in B_{r,x_m} for any $m < n$.*

12.7.5 Show that if A is a norm compact set of compact operators, and B a norm bounded and strong-$*$ compact set of bounded operators, then the collection

$$\{ST \mid S \in A, T \in B\}$$

of compact operators is norm compact.

Show also that if $s \mapsto S_s$ is a strong-$*$ continuous and norm bounded map defined on a bounded interval, then the family of maps

$$\{s \mapsto S_s S S_s^* \mid S \in A\}$$

is norm equicontinuous.

12.7.6 In part (iv) of Proposition 12.1.10 we proved a special case of the following fact: 'Let $\chi : \mathbb{R} \to [-1, 1]$ be an odd function such that $\lim_{t \to \pm\infty} \chi(t) = \pm 1$. If D_1, D_2 are essentially self-adjoint operators with the same domain S such that

$$D_1 \cdot S \subseteq S, \quad D_2 \cdot S \subseteq S,$$

and if $D_1 - D_2$ is bounded and D_1, D_2 have compact resolvent, then $\chi(D_1) - \chi(D_2)$ is compact'.

Give a different proof of this by considering the difference

$$\psi(D_1) - \psi(D_2),$$

where $\psi(x) = x/(x + i)$.

12.7.7 Let (X_n) be a sequence of finite metric spaces. Show that 'the' coarse union of the X_n only depends on the choice of metric up to coarse equivalence (see Definition A.3.9).

Moreover, show that 'the' coarse union of the sequence (X_n) coarsely embeds into Hilbert space if and only if the separated coarse union of the spaces X_n separately coarsely embeds into Hilbert space.

12.7.8 The aim of this exercise is to study analogues of Lemma 12.4.3 when the uniform continuity result is dropped. Let A be a C^*-algebra, and let $C_b([1,\infty),A)$ denote the C^*-algebra of bounded continuous functions from $[1,\infty)$ to A.

(i) Show that if A is quasi-stable, then the evaluation-at-one map $C_b([1,\infty),A) \to A$ induces an isomorphism on K-theory.
Hint: It suffices to show the kernel B of this map has trivial K-theory. Show that B fits into a pullback diagram

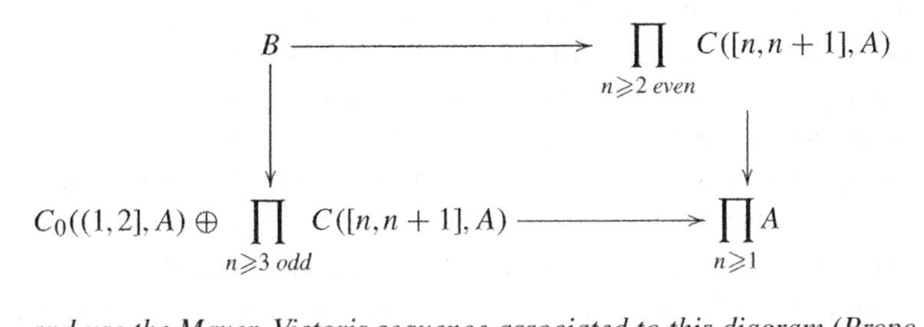

$$\begin{array}{ccc} B & \longrightarrow & \prod_{n\geqslant 2 \text{ even}} C([n,n+1],A) \\ \downarrow & & \downarrow \\ C_0((1,2],A) \oplus \prod_{n\geqslant 3 \text{ odd}} C([n,n+1],A) & \longrightarrow & \prod_{n\geqslant 1} A \end{array}$$

and use the Mayer–Vietoris sequence associated to this diagram (Proposition 2.7.15), combined with the the ability to compute products with K-theory in the presence of quasi-stability (Proposition 2.7.12).

(ii) Show that if $A = \mathbb{C}$, then the conclusion of the first part fails: one way to do this is to show that function $x \mapsto e^{ix}$ defines a non-zero class in $K_1(C_b[1,\infty))$.

12.8 Notes and References

We take the opportunity here to give a concrete description of the Bott–Dirac operator in the dimension one case, and a proof of its index-theoretic propeties. Consider the operator A on functions $u\colon \mathbb{R} \to \mathbb{C}$ defined by

$$(Au)(x) := u'(x) + xu(x).$$

This is the so-called *annihilation operator* of mathematical physics. Solving the differential equation $u' + xu = 0$, one sees that the kernel of A is spanned by $e^{-\frac{1}{2}x^2}$. More generally, if v is a Schwartz-class function on $\mathbb{R}$, then one computes that the differential equation $u' + xu = v$ has explicit solutions

$$u(x) = e^{-\frac{1}{2}x^2}\int_0^x v(t)e^{\frac{1}{2}t^2}dt + Ce^{-\frac{1}{2}x^2},$$

which are also Schwartz-class. In particular, considered as an operator $A\colon \mathcal{S}(\mathbb{R}) \to \mathcal{S}(\mathbb{R})$ from the space of Schwartz-class functions to itself, A is surjective with one-dimensional kernel, so has index one.

Our Bott–Dirac operator is given by

$$B = \begin{pmatrix} 0 & A^* \\ A & 0 \end{pmatrix},$$

where A^* is the formal adjoint of A, defined by $(A^*u)(x) = -u'(x) + xu(x)$ for Schwartz-class u. Thus B is built to be a self-adjoint operator that, when taking gradings into account, has the same index-theoretic behaviour in the graded sense as A does in the usual ungraded sense; the choice of whether to work with B or A is really just one of technical convenience.

Going into more depth, one can make the description of A even more explicit in terms of the right basis for $L^2(\mathbb{R})$: it turns out that A is a weighted shift operator. Indeed, there is an orthonormal basis of $L^2(\mathbb{R})$ given by the classical Hermite functions $(\psi_n)_{n=0}^\infty$, where each ψ_n is a degree n polynomial multiplied by $e^{-\frac{1}{2}x^2}$, so in particular ψ_0 is a constant multiple of $e^{-\frac{1}{2}x^2}$: see Proposition D.3.3. The functions $(\psi_n)_{n=0}^\infty$ are very natural and important: for example, the basis (ψ_n) diagonalises the Fourier transform. In terms of this basis, the action of A is very simple to describe: it is given by

$$A\psi_n = \sqrt{2n}\,\psi_{n-1},$$

where ψ_{-1} is interpreted as the zero function. Hence, up to a scaling factor, A behaves like the adjoint of the unilateral shift. Moreover, one can show that the Schwartz-class functions §$(\mathbb{R})$ consist precisely of series $\sum_{n=0}^\infty \lambda_n\psi_n$, where the sequence (λ_n) of complex coefficients decays faster than any polynomial in n (see for example the Appendix to V.3 of Reed and Simon [211]). In particular, this description makes it quite clear that A has index one as an operator from $\mathcal{S}(\mathbb{R})$ to itself.

The Bott–Dirac operator has been explicitly used in index theory at least since the 1980s. A version appears for example on pages 204–205 in Hörmander [142], a classical text on pseudodifferential operators, and is used as a key ingredient to prove a version of the Atiyah–Singer index theorem in sections 19.2 and 19.3 of Hörmander [142].

In terms of the material in this chapter, Theorem 12.0.2 was first proved by Yu in [272]. The original proof used a variant of the Dirac-dual-Dirac method of Kasparov [150], the infinite-dimensional Bott periodicity techniques of Higson, Kasparov and Trout [130] and the ideas that went into the proof of the Baum–Connes conjecture for a-T-menable groups due to Higson and Kasparov [129] (among other things). The Dirac-dual-Dirac method underlies many proofs of special cases of the Baum–Connes conjecture: the very rough idea is to use a (possibly very elaborate) version of Bott periodicity to replace the C^*-algebra one is interested in (something like a group C^*-algebra, or Roe algebra) with a 'proper' C^*-algebra whose K-theory can be computed using elementary techniques like Mayer–Vietoris sequences.

Our proof of Theorem 12.0.2 is still based on this idea in some sense: we use the Bott–Dirac operator that underlies one picture of Bott periodicity to replace the original Roe algebras with something much more K-theoretically tractable. However, rather than using versions of Kasparov's Dirac and dual Dirac elements, we in effect use his 'gamma element' directly to pass to something more computable; see Nishikawa [191] for a recent paper on (different cases of) the Baum–Connes conjecture that uses a philosophically similar approach. We are also able to use some tricks to reduce to the finite-dimensional case, and thus avoid any need to discuss infinite-dimensional Bott periodicity.

Before going on to discuss examples, we note that the coarse Baum–Connes conjecture fails for some metric spaces that do not satisfy the bounded geometry assumption, but do still coarsely embed into Hilbert space: the 'large spheres' of section 7 of Yu [271], as discussed in Section 13.1 below, are an example. Thus the bounded geometry assumption in Theorem 12.0.2 really is necessary.

See for example chapter 5 of Nowak and Yu [195], chapter 11 of Roe [218] or section 3 of Willett [257] for more background on coarse embeddings into Hilbert space and related issues. The class of metric spaces that coarsely embed into Hilbert space is very large. It includes for example the following classes of groups, considered as geometric objects:

- word hyperbolic groups (Sela [234]) (see also Roe [219] for a stronger result on asymptotic dimension);
- amenable groups (Bekka, Cherix and Valette [29]) (see for example Grigorchuk [109] for how wildly 'non-algebraic' this class can be);
- linear groups (Guentner, Higson and Weinberger [113]);
- relatively hyperbolic groups with suitable peripheral subgroups (Dadarlat and Guentner [74] and Ozawa [202];
- mapping class groups (Hamenstädt [120] and Kida [156]) (see also Bestvina, Bromberg and Fujiwara [31] for a stronger result on asymptotic dimension);

- outer automorphism groups of free groups (Bestvina, Guirardel and Horbez [32]);
- Thompson's group F (Farley [92]).

All but the last of these classes of groups are also known to have a stronger property called property A (section 2 of Yu [272]) (equivalently, by Guentner and Kaminker [114], Higson and Roe [134], Ozawa [201], so-called boundary amenability or C^*-exactness); property A for Thompson's group F is a well-known open problem.

Another interesting example of spaces that coarsely embed into Hilbert space, this time coming from outside the world of groups, comes from manifolds of subexponential volume growth (Gong and Yu [107]) (this class of spaces also turns out to have property A (see section 6 of Tu [241]).

Taking all of the above into account, Theorem 12.0.2 is the source of some of the most general known positive results on questions like the Novikov conjecture, and the existence of positive scalar curvature metrics.

There are, however, examples of groups that do not coarsely embed into Hilbert space: these are based on ideas around expanders as we will discuss in Chapter 13. The insight here is due to Gromov and seems to first appear at the end of Gromov [178]: the key idea is to embed expanders into groups based on random presentations. Gromov's paper [111] develops this further. See Arzhantseva and Delzant [5] and, most recently, Osajda [196] for more detailed constructions, and Sapir [229] for a construction allowing one to build fundamental groups of aspherical manifolds that do not coarsely embed into Hilbert space.

Partly inspired by the above (counter)examples, there has also been work on whether coarse embeddings into other classes of Banach spaces can have K-theoretic consequences. For example, Kasparov and Yu [153] deduces K-theoretic consequences when a space coarsely embeds into a uniformly convex Banach space, and Kasparov and Yu [154] when it coarsely embeds into a space with what is there called property (H); the latter is particularly interesting, as there are no (bounded geometry) metric spaces that are known not to embed in a property (H) Banach space. On the other hand, there are bounded geometry metric spaces that are known not to coarsely embed into any uniformly convex Banach space: certain so-called superexpanders as constructed in Lafforgue [160] or Mendel and Naor [176] have this property. It is known, however, that any bounded geometry metric space coarsely embeds into a strictly convex (in particular, reflexive) Banach space (Brown and Guentner [43]).

We should mention also that there has been a great deal of interesting work telling different classes of Banach spaces apart based on their coarse geometry, or based on which Banach spaces admit coarse embeddings into other classes. We make no attempt to give an exhaustive list of references here, just mentioning Johnson and Randrianarivony [145] and Naor and Mendel [190] for the case of L^p spaces.

Other recent results that connect to similar ideas include: Špakula and Willett [243], which uses coarse embeddability in Hilbert space to investigate the K-theory of different completions of the Roe $*$-algebra; Gong, Wu and Yu [108], studying (possibly infinite-dimensional) manifolds that are non-positively curved in some sense; and Fu and Wang [104], which proves a theorem containing both the case of Baum–Connes for a-T-menable groups, and coarse Baum–Connes for groups that coarsely embed into Hilbert space as special cases.

13

Counterexamples

Our aim in this chapter is to discuss two counterexamples to the coarse Baum–Connes conjecture, i.e. to the statement that the assembly map

$$\mu \colon KX_*(X) \to K_*(C^*(X))$$

is an isomorphism. In both cases, the counterexamples have a similar form. They consist of an infinite sequence (X_n) of bounded metric spaces (although with diameter tending to infinity) with the following property: there are operators D_n on modules over X_n that can be shown to have spectrum contained in $\{0\} \cup [c, \infty)$ for some $c > 0$, independent of n. The key point is the fact that the constant c is uniform in n: this allows one to show the existence of interesting global phenomena over the disjoint union space $X := \bigsqcup_{n=0}^{\infty} X_n$.

We give counterexamples to both the injectivity and surjectivity of the coarse Baum–Connes assembly map separately, of fairly different forms. These counterexamples are due respectively to Yu and Higson; see Section 13.5, Notes and References, at the end of the chapter for more detail.

For the injectivity counterexample, the spaces X_n are spheres (of carefully chosen dimension and radius) and the operators D_n are Dirac operators on them. The uniform spectral gap discussed above is used to show that the associated operator D on the global space $X := \bigsqcup_{n=0}^{\infty} X_n$ has essentially trivial higher index in $K_*(C^*(X))$, while Poincaré duality implies that D represents a non-trivial class in $KX_*(X)$. The injectivity counterexample we give is not coarsely equivalent to a bounded geometry space: it would be very interesting to have a bounded geometry example.

For the surjectivity counterexample, we consider a sequence (X_n) of so-called *expander graphs*: very roughly, these are graphs that are sparse in terms of not having many edges, but also highly connected in some sense. The exotic geometry of expanders allows us to show that certain combinatorial Laplacian operators Δ_n have uniform spectral gap. Putting these together gives an operator Δ in $C^*(X)$ such that the spectral projection associated to the isolated point zero in the spectrum of Δ represents a class in K-theory that cannot (at least in some special cases) be in the image of the coarse Baum–Connes assembly map. The surjectivity counterexample does have bounded geometry.

This chapter is structured as follows. In Section 13.1 we discuss the injectivity counterexample arising from large spheres. In Section 13.2 we give some background on expander graphs, and sketch a construction of such graphs based on the notion of property (τ) from representation theory of groups. Finally, in Section 13.3 we show how to construct surjectivity counterexamples from some of these expanders.

13.1 Injectivity Counterexamples from Large Spheres

The purpose of this section is to give counterexamples to injectivity of the coarse Baum–Connes assembly map coming from sequences of 'large' spheres; this example is originally due to Yu. It is worth commenting straight away that these counterexamples are not of bounded geometry, and no bounded geometry counterexamples to injectivity of the coarse Baum–Connes assembly map are known (in contrast to the surjectivity case as considered in Section 13.3 below).

Before getting to the details, we note that the proof uses differential geometry as a key ingredient. It would be very interesting to have a similar counterexample (or a different explanation of this one) that did not use any differential geometry. This is partly as understanding what is going on from a more elementary point of view might allow one to construct bounded geometry counterexamples.

The key geometric fact underlying our analysis is as follows. We will just use this as a black box.

Proposition 13.1.1 *Let*

$$S_r^d := \{(x_1, \ldots, x_{d+1}) \in \mathbb{R}^{d+1} \mid x_1^2 + \cdots x_{d+1}^2 = r^2\}$$

be the d-sphere of radius r and centre the origin in $\mathbb{R}^{d+1}$, equipped with the Riemannian metric induced[1] *and from the standard metric on $\mathbb{R}^d$. Then the scalar curvature of M is $d(d-1)/r^2$.*

[1] This is *not* the same as the restriction of the usual metric from $\mathbb{R}^{d+1}$! For example, the diameter of S_r^d is πr with the induced Riemannian metric, but $2r$ with the restricted metric.

Sketch proof The sectional curvatures of the standard round d-sphere

$$S^d := \{(x_1, \ldots, x_{d+1}) \in \mathbb{R}^{d+1} \mid x_1^2 + \cdots x_{d+1}^2 = 1\}$$

are well known to be constantly equal to one. Scaling this by r to get S_r^d multiplies the Riemannian metric by r^2, and therefore multiplies the sectional curvature by $1/r^2$, so all sectional curvatures of S_r^d are $1/r^2$. The scalar curvature of a d-manifold is $d(d-1)$ times the average (in an appropriate sense) of the sectional curvatures at that point, so we get that all scalar curvatures of $S^{d,r}$ are $d(d-1)/r^2$ as claimed. □

Now, for each n, let $X_n := S_{2n}^{2n}$. Define $X := \bigsqcup_{n=1}^{\infty} X_n$ with the coarse disjoint union metric: the metric that restricts to the given Riemannian metric on each X_n and sets the distance between distinct X_n to be infinity. Note that this space is not coarsely equivalent to a bounded geometry metric space.

We aim to prove the following theorem.

Theorem 13.1.2 *With X as above, the coarse Baum–Connes assembly map*

$$\mu\colon KX_*(X) \to K_*(C^*(X))$$

fails to be injective.

The next two lemmas complete the proof: the first says that the Dirac operators on each X_n combine to define a non-zero class in $KX_*(X)$, and the second says that this class goes to zero under the assembly map.

For each n, let D_n be the (spin) Dirac operator on X_n. Using that these operators have uniformly bounded propagation speed, a slight adaptation of the machinery of Chapter 8 gives a class $[D] \in K_*(X)$.

Lemma 13.1.3 *The natural map $c_X\colon K_*(X) \to KX_*(X)$ from the definition of $KX_*(X)$ (Definition 7.1.7) sends $[D]$ to a non-zero class in $KX_*(X)$.*

Proof It follows from Poincaré duality (Theorem 9.6.11) that for each n, the K-homology of X_n is $K_0(S^{2n}) = \mathbb{Z}[pt] \oplus \mathbb{Z}[D_n]$, where $[pt]$ is the class generated by the inclusion of any $[D_n]$ is the class of the Dirac operator. Let $Z \subseteq X$ be any net in X in the sense of Definition A.3.10 (such as exists by Lemma A.3.11). Then

$$KX_*(X) = \lim_{r\to\infty} K_*(P_r(Z))$$

by Theorem 7.2.16.

Note now that for each fixed n, as soon as r is larger than the diameter of X_n, we get that $P_r(Z \cap X_n)$ is properly homotopy equivalent to a point; on the other hand, for each fixed r and all suitably large n, $P_r(Z \cap X_n)$ is properly homotopy equivalent to X_n. Using these two facts, one computes that

$$KX_*(X) = \lim_{r\to\infty} K_*(P_r(Z)) = \frac{\prod_{n\in\mathbb{N}}(\mathbb{Z}[pt] \oplus \mathbb{Z}[D_n])}{\bigoplus_{n\in\mathbb{N}} \mathbb{Z}[D_n]}.$$

Under the canonical map, $c_X : K_*(X) \to KX_*(X)$, $[D]$ goes to the class

$$([D_1], [D_2], [D_3], [D_4], \ldots) \in \frac{\prod_{n\in\mathbb{N}}(\mathbb{Z}[pt] \oplus \mathbb{Z}[D_n])}{\bigoplus_{n\in\mathbb{N}} \mathbb{Z}[D_n]},$$

which is non-zero. □

The next lemma is a uniform version of of Lemma 10.2.5: it follows from exactly the same proof, once we have observed that Proposition 13.1.1 gives us a uniform lower bound on the scalar curvatures of each 'component' X_n.

Lemma 13.1.4 *The assembly map $\mu_X : K_*(X) \to K_*(C^*(X))$ sends the Dirac operator to zero.* □

Proof of Theorem 13.1.2 Using the definitions of assembly and the Baum–Connes assembly map (Section 7.1), we have a commutative diagram

$$\begin{array}{ccc} K_*(X) & & \\ \downarrow c_X & \searrow^{\mu_X} & \\ KX_*(X) & \xrightarrow{\mu} & K_*(C^*(X)), \end{array}$$

where the horizontal map is the Baum–Connes assembly map for X. The result follows from Lemmas 13.1.3 and 13.1.4. □

13.2 Expanders and Property (τ)

In this section, we study expander graphs, in particular, sketching a construction of such graphs coming from property (τ). This material will be used to construct surjectivity counterexamples in Section 13.3.

We will be interested in graphs considered as metric spaces; everything could be done in a purely metric language, but the combinatorial language of graphs will be convenient. For us a *graph* will consist of a set X of *vertices* equipped with a set E of *edges*, where E is a subset of the set $\{A \subseteq X \mid |A| = 2\}$ of two-element subsets of X (thus our graphs are undirected, and have no loops). The vertex set X of our graph is given the metric

$$d(x, y) := \min\{n \mid \text{there are } x = x_0, \ldots, x_n = y \text{ with } \{x_i, x_{i+1}\} \in E \text{ for all } i\}$$

(to be interpreted as ∞ if the set on the right is empty). In words, the distance between two vertices is the smallest number of edges in a path between them. A graph is *connected* if the metric above is finite-valued (or in other words, if any two vertices are connected by some edge path). The *degree* of a vertex is the number $|\{\{x,y\} \in E \mid y \in X\}|$ of edges with x as a vertex; the metric space (X,d) has bounded geometry (Definition A.3.19) if and only if there is a uniform bound on the degrees of all vertices.

Typically, we will abuse terminology and say something like 'let X be a graph' leaving the other structure implicit; we are in any case only really interested in X as a metric space.

Group theory provides an interesting class of examples.

Example 13.2.1 Let G be a finitely generated discrete group, and let $S \subseteq G$ be a finite generating set such that $S = S^{-1}$ and S does not contain the identity. Let $X = G$, and let the edge set consist of all two elements subsets $\{x,y\}$ such that $x = ys$ for some non-identity $s \in S$ (as S is symmetric, this is well-defined). The resulting (bounded geometry, connected) graph is called the *Cayley graph* of Γ with respect to S. For example, if $\Gamma = \mathbb{Z}/n$ is the cyclic group with n elements and $S = \{1, -1\}$, then the associated Cayley graph is the cycle with n vertices.

Definition 13.2.2 Let X be a finite connected graph. The *Laplacian* on X is the linear operator $\Delta = \Delta_X$ defined by

$$\Delta\colon \ell^2(X) \to \ell^2(X), \quad (\Delta u)(x) = \sum_{\{x,y\}\in E} u(x) - u(y).$$

Rewriting the formula slightly, if $d(x)$ is the degree of a vertex x, and $S_1(x)$ is the sphere of radius one about x, then

$$(\Delta u)(x) = d(x)u(x) - \sum_{y\in S_1(x)} u(y);$$

thus up to normalisation, $(\Delta u)(x)$ looks at the difference between the value of the function at x, and its average values on the sphere centred at x. This should be compared to the standard Laplacian

$$\Delta = -\sum_{i=1}^{d} \frac{\partial^2}{\partial x^2}$$

on $\mathbb{R}^d$ (the sign convention is chosen to make Δ a positive operator). In this case, one has the formula

$$(\Delta u)(x) = \lim_{r\to 0} \frac{2d}{r^2}\left(u(x) - \frac{1}{\mathrm{Vol}(S_r(x))}\int_{S_r(x)} u(y)dy\right),$$

where $S_r(x)$ is the sphere of radius r centred at x; thus the operator in Definition 13.2.2 is a sort of 'discretised version' of this, which justifies the terminology.

Here are the basic properties we will need.

Lemma 13.2.3 *The Laplacian on a finite connected graph X:*

(i) has propagation at most one (and exactly one if X has at least two points);
(ii) is positive;
(iii) has one-dimensional kernel consisting of the constant functions;
(iv) has norm bounded by four[2] *times the highest degree of any vertex in X.*

Proof The statement about propagation is immediate from the formula for Δ. Positivity follows from the computation

$$\langle u, \Delta u\rangle = \sum_{x\in X} \overline{u(x)}\left(\sum_{\{x,y\}\in E} u(x) - u(y)\right) = \sum_{\{x,y\}\in E} |u(x) - u(y)|^2.$$

From this formula, we also see that $u \in \ell^2(X)$ is in the kernel of Δ if and only if $u(x) = u(y)$ whenever $\{x, y\}$ is an edge; as we are assuming X is connected, this is equivalent to u being constant. Finally, note that the above implies that if $\|u\| = 1$ and d is an upper bound on the degrees of all vertices of X, then

$$\begin{aligned}\langle u, \Delta u\rangle &\leqslant \sum_{\{x,y\}\in E} 2|u(x)|^2 + 2|u(y)|^2 \\ &\leqslant \sum_{x\in X} 2d|u(x)|^2 + \sum_{y\in X} 2d|u(y)|^2 = 4d\|u\|^2 \\ &= 4d\end{aligned}$$

giving the norm estimate. □

We will be interested in 'heat flow' on a finite connected graph X. Let $(u_t)_{t\in[0,\infty)}$ be a smooth family of vectors in $\ell^2(X)$. Consider the *Heat equation*

$$\frac{\partial u_t}{\partial t} + \Delta u_t = 0,$$

which governs how the initial distribution u_0 'spreads out by flowing along edges' as time increases. The solution to the heat equation for a given initial distribution $u_0 \in \ell^2(X)$ is given by $u_t = e^{-t\Delta}u_0$. Note that (as one should

[2] Not optimal: we leave it as an exercise to find a better estimate.

expect), $e^{-t\Delta}$ converges in norm to the projection p_c onto the constant vectors in $\ell^2(X)$. Indeed, the functional calculus gives the

$$\|e^{-t\Delta} - p_c\| = e^{-t\lambda_1(X)}, \tag{13.1}$$

where $\lambda_1(X)$ is the first non-zero eigenvalue of Δ. Thus, the larger $\lambda_1(X)$ is, the more quickly heat flows to a constant distribution: we conclude from these heuristics that the first non-zero eigenvalue of Δ is a measure of how connected the graph X is.

Example 13.2.4 Let $X_n = \mathbb{Z}/n$ be the finite cyclic group with n elements, and consider its Cayley graph with respect to the generating set $S = \{\pm 1\}$ as in Example 13.2.1. Let $\mathbb{C}[\mathbb{Z}/n]$ denote the group algebra of $\mathbb{Z}/n$, represented on $\ell^2(\mathbb{Z}/n)$ via the regular representation, and let u be the unitary corresponding to shifting by one. Then for $n > 2$ we have

$$\Delta = 2 - u - u^*.$$

The spectrum of Δ can be computed by representation theory: indeed, as the C^*-algebra $C^*(\mathbb{Z}/n) = \mathbb{C}[\mathbb{Z}/n]$ is commutative we have

$$\operatorname{spec}(\Delta) = \{\phi(\Delta) \mid \phi\colon \mathbb{C}[\mathbb{Z}/n] \to \mathbb{C} \text{ a } *\text{-homomorphism}\}.$$

The $*$-homomorphisms from $\mathbb{C}[\mathbb{Z}/n]$ to $\mathbb{C}$ are determined by where they send u, and the possibilities are $e^{2\pi ik/n}$, $k \in \{0, \ldots, n-1\}$. Hence we have

$$\operatorname{spec}(\Delta) = \{2 - 2\cos(2\pi k/n) \mid k \in \{0, \ldots, n-1\}\}.$$

In particular, $\lambda_1(X) = 2 - 2\cos(2\pi/n)$. As this tends to zero as $n \to \infty$, we conclude from our above discussion of heat kernels that the graphs X_n get 'less and less well connected' as $n \to \infty$.

Definition 13.2.5 An *expander* is an infinite sequence (X_n) of finite connected graphs such that:

(i) $|X_n| \to \infty$ as $n \to \infty$;
(ii) there is a uniform bound on the degrees of all vertices;
(iii) there is $c > 0$ such that $\lambda_1(X_n) \geqslant c$ for all n.

The idea is that an expander consists of an infinite family of graphs which is 'uniformly well connected' in some sense, despite there being a uniform bound on all vertex degrees. Note that the computations of Example 13.2.4 show that the sequence of finite cyclic graphs is *not* an expander; indeed, it is not immediately clear that examples exist. To build examples, we will generalise Example 13.2.4 to groups with more complicated representation theory.

Construction 13.2.6 Let Γ be an infinite discrete group generated by a finite symmetric set S. Let

$$K_1 \trianglerighteq K_2 \trianglerighteq \cdots$$

be an infinite sequence of finite index nested normal subgroups of Γ such that $\bigcap_n K_n = \{e\}$ (such a sequence exists if and only if Γ is residually finite). Let $\Gamma_n := \Gamma/K_n$, and let X_n be the Cayley graph of Γ_n built with respect to the generating set given by the image of S. Then the sequence (X_n) satisfies conditions (i) and (ii) from Definition 13.2.5. Moreover, at least for all n suitably large, the Laplacian of X_n is given by the image of the operator

$$\Delta_S = \sum_{s \in S} 1 - u_s \in \mathbb{C}[\Gamma]$$

in the canonical left quasi-regular representation (see Example C.1.4) of $\mathbb{C}[\Gamma]$ on $\ell^2(X_n)$. Thus whether or not (X_n) is an expander is a property of the operator Δ_S.

Note that if $\Gamma = \mathbb{Z}$, $S = \{\pm 1\}$ and $K_n = \mathbb{Z}/2^n$, then we have a subsequence of the sequence in Example 13.2.4.

Definition 13.2.7 With notation as in Construction 13.2.6, one says that G has *property* (τ) with respect to the sequence (X_n) if (X_n) is an expander, i.e. if there exists $c > 0$ such that the image of the Laplacian Δ_S in $\mathcal{B}(\ell^2(X_n))$ has spectrum contained in $\{0\} \cup [c, \infty)$ for all n.

It is not immediately obvious, but property (τ) does not depend on the choice of generating set S: see Exercise 13.4.1 below. In general, it is a difficult problem to show that a particular group has property (τ) with respect to some sequence (X_n), and many residually finite groups (e.g. residually finite amenable groups: see Exercise 13.4.2) never have property (τ) with respect to any sequence. We will use the following theorem as a black box: see Section 13.5, Notes and References, at the end of the chapter for more information.

Theorem 13.2.8 *The following groups admit a sequence (X_n) as in Construction 13.2.6 with respect to which they have property (τ):*

(i) free groups on at least two generators;
(ii) fundamental groups of surfaces with genus at least two;
(iii) $SL(n,\mathbb{Z})$ for $n \geqslant 2$. □

Remark 13.2.9 A group G has *property (T)* if there exists $c > 0$ such that the image of $\Delta_S \in \mathbb{C}[\Gamma]$ in *any* unitary representation has spectrum contained in $\{0\} \cup [c, \infty)$. Thus a group with property (T) has property (τ) with respect to any sequence (X_n) as in Example 13.2.6, i.e. any such sequence is an expander (note that such a sequence will only exist if G is residually finite). Examples of groups with property (T) include $SL(n, \mathbb{Z})$ for $n \geqslant 3$.

On the other hand, free groups, surface groups and $SL(2, \mathbb{Z})$ all admit sequences (X_n) which are not expanders. There are also residually finite groups which have property (τ) with respect to any sequence (X_n), but do not have property (T): an example is given by $SL(2, \mathbb{Z}[1/p])$ for any prime p.

In the next section, we will show that at least some expanders are counterexamples to the coarse Baum–Connes conjecture. Here is the crucial property we will need.

Lemma 13.2.10 *Let (X_n) be an expander, and $X = \bigsqcup X_n$. Let Δ be the block-diagonal operator on $\ell^2(X) = \bigoplus_n \ell^2(X_n)$ that restricts to the Laplacian on X_n in each summand. Let p be the block diagonal operator on $\ell^2(X)$ that restricts to the projection onto the constant functions in each X_n. Then Δ is a well-defined bounded operator, and*

$$\lim_{t \to \infty} e^{-t\Delta} = p$$

in norm.

Proof The operator Δ is bounded as Lemma 13.2.3 gives a uniform bound on the norms of its restriction to each block in terms of the degrees of vertices. Let Δ_n and p_n be the restrictions of Δ and p to each block $\ell^2(X_n)$. Then

$$\|e^{-t\Delta} - p\| = \sup_n \|e^{-t\Delta_n} - p_n\| = \sup_n e^{-t\lambda_1(X_n)},$$

where the second equality follows from the functional calculus as in line (13.1) above. As there is $c > 0$ with $\lambda_1(X_n) \geqslant c$ for all n, this tends to zero as $t \to \infty$. □

The point of the proposition is that one gets *norm* convergence of the 'heat semigroup' $(e^{-t\Delta})_{t \in [0,\infty)}$ for an expander: for any bounded degree sequence (X_n) of finite connected graphs $e^{-t\Delta}$ will converge to p in the strong operator topology, but this is no use for what we need in the next section. In particular, we see that the projection p is a norm limit of operators with finite propagation, as Δ has finite propagation, whence each $e^{-t\Delta}$ is itself a norm limit of operators with finite propagation (i.e. of polynomials in Δ). This may be surprising at first: indeed, p has block-diagonal matrix representation

$$p = \begin{pmatrix} p_1 & 0 & 0 & \cdots \\ 0 & p_2 & 0 & \cdots \\ 0 & 0 & p_3 & \cdots \\ \vdots & \vdots & \vdots & \ddots \end{pmatrix},$$

where each p_n has the $|X_n| \times |X_n|$ matrix representation

$$p_n = \frac{1}{|X_n|}\begin{pmatrix} 1 & 1 & \cdots & 1 \\ 1 & 1 & \cdots & 1 \\ \vdots & \vdots & \ddots & \vdots \\ 1 & 1 & \cdots & 1 \end{pmatrix};$$

thus the matrix representation of p gets wider and wider as one goes further along the basis, and it looks like p is a long way from being finite propagation.

Definition 13.2.11 The projection p as above is called the *Kazhdan projection* associated to the expander (X_n).

13.3 Surjectivity Counterexamples from Expanders

Throughout this section, we fix a finitely generated group G and an expander (X_n) built from G as in Construction 13.2.6. Let X be the separated coarse union of (X_n) as in Definition 12.5.1: recall this means that as a set $X = \bigsqcup_{n=1}^{\infty} X_n$ is the disjoint union of the X_n, and that X is equipped with the metric that restricts to the original (graph) metric on each X_n, and that puts distinct X_n at infinite distance from each other (equivalently, we could just set X to be the graph disjoint union of the X_n and equip it with the graph metric). Our aim is to show that X is a counterexample to surjectivity of the coarse Baum–Connes assembly map, at least under additional assumptions on G.

Note that X has bounded geometry, as there is a uniform bound on the degrees of all vertices coming from the size of the fixed generating set for G. As a sanity check, let us first show that X does *not* satisfy the hypotheses of Proposition 12.6.1, and therefore it is at least possible that the coarse Baum–Connes conjecture does not hold for G.

Proposition 13.3.1 *With notation as above, X does not admit a uniform coarse embedding (see Definition 12.5.1) into Hilbert space.*

Proof Say for contradiction X does admit such a uniform coarse embedding, so there are non-decreasing functions $\rho_-, \rho_+ \colon [0,\infty) \to [0,\infty)$ with $\rho_-(t) \to \infty$ as $t \to \infty$, and functions $f_n \colon X_n \to E_n$ from each X_n to some Hilbert space E_n (which we may assume complex on tensoring with $\mathbb{C}$) such that for all n,

$$\rho_-(d(x,y)) \leqslant \|f_n(x) - f_n(y)\|_{E_n} \leqslant \rho_+(d(x,y))$$

for all $(x,y) \in X_n$. Now, for each n, consider the Hilbert space $\ell^2(X_n, E_n) = \ell^2(X_n) \otimes E_n$ and the Laplacian $\Delta_n := \Delta_{X_n} \otimes 1_{E_n}$ on this space. Note that the computations in Lemma 13.2.3 show that the kernel of Δ_n consists exactly of the constant functions from X_n to E_n. Replacing each f_n by

$$f_n - \frac{1}{|X_n|} \sum_{x \in X_n} f_n(x),$$

we may assume that each $f_n \in \ell^2(X_n, E_n)$ is orthogonal to the constant functions. Let $c \geqslant 0$ be as in the definition of an expander, whence we have that

$$c\langle f_n, f_n \rangle \leqslant \langle f_n, \Delta_n f_n \rangle.$$

Let d be an absolute bound on the degrees of all vertices in all the X_n so there are at most $d|X_n|$ edges in X_n. Expand the inner products to get

$$c \sum_{x \in X_n} \|f(x)\|^2 \leqslant \sum_{\{x,y\} \text{ an edge}} \|f_n(x) - f_n(y)\|^2 \leqslant d|X_n|\rho_+(1)^2.$$

It follows that at least half of the points in X_n must satisfy $\|f(x)\|^2 \leqslant 2d\rho_+(1)^2/c$. Combined with bounded geometry and the fact that $|X_n| \to \infty$ and $n \to \infty$, this contradicts the existence of ρ_-, and we are done. □

Now, let us move on to the coarse Baum–Connes conjecture. In order to prove that X is a counterexample to the coarse Baum–Connes conjecture, we need an additional analytic assumption on G that we now introduce.

Definition 13.3.2 A countable, proper metric space Y has the *operator norm localisation property* (ONL for short) if for any $r \geqslant 0$ and $c \in (0,1)$ there exists $s \geqslant 0$ such that for any Hilbert space H and any bounded operator T on $\ell^2(Y, H)$ with $\text{prop}(T) \leqslant r$, there exists a unit vector $u \in \ell^2(Y, H)$ such that $\text{diam}(\text{supp}(u)) \leqslant s$, and such that

$$\|Tu\| \geqslant c\|T\|.$$

The point of the definition is the fact that s only depends on the propagation bound r, and not on the specific operator T.

Remark 13.3.3 It is straightforward to check from the definition that ONL is invariant under coarse equivalences as in Definition A.3.9. Hence by Lemma A.3.13 it makes sense to speak of ONL holding for a countable group G.

Example 13.3.4 Our space X built out of an expander does not have ONL. Indeed, let p be the Kazhdan projection on $\ell^2(X)$ as in Definition 13.2.11 above, and let q be a finite propagation operator on $\ell^2(X)$ with $\|p-q\| < 1/4$. Let $r = \text{prop}(q)$, and say for contradiction there is $s \geqslant 0$ with the property in Definition 13.3.2 for this r, and for $c = 3/4$. For each N, let q_N be the restriction of q to $\bigoplus_{n\geqslant N}\ell^2(X_n)$. Then each q_N has propagation at most r, whence there is a unit vector u_N in $\ell^2(X)$ with support in a set of diameter at most s and

$$\|q_N u_N\| \geqslant (3/4)\|q_N\| \geqslant 9/16; \tag{13.2}$$

note that u_N is supported in X_n for some $n \geqslant N$. Now, the restriction of p to each $\ell^2(X_n)$ is the matrix with all entries $|X_n|^{-1}$. As $|X_n| \to \infty$ and as there is an absolute bound on the cardinality of the support of each u_N (by bounded geometry), we must have that $\|pu_N\| \to 0$ as $N \to \infty$. Hence the estimate

$$\|q_N u_N\| = \|qu_N\| \leqslant \|(q-p)u_N\| + \|pu_N\| \leqslant (1/4) + \|pu_N\|$$

implies that

$$\limsup_{N\to\infty} \|q_N u_N\| \leqslant 1/4,$$

which contradicts line (13.2) above.

On the other hand, the 'parent group' G giving rise to an expander may have ONL. We will use the next example as a black box.

Lemma 13.3.5 *The fundamental group of any closed surface of genus at least two has ONL.* □

Now, let H be an auxiliary Hilbert space. Let (X_n) be our expander built from a group G as in Construction 13.2.6 with associated separated coarse union X. Let $C^*(X)$ denote the Roe algebra of X associated to the ample X module $\ell^2(X,H)$. Fixing a unit vector $v \in H$ gives rise to an isometric inclusion

$$V\colon \ell^2(X) \to \ell^2(X,H), \quad u \mapsto u \otimes v$$

that preserves propagation. Moreover, conjugation by V sends all operators on $\ell^2(X)$ to locally compact operators on $\ell^2(X,H)$. Hence in particular conjugation by V maps the Kazhdan projection of Definition 13.2.11 to an element $p_K \in C^*(X)$, which we also call the *Kazhdan projection*.

Theorem 13.3.6 *Say G is a fundamental group of a smooth, closed manifold with contractible universal cover, that G has ONL and that (X_n) an expander constructed from G as above. Then the coarse Baum–Connes conjecture fails for X: more precisely, the class of the Kazhdan projection is not in the image of the coarse assembly map*

$$\mu\colon KX_*(X) \to K_*(C^*(X)).$$

Thanks to Example 7.4.2 part (iv), Theorem 13.2.8, Lemma 13.3.5, the hypotheses of Theorem 13.3.6 are satisfied for G the fundamental group of a surface of genus at least two. Thus there exist bounded geometry counterexamples to the coarse Baum–Connes conjecture.

The proof will occupy the remainder of this section. Fixing notation as in the statement of Theorem 13.3.6, the strategy of the proof is as follows. Let $\frac{\prod_n \mathbb{R}}{\oplus_n \mathbb{R}}$ denote the product of countably many copies of $\mathbb{R}$ divided by the direct sum in the category of abelian groups; in particular, there are no boundedness assumptions of sequences in $\prod_n \mathbb{R}$. We construct two trace-type maps

$$\mathrm{tr}, \tau \colon K_0(C^*(X)) \to \frac{\prod_n \mathbb{R}}{\oplus_n \mathbb{R}}$$

that agree on the image of μ by an analogue of Atiyah's covering index theorem (Section 10.1). We then show that for the Kazhdan projection p_K we have $\tau[p_K] \neq \mathrm{tr}[p_K]$, completing the proof.

The first of these traces tr is elementary. Indeed, as $d(X_n, X_m) = \infty$ for $n \neq m$, we have that

$$C^*(X) \subseteq \prod_n \mathcal{K}(\ell^2(X_n, H)).$$

Hence there is a map

$$K_0(C^*(X)) \to K_0\Big(\prod_n \mathcal{K}(\ell^2(X_n, H))\Big) \to \prod_n K_0(\mathcal{K}(\ell^2(X, H)), \tag{13.3}$$

where the second arrow is induced from the obvious quotient maps $\prod_n \mathcal{K}(\ell^2(X_n, H)) \to \mathcal{K}(\ell^2(X_n, H))$ onto each factor (it is an isomorphism by Proposition 2.7.12, but we do not need this).

Definition 13.3.7 Define a map

$$\mathrm{tr} \colon K_0(C^*(X)) \to \frac{\prod_n \mathbb{R}}{\oplus_n \mathbb{R}}$$

by composing the map in line (13.3) above with the canonical isomorphism

$$\prod_n K_0(\mathcal{K}(\ell^2(X, H)) \cong \prod_n \mathbb{Z},$$

and the composition

$$\prod_n \mathbb{Z} \to \prod_n \mathbb{R} \to \frac{\prod_n \mathbb{R}}{\oplus_n \mathbb{R}}$$

of the canonical inclusion and quotient map.

As the restriction of p_K to each block $\ell^2(X_n, H)$ has rank one, the following lemma follows directly from the definitions.

Lemma 13.3.8 *With notation as above, $tr(p_K)$ is the class of the constant sequence $(1, 1, 1, 1, \ldots)$ in $\prod_n \mathbb{R}/\bigoplus_n \mathbb{R}$.* □

We now look at the other trace map τ, which is more involved; in particular, it is here that we need to use that G has ONL. Let $K_n \leqslant G$ be the kernel of the quotient map $\pi_n \colon G \to X_n$. Equip G with the word metric associated to the fixed generating set S used to define the graph structure on each X_n, i.e.

$$d(g,h) = \min\{n \mid g^{-1}h = s_1 \ldots s_n \text{ with each } s_i \in S\},$$

so G becomes a proper metric space with bounded geometry. The left action of G on itself is by isometries, whence so too is the restricted action of each K_n. This action makes the ample G module $\ell^2(G, H)$ equivariant for the action of K_n, and so we may use it to define an equivariant Roe $*$-algebra $C^*(|G|)^{K_n}$ (Definition 5.2.1 – we use the notation '$|G|$' from Convention 5.1.16 instead of 'G' to avoid confusion with the group C^*-algebra) for each n. Define now a C^*-algebra

$$A := \frac{\prod_n C^*(|G|)^{K_n}}{\bigoplus_n C^*(|G|)^{K_n}}.$$

The key step in the construction of τ is the construction of a $*$-homomorphism

$$\phi \colon C^*(X) \to A,$$

which we now do.

Let then T be an element of $\mathbb{C}[X]$. As

$$\mathbb{C}[X] \subseteq \prod_n \mathcal{K}(\ell^2(X_n, H)),$$

we may write T as a sequence $(T^{(n)})$, where each $T^{(n)}$ is an element of $\mathcal{K}(\ell^2(X, H))$. We may write each $T^{(n)}$ as a matrix $(T^{(n)}_{xy})_{x,y \in X_n}$, with each $T^{(n)}_{xy}$ in $\mathcal{K}(H)$. Let $r = \text{prop}(T)$, so r is also a bound for $\text{prop}(T^{(n)})$ for all n. For each n, let $\widetilde{T^{(n)}}$ be the element of $\mathbb{C}[|G|]^{K_n}$ (see Definition 5.2.1) defined by the matrix

$$\widetilde{T^{(n)}_{xy}} := \begin{cases} T^{(n)}_{\pi_n(x)\pi_n(y)} & d(x,y) \leqslant r, \\ 0 & d(x,y) > r. \end{cases} \tag{13.4}$$

As G has bounded geometry, an easier analogue of Lemma 12.2.4 above (we leave this to the reader) shows that $\widetilde{T^{(n)}}$ is indeed a well-defined bounded operator, and that the sequence $(\widetilde{T}^{(n)})_{n=1}^{\infty}$ has uniformly bounded norms.

Definition 13.3.9 With notation as above, define

$$\phi\colon \mathbb{C}[X] \to A, \quad T \mapsto (\widetilde{T^{(n)}}).$$

The discussion above shows that ϕ is a well-defined $*$-preserving linear map. In fact, more is true.

Lemma 13.3.10 *The map $\phi\colon \mathbb{C}[X] \to A$ defined above is a $*$-homomorphism.*

Proof With K_n the kernels of the quotient maps $\pi_n\colon G \to X_n$, the fact that the K_n are nested and that $\bigcap_n K_n = \{e\}$ implies that for any s there is N such that for all $n \geqslant N$, π_n is an isometry when restricted to any ball of radius at most s (it is important here that we use the same generating set to define the metrics on G and on each X_n). However, the matricial formulas for multiplying two finite propagation operators S, T together show that $(TS)_{xy}$ depends only on information in the ball of radius $\text{prop}(T) + \text{prop}(S)$ about x. We conclude from this discussion that for operators of a fixed propagation and for suitably large n, it makes no difference whether we multiply two operators in $\mathcal{B}(\ell^2(X_n, H))$, then lift to $\mathcal{B}(\ell^2(G, H))$ via the formula in line (13.4) above; or if we lift them to $\mathcal{B}(\ell^2(G, H))$, then multiply. It follows that ϕ is multiplicative, proving the lemma. □

Here is the key use of ONL.

Lemma 13.3.11 *Say G has ONL. Then for any $T \in \mathbb{C}[X]$,*

$$\|\phi(T)\|_A = \sup_{s\geqslant 0} \limsup_{n\to\infty} \sup\{\|T^{(n)}u\| \mid u \in \ell^2(X_n, H), \|u\| = 1, diam(supp(u)) \leqslant s\}.$$

Proof Recalling that

$$A := \frac{\prod_n C^*(|G|)^{K_n}}{\bigoplus_n C^*(|G|)^{K_n}},$$

we have that with notation as in line (13.4) above,

$$\|\phi(T)\|_A = \limsup_{n\to\infty} \|\widetilde{T^{(n)}}\|.$$

It follows directly from the definition of the operator norm that $\|\phi(T)\|_A$ equals

$$\limsup_{n\to\infty} \sup_{s\geqslant 0} \sup\{\|\widetilde{T^{(n)}}u\| \mid u \in \ell^2(G, H), \|u\| = 1, \text{diam}(\text{supp}(u)) \leqslant s\}.$$

On the other hand, using that the operators $\widetilde{T^{(n)}}$ have uniformly finite propagation, we may use ONL to switch the order above to get that $\|\phi(T)\|_A$ equals

$$\sup_{s\geqslant 0}\limsup_{n\to\infty}\sup\{\|\widetilde{T^{(n)}}u\| \mid u\in \ell^2(G,H), \|u\|=1, \text{diam}(\text{supp}(u))\leqslant s\}.$$

As in the proof of Lemma 13.3.10, there is N such that for $n\geqslant N$, $\pi_n\colon G\to X_n$ is an isometry on all balls of radius $\text{prop}(T)+s$. Transferring vectors 'downstairs' from G to X_n using these isometries, we see that for any fixed $T\in\mathbb{C}[X]$ and $s\geqslant 0$,

$$\begin{aligned}\limsup_{n\to\infty}\sup\{\|\widetilde{T^{(n)}}u\| \mid u\in \ell^2(G,H), \|u\|=1, \text{diam}(\text{supp}(u))\leqslant s\}\\ =\limsup_{n\to\infty}\sup\{\|T^{(n)}u\| \mid u\in \ell^2(X_n,H), \|u\|=1, \text{diam}(\text{supp}(u))\leqslant s\},\end{aligned}$$

whence the result in the statement. □

This lemma has two crucial corollaries for our analysis.

Corollary 13.3.12 *If G has ONL, then ϕ extends to a $*$-homomorphism*

$$\phi\colon C^*(X)\to A.$$

Proof Lemma 13.3.10 gives that $\phi\colon\mathbb{C}[X]\to A$ is a $*$-homomorphism. As the right-hand side of the equation in Lemma 13.3.11 is clearly bounded above by $\|T\|$, ϕ extends by continuity to a map $\phi\colon C^*(X)\to A$. □

For the second corollary, it will be convenient to introduce the following terminology.

Definition 13.3.13 An operator $T\in C^*(X)$ is a *ghost* if for all $\epsilon>0$ there exists N such that for all $n\geqslant N$, all matrix entries $T^{(n)}_{xy}$ of the restriction $T^{(n)}$ of T to $\ell^2(X_n,H)$ have norm less than ϵ.

Note that compact operators are ghost operators. Conversely, if X has ONL then all ghost operators are compact (Exercise 13.4.3). However, note that in our case the Kazhdan projection is a ghost operator that is an infinite rank projection, so certainly not compact.

Corollary 13.3.14 *Say G has ONL. Then the kernel of $\phi\colon C^*(X)\to A$ consists exactly of the ghost operators.*

Proof The formula

$$\sup_{s\geqslant 0}\limsup_{n\to\infty}\sup\{\|T^{(n)}u\| \mid u\in \ell^2(X_n,H), \|u\|=1, \text{diam}(\text{supp}(u))\leqslant s\}$$

for $\|\phi(T)\|_A$ from Lemma 13.3.11 continues to hold for any $T\in C^*(X)$ by continuity. Using bounded geometry of X, for each s, there is a uniform bound on the number of points in any s-ball in any X_n. It follows from this that the right-hand side of the formula above is zero exactly when T is a ghost operator, giving the result. □

Finally, we are ready to construct $\tau\colon K_0(C^*(X)) \to \prod \mathbb{R}/\oplus\mathbb{R}$. For each n, let $D_n \subseteq G$ be a bounded set of coset representatives for K_n, so that $G = \bigsqcup_{g\in D_n} gK_n$. Let $\lambda\colon K_n \to \mathcal{B}(\ell^2(G,H))$ be the amplified left regular representation of G, restricted to K_n. Then Proposition 5.3.4 gives a unitary isomorphism

$$U_n\colon \ell^2(G,H) \to \ell^2(K_n)\otimes \ell^2(D_n,H), \quad u \mapsto \sum_{k\in K_n} \delta_k \otimes \chi_{D_n}\lambda_k^{-1}u \tag{13.5}$$

such that conjugation by U_n induces an isomorphism

$$C^*(|G|)^{K_n} \to C^*_\rho(K_n)\otimes \mathcal{K}(\ell^2(D_n,H)), \quad T \mapsto U_nTU_n^*.$$

Putting the various U_n together therefore gives an isomorphism

$$\psi\colon A \to \frac{\prod_n C^*_\rho(K_n)\otimes \mathcal{K}(\ell^2(D_n,H))}{\oplus_n C^*_\rho(K_n)\otimes \mathcal{K}(\ell^2(D_n,H))}. \tag{13.6}$$

On the other hand, for each n, we have a trace map

$$\widetilde{\tau_n}\colon K_0(C^*_\rho(K_n)\otimes \mathcal{K}(\ell^2(D_n,H))) \to \mathbb{R}$$

defined via the tensor product of the canonical trace on the group C^*-algebra $C^*_\rho(K_n)$ and the canonical unbounded trace on the compact operators as in Example 2.3.4. Putting all these together with the canonical map

$$K_0\Big(\prod_n C^*_\rho(K_n)\otimes \mathcal{K}(\ell^2(D_n,H))\Big) \to \prod_n K_0(C^*_\rho(K_n)\otimes \mathcal{K}(\ell^2(D_n,H)))$$

gives a trace map

$$\widetilde{\tau}\colon K_0\Big(\prod_n C^*_\rho(K_n)\otimes \mathcal{K}(\ell^2(D_n,H))\Big) \to \frac{\prod_n \mathbb{R}}{\oplus\mathbb{R}}. \tag{13.7}$$

On the other hand, the short exact sequence

$$0 \to \oplus_n C^*_\rho(K_n)\otimes \mathcal{K}(\ell^2(D_n,H)) \to \prod_n C^*_\rho(K_n)\otimes \mathcal{K}(\ell^2(D_n,H))$$
$$\to \frac{\prod_n C^*_\rho(K_n)\otimes \mathcal{K}(\ell^2(D_n,H))}{\oplus_n C^*_\rho(K_n)\otimes \mathcal{K}(\ell^2(D_n,H))} \to 0$$

gives rise to a long exact sequence on K-theory. The induced map

$$K_0(\oplus_n C^*_\rho(K_n)\otimes \mathcal{K}(\ell^2(D_n,H))) \to K_0\Big(\prod_n C^*_\rho(K_n)\otimes \mathcal{K}(\ell^2(D_n,H))\Big)$$

is injective, whence we have a canonical isomorphism

$$K_0\left(\frac{\prod_n C^*_\rho(K_n)\otimes\mathcal{K}(\ell^2(D_n,H))}{\bigoplus_n C^*_\rho(K_n)\otimes\mathcal{K}(\ell^2(D_n,H))}\right)\cong\frac{K_0(\prod_n C^*_\rho(K_n)\otimes\mathcal{K}(\ell^2(D_n,H)))}{K_0(\bigoplus_n C^*_\rho(K_n)\otimes\mathcal{K}(\ell^2(D_n,H)))}.$$

Clearly $\widetilde{\tau}$ as in line (13.7) vanishes on the image of $K_0(\bigoplus_n C^*_\rho(K_n)\otimes\mathcal{K}(\ell^2(D_n,H)))$, so the above isomorphism induces a map

$$\widetilde{\tau}\colon K_0\left(\frac{\prod_n C^*_\rho(K_n)\otimes\mathcal{K}(\ell^2(D_n,H))}{\bigoplus_n C^*_\rho(K_n)\otimes\mathcal{K}(\ell^2(D_n,H))}\right)\to\frac{\prod_n\mathbb{R}}{\bigoplus\mathbb{R}}. \tag{13.8}$$

Definition 13.3.15 Define a map

$$\tau\colon K_0(C^*(X))\to\frac{\prod_n\mathbb{R}}{\bigoplus_n\mathbb{R}}$$

by composing the map in line (13.8) with the maps on K-theory induced by the $*$-homomorphisms ϕ of Corollary 13.3.12 and the $*$-isomorphism ψ of line (13.6).

As $p_K\in C^*(X)$ is a ghost operator, the following result is immediate from the definition of τ and Lemma 13.3.14.

Lemma 13.3.16 *With notation as above, $\tau(p_K)$ is the class of the constant sequence $(0,0,0,0,\ldots)$ in $\prod_n\mathbb{R}/\bigoplus_n\mathbb{R}$.* □

The last ingredient we need for the proof of Theorem 13.3.6 is the next proposition. It is here we use the assumption that G is the fundamental group of a closed aspherical manifold. Recall that

$$\mu\colon KX_*(X)\to K_0(C^*(X))$$

denotes the assembly map.

Proposition 13.3.17 *For any $\beta\in KX_*(X)$, $\tau(\mu(\beta))=tr(\mu(\beta))$.*

Proof We first claim that for any $\beta\in KX_*(X)$, the class of $\mu(\beta)$ can be represented by a formal difference $[p]-[q]$ of idempotents in matrix algebras over the unitisation $C^*(X)^+$, which are both of finite propagation, and with the additional property that for each $x,y\in X$, $p_{xy}-q_{xy}$ is a trace class operator on H.

Indeed, to prove this, using Theorem 7.2.16 we have that $\mu(x)$ is in the image of the map

$$\mu_{P_r(X),X}\colon K_*(P_r(X))\to K_*(C^*(X))$$

for some $r \geqslant 0$. Using Theorem 6.4.20 (and the fact that the different X_n are infinitely far apart), the left-hand side splits as a product

$$K_*(P_r(X)) \cong \prod_n K_*(P_r(X_n)).$$

Let now M be a closed manifold with fundamental group G equipped with a Riemannian metric, and let $\widetilde{M}$ be its universal cover equipped with the lifted metric. Fix $x \in \widetilde{M}$ giving rise to an orbit inclusion map

$$f : G \to \widetilde{M}, \quad g \mapsto gx;$$

this is an equivariant coarse equivalence by the Svarc–Milnor theorem (Lemma A.3.14). Let $M_n := \widetilde{M}/K_n$ be the cover of M corresponding to K_n, so f induces a map $f_n : X_n \to M_n$. The sequence (f_n) is a uniform family of coarse equivalences in a natural sense. As the covering maps $\pi_n : \widetilde{M} \to M_n$ are isometries on larger and larger balls, a uniform version of the proof of Theorem 7.3.6 that we leave to the reader shows that there are $s \geqslant r$ and uniformly continuous coarse equivalences $g_n : M_n \to P_r(X_n)$ for all n suitably large such that if (β_n) is the image of β under the map

$$K_*(P_r(X)) \to K_*(P_s(X)) \cong \prod_n K_*(P_s(X_n))$$

then for all n suitably large, we have that β_n is in the image of the map

$$(g_n)_* : K_*(M_n) \to K_*(P_s(X_n)).$$

The claim follows from this and an appeal to Corollary 9.6.13 as in the proof of Theorem 10.1.8.

Having established the claim above, write $p^{(n)}$ and $q^{(n)}$ for the nth components of p and q respectively, so these are elements of $M_m(\mathcal{B}(\ell^2(X_n, H)) \cong \mathcal{B}(\ell^2(X_n, H^{\oplus m}))$. Then we have that if Tr is the canonical densely defined trace on the compact operators on $H^{\oplus m}$ (see Example 2.3.3) then $\mathrm{tr}([p]-[q])$ is the class of the sequence

$$\left(\sum_{x \in X_n} \mathrm{Tr}(p^{(n)}_{xx} - q^{(n)}_{xx})\right)$$

of integers in $\prod_n \mathbb{R} / \oplus \mathbb{R}$. On the other hand, writing $\widetilde{p^{(n)}}$ and $\widetilde{q^{(n)}}$ for the lifts of $p^{(n)}$ and $q^{(n)}$ as defined in line (13.4) (the same formula makes sense for operators in the unitisations, as long as they still have finite propagation) and using the formula in line (13.5) we have that $\tau([p]-[q])$ is given by the class of the sequence

$$\left(\sum_{x \in D_n} \mathrm{Tr}(\widetilde{p_{xx}^{(n)}} - \widetilde{q_{xx}^{(n)}}) \right)$$

in $\prod_n \mathbb{R}/\oplus\mathbb{R}$. However, using that the restriction of π_n to D_n is a bijection from D_n to X_n, this is the same as the earlier expression for $\mathrm{tr}([p]-[q])$, and we are done. □

Proof of Theorem 13.3.6 Let p_K be the Kazhdan projection. Lemma 13.3.8 implies that $\mathrm{tr}(p_K) = [1,1,1,\ldots] \in \prod \mathbb{R}/\oplus\mathbb{R}$, and Lemma 13.3.16 implies that $\tau(p_K) = [0,0,0,\ldots] \in \prod \mathbb{R}/\oplus\mathbb{R}$. However, Proposition 13.3.17 implies that tr and τ agree on the range of the assembly map, so p_K cannot be in the range of assembly. □

13.4 Exercises

13.4.1 With notation as in Example 13.2.6, show that the following are equivalent.

(i) There exists $d > 0$ and a finite subset F of G such that for all n and all $u \in \ell^2(X_n)$ in the orthogonal complement of the constant vectors, there exists $s \in F$ such that

$$\|\lambda_s u - u\| \geqslant d\|u\|$$

(here λ_s is the quasi-regular representation of G on $\ell^2(X_n)$).

(ii) There exists $c > 0$ such that for all n and all $u \in \ell^2(X_n)$ in the orthogonal complement of the constant vectors, there exists $s \in S$ such that

$$\|\lambda_s u - u\| \geqslant c\|u\|.$$

(iii) G has property (τ) with respect to (X_n).

(iv) If $C^*_{(X_n)}(G)$ denotes the completion of $\mathbb{C}[G]$ in the direct sum representation $\bigoplus_n \ell^2(X_n)$ of its quasi-regular representations on $\ell^2(X_n)$, then the image of Δ_S in $C^*_{(X_n)}(G)$ has spectrum contained in $\{0\} \cup [c,\infty)$ for some $c > 0$.

Formulate and prove similar equivalences for property (T) as defined in Remark 13.2.9.

13.4.2 Recall that a group G is *amenable* if for any $\epsilon > 0$ and finite subset F of G, there exists a unit vector $u \in \ell^2(G)$ with $\|\lambda_g u - u\| < \epsilon$ for all $g \in F$. Show that an amenable group does not have property (τ) for any sequence (X_n) as in Example 13.2.6.

13.4.3 Show that if X is a bounded geometry separated coarse union of finite metric spaces with ONL, then all ghost operators on X are compact.
Hint: compare Example 13.3.4.

13.5 Notes and References

The large sphere counterexample comes from section 8 of Yu [271]. Recently, this example has been studied in more detail: in particular, Oyono-Oyono, Yu and Zhou [200] shows that if the radii of the spheres involved increase fast enough relative to the dimensions, then the coarse Baum–Connes assembly map is injective. There is still a mysterious zone between this positive result and the existence of the counterexample in Section 13.1. Another intriguing question is whether one can deduce that this sequence of spheres is a counterexample without using any differential geometry: this would be interesting, as it might suggest methods that might be more usefully generalisable.

Expander graphs have been very widely studied due to their connections to several interesting parts of both pure and applied mathematics: Lubotzky [171] gives a beautiful tour of some of the theory of expanders and property (τ). The fact that expanders do not coarsely embed into Hilbert space was first observed by Gromov. GY learned our proof from Vincent Lafforgue, and RW learned it from Nigel Higson.

Our approach to showing that expanders give counterexamples to the coarse Baum–Connes conjecture is based on an unpublished sketch of Higson [126]. Since then, starting with Oyono-Oyono and Yu [197], there has been some study of just 'how badly' the coarse Baum–Connes conjecture fails for certain expanders, particularly if one replaces the Roe C^*-algebra with other completions of the $*$-algebra $\mathbb{C}[X]$ as first studied in Gong, Wang and Yu [106]. See also for example Chen, Wang and Yu [54], Finn-Sell [102], Finn-Sell and Wright [103], Guentner, Tessera and Yu [115] and Willett and Yu [258, 259] for some results in this direction.

Published counterexamples to various versions of the Baum–Connes conjecture appear in Higson, Lafforgue and Skandalis [131], using a somewhat different approach based on failures of exactness; this includes a different approach in the case of expanders. Exactness turns out to be closely tied to operator algebra theory and coarse geometry: in particular, Ozawa [201] shows that exactness of a group G is equivalent to property A as discussed at the end of Chapter 12. Baum, Guentner and Willett [26] surveys some of the connections between exactness and the Baum–Connes conjecture, and gives a reformulation of the Baum–Connes conjecture that obviates some counterexamples to the so-called Baum–Connes conjecture with coefficients. Nonetheless, due to phenomena related to property (T) (see Willett and Yu [260]), these methods cannot be used to obviate the counterexamples to the coarse Baum–Connes conjecture. See Brown and Ozawa [44] for an extensive discussion of exactness from an operator algebraic point of view.

Having said all of this about counterexamples, we should note that the original version of the Baum–Connes conjecture itself, as stated in Definition 7.1.11 above, is open.

The definition of ONL comes from Chen, Tessera, Wang and Yu [52]. Sako [228] showed that ONL is also equivalent to property A, and there are thus many examples of spaces with this property, including all linear groups as discussed at the end of Chapter 12. One relatively direct way to prove that higher genus surface groups have ONL as in Lemma 13.3.5 is to use that such groups are quasi-isometric to the hyperbolic plane by the Svarc–Milnor lemma (Lemma A.3.14), that the hyperbolic plane has finite asymptotic dimension (see for example corollary 9.21 in Roe [218]), and that finite asymptotic dimension implies ONL (this is essentially lemma 9.26 in Roe [218], or see remark 3.2 and proposition 4.1 in Chen, Tessera, Wang and Yu [52]).

The definition of ghost operators (Definition 13.3.13 above) is due to Yu, and appears first in section 11.5.2 of Roe [218]. It turns out that the existence of non-trivial ghost operators is actually equivalent to the failure of property A (see Roe and Willett [220]), and therefore also to the failure of ONL.

Appendices

APPENDIX A

Topological Spaces, Group Actions and Coarse Geometry

This appendix covers some background facts about topological spaces, metric and coarse geometry and group actions that are needed in the main body of the book. Results are generally only proved if we could not easily find what we need in the literature.

The structure of this appendix is as follows: Section A.1 covers background from general topology, including descriptions of the categories of topological spaces that we work with, and technical results on the existence of Borel covers with nice properties. Section A.2 covers results we need about group actions, quotient spaces, metric properties of groups and more technical results about Borel covers, this time in the presence of a group action. Finally, Section A.3 discusses ideas from coarse geometry, including basic definitions as well as some results connecting coarse geometry to group actions.

A.1 Topological Spaces

Some Results from General Topology

One of our main goals in this book is to study topological spaces. We typically work with the class of Hausdorff, locally compact, second countable topological spaces. The following foundational theorem from general topology says such spaces cannot be too wild.

Theorem A.1.1 *Let X be a locally compact, Hausdorff, second countable topological space. Then there is a metric on X that induces the topology.* □

Thus we can, and do, work with either metric or purely topological notions for this class of spaces: the choice just depends on which approach seems cleaner or more intuitive.

Another property of locally compact, second countable spaces that gets used all the time is the existence of partitions of unity.

Definition A.1.2 Let X be a topological space, and $\mathcal{U}$ an open cover of X. A *partition of unity subordinate to* $\mathcal{U}$ is a collection of functions $(\phi_i : X \to [0,1])_{i\in I}$ with the following properties:

(i) each ϕ_i is supported in some element of $\mathcal{U}$;

(ii) for any compact subset K of X, the set $\{i \in I \mid \phi_i|_K \neq 0\}$ is finite;
(iii) for all $x \in X$, $\sum_{i \in I} \phi_i(x) = 1$.

Theorem A.1.3 *Let X be a locally compact, Hausdorff, second countable space, and let $\mathcal{U}$ be an open cover of X. Then a partition of unity subordinate to $\mathcal{U}$ exists.* □

Morphisms and Duality

Here we discuss a class of maps between locally compact, Hausdorff topological spaces; these are the morphisms appropriate to our C^*-algebraic approach to topology. We first need the one-point compactification.

Definition A.1.4 Let X be a locally compact, Hausdorff, topological space. The *one-point compactification* of X is the topological space X^+ with underlying set the disjoint union $X^+ = X \sqcup \{\infty\}$ of X with a 'point at infinity', and where a subset $U \subseteq X^+$ is open if either:

(i) it is an open subset of X, or
(ii) its complement is a compact subset of X.

It is not difficult to check that the above definition makes X^+ a (Hausdorff), compact space, which is second countable if X is. Moreover, the restriction of the topology to the subset X is the original topology, and X is an open dense subset. We note that X is compact if and only if $\{\infty\}$ is an open set in X^+, i.e. if and only if ∞ is an isolated point.

We think of X^+ as a *pointed space*, i.e. a topological space with a fixed choice of basepoint, which in this case is ∞. Recall that a map between pointed spaces is itself *pointed* if it takes the basepoint to the basepoint.

Definition A.1.5 The category $\mathcal{LC}$ has objects locally compact, Hausdorff, second countable, topological spaces (possibly empty). A morphism from X to Y in $\mathcal{LC}$ is a continuous, pointed function $f\colon X^+ \to Y^+$. Composition and identities are defined using the usual composition of functions and the usual identity function.

Remark A.1.6 In topology, one often considers the category whose objects are pointed, compact, second countable, Hausdorff spaces (X, x_0), and whose morphisms are continuous pointed maps $f\colon X \to Y$. This is equivalent to our category $\mathcal{LC}$ via the functors defined on objects by

$$X \mapsto (X^+, \infty), \quad (X, x_0) \mapsto X \setminus \{x_0\}$$

and which leaves morphisms essentially unchanged. This uses the following fact, which we leave as an exercise: if X is compact, then the one-point compactification of $X \setminus \{x_0\}$ identifies as a pointed space with (X, x_0) via a homeomorphism which is the set-theoretic identity on $X \setminus \{x_0\}$, and that takes ∞ to x_0. While the category of compact pointed spaces is maybe more standard, we use the language of Definition 6.4.1 instead as we are interested primarily in non-compact spaces and it emphasises this aspect.

There is another useful description of morphisms in $\mathcal{LC}$. First we have a standard definition.

Definition A.1.7 A map $f: X \to Y$ between topological spaces is *proper* if for any compact $K \subseteq Y$, $f^{-1}(K)$ is compact in X.

Now, define a new category $\mathcal{LC}'$ as follows. A morphism from X to Y in $\mathcal{LC}'$ is a choice of an open subset U in X (possibly empty, and possibly all of X), together with a continuous and proper function $f: U \to Y$. We will write $f: (U \subseteq X) \to Y$ for such a morphism. The composition of two morphisms $f: (U \subseteq X) \to Y$ and $g: (V \subseteq Y) \to Z$ is defined to be the morphism

$$f \circ g: (f^{-1}(V) \cap U \subseteq X) \to Z$$

(the domain is allowed to be empty).

Proposition A.1.8 *The categories $\mathcal{LC}$ and $\mathcal{LC}'$ are canonically isomorphic, via an isomorphism that is the identity on objects.*

Proof Let $f: X^+ \to Y^+$ be a morphism in $\mathcal{LC}$. To get a morphism in $\mathcal{LC}'$, let $U = f^{-1}(Y)$, which is an open subset of X. Then the restriction $f|_U: U \to Y$ is proper (it is important here that the codomain is Y and not Y^+), and therefore $f|_U: (U \subseteq X) \to Y$ is a morphism in $\mathcal{LC}'$. In the other direction, let $f: (U \subseteq X) \to Y$ be a morphism in $\mathcal{LC}'$. Continuity and properness of f imply that the extension of f to X^+ defined by

$$x \mapsto \begin{cases} f(x) & x \in U, \\ \infty & x \notin U \end{cases}$$

is continuous, so gives a morphism in $\mathcal{LC}$.

We leave it to the reader to check that these processes define mutually inverse isomorphisms of categories. □

As a consequence, note that proper continuous maps $f: X \to Y$ canonically define morphisms in $\mathcal{LC}$, but that there are many others in general. An illustrative and important example is given when U is an open subset of X. Then the identity map on U defines a morphism $(U \subseteq X) \to U$ in $\mathcal{LC}'$; on the level of one-point compactifications, it corresponds to the map $X^+ \to U^+$ that collapses everything in $X^+ \setminus U$ to the point at infinity in U^+.

From the point of view of C^*-algebras, morphisms in $\mathcal{LC}'$ (equivalently $\mathcal{LC}$) are the morphisms that are dual to $*$-homomorphisms between commutative C^*-algebras. To make this precise, recall (see Definition 1.3.4) first that if A is a commutative C^*-algebra, then its spectrum $\widehat{A}$ is the space of $*$-homomorphisms $\phi: A \to \mathbb{C}$ equipped with the topology of pointwise convergence. If A^+ is the unitisation of such an A, then $\widehat{A^+}$ canonically identifies with $\widehat{A}^+$. Let $\mathcal{CC}^*$ denote the category of separable, commutative C^*-algebras and $*$-homomorphisms. Define a functor from $\mathcal{CC}^*$ to $\mathcal{LC}$ by sending a C^*-algebra A to $\widehat{A}$ on the level of objects, and a $*$-homomorphism $\phi: A \to B$ to the continuous map $\widehat{B}^+ \to \widehat{A}^+$ defined by

$$x \mapsto x \circ \phi^+,$$

where $\phi^+: A^+ \to B^+$ is the canonical unital extension of ϕ. Define a functor from $\mathcal{LC}$ to $\mathcal{CC}^*$ by sending a space X to $C_0(X)$, and a morphism $f: (U \subseteq X) \to Y$ to the $*$-homomorphism $C_0(Y) \to C_0(X)$ defined by

$$a \mapsto \widetilde{a \circ f},$$

where $\widetilde{a \circ f}$ is the extension of the function $a \circ f \in C_0(U)$ to all of X defined by setting it equal to zero outside of U.

Here is an equivalent formulation of Theorem 1.3.14 from Chapter 1 of the text.

Theorem A.1.9 *The functors defined above give a contravariant equivalence of categories between $\mathcal{LC}$ and $\mathcal{CC}^*$.* □

Borel Covers

We finish our general discussion of topological spaces with a technical lemma that gets used many times in the main text. As usual, all topological spaces considered are Hausdorff.

Lemma A.1.10 *Let X be a locally compact, Hausdorff, second countable topological space, and let $\mathcal{U}$ be an open cover of X. Then there exists a countable collection $(E_i)_{i \in I}$ of non-empty Borel subsets of X such that:*

(i) *the collection (E_i) covers X;*
(ii) *for $i \neq j$, $E_i \cap E_j = \varnothing$;*
(iii) *each E_i is contained in some element of $\mathcal{U}$;*
(iv) *each E_i has compact closure;*
(v) *for any compact subset K of X, the set $\{i \in I \mid E_i \cap K \neq \varnothing\}$ is finite;*
(vi) *each E_i is contained in the closure of its interior.*

In order to prove this, we quote a useful lemma about 'shrinking' covers; we omit the standard proof, which can be found in many texts on general topology (or makes a good exercise).

Lemma A.1.11 *Let X be a locally compact, Hausdorff, second countable topological space, and let $(U_i)_{i \in I}$ be an open cover of X. Then there is an open cover $(V_i)_{i \in I}$ such that for each $i \in I$, $\overline{V_i} \subseteq U_i$.* □

Proof of Lemma A.1.10 Using local compactness, Hausdorffness and second countability, there is a countable open cover $(A_n)_{n=1}^{\infty}$ of X such that each A_n has compact closure, and is contained in some element of $\mathcal{U}$. Use Lemma A.1.11 to produce an open cover (B_n) of X such that $\overline{B_n} \subseteq A_n$ for all n. Define $C_1 = A_1$, and for each $n > 1$ define

$$C_n := A_n \setminus \left(\bigcup_{k=1}^{n-1} \overline{B_k} \right).$$

Induction shows that $\bigcup_{k=1}^{n} C_k = \bigcup_{k=1}^{n} A_k$, and thus (C_n) is an open cover of X. Note that if $K \subseteq X$ is compact, then $K \subseteq \bigcup_{n=1}^{N} B_n$ for some N, and thus $C_n \cap K = \varnothing$ for all but finitely many n. Using Lemma A.1.11, produce a new open cover (D_n) such that $\overline{D_n} \subseteq C_n$ for all n; note that for any compact $K \subseteq X$, $K \cap \overline{D_n} = \varnothing$ for all but finitely many n.

Define now $E_1 = \overline{D_1}$ and for $n > 1$ define

$$E_n := \overline{D_n} \setminus \left(\bigcup_{k=1}^{n-1} \overline{D_k} \right).$$

Set $I := \{n \in \mathbb{N} \mid E_n \neq \varnothing\}$. We claim that $(E_i)_{i \in I}$ has the right properties. First note that each E_i is clearly Borel, and that $\bigcup_{n=1}^{\infty} E_n = \bigcup_{n=1}^{\infty} \overline{D_n} = X$, so (E_i) covers X giving property (i). Property (ii) is clear by construction. Properties (iii) and (iv) follow from the corresponding properties of (A_n), and the fact that each E_n is contained in A_n. Property (v) follows as the corresponding property holds for the collection $\{\overline{D_n} \mid n \in \mathbb{N}\}$.

Finally, for property (vi), we claim that

$$E_n \subseteq \overline{D_n \setminus \left(\bigcup_{k=1}^{n-1} \overline{D_k} \right)}.$$

As D_n is open, the set $D_n \setminus \left(\bigcup_{k=1}^{n-1} \overline{D_k} \right)$ is open, so contained in the interior of E_n; the claim will thus complete the proof. Let then x be an element of E_n, whence in particular x is in $\overline{D_n}$, so there is a sequence (x_j) in D_n that converges to x. As x is not in $\bigcup_{k=1}^{n-1} \overline{D_k}$ only finitely many elements of the sequence can be in $\overline{D_k}$ for each $k \in \{1, \ldots, n-1\}$; passing to a subsequence, we may thus assume that (x_j) is a sequence in $D_n \setminus \left(\bigcup_{k=1}^{n-1} \overline{D_k} \right)$. Hence x is in the closure of this latter set, which completes the proof. □

A.2 Group Actions on Topological Spaces

In this section we will establish our conventions on group actions, and prove some useful technical facts. If G is a group acting on a set X, and S, E are subsets of G, X respectively, then we write SE or $S \cdot E$ for the (partial) *orbit*

$$SE := \{gx \mid g \in S, \ x \in E\}$$

of E under S. If $S = \{g\}$ (respectively, $E = \{x\}$) is a singleton, this will be abbreviated to gE or $g \cdot E$ (respectively, Sx or $S \cdot x$). The *stabiliser* of $x \in X$ is defined to be $\{g \in G \mid gx = x\}$ and is denoted G_x.

The following result, which follows directly from Theorem A.1.9, gets used all the time in the main text.

Proposition A.2.1 *Say G acts by homeomorphisms on a locally compact, Hausdorff space X. Then the formula*

$$(\alpha_g(f))(x) := f(g^{-1}x)$$

defines an action α of G by $*$-automorphisms on $C_0(X)$. Moreover, every action of G on $C_0(X)$ by $*$-automorphisms arises in this way from a unique action on X. □

For the rest of this section, we will work only with proper actions as in the next definition.

Definition A.2.2 Let G be a countable discrete group acting by homeomorphisms on a locally compact, Hausdorff, second countable space X. The action is *proper* if for any compact subset K of X the set

$$\{g \in G \mid gK \cap K \neq \varnothing\}$$

is finite.

The following example is worth bearing in mind, partly for intuition.

Example A.2.3 Any action of a finite group is proper.

Proper actions share many good properties with actions of finite groups. This is illustrated by the next two lemmas.

Lemma A.2.4 *Let G act properly on a locally compact, Hausdorff, second countable space X, and let $K \subseteq X$ be compact. Then GK is closed.*

Proof Let z be in the closure of GK, and say $(g_n x_n)$ is a sequence in GK converging to z. Let U be an open neighbourhood of z with compact closure, and note that

$$\begin{aligned}\{g_n \in G \mid g_n x_n \in U\} &\subseteq \{g \in G \mid gK \cap U \neq \varnothing\} \\ &\subseteq \{g \in G \mid g(K \cup \overline{U}) \cap (K \cup \overline{U}) \neq \varnothing\};\end{aligned}$$

the last set appearing above is finite by properness, whence the first is too. Hence passing to a subsequence, we may assume that there is $g \in G$ with $g_n = g$ for all n. It follows that $gx_n \in gK$ for all n, whence $z \in gK$ as this set is closed. □

Lemma A.2.5 *Let G act properly on a locally compact, Hausdorff, second countable space X. Then when equipped with the quotient topology, X/G is locally compact, Hausdorff and second countable.*

Proof Local compactness and second countability always pass to quotient spaces; we leave the direct checks to the reader. For the Hausdorff property, let $\overline{x}$, $\overline{y}$ be distinct points in X/G with lifts $x, y \in X$. We must show that there are disjoint, G-invariant open sets containing x and y.

From Lemma A.2.4, Gy is closed. As x is in the open set $X \setminus Gy$, local compactness gives us an open set $V \ni x$ with compact closure, and which satisfies $\overline{V} \subseteq X \setminus Gy$. As $X \setminus Gy$ is G-invariant, this last inclusion implies that $G\overline{V} \subseteq X \setminus Gy$. As $\overline{V}$ is compact, $G\overline{V}$ is closed by Lemma A.2.4. Hence GV and $X \setminus G\overline{V}$ are open, disjoint, G-invariant sets containing x, y respectively. □

Example A.2.6 Let F be a finite subgroup of G, and let Y be a topological space on which F acts by homeomorphisms. Then the *balanced product* of G and Y over F, denoted $G \times_F Y$, is the quotient of $G \times Y$ by the diagonal F action defined by

$$f \cdot (g, y) = (gf^{-1}, fy), \quad f \in F, g \in G, y \in Y.$$

The space $G \times Y$ is equipped with the product topology, and $G \times_F Y$ with the quotient topology. Note that if the original topology on Y is locally compact, second countable and Hausdorff, then $G \times_F Y$ has these properties too by Lemma A.2.5, and the fact that any action of a finite group is always proper. Write $[g, y]$ for the image of $(g, y) \in G \times Y$ in $G \times_F Y$.

The formula

$$g \cdot (h, y) = (gh, y)$$

defines an action of G on $G \times Y$, which clearly passes to the quotient $G \times_F Y$. This action is proper: this follows as for any compact $K \subseteq G \times_F Y$ there is a finite subset S of G such that K is covered by $\{[g, y] \in G \times_F Y \mid g \in S\}$.

Any proper action is locally built from balanced products: this is the content of the next lemma.

Lemma A.2.7 *Let G be a discrete group acting properly by homeomorphisms on a Hausdorff, locally compact, second countable space X. Let x be a point in X, $G_x \leqslant G$ be the stabiliser of x, and let U be an open subset of X containing x. Then there is an open G_x-invariant set $V \ni x$ with compact closure contained in U such that the map*

$$\phi\colon G \times_{G_x} \overline{V} \to G \cdot \overline{V}, \quad [g, y] \mapsto gy$$

is a well-defined, equivariant homeomorphism onto its image.

Proof Local compactness gives us an open set $W_1 \ni x$ with compact closure, and which satisfies $\overline{W_1} \subseteq U$. Set

$$S := \{g \in G \mid g\overline{W_1} \cap \overline{W_1} \neq \varnothing\},$$

which is finite by properness of the action. As X is Hausdorff there is open $W_2 \ni x$ such that $W_2 \subseteq W_1$, and such that $\overline{W_2} \cap S^{-1}x = \{x\}$. Define

$$W_3 := W_2 \setminus \Bigg(\bigcup_{g \in S \setminus G_x} g\overline{W_2} \Bigg),$$

an open subset of W_2. We claim that $x \in W_3$. Indeed, if $x \in g\overline{W_2}$ for some $g \in S$, then $g^{-1}x \in S^{-1}x \cap \overline{W_2} = \{x\}$ by choice of W_2, whence $g^{-1} \in G_x$ by choice of W_2, and so $g \in G_x$; this implies that x is contained in W_3.

Choose now $W_4 \ni x$ to be any open set such that $\overline{W_4} \subseteq W_3$. We claim that $g\overline{W_4} \cap \overline{W_4} = \varnothing$ for $g \notin G_x$. Indeed, as $\overline{W_4}$ is contained in $\overline{W_1}$, $g\overline{W_4} \cap \overline{W_4} = \varnothing$ if $g \notin S$ by definition of S. On the other hand, as $\overline{W_4} \subseteq W_3$, $g\overline{W_4} \cap \overline{W_4} = \varnothing$ for $g \in S \setminus G_x$.

Finally, define $V := \bigcap_{g \in G_x} gW_4$. We claim this V has the right properties. Indeed, note first that V is an open G_x-invariant neighbourhood of x, it is contained in U, it has compact closure and for all $g \notin G_x$, $g\overline{V} \cap \overline{V} = \varnothing$ (as follows from the analogous property for W_4). Moreover, the map ϕ from the statement is clearly well defined and continuous. To complete the proof, we will construct a continuous inverse map to ϕ.

Provisionally, define

$$\psi\colon G \cdot \overline{V} \to G \times_{G_x} \overline{V}, \quad gy \mapsto [g, y], \quad g \in G,\ y \in \overline{V}.$$

To see that this is well defined, note that if $gy = hz$ for $g, h \in G$ and $y, z \in \overline{V}$, then $h^{-1}gy = z \in h^{-1}g\overline{V} \cap \overline{V}$, whence $h^{-1}g \in G_x$. Hence by definition of $G \times_{G_x} \overline{V}$,

$$[h, z] = [hh^{-1}g, g^{-1}hz] = [g, y]$$

as required. To see that ψ is continuous, say (x_n) is a sequence in $G\overline{V}$ converging to some $x \in G\overline{V}$. For definiteness, say that $x \in g\overline{V}$ for some $g \in G$. Note that as the action is proper, the set $\{n \in \mathbb{N} \mid x_n \in h\overline{V}\}$ must be infinite for some $h \in G$; moreover, as $g\overline{V}$ and $h\overline{V}$ are either disjoint or equal for all $g, h \in G$, the only $h \in G$ for which this is possible is $h = g$. Hence on throwing out finitely many elements from our sequence, we may assume that x_n is in $g\overline{V}$ for all n. Set now $y_n = g^{-1}x_n$ and $y = g^{-1}x$, which are elements of V with $y_n \to y$ as $n \to \infty$. Then

$$\psi(x_n) = [g, y_n] \to [g, y] = \psi(x) \quad \text{as} \quad n \to \infty,$$

establishing continuity. As ϕ and ψ are clearly mutual inverses, this completes the proof. □

Corollary A.2.8 *Let G act properly on a locally compact, Hausdorff, second countable space X. Let $\mathcal{U}$ be an equivariant cover of X, meaning that for all $U \in \mathcal{U}$ and $g \in G$, gU is also in U. Then there exists a partition of unity $(\phi_i)_{i \in I}$ on X that is invariant*[1] *under the G action on functions on X, and such that each ϕ_i is supported in some element of $\mathcal{U}$.*

Proof Using Lemma A.2.7, we may assume on refining that each U has the properties stated there (do this one orbit at a time). If $\pi : X \to X/G$ is the quotient map, then $\{\pi(U) \mid U \in \mathcal{U}\}$ is an open cover of X/G. The required partition of unity on X can now be constructed by taking a partition of unity as in Theorem A.1.3 on X/G for this new cover, and using the structure coming from Lemma A.2.7 to pull back. □

The following technical result provides a Borel decomposition of a metric space equipped with a proper action: it is an equivariant version of Lemma A.1.10, and is similarly important for us.

Lemma A.2.9 *Let G be a countable discrete group acting properly on a locally compact, Hausdorff, second countable space X. Let $\mathcal{U}$ be an open cover of X. Then there exists a countable collection $(E_i)_{i \in I}$ of non-empty Borel subsets of X such that:*

(i) the collection $(GE_i)_{i \in I}$ covers X;
(ii) for $i \neq j$, $GE_i \cap GE_j = \varnothing$;
(iii) each E_i is contained in some element of $\mathcal{U}$;
(iv) each E_i has compact closure;
(v) for any compact subset K of X, the set $\{i \in I \mid E_i \cap K \neq \varnothing\}$ is finite;
(vi) each E_i is contained in the closure of its interior;
(vii) for each i there is a finite subgroup $F_i \leqslant G$ such that E_i is F_i invariant and such that the function

$$\phi : G \times_{F_i} \overline{E_i} \to G\overline{E_i}, \quad [g, x] \mapsto gx$$

is an equivariant homeomorphism.

[1] More precisely, we mean that for each i there exists j such that $g\phi_i = \phi_j$, not that for each i, $g\phi_i = \phi_i$; in other words, the set $\{\phi_i \mid i \in I\}$ is invariant under the action, but it will not generally be true that each ϕ_i is fixed by G.

Proof For each $x \in X$, choose an open set $V_x \ni x$ with the properties in Lemma A.2.7, and that is contained in some $U \in \mathcal{U}$. Note that the quotient map $\pi : X \to X/G$ is open, and consider the open cover $\mathcal{V} := \{\pi(V_x) \mid x \in X\}$ of X/G. As X/G is Hausdorff, second countable and locally compact, Lemma A.2.5 gives a Borel cover $(E_{i,X/G})_{i \in I}$ of X/G with the properties in Lemma A.1.10 with respect to the open cover $\mathcal{V}$. For each i, choose $x(i)$ such that $E_{i,X/G} \subseteq \pi(V_{x(i)})$, and define $E_i = \pi^{-1}(E_{i,X/G}) \cap V_{x(i)}$. We claim that the collection $(E_i)_{i \in I}$ of Borel subsets of X has the right properties.

Note first that

$$GE_i = G(V_{x(i)} \cap \pi^{-1}(E_{i,X/G})) = GV_{x(i)} \cap \pi^{-1}(E_{i,X/G}) = \pi^{-1}(E_{i,X/G}). \tag{A.1}$$

Hence (i), (ii) and (v) follow from the corresponding properties of $(E_{i,X/G})$. On the other hand, (iii) and (iv) follow as each V_x is contained in some element of $\mathcal{U}$, and also has compact closure. For property (vii), let F_i be the stabiliser of x_i, and note that E_i is F_i invariant as it is the intersection of $V_{x(i)}$ and $\pi^{-1}(E_{i,X/G})$, and both of these sets are F_i invariant. Moreover, as $V_{x(i)}$ satisfies the conditions in Lemma A.2.7, there is a homeomorphism

$$\phi \colon G \times_{F_i} \overline{V_{x(i)}} \to G\overline{V_{x(i)}}, \quad [g,x] \mapsto gx;$$

the homeomorphism required by (vii) is then just the restriction of this to $G \times_{F_i} \overline{E_i}$.

Finally, we must show property (vi). Write Y° for the interior of a set Y. Let x be a point in E_i. As $\pi(E_i) \subseteq E_{i,X/G}$, and as $E^\circ_{i,X/G}$ is dense in $E_{i,X/G}$, there is a sequence $(y_n)_{n=1}^\infty$ in $E^\circ_{i,X/G}$ converging to $\pi(x)$. As $\pi(V_{x(i)}) \supseteq E_{i,X/G}$ by line (A.1) above, we may assume that there is a sequence (x_n) in $V_{x(i)}$ such that $\pi(x_n) = y_n$ for all n. Note that each x_n is in $V_{x(i)} \cap \pi^{-1}(E^\circ_{i,X/G})$, which is an open subset of E_i. Hence (x_n) is a sequence in E_i°. As E_i has compact closure, we may assume on passing to a subsequence that (x_n) converges to some $z \in \overline{E_i}$. We must then have that $\pi(x) = \pi(z)$, and thus there is $g \in G$ such that $gz = x$. Using part (vii), this is impossible unless $g \in F_i$. On the other hand, as F_i preserves E_i, it preserves E_i°. This shows that (gx_n) is a sequence in E_i° converging to x, so x is in the closure of E_i° as required. □

A.3 Coarse Geometry

In this section we set up conventions on the large-scale, or coarse, structure of an appropriate space.

Proper Metric Spaces

We will allow metrics to take infinite distances. As this is non-standard, we give the precise definition for the readers' convenience.

Definition A.3.1 Let X be a set. A *metric* on X is a function $d \colon X \times X \to [0,\infty]$ such that for all $x, y, z \in X$,

(i) $d(x, y) = 0$ if and only if $x = y$;

(ii) $d(x, y) = d(y, x)$;

(iii) $d(x, y) \leqslant d(x, z) + d(z, y)$ (where we adopt the usual conventions that $\infty + t = t + \infty = \infty$ and $t \leqslant \infty$ for all $t \in [0, \infty]$).

A set X equipped with a metric is called a *metric space*.

We will use the notation $B(x; r)$, or occasionally $B_d(x; r)$ or $B_X(x; r)$ if it helps to clarify what is going on, for open balls, i.e.

$$B(x; r) := \{y \in X \mid d(x, y) < r\}.$$

These balls generate a topology in the usual way (it makes no difference whether or not the metric is finite-valued). We will use other metric terminology in standard ways: for example, a subset K of X is *bounded* if it is contained in a ball of finite radius; the *diameter* of a subset $A \subseteq X$ is $\text{diam}(A) := \sup\{d(x, y) \mid x, y \in A\}$; an *isometry* is a function $f \colon X \to Y$ such that $d_Y(f(x_1), f(x_2)) = d_X(x_1, x_2)$ for all $x_1, x_2 \in X$, and so on.

There are some slight differences in the theory between metric spaces where one allows infinite distances and the usual case. Perhaps the oddest is that compact sets need no longer be bounded. Fortunately, one at least has the following.

Lemma A.3.2 *Any compact subset of a metric space is a finite union of bounded sets.*

Proof If $K \subseteq X$, the open cover $\{B(x; 1) \mid x \in K\}$ of K has a finite subcover. □

For the purposes of coarse geometry, we will almost exclusively be interested in proper metric spaces as in the next definition.

Definition A.3.3 A metric space is *proper* if every closed bounded set is compact.

For example, a finite-dimensional Euclidean space is a proper metric space, but an infinite-dimensional Hilbert space is not. Note that a closed subset of a proper metric space is itself a proper metric space with the induced metric, but that an open subset typically will not be a proper metric space.

Geodesic spaces as in the next definition are an important source of examples of proper metric spaces.

Definition A.3.4 Let X be a metric space. Let γ be a *path* in X, meaning that γ is a continuous function from $[0, 1]$ to X. The *length* of γ is

$$L(\gamma) := \sup \left\{ \sum_{i=1}^{n} d(\gamma(t_{i-1}), \gamma(t_i)) \,\middle|\, n \in \mathbb{N},\ 0 = t_0 \leqslant \cdots \leqslant t_n = 1 \right\}$$

(note that $L(\gamma)$ can be infinite). The space X is a *length space* if for all $x, y \in X$ we have

$$d(x, y) = \inf\{L(\gamma) \mid \gamma \colon [0, 1] \to X \text{ a path with } \gamma(0) = x,\ \gamma(1) = y\},$$

where we assume that $\inf(\varnothing) = \infty$, i.e. $d(x, y) = \infty$ if no path between x and y exists. The space X is a *geodesic space* if this infimum is attained (not necessarily uniquely).

Remark A.3.5 If X is a length space that is not path connected (for instance, a Riemannian manifold with more than one connected component), then there will be points at distance infinity from each other. This sort of example is one of our main reasons for allowing infinite distances.

The following is a version of the classical Hopf–Rinow theorem for general metric spaces; see Section 13.5, Notes and References, at the end of this appendix for a proof.

Theorem A.3.6 *Let (X,d) be a length space that is locally compact and complete. Then X is a proper, geodesic space.* □

In particular (this is the classical case), if X is a Riemannian manifold, then the metric it inherits from the Riemannian structure always makes it a length space. Thus if X is complete, then its metric is proper in the sense of Definition A.3.3 above.

Remark A.3.7 There is an abstract notion of coarse structure on a topological space X defined as follows. A *coarse structure* on X (that is compatible with the topology) consists of a collection of *controlled sets* $E \subseteq X \times X$ satisfying the following conditions:

(i) the collection of controlled sets is closed under finite unions, subsets, inverses[2] and compositions;[3]
(ii) there is an open set $U \subseteq X \times X$ that is controlled, and that contains the diagonal;
(iii) for all controlled sets E, the slice $E_x := \{y \in X \mid (x,y) \in E\}$ has compact closure.

The key examples of coarse structures come from proper metric spaces: one defines the controlled sets to be those on which the restriction of the metric is finite-valued. Much of what we do in this book could be carried out in the settings of abstract coarse structures, and the extra generality is sometimes useful.

However, for the reader's intuition, and as we did not have any important examples to apply the more general theory to, it seemed better to us to stick to proper metric spaces. The following metrisability theorem of Wright also says that one does not really lose any generality by keeping to the case of proper metric spaces.

Theorem A.3.8 *Let X be a second countable, locally compact, Hausdorff topological space, equipped with a countably generated[4] compatible coarse structure. Then there is a proper metric on X that induces both the topology and the coarse structure.* □

Morphisms and the Coarse Category

We now turn our attention to functions between (proper) metric spaces.

Definition A.3.9 Let X and Y be metric spaces. For a map $f: X \to Y$ and $x \in X$, the *expansion function* of f at x is the function $\omega_{f,x}: [0,\infty) \to [0,\infty]$ defined by

[2] For $E \subseteq X \times X$, the inverse of E is $E^{-1} := \{(y,x) \in X \times X \mid (x,y) \in E\}$.
[3] For $E, F \subseteq X \times X$, the composition of E and F is
$E \circ F := \{(x,z) \in X \times X \mid \text{ there is } y \in X \text{ with } (x,y) \in E \text{ and } (y,z) \in F\}$.
[4] i.e. there is a countable collection of controlled sets such that the given coarse structure is the intersection of all coarse structures that contain these sets.

$$\omega_{f,x}(r) = \sup\{d_Y(f(x_1), f(x)) \mid d_X(x_1, x) \leqslant r\}.$$

The (global) *expansion function* of f is the function $\omega_f \colon [0,\infty) \to [0,\infty]$ defined by

$$\omega_f(r) = \sup_{x \in X} \omega_{f,r}(x) \in [0,\infty].$$

A map $f \colon X \to Y$ is *uniformly expansive* if for all $r \in [0,\infty)$, $\omega_f(r)$ is finite, and is *proper* if for all compact subsets K of Y, $f^{-1}(K)$ has compact closure in X. The map f is *coarse* if it is both uniformly expansive and proper.

Two maps $f, g \colon X \to Y$ are *close* if there exists a constant $c \geqslant 0$ such that

$$d_Y(f(x), g(x)) \leqslant c$$

for all $x \in X$. A coarse map $f \colon X \to Y$ is called a *coarse equivalence* if there exists a coarse map $g \colon Y \to X$ such that $f \circ g$ and $g \circ f$ are close to the identities on Y, X respectively.

The *coarse category*, denoted $\mathcal{C}oa$, is defined to have all proper metric spaces as objects, and all closeness classes of coarse maps[5] as morphisms. Note that the isomorphisms in $\mathcal{C}oa$ are exactly the coarse equivalences.

The next lemma says that one can replace any object in $\mathcal{C}oa$ with a metric space in a way that destroys all the local structure; it is both psychologically and technically useful. First, we need a definition.

Definition A.3.10 Let X be a metric space and $r \in (0,\infty)$. A subset Z of X is *r-separated* if for all $x, y \in Z$, $d(x,y) \geqslant r$.

A *net* in X is a subset $Z \subseteq X$ such that there is $r \in (0,\infty)$ with the properties that:

(i) Z is r-separated;
(ii) for any $x \in X$ there is $z \in Z$ with $d(x,z) < r$.

If we need to specify the constant, we will say that Z is an *r-net*.

Lemma A.3.11 *For any $r \in (0,\infty)$, any proper metric space admits an r-net. Moreover, the metric on a proper metric space restricts to a proper metric on any net, and the inclusion map $i \colon Z \to X$ is a coarse equivalence.*

Proof Let $r \in (0,\infty)$. Zorn's lemma implies that there exists a maximal r-separated subset Z of X. This is an r-net, as maximality implies that every $x \in X$ must be within r of some point of Z.

Let now Z be any r-net in X. As Z is r-separated, it is closed, which implies that the restriction of the metric on X to Z is proper. It remains to show that $i \colon Z \to X$ is a coarse equivalence. We may define a map $p \colon X \to Z$ by sending each $x \in X$ to some point $p(x) \in Z$ such that $d(x, p(x)) \leqslant r$. Properness of p follows from the fact

[5] Closeness is an equivalence relation on coarse maps, and by 'closeness classes', we mean the equivalence classes for this relation. For $\mathcal{C}oa$ to be well defined, one needs to check that properties such as 'if f is close to g, then $f \circ h$ is close to $g \circ h$' hold for coarse maps; we leave it to the reader to check that this works.

that any compact subset of X must have finite intersection with Z, and the estimate $\omega_p(s) \leqslant s + 2r$ then shows that p is coarse. Finally, note that for all $x \in X$ and $z \in Z$,

$$d(x, i(p(x))) \leqslant r \quad \text{and} \quad d(y, p(i(y))) \leqslant r,$$

i.e. both compositions $p \circ i$ and $i \circ p$ are close to the identity, whence i is a coarse equivalence. □

We conclude our discussion of morphisms with a useful lemma with a similar proof to the above.

Lemma A.3.12 *Let $f\colon X \to Y$ be a coarse map between proper metric spaces. Then there exists a Borel coarse map $g\colon X \to Y$, which is close to f.*

Note that this says in particular that any morphism in *Coa* is represented by a Borel map.

Proof Let $r \in (0, \infty)$, and apply Lemma A.3.12 to the open cover $\{B(x; r) \mid x \in X\}$ of X to get a Borel cover $(E_i)_{i \in I}$ with the properties in that lemma. For each $i \in I$, choose a point x_i in the interior of E_i, and let Z be the subset $\{x_i \in X \mid i \in I\}$ of X. As the interiors of the sets E_i are all disjoint, the set Z is discrete. Let $f|_Z\colon Z \to Y$ be the map defined by restricting f to Z; this is continuous as Z is discrete.

Define a map $p\colon X \to Z$ by sending all points in each E_i to x_i; this is Borel as all the sets E_i are Borel. Finally, define $g\colon X \to Y$ to be the composition $f|_Z \circ p$. This is a composition of Borel maps, so Borel. Say $x \in X$ is in E_i for some i, whence $p(x)$ is in the same E_i, and so

$$d(f(x), g(x)) = d(f(x), f(p(x))) \leqslant \omega_f(\text{diam}(E_i)) \leqslant \omega_f(2r).$$

Hence f and g are close, which also implies that g is coarse and thus completes the proof. □

Coarse Geometry of Groups and Group Actions

We now look at groups acting on proper metric spaces, and at groups as proper metric spaces in their own right.

The next lemma says that any countable discrete group can be thought of as a proper metric space, up to canonical coarse equivalence.

Lemma A.3.13 *Let G be a countable discrete group. Then there exists a proper metric d on G, which is in addition* left invariant, *meaning that $d(gh, gk) = d(h, k)$ for all $g, h, k \in G$.*

Moreover, if d', d are any two metrics satisfying these conditions, then the identity map on G is a coarse equivalence from (G, d) to (G, d').

Proof Set $g_0 = e$, and let $g_0, g_1, g_2, \ldots$ be an ordered list of elements of G such that exactly one element from each set of the form $\{g, g^{-1}\}$, $g \in G$, occurs. Define a *length function* $l\colon G \to \mathbb{N}$ by

$$l(g) = \min\{a_1 n_1 + a_2 n_2 + \cdots + a_k n_k \mid g = g_{n_1}^{\pm a_1} \cdots g_{n_k}^{\pm a_k},\ n_i, a_i \in \mathbb{N}\}.$$

This function has the properties that $l(g) = 0$ if and only if $g = e$, $l(g) = l(g^{-1})$, and the set

$$\{g \in G \mid l(g) \leqslant r\}$$

is finite for all $r \in [0,\infty)$; we leave the checks of these properties to the reader. Define a metric on G by the formula

$$d(g,h) = l(g^{-1}h);$$

the properties of l imply that this is a proper metric, and it is clearly left-invariant.

To show the uniqueness statement, let d, d' be two proper left invariant metrics on G. It suffices to show that the identity map $\mathrm{id}\colon (G,d) \to (G,d')$ is a coarse map. As it is clearly proper, it suffices to consider its expansion function ω_{id}. Define $l(g) = d(e,g)$ and $l'(g) = d'(e,g)$, which are length functions with the properties listed above. Then

$$\begin{aligned}\omega_{\mathrm{id}}(r) &= \sup\{d'(g,h) \mid d(g,h) \leqslant r\} = \sup\{d'(e,g^{-1}h) \mid d(e,g^{-1}h) \leqslant r\} \\ &= \sup\{l'(g^{-1}h) \mid l(g^{-1}h) \leqslant r\},\end{aligned}$$

which is the supremum over a finite set, so bounded. □

We now look at *isometric* group actions on proper metric spaces, i.e. actions such that for all $x, y \in X$ and $g \in G$, $d(gx,gy) = d(x,y)$. We will assume also that the actions are proper in the sense of Definition A.2.2 above: recall this means that for any compact subset K of the space,

$$\{g \in G \mid gK \cap K \neq \varnothing\}$$

is finite. Note that a countable group G equipped with a metric d as in Lemma A.3.13 is a proper metric space, and the left translation of G on itself is a proper action by isometries.

Say now that an action of G on a proper metric space is *cobounded* if there is a bounded subset $B \subseteq X$ such that $GB = X$. The following lemma is a version of the *Svarc–Milnor lemma*, a fundamental result in the subject of geometric group theory. For the statement, if G acts on X, then an *orbit map* is any function $G \to X$ of the form $g \mapsto gx$ for some fixed $x \in X$. Note that any orbit map is equivariant.

Lemma A.3.14 *Let X be a proper metric space equipped with a proper, cobounded action of a countable discrete group G by isometric homeomorphisms. Equip G with a metric as in Lemma A.3.13. Then any orbit map $G \to X$ is a coarse equivalence, and any two such orbit maps are close.*

Proof Let $f\colon G \to X$ be any orbit map, so $f(g) = gx$ for some fixed $x \in X$ and all $g \in G$. Note that for any compact $K \subseteq Y$,

$$f^{-1}(K) = \{g \in G \mid gy \in K\} \subseteq \{g \in G \mid g(K \cup \{y\}) \cap (K \cup \{y\}) \neq \varnothing\},$$

which is finite by properness of the G-action on Y. Hence f is a proper function. Moreover, for any $r \geqslant 0$,

$$\begin{aligned}\omega_f(r) &= \sup\{d_Y(gy,hy) \mid d_G(g,h) \leqslant r\} \\ &= \sup\{d_Y(y,g^{-1}hy) \mid d_G(e,g^{-1}h) \leqslant r\} \\ &= \sup\{d_Y(y,gy) \mid d_G(e,g) \leqslant r\},\end{aligned}$$

which is the supremum over a finite set, so finite. We have thus shown that f is a coarse map.

Note now that as the G action on Y is cobounded, there exists $c \geqslant 0$ such that $d_Y(x, Gy) \leqslant c$ for all $x \in Y$. Define a map $p\colon Y \to G$ by setting $p(x)$ to be equal to any $g \in G$ which minimises $\{d_Y(x, gy) \mid g \in G\}$ (this makes sense as $Gy \cap B(x;c)$ is finite for any $x \in Y$). We claim that p is coarse. Indeed, it is proper as for any compact (i.e. finite) subset K of G, $p^{-1}(K)$ is contained in $\bigcup_{g \in K} B(gy;c)$, which has compact closure by properness of the metric on Y. On the other hand, for any $r \geqslant 0$,

$$\begin{aligned}\omega_p(r) &= \sup\{d_G(p(y_1), p(y_2)) \mid d_Y(y_1, y_2) \leqslant r\} \\ &\leqslant \sup\{d_G(g,h) \mid d_Y(gy, hy) \leqslant r + 2c\} \\ &\leqslant \sup\{d_G(e, gh^{-1}) \mid d_Y(y, g^{-1}h) \leqslant r + 2c\} \\ &\leqslant \sup\{d_G(e,g) \mid d_Y(y, gy) \leqslant r + 2c\}.\end{aligned}$$

We have that

$$\{g \in G \mid d_Y(y, gy) \leqslant r + 2c\} \subseteq \{g \in G \mid gB(y; r + 2c) \cap B(y; r + 2c) \neq \varnothing\},$$

which is finite by properness of the action and of d_Y. Hence the supremum $\sup\{d_G(e,g) \mid d_Y(y, gy) \leqslant r + 2c\}$ is over a finite set, so finite; this completes the argument that p is coarse.

We now show that $f \circ p$ and $p \circ f$ are close to the identity, which will complete the proof that f is a coarse equivalence. For $f \circ p$, note that for any $x \in Y$ and $g \in G$, $d_Y(x, f(p(x))) \leqslant c$. For $p \circ f$, note that for any $g \in G$, $p(f(g)) = p(gy)$, which by definition of p, is equal to some $h \in G$ such that $d_Y(hy, gy) = 0$ (this need not force $h = g$). Similarly to earlier arguments, note that we have

$$\{d_G(g,h) \mid d_Y(hy, gy)\} = \{d_G(e,g) \mid d_Y(y, gy) = 0\},$$

which is a finite set. Hence it is bounded by some $M > 0$ (independent of g and h). We thus have that $d_G(g, p(f(g))) \leqslant M$ for all $g \in G$.

Finally, to see that any two orbit inclusions are close, let $x, y \in X$. Then for any $g \in X$, as the action is by isometries, $d(gx, gy) = d(x, y)$. As this is independent of g, we are done. □

Corollary A.3.15 *Let X and Y be proper metric spaces, equipped with proper cobounded actions of a countable group G by isometric homeomorphisms. Then any equivariant map $f\colon X \to Y$ is a coarse equivalence.*

Proof Let $i\colon G \to X$ be any orbit inclusion map $i\colon g \mapsto gx$. Then the diagram

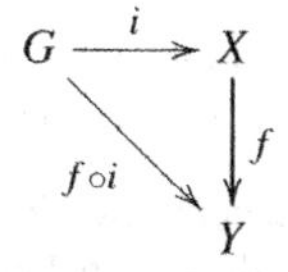

commutes. Moreover, i and $f \circ i$ are coarse equivalences by Lemma A.3.14, so f is too. □

The following category is the one we will use when considering proper metric spaces equipped with G actions.

Definition A.3.16 Let G be a countable discrete group. Let Coa^G be the category with objects given by proper metric spaces, equipped with a proper action of G by isometries. Morphisms in Coa^G are closeness classes of equivariant coarse maps. We call Coa^G the *(G-)equivariant coarse category*.

The following class of examples is important for applications to topology and geometry.

Example A.3.17 Let M be a closed Riemannian manifold. Let $\widetilde{M}$ be its universal cover with the lifted Riemannian metric. This is a proper metric space by Example A.3.4 above. Let G be the fundamental group of M, which acts properly on $\widetilde{M}$ by isometries for this metric. Thus $\widetilde{M}$ is naturally an object of Coa^G. Moreover, it is isomorphic to G itself via any orbit map $g \mapsto gx$ by Lemma A.3.14.

Finally, we give a technical lemma. The proof is essentially the same as that of Lemma A.3.12, using Lemma A.2.9 in place of Lemma A.1.10 as appropriate.

Lemma A.3.18 *Let $f: X \to Y$ be a coarse equivariant map between proper metric spaces, equipped with isometric and proper actions of a countable discrete group G. Then there exists an equivariant Borel coarse map $g: X \to Y$ which is close to f.* □

Bounded Geometry

In the last part of this section, we consider a particularly well-behaved class of discrete metric spaces.

Definition A.3.19 A metric space X has *bounded geometry* if for all $r \in (0, \infty)$, there is $n_r \in \mathbb{N}$ such that $B(x;r)$ has cardinality at most n_r.

Note that a bounded geometry metric space is proper and discrete. One of the most important classes of examples is as follows.

Example A.3.20 A countable discrete group with a proper left invariant metric as in Lemma A.3.13 is a bounded geometry metric space. This follows directly from the proof of Lemma A.3.13. As a consequence, if X is a proper metric space with a cobounded, proper, isometric action of a countable group, then one can deduce from Lemma A.3.14 that bounded geometry nets exist in X.

The following example takes much more work, but is also an important class to bear in mind for readers interested in Riemannian geometry.

Example A.3.21 Say X is a complete, connected Riemannian manifold, with injectivity radius bounded away from zero and all sectional curvatures in some bounded interval $[-M, M]$. For example, these assumptions are satisfied by the universal cover of any closed Riemannian manifold (equipped with the lifted metric).

Then any net in X is a bounded geometry metric space in the sense of Definition A.3.19. Indeed, say Z is an r-net in X for some $r \in (0,\infty)$. Then the assumptions plus some comparison results from Riemannian geometry imply the following:

(i) there is $v > 0$ such that the volume of each ball $B_X(z;r/2)$ is at least v (this uses the lower bound on the injectivity radius, and the upper bound on sectional curvature);
(ii) for each $s > 0$ there is $V(s) \in (0,\infty)$ such that for all $x \in X$, the volume of $B_X(x;s)$ is at most $V(s)$ (this uses the lower bound on sectional curvature, although actually a lower bound on Ricci curvature suffices).

Now, say the cardinality of $B_Z(z_0;s)$ is N for some $N \in \mathbb{N}$ and $z_0 \in Z$ (the cardinality must be finite, as Z is discrete and the ball $B_X(z_0;s)$ has compact closure by properness of X as in Example A.3.4). We need to find a bound for N depending only on s (and the fixed constant r). Note that the collection of balls $\{B_X(z;r/2) \mid z \in B_Z(z_0;s)\}$ is disjoint, and moreover, that

$$\bigsqcup_{z \in B_Z(z_0;s)} B_X(z;r/2) \subseteq B_X(s_0;s+r/2).$$

Taking volumes of both sides gives $Nv \leqslant V(s+r/2)$, and thus $N \leqslant V(s+r/2)/v$, completing the argument.

A.4 Exercises

A.4.1 Show that if X is a proper, geodesic metric space, then any uniformly continuous, proper map $f\colon X \to Y$ is coarse. Show that this fails if f is only assumed continuous and proper.

A.4.2 Show that a proper action of a torsion free group on a locally compact (Hausdorff) space is free.

A.4.3 Let X and Y be proper metric spaces. A coarse map $f\colon X \to Y$ is called a *coarse embedding* if there exist non-decreasing functions $\alpha_\pm\colon [0,\infty) \to [0,\infty)$ such that $\alpha_-(t) \to \infty$ as $t \to \infty$ and

$$\alpha_-(d_X(x_1,x_2)) \leqslant d_Y(f(x_1),f(x_2)) \leqslant \alpha_+(d_X(x_1,x_2)).$$

A coarse map $f\colon X \to Y$ is called a *coarse surjection* if there exists a constant $c \geqslant 0$ such that $d_Y(y,f(X)) \leqslant c$ for all $y \in Y$.

(i) Show that a map $f\colon X \to Y$ is a coarse equivalence if and only if it is a coarse embedding and a coarse surjection.
(ii) A morphism f in an abstract category is a *monomorphism* if for all morphisms g,h, $f \circ g = f \circ h \Rightarrow g = h$, and similarly f is an *epimorphism* if $g \circ f = h \circ f \Rightarrow g = h$. Show that a coarse map $f\colon X \to Y$ is a monomorphism in $\mathcal{C}oa$ if and only if it is a coarse embedding, and is an epimorphism in $\mathcal{C}oa$ if and only if it is a coarse surjection.

A.4.4 Show that if X and Y are length spaces, then any coarse equivalence $f: X \to Y$ is automatically a *quasi-isometry*: there exist constants $r > 0$ and $s, c \geqslant 0$ such that

$$r^{-1}d_X(x_1, x_2) - s \leqslant d_Y(f(x_1), f(x_2)) \leqslant r d_X(x_1, x_2) + s$$

and for all $y \in Y$, $d(y, f(X)) \leqslant c$.

A.4.5 Lemma A.2.5 states that quotient spaces by proper actions (on reasonable spaces) are Hausdorff. This can fail badly in general, as the following example shows. Let $\mathbb{Z}$ act on S^1 by $n\colon z \mapsto e^{2\pi i n\theta} z$ for θ an irrational number in $(0, 1)$. Show that the quotient space $S^1/\mathbb{Z}$ is uncountable, but has the indiscrete topology (i.e. the only open sets in the quotient are the whole thing and the empty set).

A.5 Notes and References

The facts that we need from general topology can be found (for example) in Munkres [188]. Specifically: Theorem A.1.1 follows from theorem 34.1 and exercise 2 in section 32 of Munkres [188]; Theorem A.1.3 follows from lemma 41.6, theorem 34.1 and theorem 41.4 in Munkres [188]; and Lemma A.1.11 follows from theorem 41.7, theorem 34.1 and theorem 41.4 in Munkres [188].

A proof of the Hopf–Rinow theorem as stated in Theorem A.3.6 can be found for example in section I.3 of Bridson and Haefliger [38], together with a discussion of the history. Bridson and Haefliger [38] also contains a great deal of other information about length spaces and metric geometry more generally: see for example section I.7 of Bridson and Haefliger [38] for a discussion of simplicial (and more general polyhedral) complexes as length spaces. One can also find a different version of the Svarc–Milnor lemma, and some historical references in section I.8 of Bridson and Haefliger [38].

The notion of an abstract coarse structure is due to Roe: see for example chapter 2 of Roe [218]. Wright's 'coarse metrisability theorem' stated as Theorem A.3.8 above, is proved in Wright [263]. Coarse structure is more general than our notion of proper metric space, and almost all the work in this book could have been done in that language. Nonetheless, all the examples we are interested in are in the narrower class of proper metric spaces as long as one is allowed infinite distances; we thus preferred to keep to this language as it seems more intuitive for most purposes. For readers interested in the set-up of the coarse Baum–Connes conjecture (among other things) for abstract coarse structures, a good reference is Wright's thesis (see chapter 5 of Wright [262]).

The facts from Riemannian geometry that are quoted in Example A.3.21 can be derived (for example) from theorems 103 and 107 of Berger [30]; this book is recommended in general as a source of information and inspiration on Riemannian geometry.

APPENDIX B

Categories of Topological Spaces and Homology Theories

In this appendix, Section B.1 summarises the categories of topological spaces that we work with in the main part of the text. We then briefly discuss a collection of axioms for a generalised homology theory in Section B.2: this is not used in a substantial way in the main text, but we thought it might be useful to show how K-homology fits into the sort of general framework commonly used in algebraic topology.

B.1 Categories We Work With

In this section, we will define the various categories of spaces that are used in the main text. For a locally compact, Hausdorff space X, X^+ denotes its one-point compactification as in Definition A.1.4. We think of X^+ as a pointed space with basepoint ∞; in particular a *pointed map* $f: X^+ \to Y^+$ is by definition a function from X^+ to Y^+ that takes the point at infinity in X^+ to the point at infinity in Y^+.

Definition B.1.1

(i) $\mathcal{LC}$. Objects: locally compact, second countable, Hausdorff topological spaces.
Morphisms from X to Y: continuous, pointed maps $f: X^+ \to Y^+$.

(ii) $\mathcal{C}ont$. Objects: locally compact, second countable, Hausdorff topological spaces.
Morphisms from X to Y: continuous functions $f: X \to Y$.

(iii) $\mathcal{P}ro$. Objects: proper (see Definition A.3.3) metric spaces.
Morphisms from X to Y: continuous, coarse functions $f: X \to Y$.

(iv) $\mathcal{C}oa$. Objects: proper (see Definition A.3.3) metric spaces.
Morphisms: closeness classes of coarse functions $f: X \to Y$.

(v) $\_^G$. For any of the above four categories, we allow a variant with superscript G; this means that a fixed countable discrete group G acts properly by homeomorphisms (by isometries in the metric cases) on all objects,[1] and that all morphisms are assumed equivariant.

[1] Meaning on the object X in the $\mathcal{LC}$ case, *not* on X^+. Note that such an action extends to an action on X^+. The extended action fixes ∞, and so is never proper if G is infinite.

(vi) $\mathcal{GA}$. Objects: $\mathbb{Z}/2$-graded abelian groups.
Morphisms from A to B: graded group homomorphisms.

Note that there is a canonical functor $\mathcal{P}ro \to \mathcal{LC}$ defined by forgetting the metric structure, and extending a (proper, continuous) function $f: X \to Y$ to a (pointed, continuous) function $f: X^+ \to Y^+$. In particular, a homology theory on $\mathcal{LC}$ in the sense of Section B.2 defines one on $\mathcal{P}ro$ in a natural way by compositing with the functor $\mathcal{P}ro \to \mathcal{LC}$. There is also a canonical functor from $\mathcal{P}ro$ to $\mathcal{C}oa$, defined by sending a function $f: X \to Y$ to its closeness equivalence class.

B.2 Homology Theories on $\mathcal{LC}$

For us, a $\mathbb{Z}/2$-graded[2] homology theory on $\mathcal{LC}$ is a functor from $\mathcal{LC}$ to $\mathcal{GA}$ satisfying certain axioms. To make this precise, we start with what it means for two morphisms to be homotopic.

Definition B.2.1 A *homotopy* between two morphisms f_0, f_1 from X to Y in $\mathcal{LC}$ (i.e. pointed functions $f_0, f_1: X^+ \to Y^+$) is a morphism h from $X \times [0,1]$ to Y (i.e. a pointed function $h: (X \times [0,1])^+ \to Y^+$) from $X \times [0,1]$ to Y such that the restriction of h to $(X \times \{i\})^+$ identifies with f_i for $i = 0, 1$; we say f_0, f_1 are *homotopic* if such an h exists.

Two spaces X and Y in $\mathcal{LC}$ are *homotopy equivalent* if there are morphisms f from X to Y and g from Y to X such that the compositions $f \circ g$, $g \circ f$ are homotopic to the identity morphisms on X, Y respectively.

A space X is *contractible* in $\mathcal{LC}$ if it is homotopy equivalent to the empty set.

In other words, a space X is contractible if X^+ admits a deformation retraction to the basepoint ∞ in the usual sense of homotopy theory. Thus, for example, any space of the form $X \times [0, \infty)$ is contractible in this sense.

Definition B.2.2 A *(Steenrod*[3]*) homology theory* on $\mathcal{LC}$ is a functor $H: \mathcal{LC} \to \mathcal{GA}$, written $H = H_0 \oplus H_1$, such that the following conditions hold.

(i) *Empty set*: $H(\varnothing) = 0$.
(ii) *Homotopy invariance*: if f, g are homotopic, then $H(f) = H(g)$.
(iii) *Mayer–Vietoris sequences*: if E, F are closed subsets of X, then there is a natural six-term exact sequence

$$\begin{array}{ccccc} H_0(E \cap F) & \longrightarrow & H_0(E) \oplus H_0(F) & \longrightarrow & H_0(X) \\ \uparrow & & & & \downarrow \\ H_1(X) & \longleftarrow & H_1(E) \oplus H_1(F) & \longleftarrow & H_1(E \cap F), \end{array}$$

[2] Essentially the same definition applies in the $\mathbb{Z}$-graded case: just replace six-term exact sequences with long exact sequences.
[3] The qualifier 'Steenrod' means that the cluster axiom is included.

where if

$$i_E : E \cap F \to E, \quad i_F : E \cap F \to F, \quad j_E : E \to X, \quad j_F : F \to X$$

are the respective inclusions, then the map $H(E \cap F) \to H(E) \oplus H(F)$ appearing above is $H(i_E) - H(i_F)$, and the map $H(E) \oplus H(F) \to H(X)$ is $H(j_e) + H(j_F)$.[4]

(iv) *Cluster Axiom*: if $X = \bigsqcup X_n$ is a countable disjoint union of spaces in $\mathcal{LC}$, then the inclusions $i_n : X_n \to X$ induce an isomorphism

$$\prod H(i_n) : \prod_n H(X_n) \to H(X).$$

It is perhaps more common to replace the Mayer–Vietoris axiom by an axiom assigning a six-term exact sequence to a pair (X, F) of an object in $\mathcal{LC}$ and a closed subset. This is in fact equivalent to the above definition: we prove one direction below, and leave the other to the exercises.

Proposition B.2.3 *Let (X, A) be a pair consisting of an object of $\mathcal{LC}$ and a closed subset A of X, and assume that H is a homology theory in the sense of Definition B.2.2 above, except from possibly not satisfying the cluster axiom. Then there is a natural six-term exact sequence*

$$\begin{array}{ccccc} H_0(A) & \longrightarrow & H_0(X) & \longrightarrow & H_0(X \backslash A) \\ \uparrow & & & & \downarrow \\ H_1(X \backslash A) & \longleftarrow & H_1(X) & \longleftarrow & H_1(A), \end{array}$$

where the morphisms $H(A) \to H(X)$, $H(X) \to H(X \backslash A)$ are induced by the inclusion of A into X and the 'collapse A to ∞' map $X^+ \to (X \backslash A)^+$ respectively.

Proof Let $C(X, A)$ be the space obtained by 'coning off A', i.e. by gluing a copy of $A \times [0, \infty)$ to X along $A \times \{0\}$. Let E, F be the closed subsets

$$E = X \cup (A \times [0, 1]), \quad F = A \times [1, \infty)$$

of $C(X, A)$, and note that $C(X, A) = E \cup F$. Then $E \cap F = A \times \{1\}$, E is homotopy equivalent to X, F is contractible and $C(X, A)$ is homotopy equivalent to $X \backslash A$ by collapsing the cone $A \times [0, \infty) \cup \{\infty\}$ in $C(X, A)^+$ to the point ∞. Homotopy invariance and the empty set axiom then imply that the Mayer–Vietoris sequence

$$\begin{array}{ccccc} H_0(E \cap F) & \longrightarrow & H_0(E) \oplus H_0(F) & \longrightarrow & H_0(C(X, A)) \\ \uparrow & & & & \downarrow \\ H_1(C(X, A)) & \longleftarrow & H_1(E) \oplus H_1(F) & \longleftarrow & H_1(E \cap F) \end{array}$$

[4] The exact signs here are not important as long as one is consistent: it is important for exactness, however, that they be opposite.

identifies with

$$\begin{array}{ccccc} H_0(A) & \longrightarrow & H_0(X) \oplus 0 & \longrightarrow & H_0(X \backslash A) \\ \uparrow & & & & \downarrow \\ H_1(X \backslash A) & \longleftarrow & H_1(X) \oplus 0 & \longleftarrow & H_1(A). \end{array}$$

Naturality follows from naturality of the Mayer–Vietoris sequence. A little more work to identify the morphisms involved shows that this is the desired exact sequence (possibly up to sign conventions, which we elide). □

B.3 Exercises

B.3.1 Show that if $H\colon \mathcal{LC} \to \mathcal{GA}$ satisfies axioms (i), (ii) in the above, and the existence of the long exact sequence for a pair as in Proposition B.2.3, then it satisfies the Mayer–Vietoris axiom (axiom (iii)).

B.3.2 The goal of this exercise is to show that a Steenrod homology theory as in Definition B.2.2 is (at least in some sense) determined on compact metric spaces by what it does on finite simplicial complexes. Let

$$\cdots \to X_3 \to X_2 \to X_1$$

be a sequence of continuous maps between compact metric spaces. The *inverse limit* of the sequence is the compact metrisable space defined as

$$\varprojlim X_n := \left\{ (x_n) \in \prod_n X_n \mid x_{n+1} \text{ maps to } x_n \right\} \tag{B.1}$$

(and equipped with the restriction of the product topology). It is equipped with continuous coordinate projections $p_n\colon \varprojlim X_n \to X_n$ that commute with the original maps $X_{n+1} \to X_n$, and has the universal property that if Y is any other space equipped with compatible maps $q_n\colon Y \to X_n$, then there is a unique continuous map $Y \to X$ such that the diagrams

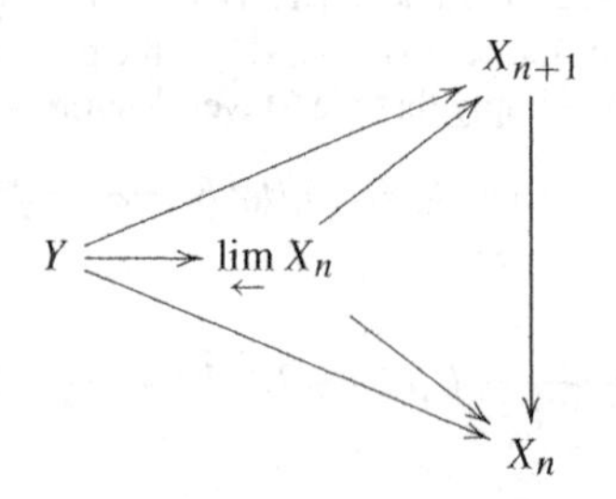

commute. Similarly, the inverse limit $\varprojlim A_n$ of the sequence of (abelian) groups

$$\cdots \to A_3 \to A_2 \to A_1$$

is defined to be the subgroup of the product group $\prod_n A_n$ defined analogously to line (B.1), and has the analogous universal property. Finally, for such a sequence of abelian groups, the $\lim^1$ *group*, denoted $\varprojlim^1 A_n$, is defined as the cokernel of the map

$$\prod_n A_n \to \prod_n A_n$$

taking a sequence (a_n) to the sequence $(a_n - (\text{image of } a_{n+1}))$.

(i) Show that if $\varprojlim X_n$ is an inverse limit as above with each X_n a compact metric space, and if H is a homology theory as in Definition B.2.2, then there is a natural *Milnor exact sequence*

$$0 \longrightarrow \varprojlim{}^1 H_{i-1}(X_n) \longrightarrow H_i(\lim_{\leftarrow} X_n) \longrightarrow \varprojlim H_i(X_n) \longrightarrow 0,$$

where the map $H_i(\varprojlim X_n) \to \varprojlim H_i(X_n)$ comes from the universal property of the inverse limit group.
Hint: define the mapping telescope *of the sequence to be*

$$\bigsqcup_n X_n \times [0,1]/\sim,$$

where $(x,1) \sim (y,0)$ whenever $x \in X_{n+1}$ maps to $y \in X_n$. Analyse this by breaking it into two infinite disjoint unions, using the Mayer–Vietoris sequence, using homotopy invariance and using the fact that disjoint unions are taken to products.

(ii) Show that any compact metric space is the inverse limit of a sequence of finite simplicial complexes.
Hint: Fix a sequence of finite covers $(\mathcal{U}_n)$ such that $sup_{U\in\mathcal{U}_n} diam(U) \to 0$ as $n \to \infty$, and such that each element of $\mathcal{U}_{n+1}$ is contained in an element of $\mathcal{U}_n$. The nerve *of such a (finite) cover $\mathcal{U}$ is defined to be the simplicial complex with a vertex for each $U \in \mathcal{U}_n$, and where a collection of vertices $U_0, \ldots, U_d$ spans a d-simplex if and only if $\bigcap_{i=0}^d U_i \neq \varnothing$. Let N_n be the nerve of $\mathcal{U}_n$, and define a continuous map $N_{n+1} \to N_n$ by sending each vertex U to some choice of vertex (open set) containing it, and extending affinely on simplices. Show that X is the inverse limit of the system*

$$\cdots \to N_3 \to N_2 \to N_1.$$

B.3.3 Show that if $\Phi\colon H \to G$ is a natural transformation of homology theories satisfying the axioms in Theorem B.2.2 such that $\Phi(pt)\colon H(pt) \to G(pt)$ is an isomorphism, then Φ is an isomorphism for all spaces in $\mathcal{LC}$.
Hint: show this first for finite simplicial complexes using the Mayer–Vietoris axiom, the homotopy invariance axiom and induction on the number of simplices. Then use limits as in Exercise B.3.2 to deduce the result for general compact metric spaces. Finally, use that for a possibly non-compact space X, the group $H(X)$ is the cokernel of the natural map $H(\{\infty\}) \to H(X^+)$ to deduce the result in general.

B.4 Notes and References

The ‘axiomatic approach’ to homology theories was developed particularly by, and is discussed in the classic text of, Eilenberg and Steenrod [87]. The Milnor exact sequence from Exercise B.3.2 (and the proof we hint at) comes from Milnor [181].

APPENDIX C

Unitary Representations

In this appendix we summarise the terminology, examples and facts that we will need about group representations. Section C.1 discusses basic properties and examples of unitary representations, and Section C.2 discusses Fell's trick, an important 'untwisting' argument that gets used a little in the main text.

Throughout this appendix, G denotes a countable discrete group.

C.1 Unitary Representations

Definition C.1.1 A *unitary representation* of G on a Hilbert space H is a group homomorphism

$$U : G \to \mathcal{U}(H), \quad g \mapsto U_g$$

from G to the unitary group of H. Two unitary representations $U : G \to \mathcal{U}(H_U)$ and $V : G \to \mathcal{U}(H_V)$ are *isomorphic* if there is an unitary isomorphism $W : H_U \to H_V$ such that $WU_g = V_g W$ for all $g \in G$.

We will also often say something like 'let H be a G representation', leaving the homomorphism, and the fact that the representation is unitary, implicit.

Example C.1.2 Let H be an arbitrary Hilbert space. The *trivial representation* on H is the homomorphism $G \to \mathcal{U}(H)$ that sends every element of G to the identity.

Example C.1.3 Let $\ell^2(G)$ denote the Hilbert space of square summable complex-values functions on G. The *left regular representation* of G on $\ell^2(G)$ is defined by $g \mapsto \lambda_g$, where

$$\lambda_g : \delta_h \mapsto \delta_{gh}$$

and the *right regular representation* by $g \mapsto \rho_g$, where

$$\rho_g : \delta_h \mapsto \delta_{hg^{-1}}.$$

These two representations are isomorphic via the unitary

$$W : \ell^2(G) \to \ell^2(G), \quad \delta_g \mapsto \delta_{g^{-1}}.$$

As they are isomorphic, it is quite common to just say 'the regular representation' and not be specific about whether one is using the left or right version.

Example C.1.4 Let H be a subgroup of G, and let $\ell^2(G/H)$ denote the square summable functions on the set $\{gH \mid g \in G\}$ of left cosets. Then the (left) *quasi-regular representation* of G on $\ell^2(G/H)$ is defined by

$$\lambda_g : \delta_{kH} \mapsto \delta_{gkH}.$$

There is similarly a right quasi-regular representation on the right coset space.

Example C.1.5 Let (X, μ) be a measure space, equipped with a measure preserving G action: this means that G acts on X, the action preserves the σ-algebra of measurable sets, and that it satisfies $\mu(gE) = \mu(E)$ for all $g \in G$ and measurable $E \subseteq X$. For each $g \in G$, define $U_g \in \mathcal{U}(L^2(X, \mu))$ by

$$(U_g u)(x) \mapsto u(g^{-1}x).$$

The map $g \mapsto U_g$ is then a representation, sometimes called the *Koopman representation*. Note that the previous example is the special case when $X = G/H$ and μ is the counting measure on X.

In addition to these examples, there are two basic constructions that build new representations out of old ones.

Definition C.1.6 Let $U : G \to \mathcal{U}(H_U)$ and $V : G \to \mathcal{U}(H_V)$ be unitary representations. Their *direct sum* is the representation defined by

$$U \oplus V : G \to \mathcal{U}(H_U \oplus H_V), \quad (U \oplus V)_g := U_g \oplus V_g,$$

and their *tensor product* is the representation defined by

$$U \otimes V : G \to \mathcal{U}(H_U \otimes H_V), \quad (U \otimes V)_g := U_g \otimes V_g.$$

Definition C.1.7 The *reduced group C^*-algebra* of G, denoted $C^*_\lambda(G)$, is the C^*-subalgebra of $\mathcal{B}(\ell^2(G))$ generated by the unitaries λ_g from the left regular representation of Example C.1.3. We also denote by $C^*_\rho(G)$ the C^*-algebra generated by the unitaries ρ_g of the right regular representation.

Note that $C^*_\lambda(G)$ and $C^*_\rho(G)$ are canonically isomorphic via conjugation by the unitary U from Example C.1.3. We will thus sometimes abuse terminology slightly and also call $C^*_\rho(G)$ the reduced group C^*-algebra of G.

Let now A be a C^*-algebra. An *action* of G on A is a homomorphism

$$\alpha : G \to \mathrm{Aut}(A), \quad g \mapsto \alpha_g$$

from G to the group of $*$-automorphisms of A; a C^*-algebra equipped with an action is called a *G-C^*-algebra*. For example, if G acts on a space X, then $C_0(X)$ comes equipped with a canonical G action by Proposition A.2.1.

Definition C.1.8 Let A be a G-C^*-algebra, and $\pi : A \to \mathcal{B}(H)$ a $*$-representation. Then a unitary representation $U : G \to \mathcal{U}(H)$ *spatially implements* the action of G on A if the *covariance relation*

$$\pi(\alpha_g(a)) = U_g \pi(a) U_g^*$$

holds for all $a \in A$ and $g \in G$. If U spatially implements π, the pair (π, U) is called a *covariant representation*.

There are always plenty of covariant representations, as exemplified by the following lemma.

Lemma C.1.9 *Let A be a G-C^*-algebra. Then there exists a faithful covariant representation of A.*

Proof Let $\pi: A \to \mathcal{B}(H)$ be a faithful representation of A (forgetting the G-action). Define an action of $\widetilde{\pi}$ of A on $\ell^2(G,H)$ by the formula

$$(\widetilde{\pi}(a)u)(g) := \pi(\alpha_{g^{-1}}(a))u(g).$$

It is not difficult to see that $\widetilde{\pi}$ is still a faithful representation of A. Moreover, if we define a representation U_g on $\ell^2(G,H)$ by

$$(U_g u)(h) := u(g^{-1}h),$$

then for any $a \in A$, $u \in \ell^2(G,H)$ and $g,h \in G$, the computation

$$\begin{aligned}\big(U_g\widetilde{\pi}(a)U_g^*u\big)(h) &= \big(\widetilde{\pi}(a)U_g^*u\big)(g^{-1}h) = \pi(\alpha_{h^{-1}g}(a))\big(U_g^*u\big)(g^{-1}h)\\ &= \big(\widetilde{\pi}(\alpha_g(a))u\big)(h)\end{aligned}$$

shows that $\widetilde{\pi}$ is covariant for U. □

C.2 Fell's Trick

In this section we introduce *Fell's trick* (also called *Fell's absorption principle*) and a consequence for representations of finite groups. Fell's trick is the next proposition: it can be summarised by saying that the regular representation 'tensorially absorbs' any other representation.

Proposition C.2.1 *Let $U: G \to \mathcal{U}(H)$ be a representation, and let H_T be the same Hilbert space as H, but equipped with the trivial representation. Let $\ell^2(G)$ be equipped with the left regular representation, and equip both tensor products $\ell^2(G) \otimes H$ and $\ell^2(G) \otimes H_T$ with the tensor product representations. Then the formula*

$$W: \ell^2(G) \otimes H \to \ell^2(G) \otimes H_T, \quad \delta_g \otimes u \mapsto \delta_g \otimes U_g^* u$$

defines an isomorphism of representations.

Proof Note that for any $\delta_g \otimes u, \delta_h \otimes v \in \ell^2(G) \otimes H$,

$$\langle W(\delta_g \otimes u), \delta_h \otimes v\rangle = \left\{\begin{array}{ll} \langle U_g^* u, v\rangle & g = h \\ 0 & g \neq h \end{array}\right\} = \langle \delta_g \otimes u, \delta_h \otimes U_h v\rangle;$$

it follows that W^* is given by the formula $\delta_h \otimes v \mapsto \delta_h \otimes U_h$, and in particular from this that WW^* and W^*W are the identity. Moreover, for any $g,h \in G$ and $u \in H$

$$W(\lambda_g \otimes U_g)(\delta_h \otimes v) = W(\delta_{gh} \otimes U_g v) = \delta_{gh} \otimes U_h^* u = (\lambda_g \otimes 1)W(\delta_h \otimes u),$$

whence W is equivariant with respect to the G actions on $\ell^2(G) \otimes H$ and $\ell^2(G) \otimes H_T$, respectively. □

We record here a consequence of Fell's trick: it says that if F is a finite group and H_∞ is an infinite-dimensional separable Hilbert space with the trivial F representation, then $\ell^2(F) \otimes H_\infty$ contains a copy of any separable F representation.

Corollary C.2.2 *Let F be a finite group and $U\colon F \to \mathcal{U}(H)$ a representation on a separable Hilbert space. Let H_∞ be an infinite-dimensional Hilbert space equipped with the trivial F representation, and $\ell^2(F) \otimes H_\infty$ the corresponding tensor product representation. Then there is an F equivariant isometry*

$$V\colon H \to \ell^2(F) \otimes H_\infty.$$

Proof Note first that there is an F equivariant isometry

$$V_1\colon H \to \ell^2(F) \otimes H, \quad u \mapsto \left(\frac{1}{\sqrt{|F|}} \sum_{g \in F} \delta_g \right) \otimes u.$$

Let H_T be the same underlying Hilbert space as H equipped with the trivial representation, so Fell's trick gives an isomorphism of representations

$$V_2\colon \ell^2(F) \otimes H \cong \ell^2(F) \otimes H_T.$$

Let $V_3\colon H_T \to H_\infty$ be any isometry (which exists as H_∞ is infinite-dimensional and H is separable). Then

$$V = (1 \otimes V_3)V_2V_1\colon H \to \ell^2(F) \otimes H_\infty$$

has the right properties. □

C.3 Notes and References

The classic reference on the C^*-algebraic approach to unitary representation theory is the second half of Dixmier [80]. A highly recommended introduction to unitary representation theory and related issues is Deitmar and Echterhoff [86]. Both of these books go much further than anything we need in this text; in particular they have a lot to say on the analytic subtleties that arise when one is dealing with non-discrete groups.

APPENDIX D

Unbounded Operators

The first two sections of this appendix summarise the facts we need about unbounded operators: Section D.1 discusses the basic definitions associated to unbounded operators and the spectral theorem, while Section D.2 discusses facts about the 'one parameter unitary group' $(e^{itD})_{t\in\mathbb{R}}$ associated to a self-adjoint operator D, and some of the Fourier theory one can thus build. Section D.3 has quite a different character: it discusses a particular concrete example in detail, and proves some important formulas.

The first two sections cover general theory, and do not really contain any proofs. Section D.3, on the other hand, has fairly detailed arguments: although this material is more or less contained in the literature, we could not find convenient references, so give details here.

D.1 Self-Adjointness and the Spectral Theorem

Definition D.1.1 Let H be a Hilbert space. An *unbounded operator* on H consists of a pair (S, D) where S is a dense subspace of H, and $D: S \to H$ is a linear operator. The subspace S is called the *domain* of the operator.

Often, we will just write D for an unbounded operator, especially if S is clear from context. However, be warned that there is often more than one 'reasonable' choice for S, and that the properties of (D, S) can be quite different if other choices are made.

Definition D.1.2 An unbounded operator (S, D) on a Hilbert space S is *closed* if its *graph*

$$\text{graph}(D) := \{(u, Du) \in H \oplus H \mid u \in S\}$$

is closed in $H \oplus H$ (equipped with the product topology). It is *closable* if the closure $\overline{\text{graph}(D)}$ of its graph is still the graph of a linear operator (equivalently, (S, D) is closeable if whenever (u, v) and (u, w) are both in $\overline{\text{graph}(D)}$, we must have $v = w$).

If (S, D) is closable, then its *closure* is the unbounded operator $\overline{D}$ with graph equal to the closure $\overline{\text{graph}(D)}$ of the graph of D (and domain equal to the projection of $\overline{\text{graph}(D)}$ onto the first coordinate).

There are several natural versions of self-adjointness for unbounded operators of varying strengths. The following is the weakest that we will consider; it may be the most intuitive definition, but unfortunately, it is also the least useful.

Definition D.1.3 An unbounded operator (S, D) on a Hilbert space H is *formally self-adjoint* if for all $u, v \in S$, $\langle Du, v\rangle = \langle u, Dv\rangle$.

A formally self-adjoint operator (S, D) on a Hilbert space H is always closeable: this follows as if (u_n) and (v_n) are sequences in S converging to some $u \in H$ (possibly not in S) and such that (Du_n) and (Dv_n) also converge in H, then for all $w \in S$,

$$\langle \lim_{n\to\infty} Du_n, w\rangle = \lim_{n\to\infty}\langle u_n, Dw\rangle = \langle u, Dw\rangle = \lim_{n\to\infty}\langle v_n, Dw\rangle = \langle \lim_{n\to\infty} Dv_n, w\rangle,$$

and as S is dense in H, this forces

$$\lim_{n\to\infty} Du_n = \lim_{n\to\infty} Dv_n.$$

Formal self-adjointness does not, however, allow one to prove a reasonable version of the spectral theorem. For this we need stronger defintions.

Definition D.1.4 Let (S, D) denote a formally self-adjoint operator on a Hilbert space H. The *minimal domain* of D is the domain of the closure $\overline{D}$. The *maximal domain* of D is the collection of all $v \in H$ such that there exists $w \in H$ such that for all $u \in S$,

$$\langle Du, v\rangle = \langle u, w\rangle.$$

The operator (S, D) is *essentially self-adjoint* if its minimal and maximal domains coincide, and it is *self-adjoint* if the minimal and maximal domains coincide with the original domain S.

Definition D.1.5 Let (S, D) be an unbounded operator on a Hilbert space H. The *resolvent set* of D is the collection of all $\lambda \in \mathbb{C}$ such that the operator $D - \lambda\colon S \to H$ is a bijection with bounded inverse. The *spectrum of D* is the complement of the resolvent set.

We will give a proof of the next result, as the argument quite nicely illustrates a point where self-adjointness is really needed, and formal self-adjointness would not be enough.

Lemma D.1.6 *If (S, D) is self-adjoint, then the spectrum of D is real.*

Proof For $v \in S$ and $\lambda \in \mathbb{C}\setminus\mathbb{R}$, we have

$$\begin{aligned}\langle (D+\lambda)v, (D+\lambda)v\rangle &= \|Dv\|^2 + 2\mathrm{Re}(\lambda)\langle Dv, v\rangle + |\lambda|^2\|v\|^2\\ &\geqslant \|Dv\|^2 - 2\mathrm{Re}(\lambda)\|v\|\|Dv\| + |\lambda|^2\|v\|^2\\ &\geqslant (\|Dv\| - \mathrm{Re}(\lambda)\|v\|)^2 + |\mathrm{Im}(\lambda)|^2\|v\|^2\\ &\geqslant \mathrm{Im}(\lambda)^2\|v\|^2. \end{aligned} \tag{D.1}$$

It follows that if $\lambda \in \mathbb{C}\setminus\mathbb{R}$, $D+\lambda\colon S \to H$ is injective with closed range.

We claim that that for any $\lambda \in \mathbb{C}$,

$$\text{Range}(D+\lambda)^{\perp} = \text{Ker}(D+\overline{\lambda}). \tag{D.2}$$

Indeed, if $u \in \text{Ker}(D+\overline{\lambda})$ then in particular $u \in S$. Let $(D+\lambda)v$ be an element of $\text{Range}(D+\lambda)$ for some $v \in S$. As both u and v are in S, we may compute

$$\langle (D+\lambda)v, u\rangle = \langle v, (D+\overline{\lambda})u\rangle = 0$$

giving the inclusion

$$\text{Range}(D+\lambda)^{\perp} \supseteq \text{Ker}(D+\overline{\lambda}).$$

For the opposite inclusion,[1] say $u \in \text{Range}(D+\lambda)^{\perp}$. Then for any $v \in S$, $\langle (D+\lambda)v, u\rangle = 0$, whence

$$\langle Dv, u\rangle = \langle v, -\overline{\lambda}u\rangle.$$

This equivalently says that u is in the maximal domain of D, whence by self-adjointness u is in the domain S of D. We then have

$$0 = \langle (D+\lambda)v, u\rangle = \langle v, (D+\overline{\lambda})u\rangle$$

for all $v \in S$, and as S is dense this forces $(D+\overline{\lambda})u = 0$ as required.

To complete the proof, we have already noted by line (D.1) that if $\lambda \in \mathbb{C}\setminus\mathbb{R}$, then $D+\overline{\lambda}\colon S \to H$ is injective. Hence line (D.2) implies that $\text{Range}(D+\lambda)^{\perp} = \{0\}$, and thus $\text{Range}(D+\lambda)$ is dense. On the other hand, line (D.1) implies that $\text{Range}(D+\lambda)$ is closed, so it equals all of H. At this point we have that $D+\lambda\colon S \to H$ is a bijection. Finally, line (D.1) implies that the inverse of $D+\lambda$ is bounded, so we are done. □

As the spectrum of a self-adjoint operator is real, the statement of the following fundamental result makes sense. It is called the *functional calculus for unbounded operators*. For the statement, let $B(\mathbb{R})$ denote the C^*-algebra of complex-valued bounded Borel functions on the real line.

Theorem D.1.7 *Let (S, D) be an unbounded self-adjoint operator on a Hilbert space H. Then there exists a unique $*$-homomorphism*

$$B(\mathbb{R}) \to \mathcal{B}(H), \quad f \mapsto f(D)$$

from the C^*-algebra of bounded Borel functions on $\mathbb{R}$ to the bounded operators on H such that for any $\lambda \in \mathbb{C}\setminus\mathbb{R}$, if $f(x) = (x-\lambda)^{-1}$, then $f(D) = (D-\lambda)^{-1}$. □

Remark D.1.8 Note that the spectral theorem extends to a version for essentially self-adjoint operators: if (S, D) is essentially self-adjoint, then its closure $(\overline{S}, \overline{D})$ as in Definition D.1.2 is self-adjoint, and one can apply the spectral theorem to this operator. We will sometimes apply the functional calculus to essentially self-adjoint operators without further comment: technically, what we are doing is applying the functional calculus to the self-adjoint closure as above.

We close this section with a useful lemma that follows from uniqueness of the functional calculus.

[1] So far the proof just uses formal self-adjointness; the next part uses self-adjointness.

Lemma D.1.9 *Let (S, D) be an essentially self-adjoint operator on H. Let f be a bounded Borel function, and let U be a unitary operator such that $U \cdot S = S$. Then $Uf(D)U^* = f(UDU^*)$.* □

D.2 Some Fourier Theory for Unbounded Operators

In this section, we collect together some useful consequences of the functional calculus. All are well explained in the literature, so we just state the results, and collect references at the end.

The basic fact here is a version of the *Stone–von Neumann theorem*, which we split into two parts.

Proposition D.2.1 *Let (S, D) be a self-adjoint operator on H. Then for any $u \in S$, the functions $\mathbb{R} \to H$ defined by*

$$t \mapsto e^{itD}u, \quad t \mapsto \cos(tD)u, \quad \text{and} \quad t \mapsto \sin(tD)u$$

take values in S, are smooth for the norm topology on H and their t-derivatives are given by

$$t \mapsto ie^{itD}Du, \quad t \mapsto -\sin(tD)Du, \quad \text{and} \quad t \mapsto \cos(tD)Du$$

respectively. □

Theorem D.2.2 *Conversely to Proposition D.2.1, let $(V_t)_{t\in\mathbb{R}}$ be a strongly continuous map from $\mathbb{R}$ to the unitary group of some Hilbert space H. Then there exists a self-adjoint operator (S, D) such that $e^{itD} = V_t$ for all $t \in \mathbb{R}$.*

Moreover, if S_0 is a dense subspace of H which is invariant under V_t and is such that for each $u \in S_0$ the map

$$\mathbb{R} \to H, \quad t \mapsto V_t u$$

is differentiable, then the operator $D_0 \colon S_0 \to H$ defined by

$$D_0 \colon u \mapsto \lim_{t\to 0} \frac{1}{i} \frac{V_t u - u}{t}$$

is essentially self-adjoint, and its closure equals the operator D above. □

The following proposition can be thought of as a version of the Fourier inversion theorem for unbounded operators.

Proposition D.2.3 *Let (S, D) be a self-adjoint operator on H, and f a Schwartz-class function on $\mathbb{R}$. Then the integral*

$$\frac{1}{2\pi} \int_{\mathbb{R}} \widehat{f}(t) e^{itD} dt$$

makes sense as a Riemann sum, and is equal to $f(D)$.

Moreover, if f is a bounded Borel function on $\mathbb{R}$ with compactly supported distributional Fourier transform, then the identity

$$f(D) = \frac{1}{2\pi}\int_{\mathbb{R}} \widehat{f}(t)e^{itD}dt$$

still holds, where now it is interpreted to mean that for every $u, v \in S$, the pairing of the distribution $\widehat{f}$ with the smooth function

$$t \mapsto \frac{1}{2\pi}\langle e^{itD}u, v\rangle$$

is equal to $\langle f(D)u, v\rangle$. □

We need one more technical result about unbounded operators whose difference is bounded.

Proposition D.2.4 *Let (S, D_1) and (S, D_2) be essentially self-adjoint operators on a Hilbert space, with the same domain; moreover, assume that*

$$D_1^n \cdot S \subseteq S \quad \text{and} \quad D_2^n \cdot S \subseteq S$$

for all n, and that $D_1 - D_2$ is a bounded operator. Let f be a bounded Borel function such that the distributional Fourier transform $\widehat{f}$ is compactly supported, and so that the function $\xi \mapsto \xi\widehat{f}(\xi)$ is smooth. Define

$$c := \frac{1}{2\pi}\int_{\mathbb{R}} |\xi\widehat{f}(\xi)|d\xi.$$

Then

$$\|f(D_1) - f(D_2)\| \leqslant c\|D_1 - D_2\|.$$ □

D.3 The Harmonic Oscillator and Mehler's Formula

This section is quite different in character from the previous two. Those sections work in broad generality and use fairly soft functional analysis techniques; here, on the other hand, we discuss some aspects of a particularly important operator in concrete detail.

Definition D.3.1 The *Harmonic oscillator* is the unbounded operator

$$H := -\frac{d^2}{dx^2} + x^2 - 1$$

on $L^2(\mathbb{R})$ with domain the Schwartz-class functions.

Define $A := x + \frac{d}{dx}$ and $A^* := x - \frac{d}{dx}$, considered as bounded operators on $L^2(\mathbb{R})$ with domain the Schwartz-class functions; A and A^* are traditionally called the *annihilation* and *creation* operators respectively. Note that integration by parts implies the relation $\langle Au, v\rangle = \langle u, A^*v\rangle$ for Schwartz-class u and v partly[2] justifying the notation. Note also that we have the relations

$$H = A^*A = AA^* - 2 \tag{D.3}$$

as operators on the Schwartz-class functions.

Define now $\psi_0(x) = \pi^{-1/4}e^{-x^2/2}$, and for $k \geqslant 0$, set

$$\psi_k = \frac{1}{\sqrt{2^k k!}}(A^*)^k\psi_0 \quad \text{and} \quad h_k(x) := \psi_k(x)e^{x^2/2}.$$

We will see below that each h_k is a polynomial, whence each ψ_k is Schwartz-class, and therefore in particular contained in $L^2(\mathbb{R})$.

Lemma D.3.2 *We have the following identities:*

(i) For all $k \geqslant 1$, $\sqrt{2k}h_k(x) = 2xh_{k-1}(x) - h_k'(x)$.
(ii) For all $k \geqslant 0$, $\sqrt{2^k k!}h_k(x) = \pi^{-1/4}(-1)^k e^{x^2}\frac{d^k}{dx^k}(e^{-x^2})$.
(iii) For all $s \in \mathbb{C}$,

$$exp(2sx - s^2) = \pi^{1/4}\sum_{k=0}^{\infty}\sqrt{\frac{2^k}{k!}}s^k h_k(x).$$

Proof For part (i), note that by definition of ψ_k we have that

$$\sqrt{2k}\psi_k = (A^*\psi_{k-1})$$

whence by definition of h_k we get

$$\sqrt{2k}h_k(x)e^{-x^2/2} = xh_{k-1}(x)e^{-x^2/2} - h_{k-1}'(x)e^{-x^2/2} + h_{k-1}(x)e^{-x^2/2},$$

so cancelling the $e^{-x^2/2}$ factors gives the result. For part (ii), note that

$$h_0(x) = \psi_0(x)e^{x^2/2} = \pi^{-1/4},$$

whence the result holds for $k = 0$. For the general case, proceed by induction on k. We have by inductive hypothesis that

$$\begin{aligned}\pi^{-1/4}(-1)^{k+1}\frac{d^{k+1}}{dx^{k+1}}e^{-x^2} &= -\sqrt{2^k k!}\frac{d}{dx}\left(e^{-x^2}h_k(x)\right)\\ &= -\sqrt{2^k k!}e^{-x^2}(2xh_k(x) - h_k'(x)).\end{aligned}$$

Applying part (i), we therefore get

$$\pi^{-1/4}(-1)^{k+1}\frac{d^{k+1}}{dx^{k+1}}e^{-x^2} = \sqrt{2^k k!}e^{-x^2}\sqrt{2(k+1)}h_{k+1}(x),$$

[2] Note, however, that there is a notion of adjoint in unbounded operator theory, and that A^* is not the adjoint of A in this sense as its domain is wrong. Notationally speaking, we are thus doing something a little non-standard here.

which gives the result on rearranging. Finally, for part (iii) consider the Taylor expansion of the analytic function e^{-z^2} for $z = s - x$ centred at x to get

$$\exp(2sx - s^2 - x^2) = \exp(-(x-s)^2) = \sum_{k=0}^{\infty} \frac{d^k}{dz^k}(e^{-z^2})\Big|_{z=x} \frac{(x-s-x)^k}{k!}.$$

Hence applying part (ii)

$$\exp(2sx - s^2) = \sum_{k=0}^{\infty} e^{x^2} \frac{d^k}{dx^k}(e^{-x^2})(-1)^k \frac{s^k}{k!} = \pi^{1/4} \sum_{k=0}^{\infty} \sqrt{2^k k!} h_k(x) \frac{s^k}{k!},$$

which implies the result. □

Proposition D.3.3 *The collection $(\psi_k)_{k=0}^{\infty}$ is an orthonormal basis for $L^2(\mathbb{R})$ consisting of eigenvectors for H with associated eigenvalue $2k$.*

Proof We show by induction that $H\psi_k = 2k\psi_k$. For $k = 0$, this is a direct computation. Assuming $H\psi_k = 2k\psi_k$, we have, using line (D.3), that

$$\begin{aligned} H\psi_{k+1} &= \frac{1}{\sqrt{2(k+1)}} A^*AA^*\psi_k = \frac{1}{\sqrt{2(k+1)}} A^*(H+2)\psi_k \\ &= \frac{2k+2}{\sqrt{2(k+1)}} A^*\psi_k = 2(k+1)\psi_{k+1}. \end{aligned}$$

Orthogonality of the ψ_k now follows as they are eigenvectors of a (formally) self-adjoint operator for distinct eigenvalues. We next show that each ψ_k has norm one by induction. Indeed, for ψ_0 this is a direct computation. Assuming $\|\psi_k\| = 1$, we see that

$$\begin{aligned} \|\psi_{k+1}\|^2 &= \frac{1}{2(k+1)} \langle A^*\psi_k, A^*\psi_j \rangle = \frac{1}{2(k+1)} \langle AA^*\psi_k, \psi_k \rangle \\ &= \frac{1}{2(k+1)} \langle (H+2)\psi_k, \psi_k \rangle = \frac{2k+2}{2k+2} \langle \psi_k, \psi_k \rangle = 1. \end{aligned}$$

It remains to show that $(\psi_k)_{k=0}^{\infty}$ has dense span. Note first that $h_0(x) = \pi^{-1/4}$, and induction on k and Lemma D.3.2 part (i) imply that h_k is a polynomial of degree k. It thus suffices to show that if $\phi_k(x) = x^k e^{-x^2/2}$ then the sequence $(\phi_k)_{k=0}^{\infty}$ has dense span. We have

$$\|\phi_k\|^2 = \int_{\mathbb{R}} x^{2k} e^{-x^2} dx = 2 \int_0^{\infty} x^{2k} e^{-x^2} dx.$$

Setting $y = x^2$, this equals $\int_0^{\infty} y^{k+1/2} e^{-y} dy$, and induction on k (or comparison to the Γ function) shows that this is bounded above by $k!$. Hence we have L^2-norm convergence of the series

$$\sum_{k=0}^{\infty} \phi_k(x) \frac{(i\lambda x)^k}{k!}$$

to the function $e^{i\lambda x - x^2/2}$ for any $\lambda \in \mathbb{R}$. It follows that if $u \in L^2(\mathbb{R})$ is orthogonal to all of the ϕ_k, then

$$\int_{\mathbb{R}} u(x) e^{-x^2/2} e^{-ix\lambda} dx = 0$$

for all $\lambda \in \mathbb{R}$. In other words, the Fourier transform of $u(x)e^{-x^2/2}$ is zero. The Plancherel formula thus forces $u(x)e^{-x^2/2}$ to be zero almost everywhere. We therefore get that u is zero almost everywhere, completing the proof. □

Lemma D.3.4 *For any $s \in \mathbb{C}$, we have*

$$\sum_{k=0}^{\infty} \frac{s^k}{\sqrt{k!\,2^k}} \psi_k(x) = e^{sx - \frac{1}{4}s^2} \psi_0(x)$$

as functions in $L^2(\mathbb{R})$ (i.e. the convergence of the series of functions on the left to the function on the right is in L^2-norm).

Proof As $(\psi_k)_{k=0}^{\infty}$ is an orthonormal basis for $L^2(\mathbb{R})$ and as the sequence $(\frac{s^k}{k!2^k})_{k=0}^{\infty}$ is square summable for any $s \in \mathbb{C}$, the series on the left-hand side converges in $L^2(\mathbb{R})$. To prove the given equality, it therefore suffices to show that we have pointwise convergence of the left-hand side to the right-hand side. Multiplying the identity in part (iii) of Lemma D.3.2 by $e^{-x^2/2}$ gives

$$\exp(2sx - s^2)\psi_0(x) = \sum_{k=0}^{\infty} \sqrt{\frac{2^k}{k!}} s^k \psi_k(x).$$

Replacing s by $s/2$ in the above gives the claimed identity. □

Consider now the operator e^{-tH} for $t \geqslant 0$.

Lemma D.3.5 *For $s \in \mathbb{R}$, set $f_s(x) = e^{isx + \frac{1}{4}s^2} \psi_0(x)$. Then for any $t \geqslant 0$,*

$$e^{-tH} f_s = f_{e^{-2t}s}.$$

Proof Using Lemma D.3.4 and that e^{-tH} is a bounded operator on $L^2(\mathbb{R})$, we have that

$$(e^{-tH} f_s) = \sum_{k=0}^{\infty} \frac{(is)^k}{\sqrt{k!\,2^k}} e^{-tH} \psi_k.$$

As ψ_k is an eigenvector for H with eigenvalue $2k$, this equals

$$\pi^{1/4} \sum_{k=0}^{\infty} \frac{(ie^{-2t}s)^k}{\sqrt{k!\,2^k}} \psi_k = f_{e^{-2t}s}$$

as claimed. □

Finally, we are ready for *Mehler's formula*, which is the main result of this section.

Theorem D.3.6 *For any $t > 0$ and $u \in L^2(\mathbb{R})$,*

$$(e^{-tH}u)(x) = \int_{\mathbb{R}} k_t(x,y)u(y)dy,$$

where

$$k_t(x,y) = \pi^{-1/2}(1-e^{-4t})^{-1/2}\exp\left(-\frac{\frac{1}{2}(1+e^{-4t})(x^2+y^2) - 2e^{-2t}xy}{(1-e^{-4t})}\right).$$

Proof For any $t > 0$, e^{-tH} has eigenvalues $\{e^{-2tn} \mid n \in \mathbb{N}\}$ with all associated eigenspaces being one-dimensional. The sequence $(e^{-2tn})_{n=0}^{\infty}$ is square-summable, so e^{-tH} is Hilbert–Schmidt. Hence there is a kernel $k_t \in L^2(\mathbb{R} \times \mathbb{R})$ with

$$(e^{-tH}u)(x) = \int_{\mathbb{R}} k_t(x,y)u(y)dy$$

for all $u \in L^2(\mathbb{R})$. Set now $l_t(x,y) = \psi_0(x)^{-1}k_t(x,y)\psi_0(y)$; this makes sense as a measurable function on $\mathbb{R}^2$, although may not be in $L^2(\mathbb{R}^2)$ any more. For fixed $s \in \mathbb{R}$, let $f_s \in L^2(\mathbb{R})$ be as in Lemma D.3.5, and note that the result of that lemma and the definitions of the kernels above give that

$$\begin{aligned}
\exp(ise^{-2t}x + \frac{1}{4}e^{-4t}s^2)\psi_0(x) &= f_{e^{-2t}s}(x) = (e^{-tH}f_s)(x) \\
&= \int_{\mathbb{R}} k_t(x,y)e^{isx+\frac{1}{4}s^2}\psi_0(y)dy \\
&= \psi_0(x)e^{\frac{1}{4}s^2}\int_{\mathbb{R}} l_t(x,y)e^{isx}dy.
\end{aligned}$$

Looking at both the first and last terms, cancelling $\psi_0(x)$ and rearranging gives that

$$\int_{\mathbb{R}} l_t(x,y)e^{isx}dy = \exp(ise^{-2t}x)\exp(-\frac{1}{4}s^2(1-e^{-4t})).$$

It follows from Fourier theory (noting that l_t is integrable in the y variable) that

$$\begin{aligned}
l_t(x,y) &= \frac{1}{2\pi}\int_{\mathbb{R}} \exp\left(-\frac{1}{4}s^2(1-e^{-4t})\right)\exp(ise^{-2t}x)e^{-isy}ds \\
&= \frac{1}{2\pi}\int_{\mathbb{R}} \exp\left(-\frac{1}{4}s^2(1-e^{-4t})\right)e^{-is(y-e^{-2t}x)}ds.
\end{aligned}$$

This is the Fourier transform of a Gaussian, so is explicitly computable. Indeed, the Fourier transform of the Gaussian $\exp(-\frac{1}{4}s^2(1-e^{-4t}))$ is

$$\xi \mapsto \sqrt{\frac{4\pi}{1-e^{-4t}}}\exp\left(\frac{-\xi^2}{1-e^{-4t}}\right).$$

The above expression is equal to this evaluated at $\xi = y - e^{-2t}x$ (and multiplied by $1/2\pi$), so we get

$$l_t(x,y) = \frac{1}{\sqrt{\pi(1-e^{-4t})}}\exp\left(\frac{-(y-e^{-2t}x)^2}{1-e^{-4t}}\right).$$

Finally, by definition of l_t, $k_t(x,y) = \psi_0(x) l_t(x,y)\psi_(y)^{-1}$, and so

$$k_t(x,y) = \frac{1}{\sqrt{\pi(1-e^{-4t})}} e^{-x^2/2} \exp\left(\frac{-(y-e^{-2t}x)^2}{1-e^{-4t}}\right) e^{y^2/2}$$

which in turn equals

$$\frac{1}{\sqrt{\pi(1-e^{-4t})}} \exp\left(\frac{-\frac{1}{2}x^2(1-e^{-4t}) - y^2 + 2e^{-2t}xy - e^{-4t}x^2 + \frac{1}{2}y^2(1-e^{-4t})}{1-e^{-4t}}\right).$$

Simplifying, we get the formula in the statement and are done. □

We can rewrite the formula above in a version that will be more convenient for us.

Corollary D.3.7 *For any* $t > 0$,

$$e^{-tH} = e^t e^{-\alpha x^2} e^{-\beta \frac{d^2}{dx^2}} e^{-\alpha x^2},$$

where

$$\alpha = \alpha(t) := \frac{\cosh(2t) - 1}{2\sinh(2t)} \quad \textit{and} \quad \beta = \beta(t) := \frac{\sinh(2t)}{2}.$$

Proof We rearrange the expression

$$-\frac{\frac{1}{2}(1+e^{-4t})(x^2+y^2) - 2e^{-2t}xy}{(1-e^{-4t})}$$

appearing in the formula for k_t in Theorem D.3.6 to get

$$-\frac{e^{-2t}(x-y)^2 + \frac{1}{2}e^{-2t}(e^{2t}+e^{-2t})(x^2+y^2) - e^{-2t}(x^2+y^2)}{e^{-2t}(e^{2t}-e^{-2t})}$$

$$= -\frac{(x-y)^2 + \left(\frac{1}{2}(e^{-2t}+e^{2t}) + 1\right)(x^2+y^2)}{e^{2t}-e^{-2t}}$$

$$= -\frac{(x-y)^2 + (\cosh(2t)+1)(x^2+y^2)}{2\sinh(2t)}.$$

Hence

$$k_t(x,y) = \pi^{-1/2}(1-e^{-4t})^{-1/2} \exp(-\alpha x^2) \exp\left(-\frac{1}{4\beta}(x-y)^2\right) \exp(-\alpha y^2).$$

Thus if we can show that

$$\pi^{-1/2}(1-e^{-4t})^{-1/2} \exp\left(-\frac{1}{4\beta}(x-y)^2\right)$$

is the integral kernel for the operator $e^t e^{-\beta \frac{d^2}{dx^2}}$ we will be done: indeed, in that case k_t will be identified with the integral kernel for the operator on the right-hand side in the formula for the statement and Theorem D.3.6 already identifies it with the integral kernel for the operator on the left-hand side.

For this, we need to be precise about Fourier transform conventions. We will use the (unitary) Fourier transform

$$U\colon L^2(\mathbb{R}) \to L^2(\mathbb{R}), \quad (Uu)(\xi) := \frac{1}{2\pi}\int_{\mathbb{R}} u(x)e^{-ix\xi}\,dx$$

with inverse given by

$$(U^*u)(x) := \frac{1}{2\pi}\int_{\mathbb{R}} u(\xi)e^{ix\xi}\,d\xi.$$

Let $f(x) = e^{-\frac{1}{4\beta}x^2}$ and let $C_f\colon L^2(\mathbb{R}) \to L^2(\mathbb{R})$ be the operator of convolution by f (which is bounded as f is integrable). We are trying to show the identity

$$\pi^{-1/2}(1-e^{-4t})^{-1/2}C_f = e^{-2t}e^{-\beta\frac{d^2}{dx^2}} \tag{D.4}$$

of operators on $L^2(\mathbb{R})$. We will first compute C_f in terms of $\frac{d^2}{dx^2}$. A standard computation shows that if M_g is the operator of multiplication by a bounded function f, then

$$UC_fU^* = \sqrt{2\pi}M_{\widehat{f}},$$

where $\widehat{f}$ is the Fourier transform of f, i.e. the function Uf. More standard computations with Gaussians (the authors suggest looking at a table of Fourier transforms – the one on Wikipedia will be sufficient for this at the time of this writing) show that

$$\widehat{f}(x) = \sqrt{2\beta}e^{-\beta x^2}.$$

Letting M_{x^2} be the self-adjoint unbounded operator on $L^2(\mathbb{R})$ with domain the Schwartz-class functions, we can think of $M_{\widehat{f}}$ as the operator $\sqrt{2\beta}e^{-\beta M_{x^2}}$ defined using the functional calculus. Hence by our computations so far and naturality of the functional calculus (Lemma D.1.9).

$$C_f = U^*\sqrt{2\pi}\sqrt{2\beta}e^{-\beta M_{x^2}}U = 2\sqrt{\pi\beta}e^{-\beta U^*M_{x^2}U}.$$

On the other hand, one computes that $U^*M_{x^2}U = \frac{d^2}{dx^2}$ and so we get that

$$C_f = \sqrt{2\pi\sinh(2t)}e^{-\beta\frac{d^2}{dx^2}}.$$

Hence

$$\pi^{-1/2}(1-e^{-4t})^{-1/2}C_f = \sqrt{\frac{2\pi\sinh(2t)}{\pi e^{-2t}2\sinh(2t)}}e^{-\beta\frac{d^2}{dx^2}},$$

which is exactly our desired formula from line (D.4). □

D.4 Notes and References

The material in the first two sections of this appendix is based on chapter VIII of Reed and Simon [211] and chapter 10 of Higson and Roe [135], with the former, in particular, being a good general introduction to unbounded operator theory. The functional calculus (Theorem D.1.7) , including the uniqueness statement used to prove Lemma D.1.9 can be found in theorem VIII.5 of Reed and Simon [211].

For a proof of what we called the Stone–von Neumann theorem (Proposition D.2.1 and Theorem D.2.2), see theorems VIII.7, VIII.8 and VIII.10 of Reed and Simon [211]. For proofs of Propositions D.2.3 and D.2.4, see propositions 10.3.5 and 10.3.7, respectively, of Higson and Roe [135].

The material in Section D.3 is based on chapter 9 of Roe [217] and section 12.9 of Cycon, Froese, Kirsch and Simon [72]. In particular, our proof of Mehler's formula is based on page 292 of Cycon, Froese, Kirsch and Simon [72].

APPENDIX E

Gradings

In this appendix, we summarise some facts about gradings on C^*-algebras and Hilbert spaces. This provides a very convenient language for some of the contents of the book. Section E.1 discusses definitions and basic examples of graded Hilbert spaces and C^*-algebras, while Section E.2 discusses graded tensor products.

E.1 Graded C^*-algebras and Hilbert Spaces

Definition E.1.1 Let A be a $*$-algebra. A *grading* on A is a $*$-automorphism $\epsilon: A \to A$ that satisfies $\epsilon^2 = \mathrm{id}$. A $*$-algebra (or C^*-algebra) equipped with a grading is called *a graded $*$-algebra* (or *graded C^*-algebra*).

If A is a C^*-algebra and there exists a self-adjoint unitary u in the multiplier algebra of A such that $\epsilon(a) = uau^*$ for all $a \in A$, then the grading is said to be *inner*.

We will often just say something like 'let A be a graded C^*-algebra', leaving the automorphism ϵ implicit. We will also often write ϵ_A for the grading on a graded $*$-algebra A.

Definition E.1.2 If A and B are graded $*$-algebras, then a $*$-homomorphism $\phi: A \to B$ is *graded* if $\phi(\epsilon_A(a)) = \epsilon_B(\phi(a))$.

Remark E.1.3 A grading on a $*$-algebra A as defined is the same thing as an action of $\mathbb{Z}/2\mathbb{Z}$ on A, and graded $*$-homomorphisms are the same thing as equivariant $*$-homomorphisms.

For computations, it will be convenient to have a notion of 'spatially induced' grading. Partly for this reason, we formalise the idea of a graded Hilbert space.

Definition E.1.4 A *graded Hilbert space* is a Hilbert space H equipped with a self-adjoint unitary operator U, called the *grading operator*. If A is a graded C^*-algebra and H a graded Hilbert space, then a representation $\pi : A \to \mathcal{B}(H)$ is *graded* if

$$\pi(\epsilon_A(a)) = U\pi(a)U^*$$

for all $a \in A$.

Remark E.1.5 Let (H, U) be a graded Hilbert space. As U is a self-adjoint operator satisfying $U^2 = 1$, H admits an orthogonal decomposition $H = H_0 \oplus H_1$ with H_i the eigenspace for the eigenvalue $(-1)^i$ of U (possibly $H_i = \{0\}$). Conversely, any orthogonal decomposition of a Hilbert space $H = H_0 \oplus H_1$ gives rise to a grading U on H: one stipulates that U acts as multiplication by $(-1)^i$ on H_i. In conclusion: gradings on Hilbert spaces are the same thing as decompositions into a pair of orthogonal subspaces. We will use either description below as convenient.

Lemma E.1.6 *Any graded C^*-algebra has a faithful graded representation.*

Proof As noted in Remark E.1.3, a grading on A is the same thing as a $\mathbb{Z}/2\mathbb{Z}$-action. Checking definitions, this lemma is just the special case of Lemma C.1.9 where the acting group is $\mathbb{Z}/2\mathbb{Z}$. □

We now turn to some examples.

Example E.1.7 Any C^*-algebra can be equipped with the *trivial* grading where ϵ is the identity. This is inner.

Example E.1.8 If A is a graded C^*-algebra, then there is a (unique) extension to a grading on the unitisation A^+ defined by

$$\epsilon_{A^+} : (a, \lambda) \mapsto (\epsilon_A(a), \lambda).$$

Example E.1.9 Let (H, U) be a graded Hilbert space. Then conjugation by U induces gradings on $\mathcal{B}(H)$ and $\mathcal{K}(H)$. As any $U \in \mathcal{B}(H)$ is in the multiplier algebra of both $\mathcal{B}(H)$ and $\mathcal{K}(H)$, these gradings are always inner.

If in the associated decomposition $H = H_0 \oplus H_1$, the subspaces H_0 and H_1 are isomorphic (i.e. have the same dimension) then this is called the *standard grading* on $\mathcal{K}(H)$. We allow the case when H is finite (necessarily even)-dimensional, in which case we get the *standard grading* on $M_{\dim(H)}(\mathbb{C})$.

The case when H is separable and infinite-dimensional is important enough that we introduce special notation for it: we write $\mathscr{K}$ for $\mathcal{K}(H)$ equipped with the standard grading.

Example E.1.10 Let $C_0(\mathbb{R})$ be equipped with the grading defined by $(\epsilon(f))(t) = f(-t)$. This is not inner, as one can see using the identification $M(C_0(\mathbb{R})) = C_b(\mathbb{R})$ of Example 1.7.7. This example is again important enough to merit its own notation: $\mathscr{S}$ denotes $C_0(\mathbb{R})$ equipped with this grading.

Note that the multiplication action of $\mathscr{S}$ on $L^2(\mathbb{R})$ is a graded representation with respect to the grading operator defined by $(Uu)(x) := u(-x)$.

Example E.1.11 Let V be a vector space over $\mathbb{R}$ equipped with an inner product. The *Clifford algebra of* V, denoted $\text{Cliff}_{\mathbb{C}}(V)$, is the unital complex algebra generated by an $\mathbb{R}$-linear copy of V and subject to the relations

$$vv = \|v\|^2$$

for all $v \in V$ (in words, 'v times v equals the norm of v squared times the identity'); one way to make this precise is sketched in Exercise E.3.2. Stipulating that each $v \in V$ is self-adjoint determines a $*$-operation on $\text{Cliff}_{\mathbb{C}}(V)$, noting that this is compatible with the relation above. The map $v \mapsto -v$ on V also preserves these relations, and thus extends to a $\mathbb{Z}/2\mathbb{Z}$ action on $\text{Cliff}_{\mathbb{C}}(V)$, making $\text{Cliff}_{\mathbb{C}}(V)$ a graded $*$-algebra.

As a concrete example let $V = \mathbb{R}$ and choose a norm one element e of $\mathbb{R}$. Then one computes that $\text{Cliff}_{\mathbb{C}}(\mathbb{R}) = \{z + we \mid z, w \in \mathbb{C}\}$ with $*$-algebra operations given by

$$(z + we)^* = \overline{z} + \overline{w}e, \quad (z_1 + w_1 e)(z_2 + w_2 e) = (z_1 z_2 + w_1 w_2) + (z_1 w_2 + w_2 z_1)e.$$

It is graded by the $*$-automorphism $z + we \mapsto z - we$. Hence as a $*$-algebra, $\text{Cliff}_{\mathbb{C}}(\mathbb{R})$ identifies with $\mathbb{C} \oplus \mathbb{C}$ via the $*$-isomorphism

$$\mathbb{C} \oplus \mathbb{C} \ni (z, w) \mapsto \frac{1}{2}(z + w) + \frac{1}{2}(z - w)e \in \text{Cliff}_{\mathbb{C}}(\mathbb{R}).$$

Under this isomorphism, the grading on $\text{Cliff}_{\mathbb{C}}(\mathbb{R})$ corresponds to the flip automorphism $(z, w) \mapsto (w, z)$, which is clearly not inner.

Having explored these examples, let us introduce some more terminology. This will be useful for certain concrete formulas and computations.

Definition E.1.12 Let A be a graded $*$-algebra. An element $a \in A$ is called *homogeneous* if $\epsilon_A(a) = a$ or $\epsilon_A(a) = -a$. If $a \in A$ is homogeneous, its *degree* is the number $\partial a \in \{0, 1\}$ defined by $\epsilon_A(a) = (-1)^{\partial a} a$.

It is common in the literature to call degree zero elements *even* and degree one elements *odd*; we will use both terminologies.

Note that as a grading is an order two automorphism of a $*$-algebra A, its eigenvalues are a subset of $\{1, -1\}$. Hence A splits as a direct sum of eigenspaces $A = A^{(0)} \oplus A^{(1)}$, with $A^{(i)}$ consisting of eigenvectors for the eigenvalue $(-1)^i$, or in other words elements of degree i. In particular, any element of A can be written uniquely as a sum of a degree zero and a degree one element; thus to determine an operation on a graded $*$-algebra, it often suffices to define it for homogeneous elements. We will use this fact without comment from now on.

Example E.1.13 Let (H, U) be a graded Hilbert space with associated decomposition $H = H_0 \oplus H_1$ as in Remark E.1.5. Then if $\mathcal{B}(H)$ has the associated inner grading from Example E.1.9, an operator on H is even (respectively, odd) if and only if when it is written as a 2×2 matrix with respect to the decomposition $H = H_0 \oplus H_1$ it is diagonal, i.e. of the form $\begin{pmatrix} T & 0 \\ 0 & S \end{pmatrix}$ (respectively, off-diagonal, i.e. of the form $\begin{pmatrix} 0 & T \\ S & 0 \end{pmatrix}$). The same descriptions apply to elements of any graded C^*-algebra that is faithfully realised as operators on H via a graded representation.

Example E.1.14 We will also use the even / odd terminology for an unbounded operator (S, D) on a graded Hilbert space (H, U) (see Section D.1 for conventions on unbounded operators). In this case, U should be assumed to preserve the domain S of the operator, whence it makes sense to consider $UDU : S \to H$. The operator D is *even* if $UDU = D$, and *odd* if $UDU = -D$. An illustrative example is the unbounded operator $D = i\frac{d}{dx}$ acting on $L^2(\mathbb{R})$ with domain $C_c^\infty(\mathbb{R})$, and with the grading defined by $(Uu)(x) := u(-x)$. This operator D is odd.

Analogously to Example E.1.13, the domain S splits into a direct sum of ± 1-eigenspaces for U as for H itself. Whether the operator is even / odd can again be thought of in terms of on / off diagonal matrices.

Example E.1.15 If $\mathscr{S}$ is as in Example E.1.10 then even and odd have their usual meanings for functions on $\mathbb{R}$.

Example E.1.16 If $\text{Cliff}_{\mathbb{C}}(V)$ is as is Example E.1.11, then an element is even if it can be written as a sum of products of an even number of elements of V, and odd if it can be written as a sum of products of an odd number of elements.

E.2 Graded Tensor Products

In this section, we discuss graded tensor products. First, we look at the purely algebraic theory, and then the spatial and maximal versions in turn.

Definition E.2.1 Let A and B be graded $*$-algebras. The *graded algebraic tensor product* $A \widehat{\odot} B$ of A and B is the algebraic tensor product over $\mathbb{C}$ equipped with the $*$-operation, multiplication and grading defined[1] on elementary tensors of homogeneous elements by the formulas

$$(a \widehat{\otimes} b)^* := (-1)^{\partial a \partial b} a^* \widehat{\otimes} b^*,$$

$$(a_1 \widehat{\otimes} b_1)(a_2 \widehat{\otimes} b_2) := (-1)^{\partial a_2 \partial b_1} a_1 a_2 \widehat{\otimes} b_1 b_2,$$

and

$$\epsilon_{A \widehat{\otimes} B}(a \widehat{\otimes} b) := \epsilon_A(a) \widehat{\otimes} \epsilon_B(b).$$

Remark E.2.2 A heuristic for the formulas above (and others in this section) is that the sign $(-1)^{\partial a \partial b}$ should be introduced whenever homogeneous elements a and b have to be moved past each other.

Remark E.2.3 The graded algebraic tensor product is both commutative and associative up to canonical isomorphism. However, there is a subtlety: the isomorphisms must be defined using the heuristic from Remark E.2.2. They are determined on elementary tensors of homogeneous elements by

$$A \widehat{\odot} B \to B \widehat{\odot} A, \quad a \widehat{\otimes} b \mapsto (-1)^{\partial a \partial b} b \widehat{\otimes} a$$

and

$$(A \widehat{\odot} B) \widehat{\odot} C \to A \widehat{\odot} (B \widehat{\odot} C), \quad (a \widehat{\otimes} b) \widehat{\otimes} c \mapsto a \widehat{\otimes} (b \widehat{\otimes} c).$$

Remark E.2.4 If A and B are $*$-algebras, recall from Remark 1.8.1 that the algebraic tensor product $*$-algebra $A \odot B$ has the following universal property: if $\phi \colon A \to C$ and $\psi \colon B \to C$ are $*$-homomorphisms with commuting images, then there is a unique $*$-homomorphism $\phi \otimes \psi \colon A \odot B \to C$ taking $a \otimes b$ to $\phi(a)\psi(b)$.

The graded tensor product $\widehat{\odot}$ has an analogous universal property, which we now describe. Define the *graded commutator* of homogeneous elements a, b in a graded $*$-algebra by

$$[a, b]_g := ab - (-1)^{\partial a \partial b} ba,$$

[1] We leave it as an exercise for the reader to show that the formulas below really do give a well-defined graded $*$-algebra structure.

and extend this to all elements by linearity. Let $\phi\colon A \to C$ and $\psi\colon B \to C$ be graded $*$-homomorphisms such that $[\phi(a), \psi(b)]_g = 0$ for all $a \in A$ and $b \in B$. Then one can check that the formula

$$a \widehat{\otimes} b \mapsto \phi(a)\psi(b)$$

uniquely determines a graded $*$-homomorphism $\phi \widehat{\otimes} \psi\colon A \widehat{\odot} B \to C$.

We will also need graded Hilbert spaces and operators on them.

Definition E.2.5 Let (H_1, U_1) and (H_1, U_1) be graded Hilbert spaces. Define the graded tensor product of H_1 and H_2, denoted $H_1 \widehat{\otimes} H_2$, to be the same underlying Hilbert space as the usual tensor product $H_1 \otimes H_2$ (see Definition 1.8.4), equipped with the grading operator $U_1 \otimes U_2$. We usually write $v \widehat{\otimes} w$ for elementary tensors in $H_1 \widehat{\otimes} H_2$ to remind us that we are in a graded setting (compare Remark E.2.6 below).

If $T_1 \in \mathcal{B}(H_1)$ are $T_2 \in \mathcal{B}(H_2)$ are bounded operators that are homogeneous for the respective gradings, then we define the operator $T_1 \widehat{\otimes} T_2$ by the formula

$$T_1 \widehat{\otimes} T_2 := T_1 U^{\partial T_2} \otimes T_2,$$

where the tensor product of operators on the right is as in Lemma 1.8.6. For non-homogeneous T_1 and T_2, we define $T_1 \widehat{\otimes} T_2$ by extending the above formula by linearity.

For $i \in \{1,2\}$, let D_i be an unbounded operator on H_i with domain S_i that is invariant under U_i as in Example E.1.8. Define $D_1 \widehat{\otimes} D_2$ to be the unbounded operator on $H_1 \widehat{\otimes} H_2$ with domain $S_1 \odot S_2$ given by the formula

$$D_1 \widehat{\otimes} D_2\colon v_1 \widehat{\otimes} v_2 \mapsto D_1 U_1^{\partial D_2} v_1 \widehat{\otimes} D_2 v_2$$

on elementary tensors. We allow the case that one[2] of D_1 and D_2 is bounded, in which case the domain of the bounded operator should just be taken to be the whole Hilbert space.

Remark E.2.6 It might help the reader's intuition to compare the formula for $T_1 \widehat{\otimes} T_2$ above with the heuristic in Remark E.2.2. Indeed, for a graded Hilbert space $H = H_0 \oplus H_1$, say v is homogeneous if v is in H_i for some i, and in this case set $\partial v = i$. The formula for $T_1 \widehat{\otimes} T_2$ is equivalent to saying that if v_1 and v_2 are homogeneous, then

$$(T_1 \widehat{\otimes} T_2)(v_1 \widehat{\otimes} v_2) = (-1)^{\partial T_2 \partial v_1} T_1 v_1 \widehat{\otimes} T_2 v_2.$$

Thus one introduces the sign $(-1)^{\partial T_2 \partial v_1}$ as the 'price' for moving T_2 past v_1.

The Spatial Graded Tensor Product

Say that $\pi_A\colon A \to \mathcal{B}(H_A)$ and $\pi_B\colon B \to \mathcal{B}(H_A)$ are graded representations, and consider the map defined by

$$\pi_A \widehat{\otimes} 1\colon A \to \mathcal{B}(H_A \widehat{\otimes} H_B), \quad a \mapsto \pi_A(a) \widehat{\otimes} 1$$

[2] If both D_1 and D_2 are bounded, the definition here does not quite agree with that for $T_1 \widehat{\otimes} T_2$ above, as the domain of $D_1 \widehat{\otimes} D_2$ would be $H_1 \odot H_2$; for two bounded operators, we always use the definition with domain all of $H_1 \widehat{\otimes} H_2$.

and similarly for $1 \widehat{\otimes} \pi_B$. One checks directly that these maps are graded $*$-homomorphisms: more concretely, with notation as in Lemma 1.8.6 they are given on homogeneous elements by the formulas $(\pi_A \widehat{\otimes} 1)(a) = \pi_A(a) \otimes 1$ and $(1 \widehat{\otimes} \pi_B)(b) = U_A^{\partial b} \otimes \pi_B(b)$. More direct checks show that with notation as in Remark E.2.4, these maps satisfy

$$[(\pi_A \widehat{\otimes} 1)(a), (1 \widehat{\otimes} \pi_B)(b)]_g = 0,$$

and therefore by that remark give rise to a graded $*$-homomorphism

$$(\pi_A \widehat{\otimes} 1) \widehat{\otimes} (1 \widehat{\otimes} \pi_B) \colon A \widehat{\odot} B \to \mathcal{B}(H_A \widehat{\otimes} H_B).$$

Definition E.2.7 With notation as above, we write $\pi_A \widehat{\otimes} \pi_B$ for $(\pi_A \widehat{\otimes} 1) \widehat{\otimes} (1 \widehat{\otimes} \pi_B)$, and call it the *graded tensor product* of π_A and π_B.

Lemma E.2.8 *With notation as above, if π_A and π_B are faithful, then $\pi_A \widehat{\otimes} \pi_B$ is also faithful. Moreover, the norm defined on $A \widehat{\odot} B$ defined by*

$$\|c\| := \|(\pi_A \widehat{\otimes} \pi_B)(c)\|$$

does not depend on the choice of π_A and π_B.

Proof The proof is very similar to that of Proposition 1.8.9. For example, let us look at faithfulness in detail. Let $\sum_{i=1}^n a_i \widehat{\otimes} b_i$ be in the kernel of $\pi_A \widehat{\otimes} \pi_B$. Rewriting, we may assume that each a_i and b_i are homogeneous, and that the set $\{b_1, \ldots, b_n\}$. is linearly independent. Then with conventions as in Remark E.2.6, for homogeneous $u, v \in H_A$ and $w, x \in H_B$, we have

$$\begin{aligned} 0 &= \left\langle v \otimes x, \left(\sum_{i=1}^n \pi_A(a_i) \widehat{\otimes} \pi_B(b_i) \right) (u \otimes w) \right\rangle \\ &= \sum_{i=1}^n (-1)^{\partial b_i \partial u} \langle v, \pi_A(a_i) u \rangle \langle x, \pi_B(b_i) w \rangle \\ &= \left\langle x, \pi_B \left(\sum_{i=1}^n (-1)^{\partial b_i \partial u} \langle v, \pi_A(a_i) u \rangle b_i \right) w \right\rangle. \end{aligned}$$

As x and w are arbitrary homogeneous elements and as π_B is injective, this forces

$$\sum_{i=1}^n (-1)^{\partial b_i \partial u} \langle v, \pi_A(a_i) u \rangle b_i = 0.$$

As $b_1, \ldots, b_n$ is a linearly independent collection, this in turn forces

$$(-1)^{\partial b_i \partial u} \langle v, \pi_A(a_i) u \rangle = 0$$

for each i. Finally, as v and u are arbitrary homogeneous elements and π_A is injective, this forces $a_i = 0$ for all i. Thus $\pi_A \widehat{\otimes} \pi_B$ is injective as claimed.

The fact that the norm $\|(\pi_A \widehat{\otimes} \pi_B)(c)\|$ does not depend on the choice of π_A and π_B can again be handled similarly to the proof of the Proposition 1.8.9: the only real difference is that it might help to assume that the net (P_i) appearing there consists only of finite rank projections that are even for the grading on $\mathcal{B}(H_B)$. We leave the details to the reader. □

Definition E.2.9 Let A and B be graded C^*-algebras, let π_A and π_B be any faithful graded representations (which always exist by Lemma E.1.6) and let $\pi_A \widehat{\otimes} \pi_B$ be the graded tensor product representation of Definition E.2.7. The *graded spatial norm* on $A \widehat{\odot} B$ is defined by

$$\|c\| := \|(\pi_A \widehat{\otimes} \pi_B)(c)\|;$$

this does not depend on the choice of π_A and π_B by Lemma E.2.8. The associated completion of $A \widehat{\odot} B$ is then a C^*-algebra denoted $A \widehat{\otimes} B$ and called the *graded spatial tensor product* of A and B.

Remark E.2.10 The canonical commutativity and associativity isomorphisms for $\widehat{\odot}$ of Remark E.2.3 extend to the spatial completions. Indeed if A, B and C are graded C^*-algebras with faithful graded representations on H_A, H_B and H_C, then the formula

$$(H_A \widehat{\otimes} H_B) \widehat{\otimes} H_C \mapsto H_A \widehat{\otimes} (H_B \widehat{\otimes} H_C), \quad (u \widehat{\otimes} v) \widehat{\otimes} w \mapsto u \widehat{\otimes} (v \widehat{\otimes} w)$$

determines a unitary isomorphism that intertwines the defining representations of $(A \widehat{\otimes} B) \widehat{\otimes} C$ and of $A \widehat{\otimes} (B \widehat{\otimes} C)$. Similarly, with conventions as in Remark E.2.6, the formula

$$H_A \widehat{\otimes} H_B \to H_B \widehat{\otimes} H_A, \quad v \widehat{\otimes} w \mapsto (-1)^{\partial v \partial w} w \widehat{\otimes} v$$

determines a unitary isomorphism that intertwines the defining representations of $A \widehat{\otimes} B$ and $B \widehat{\otimes} A$.

Remark E.2.11 Analogously to Remark 1.8.12 the graded spatial tensor product is functorial: more precisely, if $\phi \colon A \to B$ and $\psi \colon C \to D$ are graded $*$-homomorphisms, then there is a unique graded $*$-homomorphism $\phi \widehat{\otimes} \psi \colon A \widehat{\otimes} C \to B \widehat{\otimes} D$ satisfying $(\phi \widehat{\otimes} \psi)(a \widehat{\otimes} c) = \phi(a) \widehat{\otimes} \psi(c)$ on elementary tensors. This can be proved in the the same way as for the usual spatial tensor product: see the argument sketched in Exercise 1.9.16.

Example E.2.12 Let $d \geqslant 1$, and let $\text{Cliff}_{\mathbb{C}}(\mathbb{R}^d)$ be the Clifford algebra of $\mathbb{R}^d$ as in Example E.1.11, so $\text{Cliff}_{\mathbb{C}}(\mathbb{R}^d)$ is a graded $*$-algebra. Consider the bijection

$$\mathbb{R}^{d-1} \times \mathbb{R} \longleftrightarrow \mathbb{R}^d, \quad ((v_1, \ldots, v_{d-1}), v) \longleftrightarrow (v_1, \ldots, v_{d-1}, v)$$

defined by putting the $\mathbb{R}$ component on the left in the last coordinate. One can check[3] that this extends to a $*$-isomorphism

$$\text{Cliff}_{\mathbb{C}}(\mathbb{R}^{d-1}) \widehat{\otimes} \text{Cliff}_{\mathbb{C}}(\mathbb{R}) \cong \text{Cliff}_{\mathbb{C}}(\mathbb{R}^d). \quad \text{(E.1)}$$

It follows from this and the explicit description of $\text{Cliff}_{\mathbb{C}}(\mathbb{R})$ from Example E.1.11 that $\text{Cliff}_{\mathbb{C}}(\mathbb{R}^d)$ is a $*$-algebra of dimension 2^d, and moreover, that if $e_1, \ldots, e_d$ is an orthonormal basis for $\mathbb{R}^d$, then

$$\{e_{i_1} e_{i_2} \cdots e_{i_k} \mid k \in \{0, \ldots, d\}, \ i_1 < \cdots < i_k\}$$

is a basis for $\text{Cliff}_{\mathbb{C}}(\mathbb{R}^d)$ (if $k = 0$ in the above, the corresponding 'empty product' is by definition the identity of $\text{Cliff}_{\mathbb{C}}(\mathbb{R}^d)$). Note that this basis consists of homogeneous elements, with the parity of $e_{i_1} \cdots e_{i_k}$ equalling that of k.

[3] Exercise! – the unjustified claims in this example all make good practice in getting used to Clifford algebras and graded tensor products.

Stipulating that this basis is orthonormal gives $\mathrm{Cliff}_{\mathbb{C}}(\mathbb{R}^d)$ the structure of a Hilbert space, say H_d. The grading operator on $\mathrm{Cliff}_{\mathbb{C}}(\mathbb{R}^d)$ defines a self-adjoint unitary operator on H_d, turning it into a graded Hilbert space. The left multiplication of the Clifford algebra on itself gives rise to a faithful graded $*$-representation $\mathrm{Cliff}_{\mathbb{C}}(\mathbb{R}^d) \to \mathcal{B}(H_d)$. Equipped with the corresponding operator norm, $\mathrm{Cliff}_{\mathbb{C}}(\mathbb{R}^d)$ becomes a C^*-algebra. From now on, we will think of $\mathrm{Cliff}_{\mathbb{C}}(\mathbb{R}^d)$ as a graded C^*-algebra in this way.

Let us note also that the $*$-isomorphism in line (E.1) allows us to determine the structure of $\mathrm{Cliff}_{\mathbb{C}}(\mathbb{R}^d)$ as a graded C^*-algebra. Indeed, one can check that if $\{e_1, e_2\}$ is an orthonormal basis for $\mathbb{R}^2$, then the assignments

$$e_1 \mapsto \begin{pmatrix} 0 & 1 \\ 1 & 0 \end{pmatrix}, \quad e_2 \mapsto \begin{pmatrix} 0 & i \\ -i & 0 \end{pmatrix}$$

determine a (necessarily isometric) $*$-isomorphism $\mathrm{Cliff}_{\mathbb{C}}(\mathbb{R}^2) \to M_2(\mathbb{C})$; moreover, the grading corresponds to 'the' standard (inner) grading on $M_2(\mathbb{C})$ as in Example E.1.9. It follows from this, induction using the isomorphism in line (E.1), the result of Exercise E.3.3 and associativity of the graded tensor product (Remark E.2.10), that

$$\mathrm{Cliff}_{\mathbb{C}}(\mathbb{R}^d) \cong \begin{cases} M_{2^{d/2}}(\mathbb{C}), & \text{d even}, \\ M_{2^{(d-1)/2}}(\mathbb{C}) \oplus M_{2^{(d-1)/2}}(\mathbb{C}), & \text{d odd}. \end{cases} \tag{E.2}$$

The grading is the standard (inner) grading of Example E.1.9 if d is even. If d is odd, the grading is not inner: it is given by the flip $*$-automorphism $(a,b) \mapsto (b,a)$ with respect to the direct sum decomposition.

Example E.2.13 Let A be an inner graded C^*-algebra, and $\mathscr{K}$ be a standard graded copy of the compact operators on a separable infinite-dimensional Hilbert space $H = H_0 \oplus H_1$ as in Example E.1.9. The purpose of this example is to give a useful description of $A \widehat{\otimes} \mathscr{K}$: we will show that it is isomorphic as a graded C^*-algebra to $M_2(A \otimes \mathcal{K}(H_0))$ with grading given by the unitary multiplier $\begin{pmatrix} 1 & 0 \\ 0 & -1 \end{pmatrix}$. We will also show that the isomorphism we define is canonical up to homotopy equivalence. We warn the reader that we do not give the shortest possible proof of this!

Note first that by Exercise E.3.3 there is a canonical isomorphism

$$A \widehat{\otimes} \mathscr{K} \cong A \otimes \mathscr{K},$$

where both sides are equipped with the tensor product grading. Choose orthonormal bases $(e_n)_{n=1}^{\infty}$ and $(f_n)_{n=1}^{\infty}$ of H_0 and H_1 respectively. Let $\mathbb{Z}/2 = \{0,1\}$ be the group with two elements, and consider the unitary isomorphism determined as follows:

$$U \colon \ell^2(\mathbb{Z}/2) \otimes H_0 \to H_0 \oplus H_1, \quad \delta_i \otimes e_n \mapsto \frac{1}{\sqrt{2}}(e_n, (-1)^i f_n).$$

Let $\lambda_1 \otimes \mathrm{id}_{H_0}$ be the grading on $\ell^2(\mathbb{Z}/2) \otimes H_0$ determined by the tensor product of the non-trivial element λ_1 of the left regular representation associated to $\mathbb{Z}/2$ (Example C.1.3) and the identity on H_0. Then conjugation by U determines an isomorphism

$$A \otimes \mathscr{K} \cong A \otimes \mathcal{K}(\ell^2(\mathbb{Z}/2) \otimes H_0)$$

that intertwines the tensor product gradings on both sides.

Concretely represent A on some Hilbert space H_A, and let the inner grading on A be induced by some unitary V in the multiplier algebra $M(A)$ of A. Let

$$W \colon H_A \otimes \ell^2(\mathbb{Z}/2) \otimes H_0 \to \ell^2(\mathbb{Z}/2) \otimes H_0$$

be the unitary isomorphism underlying Fell's trick (Proposition C.2.1) defined by

$$W : u \otimes \delta_i \otimes v \mapsto V^i u \otimes \delta_i \otimes v.$$

It is not difficult to see that W is in the multiplier algebra of $A \otimes \mathcal{K}(\ell^2(\mathbb{Z}/2) \otimes H_0)$ and intertwines the actions of $\mathbb{Z}/2$ given by

$$i \mapsto V^i \otimes \lambda_i \otimes 1_{H_0} \quad \text{and} \quad i \mapsto 1_{H_A} \otimes \lambda_i \otimes 1_{H_0}.$$

Hence conjugating by W shows that $A \otimes \mathcal{K}(\ell^2(\mathbb{Z}/2) \otimes H_0)$ (with the tensor product grading) is isomorphic as a graded C^*-algebra to $A \otimes \mathcal{K}(\ell^2(\mathbb{Z}/2) \otimes H_0)$ equipped with the tensor product of the trivial grading on A, and the grading induced by $\lambda_1 \otimes 1_{H_0}$ on $\mathcal{K}(\ell^2(\mathbb{Z}/2) \otimes H_0)$. Using the canonical isomorphisms

$$\mathcal{K}(\ell^2(\mathbb{Z}/2) \otimes H_0) \cong M_2(\mathcal{K}(H_0)),$$

we have an isomorphism

$$A \otimes \mathcal{K}(\ell^2(\mathbb{Z}/2) \otimes H_0) \cong M_2(A \otimes \mathcal{K}(H_0)),$$

where the right-hand side is equipped with the grading induced by the unitary multiplier $\left(\begin{smallmatrix} 0 & 1 \\ 1 & 0 \end{smallmatrix}\right)$. The matrix $\left(\begin{smallmatrix} 0 & 1 \\ 1 & 0 \end{smallmatrix}\right)$ is unitarily equivalent (in $M_2(\mathbb{C})$) to $\left(\begin{smallmatrix} 1 & 0 \\ 0 & -1 \end{smallmatrix}\right)$; as such a unitary equivalence is unique up to homotopy equivalence, we get the claimed result.

The Maximal Graded Tensor Product

We now look at the maximal graded tensor product, which we will need to define some products in K-theory.

Definition E.2.14 Let A and B be graded C^*-algebras. With notations for graded commutators as in Remark E.2.4, let S be the class of all triples (ϕ, ψ, C) consisting of a graded C^*-algebra C and graded $*$-homomorphisms $\phi\colon A \to C$ and $\psi : B \to C$ that satisfy

$$[\phi(a), \psi(b)]_g = 0$$

for all $a \in A$ and $b \in B$. As in Remark E.2.4, for each triple (ϕ, ψ, C) in S we get a graded $*$-homomorphism $\phi \widehat{\otimes} \psi : A \widehat{\odot} B \to C$ that satisfies

$$(\phi \widehat{\otimes} \psi)(a \widehat{\otimes} b) = \phi(a)\psi(b)$$

on elementary tensors. We define the *maximal norm* on $A \widehat{\odot} B$ by

$$\|c\|_{\max} := \sup\{\|(\phi \widehat{\otimes} \psi)(c)\|_C \mid (\phi, \psi, C) \in S\}.$$

We define the *maximal graded tensor product*, denoted $A \widehat{\otimes}_{\max} B$, to be the associated completion of $A \widehat{\odot} B$.

Remark E.2.15 The commutativity and associativity isomorphisms of Remark E.2.3 extend to $\widehat{\otimes}_{\max}$: this follows straightforwardly from the definitions, as we leave to the reader to check.

Remark E.2.16 The maximal graded tensor product is functorial: more precisely, if $\phi\colon A \to B$ and $\psi\colon C \to D$ are graded $*$-homomorphisms, then there is a unique graded $*$-homomorphism $\phi \mathbin{\widehat{\otimes}} \psi\colon A \mathbin{\widehat{\otimes}}_{\max} C \to B \mathbin{\widehat{\otimes}}_{\max} D$ satisfying $(\phi \mathbin{\widehat{\otimes}} \psi)(a \mathbin{\widehat{\otimes}} c) = \phi(a) \mathbin{\widehat{\otimes}} \psi(c)$ on elementary tensors. This is not completely obvious from the way that we have defined $\widehat{\otimes}_{\max}$: see Exercise E.3.5 for a sketch proof.

Remark E.2.17 The maximal tensor product also has the following universal property: if A, B and C are graded C^*-algebras, and if $\phi\colon A \mathbin{\widehat{\odot}} B \to C$ is any graded $*$-homomorphism, then there is a unique extension of ϕ to a graded $*$-homomorphism $\phi\colon A \mathbin{\widehat{\otimes}}_{\max} B \to C$: see Exercise E.3.5 again.

Note that if A and B are graded C^*-algebras, then we may form both $A \mathbin{\widehat{\otimes}} B$ and $A \otimes B$, and similarly for the maximal tensor products. The norms on these C^*-algebras are actually closely related.

Lemma E.2.18 *Let A and B be graded C^*-algebras. Then the identity map on $A \odot B$ extends to Banach space isomorphisms*

$$A \mathbin{\widehat{\otimes}} B \cong A \otimes B \quad \textit{and} \quad A \mathbin{\widehat{\otimes}}_{\max} B \cong A \otimes_{\max} B.$$

Proof Consider the $*$-homomorphism $\epsilon_A \mathbin{\widehat{\otimes}} \mathrm{id}\colon A \mathbin{\widehat{\otimes}} B \to A \mathbin{\widehat{\otimes}} B$, which exists by functoriality (Remark E.2.11). For $i \in \{0,1\}$, we may thus make sense of contractive linear maps

$$\frac{1}{2}\big(\epsilon_A \mathbin{\widehat{\otimes}} \mathrm{id} + (-1)^i(\mathrm{id} \mathbin{\widehat{\otimes}} \mathrm{id})\big)\colon A\mathbin{\widehat{\otimes}}B \to A.$$

Note that if A_i denotes the subspace of A consisting of elements of degree i, then $\frac{1}{2}\big(\epsilon_A \mathbin{\widehat{\otimes}} \mathrm{id} + (-1)^i(\mathrm{id} \mathbin{\widehat{\otimes}} \mathrm{id})\big)$ is idempotent, with image the closure of $A_i \odot B$ inside $A\mathbin{\widehat{\otimes}}B$. This works similarly for ϵ_B. Thus for $i, j \in \{0,1\}$ we may define contractive linear maps $E_{ij}\colon A \mathbin{\widehat{\otimes}} B \to A \mathbin{\widehat{\otimes}} B$ by

$$E_{ij} := \frac{1}{2}\big(\epsilon_A \mathbin{\widehat{\otimes}} \mathrm{id} + (-1)^i(\mathrm{id} \mathbin{\widehat{\otimes}} \mathrm{id})\big) \circ \frac{1}{2}\big(\mathrm{id} \mathbin{\widehat{\otimes}} \epsilon_B + (-1)^j(\mathrm{id} \mathbin{\widehat{\otimes}} \mathrm{id})\big).$$

The image of E_{ij} is the closure of $A_i \odot B_j$ inside $A \mathbin{\widehat{\otimes}} B$. Moreover, note that $\sum_{i,j=0}^{1} E_{ij}$ is the identity. Putting this discussion together, we have that for any $c \in A \mathbin{\widehat{\otimes}} B$,

$$\max\{\|E_{ij}(c)\| \mid i,j \in \{0,1\}\} \leqslant \|c\| \leqslant \sum_{i,j=0,1} \|E_{ij}(c)\|. \tag{E.3}$$

In particular, the norm on $A \mathbin{\widehat{\otimes}} B$ is equivalent to the norm defined by $\|c\| := \sum_{i,j=0}^{1} \|E_{ij}(c)\|$. All this works analogously for $\widehat{\otimes}_{\max}$, $\otimes$ and $\otimes_{\max}$, using that all of these tensor products are functorial for $*$-homomorphisms (see Exercises E.3.5, 1.9.16 and 1.9.20 respectively). It follows from this discussion that to prove the desired result, it suffices to show that $\|E_{ij}(c)\|_{A \mathbin{\widehat{\otimes}} B} = \|E_{ij}(c)\|_{A \otimes B}$ for all $c \in A \odot B$, and similarly in the maximal case. It therefore suffices to prove that $\|c\|_{A \mathbin{\widehat{\otimes}} B} = \|c\|_{A \otimes B}$ for any $c \in A_i \odot B_j$ and similarly in the maximal case; this is what we will do.

Note first that the degree zero elements A_0 and B_0 in A and B respectively are C^*-subalgebras. We claim that the inclusion $A_0 \odot B_0 \to A\mathbin{\widehat{\otimes}}B$ extends to an isometric

$*$-homomorphic inclusion $A_0 \otimes B_0 \to A \widehat{\otimes} B$ and similarly in the ungraded and maximal cases: see Exercise E.3.6. Given this, let c be any element of $A_i \odot B_j$. Then

$$\|c\|^2_{A \widehat{\otimes} B} = \|c^* c\|_{A \widehat{\otimes} B}.$$

However, $c^* c$ is $A_0 \odot B_0$, and by the claim, the norm of any element here equals its norm in $A_0 \otimes B_0$. Thus we get

$$\|c\|^2_{A \widehat{\otimes} B} = \|c^* c\|_{A \widehat{\otimes} B} = \|c^* c\|_{A \otimes B} = \|c\|^2_{A \otimes B}.$$

Everything works analogously in the maximal case, so we are done. □

The following result says that for the most important examples we are interested in, it is not important whether we use $\widehat{\otimes}$ of $\widehat{\otimes}_{\max}$.

Corollary E.2.19 *Let A be a graded C^*-algebra, and B be one of: $\mathcal{K}$, $Cliff_{\mathbb{C}}(V)$ for some finite-dimensional vector space V, or a graded commutative C^*-algebra. Then the canonical map $A \widehat{\otimes}_{\max} B \to A \widehat{\otimes} B$ is an isomorphism.*

Proof If $B = \text{Cliff}_{\mathbb{C}}(V)$, $A \widehat{\odot} B$ is already a C^*-algebra for either the $\widehat{\otimes}$ or $\widehat{\otimes}_{\max}$ norms, so these norms are the same by uniqueness of C^*-algebra norms (Corollary 1.3.16). If $B = \mathcal{K}$, then one can use that $\mathcal{K}$ is the closure of the union of graded subalgebras $M_{2n}(\mathbb{C})$ (with the standard even grading) and that each tensor product $A \widehat{\odot} M_{2n}(\mathbb{C})$ has a unique C^*-algebra norm: compare Exercise 1.9.18.

In the commutative case, we use Lemmas 1.8.13 and E.2.18. □

E.3 Exercises

E.3.1 Show that if A is a graded C^*-algebra, then A has an approximate unit consisting of even elements.
Hint: if (h_i) is an arbitrary approximate unit, then $\frac{1}{2}(h_i + \epsilon(h_i))$ works.

E.3.2 Let V be a real vector space, and let $\odot_{\mathbb{R}}$ denote the algebraic tensor product over $\mathbb{R}$. The *tensor algebra* $T(V)$ is defined to be the direct sum

$$T(V) := \bigoplus_{n=0}^{\infty} V^{\otimes n},$$

where

$$V^{\otimes n} := \underbrace{V \odot_{\mathbb{R}} \cdots \odot_{\mathbb{R}} V}_{n \text{ times}}$$

and $V^{\otimes 0} := \mathbb{R}$. The vector space $T(V)$ is equipped with the multiplication defined on elementary tensors $v_1 \otimes \cdots \otimes v_n \in V^{\otimes n}$ and $w_1 \otimes \cdots \otimes w_m \in V^{\otimes m}$ by

$$(v_1 \otimes \cdots \otimes v_n)(w_1 \otimes \cdots \otimes w_m) := v_1 \otimes \cdots \otimes v_n \otimes w_1 \otimes \cdots \otimes w_m \in V^{\otimes(n+m)}$$

and made into a $*$-algebra via the operation defined on elementary tensors by

$$(v_1 \otimes \cdots \otimes v_n)^* := v_n \otimes \cdots \otimes v_1.$$

Let $\iota \colon V \to T(V)$ be the linear map defined via the tautological inclusion of $V = V^{\otimes 1}$ in $\bigoplus_{n=0}^{\infty} V^{\otimes n}$.

Let $T_{\mathbb{C}}(V)$ denote the tensor product $*$-algebra $T(V) \odot_{\mathbb{R}} \mathbb{C}$. Moreover, if V is equipped with an inner product, let I be the ideal in $T_{\mathbb{C}}(V)$ generated by all elements of the form $v \otimes v - \|v\|^2 1_{T_{\mathbb{C}}(V)}$, where $\|\cdot\|$ is the norm associated to the inner product on V.

(i) Show that $T(V)$ has the following universal property: for any linear map $\phi\colon V \to A$ from V into a real $*$-algebra A that satisfies $\phi(v)^* = \phi(v)$ for all $v \in V$, there is a unique algebra homomorphism filling in the dashed arrow below

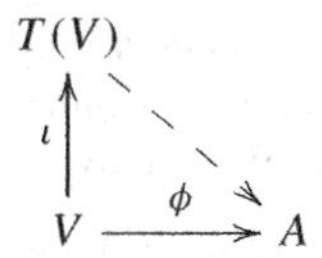

so the diagram commutes. Show that $T_{\mathbb{C}}(V)$ has the analogous universal properties for (real) linear maps from V into a complex $*$-algebra.

(ii) Show that if $\phi\colon V \to A$ is any real-linear map to a unital complex $*$-algebra such that $\phi(v)^2 = \|v\|^2 1_A$ and $\phi(v)^* = \phi(v)$ for all $v \in V$, then there is a unique unital $*$-algebra homomorphism making the following diagram commute

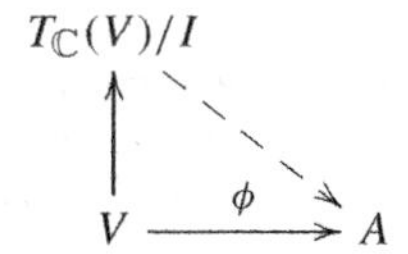

(here the vertical map is the composition

$$V \to T(V) \to T_{\mathbb{C}}(V) \to T_{\mathbb{C}}(V)/I,$$

where the first map is the canonical inclusion $\iota\colon V \to T(V)$, the second is the map $v \mapsto v \otimes 1_{\mathbb{C}}$ and the third is the canonical quotient).

From this exercise, we conclude that $T_{\mathbb{C}}(V)/I$ is one way to make rigorous sense of the 'generators and relations' description of the Clifford algebra from Example E.1.11: more precisely, we may reasonably define $\mathrm{Cliff}_{\mathbb{C}}(V) := T_{\mathbb{C}}(V)/I$.

E.3.3 Let A and B be graded C^*-algebras, with at least one of the gradings inner. Show that there is an isomorphism of C^*-algebras $A \mathbin{\widehat{\otimes}} B \cong A \otimes B$, that is also compatible with the gradings where $A \otimes B$ is given the canonical tensor product grading determined by $\epsilon_{A\otimes B}(a \otimes b) := \epsilon_A(a) \otimes \epsilon_B(b)$.
Hint: if, say, B is inner graded by u, the map $A \mathbin{\widehat{\otimes}} B \to A \otimes B$ determined on elementary tensors of homogeneous elements by the formula $a \mathbin{\widehat{\otimes}} b \mapsto a \otimes u^{\partial a} b$ works. Show this.

E.3.4 More generally than the computations in Example E.2.12, show that if V and W are finite-dimensional real vector spaces, then there is a canonical isomorphism $\mathrm{Cliff}_{\mathbb{C}}(V \oplus W) \cong \mathrm{Cliff}_{\mathbb{C}}(V) \mathbin{\widehat{\otimes}} \mathrm{Cliff}_{\mathbb{C}}(W)$.

E.3.5 (i) Let $\pi : A \mathbin{\widehat{\odot}} B \to \mathcal{B}(H)$ be a non-degenerate graded representation. Show that there are graded representations $\pi_A : A \to \mathcal{B}(H)$ and $\pi_B : B \to \mathcal{B}(H)$ such that (with notation as in Remark E.2.4) $[\pi_A(a), \pi_B(b)]_g = 0$ for all $a \in A$ and $b \in B$, and $\pi = \pi_A \mathbin{\widehat{\otimes}} \pi_B$.

Hint: analogously to Exercise 1.9.20, choose an even approximate unit (h_i) for B as in Exercise E.3.4. Show that the net $\pi(a \widehat{\otimes} h_i)$ strongly converges for all $a \in A$, and define $\pi_A(a)$ to be the associated limit. Proceed similarly for π_B.

(ii) Use the previous part to prove the functoriality claim for $\widehat{\otimes}_{\max}$ in Remark E.2.16. *Hint: with notation as in Remark E.2.16, fix a faithful non-degenerate representation $\pi : C \widehat{\otimes}_{\max} D \to \mathcal{B}(H)$. Let π_C and π_D be representations of C and D as in the first part of the exercise. Using Remark E.2.4, we then get a $*$-homomorphism*

$$(\pi_C \circ \phi) \widehat{\otimes} (\pi_D \circ \psi) \colon A \widehat{\odot} B \to \mathcal{B}(H),$$

which extends to $A \widehat{\otimes}_{\max} B$ by definition of the maximal norm. Show that this takes image in $C \widehat{\otimes}_{\max} D$, and is the required $*$-homomorphism.

(iii) Use a similar argument to justify the universal property in Remark E.2.17.

E.3.6 With notation as in the proof of Lemma E.2.18, prove that the natural $*$-homomorphic inclusion $A_0 \odot B_0 \to A \widehat{\odot} B$ extends to isometric inclusions

$$A_0 \otimes B_0 \to A \widehat{\otimes} B \quad \text{and} \quad A_0 \otimes_{\max} B_0 \to A \widehat{\otimes}_{\max} B,$$

and similarly there are inclusions

$$A_0 \otimes B_0 \to A \otimes B \quad \text{and} \quad A_0 \otimes_{\max} B_0 \to A \otimes_{\max} B$$

in the ungraded case.

Hint: the spatial case is the easier of the two, and we leave it to the reader. The maximal case is trickier: one way to proceed is as follows. Choose a faithful representation $\pi : A_0 \otimes_{\max} B_0 \to \mathcal{B}(H)$. Define a form on $(A \widehat{\odot} B) \odot H$ by the formula

$$\langle c \otimes u, d \otimes v \rangle := \langle u, \pi(E_{00}(c^* d))v \rangle$$

on elementary tensors. Show that the separated completion is a Hilbert space $\widetilde{H}$. For $c \in A \widehat{\odot} B$ and $d \otimes v \in (A \widehat{\odot} B) \odot H$, show that the formula

$$\widetilde{\pi}(c) \colon d \otimes v \mapsto cd \otimes v$$

determines a well-defined representation $\widetilde{\pi}$ of $A \widehat{\otimes}_{\max} B$ on $\widetilde{H}$. Finally, show that the restriction of $\widetilde{\pi}$ to $A_0 \odot B_0$ is isometric.

E.4 Notes and References

The introduction of Clifford algebras into the study of K-theory is due to Atiyah, Bott and Shapiro [9]. Since then, gradings have been a useful tool in K-theory, particularly in the setting of Kasparov's bivariant theory (see Kasparov [150]). Gradings and Clifford algebras become particularly important in the setting of real K-theory: we will not touch on this here, but the theory is important for applications to topology and geometry. The interested reader can see chapter 1 of Lawson and Michelsohn [164] for a detailed study.

Graded tensor products of C^*-algebras are (fairly briefly) discussed in section 1.2 of Higson and Guentner [128] and section 14.4 of Blackadar [33].

References

[1] J. F. Adams. *Stable Homotopy and Generalised Homology*. University of Chicago Press, 1974.

[2] P. Antonini, S. Azzali and G. Skandalis. The Baum–Connes conjecture localised at the unit element of a discrete group. arXiv:1807.05892, 2018.

[3] P. Ara, K. Li, F. Lledó and J. Wu. Amenability and uniform Roe algebras. J. Math. Anal. Appl. 459(2):257–306, 2018.

[4] W. Arveson. *A Short Course in Spectral Theory*. Springer, 2002.

[5] G. Arzhantseva and T. Delzant. Examples of random groups. Available at www.mat.univie.ac.at/%7Earjantseva/Abs/random.pdf, 2008.

[6] M. Atiyah. *K-theory*. W. A. Benjamin, 1967.

[7] M. Atiyah. Global theory of elliptic operators. In *Proceedings of the International Conference on Functional Analysis and Related Topics, Tokyo, April 1969*. University of Tokyo Press, pp. 21–30.

[8] M. Atiyah. Elliptic operators, discrete groups and von Neumann algebras. *Asterisque*, 32–33:43–72, 1976.

[9] M. Atiyah, R. Bott and A. Shapiro. Clifford modules. *Topology*, 3:3–38, 1964.

[10] M. Atiyah and W. Schmid. A geometric construction of the discrete series for semisimple Lie groups. *Invent. Math.*, 42:1–62, 1977.

[11] M. Atiyah and G. Segal. The index of elliptic operators II. *Ann. Math.*, 87(3):531–545, 1968.

[12] M. Atiyah and I. Singer. The index of elliptic operators I. *Ann. Math.*, 87(3): 484–530, 1968.

[13] M. Atiyah and I. Singer. The index of elliptic operators III. *Ann. Math.*, 87(3):546–604, 1968.

[14] A. Bartels. Squeezing and higher algebraic K-theory. *K-Theory*, 28(1):19–37, 2003.

[15] A. Bartels. K-theory and actions on Euclidean retracts. In *Proceedings of the International Congress of Mathematicians—Rio de Janeiro 2018. Vol. II. Invited lectures*. World Scientific Publishing, 2018, pp. 1041–1062.

[16] A. Bartels and M. Bestvina. The Farrell–Jones conjecture for mapping class groups. Invent. Math. 215(2):651–712, 2019.

[17] A. Bartels, T. Farrell, L. Jones and H. Reich. On the isomorphism conjecture in algebraic K-theory. *Topology*, 43:157–213, 2004.

[18] A. Bartels and W. Lück. The Borel conjecture for hyperbolic and CAT(0)-groups. *Ann. Math.*, 175(2):631–689, 2012.

[19] A. Bartels, W. Lück and H. Reich. The K-theoretic Farrell–Jones conjecture for hyperbolic groups. *Invent. Math.*, 172(1):29–70, 2008.

[20] P. Baum and A. Connes. Chern character for discrete groups. In Y. Matsumoto, T. Mizutani and S. Morita, eds., *A Fête of Topology*. Academic Press, 1988, pp. 163–232.

[21] P. Baum and A. Connes. Geometric K-theory for Lie groups and foliations. *Enseign. Math. (2)*, 46:3–42, 2000 (first circulated 1982).

[22] P. Baum, M. Connes and N. Higson. Classifying space for proper actions and K-theory of group C^*-algebras. American Mathematical Society, 1994, pp. 241–291 (Contemporary Mathematics, vol. 167).

[23] P. Baum and R. G. Douglas. K-homology and index theory. In R. V. Kadison, ed., *Operator Algebras and Applications, Part I*. American Mathematical Society, 1982, pp. 117–173 (Proceedings of Symposia in Pure Mathematics, vol. 38).

[24] P. Baum and R. G. Douglas, Index theory, bordism, and K-homology. In R. G. Douglas and C. Schochet, eds., *Operator Algebras and K-theory*. American Mathematical Society, 1982, pp. 1–31 (Contemporary Mathematics, vol. 10).

[25] P. Baum, E. Guentner and R. Willett. Expanders, exact crossed products, and the Baum–Connes conjecture. *Ann. K-theory*, 1(2):155–208, 2015.

[26] P. Baum, E. Guentner and R. Willett. Exactness and the Kadison–Kaplansky conjecture. In R. S. Doran and E. Park, eds., *Operator Algebras and Their Applications*. American Mathematical Society, 2016, pp. 1–33 (Contemporary Mathematics, vol. 671).

[27] P. Baum, N. Higson and T. Schick. On the equivalence of geometric and analytic K-homology. *Pure Appl. Math. Q.*, 3(1):1–24, 2007.

[28] P. Baum, N. Higson and T. Schick. A geometric description of equivariant K-homology for proper actions. In E. Blanchard et al., eds., *Quanta of Maths*. American Mathematical Society, 2010, pp. 1–22 (Clay Mathematics Proceedings, vol. 11).

[29] B. Bekka, P.-A. Cherix and A. Valette. Proper affine isometric actions of amenable groups. In S. C. Ferry, A. Ranicki, and J. M. Rosenberg, eds., *Novikov Conjectures, Index Theory Theorems and Rigidity, vol. 2*. Cambridge University Press, 1993, pp. 1–4 (London Mathematical Society Lecture Note Series; vol. 227).

[30] M. Berger. *A Panoramic View of Riemannian Geometry*. Springer, 2003.

[31] M. Bestvina, K. Bromberg and K. Fujiwara. Constructing group actions on quasi-trees and applications to mapping class groups. *Publ. Math. Inst. Hautes Études Sci.*, 122:1–64, 2015.

[32] M. Bestvina, V. Guirardel and C. Horbez. Boundary amenability of $Out(F_n)$. arXiv:1705.07017, 2017.

[33] B. Blackadar. *K-Theory for Operator Algebras*, 2nd ed. Cambridge University Press, 1998 (Mathematical Sciences Research Institute Publications, vol. 5).

[34] B. Blackadar. *Operator Algebras: Theory of C^*-Algebras and Von Neumann Algebras*. Springer, 2006.

[35] M. Bökstedt, W. C. Hsiang and I. Madsen. The cyclotomic trace in algebraic K-theory of spaces. *Invent. Math.*, 111(3):465–539, 1993.

[36] R. Bott and L. Tu. *Differential Forms in Algebraic Topology*. Springer-Verlag, 1982.

[37] B. Braga and I. Farah. On the rigidity of uniform Roe algebras over uniformly locally finite coarse spaces. arXiv:1805.04236, 2018.

[38] M. Bridson and A. Haefliger. *Metric Spaces of Non-Positive Curvature. Springer*, 1999. (Grundlehren der Mathematischen Wissenschaft; vol. 319.)

[39] K. S. Brown. *Cohomology of Groups*. Springer, 1982. (Graduate Texts in Mathematics; vol. 87.)

[40] L. G. Brown. Stable isomorphism of hereditary subalgebras of C^*-algebras. *Pacific J. Math.*, 71(2):335–348, 1977.

[41] L. G. Brown. The universal coefficient theorem for Ext and quasidiagonality. In G. Arsene, ed., *Operator Algebras and Group Representations, vol. I* (Neptun, 1980). Pitman, 1984, pp. 60–64 (Monographs and Studies in Mathematics; vol. 17).

[42] L. G. Brown, R. G. Douglas and P. Fillmore. Extensions of C^*-algebras and K-homology. *Ann. Math.*, 105:265–324, 1977.

[43] N. Brown and E. Guentner. Uniform embeddings of bounded geometry spaces into reflexive Banach spaces. *Proc. Amer. Math. Soc.*, 133(7):2045–2050, 2005.

[44] N. Brown and N. Ozawa. *C^*-Algebras and Finite-Dimensional Approximations*. American Mathematical Society, 2008. (Graduate Studies in Mathematics; vol. 88.)

[45] A. Buss, S. Echterhoff and R. Willett. Exotic crossed products and the Baum–Connes conjecture. *J. Reine Angew. Math.*, 740:111–159, 2018.

[46] G. Carlsson and E. Pedersen. Controlled algebra and the Novikov conjecture for K- and L-theory. *Topology*, 34(3):731–758, 1995.

[47] J. Carrión and M. Dadarlat. Almost flat K-theory of classifying spaces. J. Noncommut. Geom. 12(2):407-438, 2018.

[48] J. Chabert and S. Echterhoff. Permanence properties of the Baum–Connes conjecture. *Doc. Math.*, 6:127–183, 2001.

[49] J. Chabert, S. Echterhoff and R. Nest. The Connes–Kasparov conjecture for almost connected groups and for linear p-adic groups. *Publ. Math. Inst. Hautes Études Sci.*, 97(1):239–278, 2003.

[50] J. Chabert, S. Echterhoff and H. Oyono-Oyono. Going-down functors, the Künneth formula, and the Baum–Connes conjecture. *Geom. Funct. Anal.*, 14(3):491–528, 2004.

[51] S. Chang, S. Weinberger and G. Yu. Positive scalar curvature and a new index theory for noncompact manifolds. *J. Geom. Phys.*, 149:22pp., 2020.

[52] X. Chen, R. Tessera, X. Wang and G. Yu. Metric sparsification and operator norm localization. *Adv. Math.*, 218(5):1496–1511, 2008.

[53] X. Chen and Q. Wang. Ideal structure of uniform Roe algebras of coarse spaces. *J. Funct. Anal.*, 216(1):191–211, 2004.

[54] X. Chen, Q. Wang and G. Yu. The maximal coarse Baum–Connes conjecture for spaces which admit a fibred coarse embedding into Hilbert space. *Adv. Math.*, 249:88–130, 2013.

[55] X. Chen and S. Wei. Spectral invariant subalgebras of reduced crossed product C^*-algebras. *J. Funct. Anal.*, 197(1):228–246, 2003.

[56] L. A. Coburn, R. G. Douglas, D. G. Schaeffer and I. Singer. C^*-algebras of operators on a half-space II: Index theory. *Publ. Math. Inst. Hautes Études Sci.*, 40(1):69–79, 1971.

[57] A. Connes. A survey of foliations and operator algebras. In R. V. Kadison, ed., *Operator Algebras and Applications, Part I*. American Mathematical Society, 1982, pp. 521–628 (Proceedings of the Symposium in Pure Mathematics; vol. 38).

[58] A. Connes. Non-commutative differential geometry. *Publ. Math. Inst. Hautes Études Sci.*, 62(2):41–144, 1985.

[59] A. Connes. Cyclic cohomology and the transverse fundamental class of a folitation. In H. Araki and E. G. Effros, eds., *Geometric Methods in Operator Algebras*. Longman Scientific & Technical, 1986, pp. 52–144 (Pitman Research Notes in Mathematics; vol. 123).

[60] A. Connes. *Noncommutative Geometry*. Academic Press, 1994.

[61] A. Connes and N. Higson. Déformations, morphismes asymptotiques et K-théorie bivariante. *C. R. Acad. Sci. Paris Sér. I Math.*, 311:101–106, 1990.

[62] A. Connes, Mikhael Gromov and H. Moscovici. Conjecture de Novikov et fibrés presque plats. *C. R. Acad. Sci. Paris Sér. I Math.*, 310(5):273–277, 1990.

[63] A. Connes, Mikhael Gromov and H. Moscovici. Group cohomology with Lipschitz control and higher signatures. *Geom. Funct. Anal.*, 2(1):1–78, 1993.

[64] A. Connes and H. Moscovici. The L^2-index theorem for homogeneous spaces of Lie groups. *Ann. Math.*, 115(2):291–330, 1982.

[65] A. Connes and H. Moscovici. Cyclic cohomology, the Novikov conjecture and hyperbolic groups. *Topology*, 29(3):345–388, 1990.

[66] A. Connes and G. Skandalis. The longitudinal index theorem for foliations. *Publ. Math. Inst. Hautes Études Sci.*, 20:1139–1183, 1984.

[67] G. Cortiñas. Algebraic v. topological K-theory: A friendly match. In P. Baum and G. Cortiñas, eds., *Topics in Algebraic and Topological K-theory*, Springer Lecture Notes, 2008.

[68] G. Cortiñas and G. Tartaglia. Operator ideals and assembly maps in K-theory. *Proc. Amer. Math. Soc.*, 142:1089–1099, 2014.

[69] J. Cuntz. Generalized homomorphisms between C^*-algebras and KK-theory. In Ph. Blanchard and L. Streit, eds., *Dynamics and Processes (Bielefeld 1981)*. Springer, 1983, pp. 31–45 (Lecture Notes in Mathematics, vol. 1031).

[70] J. Cuntz. Noncommutative simplicial complexes and the Baum–Connes conjecture. *Geom. Funct. Anal.*, 12(2):307–329, 2002.

[71] J. Cuntz, R. Meyer and J. Rosenberg. *Topological and Bivariant K-theory*. Birkhäuser, 2007 (Oberwolfach Seminars; vol. 36).

[72] H. Cycon, R. Froese, W. Kirsch and B. Simon. *Schrödinger Operators with Applications to Quantum Mechanics and Global Geometry*. Springer, 1987.

[73] M. Dadarlat. Group quasi-representations and index theory. *J. Topol. Anal.*, 4(3):297–319, 2012.

[74] M. Dadarlat and E. Guentner. Uniform embeddability of relatively hyperbolic groups. *J. Reine Angew. Math.*, 612:1–15, 2007.

[75] M. Dadarlat, R. Willett and J. Wu. Localization C^*-algebras and K-theoretic duality. *Ann. K-theory*, 3(4):615–630, 2018.

[76] K. Davidson. *C^*-algebras by Example*. Fields Institute monographs, 1996.

[77] J. Davis and W. Lück. Spaces over a category and assembly maps in isomorphism conjectures in K- and L-theory. *K-Theory*, 15:201–252, 1998.

[78] M. Davis. Groups generated by reflections and aspherical manifolds not covered by Euclidean space. *Ann. Math.*, 117(2):293–324, 1983.

[79] R. Deeley and M. Goffeng. Realizing the analytic surgery group of Higson and Roe geometrically, part I: The geometric model. *J. Homotopy Relat. Str.*, 12:109–142, 2017.

[80] J. Dixmier. *C^*-Algebras*. North Holland Publishing Company, 1977.

[81] R. G. Douglas. *Banach Algebra Techniques in Operator Theory*. Academic Press, 1972.

[82] R. G. Douglas. *C^*-algebra Extensions and K-homology*. Princeton University Press, 1980 (Annals of Mathematics Studies, vol. 95).

[83] A. Dranishnikov, S. Ferry and S. Weinberger. Large Riemannian manifolds which are flexible. *Ann. Math.*, 157:919–938, 2003.

[84] A. Dranishnikov, S. Ferry and S. Weinberger. An étale approach to the Novikov conjecture. *Comm. Pure Appl. Math.*, 61(2):139–155, 2007.

[85] A. Dranishnikov, J. Keesling and V. V. Uspenskii. On the Higson corona of uniformly contractible spaces. *Topology*, 37(4):791–803, 1998.

[86] A. Deitmar and S. Echterhoff. *Principles of Harmonic Analysis*. Springer, 2009.

[87] S. Eilenberg and N. Steenrod. *Foundations of Algebraic Topology*. Princeton University Press, 1952.

[88] G. Elliott, T. Natsume and R. Nest. The Atiyah–Singer index theorem as passage to the classical limit in quantum mechanics. *Comm. Math. Phys.*, 182(3):505–533, 1996.

[89] H. Emerson. Noncommutative Poincaré duality for boundary actions of hyperbolic groups. *J. Reine Angew. Math.*, 564:1–33, 2003.

[90] H. Emerson and R. Meyer. Dualizing the coarse assembly map. *J. Inst. Math. Jussieu*, 5:161–186, 2006.

[91] H. Emerson and R. Meyer. A descent principle for the Dirac dual Dirac method. *Topology*, 46:185–209, 2007.

[92] D. Farley. Proper isometric actions of Thompson's groups on Hilbert space. *Int. Math. Res. Not.*, 45:2409–2414, 2003.

[93] T. Farrell and W. C. Hsiang. On Novikov's conjecture for non-positively curved manifolds. *Ann. Math.*, 113:199–209, 1981.

[94] T. Farrell and L. Jones. K-theory and dynamics I. *Ann. Math.*, 124(2):531–569, 1986.

[95] T. Farrell and L. Jones. A topological analogue of Mostow's rigidity theorem. *J. Amer. Math. Soc.*, 2:257–370, 1989.

[96] T. Farrell and L. Jones. Classical aspherical manifolds. In *CBMS Regional Conference Series in Mathematics*, volume 75, 1990.

[97] T. Farrell and L. Jones. Topological rigidity for compact non-positively curved manifolds. In R. Greene and S. T. Yau, eds., *Differential Geometry: Riemannian Geometry*. American Mathematical Society, 1993, pp. 229–274 (Proceedings of the Symposium in Pure Mathematics, vol. 54).

[98] T. Farrell and L. Jones. Rigidity for aspherical manifolds with $\pi_1 \subset \mathrm{GL}_m(\mathbb{R})$. *Asian J. Math.*, 2:215–262, 1998.

[99] S. Ferry, A. Ranicki and J. Rosenberg. A history and survey of the Novikov conjecture. In S. Ferry, A. Ranicki and J. Rosenberg, eds. *Novikov Conjectures, Index Theorems and Rigidity, Volume I*. Cambridge University Press, 1993, pp. 7–66 (London Mathematical Society Lecture Note Series, vol. 226).

[100] S. Ferry and S. Weinberger. Curvature, tangentiality, and controlled topology. *Invent. Math.*, 105(2):401–414, 1991.

[101] S. Ferry and S. Weinberger. A coarse approach to the Novikov conjecture. In S. Ferry, A. Ranicki and J. Rosenberg, eds., *Novikov Conjectures, Index Theory Theorems and Rigidity, Volume I*. Cambridge University Press, 1993, pp. 147–163 (London Mathematical Society Lecture Note Series, vol. 226).

[102] M. Finn-Sell. Fibred coarse embeddings, a-T-menability and the coarse analogue of the Novikov conjecture. *J. Funct. Anal.*, 267(10):3758–3782, 2014.

[103] M. Finn-Sell and N. Wright. Spaces of graphs, boundary groupoids and the coarse Baum–Connes conjecture. *Adv. Math.*, 259:306–338, 2014.

[104] B. Fu and Q. Wang. The equivariant coarse Baum–Connes conjecture for spaces which admit an equivariant coarse embedding into Hilbert space. *J. Funct. Anal.*, 271(4):799–832, 2016.

[105] E. Ghys and P. de la Harpe, editors. *Sur les groupes hyperboliques d'après Mikhael Gromov*. Birkhäuser, 1990 (Progress in Mathematics, no. 83).

[106] G. Gong, Q. Wang and G. Yu. Geometrization of the strong Novikov conjecture for residually finite groups. *J. Reine Angew. Math.*, 621:159–189, 2008.

[107] G. Gong and G. Yu. Volume growth and positive scalar curvature. *Geom. Funct. Anal.*, 10:821–828, 2000.

[108] S. Gong, J. Wu and G. Yu. The Novikov conjecture, the group of volume preserving diffeomoprhisms and Hilbert–Hadamard spaces. arXiv:1811.02086, 2018.

[109] R. Grigorchuk. Degrees of growth of finitely generated groups and the theory of invariant means. *Izv. Akad. Nauk SSSR*, 48(2):939–985, 1984.

[110] M. Gromov. Asymptotic invariants of infinite groups. In G. Niblo and M. Roller, eds., *Geometric Group Theory*, volume 2, pp. 1–295. Cambridge University Press, 1993. London Math Society Lecture Notes vol. 182.

[111] M. Gromov. Random walks in random groups. *Geom. Funct. Anal.*, 13(1): 73–146, 2003.

[112] E. Guentner, N. Higson and J. Trout. Equivariant E-theory. *Mem. Amer. Math. Soc.*, 148(703), 2000.

[113] E. Guentner, N. Higson and S. Weinberger. The Novikov conjecture for linear groups. *Publ. Math. Inst. Hautes Études Sci.*, 101:243–268, 2005.

[114] E. Guentner and J. Kaminker. Exactness and the Novikov conjecture. *Topology*, 41(2):411–418, 2002.

[115] E. Guentner, R. Tessera and G. Yu. Operator norm localization for linear groups and its applications to K-theory. *Adv. Math.*, 226(4):3495–3510, 2010.

[116] E. Guentner, R. Tessera and G. Yu. A notion of geometric complexity and its application to topological rigidity. *Invent. Math.*, 189(2):315–357, 2012.

[117] E. Guentner, R. Willett and G. Yu. Finite dynamical complexity and controlled operator K-theory. arXiv:1609.02093, 2016.

[118] E. Guentner, R. Willett and G. Yu. Dynamic asymptotic dimension: relation to dynamics, topology, coarse geometry, and C^*-algebras. *Math. Ann.*, 367(1): 785–829, 2017.

[119] U. Haagerup and J. Kraus. Approximation properties for group C^*-algebras and group von Neumann algebras. *Trans. Amer. Math. Soc.*, 344:667–699, 1994.

[120] U. Hamenstädt. Geometry of the mapping class groups. I. Boundary amenability. *Invent. Math.*, 175(3):545–609, 2009.

[121] B. Hanke and T. Schick. Enlargeability and index theory. *J. Differential Geometry*, 74(2):293–320, 2006.

[122] B. Hanke and T. Schick. The strong Novikov conjecture for low degree cohomology. *Geometriae Dedicata*, 135:119–127, 2008.

[123] A. Hatcher. *Algebraic Topology*. Cambridge University Press, 2002.

[124] N. Higson. C^*-algebra extension theory and duality. *J. Funct. Anal.*, 129: 349–363, 1995.

[125] N. Higson. The Baum–Connes conjecture. In *Proceedings of the International Congress of Mathematicians, Volume II (Berlin 1998)*. Doc. Math., Extra Vol. II, 637–646, 1998.

[126] N. Higson. Counterexamples to the coarse Baum–Connes conjecture. Unpublished paper, Pennsylvania State University. Available at http://www.personal.psu.edu/ndh2/math/Unpublished_files/Higson%20-%201999%20-%20Counterexamples%20to%20the%20coarse%20Baum-Connes%20conjecture.pdf

[127] N. Higson. Bivariant K-theory and the Novikov conjecture. *Geom. Funct. Anal.*, 10:563–581, 2000.

[128] N. Higson and E. Guentner. Group C^*-algebras and K-theory. In S. Doplicher and R. Longo, eds., *Noncommutative Geometry*. Springer, 2004, pp. 137–252 (Springer Lecture Notes, no. 1831).

[129] N. Higson and G. Kasparov. E-theory and KK-theory for groups which act properly and isometrically on Hilbert space. *Invent. Math.*, 144:23–74, 2001.

[130] N. Higson, G. Kasparov and J. Trout. A Bott periodicity theorem for infinite dimensional Hilbert space. *Adv. Math.*, 135:1–40, 1999.

[131] N. Higson, V. Lafforgue and G. Skandalis. Counterexamples to the Baum–Connes conjecture. *Geom. Funct. Anal.*, 12:330–354, 2002.

[132] N. Higson, E. Pedersen and J. Roe. C^*-algebras and controlled topology. *K-Theory*, 11:209–239, 1997.

[133] N. Higson and J. Roe. On the coarse Baum–Connes conjecture. In S. C. Ferry, A. Ranicki, and J. M. Rosenberg, eds., *Novikov Conjectures, Index Theory Theorems and Rigidity, vol. 2*. London Mathematical Society, 1995, pp. 227–254 (London Mathematical Society Lecture Note Series, vol. 227).

[134] N. Higson and J. Roe. Amenable group actions and the Novikov conjecture. *J. Reine Angew. Math.*, 519:143–153, 2000.

[135] N. Higson and J. Roe. *Analytic K-homology*. Oxford University Press, 2000.

[136] N. Higson and J. Roe. Mapping surgery to analysis I: Analytic signatures. *K-theory*, 33:277–299, 2005.

[137] N. Higson and J. Roe. Mapping surgery to analysis II: Geometric signatures. *K-theory*, 33:301–324, 2005.

[138] N. Higson and J. Roe. Mapping surgery to analysis III: Exact sequences. *K-theory*, 33:325–346, 2005.

[139] N. Higson, J. Roe and G. Yu. A coarse Mayer–Vietoris principle. *Math. Proc. Cambridge Philos. Soc.*, 114:85–97, 1993.

[140] M. Hilsum and G. Skandalis. Invariance par homotopie de la signature á coefficients dans un fibré presque plat. *J. Reine Angew. Math.*, 423:73–99, 1992.

[141] N. Hitchin. Harmonic spinors. *Adv. Math.*, 14:1–55, 1974.

[142] L. Hörmander. *The Analysis of Linear Partial Differential Operators III: Pseudo-Differential Operators*. Springer, 1994 (Grundlehren der Mathematischen Wissenschaften, vol. 274).

[143] W. C. Hsiang. Geometric applications of algebraic K-theory. In *Proceedings of the International Congress of Mathematics, Volumes 1,2 (Warsaw 1983)*. PWN, 1984, pp. 98–118, 1983.

[144] B. Hunger. Almost flat bundles and homological invariance of infinite K-area. arXiv:1607.07820, 2016.

[145] W. B. Johnson and N. L. Randrianarivony. l^p $(p > 2)$ does not coarsely embed into a Hilbert space. *Proc. Amer. Math. Soc.*, 4(4):1045–1050, 2006.

[146] J. Kaminker and J. G. Miller. Homotopy invariance of the analytic index of signature operators over C^*-algebras. *J. Operat. Theor.*, 14(1):113–127, 1985.

[147] M. Karoubi. *K-theory*. Springer, 1978.

[148] G. Kasparov. Topological invariants of elliptic operators I: K-homology. *Math. USSR-Izv.*, 9(4):751–792, 1975.

[149] G. Kasparov. Equivariant KK-theory and the Novikov conjecture. *Invent. Math.*, 91(1):147–201, 1988.

[150] G. Kasparov. K-theory, group C^*-algebras, and higher signatures (conspectus). In S. Ferry, A. Ranicki and J. Rosenberg, eds., *Novikov Conjectures, Index Theory Theorems and Rigidity, Volume I*. Cambridge University Press, 1993, pp. 101–147 (London Mathematical Society Lecture Note Series, vol. 226).

[151] G. Kasparov and G. Skandalis. Groups acting on buildings, operator K-theory and Novikov's conjecture. *K-theory*, 4:303–337, 1991.

[152] G. Kasparov and G. Skandalis. Groups acting properly on "bolic" spaces and the Novikov conjecture. *Ann. Math.*, 158(1):165–206, June 2003.

[153] G. Kasparov and G. Yu. The coarse geometric Novikov conjecture and uniform convexity. *Adv. Math.*, 206(1):1–56, 2006.

[154] G. Kasparov and G. Yu. The Novikov conjecture and geometry of Banach space. *Geom. Topol.*, 16:1859–1880, 2012.

[155] J. Keesling. The one-dimensional Čech cohomology of the Higson compactification and its corona. *Topology Proc.*, 19:129–148, 1994.

[156] Y. Kida. The mapping class group from the viewpoint of measure equivalence. *Mem. Amer. Math. Soc.*, 916, 2008.

[157] T. Kondo. $CAT(0)$ spaces and expanders. *Math. Z.*, 271(1–2):343-355, 2012

[158] Y. Kubota. The realtive Mischenko–Fomenko higher index and almost flat bundles. arXiv:1807.03181, 2018.

[159] V. Lafforgue. Banach KK-theory and the Baum–Connes conjecture. In D. Li, ed., *Proceedings of the International Congress of Mathematics, Vol. II (Beijing, 2002)*. Higher Ed. Press, 2002, pp. 795–812.

[160] V. Lafforgue. Un renforcement de la propriété (T). *Duke Math. J.*, 143(3): 559–602, 2008.

[161] V. Lafforgue. La conjecture de Baum–Connes à coefficients pour les groupes hyperboliques. *J. Noncommut. Geom.*, 6(1):1–197, 2012.

[162] V. Lafforgue and M. de la Salle. Non commutative L^p without the completely bounded approximation property. *Duke Math. J.*, 160:71–116, 2011.

[163] E. C. Lance. *Hilbert C^*-modules (A Toolkit for Operator Algebraists)*. Cambridge University Press, 1995.

[164] H. B. Lawson and M.-L. Michelsohn. *Spin Geometry*. Princeton University Press, 1989.

[165] E. Leichtnam and P. Piazza. Spectral sections and higher Atiyah–Patodi–Singer index theory on Galois coverings. *Geom. Funct. Anal.*, 8(1):17–58, 1998.

[166] K. Li and R. Willett. Low-dimensional properties of uniform Roe algebras. *J. London Math. Soc.*, 97:98–124, 2018.

[167] X. Li and J. Renault. Cartan subalgebras in C^*-algebras. Existence and uniqueness. Trans. Amer. Math. Soc. 372:1985–2010, 2019.

[168] A. Lichnerowicz. Spineurs harmoniques. *C. R. Acad. Sci. Paris*, 257:7–9, 1963.

[169] J.-L. Loday. K-théorie algébrique et représentations de groupes. *Ann. Sci. École Norm. S.*, 9(3):309–377, 1976.

[170] J. Lott. Higher eta-invariants. *K-theory*, 6(3):191–233, 1992.

[171] A. Lubotzky. *Discrete Groups, Expanding Graphs and Invariant Measures*. Birkhäuser, 1994.

[172] W. Lück. The relationship between the Baum–Connes conjecture and the trace conjecture. *Invent. Math.*, 149:123–152, 2002.

[173] W. Lück and H. Reich. The Baum–Connes and Farrell–Jones conjectures in K- and L-theory. In E. Friedlander and D. Grayson, eds., *Handbook of K-theory*, volume 2. Springer, 2005, pp. 703–842.

[174] G. Lusztig. Novikov's higher signature and families of elliptic operators. *J. Differential Geometry*, 7:229–256, 1972.

[175] V. Mathai. The Novikov conjecture for low degree cohomology classes. *Geometriae Dedicata*, 99:1–15, 2003.

[176] M. Mendel and A. Naor. Non-linear spectral caculus and super-expanders. *Publ. Math. Inst. Hautes Études Sci.*, 119(1):1–95, 2014.

[177] R. Meyer and R. Nest. The Baum–Connes conjecture via localisation of categories. *Topology*, 45(2):209–259, March 2006.

[178] Mikhael Gromov. Spaces and questions. *Geom. Funct. Anal.*, 2000, Special Volume, Part I, 118–161.

[179] Mikhael Gromov and H. Blaine Lawson Jr. Spin and scalar curvature in the presence of a fundamental group. *Ann. Math.*, 111(2):209–230, 1980.

[180] Mikhael Gromov and H. Blaine Lawson, Jr. Positive scalar curvature and the Dirac operator on complete Riemannian manifolds. *Publ. Math. Inst. Hautes Études Sci.*, 58(1):295–408, 1983.

[181] J. Milnor. On axiomatic homology theory. *Pacific J. Math.*, 12(1):337–341, 1962.

[182] J. Milnor. *Morse Theory*. Princeton University Press, 1963.

[183] J. Milnor. *Introduction to Algebraic K-theory*. Princeton University Press, 1971 (Annals of Mathematics Studies, vol. 72).

[184] I. Mineyev and G. Yu. The Baum–Connes conjecture for hyperbolic groups. *Invent. Math.*, 149(1):97–122, 2002.

[185] A. S. Miščenko. Infinite dimensional representations of discrete groups and higher signatures. *Izv. Akad. Nauk. SSSR Ser. Mat.*, 8(1):81–106, 1974.

[186] C. Moore and C. Schochet. *Global analysis on foliated spaces*, 2nd ed. New York: Cambridge University Press, 2006 (Mathematical Sciences Research Institute Publications, vol. 9).

[187] H. Moriyoshi and P. Piazza. Eta cocycles, relative pairings and the Godbillon–Vey index theorem. *Geom. Funct. Anal.*, 22(6):1708–1813, 2012.

[188] J. Munkres. *Topology*, 2nd ed. Prentice-Hall, 2000.

[189] G. Murphy. *C^*-algebras and Operator Theory*. Academic Press, 1990.

[190] A. Naor and M. Mendel. Metric cotype. *Ann. Math.*, 168(1):247–298, 2008.

[191] S. Nishikawa. Direct splitting method for the Baum–Connes conjecture. arXiv:1808.08298, 2018.

[192] F. Noether. Über eine klasse singulärer integralgleichungen. *Math. Ann.*, 82(1-2): 42–63, 1920.

[193] S. Novikov. Topological invariance of rational classes of Pontrjagin (in Russian). *Dokl. Akad. Nauk SSSR*, 163:298–300, 1965.

[194] S. Novikov. Algebraic construction and properties of Hermitian analogues of K-theory over rings with involution from the point of view of Hamiltonian formalism: Some applications to differential topology and to the theory of characteristic classes. *Izv. Akad. Nauk SSSR*, 34:253–288, 475–500, 1970.

[195] P. Nowak and G. Yu. *Large scale geometry*. European Mathematical Society, 2012 (EMS Textbooks in Mathematics).

[196] D. Osajda. Small cancellation labellings of some infinite graphs and applications. arXiv:1406.5015, 2014.

[197] H. Oyono-Oyono and G. Yu. K-theory for the maximal Roe algebra of certain expanders. *J. Funct. Anal.*, 257(10):3239–3292, 2009.

[198] H. Oyono-Oyono and G. Yu. On quantitative operator K-theory. *Ann. Inst. Fourier (Grenoble)*, 65(2):605–674, 2015.

[199] H. Oyono-Oyono and G. Yu. Quantitative K-theory and Künneth formula for operator algebras. *J. Funct. Anal.*, 277(7):2003–2091, 2019.

[200] H. Oyono-Oyono, G. Yu and D. Zhou. Quantitative index, Novikov conjecture, and coarse decomposability. Preprint, 2019.

[201] N. Ozawa. Amenable actions and exactness for discrete groups. *C. R. Acad. Sci. Paris Sér. I Math.*, 330:691–695, 2000.

[202] N. Ozawa. Boundary amenability of relatively hyperbolic groups. *Topol. Appl.*, 153:2624–2630, 2006.

[203] W. L. Paschke. K-theory for commutants in the Calkin algebra. *Pac. J. Math.*, 95:427–437, 1981.

[204] G. K. Pedersen. *C^*-Algebras and their Automorphism Groups*. Academic Press, 1979.

[205] M. Pflaum, H. Posthuma and X. Tang. The transverse index theorem for proper cocompact actions of Lie groupoids. *J. Differ. Geom.*, 99(3):443–472, 2015.

[206] P. Piazza and T. Schick. Rho-classes, index theory, and Stolz' positive scalar curvature sequence. *J. Topol.*, 4:965–1004, 2014.

[207] M. Pimsner and D.-V. Voiculescu. Exact sequences for K-groups and Ext-groups of certain crossed product C^*-algebras. *J. Operat. Theor.*, 4(1):93–118, 1980.

[208] Y. Qiao and J. Roe. On the localization algebra of Guoliang Yu. *Forum Math.*, 22(4):657–665, 2010.

[209] I. Raeburn and D. Williams. *Morita Equivalence and Continuous Trace C^*-algebras*. American Mathematical Society, 1998.

[210] D. A. Ramras, R. Tessera and G. Yu. Finite decomposition complexity and the integral Novikov conjecture for higher algebraic K-theory. *J. Reine Angew. Math.*, 694:129–178, 2014.

[211] M. Reed and B. Simon. *Methods of Modern Mathematical Physics I: Functional Analysis*, revised and enlarged edition. Academic Press, 1980.

[212] J. Roe. An index theorem on open mainfolds, I. *J. Differ. Geom.*, 27:87–113, 1988.

[213] J. Roe. An index theorem on open mainfolds, II. *J. Differ. Geom.*, 27:115–136, 1988.

[214] J. Roe. Coarse cohomology and index theory on complete Riemannian manifolds. *Mem. Amer. Math. Soc.*, 104(497), 1993.

[215] J. Roe. From foliations to coarse geometry and back. In X. Masa and J. A. Alvarez López, eds., *Analysis and geometry in foliated manifolds (Santiago de Compostela, 1994)*. World Scientific, 1995, pp. 195–205.

[216] J. Roe. *Index Theory, Coarse Geometry and Topology of Manifolds*. American Mathematical Society, 1996 (CBMS Conference Proceedings, vol. 90).

[217] J. Roe. *Elliptic Operators, Topology and Asymptotic Methods*, 2nd edition. Chapman and Hall, 1998.

[218] J. Roe. *Lectures on Coarse Geometry*. American Mathematical Society, 2003 (University Lecture Series, vol. 31).

[219] J. Roe. Hyperbolic groups have finite asymptotic dimension. *Proc. Amer. Math. Soc.*, 133(9):2489–2490, 2005.

[220] J. Roe and R. Willett. Ghostbusting and property A. *J. Funct. Anal.*, 266(3):1674–1684, 2014.

[221] M. Rørdam, F. Larsen and N. Laustsen. *An Introduction to K-Theory for C^*-Algebras*. Cambridge University Press, 2000.

[222] J. Rosenberg. C^*-algebras, positive scalar curvature, and the Novikov conjecture. *Publ. Math. Inst. Hautes Études Sci.*, 58(1):197–212, 1983.

[223] J. Rosenberg. Analytic Novikov for topologists. In S. Ferry, A. Ranicki, and J. Rosenberg, eds., *Novikov Conjectures, Index Theory Theorems and Rigidity, Volume I*. Cambridge University Press, 1993, pp. 338–368 (London Mathematical Society Lecture Note Series, vol. 226).

[224] J. Rosenberg. *Algebraic K-theory and its applications*. Springer, 1994.

[225] J. Rosenberg. Manifolds of positive scalar curvature: a progress report. In J. Cheeger and K. Grove, eds., *Surveys in Differential Geometry XI: Metric and Comparison Geometry*. International Press, 2007.

[226] J. Rosenberg and C. Schochet. The Künneth theorem and the universal coefficient theorem for Kasparov's generalized K-functor. *Duke Math. J.*, 55(2): 431–474, 1987.

[227] J. Rosenberg and S. Stolz. Manifolds of positive scalar curvature. In G. E. Carlsson, R. L. Cohen, W. C. Hsiang and J. D. S. Jones, eds., *Algebraic Topology and Its Applications*. Springer, 1994, pp. 241–267 (Mathematical Sciences Research Institute Publications, vol. 27).

[228] H. Sako. Property A and the operator norm localization property for discrete metric spaces. *J. Reine Angew. Math.*, 690:207–216, 2014.

[229] M. Sapir. A Higman embedding preserving asphericity. *J. Amer. Math. Soc.*, 27:1–42, 2014.

[230] T. Schick. A counterexample to the (unstable) Gromov–Lawson–Rosenberg conjecture. *Topology*, 37(6):1165–1168, 1998.

[231] R. Schoen and S.-T. Yau. Existence of incompressible minimal surfaces and the topology of three-dimensional manifolds of non-negative scalar curvature. *Ann. Math.*, 110:127–142, 1979.

[232] R. Schoen and S.-T. Yau. On the structure of manifolds with positive scalar curvature. *Manuscripta Math.*, 28:159–183, 1979.

[233] L. Schweitzer. A short proof that $M_n(A)$ is local if A is local and Fréchet. *Internat. J. Math.*, 4(2), 1992.

[234] Z. Sela. Uniform embeddings of hyperbolic groups in Hilbert space. *Isr. J. Math.*, 80(1-2):171–181, 1992.

[235] L. Shan and Q. Wang. The coarse geometric Novikov conjecture for subspaces of non-postively curved manifolds. *J. Funct. Anal.*, 248(2):448–471, 2007.

[236] B. Simon. *Trace ideals and their applications*, 2nd edition. American Mathematical Society, 2005.

[237] S. Stolz. Simply connected manifolds of positive scalar curvature. *Ann. Math.*, 136(3):511–540, 1992.

[238] M. Takesaki. On the cross-norm of the direct product of C^*-algebras. *Tohoku Math. J.*, 16(1):111–122, 1964.

[239] X. Tang, R. Willett and Y.-J. Yao. Roe C^*-algebra for groupoids and generalized Lichnerowicz vanishing theorem for foliated manifolds. Math. Z. 290(3-4):1309–1338, 2018.

[240] J. Trout. On graded K-theory, elliptic operators, and the functional calculus. *Illinois J. Math.*, 44(2):194–309, 2000.

[241] J.-L. Tu. Remarks on Yu's property A for discrete metric spaces and groups. *B. Soc. Math. Fr.*, 129:115–139, 2001.

[242] J. von Neumann. *Continuous Geometry*. Princeton University Press, 1998 (Princeton Landmarks in Mathematics and Physics).

[243] J. Špakula and R. Willett. Maximal and reduced Roe algebras of coarsely embeddable spaces. *J. Reine Angew. Math.*, 678:35–68, 2013.

[244] J. Špakula and R. Willett. On rigidity of Roe algebras. *Adv. Math.*, 249:289–310, 2013.

[245] C. Wahl. Higher ρ-invariants and the surgery structure set. *J. Topol.*, 6(1):154–192, 2013.

[246] C. T. C. Wall. *Surgery on Compact Manifolds*, 2nd edition. American Mathematical Society, 1999 (Mathematical Surveys and Monographs, vol. 69).

[247] B.-L. Wang and H. Wang. Localized index and L^2-Lefschetz fixed point theorem for orbifolds. *J. Differ. Geom.*, 102(2):285–349, 2016.

[248] N. E. Wegge-Olsen. *K-Theory and C^*-Algebras (A Friendly Approach)*. Oxford University Press, 1993.

[249] S. Wei. On the quasidiagonality of Roe algebras. *Sci. China Math.*, 54: 1011–1018, 2011.

[250] C. Weibel. *The K-book: An Introduction to Algebraic K-theory*. American Mathematical Society, 2013 (Graduate Studies in Mathematics, vol. 145).

[251] S. Weinberger. Aspects of the Novikov conjecture. In J. Kaminker, ed., *Geometric and Topological Invariants of Elliptic Operators*. American Mathematical Society, 1990, pp. 281–297 (Contemporary Mathematics, vol. 105).

[252] S. Weinberger. *The Topological Classification of Stratified Spaces*. University of Chicago Press, 1994.

[253] S. Weinberger. Variations on a theme of Borel. Available at http://math.uchicago .edu/~shmuel/VTBdraft.pdf, 2017.

[254] S. Weinberger, Z. Xie and G. Yu. Additivity of higher rho invariants and nonrigidity of topological manifolds. arXiv:1608.03661, 2016. To appear, Comm. Pure. Appl. Math.

[255] S. Weinberger and G. Yu. Finite part of operator K-theory for groups finitely embeddable into Hilbert space and the degree of nonrigidity of manifolds. *Geom. Topol.*, 19(5):2767–2799, 2015.

[256] S. White and R. Willett. Cartan subalgebras in uniform Roe algebras. arXiv:1808.04410, 2018. To appear, Groups, Geom, Dyn.

[257] R. Willett. Some notes on property A. In G. Arzhantseva and A. Valette, eds., *Limits of Graphs in Group Theory and Computer Science*. EPFL Press, 2009, pp. 191–281.

[258] R. Willett and G. Yu. Higher index theory for certain expanders and Gromov monster groups I. *Adv. Math.*, 229(3):1380–1416, 2012.

[259] R. Willett and G. Yu. Higher index theory for certain expanders and Gromov monster groups II. *Adv. Math.*, 229(3):1762–1803, 2012.

[260] R. Willett and G. Yu. Geometric property (T). *Chinese Ann. Math. Ser. B*, 35(5):761–800, 2014.

[261] W. Winter and J. Zacharias. The nuclear dimension of C^*-algebras. *Adv. Math.*, 224(2):461–498, 2010.

[262] N. Wright. C_0 coarse geometry. PhD thesis, The Pennsylvania State University, 2002.

[263] N. Wright. Simultaneous metrizability of coarse spaces. *Proc. Amer. Math. Soc.*, 139(9):3271–3278, 2011.

[264] C. Wulff. Coarse co-assembly as a ring homomorphism. J. Noncommut. Geom. 10(2):471–514, 2016.

[265] Z. Xie and G. Yu. Positive scalar curvature, higher rho invariants and localization algebras. *Adv. Math.*, 262:823–866, 2014.

[266] Z. Xie and G. Yu. Higher rho invariants and the moduli space of positive scalar curvature metrics. *Adv. Math.*, 307:1046–1069, 2017.

[267] Z. Xie and G. Yu. Delocalized eta invariants, algebraicity, and K-theory of group C^*-algebras. arXiv:1805.07617, 2018. To appear, Int. Math. Res. Not.

[268] G. Yu. Baum–Connes conjecture and coarse geometry. *K-theory*, 9(3):223–231, 1995.

[269] G. Yu. Coarse Baum–Connes conjecture. *K-theory*, 9:199–221, 1995.

[270] G. Yu. Localization algebras and the coarse Baum–Connes conjecture. *K-theory*, 11(4):307–318, 1997.

[271] G. Yu. The Novikov conjecture for groups with finite asymptotic dimension. *Ann. Math.*, 147(2):325–355, 1998.

[272] G. Yu. The coarse Baum–Connes conjecture for spaces which admit a uniform embedding into Hilbert space. *Invent. Math.*, 139(1):201–240, 2000.

[273] G. Yu. The algebraic K-theory Novikov conjecture for group algebras over the ring of Schatten class operators. *Adv. Math.*, 307:727–753, 2017.

[274] J. Zacharias. On the invariant translation approximation property for discrete groups. *Proc. Amer. Math. Soc.*, 134(7):1909–1916, 2006.

[275] R. Zeidler. Secondary large-scale index theory and positive scalar curvature. PhD thesis, Georg-August-Universität, Göttingen, 2016.

[276] V. F. Zenobi. Mapping the surgery exact sequence for topological manifolds to analysis. *J. Topol. Anal.*, 9(2):329–361, 2017.

[277] W. Zhang. Positive scalar curvature on foliations. *Ann. Math.*, 185:1035–1068, 2017.

Index of Symbols

Index